Complete Brake Systems

Complete Brake Systems

Edited by Robert Scharff

Delmar Publishers Inc.®

NOTICE TO THE READER

Publisher does not warrant or guarantee any of the products described herein or perform any independent analysis in connection with any of the product information contained herein. Publisher does not assume, and expressly disclaims, any obligation to obtain and include information other than that provided to it by the manufacturer.

The reader is expressly warned to consider and adopt all safety precautions that might be indicated by the activities described herein and to avoid all potential hazards. By following the instructions contained herein, the reader willingly assumes all risks in connection with such instructions.

The publisher makes no representations or warranties of any kind, including but not limited to, the warranties of fitness for particular purpose or merchantability, nor are any such representations implied with respect to the material set forth herein, and the publisher takes no responsibility with respect to such material. The publisher should not be liable for any special, consequential or exemplary damages resulting, in whole or in part, from the readers' use of, or reliance upon, this material.

Text, Design, and Production by
Scharff Associates, Ltd.

Scharff Staff
Production Manager: Marilyn Strouse-Hauptly
Staff Writer: Dave Caruso
Cover Design: Eric Schreader
Logo Design: Ed Foulk

Delmar Staff
Editor-in-Chief: Mark W. Huth
Associate Editor: Joan Gill

For information, address Delmar Publishers Inc.
3 Columbia Circle, PO Box 15015
Albany, New York 12212-5015

Printed in the United States of America
Published simultaneously in Canada
by Nelson Canada,
A division of The Thomson Corporation

10 9 8

ISBN: 0-8273-3539-3

Contents

Preface

Not long ago, almost any mechanic could completely overhaul a brake system and be fairly confident of the results, but not so today. The automotive braking system has become vastly complex. A confounding array of values and control mechanisms to monitor, and even modulate hydraulic line pressures have appeared. The fluids have changed, the tolerances have become tighter, and the components have become more intricate. Now, even electronic control systems are used to control an automobile's brakes. In order to repair the braking systems of today, and those of the future, you must understand them. How do they operate? How do you diagnose problems? How do you repair and replace components?

The title of this book—*Complete Brake Systems*—was selected because of the first rule of brake service work, as explained in later chapters, is to do a **complete** job. For example, if new linings are installed without regard to the condition of the hydraulic system, the presence of a leaking or frozen caliper or wheel cylinder will cause the customer to come back complaining that you cannot be trusted to do good work, and your reputation and schedule will be damaged. Worse yet, if you overlook a weak brake hose or a condition that results in severe pulling or poor stopping power, your work could be the cause of an accident.

To accomplish the task of providing the information to do a complete brake job, the book is divided into fourteen chapters plus a glossary. Chapter 1 explains the theory and workings of the overall brake system, and lays the groundwork for the service instructions in the succeeding chapters. Chapters 2, 3, and 4 discuss in detail the various components of the hydraulic system (the heart of any modern passenger or light-duty brake system), how the brakes are bled, and how to maintain this important system.

Chapters 5 and 6 cover the theory and servicing of drum brakes, while disc brake theory and service are discussed in Chapters 7 and 8. Parking brakes are detailed in Chapter 9, with Chapter 10 covering power brakes—both vacuum and hydro-boost types.

Chapter 11 describes the operation and service procedures relating to the anti-lock braking system used on passenger cars so equipped. Following in Chapters 12 and 13 are descriptions and servicing procedures of light-duty truck and import vehicle brake systems.

Chapter 14 contains concise troubleshooting charts for drum, disc, power, and electric brakes as well as those light- and medium-duty trucks that contain air brakes. This chapter also includes a detailed approach to diagnosing the complete braking system, plus a comprehensive inspection report. The brake technician, whether experienced or inexperienced, is encouraged to use this table when performing complete brake system service. The glossary at the back of the book includes the most commonly used brake terms.

Complete Brake Systems contains various cautions and warnings that should be carefully noted to avoid personal injury or vehicle damage when servicing. It is important to understand that these cautions and warnings could not possibly cover all conceivable ways in which service might be done or of all the possible hazards in each way, nor can the contributors, the publisher, or the editors and their staff possibly know or investigate all such ways. It is therefore the responsibility of anyone using service procedures or tools, whether or not recommended in this book, to satisfy himself/herself thoroughly that neither personal safety nor vehicle safety will be jeopardized by the service procedures and tools selected.

ACKNOWLEDGMENTS

I would like to thank the following companies for permission to use information from their training programs and manuals, which are protected by copyright.

- *Domestic and Import Car Brake Service Manual*—Allied Aftermarket Division, Allied Signal Corporation.
- *Disc Brake Manual and Various Brake Service Bulletins*—Ammco Division, and Hennessy Industries, Inc.
- *Passenger Car Brake Service Manual*—Raybestos Division, Brake System, Inc.
- *Service Manual #799*—EIS Brake Parts, Inc.
- *Diagnostic Analysis and Reconditioning Manual*—Wagner Division, Cooper Industries, Inc.

Grateful acknowledgment is also made to the following companies, which provided brake data and other reference material used in the preparation of this book.

AC-Delco (General Motors Corp.)
Airtex Automotive Division
American Honda Motor Company, Inc.
APS, Inc.
Arrow Automotive Industries
John Bean Division, FMC Corporation
Robert Bosch Corp.
Buick Division, General Motor Corporation
Brake-Pro Systems Division, Tenneco, Inc.
Cadillac Motor Division, General Motors Corp.
Chevrolet Motor Division, General Motors Corp.
Chrysler Corp.
CR Industries, Inc.
Dana Corp.
Echlin Inc.
E. Edelmann and Company
Ford Motor Company
Gibson Products, Division of Rolero, Inc.
GKN Aftermarket Import Parts, Inc.
Hunter Engineering Company
International Precision Industries, Inc.
Isuzu Motors, Ltd.
Kelsey-Hayes Company
Lee Manufacturing Company
Lisle Corp.
Lucas Girling, Ltd.
Maremont Corporation
Midland Brake, Inc.
National Institute for Auto Service Excellence (ASE)
Nissan Motor Corporation
Oldsmobile Motor Division, General Motors Corp.
Standard Motor Products, Inc.
Subaru of America, Inc.
Toyota Motor Sales, Inc.
TRW Automotive Aftermarket, Inc.
United Brake Parts Company
Vacula Automotive Products, Inc.

I would like to thank the following for reviewing the manuscript at various stages and for their helpful comments:

Mr. Douglas Gossett
Jefferson State Vo-Tech
Louisville, Kentucky

Mr. Eugene Scotthoefer
Henry Ford College
Dearborn, Michigan

Mr. R. Ashby Melton
Central Texas College
Killeen, Texas

Mr. Richard Proulx
Mid-Florida Technical Institute
Orlando, Florida

Mr. Bill J. Steen
Citrus College
Glendora, California

CHAPTER 1

BRAKE PRINCIPLES AND FUNDAMENTALS

OBJECTIVES

Upon completing this chapter, you should be able to

- Explain how the principle of friction is involved in braking action.
- Describe how a hydraulic system transfers foot-pressure into braking power.
- Identify the factors involved in controlled braking.
- Explain the development of automotive brake systems.
- Recognize the two basic types of hydraulic brakes.
- Name the components used in various brake systems.
- Explain why most cars made today use a combination brake system.
- Describe proper brake service and safety techniques.
- Identify the basic brake tools.
- Use vehicle manuals and aftermarket literature.
- Explain ASE certification.

How do you stop a four thousand pound object hurtling 55 miles per hour down a steep hill—especially if you are inside of it? Do you drag your feet? Do you grab the next street light or tree limb? Do you jump?

If the object is a modern automobile, you simply apply the brakes. The brakes on a vehicle are designed to absorb the tremendous energy of a moving car or truck and convert the momentum into another form of energy: heat. Brakes do this by using friction.

FRICTION

Friction is the resistance to movement exerted by two objects in contact with each other. Two forms of friction play a part in controlling a vehicle (Figure 1-1): *kinetic,* or moving, *friction* and *static,* or stationary, *friction.* The amount of friction, or resistance to movement, depends on the type of materials in contact, the smoothness of their rubbing surfaces, and the pressure holding them together (often gravity or weight). For exam-

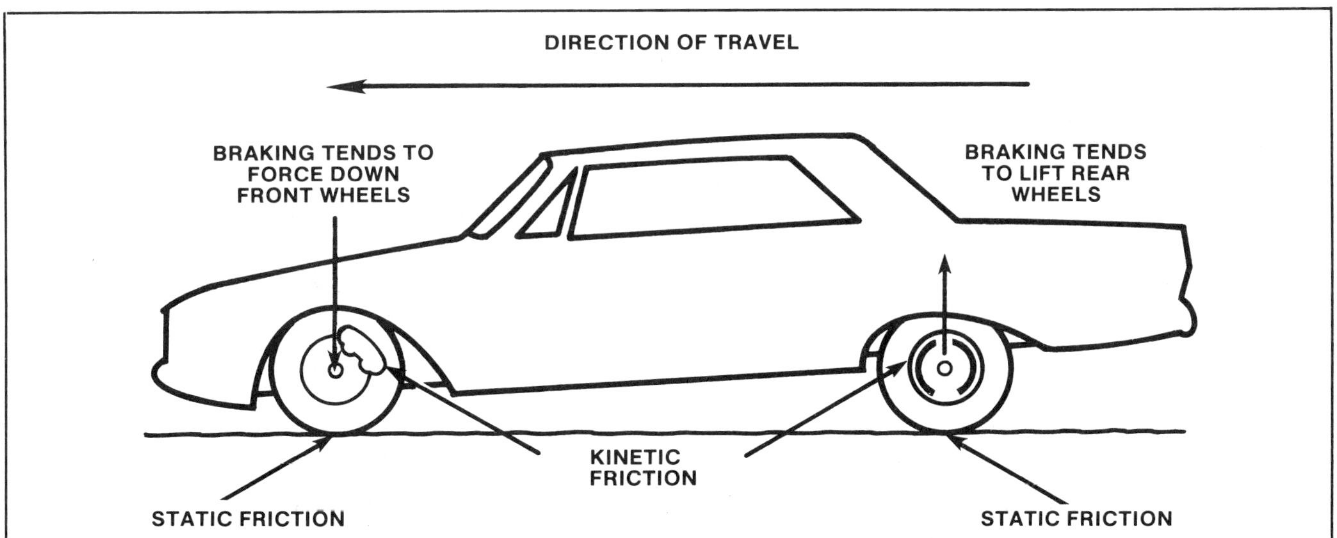

Figure 1-1 The braking action creates kinetic friction in the brakes and static friction between tire and road to slow the vehicle. As shown, when brakes are applied, the vehicle weight is transferred to the front wheels and is unloaded on the rear wheels.

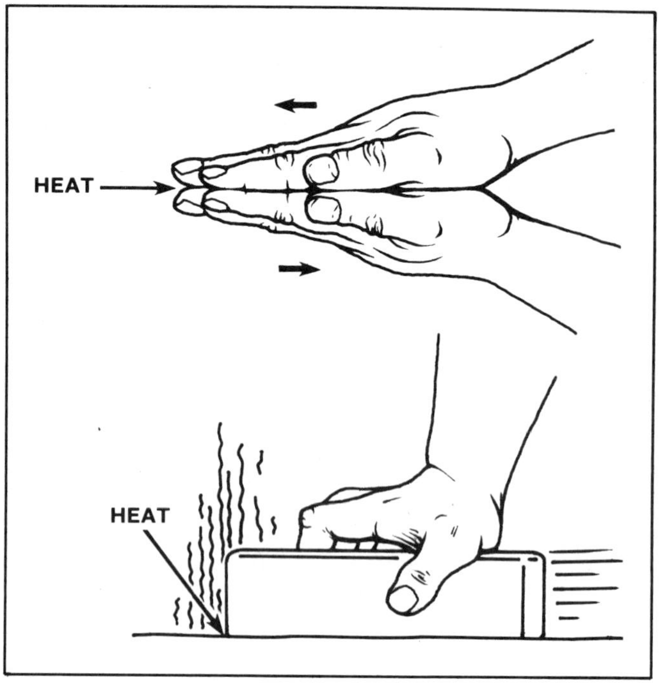

Figure 1-2 Friction produces heat.

ple, slide your hand lightly across a polished table top. Very little effort is required because of the smooth texture of the rubbing surfaces and the small amount of pressure exerted. Now rub your hand over a piece of sandpaper. Feel the difference in friction? The rough texture of the sandpaper resists the movement of your hand. Rub any surface long enough or hard enough and you will begin to feel heat (Figure 1-2). Friction always converts moving, or kinetic, energy into heat. The greater the friction between two moving surfaces, the greater the amount of heat produced.

Kinetic Friction

As you apply the brakes on a moving automobile, rough-textured pads or shoes are pressed against rotating parts of the vehicle—either rotors or drums. The **kinetic energy,** or momentum, of the vehicle is then converted into heat energy by the kinetic friction of rubbing surfaces. The car or truck slows down.

Static Friction

When the vehicle comes to a stop, it is held in place by static friction. The friction between the surfaces of the brakes, as well as the friction between the tires and the pavement, resist any movement. To overcome the static friction that holds the car motionless, the brakes are released; the heat energy of combustion in the engine crankcase is converted into kinetic, or moving, energy by the transmission and drivetrain; and the vehicle moves forward.

Static friction also plays an important part in controlling a moving vehicle. The rotating tires grip the road, and the static friction between these two surfaces enables the driver to control the speed and direction of the car. When the brakes are applied, the kinetic friction of the rubbing brake components slows the rotation of the tires. This increases the static friction between the tires and the road, decreasing the forward motion of the car. If the kinetic or sliding friction of the brake components overcomes the static friction between the tires and road, the wheels will lock up and the car will begin to slide. Static friction then exists between the components in the brakes and kinetic friction between the skidding tires and the road—the car is out of control. Obviously, the most effective braking effort is achieved just before wheel lockup occurs.

FACTORS IN BRAKING

There are three principal factors governing the amount of friction developed in a braking system and a fourth factor, which is the result of friction. These are pressure, coefficient of friction, frictional contact surface, and heat dissipation.

Pressure

Less pressure is required to fully activate drum brakes than is required by disc brakes, for reasons we shall presently see. On drum brakes, the frictional contact surface is inside the drum. Brake shoes are pushed outward. After initial contact, a frictional drag develops and pulls the shoe against the drum. This braking action energizes the shoe, actually assisting in the braking action. Thus, the hydraulic system needs to generate only a part of the pressure needed to stop the car. On disc brakes, the frictional contact surfaces consist of pads pushed inward against both sides of the rotor or disc (Figure 1-3). Much more pressure is required with this method of braking because disc brakes are non-energized. That is, all the braking force must come through the hydraulic system. For this reason, a power brake unit is often used to assist in applying disc brakes, especially in larger vehicles.

Coefficient of Friction

The amount of friction between two surfaces is expressed as a *coefficient of friction (C.O.F.).* The

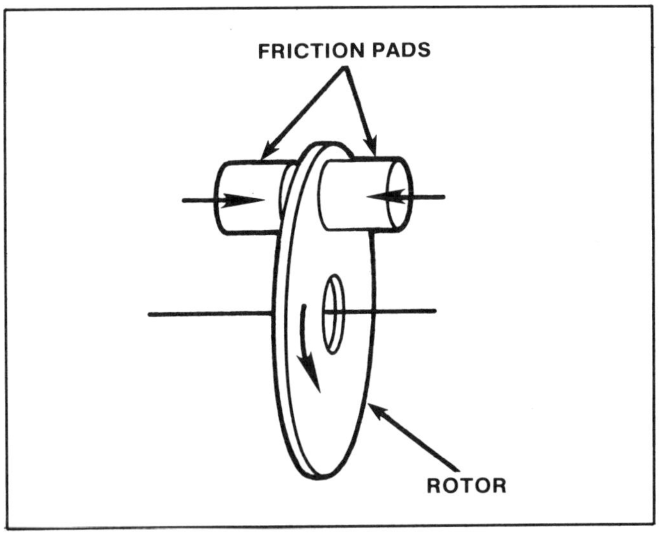

Figure 1-3 Disc brake operation

coefficient of friction is determined by dividing the force required to pull an object across a surface by the weight of the object (Figure 1-4). For example, if it requires 60 pounds of pull to slide a 100-pound metal part across a concrete floor, the coefficient of friction is 100 ÷ 100 or 1.00. To pull a 100-pound block of ice across the same floor might require only 2 pounds of pull. The coefficient of friction then would be only 0.02. As it applies to automotive brakes, the C.O.F. expresses the frictional relationship between pads and rotors or shoes and drums and is carefully engineered to ensure optimum performance. Therefore, when replacing pads or shoes, it is important to use replacement parts with similar C.O.F. If, for example, the C.O.F. is too high, the brakes will be too "sticky" to stop the car smoothly. Premature wheel lockup or grabbing would result. If the coefficient is too low, the friction material will tend to slide over the machined surface of the drum or rotor rather than slowing it down. Most automotive friction materials are thus engineered with a C.O.F. of between .25 and .55, depending upon their intended application.

Frictional Contact Surface

The third factor is the amount of surface, or area, which is in contact. Simply put, bigger brakes will stop a car more quickly than smaller brakes used on that same car. Similarly, brakes on all four wheels will slow or stop a moving vehicle faster than will brakes on only two wheels, assuming the vehicles are equal in size.

Heat Dissipation

Obviously, another important factor in brake design is heat dissipation. The tremendous heat created by the rubbing brake surfaces must be conducted away from the pad and rotor (or shoe and drum) and be absorbed by the air. This is done by the components of the brake mechanism. The greater the surface areas of the brake components, the faster the heat can be dissipated. Thus, the weight and potential speed of the vehicle determines the size of the braking mechanism and the friction surface area of the pad or shoe. Brakes that do not effectively dissipate heat will experience **brake fade** during hard, continuous braking. The linings of the pad and shoe become glazed, and the rotor and drum become hardened. Therefore, the coefficient of friction is reduced and excessive foot

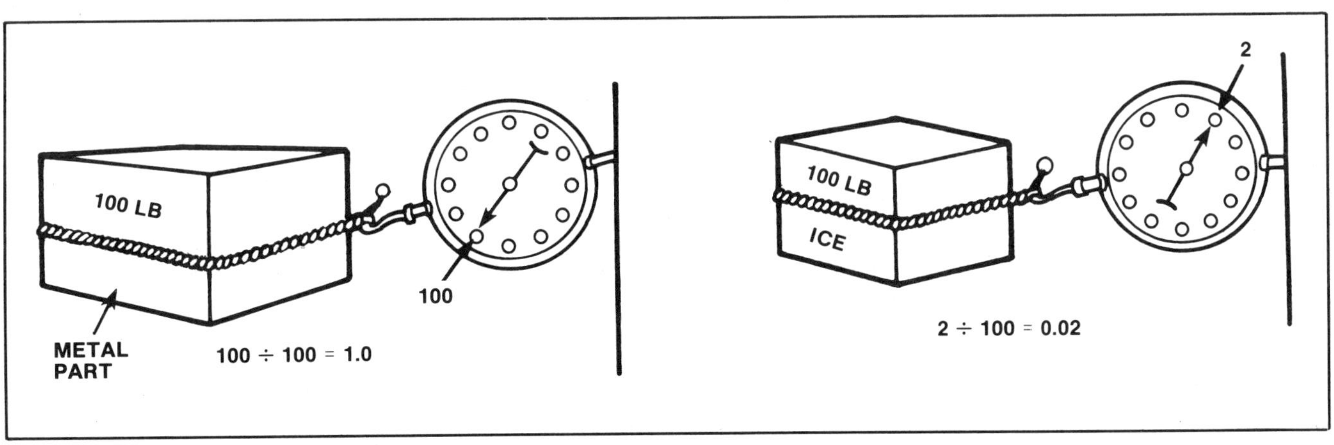

Figure 1-4 Coefficient of friction is the weight of the object divided by the pounds of pull.

pressure must be applied to the brake pedal to produce the desired braking effect. In other words, brake fade is the loss of braking friction.

BRAKING ABILITY OF A VEHICLE

The effectiveness of brakes depends upon a number of combined elements, any one of which could have an effect on performance. The size of the pads and shoes and the area that they contact as well as the force applied to them are some of these factors. The larger the radius of the drum and wheel, the better the leverage and the more effective the braking. The coefficient of friction of the braking surfaces and the contact between the tire and the road also affect vehicle braking distance.

Other factors that affect a vehicle's braking ability include:

1. **Weight.** When the brakes are applied, the weight of the vehicle is one of the factors that determines stopping ability. If the weight is doubled, the energy of motion that must be converted into heat energy is also doubled. Brakes designed to control a vehicle with a gross weight of 4,500 pounds, for instance, cannot properly absorb and dissipate the amount of heat produced if the same vehicle is overloaded to a gross weight of 9,000 pounds.
2. **Vehicle speed.** When the brakes are applied, vehicle speed decreases as the energy of motion is converted into heat energy and the vehicle finally stops. The effect of higher speed is even more serious than the effect of the greater weight. When the speed of a vehicle is doubled, four times as much energy of motion is produced. Consequently, at twice the speed, the brakes must absorb and dissipate four times as much heat. If the weight is doubled, too, the brakes must handle eight times as much heat.
3. **Tires and road.** The contact between the tires and the road occurs as rolling friction and it is this friction that must actually stop the vehicle. If the brakes overcome rolling friction and lock a wheel on the vehicle, the heat must be produced by the small skid area of the tire. Tire diameter is also a factor since leverage between the brake and the wheel is reduced by increases in the rolling radius. All other things being equal, the larger the tire rolling radius, the harder the vehicle is to stop.

FRICTION MATERIALS

Linings are the most important part of any brake. They vary in composition and quality levels, but all linings should have the following characteristics for optimum brake performance:

- *Resistance to Heat Fade*—The ability to maintain effectiveness at higher temperatures.
- *Quick Recovery*—A lining's ability to return to its prefade friction level under cooling conditions.
- *Good Speed Sensitivity*—The ability to maintain effectiveness against drum and disc surfaces at varying speeds.
- *Resistance to Water Fade*—The speed at which a lining recovers from loss of friction due to moisture from rain or puddles. Moisture in the brake lining makes brakes grab, while a water film causes slippage.
- *Long Lining Life*—The ability to give maximum performance and longer wear without damaging disc and drum surfaces.

For many decades, asbestos has been the standard brake lining material. It offers good friction qualities, long wear, and low noise. But new materials are beginning to be used because of two important considerations: the health hazards of breathing asbestos dust and the need for extra fade resistance in the brakes used on late-model cars, especially those with down-sized brakes and front-wheel drive where the front brakes do up to 80 percent of the work.

Fully metallic linings of sintered iron have been used for years in heavy-duty and racing applications because they have great fade resistance. But they require very high pedal pressure and tend to wear out drums and rotors. So, semi-metallic linings were developed. This friction material is made of iron fibers molded with an adhesive matrix. It offers excellent fade resistance that meets the needs of today's cars, yet has good friction characteristics so that a reasonable amount of pedal pressure is required. However, these characteristics do not wear out drums and rotors excessively (some early types and off-brands, however, have been known to destroy drums and rotors). These modern linings are generally referred to as *semimet.*

A proper rotor finish is important with semimetallic brake linings in order to prevent noise and

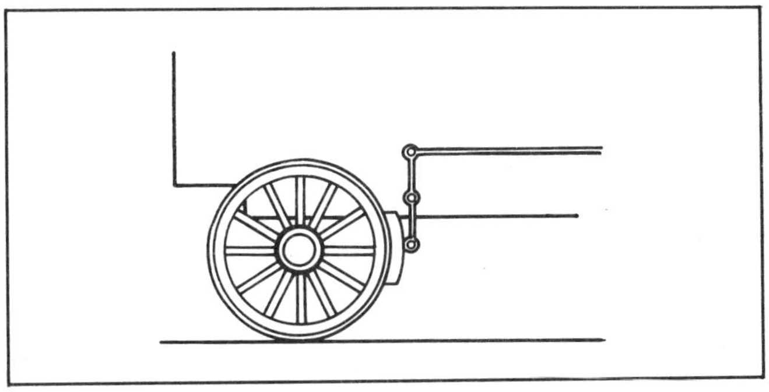

Figure 1-5 Wagon brake

achieve good break-in and performance. The directional finish left by a lathe should not be used. Instead, the rotor should be ground using the right equipment in order to provide a non-directional, swirl-like finish. The correct grit to use is #120, which is finer than the #80 normally recommended with asbestos linings.

If grinding equipment is not available, an acceptable temporary finish can be achieved by sanding the rotor in a non-directional pattern with #120 grit sandpaper.

Other new lining materials are being developed, also. Neither asbestos nor semi-met, these new lining materials are called **non-asbestos** and are made of modern synthetic substances, sometimes reinforced with Aramid fiber. The major brake lining manufacturers are constantly experimenting with new materials, but brand-name manufacturers will not bring a lining to market until they are satisfied that it meets all established criteria for long life, friction characteristics, and drum and rotor wear.

Never replace semi-mets with ordinary asbestos linings or any other lining material that is not recommended by the vehicle manufacturer. All linings sold today have edge codes, which identify the manufacturer, the specific formulation, and the coefficient of friction tested under a specific set of conditions. However, the test is simply for a uniform means of identification. It measures only initial friction and does not indicate significant characteristics that can better determine brake lining quality. Often, the best way to determine quality is to go by the reputation of the manufacturer.

The friction surface, in addition to giving a smooth lining contact surface, also conducts most of the heat away from the linings and radiates it into the air. Some of the heat also passes through the linings and brake shoes to the backing plates, which also radiate the heat into the air. Friction surfaces such as brake drums and rotors must have the following qualities to perform satisfactorily:

- Smooth contact surface
- Ability to conduct and radiate heat
- Strength and light weight
- Resistance to scoring

BRAKE SYSTEMS

Since the first horseless carriage roamed the cobbled streets of America, many different mechanisms have been used to brake its momentum. The earliest was the wagon brake (Figure 1-5). It was simply a long lever with a friction block of leather or wood on one end. When the lever was pulled, the friction block was brought to bear against the rotating rim of the wheel. The wagon brake, however, was better suited to the plodding pace of a horse drawn carriage. When the pneumatic tire was introduced, the wagon brake became impractical.

Another braking mechanism was introduced: the external band brake (Figure 1-6). The external band brake consisted of flexible steel bands lined with a friction material, such as leather. The bands were wrapped around a drum or hub and were tightened by application of a lever or pedal. The effectiveness of this design was limited, though, by the moisture and dirt the brake was exposed to.

To lengthen the service life of the brake and to overcome the negative effects moisture and dirt had on braking power, the internal drum brake (Figure 1-7) was developed. Shoes lined with a

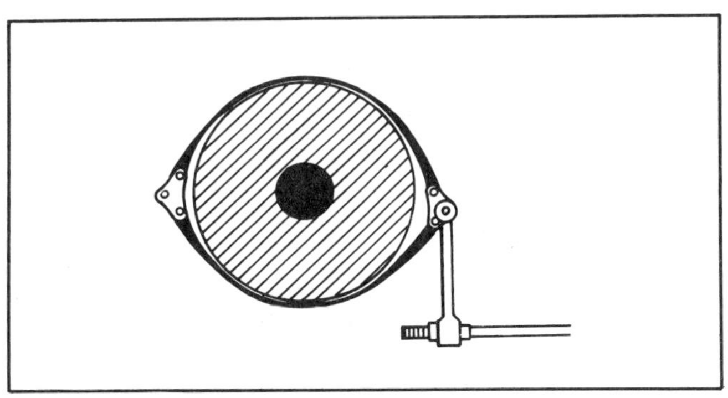

Figure 1-6 External band brake

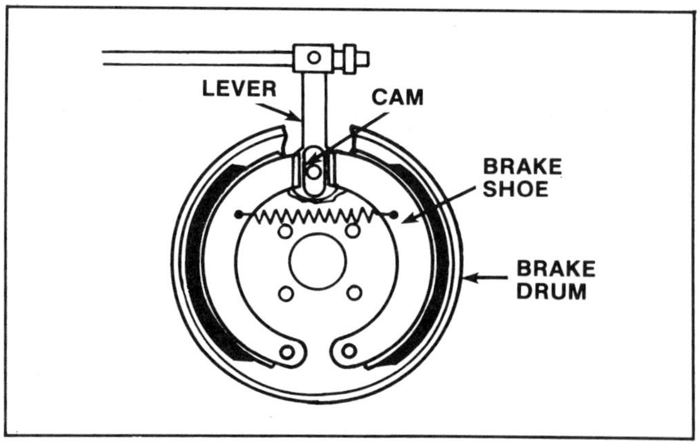

Figure 1-7 Internal drum brake

friction material, such as asbestos, were housed inside a rotating drum. When mechanically actuated, a lever and cam mechanism expanded the brake shoes against the rotating drum. The drum protected the friction linings from dirt and moisture and radiated heat into the air.

These early braking mechanisms were first used on rear wheels. Later, through a complex system of cables, levers, rods, and linkages, brakes were applied to all four wheels. But a mechanical actuating system required constant adjustment and lubrication and had many moving parts to wear and brake. Worse yet it was impossible to engage all four brakes simultaneously and with equal force. The result was pulling, jerking, and uncertain stops.

THE HYDRAULIC BRAKE SYSTEM

These problems were overcome with the introduction of *hydraulics*. A hydraulic system (Figure 1-8) uses a liquid (brake fluid) to transfer pressure from the brake pedal to the pads or shoes. This transfer of pressure is reliable and consistent because liquids are not compressible. That is, pressure applied to a liquid in a closed system is transmitted by that liquid equally to every other part of that system. This includes the hydraulic brake system of a typical passenger car or truck. Apply a force of 5 pounds per square inch (PSI) through the master cylinder and you can measure 5 PSI anywhere in the lines and at each wheel where the brakes operate.

For example, move piston 1 a distance of 1 inch and piston 2 will also move 1 inch (Figure 1-9). The force can be increased at output (that is, at the wheel) by increasing the size of the wheel piston, though piston travel will be *decreased*. The force at output can be decreased by decreasing the size of the wheel piston, but the piston travel will *increase*.

Thus, to double the output force of the 5 PSI at the master cylinder to 10 PSI at the wheels, simply use a wheel cylinder piston with a surface area of

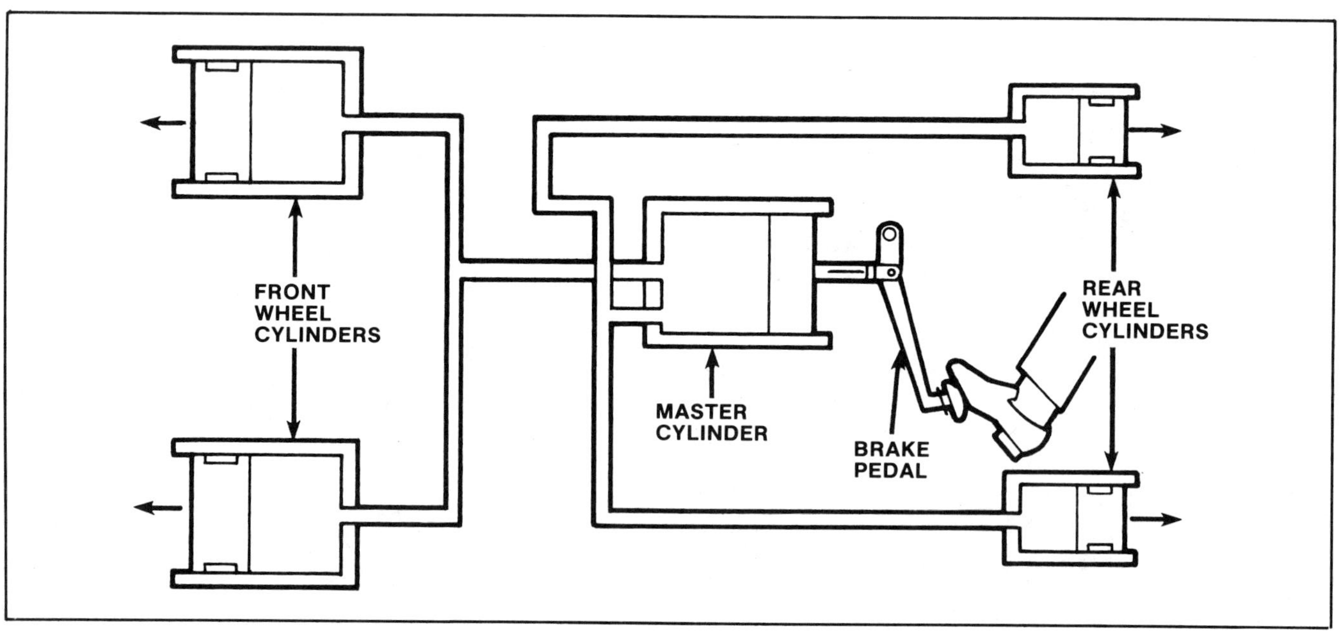

Figure 1-8 Schematic of automotive hydraulic system

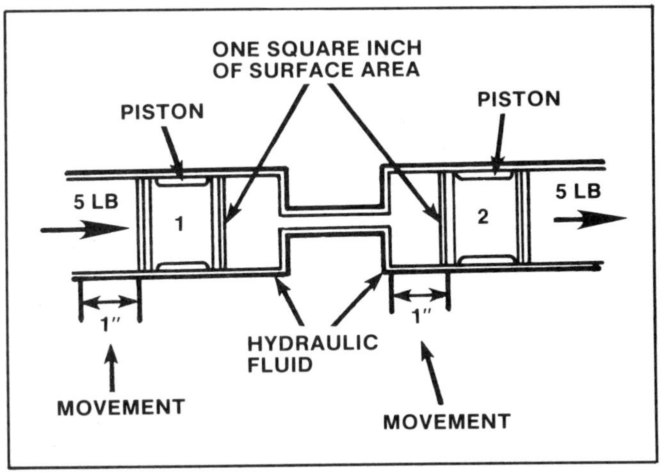

Figure 1-9 Force and movement are transmitted equally.

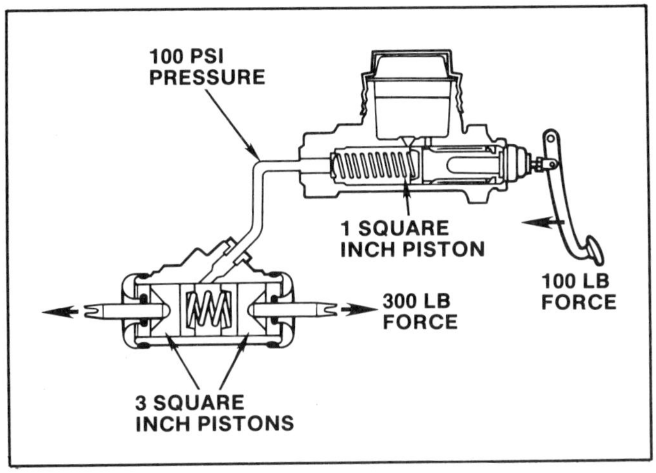

Figure 1-11 Output force increases with piston size.

2 square inches (Figure 1-10). To triple the force of 100 PSI, use a piston with 3 square inches and 300 pounds of output will result (Figure 1-11). No matter what the fluid pressure is, the output force can be increased with a larger piston, though piston travel will decrease proportionately. In actual practice, however, fluid movement in an automotive hydraulic brake system is very slight. In an emergency, when the pedal goes all the way to the floor, the volume of fluid displaced only amounts to about 20 cubic centimeters. About 15 cubic centimeters goes to the front discs and 5 cubic centimeters goes to the rear drums. And even under these conditions, the wheel cylinder and caliper pistons move only very slightly.

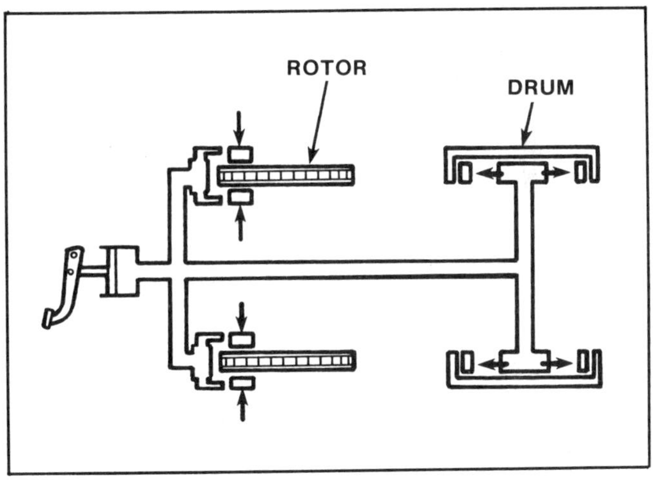

Figure 1-12 Typical hydraulic brake layout

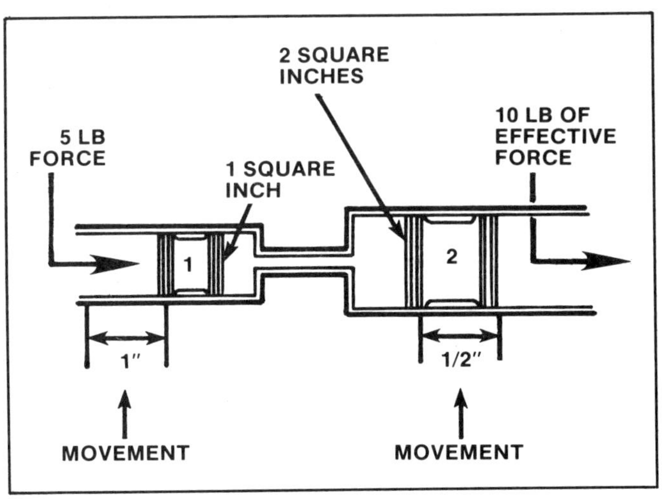

Figure 1-10 Force increases—movement decreases

Of course, the hydraulic system does not stop the car all by itself. In fact, it really just transmits the action of the driver's foot on the brake pedal out to the wheels (Figure 1-12). In the wheels, sets of friction pads are forced against rotors or drums to slow their turning and bring the car to a stop. What happens is that mechanical force (the driver stepping on the brake pedal) is changed into hydraulic pressure, which is changed back into mechanical force (brake shoes and disc pads contacting the drums and rotors). The amount of force acting upon the friction pads and shoes is equivalent to the PSI applied to the pedal multiplied by the area of the piston affected. A force of 25 PSI applied to the pedal times 4 square inches of piston area equals 100 pounds of pressure in the system (Figure 1-13).

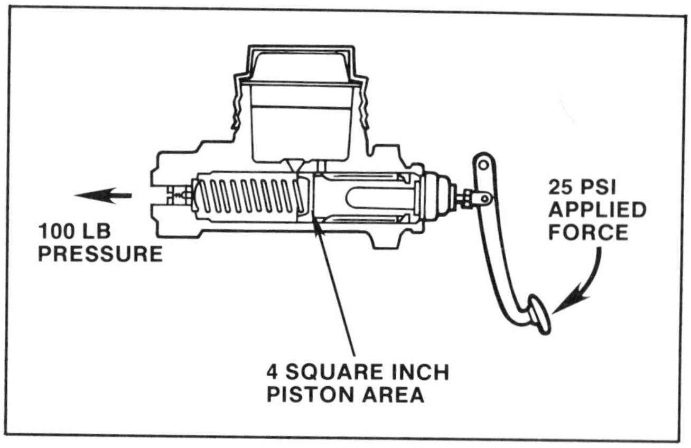

Figure 1-13 Multiplied force in a master cylinder

Advantages of a Hydraulic System

Using the principles of hydraulic pressure, brake manufacturers are able to design brakes with very precise piston strokes and braking pressures. Unlike the old mechanical actuating systems, a hydraulic system provides equal pressure to all four brakes simultaneously, enabling a smooth, straight line stopping action.

Not only does the hydraulic system provide smoother braking than a mechanical system, but it has several other advantages as well. One of these is few moving parts to wear and break. There is actually very little movement in a hydraulic brake system. Less than an ounce of fluid is pumped through the system even during hard braking. The hydraulic actuating system is also a sealed system. Moisture and dirt do not interfere with its operation. Adjustments are seldom necessary and no lubrication is required.

TYPES OF HYDRAULIC BRAKES

As already mentioned, there are two basic types of hydraulic brakes: drum and disc.

Drum Brakes

A drum brake assembly consists of a cast-iron drum, which is bolted to and rotates with the vehicle wheel, and a fixed backing plate to which are attached the shoes and other components—wheel cylinders, automatic adjusters, linkages, and so on (Figure 1-14). Additionally, there might be some extra hardware for parking brakes. The shoes are surfaced with frictional linings, which contact the

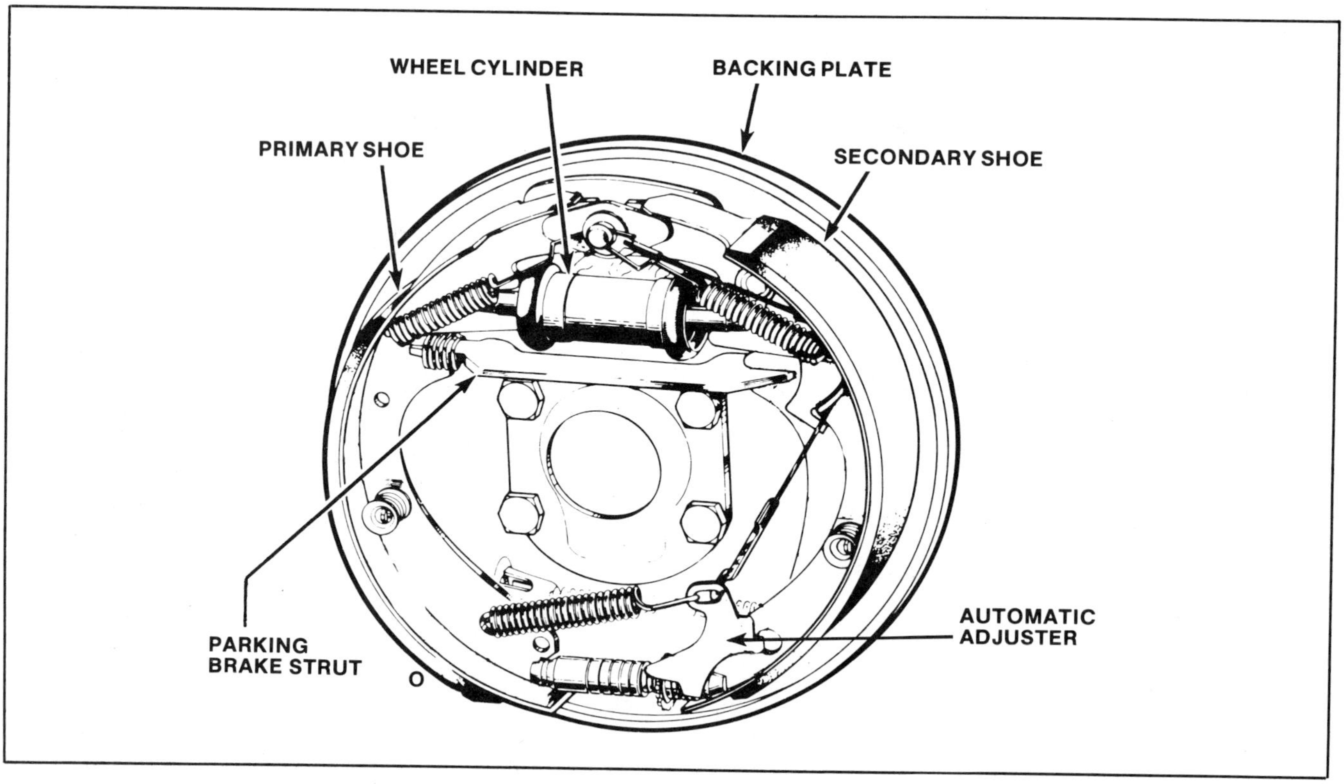

Figure 1-14 Typical drum brake

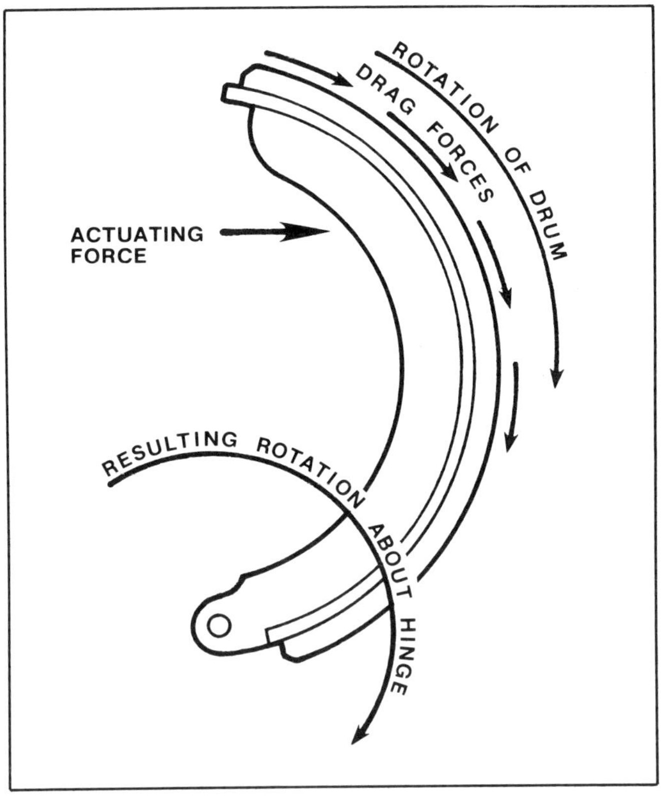

Figure 1-15 How a shoe became self-energized

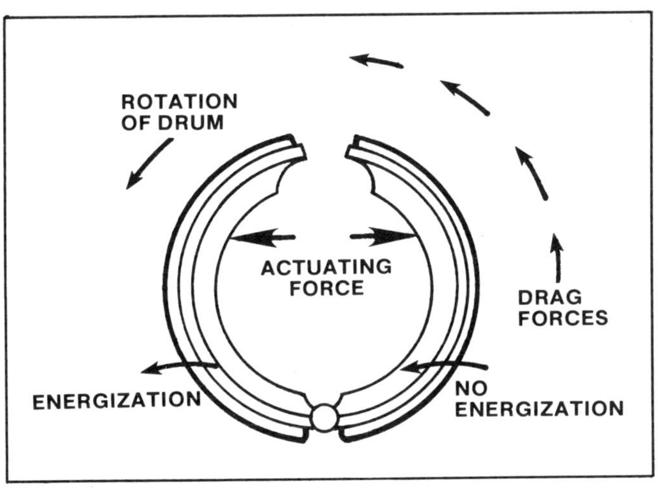

Figure 1-16 Typical non-servo type brake action

inside of the drum when the brakes are applied. The shoes are forced outward, against the action of the return springs, by pistons which are actuated by hydraulic pressure. Just how this pressure is brought to the pistons will be covered later. As the drum rubs against the shoes, the energy of the moving drum is transformed into heat, and this heat energy is passed into the atmosphere.

When the brake shoe is engaged, the frictional drag acting around its circumference tends to rotate it about its hinge point, the brake anchor (Figure 1-15). If the rotation of the drum corresponds to an outward rotation of the shoe, then the drag will pull the shoe tighter against the inside of the drum, and the shoe will be self-energizing.

Non-Servo Drum Brakes. In most brake designs, shoe energization is usually supplemented by another feature called "servo" action (Figure 1-16), but for now let us consider a simple non-servo drum brake and how the energization figures in. Each shoe is separately anchored at its lower end, the heel; the upper end, the toe, is where the actuating force is applied. When the brakes are applied in normal vehicle motion, the forward, or leading, shoe is energized, while the trailing shoe receives no energization. If the brakes are applied when the car is in reverse—remember, the drums are rotating the other way now—the leading and trailing shoes exchange roles; that is, what was previously the trailing shoe now becomes energized. Since most of the time it is the front shoes that receive the energization in a non-servo brake system, they will tend to wear faster.

Servo Type Drum Brakes. A servo brake is termed such because the action of one shoe is governed by an input from the other. There is an anchor, but it only serves as an anchor for the secondary shoe, while the secondary shoe serves as an anchor for the primary shoe.

When the brakes are applied, the primary shoe reacts first—it has a weaker return spring. The shoe lifts off the anchor (which at this point is only acting as a stop) and contacts the drum surface. Since the shoe is pivoted at the bottom, the drag forces tend to rotate it deeper into the drum; it is energized. Now take a look at what happens to the secondary shoe (Figure 1-17): it is being acted upon at its heel, via a connecting link, by the primary shoe. This forces the secondary shoe in the direction of the drum rotation—but it cannot rotate, because its upper end is butted against the fixed anchor. Now, since the drag forces (from contact with the drum) try to rotate it about the anchor, it too becomes energized. In the applied position, the servo brake behaves as if it were one continuous shoe—with the actuating force pushing on one end, and the other end held fast by the fixed anchor—and with drag forces causing the whole affair to be energized.

Servo brakes exist in different configurations. There is usually some form of piston pressure

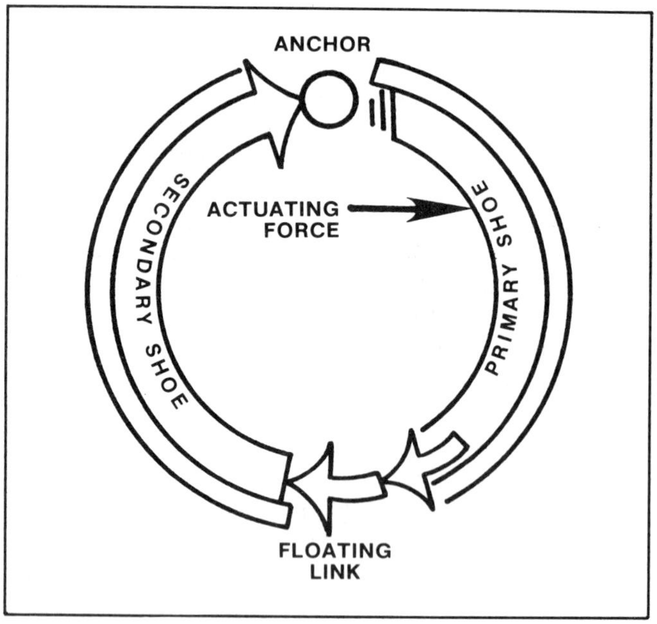

Figure 1-17 Typical servo type brake action

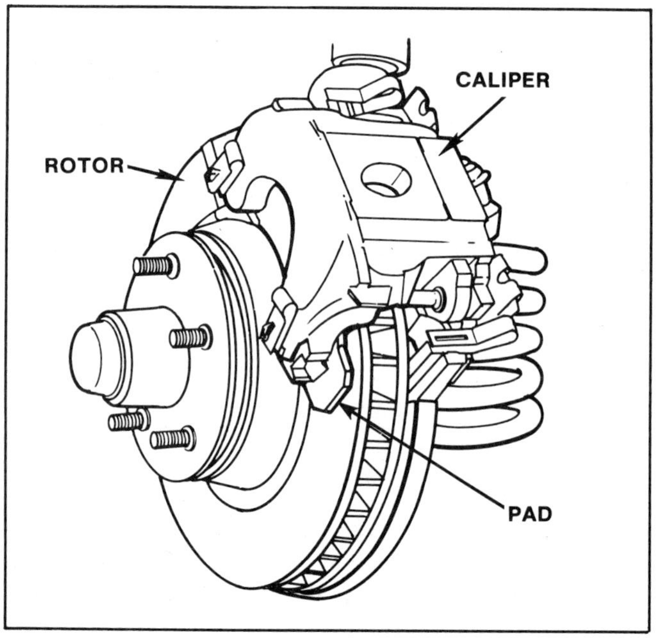

Figure 1-18 A typical disc brake

against the secondary shoe, but this is incidental; the principal of operation remains the same.

Disc Brakes

Disc brakes resemble the brakes on a bicycle: the friction elements are in the form of pads, which are squeezed or clamped about the edge of a rotating wheel. With automotive disc brakes, this wheel is a separate unit, called a *rotor,* inboard of the vehicle wheel (Figure 1-18). The rotor is made of cast iron, and since the pads clamp against both sides of it, both sides are machined smooth. Usually the two surfaces are separated by a finned center section for better cooling. The pads are attached to metal shoes, which are actuated by pistons, the same as with drum brakes. The pistons are contained within a caliper assembly, a housing that wraps around the edge of the rotor. The caliper is kept from rotating by way of bolts holding it to the car's suspension framework.

The caliper is a housing containing the pistons and related seals, springs, and boots, as well as the cylinder(s) and fluid passages, necessary to force the friction linings or pads against the rotor. The caliper resembles a hand in the way it wraps around the edge of the rotor. It is attached to the steering knuckle. Some models employ light spring pressure to keep the pads close against the rotor; in other caliper designs this is achieved by a unique type of seal that allows the piston to be pushed out the necessary amount, then retracts it just enough to pull the pad off the rotor.

Unlike shoes in a drum brake, the pads act perpendicular to the rotation of the disc when the brakes are applied. This effect is different from that produced in a brake drum, where frictional drag actually pulls the shoe into the drum. Disc brakes are said to be non-energized, and so require more force to achieve the same braking effort. For this reason, they are ordinarily used in conjunction with a power brake unit.

Disc brake calipers fall into two categories (Figure 1-19): fixed and moving designs. The latter includes both sliding and floating caliper types.

Disc brakes offer four major advantages over conventional drum brakes:

1. *Resistance to Heat Fade—* Disc brakes are generally more resistant to heat fade during high-speed brake stops or repeated stops. The design of the disc brake rotor exposes more surface to the air and thus dissipates heat more efficiently.
2. *Resistance to Water Fade—* Disc brakes also are affected very little by water fade because the rotation of the rotor tends to throw off moisture. That is, the squeeze of the sharp edges of brake pads clears the surface of water.
3. *More Straight-Line Stops—* Due to their clamping action, disc brakes are less apt to pull and generally produce more uniform straight-line stops.
4. *Automatic Adjusting—* Disc brakes automatically adjust as pads wear.

Because of the advantages that the disc brake offers over the drum brake, most cars manufactured in the United States since the 1960s employ a *combination brake system.* That is, disc brakes are used on the front wheels and drum brakes on the rear wheels.

One of the main reasons for using a combination brake system is the increased popularity of front-wheel drive vehicles. Front engine/rear drive vehicles have their weight load reasonably distributed from front to rear while front-wheel drive vehicles exhibit extreme forward weight bias. The engine, transmission, and other drivetrain components are placed directly over the front wheels and contribute to the forward weight bias. The actual distribution varies from model to model, but in general, it is distributed forward of center so as to give the vehicle good handling and road-holding characteristics. (Light trucks, of course, have more than 80 percent of their weight forward of center when unloaded.) In addition, during braking as much as 75 percent of the vehicle weight can be transferred forward. Because of this, more braking force must be applied to the front wheels than to the rear. (If under hard-braking conditions, the brakes were applied equally to the front and rear, the rear wheels might lock up prematurely, causing skidding.)

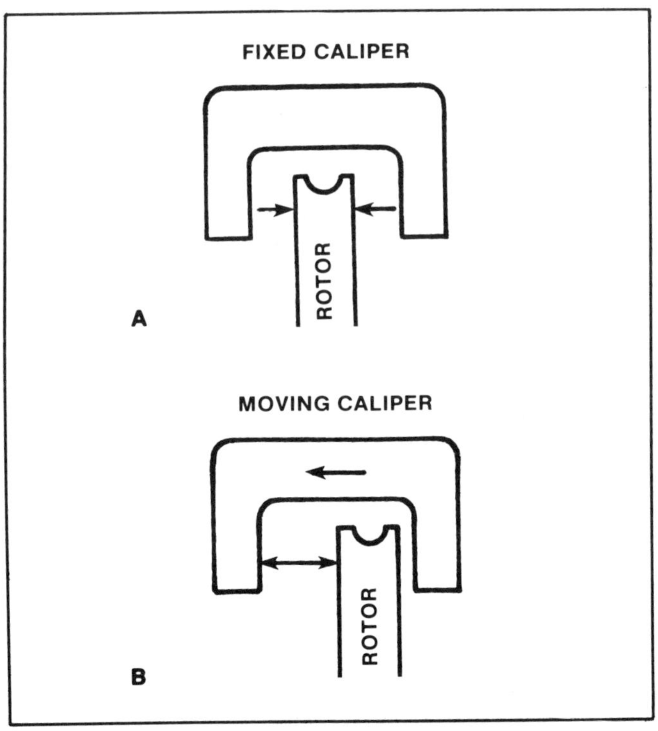

Figure 1-19 Disc brake caliper designs: (A) fixed and (B) moving. The moving caliper includes the sliding and floating types.

Since the front wheels provide at least three-fourths of the braking effort (as evidenced by the way the front end of a car dips down during braking, Figure 1-1), the demands placed upon the rear wheels are correspondingly light. Because of some of the reasons discussed previously, and because most existing parking brake designs still employ drums, rear brakes are often of the drum type even though the front employs discs. Along with the increase of front-wheel drive vehicles, those with the four-wheel disc brake system and the introduction of rear disc parking brakes (see Chapter 12) has surfaced. However, this might change in the future.

OTHER HYDRAULIC BRAKE COMPONENTS

The modern hydraulic brake system (Figure 1-20) is a complex system with many hydraulic components. Since 1967, all cars sold in the United States have had two hydraulic systems. Most rear drive vehicles have one system for the front brakes and one for the rear brakes. Front wheel drive vehicles have a diagonal split system—one front and one rear brake in each system. A tandem-type master cylinder provides pressure to both systems simultaneously. This arrangement ensures that if one system fails, at least 50 percent of the braking power will remain.

Rear wheel lockup has been greatly reduced by the invention of the proportioning valve. The proportioning valve acts to regulate the hydraulic pressure available to the rear wheels. During hard braking, the back end of the car tends to lift as the weight of the car shifts forward. This decreases the pressure holding the rear wheels to the road. By limiting the pressure applied to the rear brakes, the proportioning valve reduces the likelihood of the rear wheels locking and skidding.

Other hydraulic devices such as metering valves, combination valves, and pressure sensors are discussed fully in Chapters 2, 3, and 4. Properly servicing a brake system requires not only knowing how to replace pads or linings, but a thorough knowledge of how the various hydraulic components affect the brake system's operation—an essential for satisfactorily completing a brake job.

Power Brakes. Power brakes on passenger cars are used in conjunction with a basic hydraulic braking system. As described in Chapter 9, all of the systems produced today employ some sort of a booster unit (Figure 1-21), which acts like an extra strong foot on the brake pedal. The booster unit can derive its force from vacuum or from the power

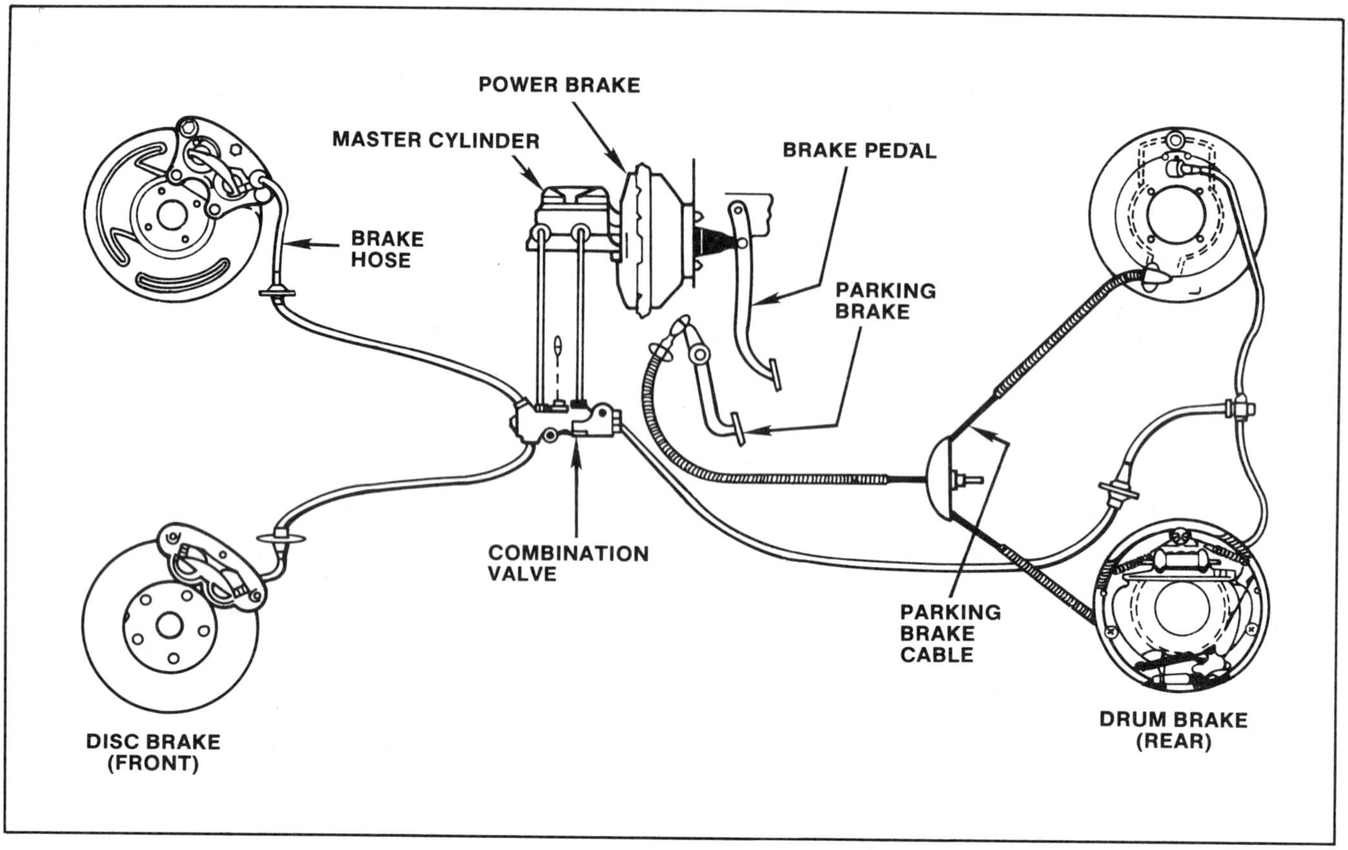

Figure 1-20 Typical brake components of hydraulic system

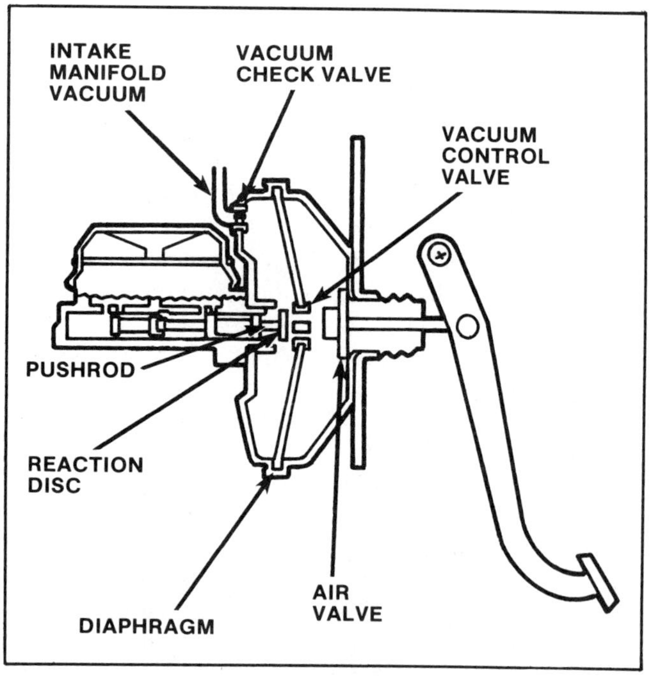

Figure 1-21 Vacuum booster components

steering pump. Either way, it is located between the brake pedal and the master cylinder. When the brake pedal is depressed, the booster unit multiplies the pedal force to the master cylinder. From here on, the system is identical to a basic hydraulic system.

Anti-Lock Brakes. This system is another control arrangement that is used in conjunction with a basic hydraulic braking operation. During hard braking conditions, it is possible for the wheels of a vehicle to lock, resulting in reduced steering as well as braking. On vehicles equipped with the anti-lock brake system, however, an electronic sensor constantly monitors wheel rotation (Figure 1-22). If one or more of the wheels begins to lock, the system opens and closes solenoid valves, cycling up to 10 times per second. This applies and releases the brakes rapidly and repeatedly, so that the front wheels alternately steer and brake. This makes it possible for vehicles equipped with the anti-lock brake system to avoid skidding under conditions that might cause vehicles not so equipped to handle differently.

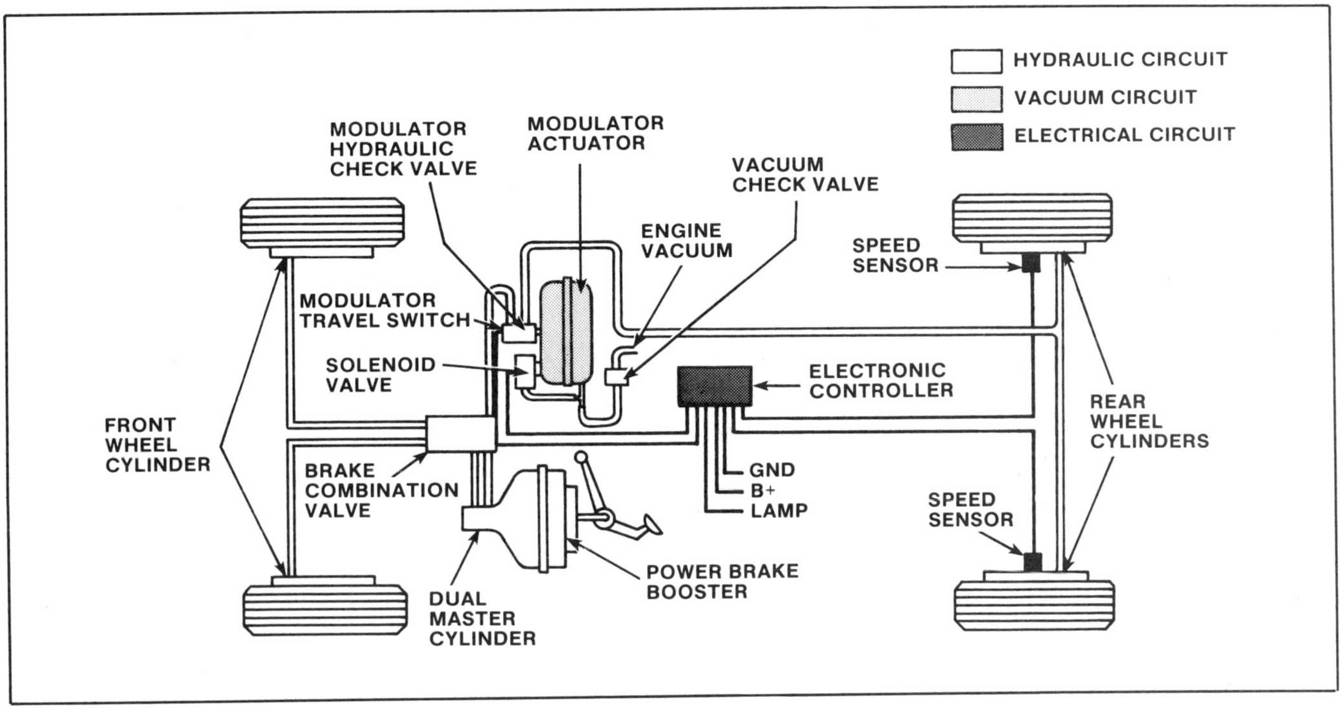

Figure 1-22 Control circuits of a typical anti-lock system.

The anti-lock or anti-skid brake system has a controller that senses rotation at each of the wheels through wheel sensors. It can apply the anti-lock brake system to each of the front wheels independently, to the rear wheels as a pair, or to any combination of these three, as the need arises. Complete details on the operation and servicing of this system can be found in Chapter 10.

Parking Brakes. The rear wheel brakes act to hold the car stationary for parking (Figure 1-23). Parking brakes, however, are not actually a part of the hydraulic brake service system. They are actuated mechanically, rather than by hydraulic pressure. The subject of parking brakes is discussed in Chapter 11.

BRAKE SAFETY PROCEDURES AND TOOLS

Because studies have indicated that exposure to excessive amounts of asbestos dust might be a potential health hazard, OSHA has set maximum limits of asbestos dust to which workers can be exposed. Since most automotive friction materials normally contain a sizable amount of asbestos, it is important that people who handle brake linings and clutch facings understand the nature of the problem and know the precautions to be taken.

Areas where brake work is done should be set aside, if possible, and entrances should be posted with an asbestos exposure sign that reads:

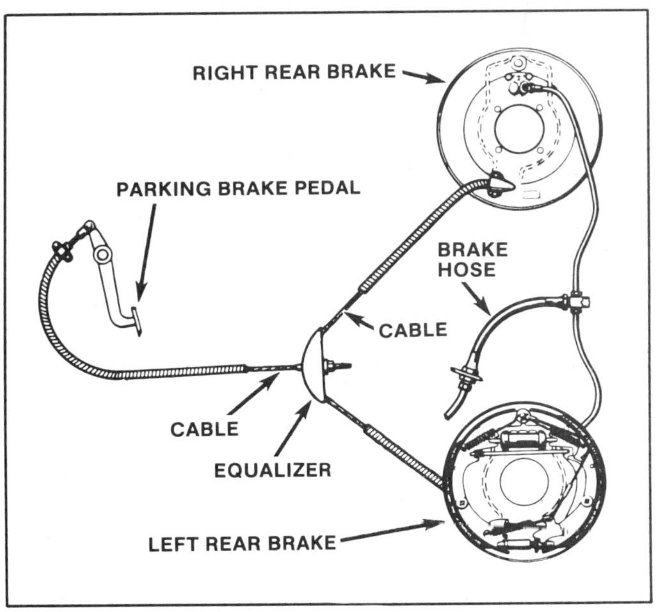

Figure 1-23 Typical integral parking brake

ASBESTOS DUST HAZARD

1. **Avoid breathing dust.**
2. **Wear assigned protective equipment.**
3. **Do not remain in area unless your work requires it.**
4. **Breathing asbestos dust may be hazardous to your health.**

The amount of asbestos in the dust from brake lining wear is normally at an extremely low level because of chemical breakdown during use. If machining of friction material does not take place, simple procedures will minimize exposure. During brake servicing, the mechanic should wear a respirator approved by NIOSH for asbestos dust. It should be worn during all procedures starting with the removal of the wheels and including reassembly.

When removing worn friction materials, remove the accumulated dust in the assemblies with an industrial vacuum cleaner equipped with a high-efficiency filter system (Figure 1-24). If such equipment is not available, dust can be removed with a damp cloth. Do not use compressed air or dry brushing for cleaning unless the assembly is enclosed and properly exhausted.

Whenever possible, purchase friction materials pre-ground and ready for installation. If machining is necessary, the precautions that must be taken are extremely important. This is the operation in brake servicing when exposure to asbestos dust can be at its highest. In addition to the approved respirator (Figure 1-25), there must be local exhaust ventilation such that the worker exposures are maintained at least below the OSHA asbestos standards. If there is any question as to the efficiency of asbestos dust removal by the machine, the manufacturer should be contacted.

Industrial vacuum cleaner bags containing asbestos dust and cloths used for wiping brake assemblies should be sealed in plastic bags and labeled with the following warning printed in letters of sufficient size and contrast to be readily visible and legible:

CAUTION

1. **Contains asbestos fibers.**
2. **Avoid creating dust.**
3. **Breathing asbestos dust may cause serious bodily harm.**

All asbestos waste should be disposed of in accordance with OSHA and EPA asbestos regulations. During removal of vacuum bags, an approved respirator should be worn.

Good housekeeping is essential in a workplace where materials containing asbestos are handled. An industrial vacuum equipped with multiple-

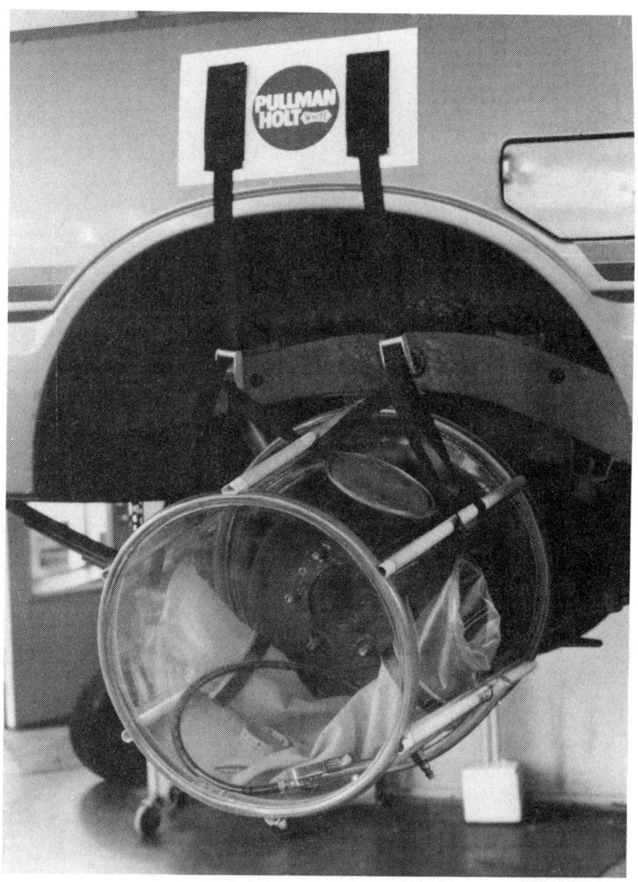

Figure 1-24 Brake bubble enclosure system with asbestos vacuum *(courtesy of Pullman, White, Holt, Inc.)*

Figure 1-25 Wear a respirator when working on brakes.

stage, high-efficiency filters should be used for removing accumulations of asbestos dust and waste. Never use compressed air or dry sweeping for cleaning. Water or other dust suppressants should be applied if brooms are used.

Good personal hygiene practices are important in minimizing asbestos dust exposure.

1. Do not smoke.
2. Wash before eating.
3. Shower after work.
4. Change to work clothes upon arrival at work, and change from work clothes after work. Work clothing should not be taken home. Laundering of asbestos-contaminated clothing should be done so as to prevent the release of airborne asbestos fibers in excess of the exposure limits.

WARNING: Do not breathe asbestos dust.

In the following chapters of this book, the text and procedures contain special notations labeled SHOP TALK, CAUTIONS, AND WARNINGS. Each one is there for a specific purpose. SHOP TALKS give you added information that will help you to complete a particular procedure. CAUTIONS are given to prevent you from making an error that could damage the vehicle. WARNINGS remind you to be especially careful in those areas where carelessness can cause personal injury. The following list contains some general WARNINGS that you should follow when you work on a vehicle.

- Always wear safety glasses for eye protection.
- Use safety stands (Figure 1-26) whenever a procedure requires you to be under the vehicle.
- Be sure that the ignition switch is always in the OFF position, unless otherwise required by the procedure.
- Set the parking brake when working on the vehicle. If you have an automatic transmission, set it in PARK unless instructed otherwise for a specific service operation. If you have a manual transmission, it should be in REVERSE (engine OFF) or NEUTRAL (engine ON) unless instructed otherwise for a specific service operation.
- Operate the engine only in a well-ventilated area to avoid the danger of carbon monoxide.
- Keep yourself and your clothing away from moving parts when the engine is running, especially the fan and belts.

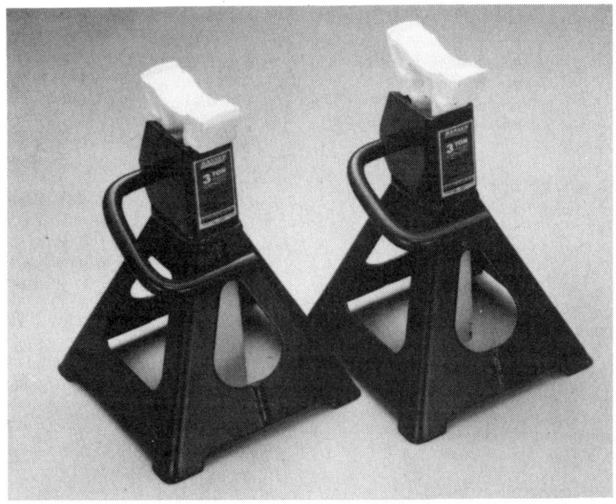

Figure 1-26 Whenever working under a vehicle, use safety stands.

- To prevent serious burns, avoid contact with hot metal parts such as the radiator, exhaust manifold, tail pipe, and catalytic converter.
- Do not smoke while working on the vehicle.
- To avoid injury, always remove rings, watches, loose hanging jewelry, and loose clothing before beginning to work on a vehicle. Tie long hair securely behind the head.
- Keep hands and objects clear of the radiator fan blades. Electric cooling fans can start to operate at any time by an increase in underhood temperatures, even though the ignition is in the OFF position. Therefore, care should be taken to ensure that the electric cooling fan is completely disconnected when working under the hood.
- Special attention should be exercised when working with spring or tension-loaded fasteners and devices such as E-clips, C-clips, snap rings, or circlips because careless removal can cause personal injury.

Brake Tools

Some basic special tools are needed to recondition brakes. These are inexpensive and easy to get. Of course, additional tools can be obtained to make the job easier. Expensive brake machining equipment is not a good investment unless you do a lot of brake work. An alternative is to have an automotive parts jobber do the measuring and machining of drums and rotors, or even the rebuilding of calipers and master cylinders. Many of the tools listed here are illustrated in Figure 1-27.

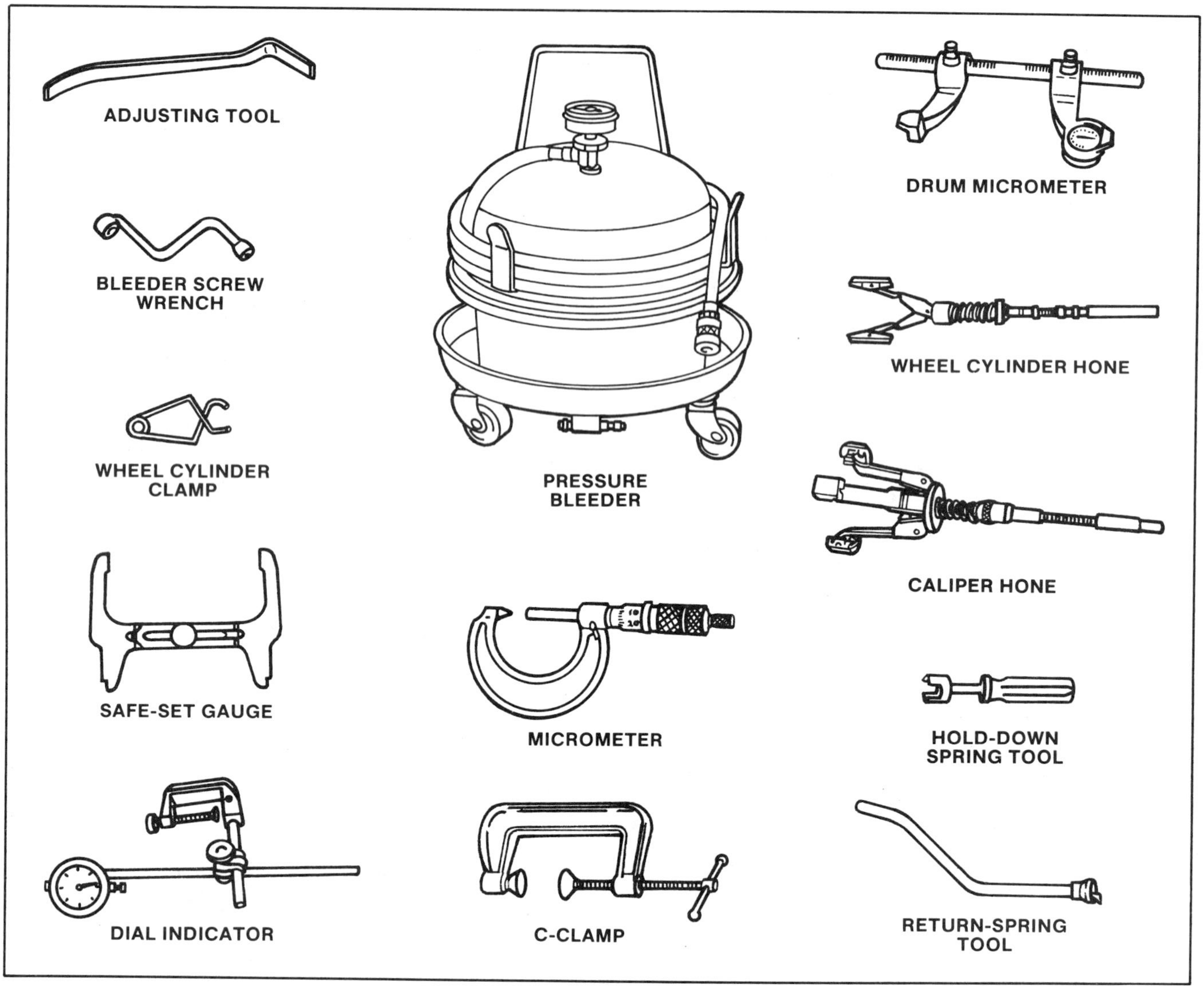

Figure 1-27 Common brake tools

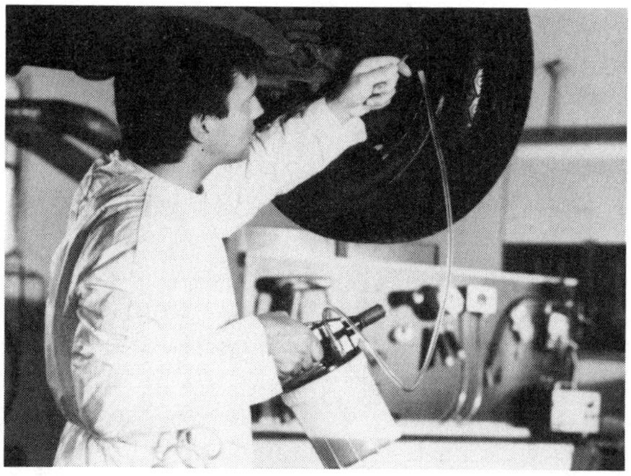

Figure 1-28 Small one-man vacuum bleeder in operation *(courtesy of Vacula Automotive Products Inc.)*

BASIC TOOLS

Return spring remover-installer
Hold-down spring remover-installer
Brake adjuster
Large sized C-clamp
Wheel cylinder hone
Electric drill

MEASURING TOOLS

Micrometer for checking brake rotor thickness
Brake drum micrometer
Dial indicator and bracket

ADDITIONAL BRAKE TOOLS

Wheel cylinder clamps
Bleeder screw wrenches
Brake line wrenches

Drum and shoe gauges
Drum brake backing plate washer or industrial vacuum cleaner
Brake pressure bleeder unit (Figure 1-28)
Brake drum and rotor lathe
Flaring tool
Bleeder wrench
Brake pedal depressor (Figure 1-29)
Seal installer
Piston installer
Pressure bleeder adapter
Combination valve bleeder
Steering wheel holder
Brake vacuum

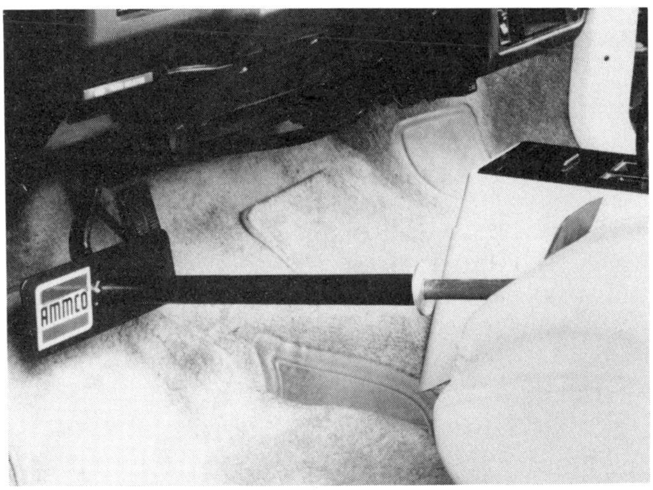

Figure 1-29 Brake pedal depressor *(courtesy of Hennessy Industries, Inc.)*

SERVICE MANUALS AND MANUFACTUER CATALOGS

Two of the best tools of a brake technician are the vehicle's service manual and the aftermarket brake catalogs and data sheets. The main source of all repair and specifications for any car, van, or truck is the vehicle manufacturer. All service manuals have a specific section in the manual for that vehicle's brake (Figure 1-30).

Many of the larger aftermarket brake manufacturers and suppliers have excellent catalogs (Figure 1-31) that are a fine source of information for the brake technician. Some of the companies also provide updated service bulletins on their brake products (Figure 1-32).

— ASE CERTIFICATION —

Just as doctors, nurses, accountants, electricians, and other professionals are licensed or certified in order to practice their profession, so the mechanic can be certified. Certification protects the general public and the practitioner or professional. It assures the general public and the prospective employer that certain minimum standards of performance have been met. Many employers now expect their mechanics or technicians

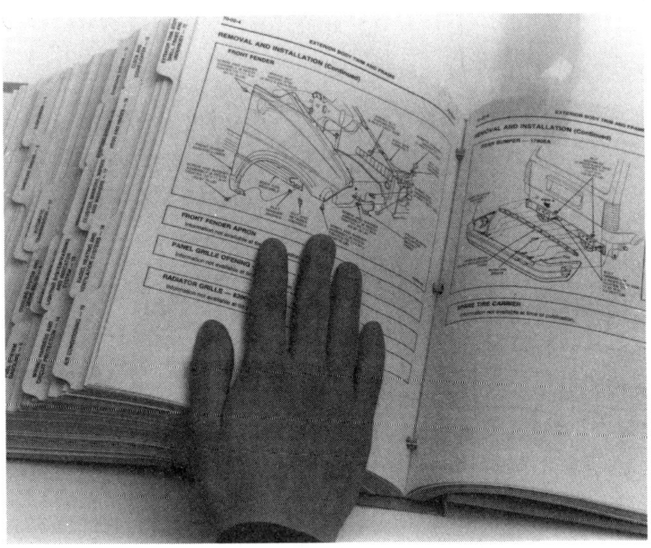

Figure 1-30 Typical page from a manufacturer's vehicle service manual

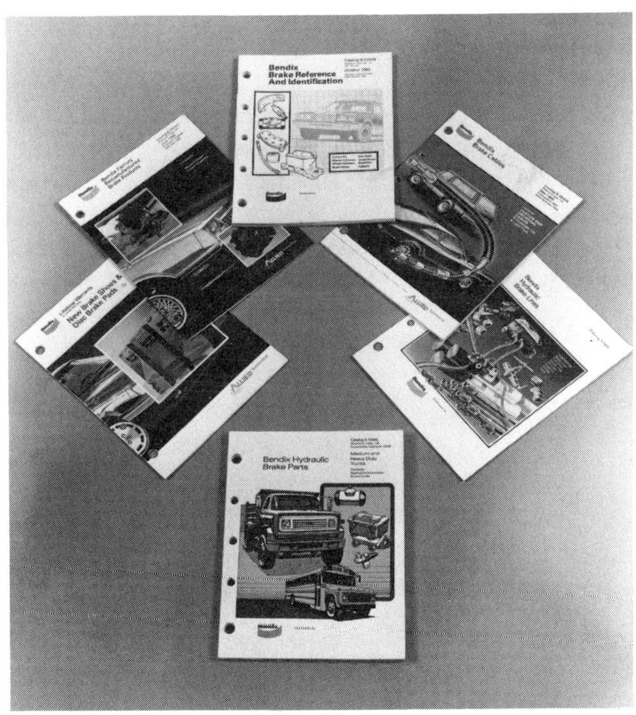

Figure 1-31 Typical brake aftermarket furnished by the manufacturers

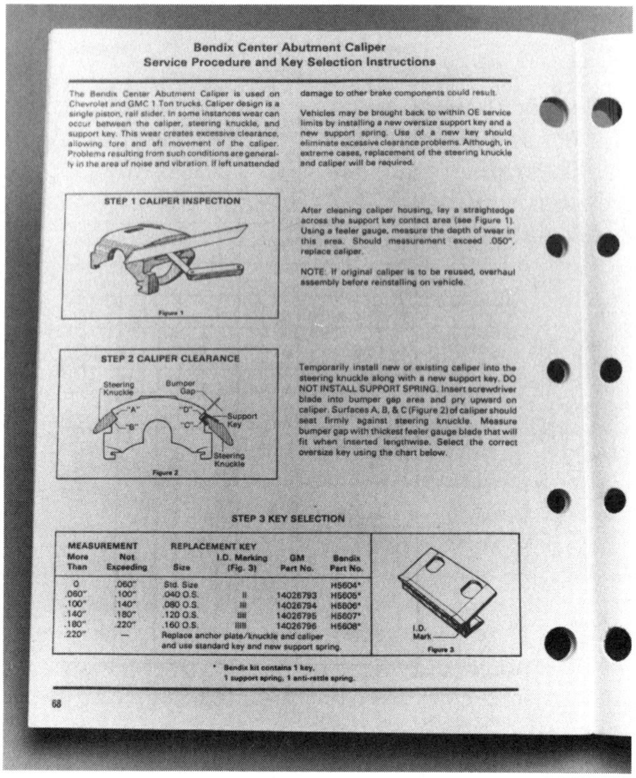

Bendix Center Abutment Caliper
Service Procedure and Key Selection Instructions

The Bendix Center Abutment Caliper is used on Chevrolet and GMC 1 Ton trucks. Caliper design is a single piston, rail slider. In some instances wear can occur between the caliper, steering knuckle, and support key. This wear creates excessive clearance, allowing fore and aft movement of the caliper. Problems resulting from such conditions are generally in the area of noise and vibration. If left unattended damage to other brake components could result.

Vehicles may be brought back to within OE service limits by installing a new oversize support key and a new support spring. Use of a new key should eliminate excessive clearance problems. Although, in extreme cases, replacement of the steering knuckle and caliper will be required.

STEP 1 CALIPER INSPECTION

Figure 1

After cleaning caliper housing, lay a straightedge across the support key contact area (see Figure 1). Using a feeler gauge, measure the depth of wear in this area. Should measurement exceed .050", replace caliper.

NOTE: If original caliper is to be reused, overhaul assembly before reinstalling on vehicle.

STEP 2 CALIPER CLEARANCE

Steering Knuckle
Bumper Gap
"A"
"B"
"C"
"D"
Support Key
Steering Knuckle

Figure 2

Temporarily install new or existing caliper into the steering knuckle along with a new support key. DO NOT INSTALL SUPPORT SPRING. Insert screwdriver blade into bumper gap area and pry upward on caliper. Surfaces A, B, & C (Figure 2) of caliper should seat firmly against steering knuckle. Measure bumper gap with thickest feeler gauge blade that will fit when inserted lengthwise. Select the correct oversize key using the chart below.

STEP 3 KEY SELECTION

MEASUREMENT More Than	Not Exceeding	REPLACEMENT KEY Size	I.D. Marking (Fig. 3)	GM Part No.	Bendix Part No.
0	.060"	Std. Size			H5604*
.060"	.100"	.040 O.S.	II	14026793	H5605*
.100"	.140"	.080 O.S.	III	14026794	H5606*
.140"	.180"	.120 O.S.	IIII	14026795	H5607*
.180"	.220"	.160 O.S.	IIIII	14026796	H5608*
.220"	—	Replace anchor plate/knuckle and caliper and use standard key and new support spring.			

I.D. Mark

Figure 3

* Bendix kit contains 1 key, 1 support spring, 1 anti-rattle spring.

68

Figure 1-32 Brake updates furnished by the manufacturers

to be certified. The certified technician is recognized as a professional by the public, employer, and peers. For this reason, the certified technician usually receives higher pay than the noncertified operator.

Mechanics can get certified in one or more technical areas by taking and passing a written mechanic certification test of either 40 or 80 questions on diagnosis and repair techniques. The National Institute for Automotive Service Excellence (ASE) offers a voluntary certification program that is recommended by the major vehicle manufacturers in the United States. This program has certification tests that fall into eight categories or specialties as follows:

1. Engine repairs (80 questions)
2. Engine performance (80 questions)
3. Electrical systems (40 questions)
4. Suspension and steering—front end (40 questions)
5. Brakes (40 questions)
6. Manual drivetrain and rear axle (40 questions)
7. Automatic transmissions/transaxle (40 questions)
8. Heating and air-conditioning (40 questions)

Figure 1-33 Certified ASE Automobile Technician shoulder patch

The technician must have at least two years of experience in order to take the test and become certified. Those who wish to be certified as a Specialist Technician need to pass only the examination that pertains to the area of specialty. To be certified as a General Technician requires passing examinations in all eight areas. A nominal fee is charged for each test taken.

A technician who passes one examination receives an Automobile Technician shoulder patch (Figure 1-13). A Master Automobile Technician meeting minimum standards in all eight categories receives the shoulder patch shown in Figure 1-14.

Figure 1-34 Certified ASE Master Automobile Technician shoulder patch

To help prepare for the Suspension and Steering program, the test questions at the end of each chapter are similar in design and content to that used by the ASE. For further information on the ASE certification program, write: National Institute for Auto Service Excellence, 1920 Association Drive, Reston, Virginia 22091.

REVIEW QUESTIONS

Review the information in this chapter by answering the following questions.

1. Kinetic energy is
   a. stationary energy.
   b. heat energy.
   c. energy in motion.
   d. static energy.
2. Friction of motion
   a. depends only on the type of materials in contact.
   b. produces heat.
   c. prevents heat dissipation.
   d. none of the above.
3. A sign of brake fade is
   a. a hardened rotor and drum.
   b. glazed pad and shoe linings.
   c. both a and b.
   d. neither a nor b.
4. When force is applied to a liquid confined in a system, which of the following is *not true?*
   a. Pressure is produced.
   b. Equal pressure is exerted on all equal areas.
   c. The liquid is compressed.
   d. The pressure acts at right angles to all surfaces in the system.
5. Which of the following affects a vehicle's braking ability?
   a. vehicle weight.
   b. radius of the drum and wheel
   c. coefficient of friction of the braking surfaces
   d. all of the above
6. Which of the following is not an advantage of disc brakes over conventional drum brakes?
   a. resistance to water fade
   b. more straight-line stops
   c. less rear wheel lockup
   d. automatic adjusting
7. Mechanic A replaced worn semi-metallic linings with asbestos linings. Mechanic B replaced worn semi-metallic linings with non-asbestos linings. Who is right?
   a. Mechanic A
   b. Mechanic B
   c. both A and B
   d. neither A nor B
8. Internal drum brake shoes are expanded against a rotating drum in which type of brake?
   a. wagon brake
   b. disc brake
   c. band brake
   d. drum brake
9. The caliper in a disc brake
   a. clamps the edge of a rotor.
   b. tightens around a drum.
   c. presses against a rim.
   d. houses the pistons.
10. The science of transferring pressure with liquids is called
    a. hydraulics.
    b. kinetics.
    c. coefficient of friction.
    d. all of the above.
11. Which of the following factors affect a vehicle's braking ability?
    a. vehicle weight
    b. vehicle speed
    c. none of the above
    d. all of the above
12. When 100 PSI of pressure is exerted on a piston with a surface area of 1 square inch, the output force of the piston is
    a. 50 pounds.
    b. 100 pounds.
    c. 200 pounds.
    d. none of the above.
13. A hydraulic device that limits the pressure available to the rear wheels is
    a. a master cylinder.
    b. a proportioning valve.
    c. a wheel cylinder.
    d. none of the above.
14. Mechanic A uses compressed air to remove asbestos dust from work areas. Mechanic B uses an industrial vacuum equipped with filters to remove asbestos dust from work areas. Who is right?
    a. Mechanic A
    b. Mechanic B
    c. both A and B
    d. neither A nor B

15. The booster unit in a power brake system
    a. monitors wheel rotation.
    b. houses the pistons and related seals.
    c. multiplies the pedal force to the master cylinder.
    d. employs light spring pressure to keep the brake pads close against the rotor.
16. The amount of friction between two surfaces is expressed as
    a. coefficient of friction.
    b. heat dissipation.
    c. frictional contact surface.
    d. brake fade.
17. Which of the following is a necessary characteristic of brake lining?
    a. resistance to heat fade
    b. resistance to water fade
    c. good speed sensitivity
    d. all of the above
18. Which of the following is a type of hydraulic brake?
    a. drum
    b. disc
    c. both a and b
    d. neither a nor b
19. What type of drum brake has the action of one shoe governed by an input from the other?
    a. servo
    b. nonservo
    c. disc
    d. all of the above
20. Rear wheel lockup has been greatly reduced by the invention of
    a. power brakes.
    b. the proportioning valve.
    c. solenoid.
    d. the diagonal split system.
21. With automotive disc brakes, pads are squeezed or clamped against the surfaces of a rotating wheel called a
    a. rotor.
    b. disc.
    c. cylinder.
    d. drum.
22. Which of the following is *not* actually a part of a typical hydraulic system?
    a. combination valve
    b. master cylinder
    c. brake pedal
    d. parking brake
23. Which of the following is a basic brake tool?
    a. wheel cylinder hone
    b. dial indicator and bracket
    c. return spring remover-installer
    d. all of the above

CHAPTER 2

HYDRAULIC SYSTEM—COMPONENT OPERATIONS

OBJECTIVES

Upon completing this chapter, you should be able to

- Explain the principles of the dual braking system.
- Identify the differences between a front/rear split system and a diagonally split system.
- Explain the importance of the dual master cylinder to the hydraulic system, and describe the function of each of its main components.
- Recognize other master cylinders and their applications, including the combination brake, dual-bore, vertical, step-bore dual, quick take-up, Bendix mini, and aluminum/composite master cylinders.
- Explain the importance of hydraulic tubes and hoses, wheel cylinders, and calipers to the hydraulic system.
- Describe the function of the various safety switches and valves, including the brake warning light, metering and proportioning valves, combination valves, stoplight switch, and brake fluid level sensor.
- Identify the three basic types of hydraulic brake fluid, as well as the precautions that should be taken to minimize brake fluid contamination.

The hydraulic system, as previously mentioned, controls a car's braking action and consists of a master cylinder, fluid lines and hoses, control valves, and calipers and/or wheel cylinders. When the brake pedal is depressed, the master cylinder creates pressure that is transmitted to the calipers and wheel cylinders via lines and hoses (Figure 2-1A). Sliding rubber seals contain the fluid and prevent leakage.

Return springs in the master cylinder help the pedal return to its unapplied position (Figure 2-1B). Other valves, such as the metering valve, proportioning valve, or combination valve, regulate the application of fluid pressure to the calipers and wheel cylinders to achieve efficient, balanced braking. With drum brakes, residual pressure valves can be used to retain a small amount of pressure in the lines and wheel cylinders.

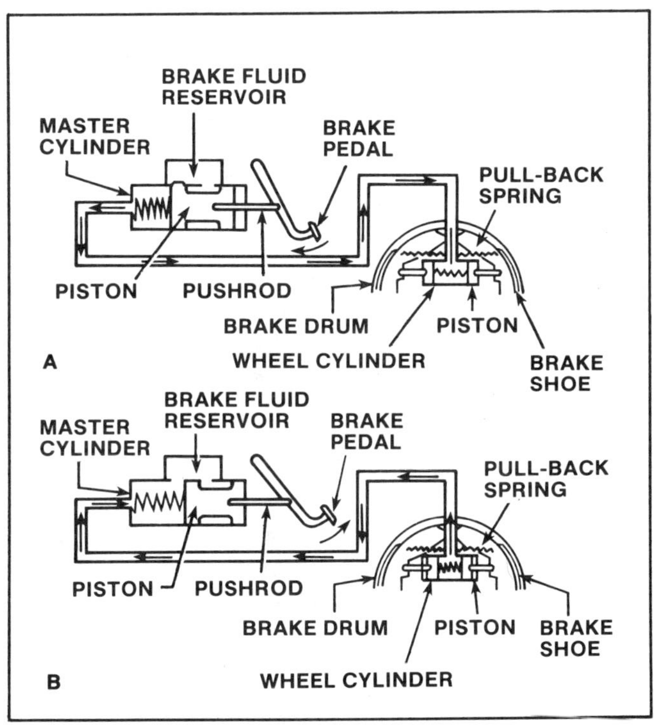

Figure 2-1 (A) Applied and (B) released brake pedal and the movement of components by hydraulic pressure

SINGLE BRAKING SYSTEMS

With very few exceptions, the single circuit hydraulic system (Figure 2-2) was used on cars made through 1966. It had a master cylinder with one piston that operated the brakes on all four wheels. Later, its use was confined to trucks, and industrial and construction equipment.

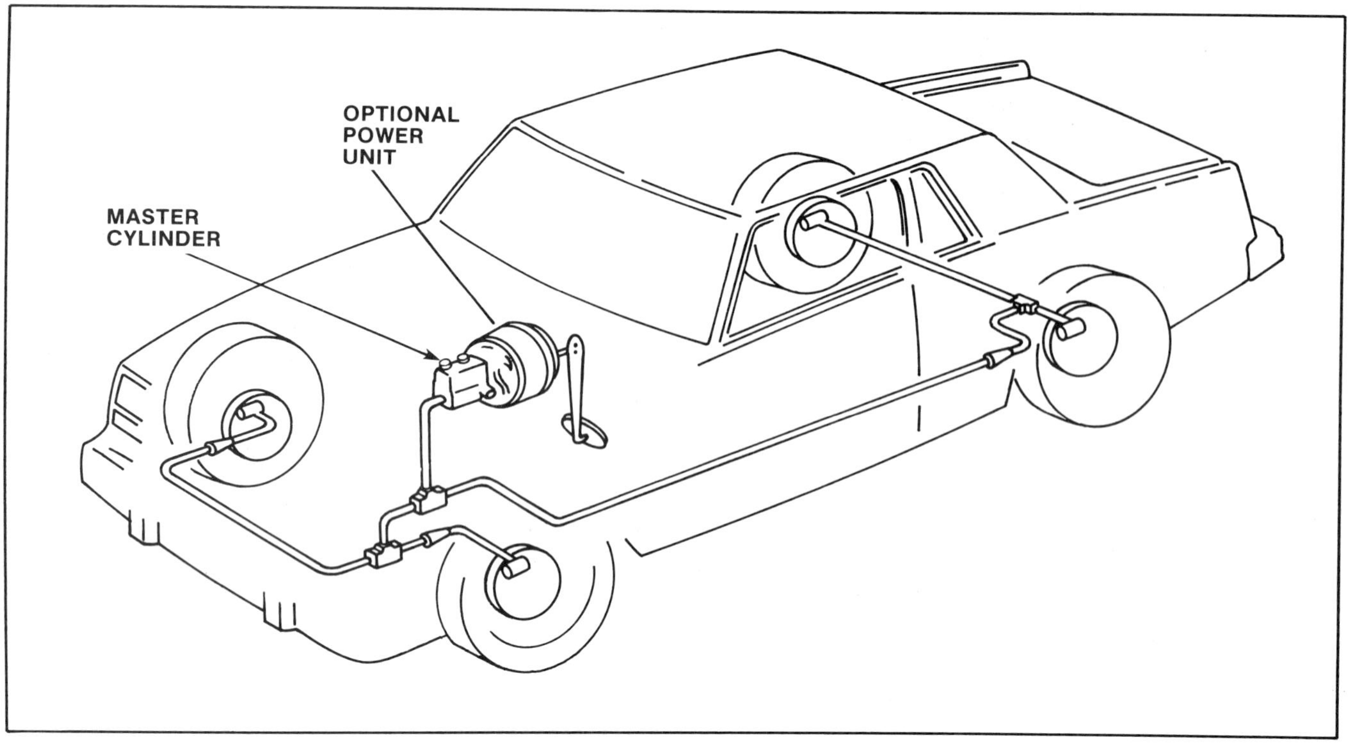

Figure 2-2 Single system braking

DUAL BRAKING SYSTEMS

The dual system differs from the single system by employing a *tandem* master cylinder, which is essentially two master cylinders that are formed by installing two separate pistons and fluid reservoirs into one cylinder bore. Dual brake lines split the wheel cylinders into two groups; each are actuated by a separate master cylinder piston.

Whether they were called dual, split, or safety systems, these braking systems were introduced in 1962 on the Cadillac and the Rambler, and in 1963 on the Studebaker Lark. Since 1967, Federal law has required that all cars be equipped with two separate brake systems. If one circuit fails, the other will provide enough braking power to safely stop the car.

The dual system includes a red warning lamp on the instrument panel and, to activate it, a pressure differential switch connected to both sides of the system. The switch is sensitive to any loss of hydraulic pressure that might result from a leak in either side of the system and alerts the driver by turning on the warning lamp. With a few braking exceptions, low-fluid warning switches located in the master cylinder reservoir can be used in place of the pressure differential switch. On some other vehicles the same lamp performs the function of indicating that the parking brake is applied.

Front/Rear Split System

From the beginning of dual systems in domestic cars, the hydraulic circuits have been separated front and rear (Figure 2-3). Both front wheels are on one hydraulic circuit and both rear wheels on another. If a failure of a rear (secondary) brake system should occur, initial brake pedal movement causes the unrestricted secondary piston to bottom in the master cylinder bore. Primary piston movement displaces hydraulic fluid in the primary section of the dual master cylinder to actuate the front brake system.

Should the front (primary) brake system fail, initial brake pedal movement causes the unrestricted primary piston to bottom out against the secondary piston. Continued downward movement of the brake pedal moves the secondary piston to displace hydraulic fluid in the rear brake system to actuate the rear brakes.

The increased pedal travel and the increased pedal effort required to compensate for the loss of the failed portion of the brake system provides a warning that a partial brake system failure has occurred. When the ignition switch is turned on, a brake warning light on the instrument panel provides a visual indication that one of the dual brake systems has become inoperative.

Should a failure of either the front or rear brake hydraulic system occur, the hydraulic fluid pressure differential resulting from pressure loss of the failed brake system forces the valve toward the low pressure area to light the brake warning lamp.

The axle-to-axle brake system (as the front/rear split arrangement is frequently called) is still used on several larger conventional front engine/rear drive vehicles.

Diagonally Split System

This system operates on the same principles as conventional front and rear split systems. It uses primary and secondary master cylinders that move simultaneously to exert hydraulic pressure on their respective systems.

The hydraulic brake lines on this system, however, have been diagonally split front to rear (left front to right rear and right front to left rear) in place of separate lines to the front and rear wheels (Figure 2-4). The circuit split can occur within the master cylinder, or externally at the proportioning valves or pressure differential switch. A dual master cylinder pressurizes the circuits. Since both rear brakes and both front brakes are on separate circuits, a proportioning or metering valve must be used in each circuit if required in the system.

In the event of a system failure this would cause the remaining good system to do all the braking on one front wheel and the opposite rear wheel, thus maintaining 50 percent of the total braking force. The hydraulic pressure loss would result in a pressure differential in the system and cause a warning light on the dashboard to glow as in front and rear split systems.

During a half-system stop, with only one front brake operational, the vehicle would have a tendency to pull toward the direction of the functional

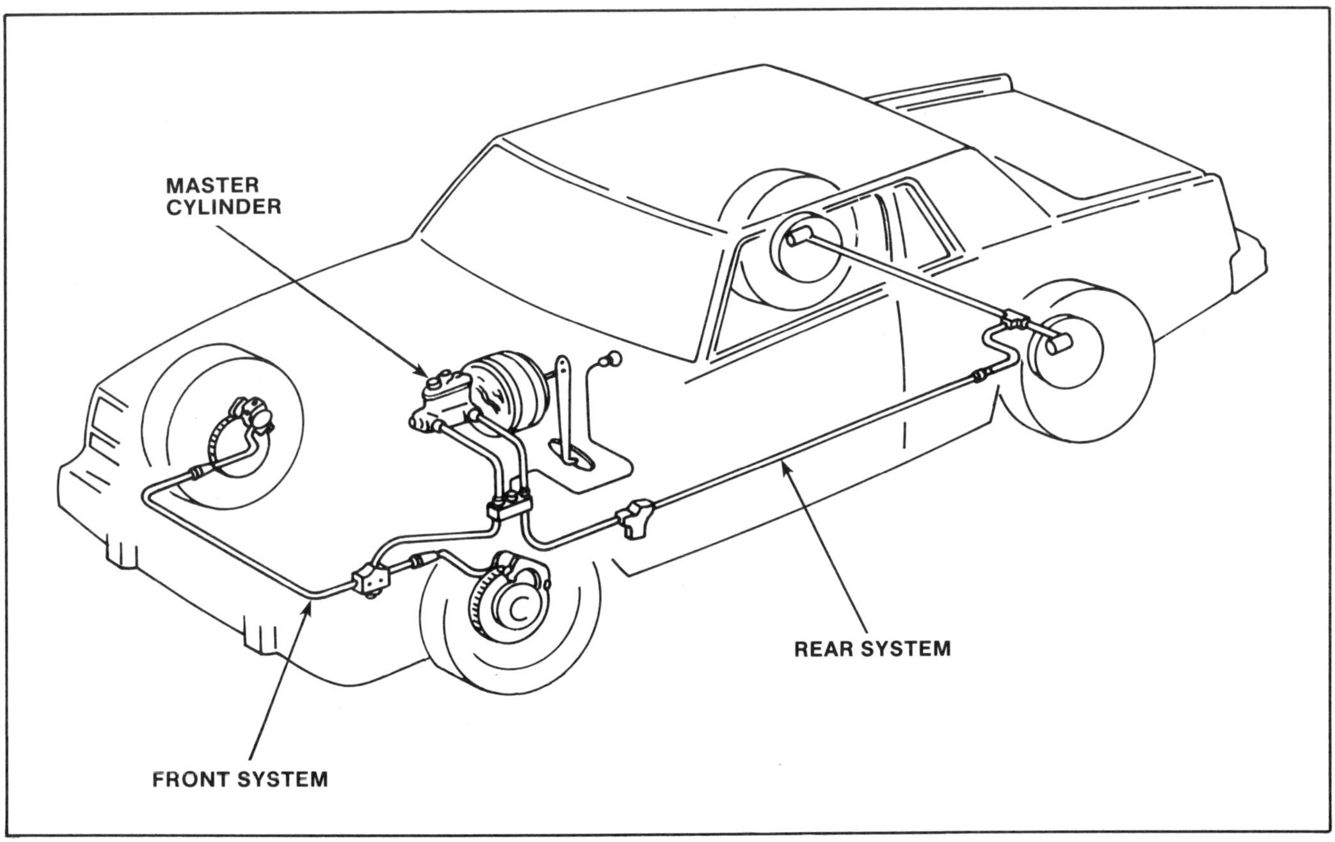

Figure 2-3 Front/rear split hydraulic system

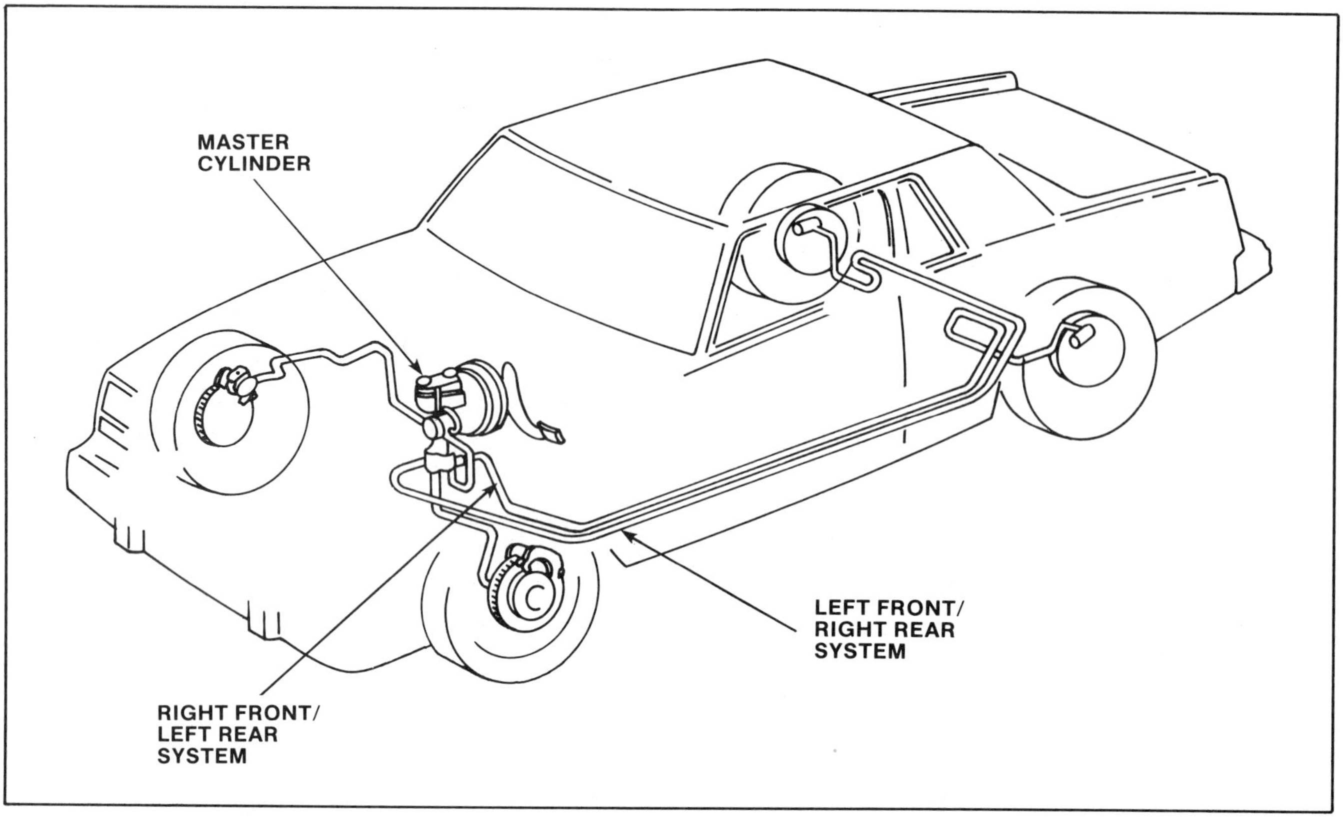

Figure 2-4 Dual diagonally split hydraulic system

brake. To compensate for this problem, the front-end geometry is designed with a negative scrub radius. This also allows greater vehicle control in the event of a front tire blowout.

Most front-wheel drive vehicles use a diagonal split brake system as opposed to the axle-to-axle system. Due to the extreme forward weight bias of front-wheel drive vehicles, as mentioned in Chapter 1, the rear brakes provide less assistance in braking. Because of this, a front-wheel drive vehicle does not stop well with the rear brakes alone, a situation that could arise with an axle-to-axle system. Incorporating the diagonal split brake system on front-wheel drive vehicles assures safe stops and controlled braking under half-system operation.

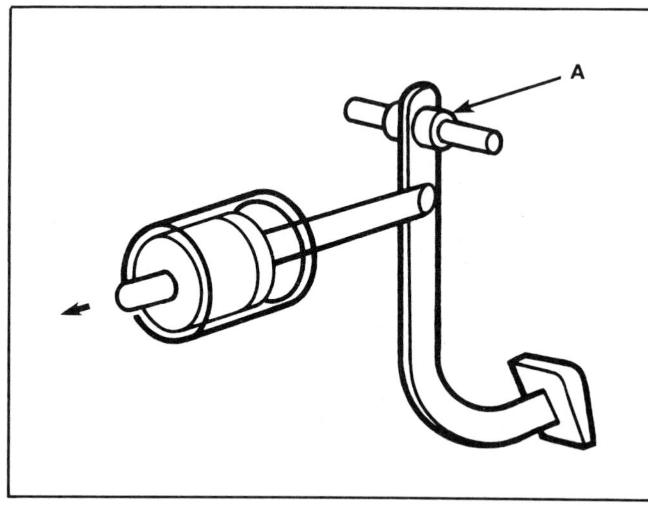

Figure 2-5 Brake pedal action

BRAKE PEDALS

The brake pedal is where the brake's hydraulic system gets its start. When the brake pedal is depressed, force is applied to the master cylinder. On a basic hydraulic brake system (where there is no power assist), the force applied is transmitted mechanically. As the pedal in Figure 2-5 pivots at point A, the force applied to the brake pedal is multiplied mechanically. The force that the pushrod applies to the master cylinder piston is, therefore, much greater than the force applied to the brake pedal. As was stated earlier in the chapter, this force is transmitted by hydraulic pressure from the master cylinder to each of the wheel cylinders. This pressure is maintained as long as the brake pedal remains depressed. A brake stoplight switch is usually mounted on the pedal and whenever the pedal is depressed the rear brake lights are illuminated.

Brake pedals are of either the suspended or frame-mounted type. Since manual brakes need more force multiplication by leverage than power brakes to generate sufficient hydraulic pressure, the foot pad is high off the floor (Figure 2-6). Power brakes, as mentioned in Chapter 1, help the driver to produce the necessary hydraulic pressure and, therefore, the leverage required is less. The power brake foot pad, as a result, might be closer to the floor (Figure 2-7).

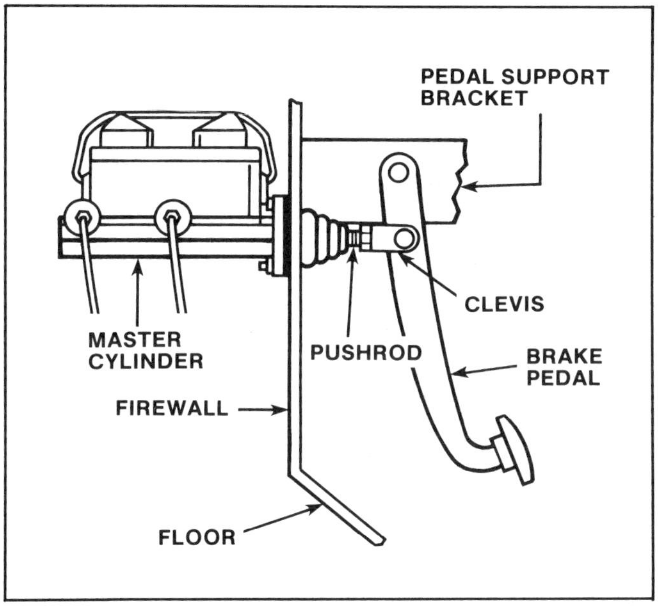

Figure 2-6 Suspended pedal—manual brake

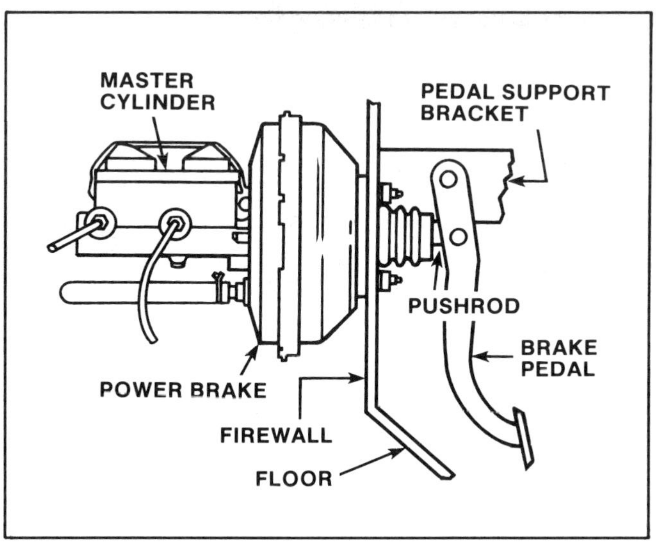

Figure 2-7 Suspended pedal—power brake

In many cases, master cylinders are tilted on the firewall to achieve the desired mechanical relationship with the pedal linkage.

Some older model passenger cars and light trucks use a frame-mounted brake pedal as shown in Figure 2-8. The adjustable pushrod length is pre-set at the factory and usually does not need to be changed. The length should provide 1/4 to 1/2 inch (Figure 2-9) free-play at the foot pad from the fully released position to close the master cylinder replenishing port(s). Some movement or a spurt in the brake fluid in the reservoir should be noticeable when the ports close.

MASTER CYLINDERS

The heart of the hydraulic brake system is the master cylinder. It converts the mechanical force of the driver's foot on the brake pedal to hydraulic pressure. The master cylinder has a bore that contains a piston assembly and return spring. Seals on the piston prevent fluid leakage. A cup seal on the forward end keeps the brake fluid ahead of the piston when it is put under pressure. Although master cylinders differ in number of pistons, reservoir design, and integrated hydraulic components, the operation of all master cylinders is basically the same.

Master cylinder construction will generally be cast iron (Figure 2-10A) with an integral fluid res-

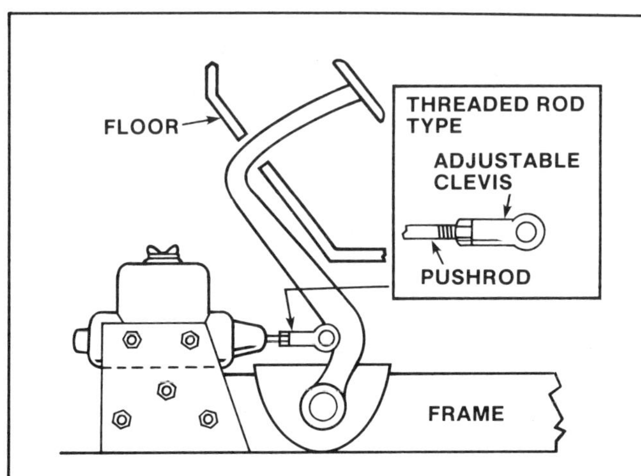

Figure 2-8 Frame-mounted pedal

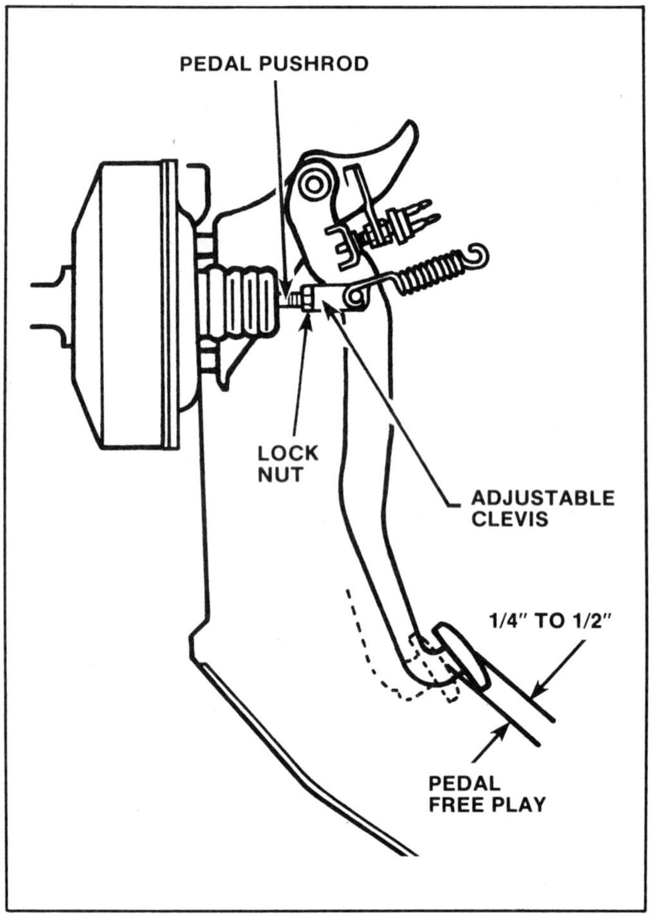

Figure 2-9 Firewall-mounted booster showing free pedal play

ervoir or weight-saving aluminum with a separate molded nylon or fiberglass reinforced plastic reservoir. Aluminum body master cylinders feature an anodized body (Figure 2-10B) to protect against corrosion and to extend bore life.

CAUTION: It is recommended that aluminum master cylinders not be rebuilt if pitting or scoring of the cylinder bore is evident. A new unit should be installed.

SINGLE SYSTEM MASTER CYLINDERS

To understand how a master cylinder works, it is necessary to examine a single system master cylinder. Single piston master cylinders were used on most passenger cars before 1967. A cross section of a typical single piston cylinder is shown in Figure 2-11. The cylinder and the fluid reservoir are an integral unit, usually made of cast iron. A vent port and compensating or replenishing port pass from the reservoir into the piston chamber. The chamber contains a piston, a return spring, and a residual pressure check valve. On the brake pedal end of the cylinder, a pushrod connects the cylinder to the power booster or brake pedal clevis. The opposite end of the cylinder contains the threaded fluid outlet to which the hydraulic line is connected.

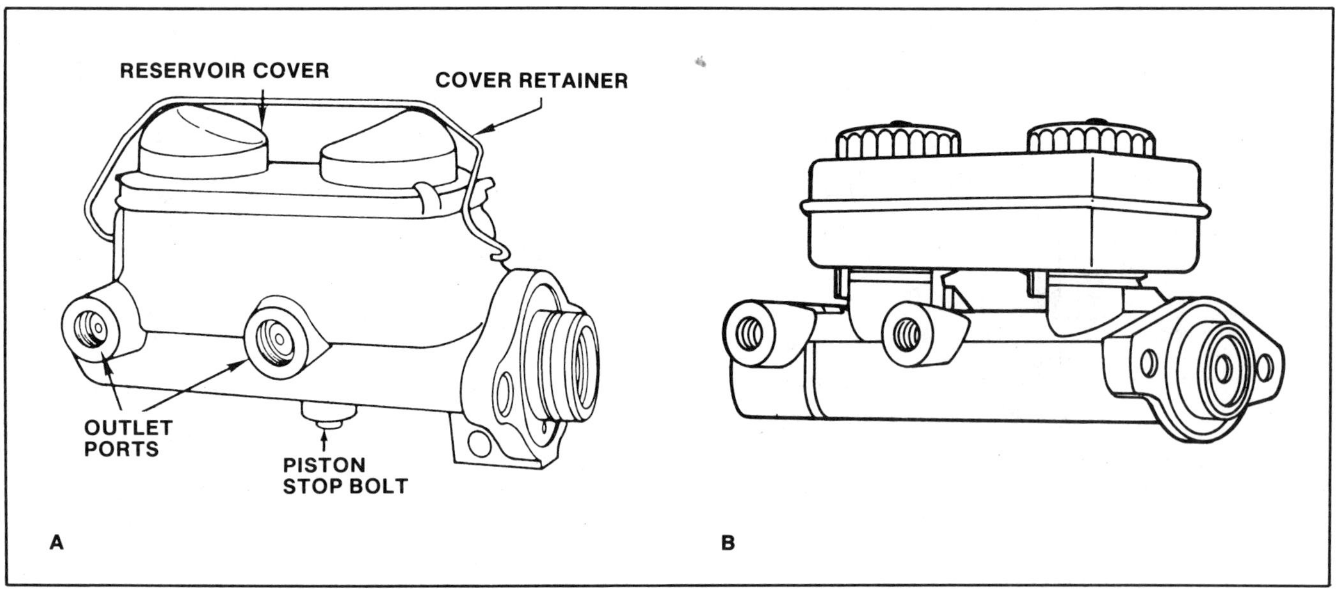

Figure 2-10 (A) Typical cast-iron dual master cylinder and (B) typical aluminum/composite dual master cylinder

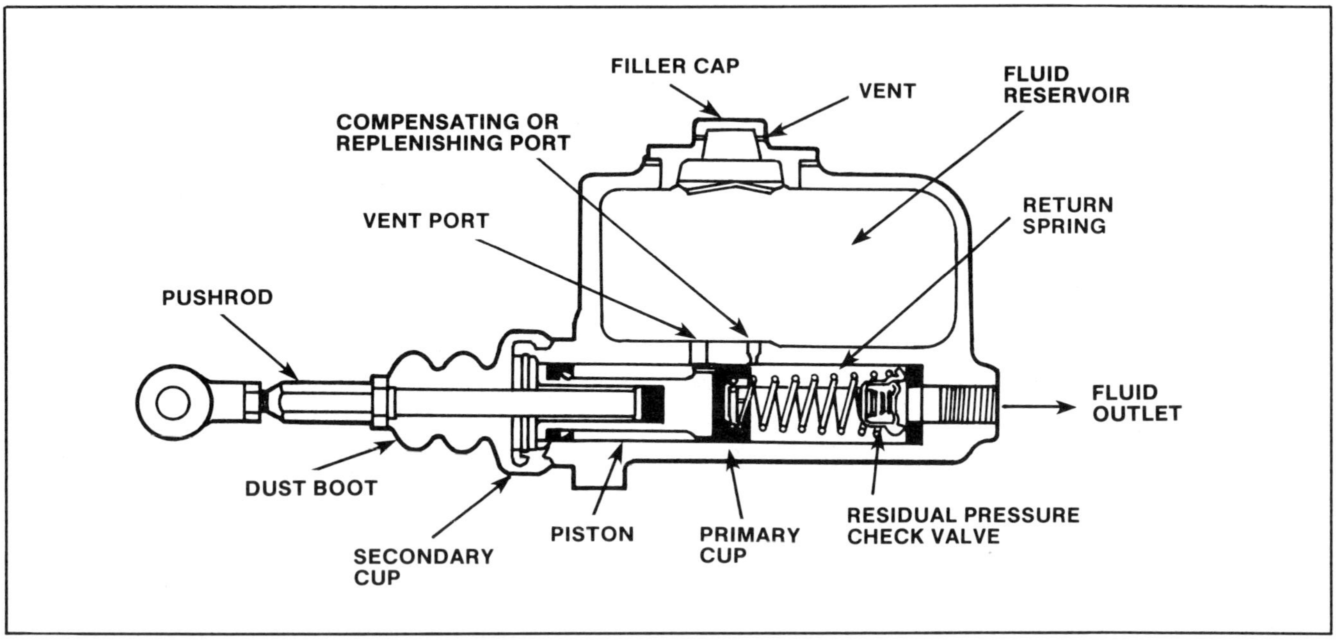

Figure 2-11 Typical single system master cylinder

When the brake pedal is depressed, the force from the brake pedal is transferred to the piston by a pushrod. The piston exerts pressure on the fluid, which flows out through the fluid outlet port into a single hydraulic system, serving both front and rear brakes. When the brake pedal is released, the return spring in the cylinder forces the piston back to its original position. If the piston returns too quickly, vacuum will be generated in the cylinder. Fluid from the reservoir then enters the cylinder through the vent port to fill the void. As the brakes are fully released, excess fluid is returned to the reservoir through the replenishing or compensating port to relieve pressure in the system.

SHOP TALK

Master cylinder terminology is often confusing. For example, replenishing and compensating ports are synonymous terms. That is, they serve the same function but are called different names by different manufacturers and in different parts of the country. The same is true of the master cylinder's vent port, by-pass port, or relief passage. All these names actually perform the same function and in basically the same manner, but are called by different names.

DUAL MASTER CYLINDERS

In order to protect against a failure of the system, all passenger cars—and most light-duty trucks—are now required to have two (dual) hydraulic systems. The dual master cylinder arrangement provides for the use of either the front/rear or diagonally split system previously mentioned. If one system fails, the other system remains operational, thus providing much greater protection against the total loss of stopping power.

The dual master cylinder performs in the general following manner:

- *Rear Wheel Brakes*—The rear wheel brakes are connected to the secondary fluid outlet port and are actuated by the secondary piston assembly.
- *Front Wheel Brakes*—The front wheel brakes are connected to the primary fluid outlet port (farthest from the dash panel) and are actuated by the primary piston assembly.

The important parts of a typical dual master cylinder are shown in Figures 2-12 and 2-13 and their functions are described here:

- *Cap (or Bail, Cover, and Gasket)*—These parts keep an airtight seal on the master cylinder while letting it "breathe." They keep moisture from entering the system. Moisture breaks down the ability of brake fluid to lubricate properly. It also causes corrosion. The cap, or cover gasket, acts like a diaphragm. It expands and contracts as the fluid pressure rises or falls. When the brakes are applied, the de-

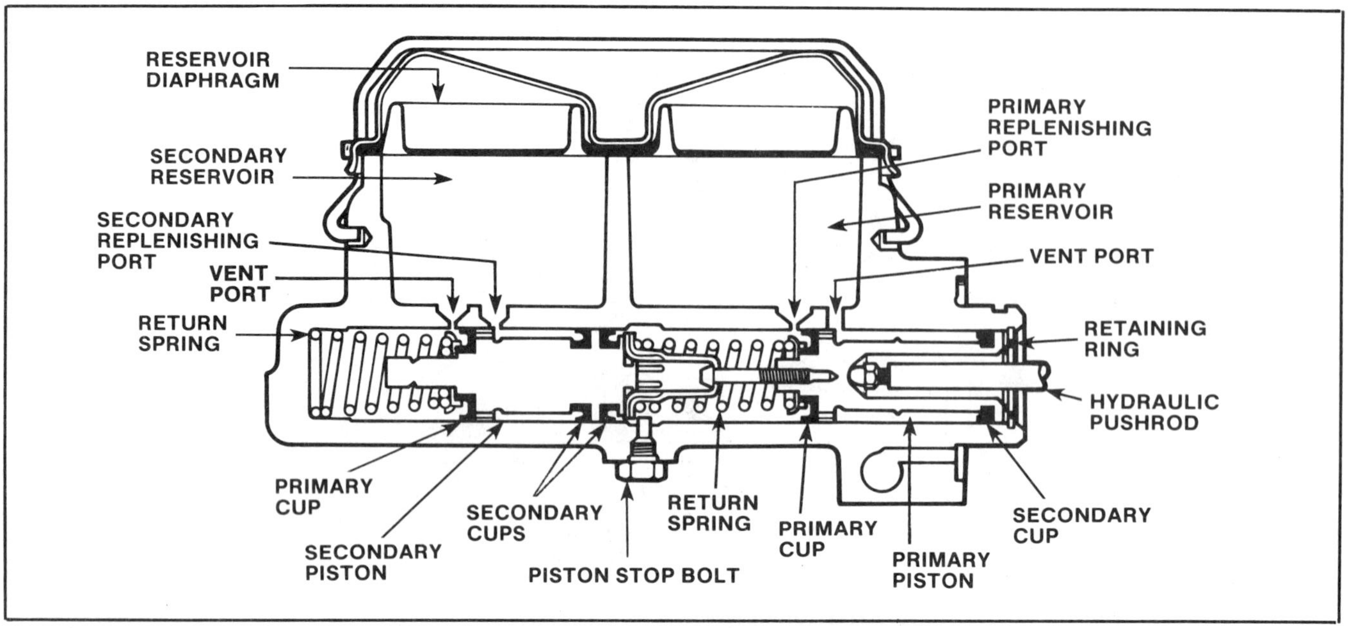

Figure 2-12 Cross section of a typical dual system master cylinder

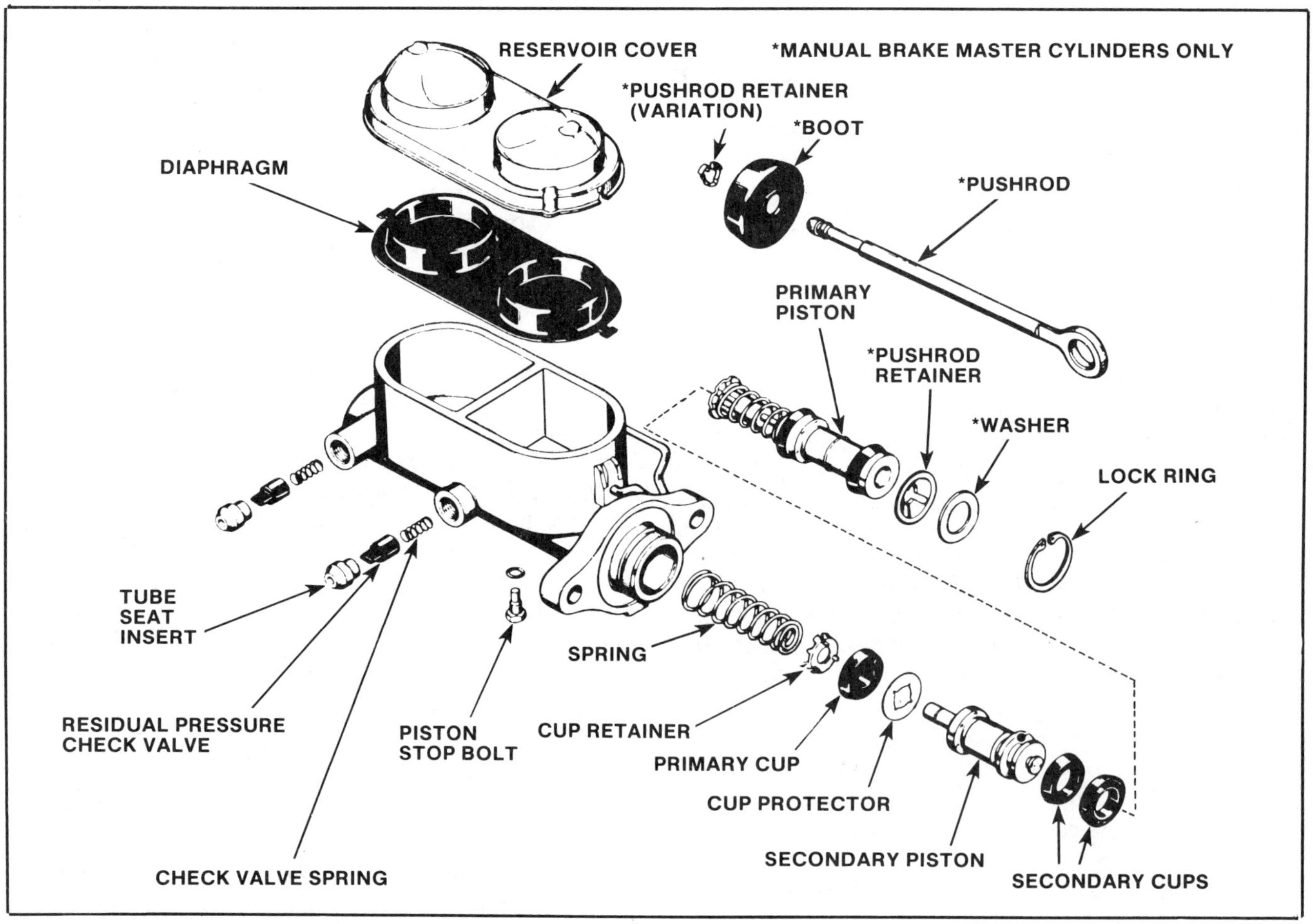

Figure 2-13 Typical Bendix dual system master cylinder—disassembled

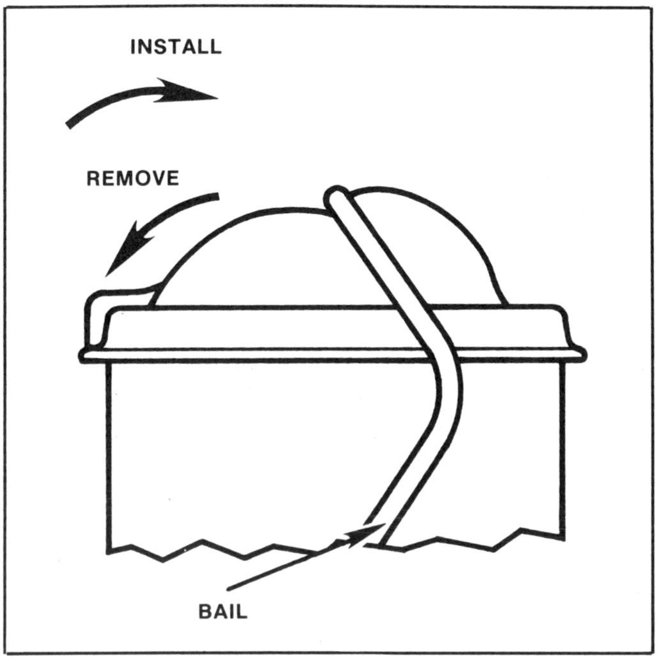

Figure 2-14 Cover and retainer bail wire

mand for brake fluid could create a vacuum below the diaphragm. This will keep fluid from flowing out of the reservoir. For the gasket to expand or contract, the space between the gasket and the cover must be vented. The venting is accomplished with small grooves or drilled holes in the cap or cover. These holes allow air above the seal and below the cap or cover. (Most seals have a bellow design. As the fluid goes down, the space above the gasket fills with air. When the brakes are released, the fluid returns to the master cylinder reservoirs and the air is forced out the vent holes.) The wire bail-type retainer (Figure 2-14) or a bolt holds the cap or cover on.

- *Reservoirs*—Extra brake fluid is stored in two separate reservoirs. Made of either cast iron or specially-blended plastic, they are designed to function independently as a protection against lost brake fluid. Without this reserve brake fluid available, a driver can run low due to evaporation, etc., and such a condition could cause the brake system to function improperly and/or to fail. It is important to keep the reservoir filled to its proper level as described in Chapter 3.
- *Brake Fluid*—A specially blended liquid that, under pressure, acts as a link between the master cylinder and the wheel cylinders and causing the wheel cylinder mechanism to engage the brake shoes or pads with the brake drums or disc (rotors).

CAUTION: Brake fluid can damage the finish of a vehicle. Be sure to flush with water immediately if any brake fluid is spilled on a painted surface.

- *Cylinder Bore and Pistons*—The dual master cylinder reservoir has a single piston bore that contains the two pistons (primary and secondary) lined-up in tandem. Each piston has a return spring in front of it. This holds the piston cup slightly behind the replenishing port for each reservoir. In addition to keeping the replenishing ports uncovered, the return springs also return the brake pedal when the force has been removed from it. As the brakes are applied, the stiffer primary piston spring pushes the secondary piston and spring slightly. Then, the cup at the front end of the secondary piston passes and closes off the primary replenishing port on the secondary side of the master cylinder. Some master cylinders have a piston stop bolt in a threaded hole in the side of the bore or in the bottom of the front reservoir. Others do not have piston stop bolts at all. Do not install a stop bolt in the reservoir of a master cylinder if one was not there originally. Some cylinders have a tapped hole, but no bolt was installed in production. This is done on purpose, and is not an error. However, it should be plugged to prevent brake failure.
- *Retaining Ring*—A retaining ring fits into a groove near the end of the bore and holds the piston.
- *Primary and Secondary Ports or Outlets*—The primary outlet (or port) is the one closest to the cowl panel. It is off of the primary piston (the one which receives the initial, or primary movement).
- *Residual Check Valves*—The function of the residual check valve is to keep light (or residual) pressure in the brake lines and also at the wheel cylinders. Without residual pressure, air can be sucked past the wheel cylinder cups and into the wheel cylinders. This could occur when the brake pedal is released quickly. Residual check valves are designed to permit hydraulic fluid and pressure to flow from the master cylinder into the hydraulic lines, when the pressure in the master cylinder becomes

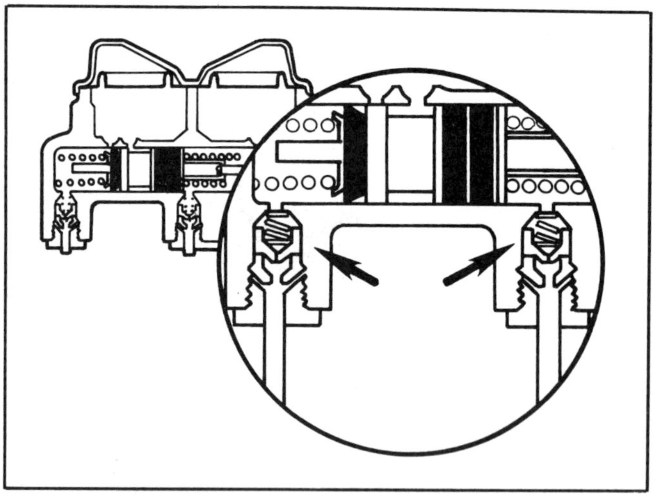

Figure 2-15 Outlet check valve springs

great enough to open the valve. After allowing the fluid to flow back into the master cylinder when the pedal is released, the outlet check valve spring (Figure 2-15) takes over and keeps the hydraulic lines and wheel cylinders full of fluid, ready for the next application of the brakes. It should be remembered that a certain amount of residual line pressure is necessary even with complete pedal release. This helps to prevent air from entering the wheel cylinders as they retract. This slight pressure is maintained by a residual pressure check valve built into the master cylinder (Figure 2-16). On most of the newer car models, automobile manufacturers have eliminated residual pressure check valves from the brake system by equipping the wheel cylinders with cup expanders. The cup expanders keep the cup seals tightly pressed against the cylinder walls to prevent air from entering the system.

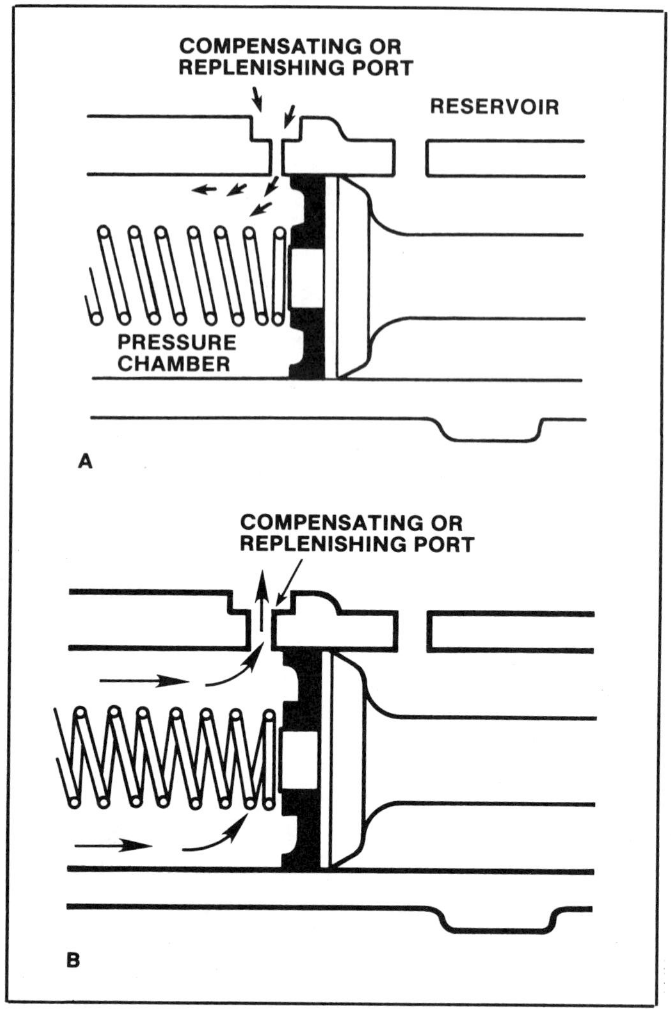

Figure 2-17 (A) Reservoir replenishing pressure chamber; (B) fluid returns to the reservoir

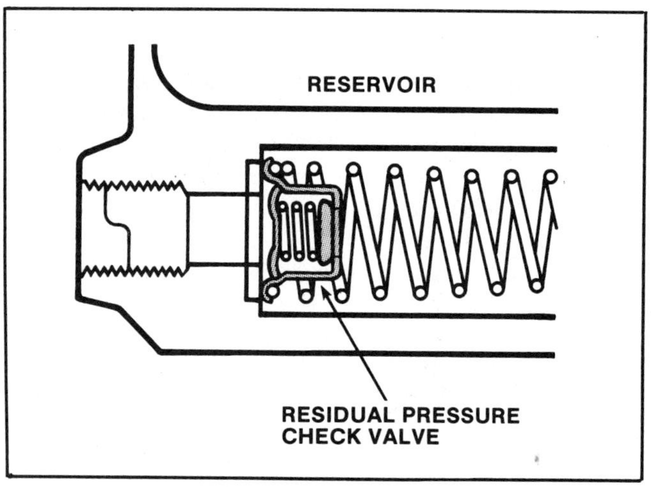

Figure 2-16 Residual check valve holds pressure in the line.

- *Replenishing or Compensating Ports*—Light residual pressure is needed for operating the drum brakes. However, when you use the brakes repeatedly, too much pressure can build up. The replenishing or compensating port (Figure 2-17) takes care of this problem. After many stops, like you experience in city driving, the brake drums and wheel cylinders get very hot. This causes the brake fluid to heat up and expand. Such a pressure buildup can become so strong that it keeps the brake shoes from returning after releasing the brake pedal. If the brake shoes do not return properly, they might drag against the drums. This causes more heat to build up and pressure is further increased.

SHOP TALK

Extra brake fluid is forced into the hydraulic system by pumping the brakes. Pressure builds when the brake pedal is released and the piston moves back far enough to uncover the replenishing ports. This removes the excess pressure from the brake lines and wheel cylinders and allows the fluid to return to the brake master cylinder reservoir.

- *Vent Ports*—Vent ports help to pump up a low brake pedal by allowing more fluid in the lines. There are several reasons for a brake pedal to be lower than normal. One is that the clearance between the brake lining and the drum is too large. This can happen when the automatic brake adjusters, if the vehicle is so equipped, are not working properly. Another reason could be the result of air leaking into the brake system. If a brake pedal is pumped too rapidly, the return springs in the master cylinder might move the pistons back very fast. At the other end of the system, the brake shoe return springs move the wheel cylinder pistons much more slowly. The flow of brake fluid from the wheel cylinders is then delayed. It cannot match the fast return of the master cylinder pistons. As a result, pressure drops in the master cylinder. The pressure in the master cylinder actually becomes lower than the pressure in the reservoir. Fluid then rushes through the vent ports; through holes in the piston; past the piston cup; and, back into the area ahead of the piston cups. Vent ports are important because the replenishing ports are not large enough to handle the extra flow of fluid needed when the pedal is pumped.

Operation of the Dual System Master Cylinder. To understand the operation of a dual master cylinder, it is important to know how the various parts of the unit work together. Actually, the operation of the dual master cylinder is like that of two single piston types placed end to end in the same bore. The two pistons in the master cylinder are not rigidly connected. Each half really operates like a single system. As just mentioned, each piston has its own reservoir with vent and replenishing ports. Stepping on the brake pedal moves a pushrod forward, causing the first or primary piston to move forward. The fluid ahead of it cannot be compressed, so the secondary piston moves. As the pistons progress deeper into the cylinder bore, the brake fluid that is put under pressure transmits this force through both systems to friction pads in the wheels (Figure 2-18).

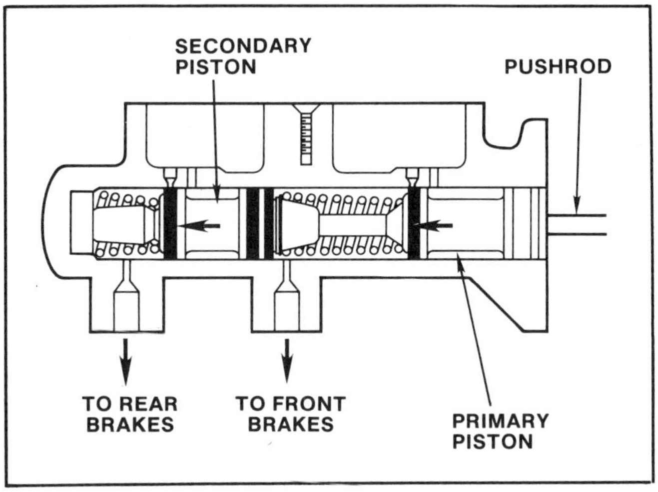

Figure 2-18 Master cylinder—brakes applied

The pistons have seals at both ends: a cup seal with a one-way sealing property at the forward end and a secondary cup at the rear. Each piston has a return spring with the primary piston spring between the two pistons. On the forward stroke, the lip of each cup seal is forced against the bore wall. When the piston springs quickly back to place, the brake fluid cannot return through the lines fast enough to avoid creating a low-pressure condition ahead of it. Fluid must reach the low-pressure area in time for another stroke of the cylinder. On the return stroke, the edges of the cup pull away from

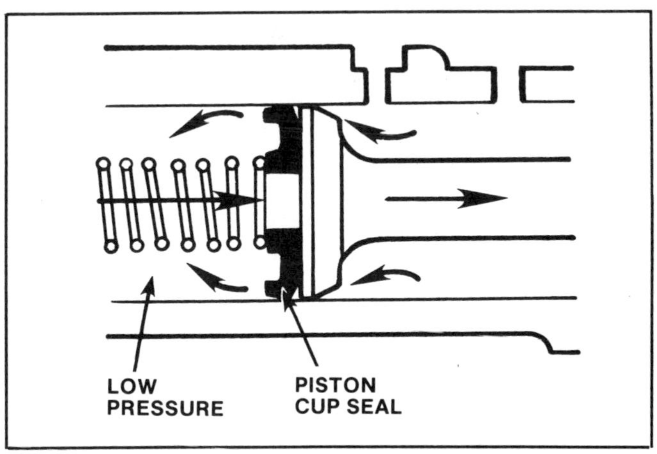

Figure 2-19 Brake fluid flows around piston cup.

the bore enough to allow fluid to pass around the piston assembly to the area of low pressure (Figure 2-19).

At the same time, the reservoir returns the small amount of fluid it had taken in during piston action (Figure 2-20). This way, the master cylinder returns to normal before all the fluid returns from the lines. Finally, the pads and shoes return to position. That is, once the brake pedal is released and the master cylinder piston has returned to the rest position, shoe return springs (in the drum brakes) and piston seals (in the disc brakes) cause

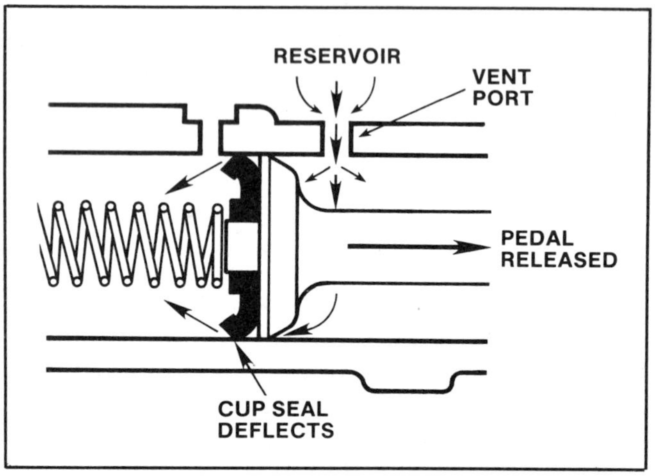

Figure 2-20 The reservoir returns fluid to the bore.

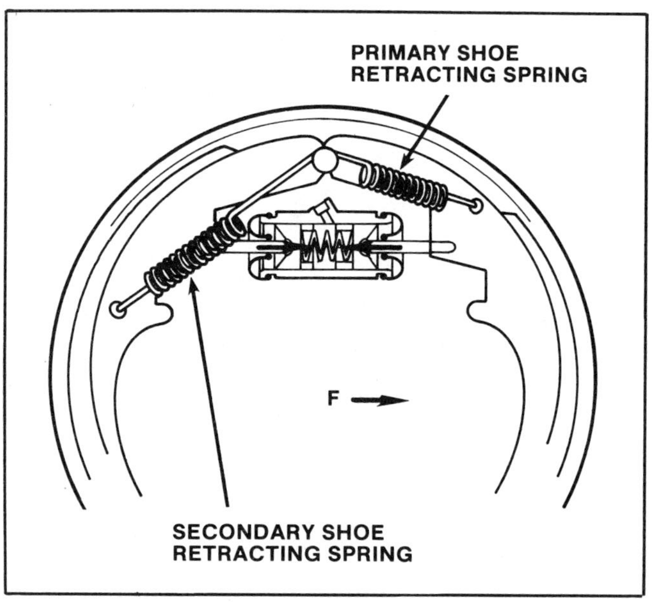

Figure 2-21 Shoe return springs retracting wheel cylinder in brake drum operation

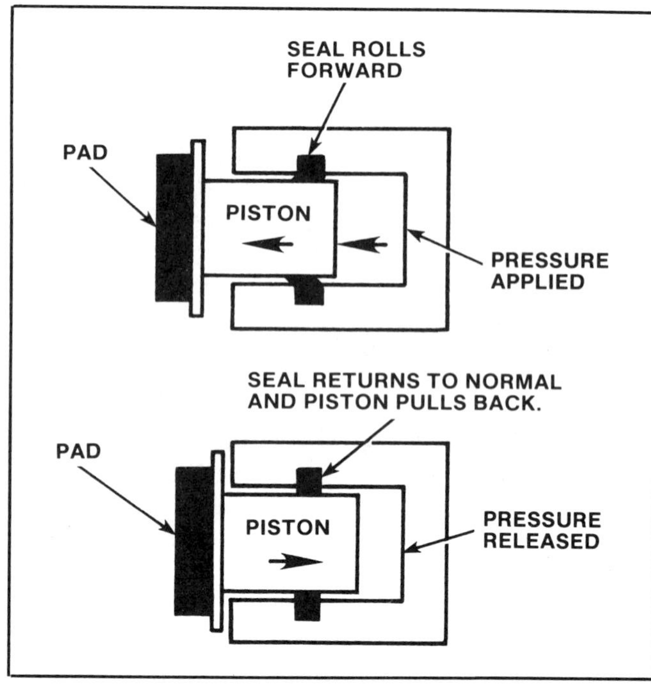

Figure 2-22 Disc brake piston seal operation

these pistons to retract, as shown in Figures 2-21 and 2-22.

Partial System Failure. As mentioned earlier in the chapter, the major advantage of the dual master cylinder over the single type is that if there is a failure in one of the two independent hydraulic systems, the other one will still function.

If a failure occurs in the hydraulic system served by the primary piston, this piston will move forward but will not develop pressure. The piston extension contacts the secondary piston and pedal effort is transmitted directly to that piston to build hydraulic pressure to operate the brakes in the secondary system (Figure 2-23).

If the secondary system suffers a leak, both pistons move forward until the secondary piston bottoms out at the end of the master cylinder bore. Then, the primary piston develops hydraulic pressure to operate the brakes in the primary system (Figure 2-24). When a failure occurs, in half of the system, the pedal will move noticeably farther before braking occurs.

Combination Brake System Master Cylinder

Combination brake system dual master cylinders are much the same as the dual-system units just described. The main difference is the unequal size of the two reservoirs (Figure 2-25). The larger

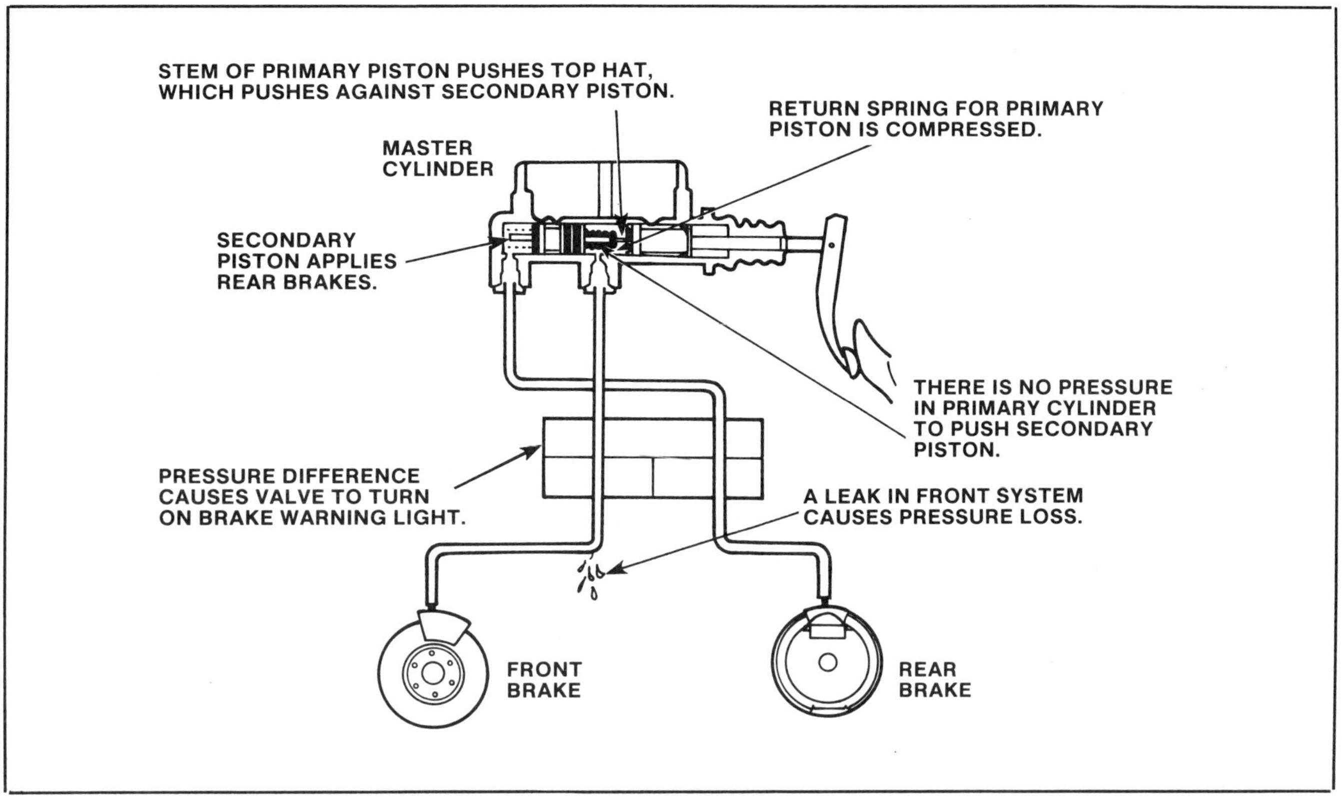

Figure 2-23 Leak in primary hydraulic system

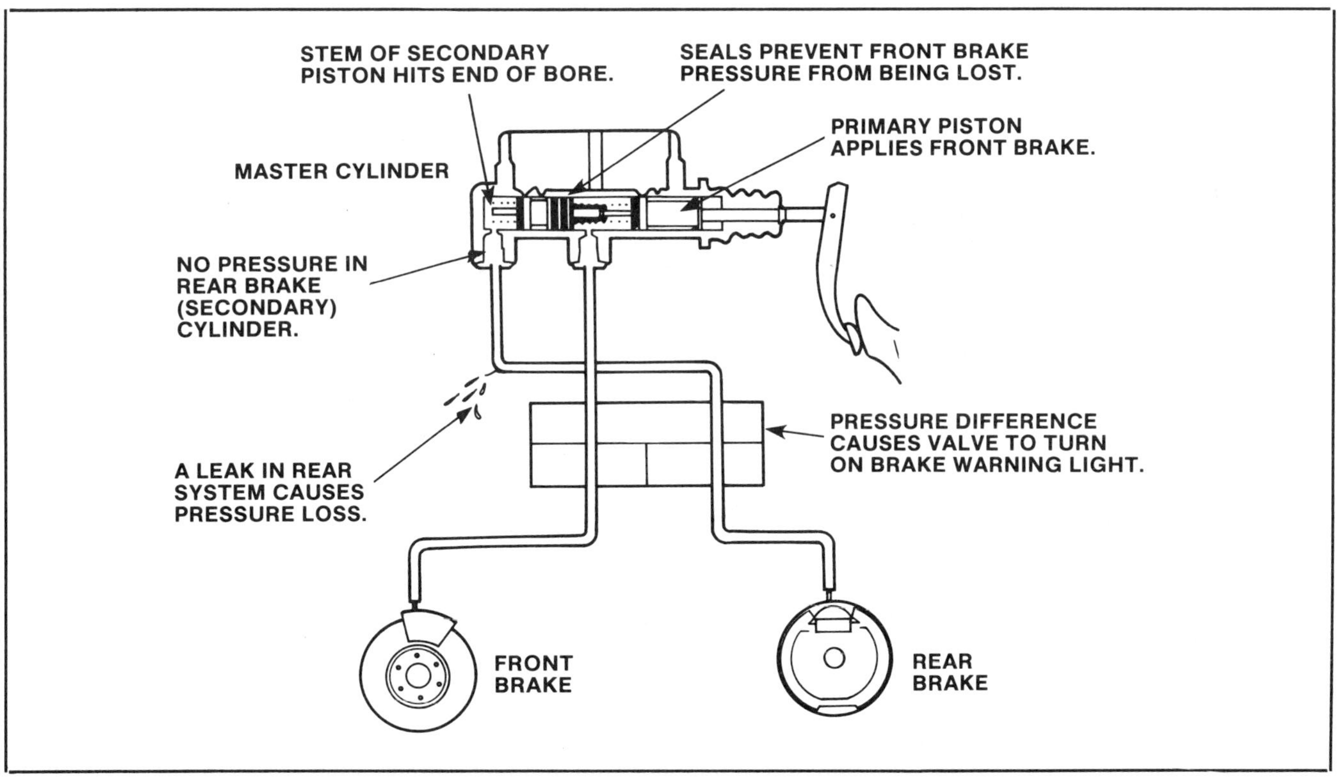

Figure 2-24 Leak in secondary hydraulic system

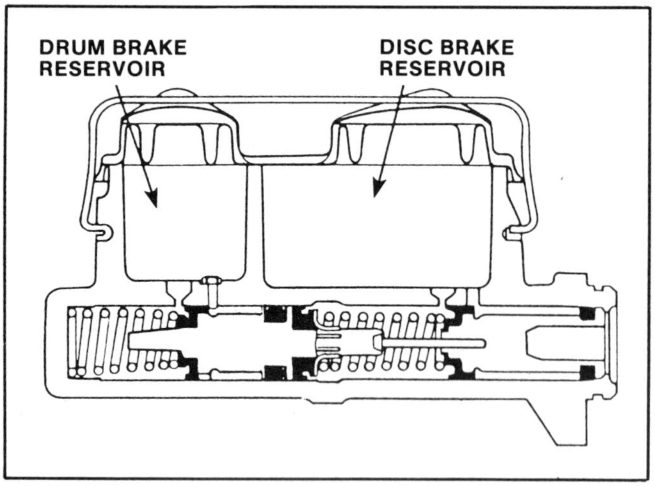

Figure 2-25 Combination brake system master cylinder

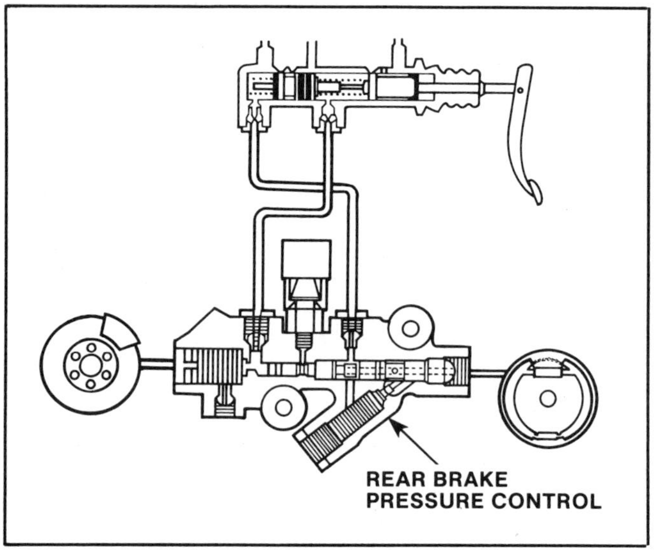

Figure 2-26 Brake proportioning valve

reservoir is connected to the disc brakes to provide the greater fluid reserve required for disc brakes.

Another difference is in the usage of residual pressure check valves. The outlet connected to the disc brakes should not contain a check valve. When using a no-check valve master cylinder, wheel cylinders must contain cup expanders to prevent cup collapse and air from entering the system. (Master cylinders used on four-wheel disc systems are very similar to the combination master cylinder just described.)

As shown in Figure 2-26, some combination brake system master cylinders contain a rear brake pressure control. As can be noted, the master

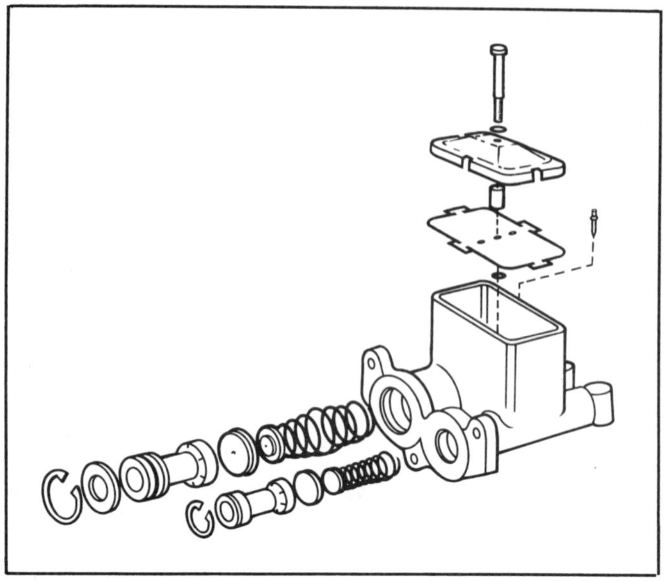

Figure 2-27 Dual bore type master cylinder

cylinder has a primary and secondary cylinder. The primary cylinder actuates the front wheel disc brakes, while the secondary cylinder is for the rear wheel drum brakes. Fluid flow to the front brakes is not restricted. However, fluid flow to and from the rear brakes goes through the pressure control valve. This valve provides balanced braking action between front and rear brakes under a wide range of braking conditions. As an example, at the higher master cylinder pressures, the control valve reduces the pressure to the rear brakes. This reduced pressure balances the self-energizing rear brake action with that of the front disc brakes, which are not self-energizing.

Dual-Bore Master Cylinders

The housing of this type of master cylinder accommodates two cylinders side by side (Figure 2-27). Only one cylinder functions in the brake system, the other is used for hydraulic clutch actuation. Also, a common fluid reservoir with an internal partition is used. The clutch system does not have a residual pressure check valve. Each hydraulic system must be bled separately.

Hydraulic Clutch Linkage

Although hydraulic clutches are not directly related to brakes, they bear mentioning here because their components and service requirements are similar to those of brake systems (Figure 2-28). Due to the design of the frame and body on some cars and many trucks, it is impractical to run mechani-

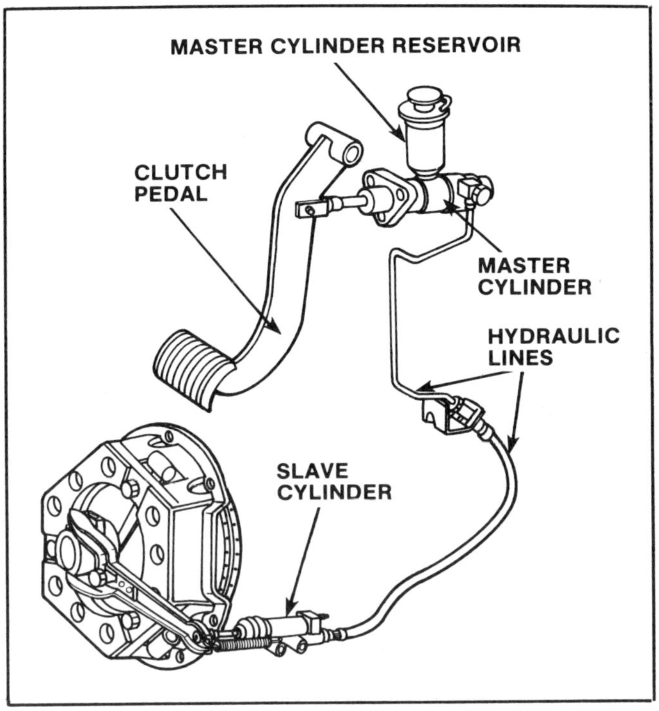

Figure 2-28 Hydraulic clutch control

cal linkage from the clutch pedal to the clutch release arm, so clutch operation is done by hydraulics. A master cylinder (Figure 2–29) is connected to the clutch pedal (very similar to the single system brake master cylinder) and a slave cylinder is connected to the clutch release arm (similar to the single-end wheel cylinder used in some drum brakes). A hydraulic line connects the clutch master and slave cylinders (Figure 2–30).

As the pedal is depressed, the piston moves forward, pressurizing the fluid. This forces the slave cylinder piston to move, depressing the clutch release arm.

SHOP TALK

If you should encounter one of these systems, be sure parts are available before you start any disassembly operations.

Vertical Master Cylinders

Vertical cylinders are used on trucks that have tilt, forward control, or other unconventional cab arrangements requiring cylinder and brake pedal

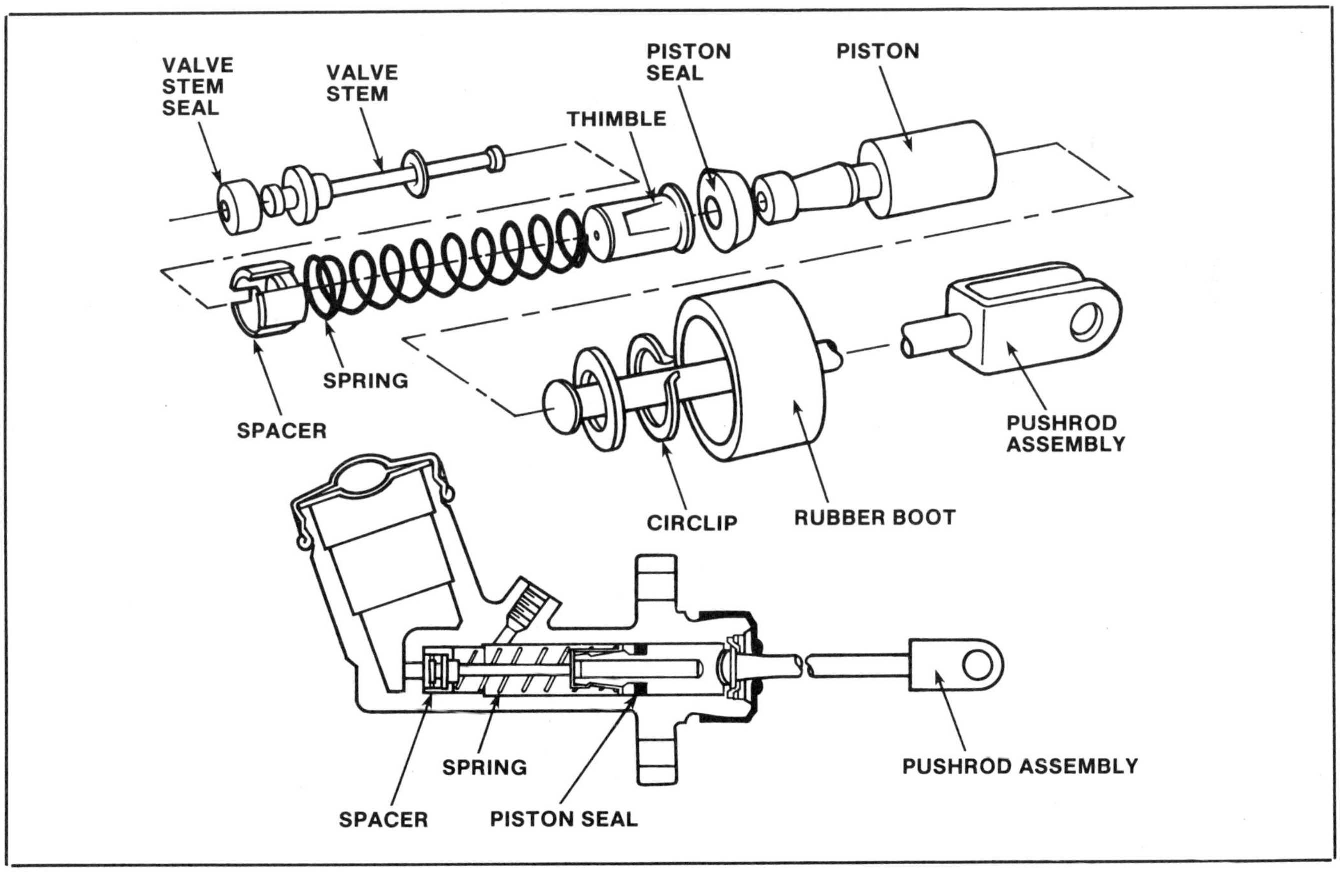

Figure 2-29 Clutch master cylinder

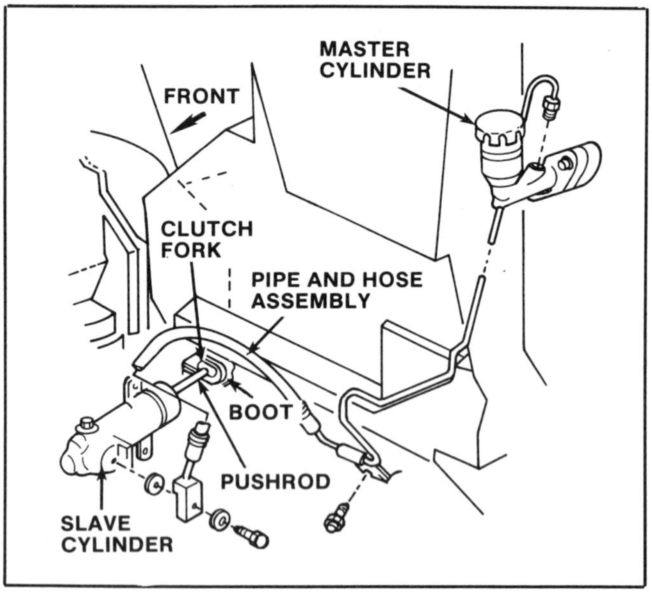

Figure 2-30 Hydraulic clutch linkage

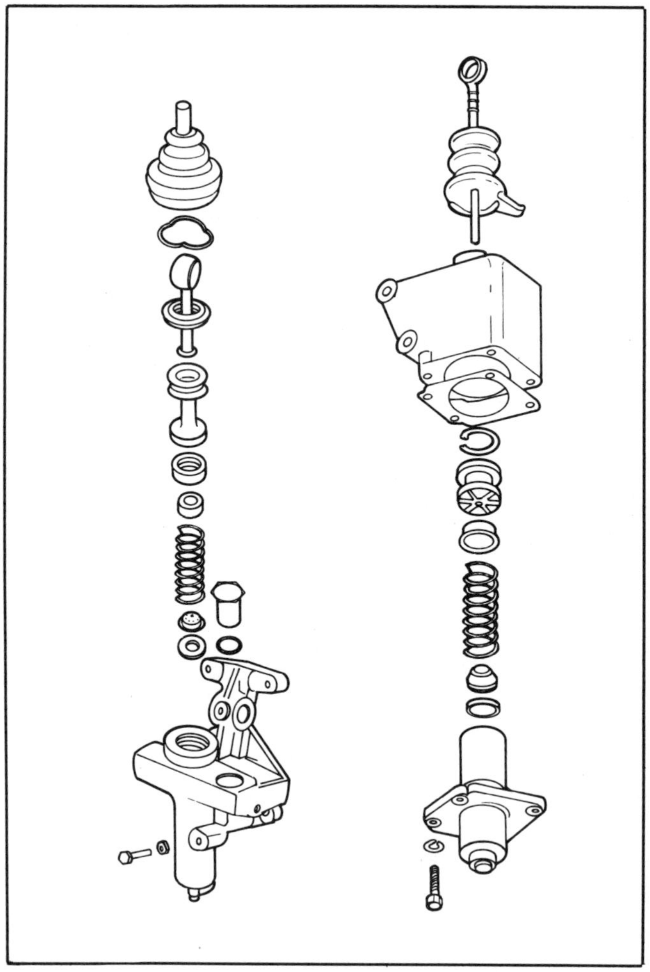

Figure 2-31 Vertical type master cylinder

mountings peculiar to the cab design (Figure 2-31). Vertical units differ from conventional cylinders in the relation of the fluid reservoir to the cylinder proper, and in the mounting. Some cylinders have a two-piece housing: one for the reservoir, and the other for the cylinder proper, which are bolted together. Others are mounted on a pivot and have flexible hose connections, permitting a tilt cab to swing up on its hinge.

Step-Bore Dual-System Master Cylinder

As shown in Figure 2-32, this cylinder is the same as other cylinders, except that the forward part of the bore is slightly smaller in diameter than the rear part. This difference contributes to correct brake balance.

As described in Chapter 13, Volvo has incorporated a step-bore master cylinder into a sophisticated brake system. It provides automatic brake proportioning, no pedal loss, and automatically increased pressure on the remaining circuit during partial system failure.

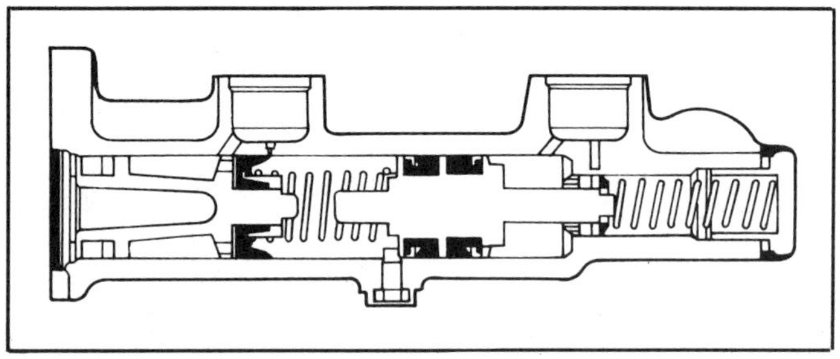

Figure 2-32 Step bore master cylinder

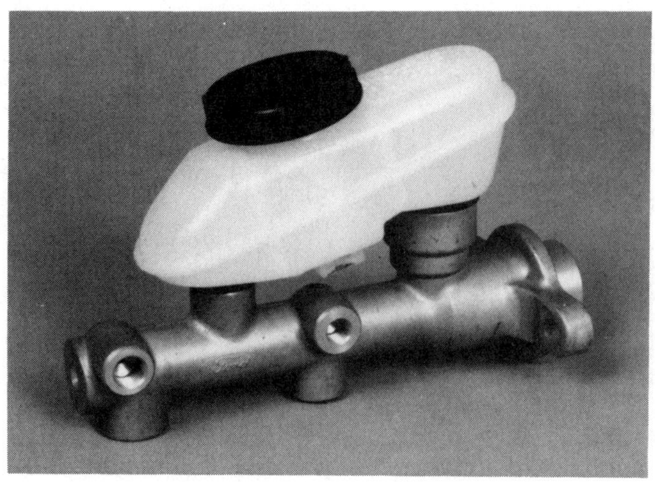

Figure 2-33 Typical Ford "take-up" master cylinder

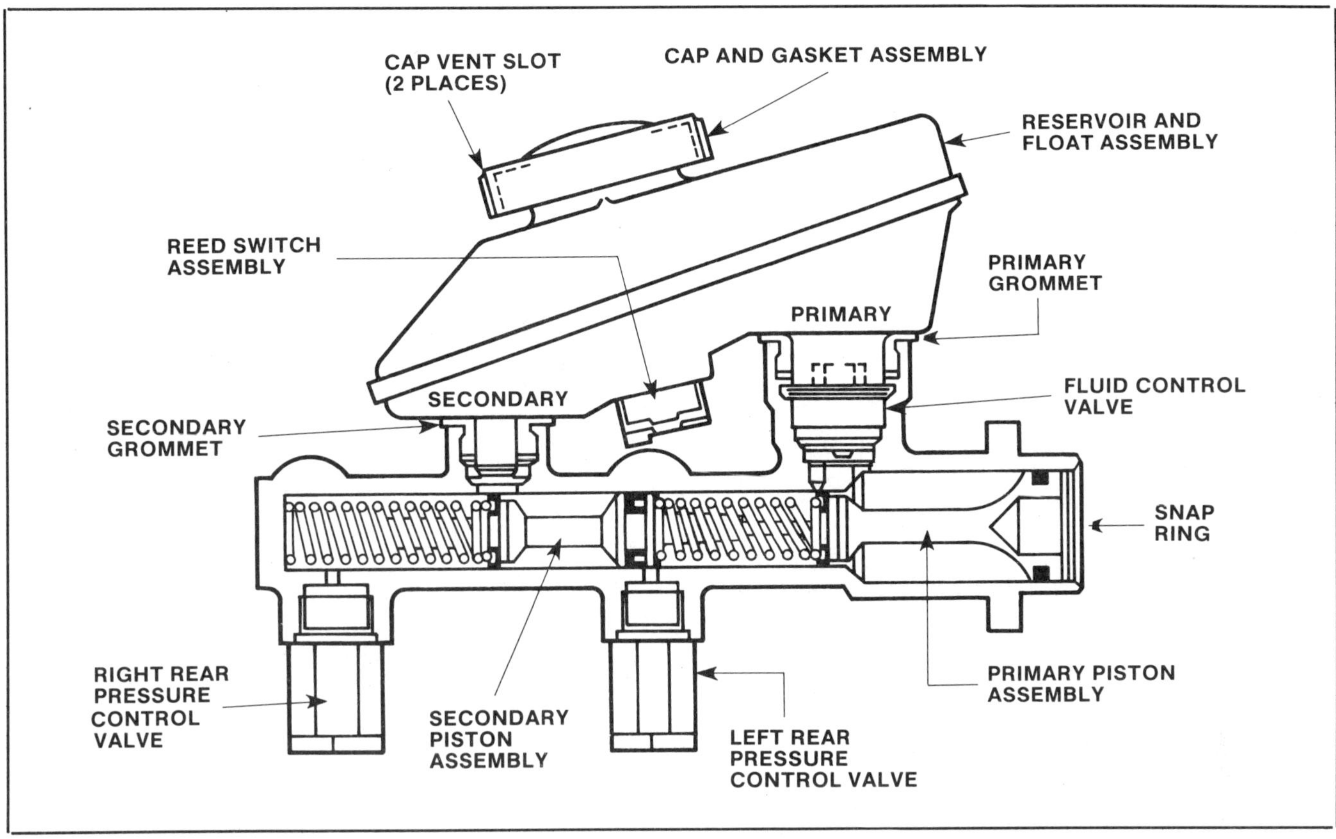

Figure 2-34 Parts of a Ford dual "quick take-up" master cylinder

GM "Quick Take-Up" Master Cylinders

In 1980, General Motors and Delco-Moraine introduced the "quick take-up" step-bore type master cylinder. Two years later, Ford developed their version of the quick take-up master cylinder (Figure 2-33). Both master cylinders are designed to be used with front-wheel drive (FWD), diagonally split brake systems.

These master cylinders (Figure 2-34) provide a large volume of fluid at low pressure with only a small amount of pedal movement. This feature allows it to work well with the "zero-drag" disc brake calipers that manufacturers employ as a means of reducing parasitic losses and, hence, raising fuel mileage. These calipers use piston seals that retract the pads completely from the rotor when the pedal is in the released position. This increased clearance between linings and disc must be compensated for or brake pedal travel would be excessive before friction was generated, so the high volume displacement of the quick take-up master cylinder becomes necessary.

Although both designs are similar, the GM unit needs further inspection. As can be noted in Figure 2-35, it has a stepped bore; the primary low-pressure chamber is larger than the primary high-pressure chamber. Between the reservoir and the primary low chamber on the primary side of the master cylinder is a *quick take-up valve,* or a two-piece plastic component with a spring-loaded check valve (normally in the closed position). There is a bypass groove along the side with a large lip seal around the valve with a series of small holes around the outer edge.

The GM quick take-up master cylinder operates as follows: In the initial brake application stage (Figure 2-36A), more fluid is displaced in the primary low-pressure chamber than in the high-pressure chamber because of the larger diameter of the low-pressure chamber and piston. The excess volume of fluid must go somewhere, so it goes around the primary piston lip seal into the primary high-pressure chamber. From there it flows to two wheels. Since the pressure and displacement is equal in both the primary and secondary systems, the secondary piston must move farther because its bore is smaller. Flow and pressure to the two wheels it serves is the same as that to the two wheels served by the primary piston. This large

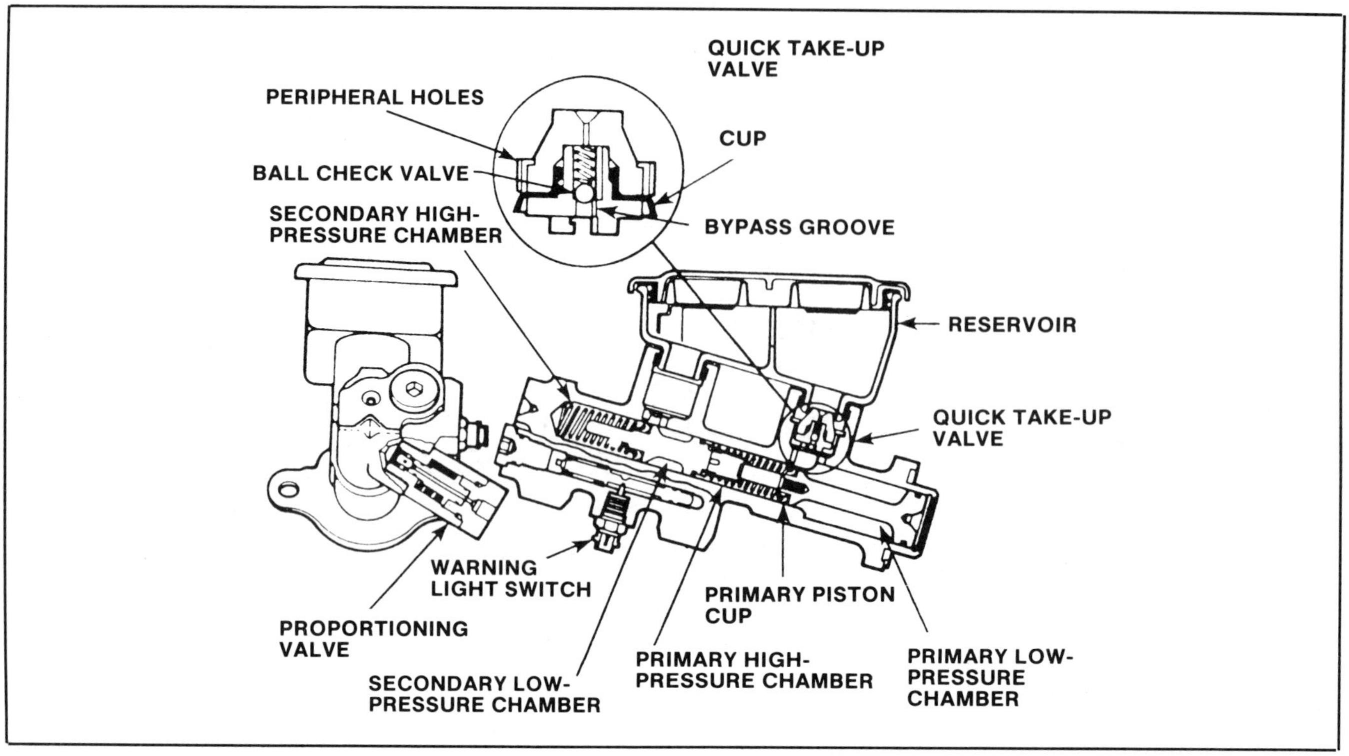

Figure 2-35 GM type quick take-up master cylinder

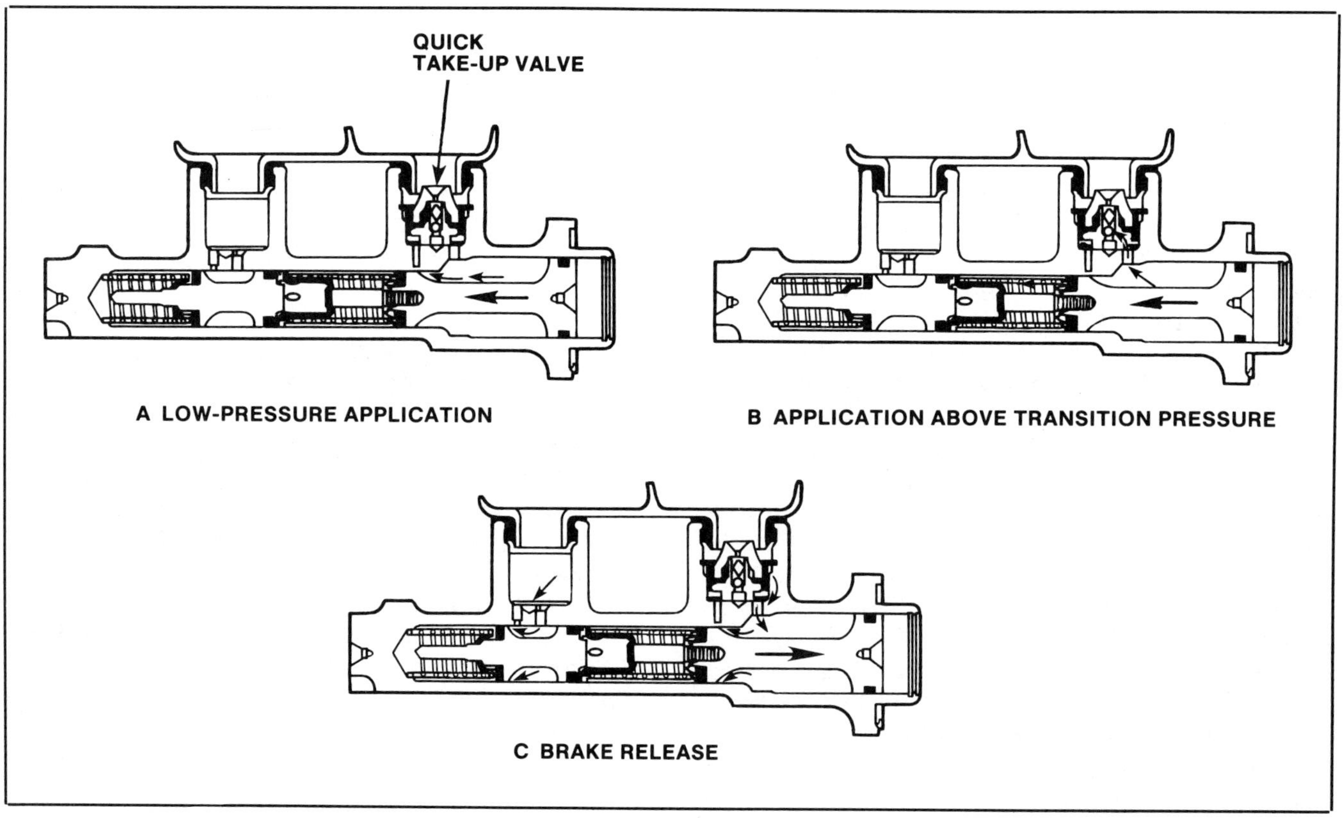

Figure 2-36 GM type quick take-up master cylinder in operation

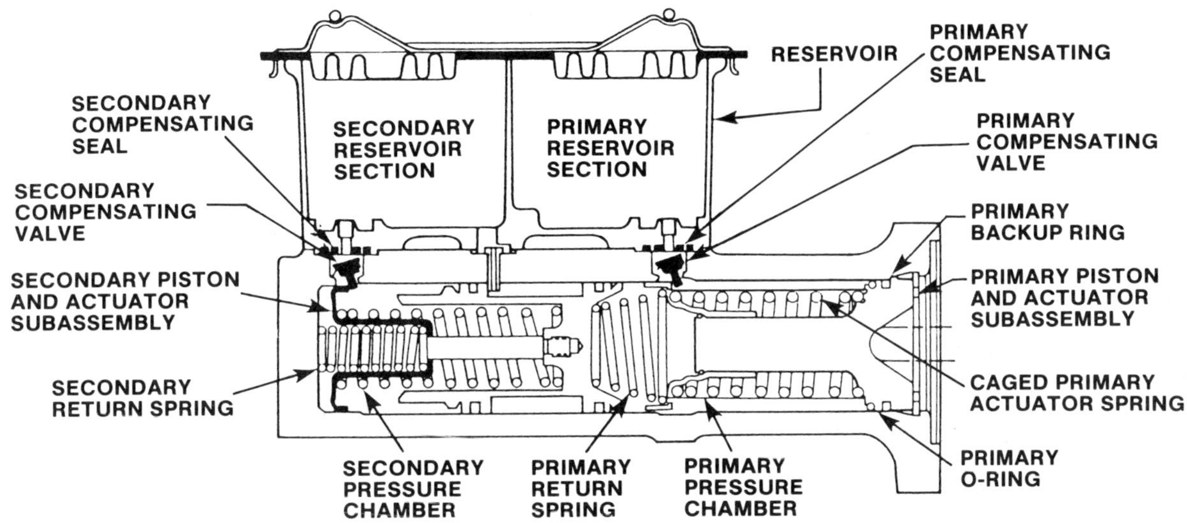

Figure 2-37 Parts of Bendix mini-master cylinder

initial displacement has now taken up all clearances in the brake system and applied a light pressure throughout the system.

The quick take-up valve contains a spring-loaded check valve that holds pressure in the primary low-pressure chamber up to a specified value. When that pressure is exceeded in the cylinder, this spring-loaded valve opens. Then, fluid in the primary low-pressure chamber will bypass to the reservoir. The primary and secondary pistons function as a normal dual master cylinder (Figure 2-36B). These pistons operate in the smaller bore to develop the higher brake operating pressures required.

When the brakes are released, fluid replenishment occurs through a normal replenishing port for the secondary piston. (Remember, the piston moves back faster than the fluid returns, and fluid must flow around the cup to the front of the piston to balance the pressures.) Fluid makeup for the primary piston occurs through the peripheral holes in the quick take-up valve, around the quick take-up cup, and through the replenishing port to the low-pressure chamber of the primary piston. The fluid then flows around the primary piston cup into the high-pressure chamber (Figure 2-36C).

A bypass groove in the quick take-up valve accommodates fluid expansion in the primary chambers through the vent and replenishing ports. The secondary chambers handle expansion and contraction through the normal vent and replenishing ports directly connected to the front reservoir.

When servicing the quick take-up valve, remember that the GM type cannot be replaced. However, the Ford type can be removed by rotating the reservoir and removing it with a socket.

Bendix Mini-Master Cylinder

The Bendix mini-master cylinder (Figure 2-37) is a lighter, shorter dual-hydraulic brake system master cylinder. It is designed for use in the medium truck industry and a variety of off-highway systems. The mini-master cylinder uses tilt type compensation valves and telescoping actuators to reduce weight and length. When used with the Hydro-Max (see Chapter 9) or equivalent hydraulic power brake booster, the mini-master cylinder provides fluid displacement for most vehicles through 46,000 gvw (gross vehicle weight).

Each pressure chamber has a piston/actuator subassembly containing a pre-loaded (caged) spring and a return spring. In the released position, actuators of both the primary and secondary pistons are in contact with their respective compensating valve stems, which project into the cylinder bore. This maintains the valves in an open position, which allows hydraulic fluid in the reservoir to replenish any fluid displaced from the cylinder bore.

Initial forward travel of the primary piston moves the primary actuator away from its compensating valve counterpart, permitting the valve to seat. Closure of this valve shuts off the passage

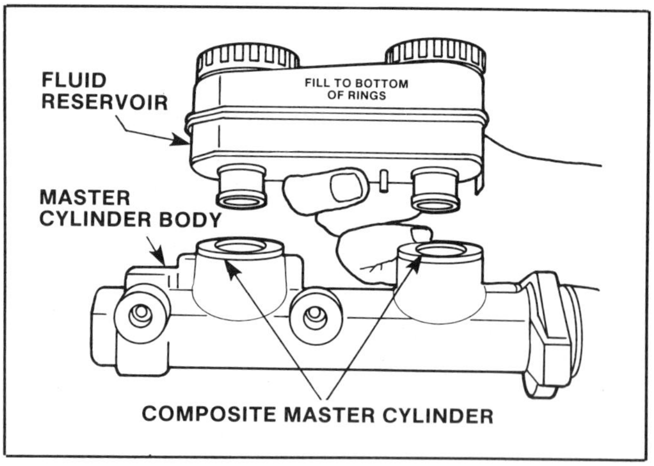

Figure 2-38 Design of a typical aluminum/composite master cylinder

between the primary pressure chamber and the reservoir section serving the primary chamber.

Further movement of the primary piston creates a pressure in the primary pressure chamber, causing the secondary piston and actuator to move. As the secondary piston and actuator move, the secondary compensating valve closes, shutting off the passage between the secondary pressure chamber and the reservoir section serving the secondary chamber. Additional movement of the primary piston causes both chambers to build pressure.

When the load on the primary piston is removed, fluid pressure in each chamber, combined with return spring force, causes the primary and secondary pistons to return to their initial released positions. Each actuator opens its respective compensating valve, which reopens the passage between the individual reservoir sections and their associated pressure chamber.

Should the rate of release be great enough to cause a partial vacuum in a chamber, the compensating valve will open to allow replenishment of fluid in the cylinder bore.

Any excess fluid remaining at the end of the stroke due to pumping and/or volume change due to temperature fluctuation is released as the compensating valve ports open.

Aluminum/Composite Master Cylinders

Master cylinders with aluminum bodies have been used on special applications for many years. Today, as mentioned earlier in the chapter, their use is becoming more widespread due to the weight advantage over cast iron. The machined aluminum bore is usually anodized to resist corrosion, prevent galling, and provide better wear resistance. Aluminum master cylinder bores are not serviced if scored or pitted, but must be replaced. A composite master cylinder has a plastic reservoir attached to a metal barrel. That is, the cylinder itself is made of metal (usually aluminum), and the reservoir is made of nylon or fiberglass-reinforced plastic (Figure 2-38). This design was adopted because it is lighter than the one-piece, cast-iron type and has the advantage of a translucent reservoir, which makes it possible to check fluid level without removing the cap.

To avoid damage and leaks, care must be taken when disassembling a composite master cylinder. To remove the reservoir from the cylinder, pry carefully between the reservoir outlets using a wide pry bar. To assemble, first inspect the rubber grommets that the reservoir outlets fit into, and replace them if they are cracked or torn. Then, lubricate the grommets with special silicone brake assembly lubricant or fresh brake fluid, and insert the reservoir outlets into them slowly until they seat.

One of the newer aluminum/composite master cylinders features a plastic reservoir and an integral proportioning valve/warning switch assembly. Two outlet ports are provided for the front brakes, which are served by the front reservoir and secondary piston. One outlet port is provided for the rear brakes, which is served by the rear reservoir and primary piston.

The integral proportioning valve balances outlet pressure to the rear brakes to prevent rear wheel lock-up under light rear wheel load conditions. After a pre-determined rear inlet pressure has been reached, hydraulic outlet pressure to the rear brake will be shared. A bypass feature is designed into the valve to assure full system pressure to the rear brakes in the event of a front system failure or full system pressure and to the front brakes in the event of a rear brake system failure. The pressure differential warning switch monitors front and rear brake system hydraulic pressures from the master cylinder. Should the front or rear brake system fail, the switch will latch in the warning position, causing a brake warning lamp on the instrument panel to light. The switch will remain in the warning position until the failure has been repaired.

Some aluminum/composite master cylinders have an electric fluid level indicator as an integral part of the fluid reservoir (Figure 2-39). It is serviced as a part of the plastic reservoir assembly. It consists of a float and magnet assembly and a reed switch. When the fluid in the reservoir gets to a pre-determined level, the magnet—which is attached to the float—actuates the reed switch, causing the warning lamp to light. Loss of fluid from either the primary (front) or secondary (rear) system will cause this system to actuate. If the fluid level indicator is deemed to be inoperative, the plastic reservoir must be replaced.

HYDRAULIC TUBES AND HOSES

Hydraulic tubes and flexible hoses (Figure 2-40) serve as the "arteries" and "veins" of the hydraulic system because they transmit fluid pressure (the blood) from the master cylinder (the heart) to the wheel cylinders and calipers (the body and working parts of the brake system).

Figure 2-41 shows where hydraulic tubes and flexible hoses are found in a typical vehicle. Early brake lines were made of rubber. However, the chemical materials in the brake fluid "ate away" at the rubber and caused the material to disintegrate. This caused brake lines to leak or split. Other types of tubing were experimented with in order to find the most reliable material. Not only did the brake lines have to resist the chemical actions of the ingredients in the brake fluid, they also had to be strong enough to handle extremes in pressure changes. It is important to keep in mind that the pressure in the brake lines approaches 1500 PSI during a strong panic stop. Also remember that even a partial blockage in the lines is detrimental to braking action because it restricts the flow of fluid. Although applying pressure can force the fluid past the blockage, the affected brake cannot release, or releases slowly and drags.

HYDRAULIC LINE TUBES

Most brake line tubing is made of double-walled steel with a rust-resistant lead alloy coating. Diameters vary and should be checked when replacing any old tubing with new. Tubing can be found

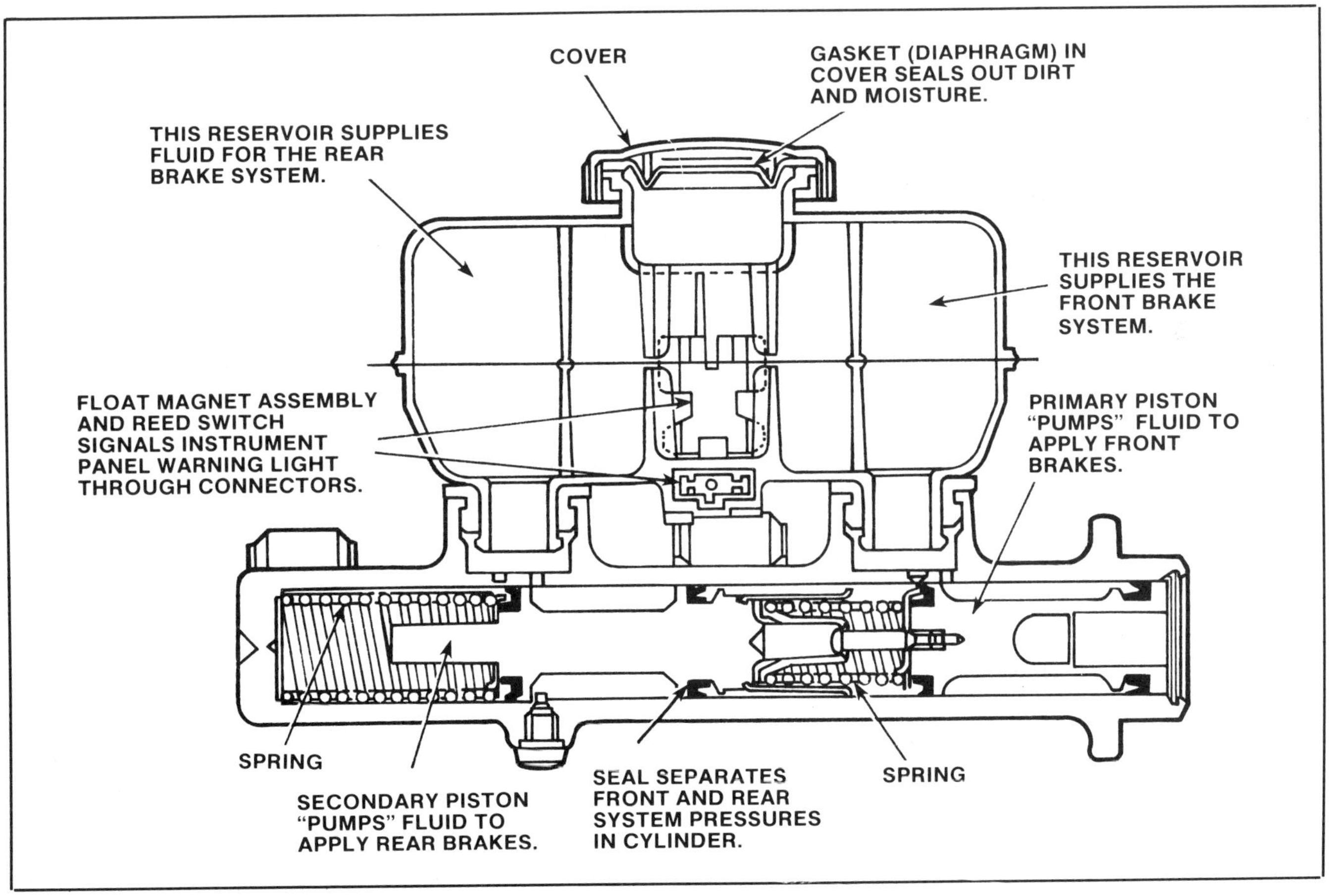

Figure 2-39 Cutaway view of a typical aluminum/composite master cylinder

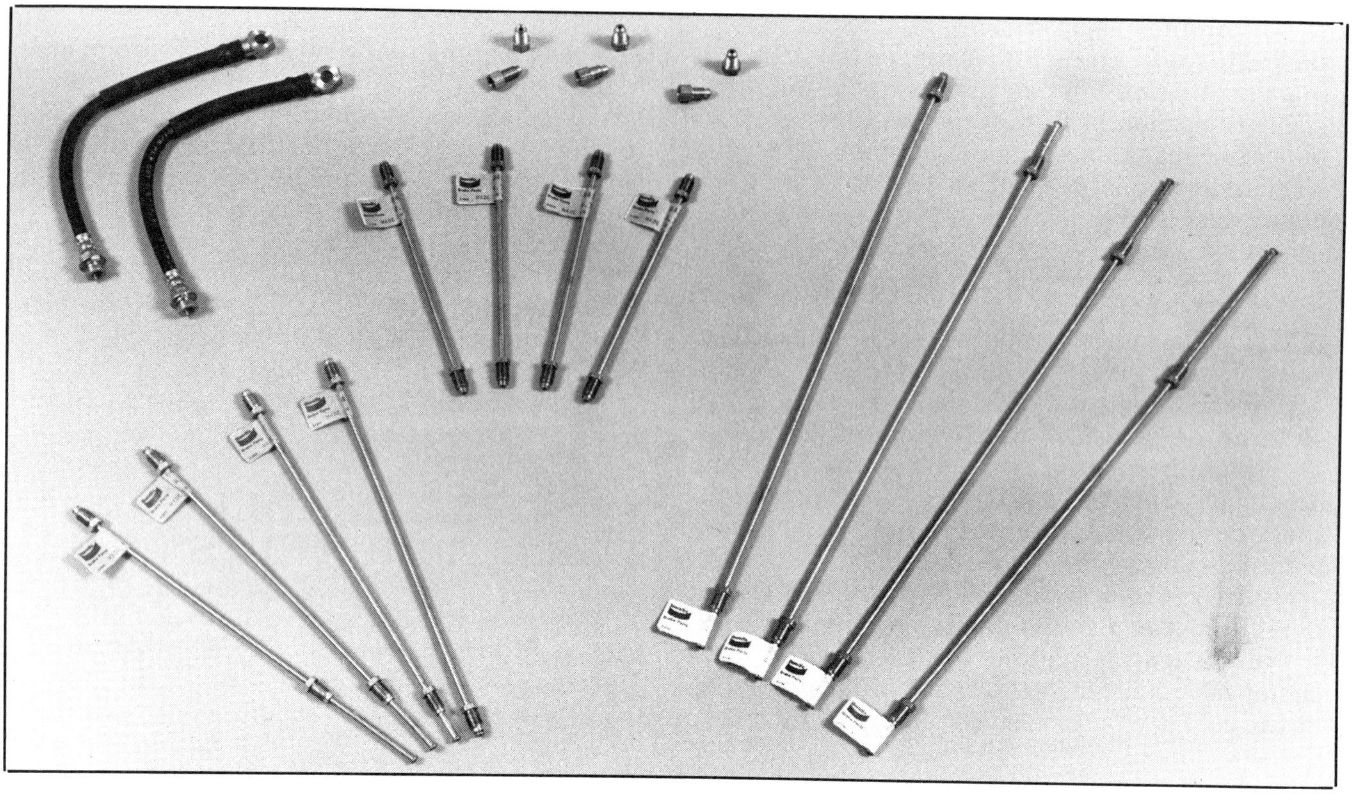

Figure 2-40 Typical hydraulic lines, and hoses and fittings

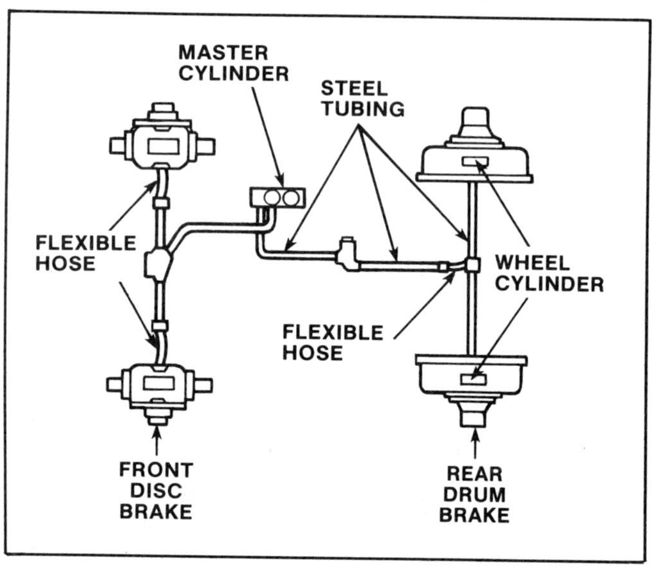

Figure 2-41 Use of brake lines and hoses

in varying lengths already flared and with the fittings attached. Always use pre-fabricated tubing whenever possible. But it might be necessary to cut a piece of tubing to a needed length and flare it as described in Chapter 4.

CAUTION: The use of copper or other forms of unapproved tubing will not be adequate to hold up under the extreme pressure exerted on the brake lines and should never be used.

HYDRAULIC HOSES

Hoses form a flexible link between the brakes (or axle) and the car frame or body. They must withstand high fluid pressures without leakage and must be free to flex during chassis deflection and wheel turns without damage. Hoses come in different sizes, lengths, materials, and end fittings (Figure 2-42). It is important to use the correct size hose for a given application. A hose that is too long will rub against the chassis or other part and eventually wear its cover through. In extreme cases, this abrasion can wear all the way through the hose and cause a hydraulic failure. A hose that is too short can break in tension. Aging and constant exposure to weather can result in hardening and cracking of a hose. The connections at the metal couplings are especially critical. Blisters in a hose suggest seepage in the hose joint that has penetrated underneath the outer hose covering and could indicate a torn liner, which can act as a check valve in the liner (Figure 2-43). Never take a chance with a faulty hose; replace it.

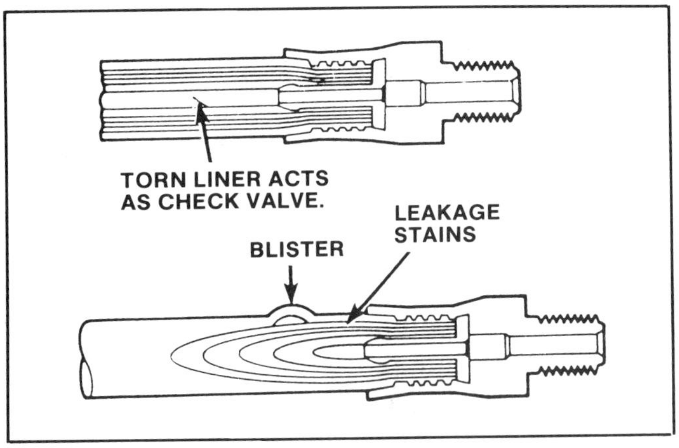

Figure 2-43 Possible internal defects

WHEEL CYLINDERS AND CALIPERS

The wheel cylinders and calipers are at the opposite end of the hydraulic system from the master cylinder and serve the opposite purpose. As the master cylinder changes mechanical force into hydraulic pressure, the wheel cylinder or caliper converts the hydraulic pressure back into mechanical force. Wheel cylinders are used with drum brakes, while calipers are employed with disc brakes.

WHEEL CYLINDERS

The wheel cylinder (Figure 2-44) performs in response to the master cylinder. It receives fluid pressure from the hydraulic line through its inlet port. As the pressure increases, the wheel cylinder cups and pistons are forced apart (Figure 2-45A). As a result, the hydraulic pressure is converted into mechanical force acting on the brake shoes.

When the brake pedal is released, the pressure inside the wheel cylinder decreases. The brake shoe retracting springs are then able to overcome the residual pressure and the spring in the cylinder. This pushes the pistons back together (Figure 2-45B) and forces fluid out of the wheel cylinder.

Figure 2-44 Typical wheel cylinder

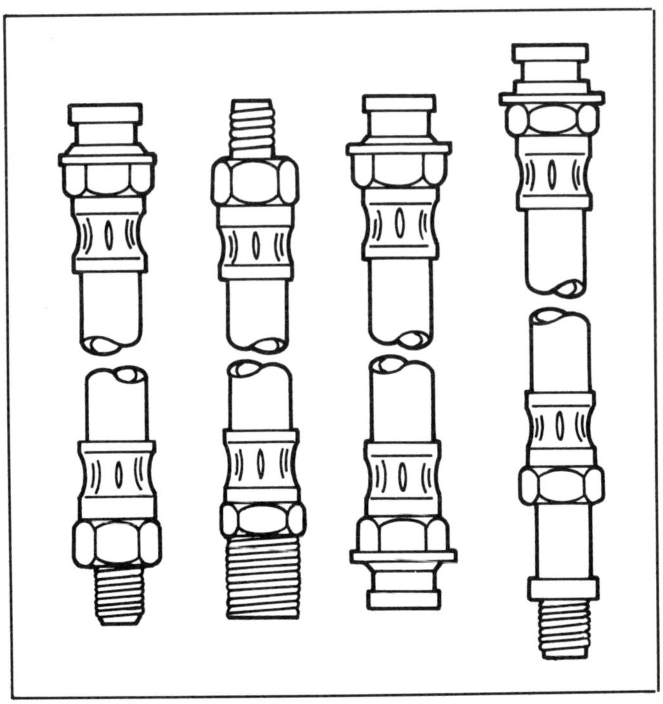

Figure 2-42 Typical hose ends

The wheel cylinder size can vary from front to rear. The variation in wheel cylinder size (diameter) is one of the factors controlling the distribution of braking force in a vehicle. Larger diameter wheel cylinders are normally specified for the front brakes of front engine passenger cars equipped with drum brakes. Bleeder screws are provided to remove air trapped in the system.

Four types of wheel cylinders are normally used with drum brakes:

- *Double Piston with Straight Bore*—This cylinder is the type most commonly used on modern automobile brakes. A cross section of a typical double-piston straight-bore cylinder is

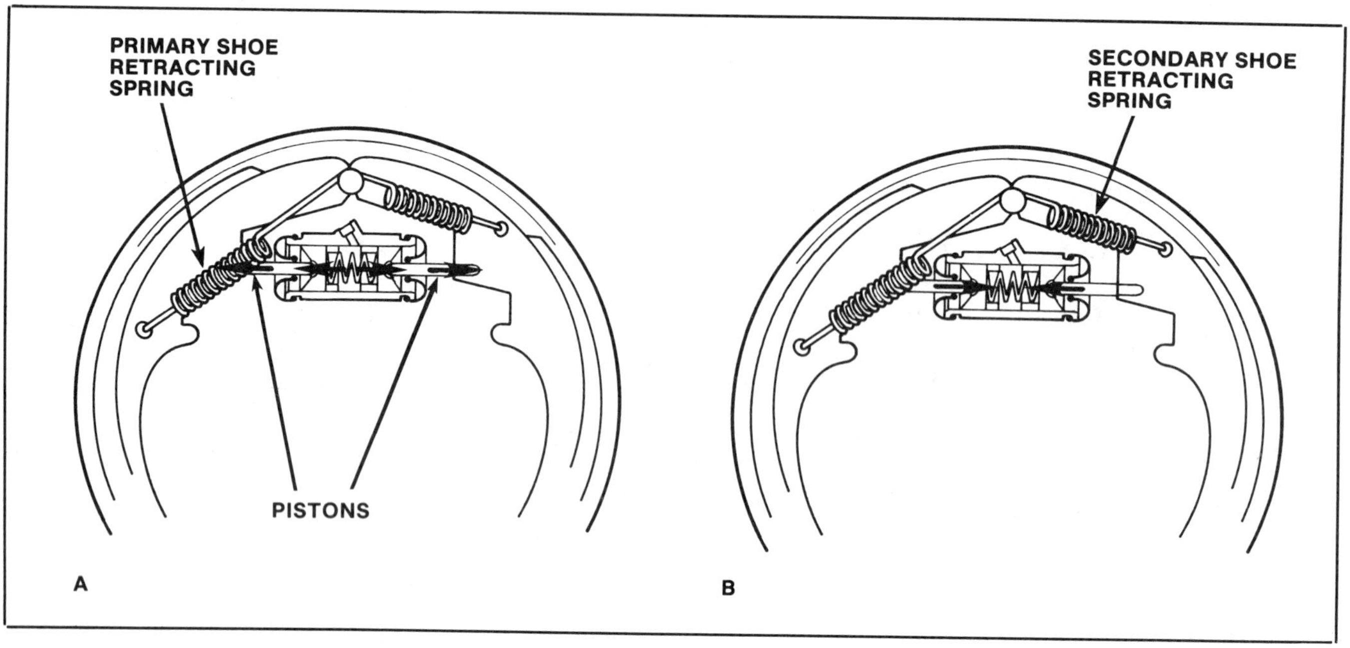

Figure 2-45 (A) Hydraulic pressure providing outward forces on brake shoes; (B) spring tension providing inward force on pistons.

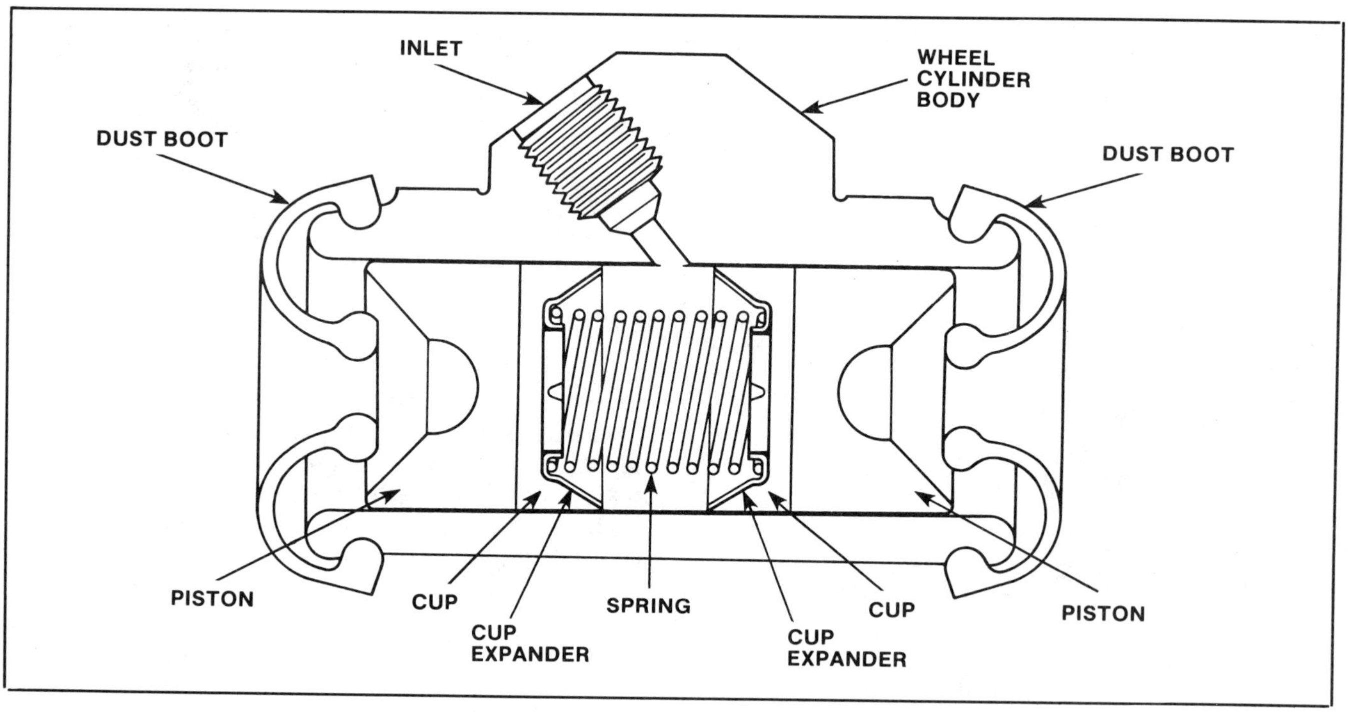

Figure 2-46 Cross section of typical wheel cylinder

shown in Figure 2-46. All of the components of a typical cylinder are shown except a bleeder screw, which is not illustrated. Figure 2-47 shows two double-piston straight-bore cylinders, but with variations in the method of attaching the dust boots to the cylinder body. The dust boots on the end or ends of the wheel cylinder seal the bore from contamination that would adversely affect its performance. While serving as a dust seal, these boots also hide minor fluid leaks. It is recommended, therefore, that the dust boot be pulled back during inspection to check for low-volume leaks.

- *Double Piston with Step Bore*—This type is used on some non-servo brakes and has the same components as the straight bore type. As shown in Figure 2-48, two different sized dust boots, cups, and pistons are used. Opposed pistons of different diameters exert different amounts of force.
- *Single Piston*—A single piston wheel cylinder has only one cup, piston, and dust boot as shown in Figure 2-49. There are single piston, fixed wheel cylinders, generally used in pairs, and single piston, floating wheel cylinders (mostly import application). On a duo-servo brake, only the primary shoe is activated. On a non-servo brake, two single piston wheel cylinders are used. When this is the case, the toe end of both shoes are forced against the drum.
- *Baffle Type Piston*—This type of cylinder (Figure 2-50) has been used in some cases to eliminate a "clunking" noise when the direction of the braking is changed from forward to reverse or reverse to forward. The baffle is drilled with a very small orifice (metering hole) at the top of the bore to restrict fluid flow from the front half to the rear half of the cylinder. This causes the primary shoe to actuate first, followed by the secondary. Even though this hole is very small, the pressure on either side of the baffle is balanced after a short period of time.

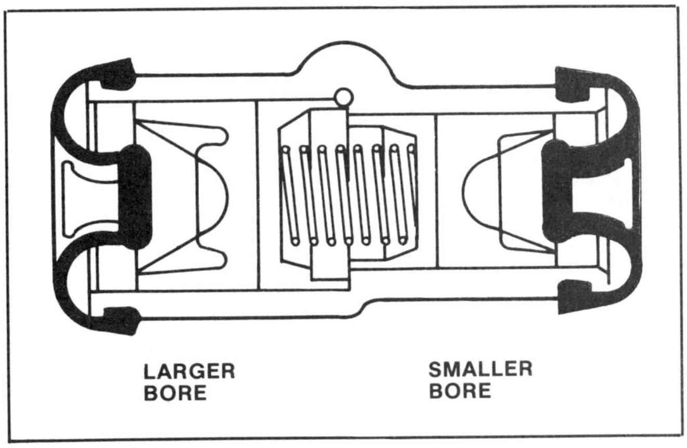

Figure 2-48 Step bore wheel cylinder

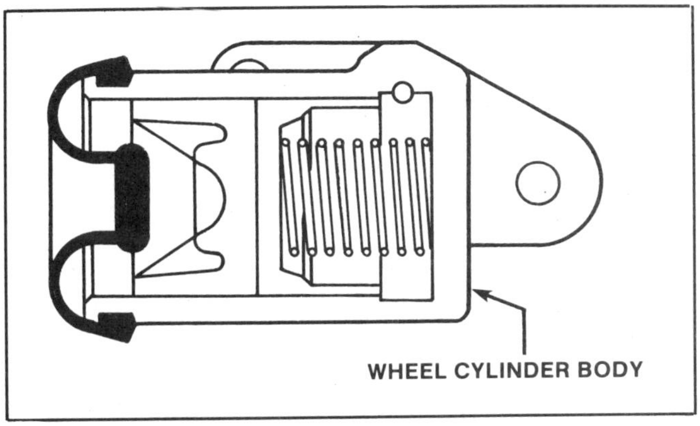

Figure 2-49 Single piston fixed wheel cylinder

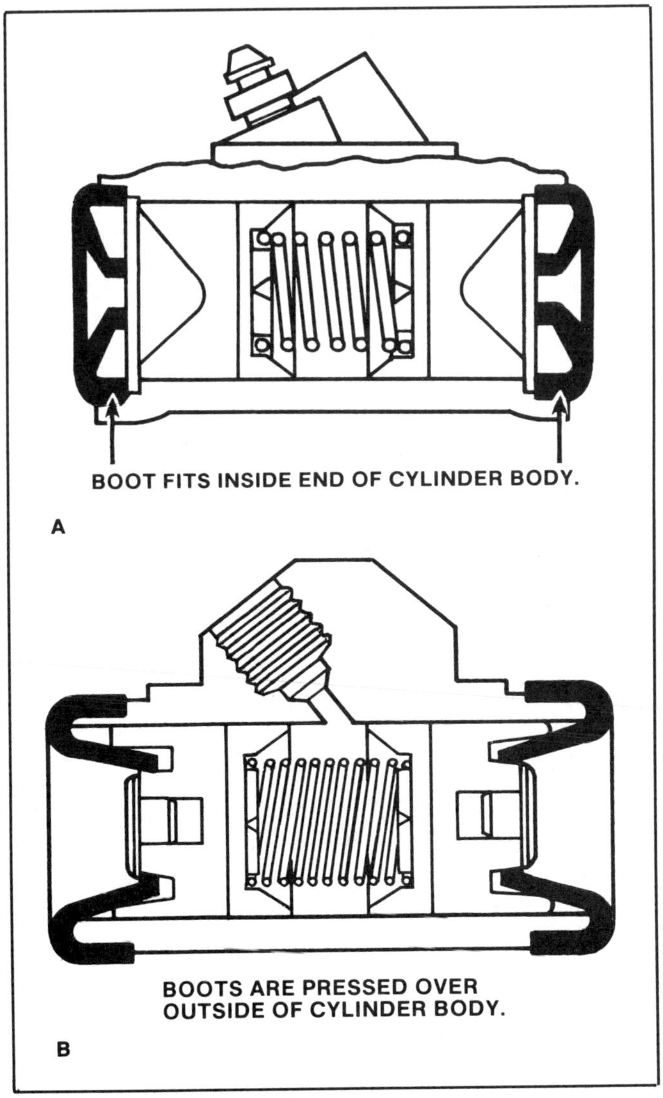

Figure 2-47 (A) Wheel cylinder with internal boots and (B) with external press-on boots

Operation of Wheel Cylinders

The space in the cylinder bore between the cups must remain filled with fluid at all times. When the brake pedal is depressed, additional brake fluid is

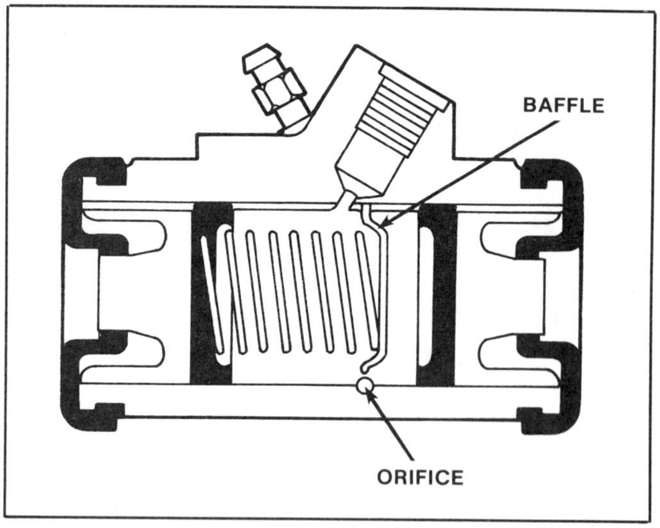

Figure 2-50 Wheel cylinder with baffle in bore

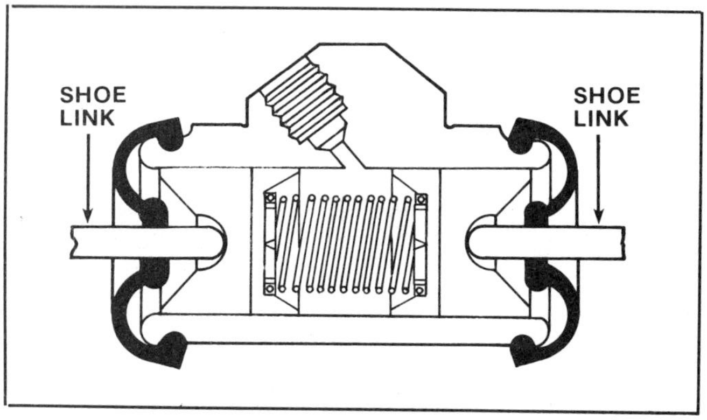

Figure 2-51 Wheel cylinder in applied position

forced into the cylinder bore. The additional fluid, which is under pressure, moves the cups and pistons outward in the bore (Figure 2-51). This piston movement forces the shoe links and brake shoes outward to contact the drum and thus apply the brakes.

SHOP TALK

On some brakes, the end of the shoe web bears directly against the pistons, and shoe links are not used.

The wheel cylinder casting is manufactured so as to produce a very exacting finish on the cylinder bore wall. This is done by means of "bearingizing" whereby the bore is sized and finished by drawing a hardened steel ball over the surface at high pressures. This produces a finish that is neither too smooth nor too coarse. A finish that is too smooth will cause the sealing cups to operate on a dry surface without lubrication and, if too coarse, will cause rapid wear of the sealing lip of the rubber cup. The "bearingized" surface provides good lubrication and wear characteristics between the cup lip and the cylinder bore.

A spring is compressed between the two rubber sealing cups to keep these cups in contact with the piston under all conditions. The cup seals have flanged edges with cup expanders to flatten them against the cylinder walls to prevent fluid leakage. A lightweight coil spring sits between them. Pushrods extend out the ends through rubber dust boots. The cylinder has a bleeder screw and brake line connection. It is usually attached to the brake drum backing plate.

DISC BRAKE CALIPERS

Disc brake calipers respond to hydraulic pressure differently from drum brake wheel cylinders. Instead of spreading shoes apart, they force the pads against the rotor, so there is no servo action to multiply stopping force. Calipers can have one, two, or four cylinders and pistons, although the single piston design (Figure 2-52) is the most popular. In two and four piston configurations, the caliper is rigidly mounted to the wheel spindle or steering knuckle, and one or a pair of pistons is found on each side of the rotor. When the brakes are applied, the pistons move outward in their bores, pushing the pads directly against the rotor. In the single piston type, the caliper is free to move from side to side, so as the piston pushes on the pad on its side of the caliper, the action/reaction principle works to transfer force to the other side, pulling the opposite pad against the rotor.

There are three different types of caliper designs, but all are built to straddle the rotor. All use one or more pistons in a cylinder bore.

1. **Fixed Caliper.** A fixed caliper design is rigidly mounted to the wheel spindle bracket (Figure 2-53). Most types are equipped with two pistons and cylinders on each side. A crossover tube connects the caliper halves. One or both sides might have bleeder screws located in the top of the caliper. When the brakes are applied, the caliper stays in place. The pistons move inward from both sides of the caliper and press the pads against the rotor. Each piston has a groove to fit a dust boot and to protect the bore.

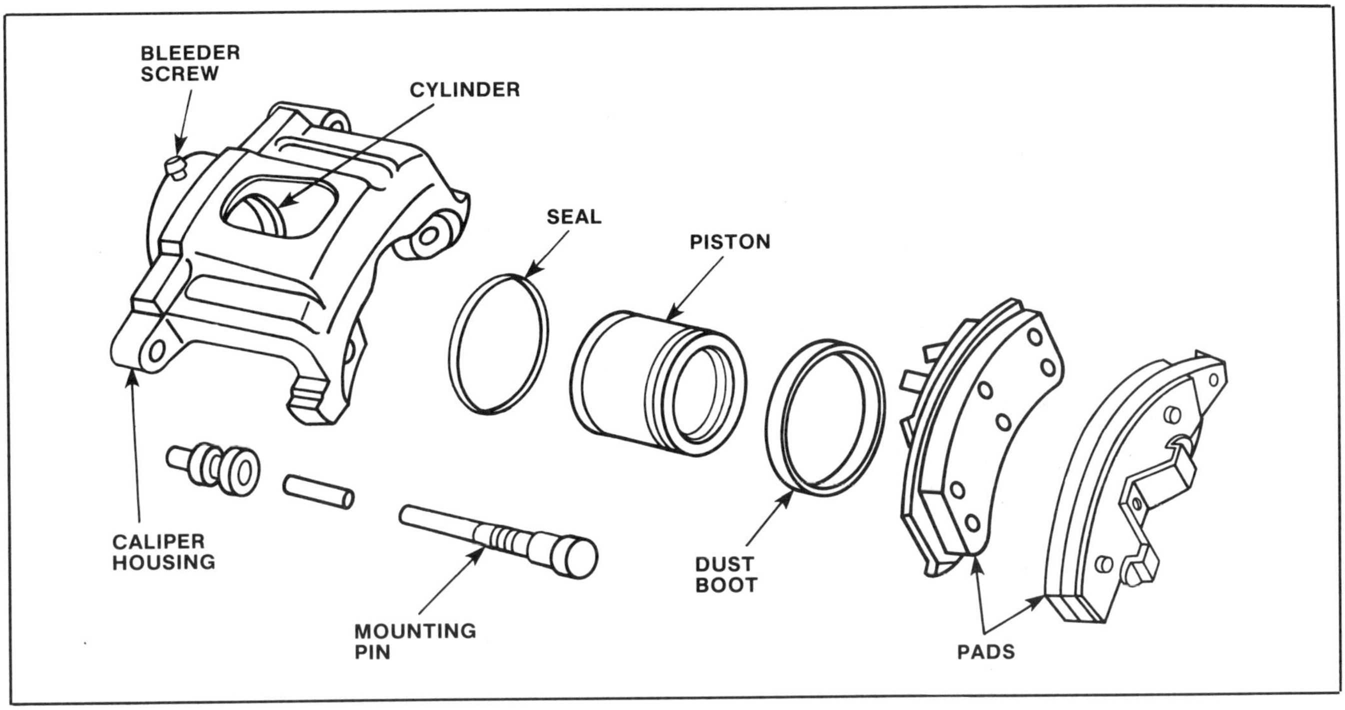

Figure 2-52 Typical single piston disc brake caliper—disassembled

2. **Sliding Caliper.** A sliding caliper moves freely from side to side in slots or machined "ways" on an anchor plate, and held by a caliper support key (Figure 2-54). Both sides have pads, but the pistons and bores are only on the inboard half of the caliper. The inboard side is also where both the inlet port and bleeder screw are. There is a dust boot at the end of the piston.

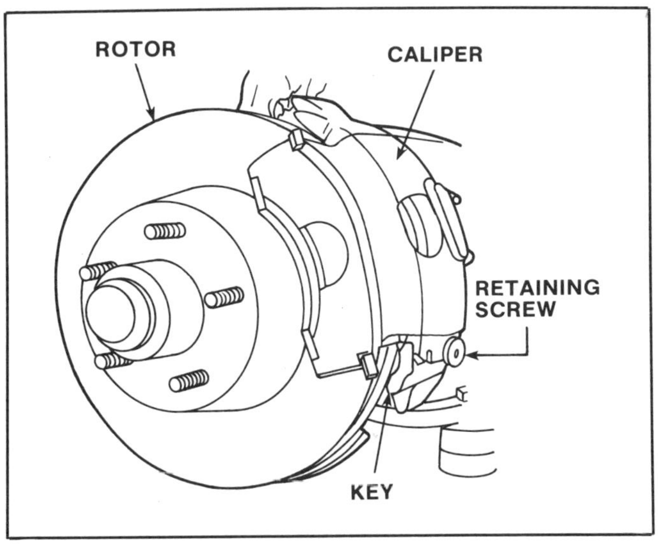

Figure 2-54 Typical sliding caliper disc brake—one piston

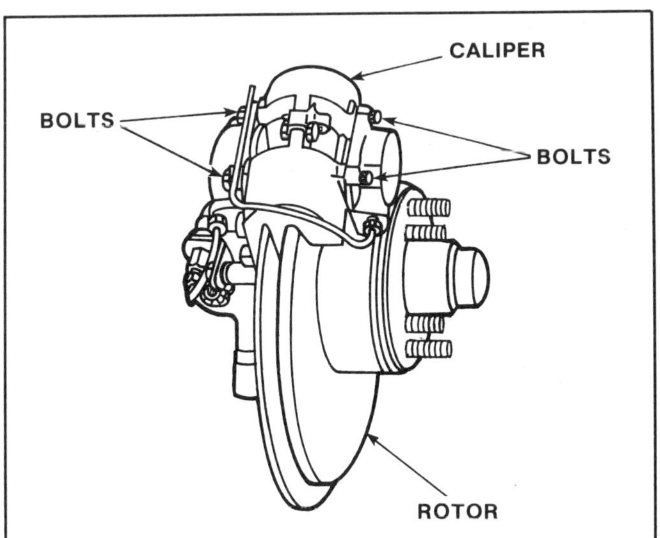

Figure 2-53 Typical fixed caliper

3. **Floating Caliper.** Floating calipers operate principle as the sliding type, but the caliper rides back and forth on guide pins held by bushings and O-rings. The single piston floating caliper is a common design. As with

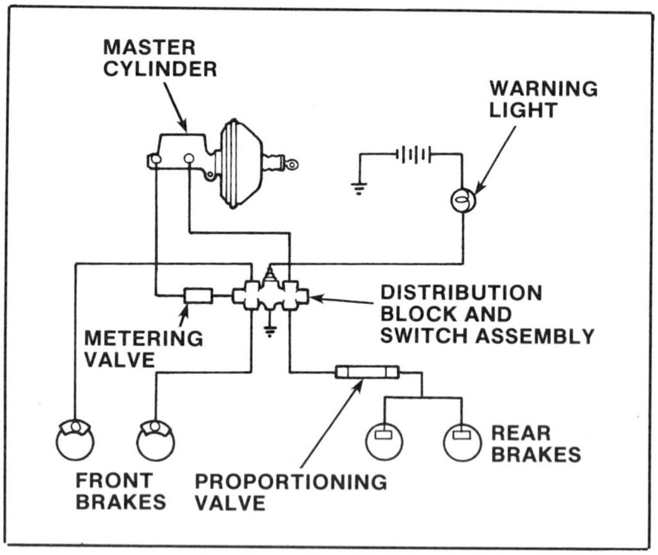

Figure 2-55 Brake system with safety equipment

the sliding caliper, pads rest lightly against the rotor. They grip with quite a small movement of the piston. When released, the piston does not retract very far. By being so close to the rotor there is little delay in brake application.

Complete details on the operation and servicing of fixed, sliding, and floating disc caliper brakes are given in Chapters 7 and 8.

BRAKE SAFETY SWITCHES AND VALVES

Switches and valves installed in the vehicle's brake system are the principal safety devices (Figure 2-55). For example, if failure should occur in either brake system, the pressure loss differential activates a brake failure warning switch to light the brake warning light in the instrument panel

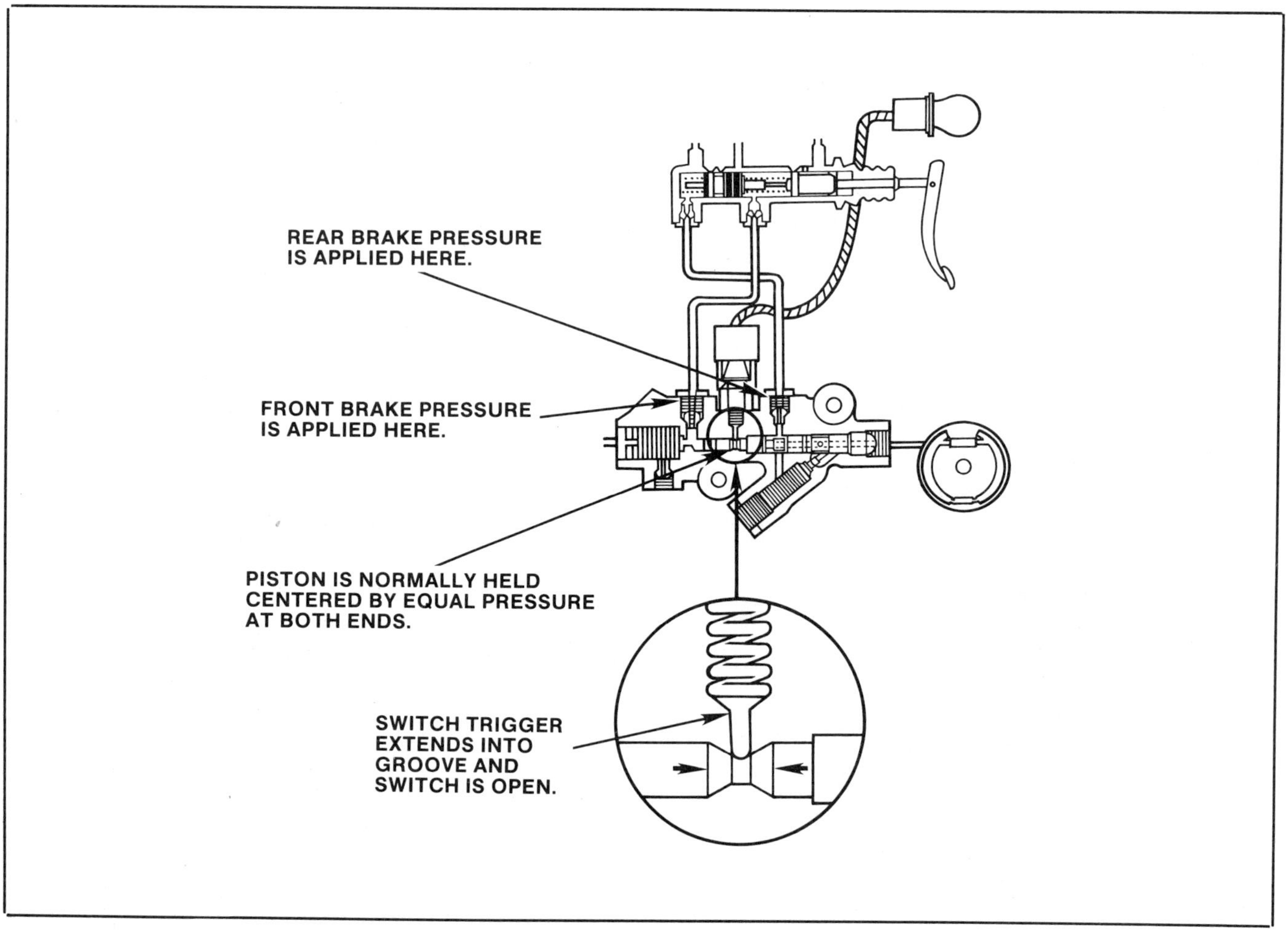

Figure 2-56 Pressure differential valve under normal conditions

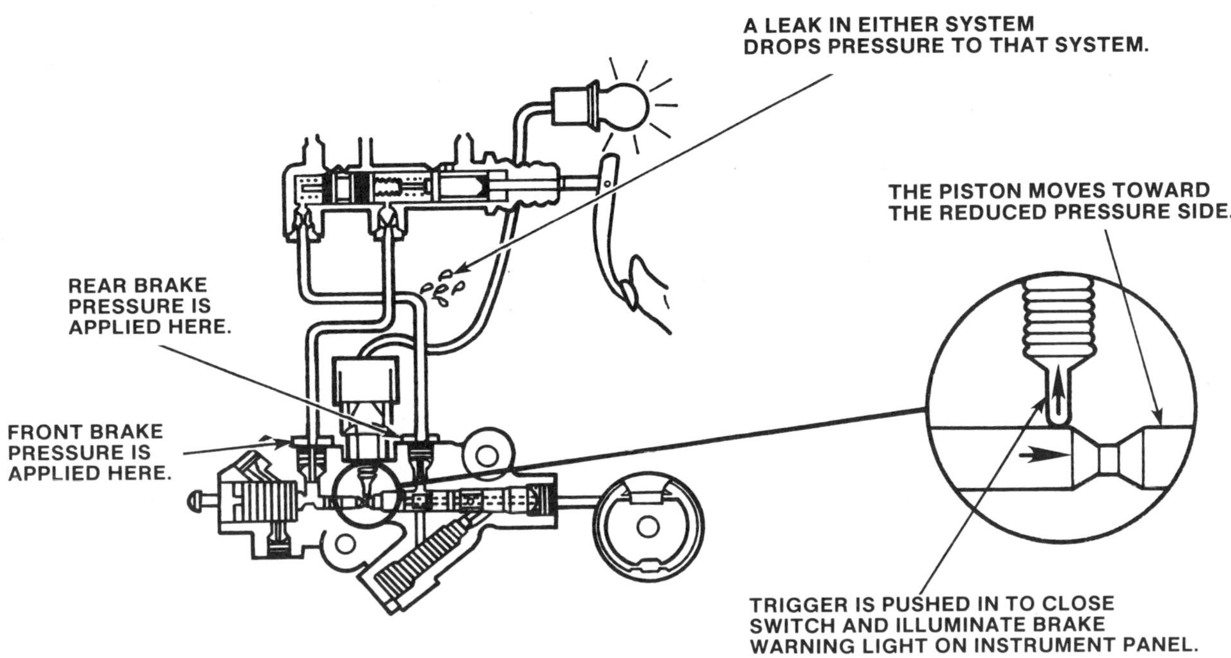

Figure 2-57 Pressure differential valve unseated

and alert the driver to the failure. Apart from the hydraulic system, the brake warning light serves a second purpose. It signals that the parking brake is set. This occurs when the ignition switch is turned on and the parking brake is applied.

Most modern auto systems have discs on the front and drums on the rear. Both types of brakes work differently and activate at different levels of pressure. Discs require higher hydraulic pressure than drums and respond quickly. Drums must overcome the tension of return springs before they activate, but require lower line pressure. Because the two types of brakes differ in action and hydraulic requirements, metering (hold-off) valves are used to balance pressure in the system. A metering valve momentarily delays front disc braking action until the rear drum brake shoes are forced against the brake drums. The proportioning valve (or P-valve) balances front-to-rear braking action as the car comes to quick stops; and, it prevents premature locking of the rear wheels.

BRAKE WARNING LIGHT (PRESSURE DIFFERENTIAL SWITCH)

A pressure differential valve is used in all dual brake systems to operate a warning light switch. Its main purpose is to tell the drive if pressure is lost in either of the two hydraulic systems. Since each brake hydraulic system functions independently, it is possible that the driver might not notice right away that pressure and braking are lost. When a pressure loss occurs, brake pedal travel will increase and a more-than-usual effort will be needed for braking. Should the driver not notice the extra effort needed, the warning light actuated by the hydraulic system safety switch serves to alert him/her.

Under normal conditions the hydraulic pressure on each side of the pressure differential valve piston is balanced (Figure 2-56). The piston is located at its center point, so that the spring-loaded warnings switch plunger fits into the tapered groove of the piston. This leaves the contacts of the warning switch open; the brake warning light stays off.

If there is a leak in the front or rear braking system, the hydraulic pressure in the two systems will be unequal. For example, if there is a leak in the system supplying the front brakes, there will be lower pressure in the front system when the brake pedal is applied. The hydraulic pressure in the rear system then pushes the piston toward the front side, where the pressure is lower. As the piston moves the plunger is pushed out. This closes the switch and illuminates the brake warning light, as shown in Figure 2-57.

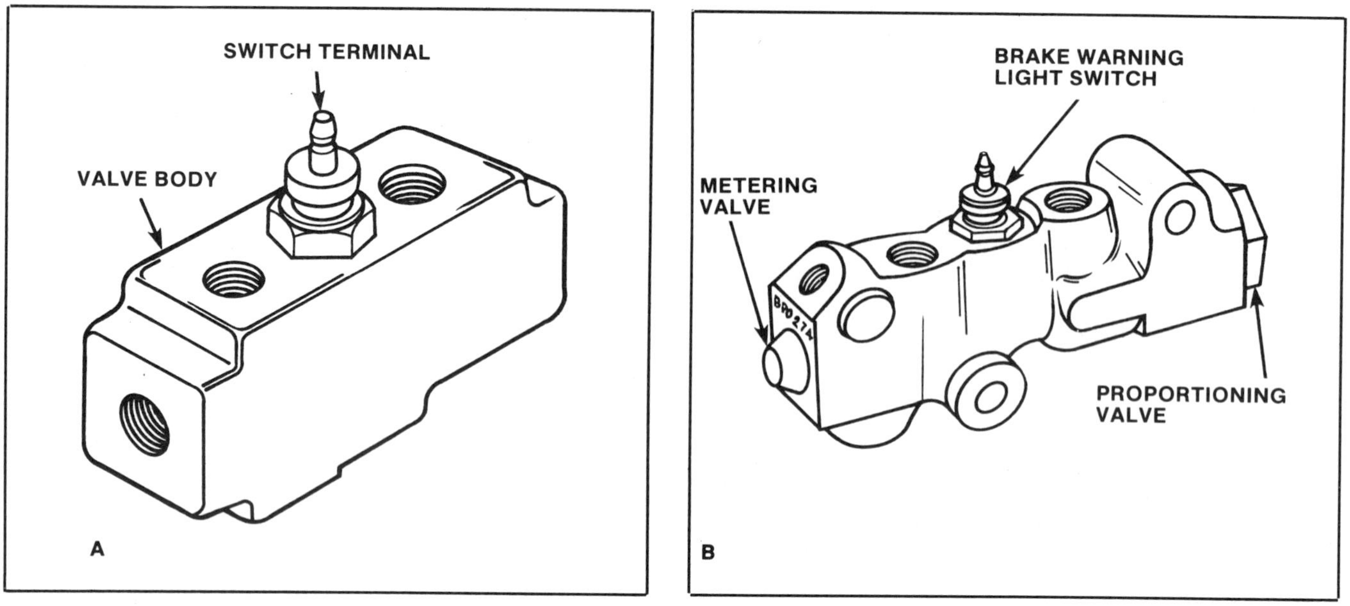

Figure 2-58 (A) Typical brake warning light switch; (B) three-function combination valve

While all brake warning light switches serve the same function, there are three common variations in the design of these switches. These variations include switches with centering springs, without centering springs, and with centering springs and two pistons. Separate pressure differential valves and brake warning light switches (Figure 2-58A) are sometimes found in vehicles; however, the trend today is toward the combination valves (Figure 2-58B) discussed later in this chapter. The brake warning light can also be integrated with the master cylinder.

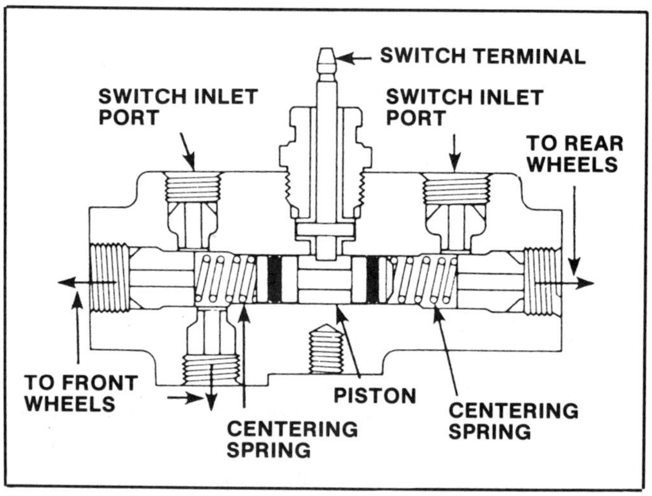

Figure 2-59 Cross section of switch with centering springs

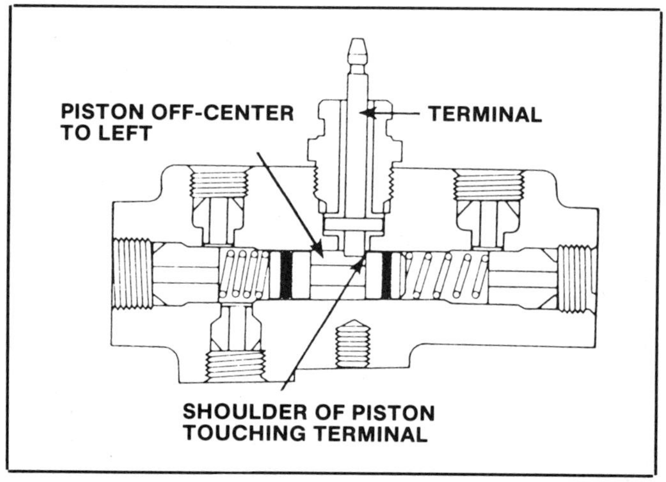

Figure 2-60 Loss of pressure in front hydraulic system

Switch with Centering Springs

A typical switch of this type consists of a valve body, a piston, two centering springs, a switch terminal, and the inlets and outlets. The hydraulic lines from the master cylinder are connected to the two switch inlet ports as shown in Figure 2-59. Under normal conditions, the pressure is equal on each end of the piston in the switch bore. With the pressure equal, the springs hold the piston in the centered position, and the piston does not touch the switch that projects into the valve bore.

If there is a failure in one of the hydraulic systems, the pressure is no longer equal on each end of the piston. The higher hydraulic pressure in the operative side moves the piston toward the inoperative side of the switch.

Figure 2-60 illustrates what takes place if the brakes are applied and there is a leak or loss of pressure in the front hydraulic system. The higher pressure in the rear hydraulic system has moved the piston toward the front hydraulic side of the switch. The shoulder of the piston touches the switch terminal. This provides a ground for the brake warning light circuit and causes the warning light to come on.

When the brakes are released and the hydraulic pressure again is equal on each end of the piston, the springs move the piston to the centered position. This breaks the electrical contact, and the warning light should go out.

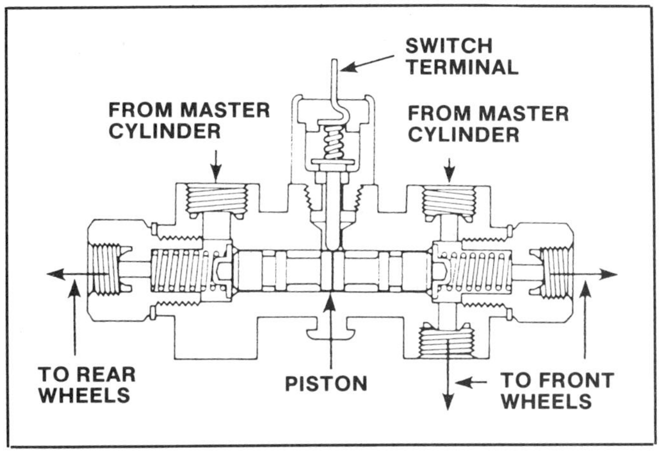

Figure 2-63 Cross section of switch with centering springs and two pistons

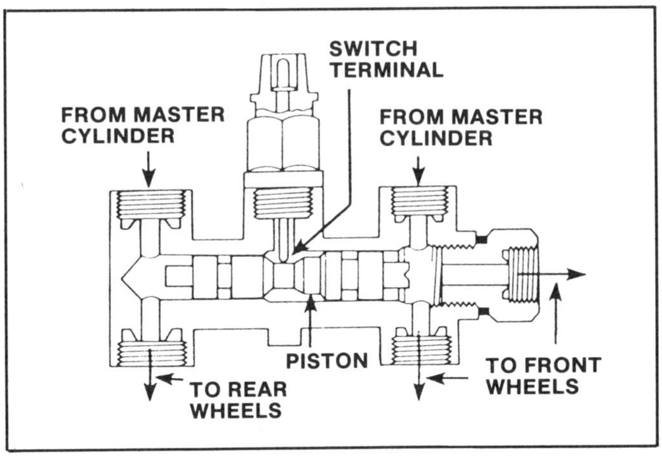

Figure 2-61 Cross section of switch without centering springs

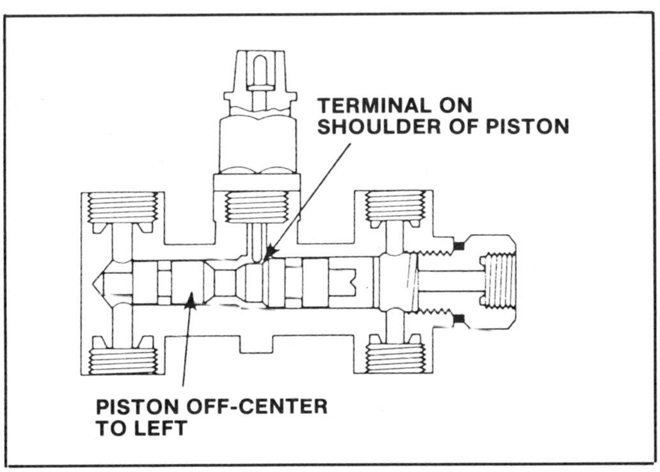

Figure 2-62 Loss of pressure in rear hydraulic system

Switch without Centering Springs

Except for the springs, this switch includes the same parts as the switch with centering springs. The appearance of the piston and the switch terminal is different (Figure 2-61).

Figure 2-62 illustrates the movement that would take place if the brakes were applied and there was a leak or loss of pressure in the rear hydraulic system. The higher pressure in the front hydraulic system has moved the piston toward the rear hydraulic side of the switch. The switch terminal has been moved upward by the shoulder of the piston. This closes the circuit and causes the brake warning light to come on.

With this type of switch, the warning light will remain on until the piston is centered in the bore of the switch.

Switch with Centering Springs and Two Pistons

This switch contains the same functional parts as the other two switches, except that it has two pistons (Figure 2-63). In the normal position, the shoulders on the end of the pistons hold the switch terminal in the UP position. This keeps the electrical circuit open.

Figure 2-64 illustrates the movement that would occur if the brakes were applied and there was a leak or loss of pressure in the front hydraulic system. The higher pressure in the rear hydraulic system has moved the two pistons toward the front hydraulic side of the switch. This permits the

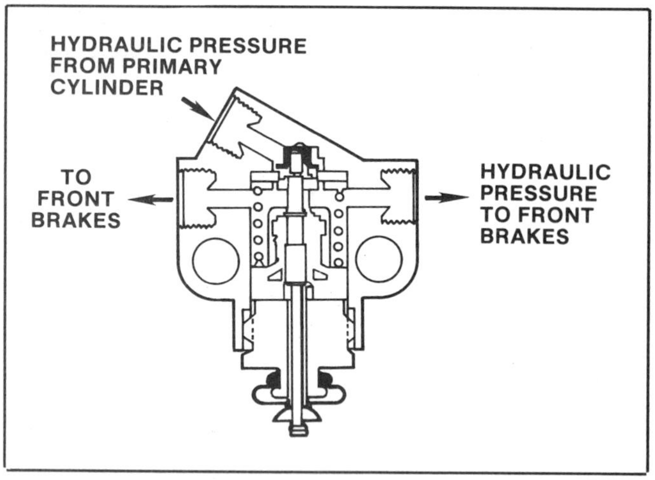

Figure 2-64 Metering valve

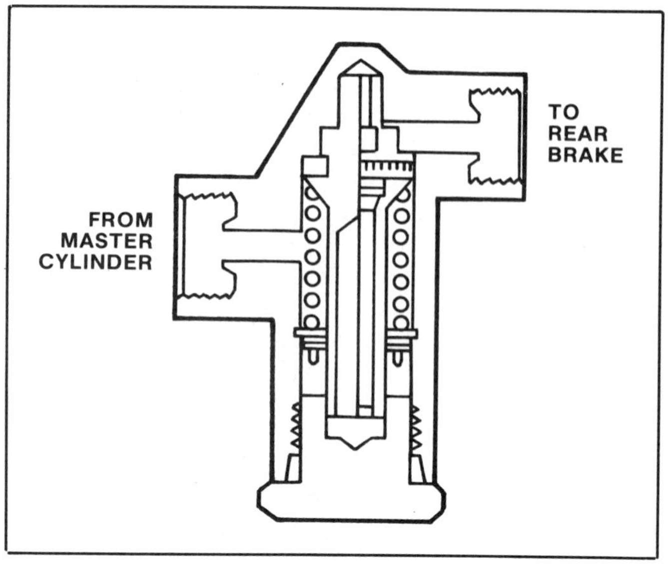

Figure 2-65 Typical proportioning valve—cross section and disassembly

switch terminal to move downward and close the electrical circuit to the brake warning light.

The pistons will remain off-center and the warning light will remain on until the switch terminal is removed from the switch body. Removing the terminal permits the centering springs to center the pistons in the switch bore.

> **CAUTION:** With this type switch on certain cars and trucks, the switch terminal must be removed before bleeding the hydraulic system to prevent damage to the unit. Always check the vehicle's service manual before bleeding a hydraulic system that contains a switch with centering springs and two pistons.

METERING AND PROPORTIONING VALVES

Along with the typical brake warning switch, most vehicles have some type of hydraulic control valve in the brake system. Metering and proportioning valves are used to balance the braking characteristics of disc and drum brakes. Some cars, in particular earlier models, use similar components separately to suit the type of brake system they employ. Newer cars with front disc/rear drum brake systems combine the three functions into a single component that will be explained later.

The braking response of disc brakes is immediate when the brake pedal is applied. It is directly proportional to the effort applied at the pedal. Drum brake response is delayed while rear brake hydraulic pressure moves the wheel cylinder pistons to overcome the force of their return springs and force the brake shoes to contact the drum. Their actions are self-energizing and tend to multiply the pedal effort.

Metering Valves

A metering valve in the front brake line holds off pressure going from the master cylinder to the front disc calipers. This delay allows pressure to build up in the rear drums first. When the rear brakes begin to take hold, the hydraulic pressure builds to the level needed to open the metering valve (Figure 2-65). When the metering valve opens, line pressure is high enough to operate the front discs. This process provides for better balance of the front and rear brakes. It also prevents lockup of the front brakes by keeping pressure from them until the rear brakes have started to operate. The metering valve has the most effect at the start of each brake operation and all during light braking conditions.

Proportioning Valves

The self-energizing action of the delayed response rear drum brakes can cause them to lock the rear wheels at a lower hydraulic pressure than the front brakes, which react immediately to pressure upon brake pedal application. The proportioning valve is used to control rear brake pressures, particularly during hard stops. When the pressure to the rear brakes reaches a specified level, the proportioning valve overcomes the force of its

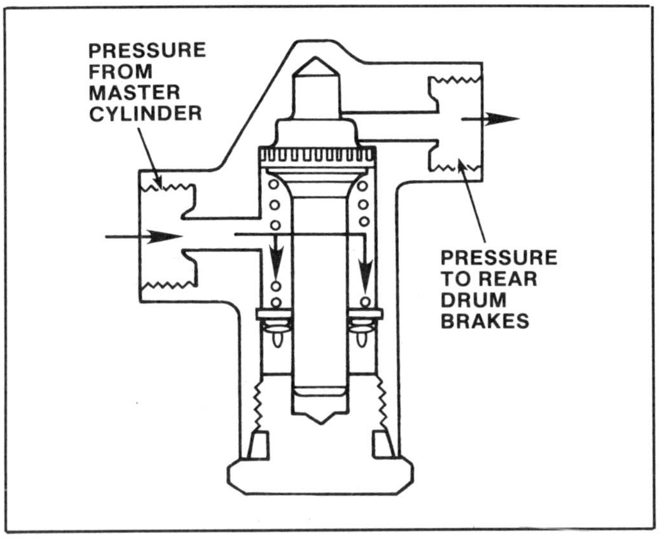

Figure 2-66 Proportioning valve

spring-loaded piston. By doing so, it regulates rear brake system pressure and adjusts for the difference in pressure between front and rear brake systems. This keeps front and rear braking forces in balance.

The proportioning unit delivers full pressure to the rear brakes up to a certain level. This is called the *split point*. Beyond this point, increasing pressure to the brakes is reduced according to a certain ratio. Thus, pressure transmitted to both front and rear brakes under light brake pedal application is approximately equal.

The proportioning valve (Figure 2-66) does not function during normal braking, when low output pressures are sufficient to actuate the rear brakes. Fluid flows into the proportioner. It moves around the insert, through the angular hole, and into the proportioning valve. It continues past the open seat into the center hole of the insert and, then, out to the rear brakes. The proportioner plunger is spring loaded to keep the valve open until predetermined split point pressures are exceeded.

Higher pressures resulting from hard brake pedal application overcome the spring's resistance. The plunger closes the valve. Output pressure to the rear brakes is kept at the relatively low split point pressure. The valve must be unseated before further fluid flow can reach the rear brakes. This occurs when additional pressure from the master cylinder, due to hard braking, creates an imbalance in pressure in the annular space between the valve seat and the O-ring. This condition causes the plunger to open the valve. Pressure to the rear brakes can then continue to increase at a ratio determined by the cross-sectional area of the annular space and the area of the exposed valve plate. For example, increasing pressure to the rear brakes can take place in some valve designs at a ratio of 3:7 when split point pressures from 220 to 470 PSI are exceeded.

A proportioning valve is hooked into either the rear brake line or the rear outlet ports of the master cylinder. Increasing pedal force sends more pressure to the front brakes, while the proportioning valve limits the pressure sent to the rear brakes. At a certain pressure level, the proportioner allows increases at a slower rate than the front. This way, more braking capacity shifts to the front at the same time the load shifts forward. The possibility of the car skidding is reduced.

SHOP TALK

Distribution of weight on some vehicles, such as station wagons, is such that a rear-to-front weight transfer does not present a similar problem. For this reason, standard proportioning valves are not required on all station wagon models.

Dual-Body Proportioning Valve. In some applications, such as diagonal split brake systems, a dual-body proportioning valve is used. The valve body is generally located in the engine compartment mounted to the firewall or fenderwell. The brake lines feed directly from the master cyl-

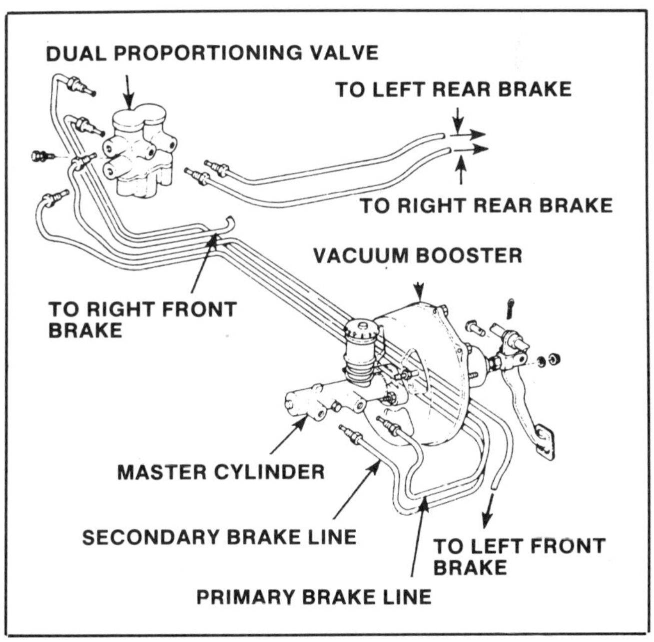

Figure 2-67 Dual-body proportioning valve

inder to the proportioning valve where the system is divided diagonally (Figure 2–67).

Height-Sensing, Brake-Proportioning Valve. The previously mentioned valves limit rear brake pressure when a pre-set value is reached. Therefore, the pre-set proportioning valve (Figure 2-68) cannot accurately compensate for variations in weight load and transfer in a panic stop. On certain light trucks there is a height-sensing, brake-proportioning (or regulated-proportioning) valve. This valve is located on a rear cross member and is activated by a linkage connected to the rear

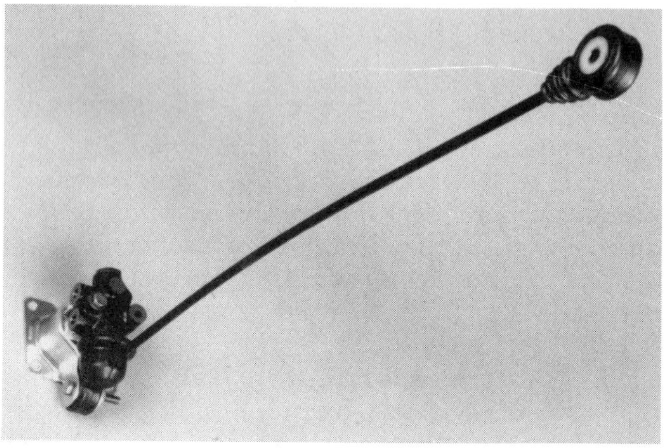

Figure 2-68 Typical load-sensing proportioning valve

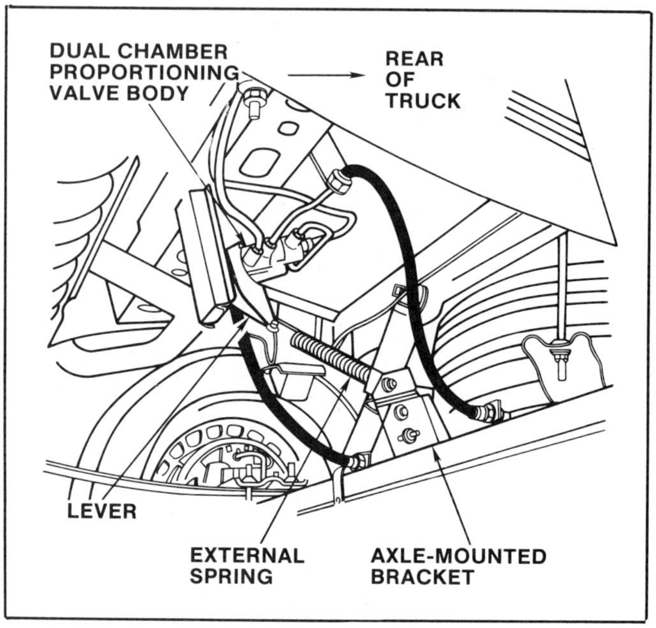

Figure 2-69 Height-sensing, brake-proportioning valve

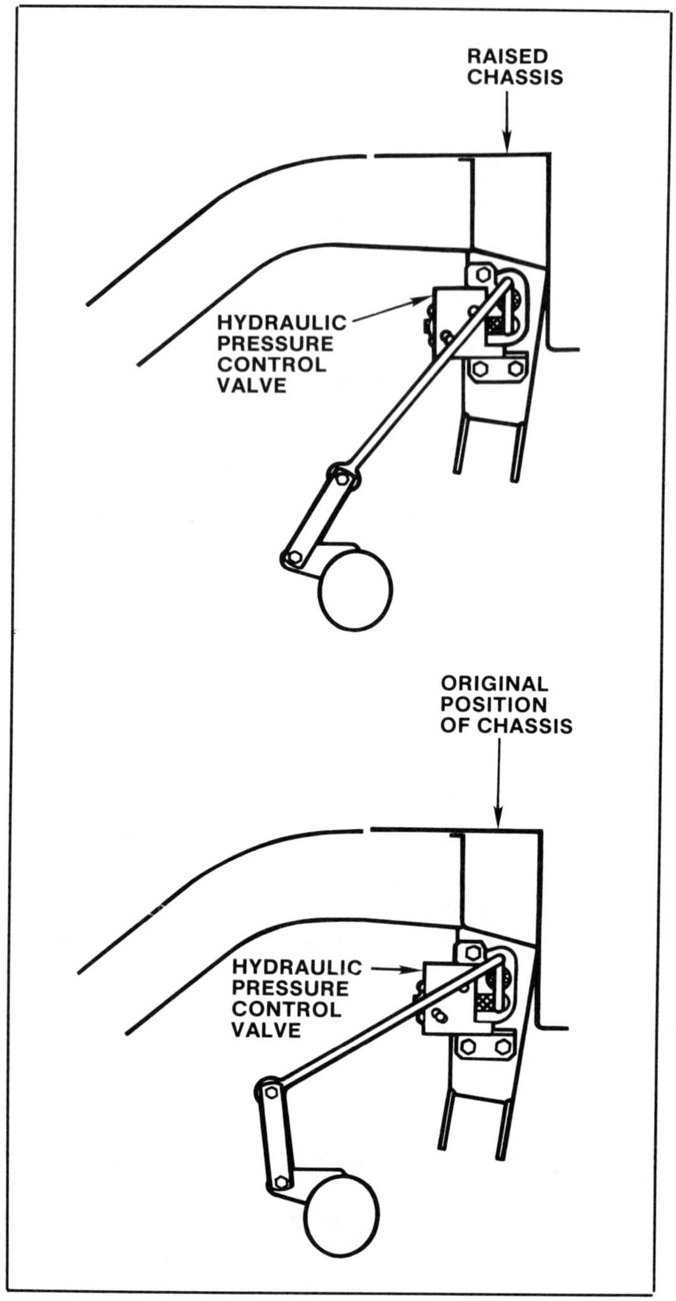

Figure 2-70 Hydraulic pressure control valve actuation

axle housing cover, as shown in Figure 2–69. On passenger cars, the height-sensing, brake-proportioning valve is attached to the vehicle underbody rear floor crossmember on the left side between the rear suspension arms. This valve functions in the following manner (Figure 2–70).

When the vehicle is unloaded, the valve closes partially, reducing the hydraulic pressure to the rear brakes. As the vehicle is loaded, the weight

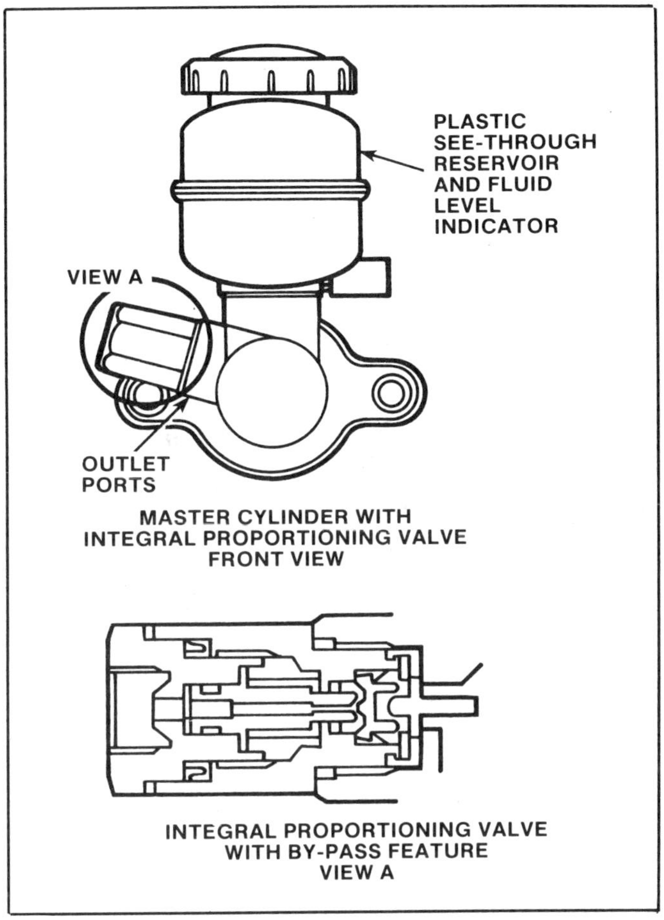

Figure 2-71 Master cylinder assembly and integral proportioning valve

presses the back end of the vehicle down on the springs. This activates the valve, which opens, supplying more hydraulic pressure to the rear wheels. When the vehicle is in the fully loaded condition, the valve is fully open providing full hydraulic pressure to the rear brakes.

CAUTION: Use of lift kits or altering vehicle riding height is not recommended on vehicles equipped with a height-sensing, brake-proportioning valve.

The proportioning valve should be inspected whenever the brakes are serviced. If the valve is leaking, it is defective and should be replaced. Proportioning valves are non-adjustable and non-repairable. If a valve is defective, it should be replaced as an assembly.

Integral Proportioning Valve. The proportioning valve is integral to the master cylinder on many of the newer light trucks and vans. The valve functions to recirculate rear brake system hydraulic pressure. A plug and tube seat is pressed into the rear end of the central bore. A hole through the center of the plug and tube seat directs the hydraulic fluid down the valve to the rear brake system outlet port. In case of the front brake system malfunction, the proportioning valve with a bypass feature allows full hydraulic pressure to the rear brake system (Figure 2-71).

COMBINATION VALVES

Most newer cars have a combination valve in their hydraulic system (Figure 2-72). This valve is simply a single unit that combines the metering and proportioning valves with the pressure differential valve. Combination valves are described as three-function or two-function valves, depending on the number of functions they perform in the hydraulic system.

Three-Function Valves. This type of valve (Figure 2-73) performs the functions of the

- Metering Valve
- Brake Warning Light Switch
- Proportioning Valve

Two-Function Valves. There are two variations of the two-function combination valve. One variation (Figure 2-74) performs the proportioning

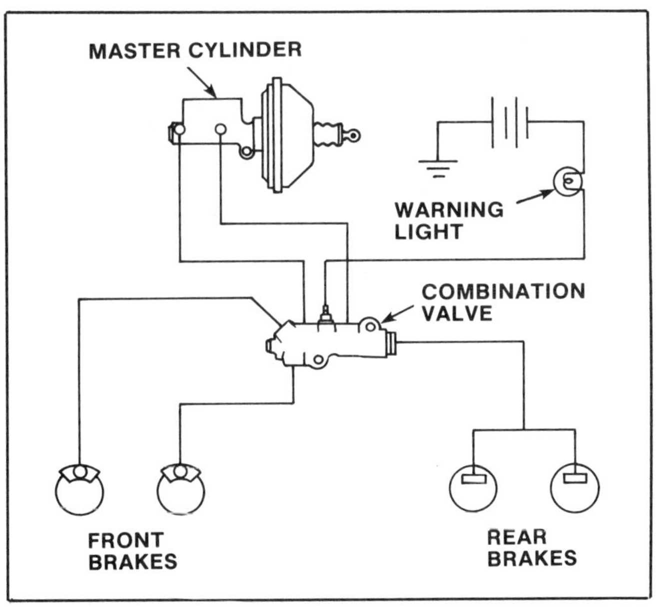

Figure 2-72 Brake system with combination valve

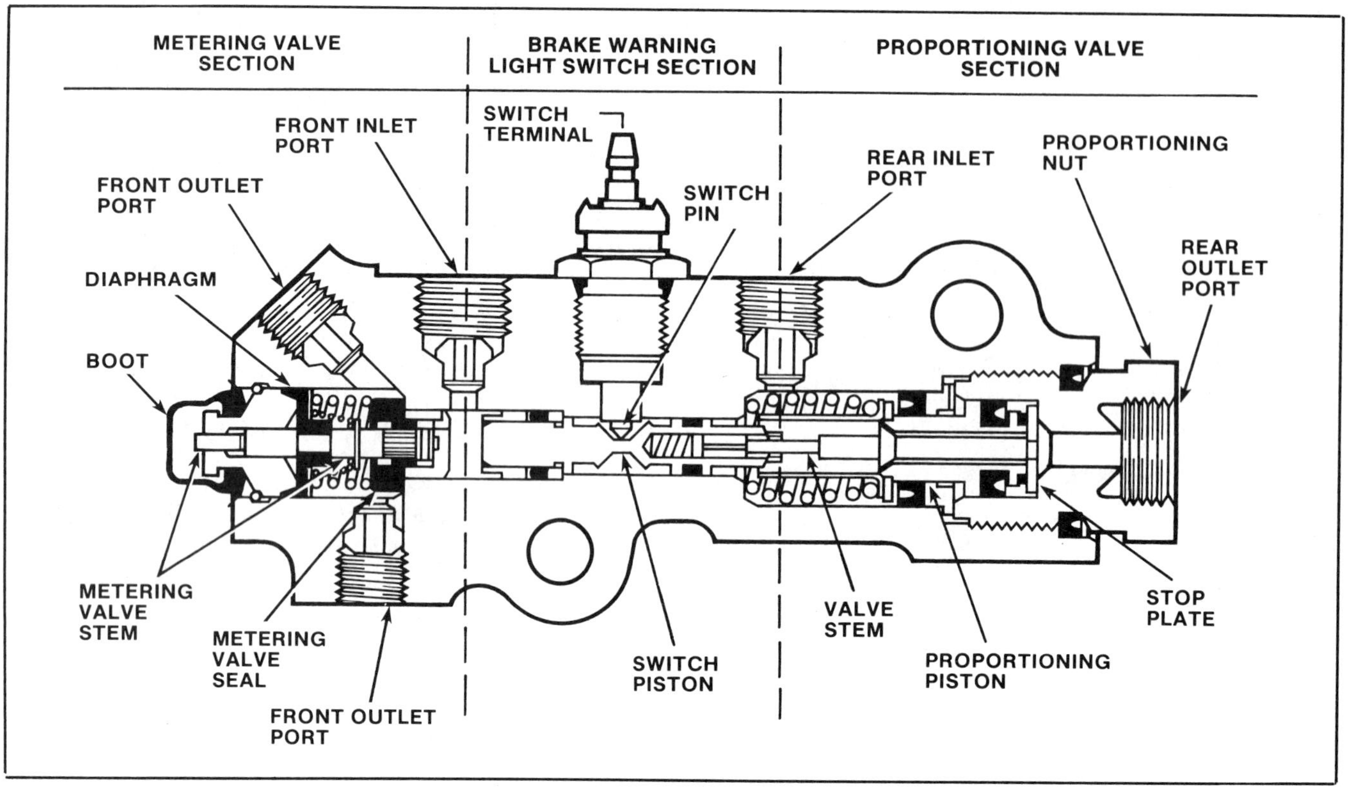

Figure 2-73 A three-function combination valve

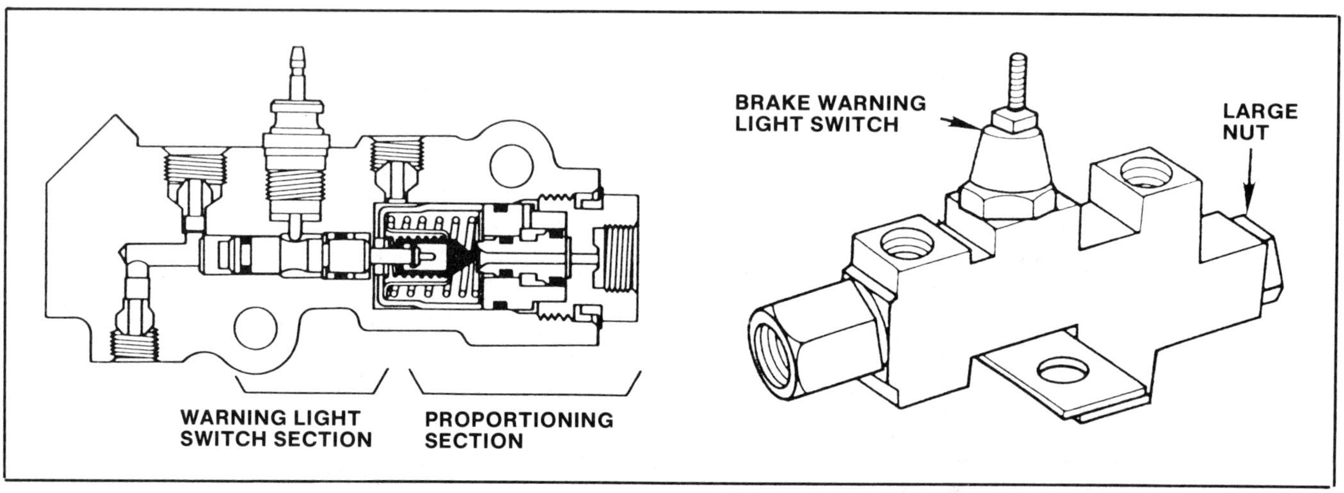

Figure 2-74 Two-function valve (proportioning/brake warning light)

valve and brake warning light switch functions. The other variation, shown in Figure 2-75, performs the metering valve and brake warning light switch function.

If any one of its several operations fails, the entire combination valve must be replaced, because these units are not repairable.

STOPLIGHT SWITCH

The stoplight (stoplight/speed control) switch and mounting bracket assembly (Figure 2-76) is attached to the brake pedal bracket and is activated by pressing the brake pedal. There are two general types of stoplight switches:

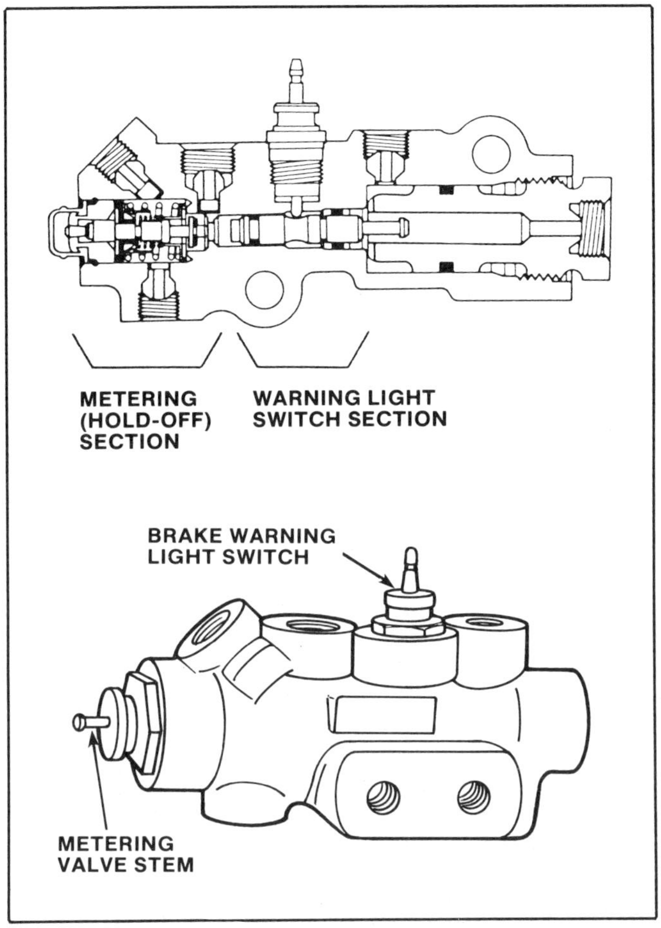

Figure 2-75 Two-function valve (metering/ brake warning light)

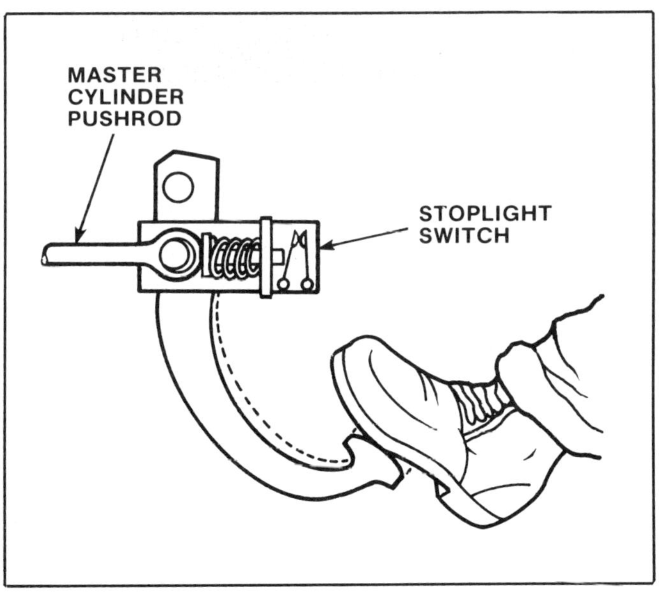

Figure 2-76 Mechanically activated stoplight switch

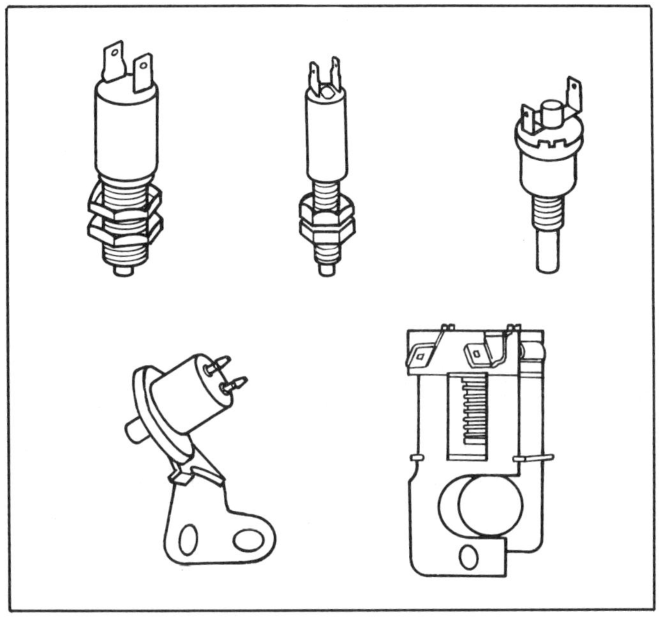

Figure 2-77 Mechanically operated switches

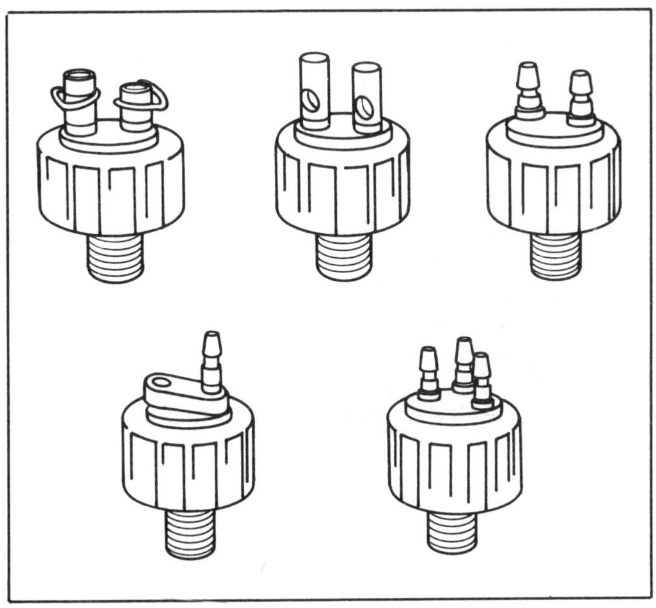

Figure 2-78 Hydraulically operated switches

- *Mechanically operated switches (Figure 2-77)*—Used on most of the recent model cars.
- *Hydraulically operated switches (Figure 2-78)*—Used on some older model cars.

These switches, as can be seen in the illustrations, are supplied in many different shapes and sizes to comply with the car manufacturer's requirements.

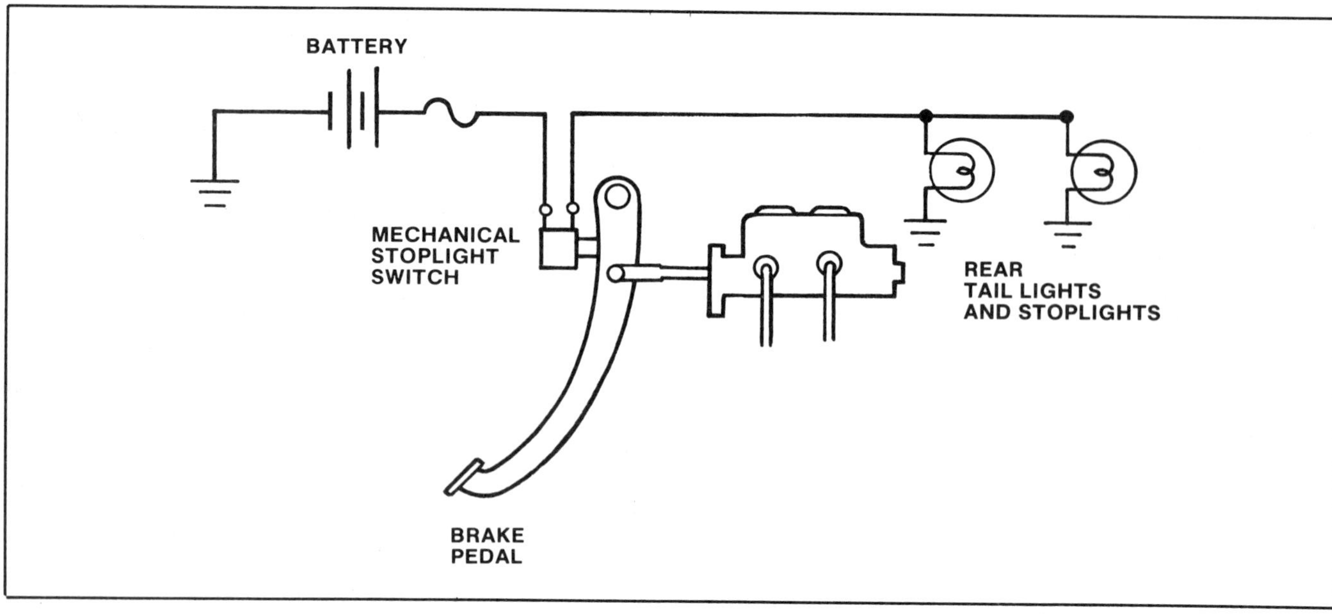

Figure 2-79 Stoplight switch layout

The mechanical stoplight switch is operated by contact with the brake pedal or with a bracket attached to the pedal. The hydraulic switch is operated by hydraulic pressure developed in the master cylinder. In both types, the circuit through the switch is *open* when the brake pedal is released. When the brakes are applied, the circuit through the switch *closes* and causes the stoplights to come on (Figure 2-79).

BRAKE FLUID

Brake fluid is the lifeblood of any hydraulic brake system. It is what makes the system operate properly.

Modern brake fluids are specially blended to enable them to perform a variety of functions.

- *Temperature Extremes*—Brake fluids must be able to flow freely at extremely high temperatures (500 degrees F) and at very low temperatures (-104 degrees F).
- *Lubricating Ability*—Brake fluid must serve as a lubricant to the many parts with which it comes into contact to insure smooth and even operation.
- *Anti-Corrosion, Anti-Rust*—In addition, brake fluid must fight corrosion and rust in the brake lines and various assemblies and components it services.
- *Resist Evaporation*—Another important property of brake fluid is that it must resist evaporation.
- *Compatible with Rubber*—Some of the earliest brake fluids had chemicals in them that ate away at the rubber components in the brake system (i.e., cups and seals). Modern brake fluid must be compatible with rubber to avoid damage to the cups and seals in the system.
- *Controlled Amount of Swell*—Brake fluids must provide a controlled amount of swell to the brake system cups and seals. There must be just enough swell to form a good seal. However, the swell cannot be too great, because if it is, it will cause drag and poor brake response.
- *SAE/DOT*—Every can of brake fluid carries the identification letters of SAE and DOT. These letters (and corresponding numbers) indicate the nature, blend, and performance characteristics of that particular brand of brake fluid. SAE comes from the Society of Automotive Engineers. In 1945, the Society of Automotive Engineers drew up some basic specifications for the manufacture of brake fluids to insure quality, safety, and performance. In 1963, the US Department of Transportation (DOT) adopted the SAE's standards. These standards were updated again in 1972.

There are three basic types or classifications of hydraulic brake fluids:

- *DOT 3*—A conventional type brake fluid with a minimum dry ERBP (Equilibrium Reflux Boiling Point) of 401 degrees F (205 degrees C) and a minimum wet ERBP of 284 degrees F (140 degrees C).
- *DOT 4*—This is a conventional type brake fluid with a minimum dry ERBP of 446 degrees F (230 degrees C) and a minimum wet ERBP of 356 degrees F (180 degrees C).
- *DOT 5*—A unique silicone-based brake fluid with a minimum dry ERBP of 500 degrees F (260 degrees C) and a minimum wet ERBP of 356 degrees F (180 degrees C).

Dot 4 is the most commonly used brake fluid for most conventional brake systems, while DOT 3 is generally recommended for most ABS systems and some power brake setups. In the last few years, DOT 5 has lost its demand by brake servicing experts. However, remember that it is best to follow the vehicle manufacturer's recommendations.

CAUTION: Only use approved brake fluid in a brake system. Any other lubricant, such as power steering fluid, automatic transmission fluid, or engine oil, which has a petroleum base, *must never be used in the brake system.* Petroleum-based fluids will attack the rubber components in the brake system, like the piston cups and seals, and cause them to swell and disintegrate. If you open a brake fluid reservoir and find that the diaphragm is swelled out of shape and it is larger than it should be, you can be sure someone has previously put some kind of petroleum-based fluid in it. Such a discovery will be costly to the vehicle's owner as he/she will have to replace all rubber components in the entire brake system.

Brake fluid is adversely affected by dirt, rust, oil, gas, grease, water—practically everything that is *not* brake fluid. Therefore, be extremely careful whenever removing the cap of a master cylinder—or when opening up any other portion of a system for any reason. Never use brake fluid from anything other than its original container or a closed bleeder tank. Even if all these precautions are taken, the hydraulic system can still become contaminated—by moisture in the air. Here is how to minimize contamination:

1. Keep the master cylinder tightly covered. Always recap it immediately after filling.
2. Use the smallest possible can of fluid, and use it all if possible. If, for instance, there is a choice of using two small cans or a portion of a large can, use the two small ones.
3. Tightly cap the fluid container after use.
4. When using a pressure brake bleeder, keep its fluid reservoir tightly closed—just like you would a master cylinder.
5. If you suspect any fluid might have become contaminated, *throw it out.*
6. Do not re-use old brake fluid.
7. Do not re-use an old brake fluid container since you do not know what else could have been in the can.
8. Do not transfer brake fluid from its original container to anything other than a container specifically designed to hold brake fluid, such as a pressure bleeder.

CAUTION: Some import vehicle manufacturers employ, as original equipment, a brake fluid that is petroleum (mineral oil) based. These vehicles use a special type of rubber that is not affected by petroleum-based materials. To conform with domestic requirements, these fluids are color coded green, but should not be used unless recommended by the vehicle manufacturer. Never mix a petroleum-based fluid with any other type of brake fluid. If this precaution is not taken, all rubber components in the system could be damaged.

Brake Fluid Level

The master cylinder should be filled to the following levels, depending on type, with clean specified brake fluid:

- *Cast-Iron Master Cylinders*—Cast-iron master cylinders should be filled to within 1/4 inch of the top of the reservoir to allow for brake fluid expansion when heated.
- *Aluminum Master Cylinders*—Aluminum master cylinders with transparent plastic reservoirs are marked with "FILL" to " " to show the proper level for the brake fluid (Figure 2-80).

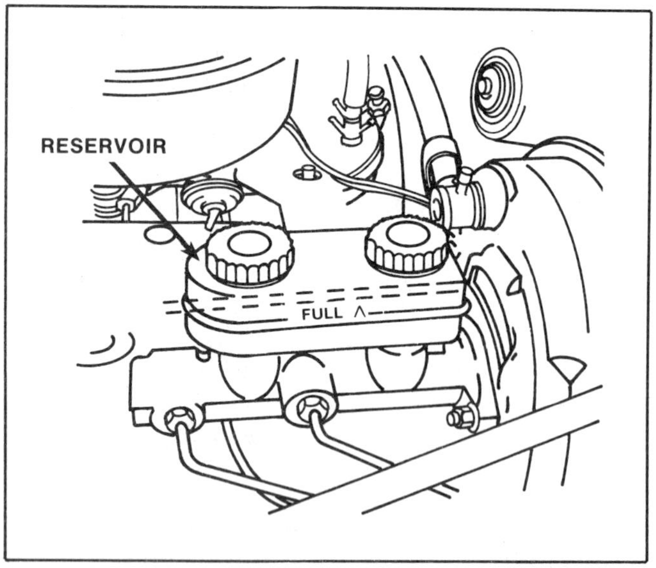

Figure 2-80 "Full" marking on transparent plastic reservoir

> **CAUTION:** Overfilling can cause the front pads to drag when the cover and diaphragm assembly is reinstalled. Underfilling can introduce air into the system.

Most auto manufacturers recommend that the brake fluid be checked at least twice each year. Remember that the brake fluid level will drop as the lining wears. A low level is a safety hazard if it falls to a point where it allows air into the master cylinder system.

BRAKE FLUID LEVEL SENSORS

Some vehicles have brake fluid level sensors as a component in the visual electronic message center. The purpose of this sensor (Figure 2-81) is to provide the driver with an early warning message that the brake fluid in the master cylinder reservoir has dropped below the normal level.

As the brake fluid in the master cylinder reservoir drops below the designated level, the sensor closes the warning message circuit. About 15 seconds later the message "BRAKE FLUID LOW" appears on the instrument panel. At this time, the master cylinder reservoir should be checked and filled to the bottom of the ring with the specified brake fluid.

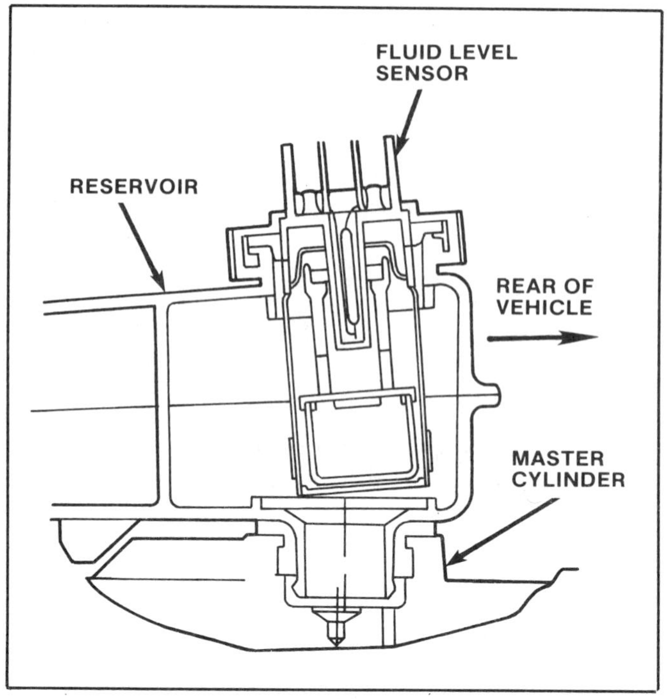

Figure 2-81 Reservoir mounted fluid level sensor

REVIEW QUESTIONS

Review the information in this chapter by answering the following questions.

1. Since 1967, federal law has required all cars to be equipped with
   a. safety braking systems.
   b. dual braking systems.
   c. split braking systems.
   d. all of the above.
2. Most front-wheel drive vehicles use a
   a. diagonally split brake system.
   b. axle-to-axle brake system.
   c. front/rear split brake system.
   d. left/right split brake system.
3. Where does the hydraulic brake system get its start?
   a. the master cylinder
   b. the wheel cylinders
   c. the brake lines
   d. the brake pedal
4. What is known as the heart of the hydraulic brake system?
   a. the master cylinder
   b. the wheel cylinders

c. the brake lines
d. the brake pedal

5. When the brake pedal is depressed, the force is transferred to the piston by a
   a. return spring.
   b. metering valve.
   c. replenishing or compensating port.
   d. pushrod.
6. On a given aluminum master cylinder, the cylinder bore has obviously been scored. Mechanic A decides to rebuild the unit. Mechanic B decides to install an entirely new unit. Who is right?
   a. Mechanic A
   b. Mechanic B
   c. both A and B
   d. neither A nor B
7. What part of a dual master cylinder prevents moisture from entering the system?
   a. the residual check valves
   b. the replenishing or compensating port
   c. the retaining ring
   d. the cap
8. On a dual master cylinder, what is the function of the replenishing ports?
   a. to keep light pressure in the brake lines and at the wheel cylinders
   b. to store extra brake fluid
   c. to help pump up the brake pedal when it is low
   d. to help manual brakes generate sufficient hydraulic pressure
9. Mechanic A decides to install a GM "quick take-up" master cylinder on a front-wheel drive car with a diagonally split braking system. Mechanic B decides to install a GM "quick take-up" master cylinder on a rear-wheel drive car with front/rear split braking system. Who is right?
   a. Mechanic A
   b. Mechanic B
   c. both A and B
   d. neither A nor B
10. Mechanic A uses copper brake line tubing. Mechanic B uses steel brake line tubing. Who is right?
    a. Mechanic A
    b. Mechanic B
    c. both A and B
    d. neither A nor B
11. Calipers are used with
    a. disc brakes only.
    b. disc brakes and drum brakes.
    c. drum brakes only.
    d. none of the above.
12. The pressure differential valve
    a. operates a warning light switch.
    b. is used in all dual brake systems.
    c. alerts the driver if pressure is lost in either of the two systems.
    d. all of the above.
13. Which valve has the most effect at the start of each brake operation and all during light braking conditions?
    a. the proportioning valve
    b. the dual-body proportioning valve
    c. the metering valve
    d. the integral proportioning valve
14. Which type of wheel cylinder is most often used on today's cars?
    a. double piston with step bore
    b. double piston with straight bore
    c. baffle-type piston
    d. single piston
15. Which of the following statements concerning brake fluid is incorrect?
    a. Brake fluid is adversely affected by gasoline.
    b. An important property of brake fluid is that it must resist evaporation.
    c. Every can of brake fluid carries the identification letters of SAE and DOT.
    d. Most auto manufacturers recommend that the brake fluid be checked at least once every two years.
16. In the typical hydraulic system, the sliding rubber seals
    a. prevent leakage.
    b. contain the brake fluid.
    c. both a and b
    d. neither a nor b
17. The front/rear split brake system is also known as the
    a. axle-to-axle system.
    b. tandem system.
    c. diagonally split system.
    d. horizontally split system.
18. What are the two types of brake pedals?
    a. suspended and actuated
    b. suspended and frame-mounted
    c. integral and frame-mounted
    d. free-play and adjustable

19. In a single system master cylinder, the brake fluid from the reservoir enters the cylinder through the
    a. compensating port.
    b. replenishing port.
    c. both a and b.
    d. neither a nor b.

20. Which of the following statements concerning the dual system master cylinder is incorrect?
    a. If failure occurs in one of the independent hydraulic systems, the other one still functions.
    b. Each piston has its own reservoir with vent and replenishing ports.
    c. The pistons are rigidly connected.
    d. The cap, or cover gasket, acts like a diaphragm.

21. What type of master cylinder is used on trucks with tilt, forward control, or other unconventional cab arrangements?
    a. vertical
    b. dual-bore
    c. GM quick take-up
    d. none of the above

22. During a panic stop, the pressure in the brake lines can reach up to
    a. 150 PSI.
    b. 1500 PSI.
    c. 1000 PSI.
    d. 750 PSI.

23. In which piston configuration is the caliper rigidly mounted to the wheel spindle or steering handle?
    a. single-piston
    b. two-piston
    c. four-piston
    d. both b and c

CHAPTER 3

HYDRAULIC SYSTEM—BLEEDING

OBJECTIVES

Upon completing this chapter, you should be able to

- Explain the importance of properly bleeding a hydraulic brake system.
- Describe the bleeding sequence, and identify common bleeding locations.
- Explain the various techniques of manual bleeding, and identify the differences between them.
- Explain the techniques of pressure bleeding and surge bleeding.
- Describe the procedure for bench bleeding a master cylinder.
- Recognize when it is necessary to flush a hydraulic system, and describe how it is done.
- Explain the procedure for final touch bleeding.

The brake hydraulic system must be free of air to function properly. Any air in the system is compressed when pressure is applied, resulting in greater pedal travel and a spongy brake pedal. Air is lighter than brake fluid, so air bubbles will float on top of the fluid in wheel cylinders or work back into the master cylinder. Air can enter the system if any of the hydraulic parts are disconnected or if the brakes are operated with the master cylinder fluid level very low.

Bleeding is a service procedure that introduces fresh brake fluid into the system and consequently expels the air bubbles and the old aerated fluid.

To completely drain the system, open the bleeder screws and pump the pedal slowly until only air is expelled. If the fluid reservoir has a diaphragm type seal, first remove it. If the vehicle is equipped with a pressure metering valve, hold it open. To catch the fluid, slip a hose over each bleeder screw and drain into containers.

BLEEDING LOCATIONS

Depending upon the point at which the hydraulic system was opened to air, bleeding might be required at all four wheels and the master cylinder; or it might be required at only a wheel or two (Figure 3-1). If both tubes are disconnected from the master cylinder (one tube on a single system master cylinder), or if air was introduced into the system through low fluid level in both master cylinder reservoirs (one reservoir on a single system master cylinder), bleeding is then required at all four wheels and the master cylinder.

On the other hand, only the rear wheels or the front wheels must be bled, as appropriate, if the tube serving that set of wheels was disconnected at the master cylinder, or if air was introduced into the system through low fluid level in only the master cylinder reservoir serving that set of wheels.

Bleeder screws are located strategically at high points on the wheel cylinders, and it is through them that the air and fluid are actually expelled. Normally, a bleeder screw is installed in each drum brake wheel cylinder, in each disc brake caliper, adjacent to the outlet port of some master cylinders, and on some combination valves on disc brake systems. Some bleeder screws have a threaded passage and a protective dust cap screw that must be removed before a drain hose is attached to the bleeder screw.

SHOP TALK

On some vehicles, certain Chrysler products for example, it is possible to install a right-hand caliper on the left side and vice versa. When this is done, the bleeder screw will be at the bottom of the caliper's cylinder, making it impossible to bleed out the air. In the manufacturer's service manual, check whether or not this is necessary.

BLEEDING PROCEDURE

The bleeding operation itself is fairly well standardized. The first step in all cases is cleaning the

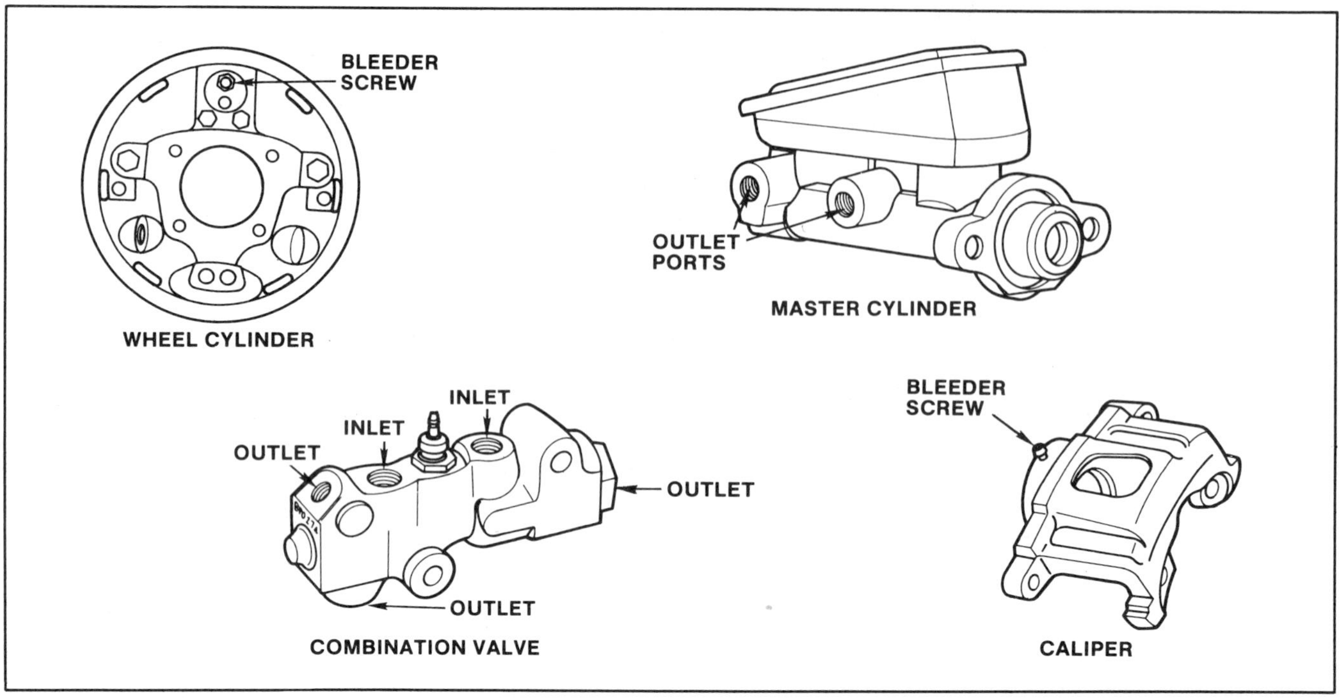

Figure 3-1 Typical bleeder points

dirt from the filler cap before removing it from the master cylinder. This should be done thoroughly. Anytime the fluid in the reservoir falls below the level of the intake port(s), air will enter the system. If this has happened, all you will usually have to do is refill the reservoir and slowly pump the pedal a few times to purge the air. If this does not work, bleed the system.

CAUTION: Be careful not to spill or drip any brake fluid on the car finish. If you do, *flush it immediately with water.*

Air bubbles will float to the top of the fluid in the wheel cylinders, or work their way back into the master cylinder. Bleeder screws are located strategically at high points on each wheel cylinder or caliper, near the outlet ports of some master cylinders, and on some combination valves. A bleeder screw normally is installed

- In each drum brake wheel cylinder.
- In each disc brake caliper. Some calipers include two or more bleeder screws.
- Adjacent to the outlet port of some master cylinders, particularly those that are mounted on an angle.
- On some combination valves used with disc brakes.

General practice, with any type of bleeding, is to slip a hose over the end of the bleeder screw. Put the free end of the hose into a jar half filled with brake fluid (Figure 3-2). Always keep the end of the hose submerged in brake fluid.

BLEEDING SEQUENCE

The following bleeding sequence can be considered typical for *most* hydraulic systems:

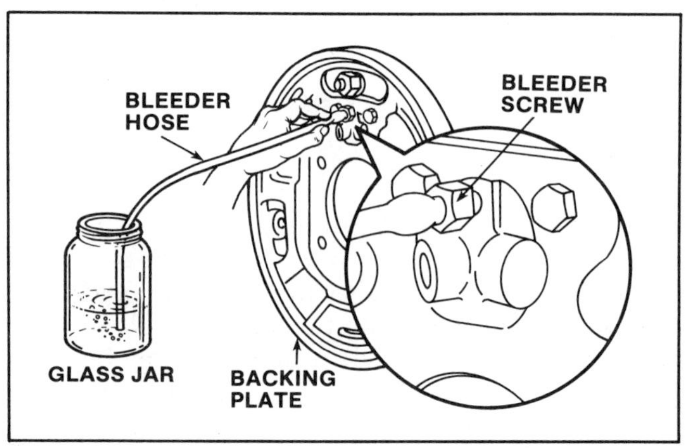

Figure 3-2 Brake bleeding

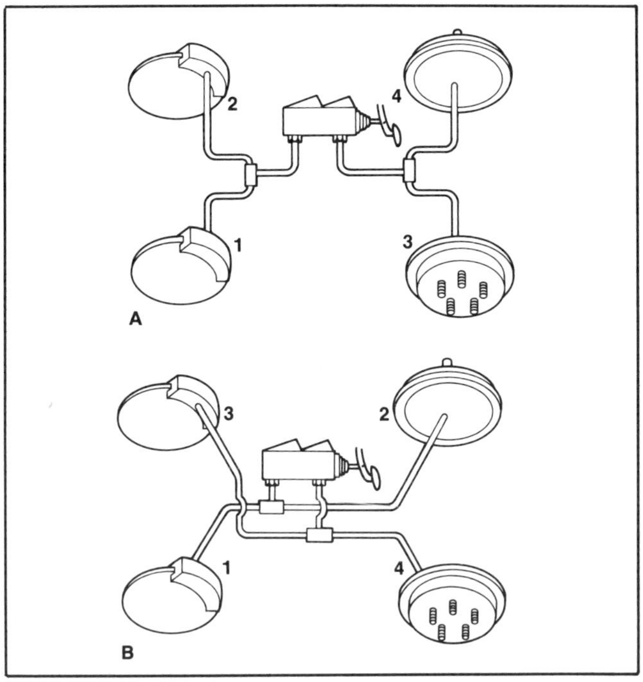

Figure 3-3 Bleeding sequence for each wheel cylinder

1. Bleed at the master cylinder bleeder screws, if so equipped, or by cracking the outlet tube nuts at the master cylinder while a helper applies medium pressure to the brake pedal.
2. Bleed at the bleeder screw on the combination valve, if so equipped.
3. Bleed at each wheel cylinder and caliper in succession. Start with the wheel cylinder or caliper farthest from cylinder and work back to the master cylinder. This procedure will help prevent trapping air in brake lines between the two units (Figure 3-3A). If there are two bleeder screws per wheel, do the lower one first. If it is a diagonally split system (Figure 3-3B), do one-half of the system at a time.
4. On drum brakes with two wheel cylinders, bleed the upper cylinder first.

It is important to always check with the vehicle's service manual to be sure that the proper sequence for that vehicle is followed. With a GM quick take-up master cylinder, for example, it is necessary to bleed all wheel cylinders and calipers in the following sequence:

1. Right rear wheel cylinder
2. Left front caliper
3. Left rear wheel cylinder
4. Right front caliper

If bled in the typical or conventional sequence, you might chase air throughout the system.

METHODS OF BLEEDING

Any time the hydraulic system is opened for service, it will be absolutely necessary to bleed out any air that has entered. It is also good practice to bleed the brakes when a simple reline is done.

The procedure is to open bleeder screws at individual wheels, one at a time. Then, apply pressure to the system either by stroking the pedal or through the use of a pressure bleeder. This will force a solid column of fluid through the lines and the wheel cylinders and calipers, expelling any air bubbles that might be present. This is important because air, unlike fluid, is compressible, so any amount of it in the system will cause a spongy pedal, or require that the pedal be pumped in order to generate sufficient pressure to stop the vehicle. If the rear calipers are difficult to bleed, lightly tap them with a hammer to regroup the air bubbles so they can be expelled.

The various methods of bleeding a brake hydraulic system are as follows:

Manual Bleeding

Manual bleeding (Figure 3–4) requires the use of the brake pedal and master cylinder as a hydraulic pump to expel air and brake fluid from the system when a bleeder screw is opened. This method of bleeding is normally a two-man operation—one to pump the brake pedal and another to operate the bleeder valves.

Starting at the wheel that is determined by the proper bleeding sequence for the hydraulic system being worked on, the technician opens (during pedal application) and closes (prior to pedal release) the bleeder screws and observes the emission of air bubbles while an assistant pumps the brake pedal slowly in order to move the fluid through the system and expel trapped air. A piece of hose is slipped over the bleeder screw and the free end is inserted into a jar partially filled with brake fluid. The technician can determine accurately when all air has been purged from the system by observing when bubble-free fluid flows from the hose. Air in the system is visible when bubbles emerge from the end of the hose inserted in the fluid as the pedal is depressed. Keeping the end of the hose inserted in fluid will prevent the accidental admission of air to the system when the bleeder screw is open. Pedal pumping action must be done *slowly* to prevent any undue buildup of pressure and increase in turbulence.

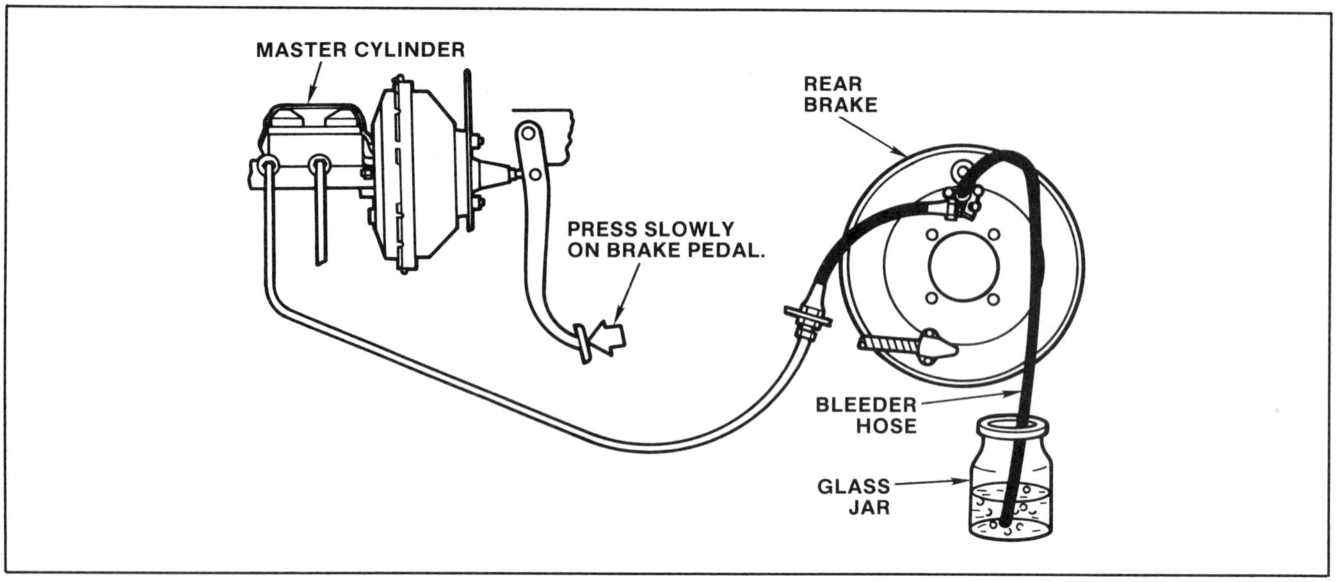

Figure 3-4 Manual bleeding

Fill the master cylinder to the proper level specified in Chapter 2 with clean brake fluid before starting and after completing the bleeding. Also check the master cylinder fluid level frequently during bleeding and keep the reservoirs at least half full. Manual bleeding requires that the fluid level in the master cylinder be checked and replenished periodically. If the level drops below the intake port, air will enter the system. The bleeder screw should be open only when the pedal is depressed and closed before the pedal is released. Residual pressure valves in the system will not hinder bleeding. The metering valve (if so equipped) should be deactivated to properly bleed front disc brakes.

The following steps apply to one bleeder screw and should be repeated at all other bleed points:

1. Attach a bleeder hose to the bleeder screw. Place the other end of the hose in a glass jar partially filled with brake fluid.

SHOP TALK

Be sure that the free end of the hose is submerged in brake fluid. This helps to show air bubbles as they come out of the system and prevents air from being accidentally sucked into the system through the bleeder screw.

2. Apply moderate (40 to 50 pounds), steady pressure on the brake pedal and open the bleeder screw.
3. When fluid coming from the submerged end of the hose is clear and free from air bubbles, close the bleeder screw and release the brake pedal.
4. If the brake pedal goes to the toe board without removing all of the air bubbles, close the bleeder screw and release the brake pedal slowly. Then repeat steps 2 and 3.

CAUTION: In order to insure that no air can be sucked back into the system, the bleeder valve must be closed at the end of each stroke, before the brake pedal is released.

Manual Bleeding with a One-Man Bleeder Hose. The one-man bleeder hose containing a one-way valve is widely available. This is attached to the bleeder screw and only allows fluid to flow out of the bleeder, not in the reverse direction, so it eliminates the possibility of air being drawn back into the system when the brake pedal is released. Open the bleeder screw, attach the hose, then place the end of the hose in a container half filled with fluid. Pump the pedal slowly several times. If the hose is transparent, check it for the presence of bubbles. Continue pumping until a clear stream exits from the bleeder, then tighten the bleeder screw, and move on to the next wheel. Do not allow the master cylinder reservoir to run dry.

Bleeding Slave Cylinders. A slave cylinder, such as is used to actuate a clutch, usually has a bleeder screw on the housing. Bleed a slave cylinder in the same manner as a wheel cylinder.

Manual Bleeding Techniques for Quick Take-Up Master Cylinder. Some GM cars 1980 to present are equipped with a quick take-up master cylinder that requires a special manual bleeding procedure. To bleed this type of master cylinder, proceed as follows:

1. Disconnect the brake line to the left front wheel and add brake fluid to the master cylinder until fluid flows from the port.
2. Reconnect and tighten the brake line.
3. Slowly depress the brake pedal and hold. Again loosen the left front line to purge any air from the system. Tighten the fitting and slowly let up on the pedal. Wait 15 seconds.
4. Repeat this sequence on the left front line until all air is removed. Remember to include the wait period. When all air has been removed from the left front connection, bleed the right front connection following the same procedure.

Individual calipers and wheel cylinders are bled only after all air is expelled from the master cylinder. The actual sequence for bleeding the individual wheel cylinder and caliper for vehicles equipped with a quick take-up cylinder was given earlier in the chapter. To perform the actual bleeding, slowly depress the brake pedal and hold. Loosen the bleeder screw to release the air. Then, tighten the screw and slowly let up on the brake pedal. Wait 15 seconds and repeat until all the air is removed.

SHOP TALK

It might be necessary to repeat this operation 10 or more times per wheel to expel all the air from the wheel cylinders and calipers.

Other Methods of Manual Bleeding. Other methods of manual bleeding are sometimes employed to purge air from certain hydraulic systems. One of those methods is *gravity,* or vacuum, *bleeding.* It allows the brake fluid, by its own weight, to push the air out through a bleeder valve located low in the system. Another method is *reverse bleeding.* It involves pumping fluid into the system through a low point, such as the wheel cylinder, to force air out through the master cylinder. Both of these methods are not frequently employed today. If they are required (mainly in light-duty truck applications), they are specified and described in the manufacturer's service manual.

Pressure Bleeding

Pressure bleeding is fastest because the master cylinder does not have to be refilled several times, and the job can be done by one man. This method of bleeding, of course, eliminates the need for pedal pump. For dispensing hydraulic brake fluid under pressure, use a modern pressure bleeder system as shown in Figure 3-5.

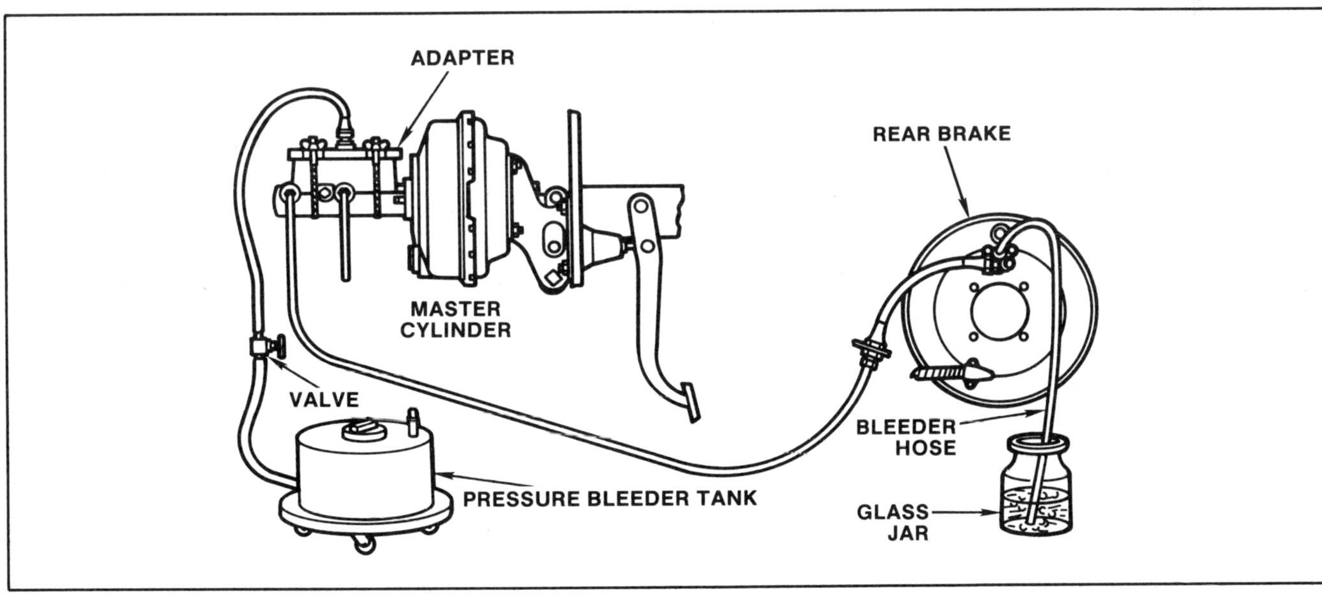

Figure 3-5 Pressure bleeding

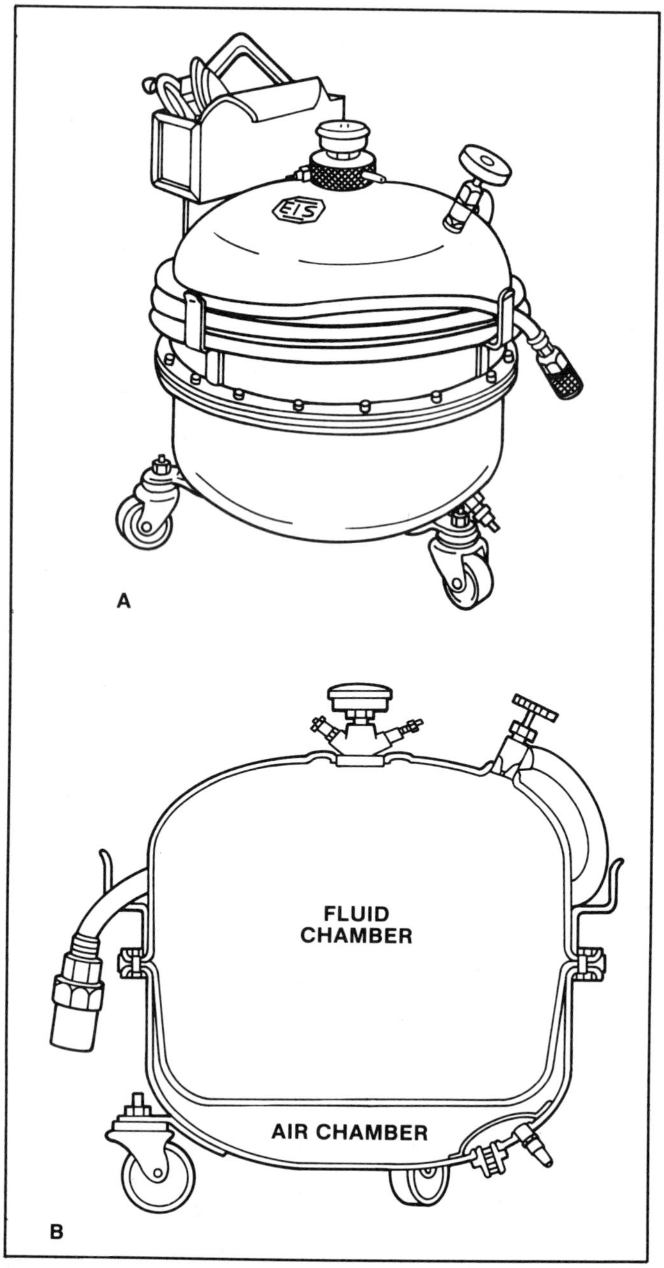

Figure 3-6 (A) Typical pressure bleeder; (B) cross section of a typical pressure bleeder

The pressure bleeder (Figure 3-6A) is a tank that is separated into two sections by a flexible diaphragm. The top section is filled with brake fluid and air pressure is admitted to the bottom section (Figure 3-6B). As the pressure pushes on the diaphragm, the fluid above it is pressurized. A supply hose runs from the top of the tank to the master cylinder.

The hose is connected to the master cylinder by means of an adapter fitting that fits over the reservoir, taking the place of the reservoir cap. These adapters (Figure 3-7) exist in different configurations, for the different types of reservoirs, including the plastic reservoirs on some of the newer vehicles.

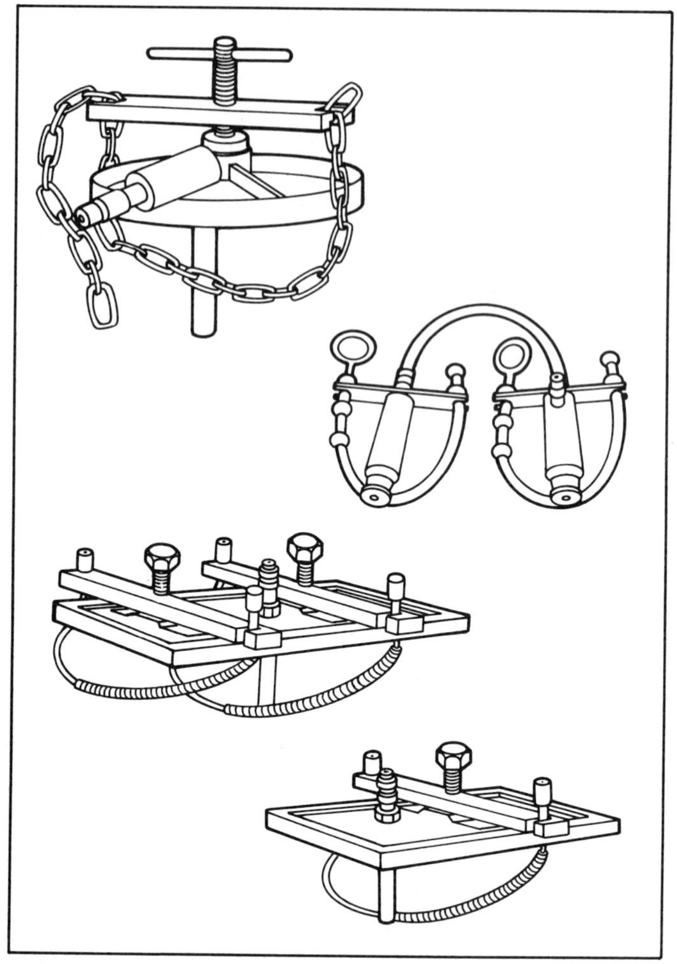

Figure 3-7 Typical adapters for pressure bleeders

The procedure for pressure bleeding is as follows:

1. Bring the unit to a working pressure of 15 to 20 pounds per square inch (PSI). Be sure that there is enough brake fluid in the pressure bleeder to complete the bleeding operation.
2. Clean the master cylinder cover and remove it. Remove the reservoir diaphragm gasket if there is one. Clean the gasket seat and fill the reservoir.
3. Attach the adapter to the reservoir.

SHOP TALK

A special adapter for master cylinders with a nylon fluid reservoir is shown in Figure 3-8.

4. Connect the hose to the adapter fitting, making sure that the coupling sleeve is fully engaged.
5. Open the fluid supply valve.
6. Bleed the system in the normal sequence, beginning with the master cylinder.
7. Disconnect the pressure bleeder: Close the supply valve. Then, with the coupling wrapped in a rag to protect the car finish from drips, undo the coupler sleeve above the master cylinder. Make sure that the reservoir fluid level is within 1/4 inch of the top, then replace the gasket and cap.

CAUTION: To prevent air in the pressure tank from getting into the lines, do not shake the tank while air is being added to the tank or after it has been pressurized. Set the tank in the required location, bring the air hose to the tank, and do not move it during the bleeding operation. The tank should be kept at least one-third full.

The hydraulic pressure generated by manual bleeding is sufficent to open the metering or combination valve and allow fluid to flow to the front wheels. When pressure bleeding, it will be necessary to hold the valve stem open manually.

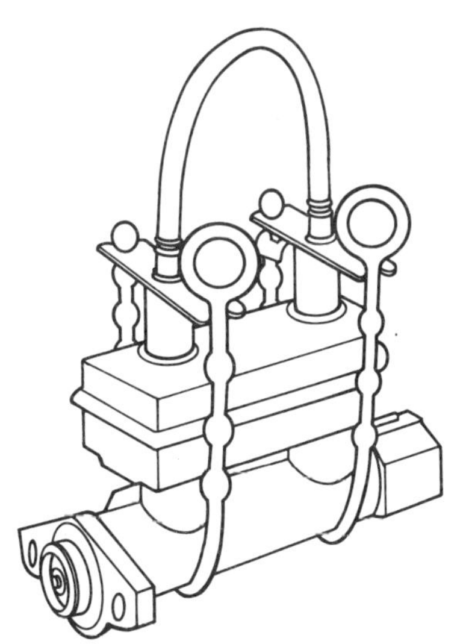

Figure 3-8 Pressure bleeder adapter for nylon reservoir

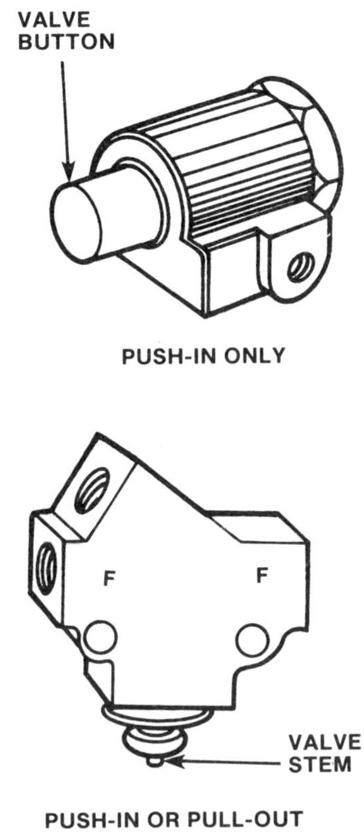

Figure 3-9 Types of valve stems

To hold the metering or combination valve open, either push the valve stem in or pull it out (Figure 3-9), depending on the valve type. Do not use more than 25 pounds pressure to push the stem in or pull it out. Otherwise, the valve can be damaged. To hold the metering valve open, use the special tool available for this task (Figure 3-10). Never use a screw clamp, wedge, or block that can put excessive pressure on the valve stem. Also, be sure to remove the special tool when the bleeding operation is completed.

Surge Bleeding

This method includes both manual and pressure bleeding, and deliberately creates a churning, higher pressure turbulence in the wheel cylinders so that any trapped air can be drawn off in the form of aerated fluid. It is important to remove all possible air before surging; thus, this method is never used unless the routine manual or pressure bleeding method proves inadequate.

The following simple procedure is recommended for surge bleeding:

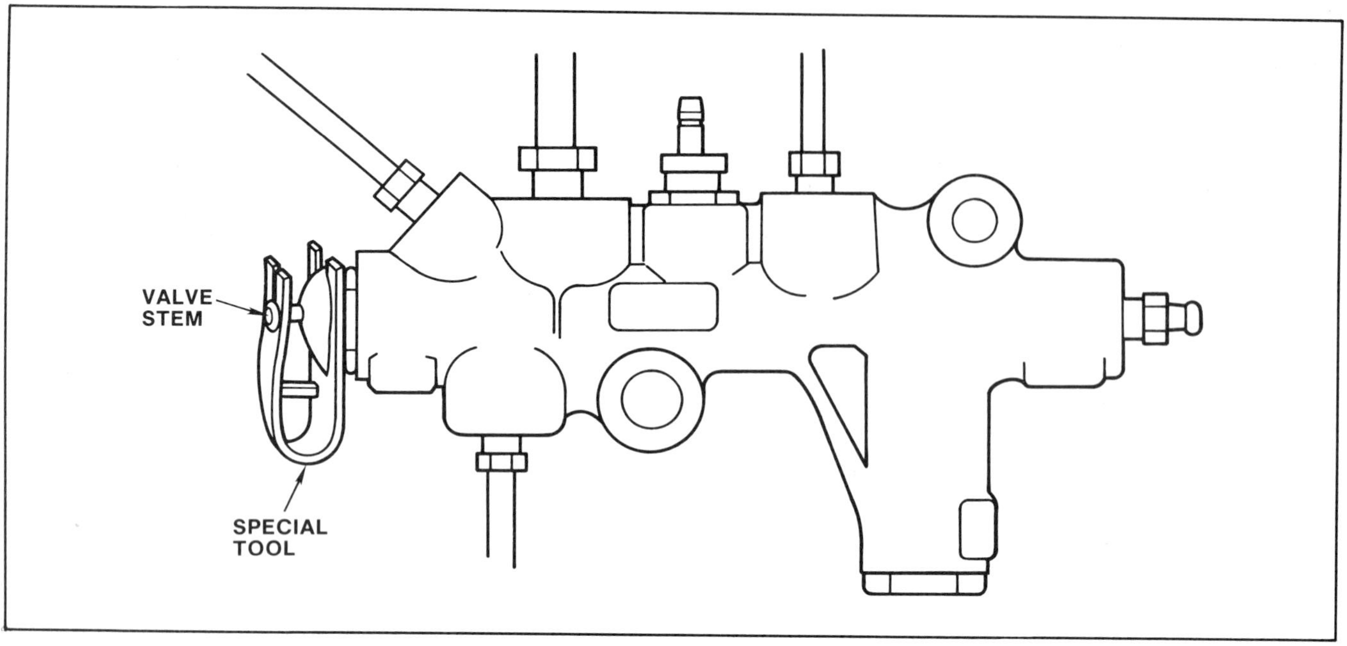

Figure 3-10 Special tool to hold open metering valve

1. Bleed the brakes at all wheels in the usual manner.
2. At each wheel cylinder, in turn, open the bleeder screw and press the brake pedal down sharply several times. Close the bleeder screw. The action creates a turbulence in each cylinder, forcing out practically all of the remaining trapped air.

BENCH BLEEDING MASTER CYLINDERS

The only sure way to extract all the air from a new or reconditioned master cylinder is to bench bleed it before installation. This is especially true with modern dual-piston cylinders, and those such as found on GM cars that are mounted at an angle from the horizontal. In fact, to expel all air trapped in the cylinder, tandem master cylinders must be bench bled before they are installed on the vehicle. Bench bleeding reduces the possibility of air getting into the brake lines.

Follow this simple procedure for bench bleeding:

1. Mount the master cylinder in a vise (Figure 3-11A). Insure that bore is horizontal. Do not apply excessive pressure to the casting.
2. Connect short lengths of tubing to the outlet ports and bend as shown (Figure 3-11B).

SHOP TALK

Bench bleeding kits are available that contain an assortment of fittings and tubing that will handle most domestic vehicles. Insure that the tube nuts are tightened securely to guard against air intake on the return stroke.

3. Fill the reservoirs with fresh brake fluid (Figure 3-11C). Make sure the level is above the ends of the tubes.
4. Using a wooden dowel or a smooth, round-tipped rod, slowly pump the master cylinder until all bubbles stop appearing at the tube ends (Figure 3-11D). On application, make sure that both pistons are completely bottomed in the bore. On release, allow the pistons to return completely to the stop.

SHOP TALK

Slow piston return is normal on some master cylinders during bench bleeding.

CAUTION: Be careful to avoid spraying brake fluid. Do not hold face directly above reservoirs.

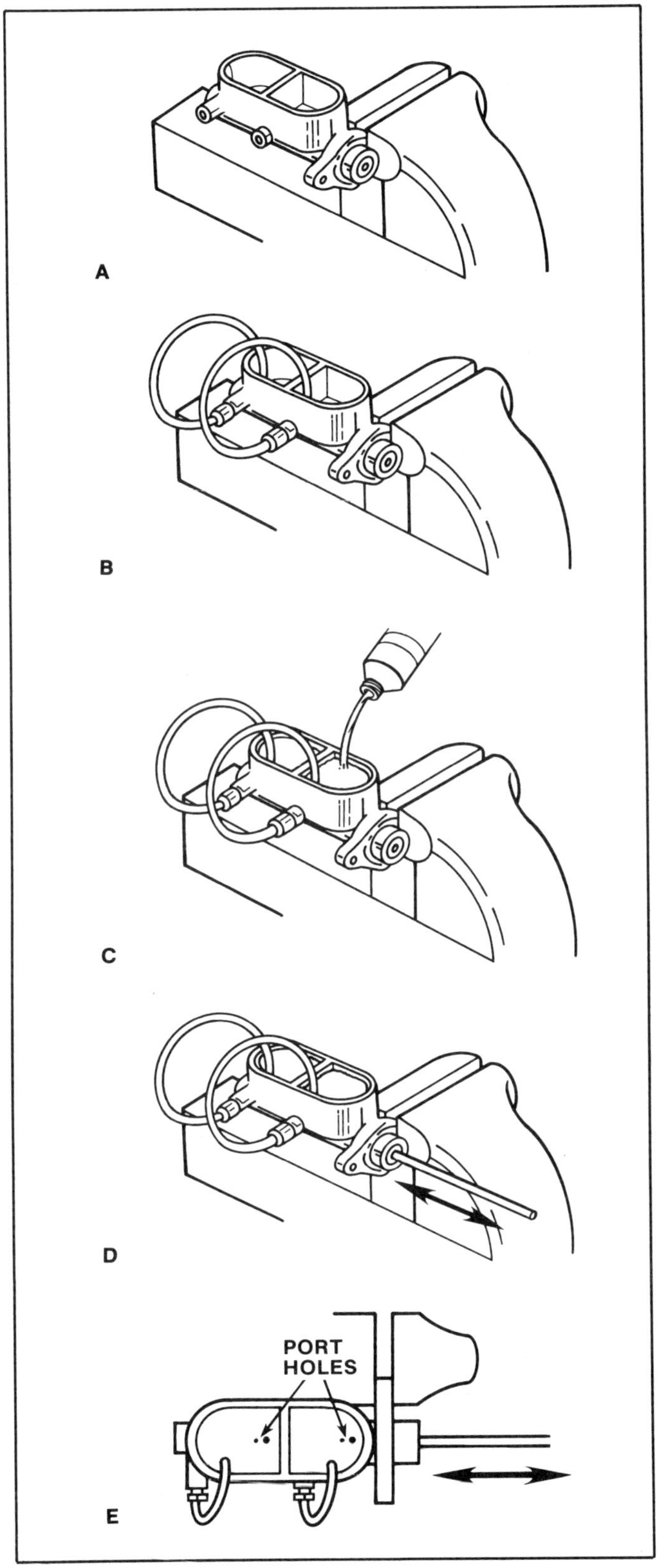

Figure 3-11 Bench bleeding tandem master cylinder

5. Using the tool from the previous step, push the piston approximately 1/4 to 1/2 inch in from the fully released position. Allow the piston to return about 1/8 inch, then push it back to the initial 1/4 to 1/2 inch position. Rapidly move the piston back and forth in this manner until air bubbles stop coming from the ports located inside the reservoirs (Figure 3-11E).

SHOP TALK

Bleeding the master cylinder can be performed on the vehicle if the brake lines are not connected and the cylinder bore is horizontal.

Use the above procedure, but use the brake pedal to stroke the master cylinder. (This procedure will require two people, one to pump the pedal and one to watch for air bubbles coming from the tubes and ports.)

6. When all air has been purged from the master cylinder, remove the tubes. Refill the reservoirs and securely install the master cylinder cap.
7. Install the master cylinder on the vehicle. Attach the lines, but do not tighten the tube connections.
8. Force out any air that might have been trapped in the connections by slowly depressing the pedal several times. Tighten the nut slightly before releasing the pedal, and loosen before depressing it each time. Catch the fluid in a rag to avoid damaging the car finish. Tighten the connections to manufacturer's specifications when air bubbles are no longer present in the fluid. Make sure the master cylinder reservoirs are adequately filled with brake fluid.

Another bench bleeding technique for the master cylinder utilizes a specially designed bleeder syringe. The procedure is as follows:

1. Plug outlet holes of master cylinder and gently clamp it in a vise with the pushrod end slightly elevated (Figure 3-12A). Do not clamp the cylinder by the bore. Do not clamp reservoirs tightly or damage can result.
2. Pour brake fluid into master cylinder until half full.
3. Remove a plug from the cylinder. Depress the syringe plunger fully and place its rubber tip firmly against the outlet hole to form

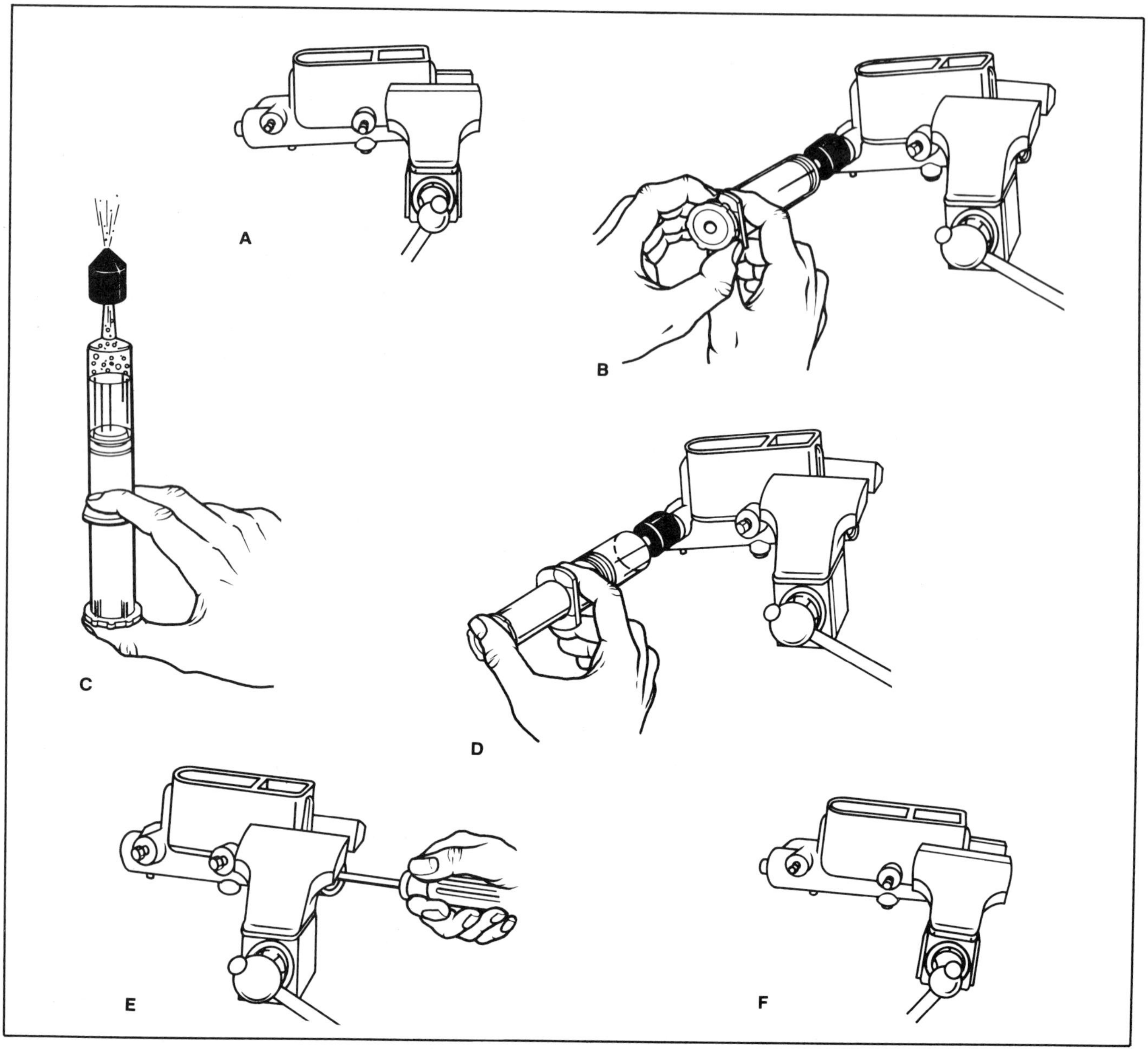

Figure 3-12 Bleeding the power brake unit

a seal (Figure 3-12B). Slowly pull out the plunger until the syringe is half full of fluid, and remove the syringe from the outlet.

4. Invert the syringe and slowly depress its plunger until all air is expelled (Figure 3-12C).
5. Now hold the syringe (with fluid) against the same outlet and slowly depress the plunger (Figure 3-12D). Note air bubbles in the reservoir. When these bubbles stop, remove the syringe and plug the outlet. Repeat this procedure at the other outlet(s). Plug outlets tightly.
6. Reclamp the master cylinder in the vise with the pushrod end tilted downward slightly.
7. Slowly slide the master cylinder pushrod back and forth about 1/8 inch until no air bubbles are seen in the reservoirs (Figure 3-12E).
8. Again mount the master cylinder with the pushrod end up (Figure 3-12F). Fill the syringe with brake fluid and expel air as in

step 4. Remove one outlet plug at a time and repeat steps 4 and 5. The master cylinder is now fully bench bled.

SOME UNIQUE BLEEDING SITUATIONS

There are some special bleeding procedures that are required on some vehicles. The most common of these are:

Tandem Master Cylinder with Two Filler Caps (Figure 3-13). Some earlier car and truck models have the fluid reservoir separated into two compartments by a baffle that is open across the top. Insert a finger into one of the filler openings to detect this baffle.

A pressure bleeder can be connected to either opening; however, the other filler opening must be sealed with a blind, non-vented cap. This can be achieved by plugging the air vent in a standard cap. Completely fill the reservoir with fluid before installing the adapter, as fluid must flow over the baffle. Other models have two entirely separate reservoirs (Figure 3–14).

The pressure bleeder adapter should be installed over the rear filler cap opening first to bleed one axle system; transfer the adapter to the front opening in order to bleed the remaining axle system.

Dual Bore Master Cylinder (Some Chevrolet, GMC, and IHC trucks). The brake and clutch systems must be bled separately. The clutch system has no residual pressure check valve, so,

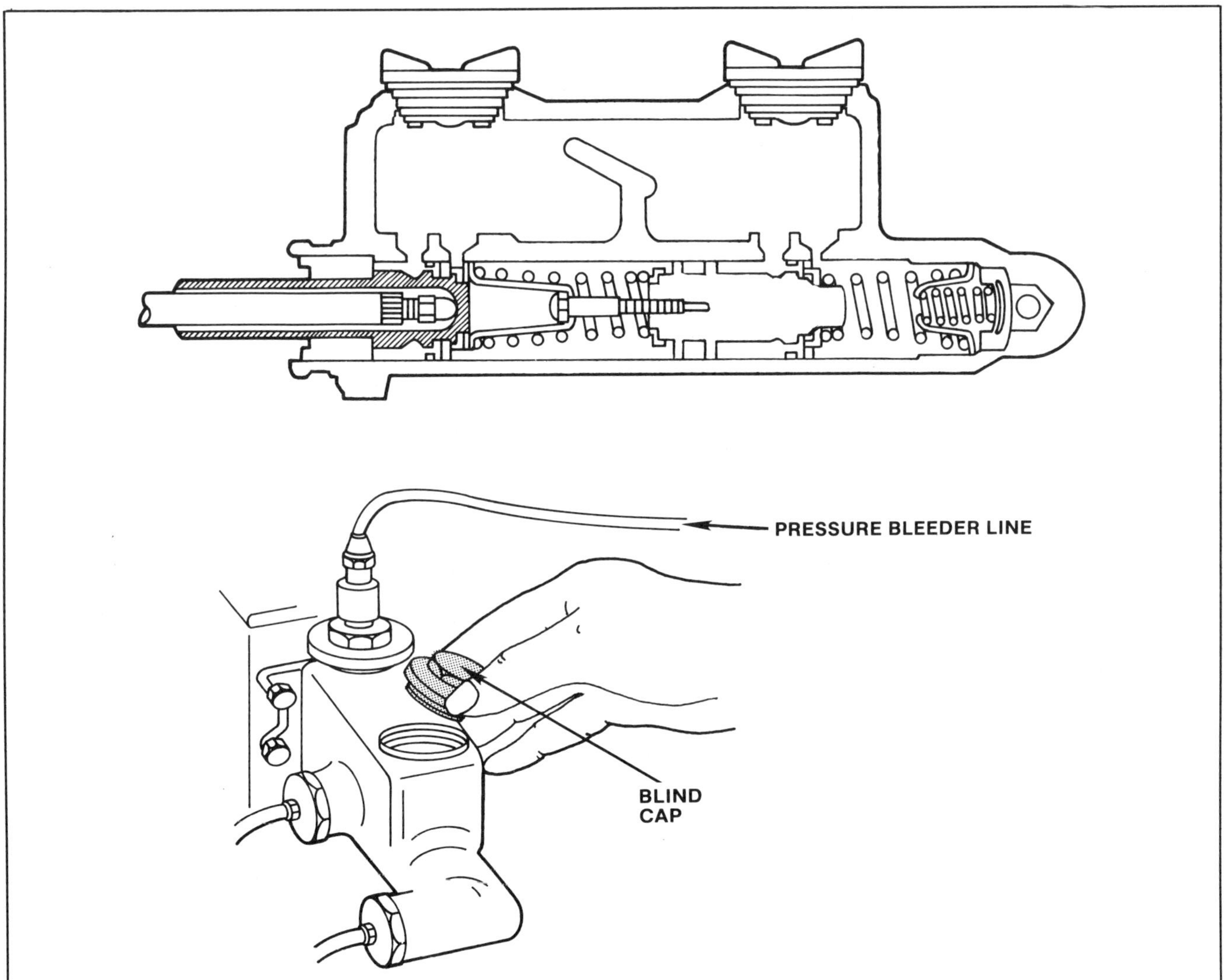

Figure 3-13 Bleeding tandem master cylinder with two filler caps

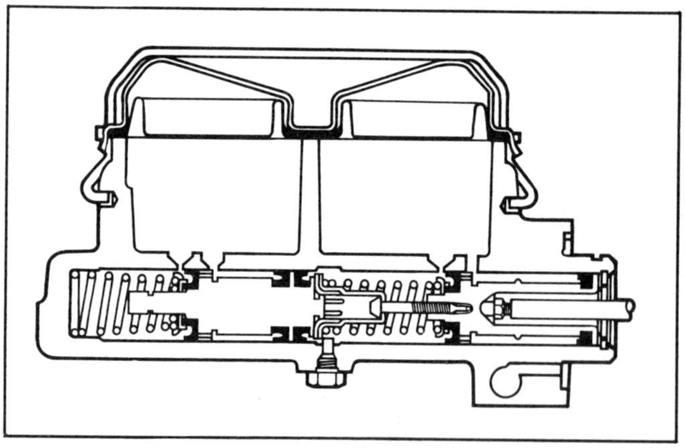

Figure 3-14 Tandem master cylinder with separate reservoirs

when bleeding manually, the bleeder screws must be closed before the pedal is released.

If a pressure bleeder cannot be attached to the fluid reservoir (some trucks), disconnect the brake line at the master cylinder and attach the bleeder supply hose directly to it. After bleeding the system, connect the master cylinder line and manually bleed the master cylinder, through a loosened tube nut, catching the fluid in a rag.

Midland Low-Input Power Unit (Some light-duty trucks). This type of power unit will not permit a low-bleeding pressure to pass through it.

1. If the master cylinder and the tubing upstream from the power unit require bleeding, loosen the tube connection at the power unit and bleed at this point. Retighten tube nut.
2. Charge the pressure bleeder with compressed air to its maximum pressure capacity. Start the engine.
3. Have an assistant quickly pump the brake pedal to its greatest height, at least three strokes, and maintain foot pressure on the pedal. While the assistant depresses the pedal only 3/4 stroke, open and close the bleeder screw on the power unit. Repeat the action until clean bubble-free fluid flows from the screw.
4. Bleed the wheel cylinders with combined pressure and manual bleeding but, unlike surge bleeding, operate the pedal smoothly to avoid increasing fluid turbulence.

Two Cylinder Brakes. Figure 3-15 shows some two wheel cylinder brake configurations.

Make sure that the wheel cylinders and the manifold tube connections are installed in the correct positions. If two bleeder screws per brake are provided, they should be located in the upper ports of the cylinders; tube connections should be in the lower ports. If only one bleeder screw per brake is provided, it is located in the upper port of the upper cylinder. The lead-in pipe or hose should be connected to the lower port of the opposite cylinder. If the cylinders are piped to opposite sides of a dual system (two lead-ins), determine the correct bleeding rotation by tracing the pipes back to the master cylinder.

Non-Horizontal Cylinders. When the brakes are mounted so that the wheel cylinder bore is not level (Figure 3-16), cup fillers are usually provided to minimize air entrapment. Should the cylinder trap air in the "blind" corner higher than the bleeder screw, surge bleeding might be necessary to dislodge it.

The procedures for bleeding power brakes and anti-lock systems are covered in Chapters 9 and 10.

FLUSHING A HYDRAULIC SYSTEM

It might sometime become necessary to flush out the system due to the presence of mineral oil, kerosene, gasoline, etc., which will cause swelling of rubber piston cups and valves so they become inoperative. Also whenever new brake components are installed in the hydraulic system, it is recommended that the entire hydraulic system be thor-

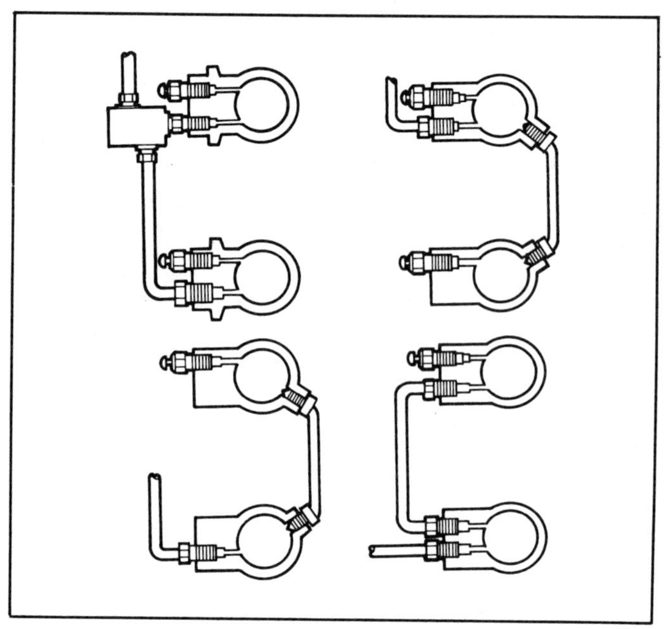

Figure 3-15 Bleeding two-cylinder brakes

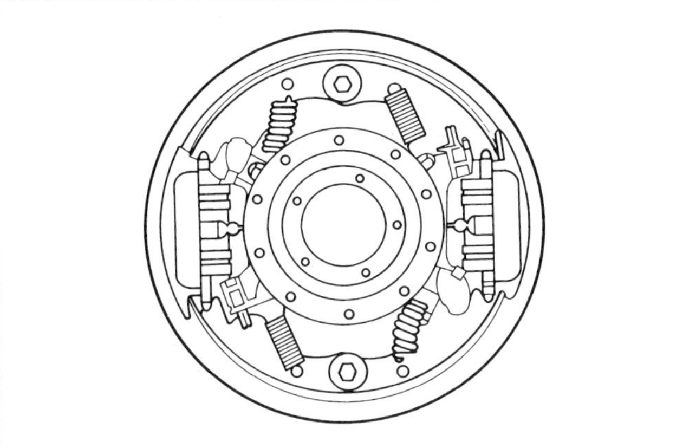

Figure 3-16 Typical bleeding non-horizontal wheel cylinders

oughly flushed with clean brake fluid. The flushing procedure is as follows:

Flushing is performed at each wheel in the same manner as the bleeding operation, except that the bleeder valve is opened 1-1/2 turns and the fluid is forced through the lines and bleeder valve into a container (Figure 3–17). Flushing should be done with a pressure bleeder since you must provide a constant supply of fresh fluid in the reservoir to replace the old fluid that gets drained out. If the vehicle has a pressure metering valve, it must be held open. Then, working one wheel at a time with

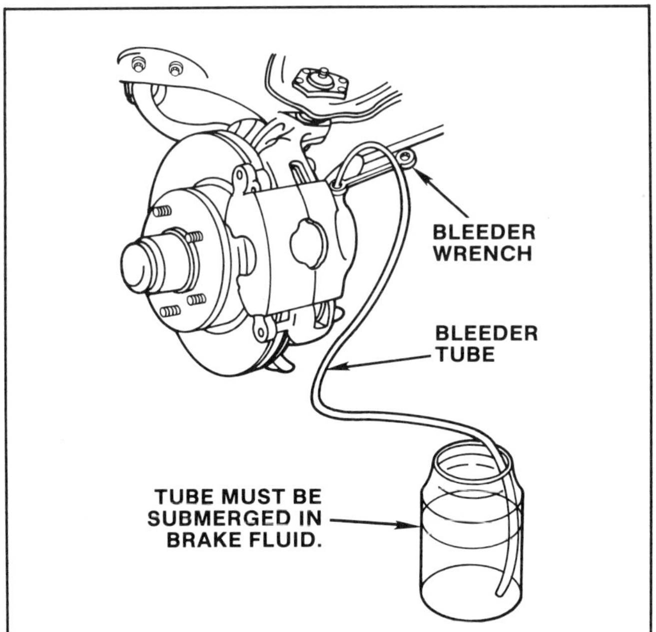

Figure 3-17 Flushing each wheel is similar to bleeding.

the free end of the bleeder hose in the bleeder jar, open the bleeder screw until only clean new fluid comes out. Approximately one quart of clean brake fluid is required to flush the hydraulic system. After completing the flushing operation at all bleeder valves, check to ensure the master cylinder is filled to the proper level.

Do not flush the hydraulic system with any alcohol solvents or any other solutions besides brake fluid. Denatured alcohol is fine for cleaning component parts on the bench, but if any of it gets trapped in the hydraulic system it will reduce the boiling point of the fluid even more so than water.

FINAL TOUCH BLEEDING

Many late-model master cylinders tend to trap air in the bore because the outlets are slightly below the top of the bore. To remove this air from the system after pressure bleeding, follow these steps.

1. Position the car so the master cylinder is slightly nose down.
2. Park on a slope or raise the rear on a lift.
3. Set the parking brake and block the wheels.
4. Remove the master cylinder cover and lightly tap (jiggle) the brake pedal several times. The trapped air will come up through the replenishing port in fine bubbles.
5. When bleeding is complete, you will see a spurt of fluid and the bubbles will stop.

As a final test, make two or three severe applications and hold pedal down. Let the reservoir settle and then release the pedal. If nothing happens, the system is bled and tight. If there is a surge in the reservoir on the return stroke, you have an air or adjustment problem. Find and fix it.

On some vehicles, the warning light valve must be manually reset after the hydraulic system is bled. If this applies, refer to Chapter 4 for the procedure.

Finally, after the brake system is bled and a good firm pedal established, then road test the vehicle as described in Chapter 14 to insure proper operation of the brake system.

REVIEW QUESTIONS

Review the information in this chapter by answering the following questions.

1. In a given hydraulic system, both tubes are disconnected from the master cylinder. Me-

chanic A says to bleed all four wheels and the master cylinder. Mechanic B says only the front wheels must be bled. Who is right?
   a. Mechanic A
   b. Mechanic B
   c. both A and B
   d. neither A nor B
2. When bleeding a hydraulic system, the air and fluid are expelled through the
   a. outlet ports.
   b. bleeder screws.
   c. inlet ports.
   d. metering valve.
3. What is the first location in the bleeding sequence for most hydraulic systems?
   a. Bleed at the bleeder screw on the combination valve.
   b. Bleed at each wheel cylinder and caliper in succession, starting with the wheel closest to the master cylinder.
   c. Bleed at the master cylinder bleeder screws.
   d. None of the above.
4. In order to ensure that no air can be sucked back into the system during manual bleeding, the bleeder valve
   a. must remain closed.
   b. must be closed at the beginning of each stroke.
   c. must be closed at the end of each stroke.
   d. must be deactivated before beginning.
5. Gravity and reverse bleeding are used mainly in
   a. light-duty truck applications.
   b. passenger car applications.
   c. conjunction with pressure bleeding.
   d. all of the above.
6. While pressure bleeding, Mechanic A uses a screw clamp to hold the metering valve open. While pressure bleeding, Mechanic B uses wedges to hold the metering valve open. Who is right?
   a. Mechanic A
   b. Mechanic B
   c. both A and B
   d. neither A nor B
7. When bench bleeding a master cylinder, which of the following should not be done?
   a. Mount the master cylinder in a vise.
   b. Pump the master cylinder rapidly for at least 5 minutes.
   c. Make sure that the master cylinder bore is horizontal.
   d. Fill the reservoirs with fresh brake fluid.
8. A bleeder screw is normally installed
   a. in each disc brake caliper.
   b. on some combination valves used with disc brakes.
   c. in each drum brake wheel cylinder.
   d. all of the above.
9. When working with quick take-up master cylinders, individual calipers and wheel cylinders are bled
   a. first.
   b. only after adding sufficient brake fluid to the master cylinder.
   c. only after all air is expelled from the master cylinder.
   d. whenever convenient.
10. Which method of bleeding is fastest?
    a. pressure
    b. gravity
    c. surge
    d. both b and c
11. Flushing the hydraulic system should always be done with
    a. alcohol solvents.
    b. mineral oil.
    c. a pressure metering valve.
    d. a pressure bleeder.
12. New brake components have just been installed in a hydraulic system. Mechanic A says that it is not necessary to flush the entire system with clean brake fluid. Mechanic B says that the entire system should be thoroughly flushed. Who is right?
    a. Mechanic A
    b. Mechanic B
    c. both A and B
    d. neither A nor B
13. Manual bleeding is normally a
    a. two-man operation.
    b. three-man operation.
    c. one-man operation.
    d. four-man operation.
14. When surging, it is important to
    a. create as little turbulence as possible.
    b. bleed the brakes only at the rear wheels.
    c. both a and b.
    d. remove all possible air beforehand.
15. On dual-bore master cylinder vehicles, the brake and clutch systems
    a. cannot be bled manually.
    b. can only be bled using the reverse method.
    c. must be bled separately.
    d. all of the above.

CHAPTER 4

HYDRAULIC SYSTEM—SERVICING

OBJECTIVES

Upon completing this chapter, you should be able to

- Perform a complete inspection of the basic components of a hydraulic brake system.
- Inspect a brake pedal for free height and travel, and remove a typical brake pedal assembly.
- Service stoplight switches, brake fittings, lines, and hoses.
- Recognize conditions that are considered normal for a master cylinder and therefore do not require servicing.
- Perform basic tests on a master cylinder, including testing for trapped air, checking for external leaks, testing for a bypass, testing for open replenishing ports, and checking for brake pedal reserve.
- Service the various types of master cylinders, including the Bendix dual system, the Delco-Moraine dual system, the quick take-up, the dual aluminum, and the Bendix mini.
- Inspect the various safety valves and switches to determine if they are operating properly, and replace as needed.

The brake hydraulic must function perfectly for safe vehicle operation. That means *no* leaks or failures can be tolerated. Although it is a tall order, modern brake systems are up to the task if the components are inspected periodically and maintained. Whenever the brake pads or linings are changed, the entire hydraulic system should be inspected carefully. Any signs of trouble should be corrected immediately.

Two major problems can occur in hydraulic brake systems: internal and external leaks. Hydraulic brake systems and their components (Figure 4–1), the master cylinder, and valves do not cause the vehicle to vibrate or pull, the brakes to grab or squeal, or the brake pedal to pulse. In most cases, they will not even make the brakes drag. The sources of these problems are found elsewhere in the brake system (see Chapter 14), but it is always a good policy to first check the entire hydraulic system for leaks before continuing to diagnose any other brake problems.

To carefully inspect the basic components of the hydraulic system, proceed as follows:

1. Depress and release the brake pedal several times (engine running for power brakes), and check for friction and noise. Pedal movements should be smooth, and the pedal should return with no lag or noise.
2. Move the brake pedal from side to side. Excessive side movement indicates worn pedal mounting parts.
3. Depress the pedal with heavy foot pressure (engine running for power brakes) to check for sponginess and pedal reserve. The pedal should remain firm and the foot pad should be at least 2 inches from the floor for manual brakes, and 1 inch for power brakes.
4. Hold the pedal depressed with medium foot pressure (25 to 36 pounds) for about 15 seconds to make sure that the pedal does not drop under steady pressure. If it does, the master cylinder might have an internal leak.
5. Depress and release the brake pedal several times to check stoplight operation. The lights should come on each time the pedal is depressed and go off each time it is released. The dashboard warning light should not come on when ignition is switched to the ON position. Be sure to check that the warning light works.

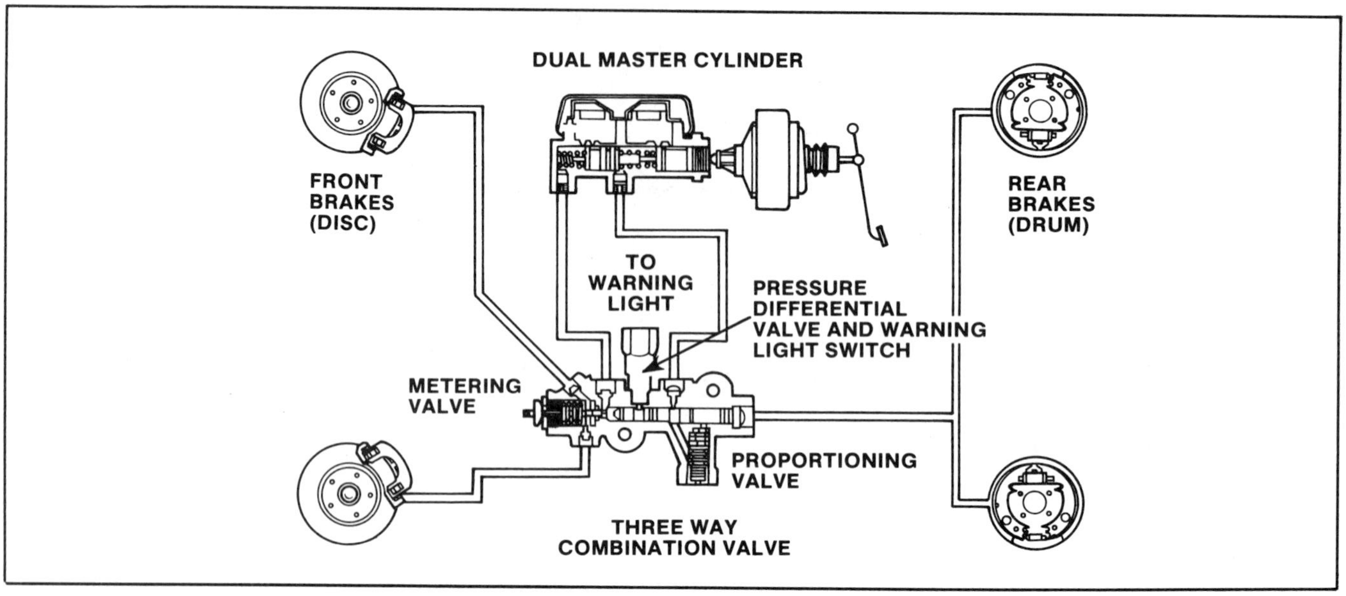

Figure 4-1 A typical power braking system

6. Brake lines should not be kinked, dented, or otherwise damaged, and there should be no leakage.
7. Brake hoses should be flexible, and free of leaks, cuts, cracks, and bulges.
8. Backing plates should be free of brake fluid and grease, and any parts attached to it should be tight.
9. Check all hydraulic connections for leaks.
10. Carefully inspect the master cylinder and the various safety valves and switches for leaks and proper operation.

BRAKE PEDALS AND STOPLIGHTS

Changes in brake pedal feel or travel are usually the first indicators that something could be wrong in the brake system. Normally, the brake pedal (on vehicles that do not have power assist) is firm, not spongy. A low pedal can indicate a leak if the brake warning light comes on. Or, it can indicate that the brakes need adjustment. Driving in reverse and stopping firmly, several times, should actuate the automatic adjustment. The height of brake pedals that have power-assisted brakes is normally lower.

To measure brake pedal free height and to check the brake pedal travel, release the parking brake and while the brakes are cold measure as follows:

Insert a slender, sharp-pointed prod through the carpet and sound deadener to the dash panel metal. Measure the distance to the center on top of the brake pedal pad (Figure 4-2). This dimension is the pedal free height. The pedal travel is measured parallel to the vertical centerline of the steering column with about a 25-pound load applied to the centerline of the brake pedal pad. (Checks on the power brake vehicle should be made with the engine running.) These measurements should be approximately:

	Pedal Free Height	Pedal Travel
Manual Brakes	7″ to 8″	3-1/2″ to 4-1/2″
Power Brakes	7″ to 8″	2″ to 3″

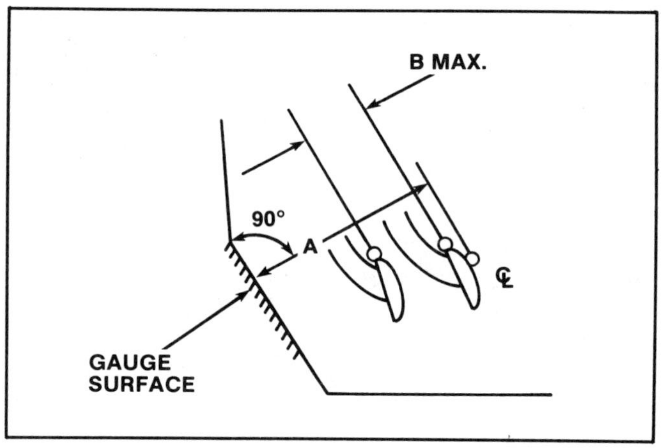

Figure 4-2 "A" is the pedal free height; "B" is the pedal travel.

Exact pedal free height and travel can be found in the vehicle's service manual.

SHOP TALK

When adjusting brakes, it is important to obtain a small amount of free play at the pedal for proper operation of the brake system.

If the position of the pedal is not within specification, check the brake pedal for missing, worn, or damaged bushings, or loose attaching bolts, and replace them, if required. If the pedal free height is still out of specification, check the brake pedal, booster, or master cylinder to be sure the correct parts are installed. Replace the worn or damaged parts as necessary.

Should the brake pedal be found faulty, replace or repair the assembly. To remove a typical brake pedal assembly (Figure 4-3), proceed as follows:

1. Disconnect the negative battery cable.
2. Disconnect the stoplight switch wire connector from the switch.
3. Remove the clutch cable from the clutch pedal (manual transmission vehicles). In the case of a vehicle with an automatic transmission, follow instructions in the manufacturer's service manual.
4. Loosen the four pedal support attaching nuts at the dash panel. Remove the pushrod retainer and nylon washer. Slide the stoplight switch outboard along the brake pedal pin just far enough for the outer hole of the switch frame to clear the pin. Remove the switch by sliding it downward. Remove the black, stoplight switch bushing from the pushrod.
5. Remove the locknut from the pivot bolt. Pull the pivot bolt out, and remove the brake pedal.

To install a new or repaired brake pedal assembly, reverse the removal procedure. But before reinstalling the unit, apply a coating of SAE 10W-40 engine oil to the bushings. Locate the bushing and pivot spacer in the brake pedal hub. Be sure to tighten the nuts to the vehicle manufacturer's specifications.

If the height is correct, but the brake pedal feels spongy, this might indicate that there is air in the system. Bleeding should correct this. A brake pedal that feels somewhat soft can, however, be nor-

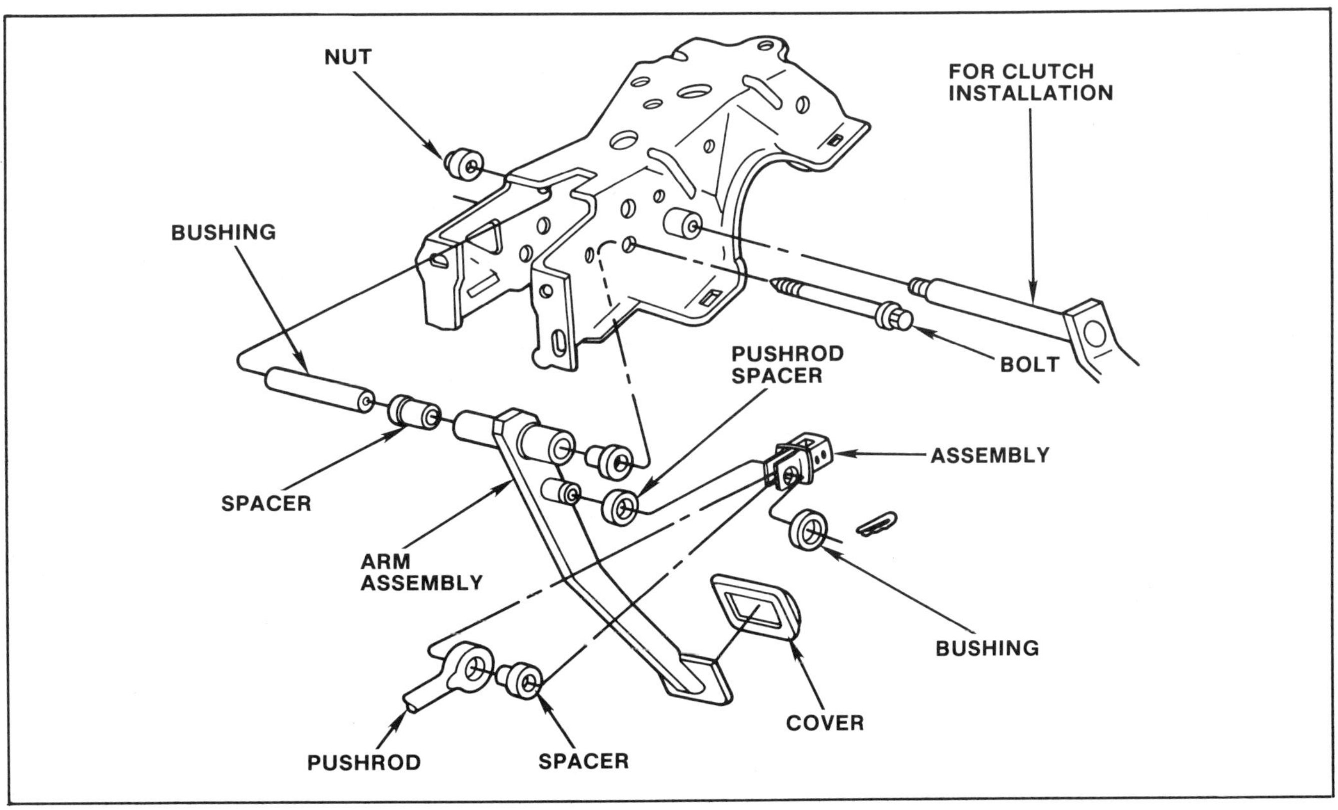

Figure 4-3 Typical brake pedal assembly

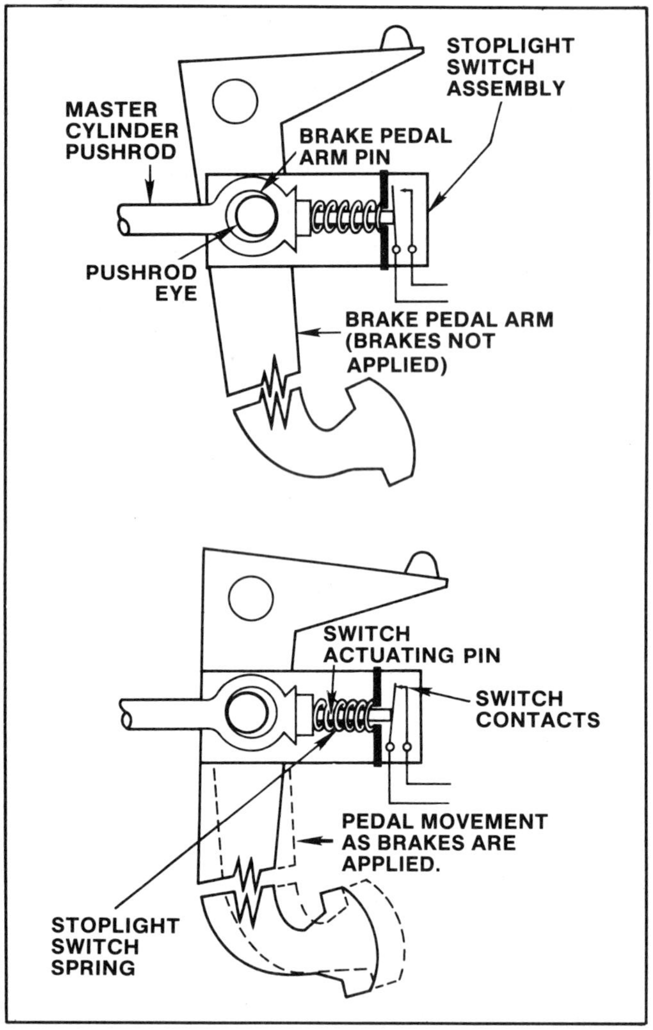

Figure 4-4 Mechanical stoplight switch operation. Forward movement of the brake pedal arm pin releases the switch-actuating pin and closes the switch contacts.

mal. It should be compared with a similar vehicle that has a brake system known to be in good order.

STOPLIGHT SWITCHES

Stoplight switches can become faulty for a number of reasons: wear, breakage, corrosion, or leakage (hydraulic switch). Most mechanical stoplight switches are adjustable (Figure 4-4).

SHOP TALK

If one or both turn signals will not flash, it might mean that the brake (stop) lights are not working.

Before replacing a stoplight switch, use a jumper wire to bypass the switch and check the electrical circuit for continuity. Replace any burned out bulbs or defective sockets, connectors, or wires. Also before replacing a mechanical stoplight switch, be sure that the switch is properly adjusted. The switch should be adjusted so that the circuit closes when the pedal is depressed approximately 1/4 to 1/2 inch. If the stoplights come on when using a jumper wire but do not come on when the pedal is depressed, the switch is faulty and should be replaced.

Figure 4-5 Check the rear brake boiling plate on each rear wheel for signs of brake fluid leakage or oil spray.

BRAKE FITTINGS, LINES, AND HOSES

Exposure to the elements, neglected maintenance, and component design all contribute to rusting and corrosion of brake fittings, lines, and hardware (Figure 4-5). Nowhere else in the system are these problems more troublesome than when bleeder screws are frozen fast to the cylinder or caliper, and flare nuts to a variety of components (Figure 4-6).

The bleeder screw, of course, must be opened to bleed air from cylinders or calipers (Figure 4-7). Usually, this small screw corrodes fast to the housing and is not easily removed. Unless a six-sided wrench of the correct size is used, rounding off of the shoulders is the first problem the mechanic faces.

Holding the correct wrench in position, the brake technician must be careful not to exert too much force in attempting to back out the bleeder screw. Unless this work is done carefully, the small screw will break off in the housing and require drilling and tapping, or replacement of the entire cylinder or caliper.

Before performing caliper or cylinder service, always check the bleeder screw to see if it can be removed. If the caliper or cylinder is otherwise serviceable, there are two methods to remove a frozen bleeder screw:

- Use a special wrench that will allow you to exert pressure on the screw and strike it with a hammer. The shocks to the housing from the hammer blows and the tension against it might free the screw.
- Use heat from a welding torch to bring the housing around the screw to a dull red. Then, apply light pressure with the correct size wrench. When this method is used, however, the rubber and plastic components in the caliper or piston must be removed to prevent damage from the heat.

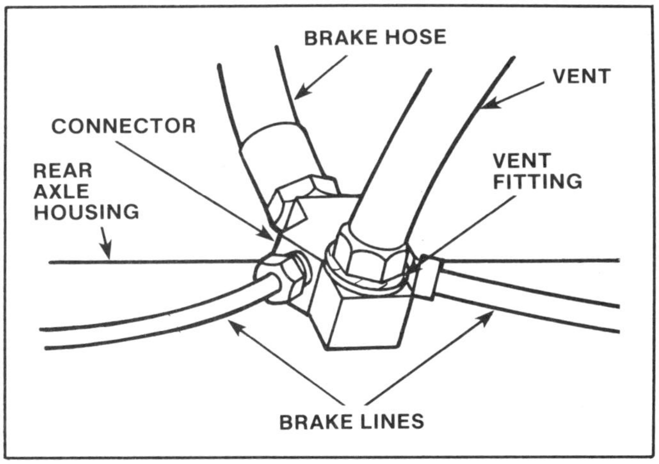

Figure 4-6 Rear brake tube connector

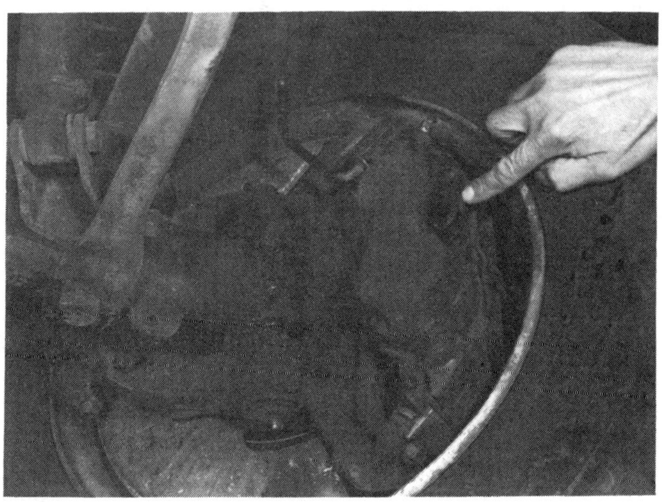

Figure 4-7 Visually inspect the brake caliper area for signs of brake fluid leakage.

BRAKE LINES

Flare nuts, themselves, do not normally create a problem, but the line that passes through them does. The line corrodes to the nut and as the nut is backed out with the flare-nut wrench, the line twists. If liberal applications of penetrating oil do not permit the nut to turn on the line, heat or replacement of the line is recommended. Again, when using heat, plastic and rubber components must be removed to prevent damage. Apply a pencil-thin flame to all sides of the flare nut, never to the line. The steel nut will glow when it is hot enough, then the nut should be backed out with the flare wrench.

Tubing should be inspected from time to time for kinks, cracks (ruptures), dents (Figure 4-8), and corrosion or rust. It is important that corroded or rusted lines be replaced since they can burst under a panic stop situation and high pressures. Whenever brake lines are replaced, the mechanic will have to do a certain amount of fabrication. Because of the many bends and lengths used, the lines are supplied in various dimensions and lengths. All brake lines are steel and should only be replaced with steel.

SHOP TALK

Brake lines and flare nuts (Figure 4-9) are available in both metric and conventional sizes. Care should be exercised not to confuse or mix the two.

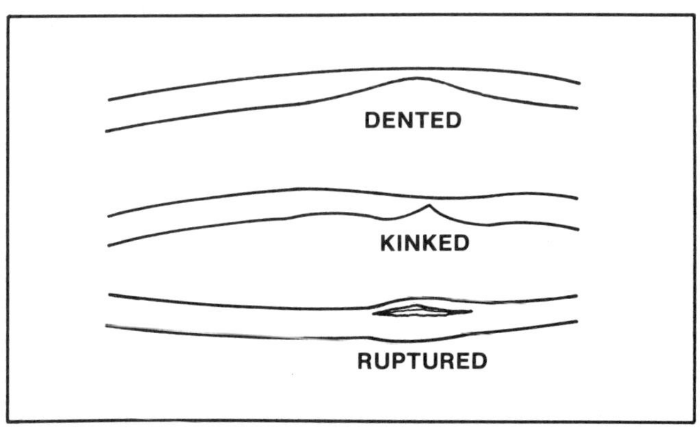

Figure 4-8 Brake lines should be inspected periodically for damage.

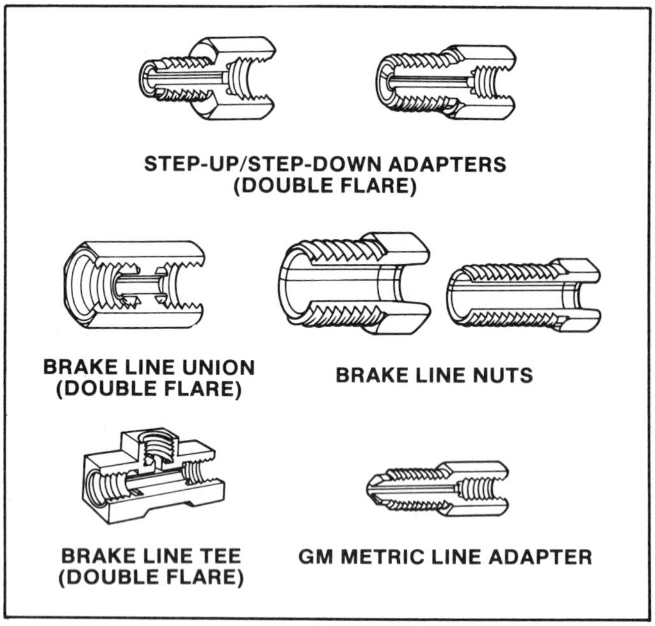

Figure 4-9 Various brake line fittings

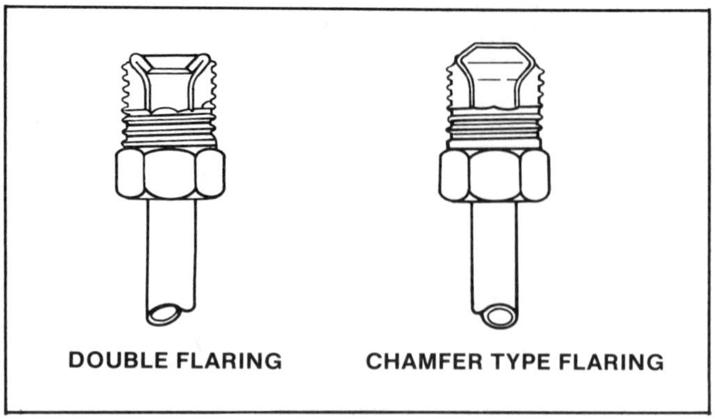

Figure 4-10 Two types of flares

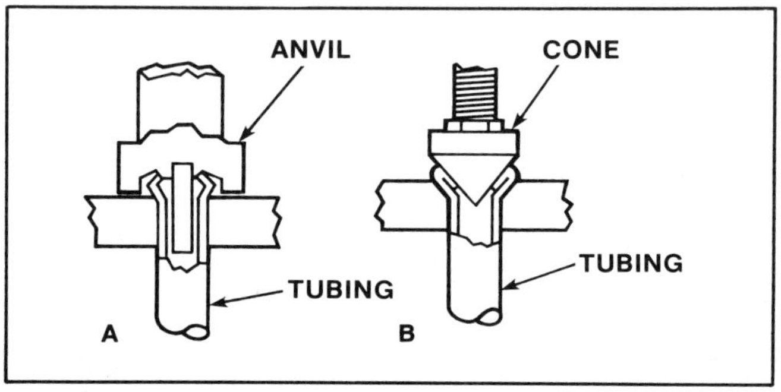

Figure 4-11 (A) Anvil folds over tubing; (B) cone creates double thickness

Flaring Brake Lines

A common task during brake system service is to replace the brake lines. To fabricate a brake line, select the correct tube and fitting dimension, and start with a standard length that is slightly longer than the old line. With the old line as a reference, use a tubing bender to form the same bends in the new line as exist in the old. Each end of the new line must be flared to provide the seal and the strength required by the high hydraulic pressures. A tubing cutter should be used to cut the ends of the tubing perfectly square and to keep the tubing itself perfectly round. A tubing cutter does this by supporting the tubing with a pair of rollers while a sharp cutting wheel is rolled around the tubing. After the tubing has been cut to size, any burrs that might have formed inside the edge of the tubing must be removed. A reamer attached to the tool cutter will usually do this job nicely.

When replacing brake lines, it might be necessary to bend the new line to duplicate the shape of the one being removed. Although steel tubing can be bent by hand to obtain a gentle curve, any attempt to bend a tight curve by hand will usually kink the tubing. Since a kink in a brake line will weaken the line, a kinked line should never be used. To avoid kinking, always use a bending tool. There are several types of tube bending tools available; therefore, be sure to follow the instructions carefully for the type your shop employs. Once the tubing is bent into the proper shape, assemble the flare nuts on the tubing before undertaking the flaring operation.

There are two types of flares: the *double flare* and *ISO flare* (Figure 4-10). Each uses a particular style of fitting, and they are not interchangeable. A double flare is made with a special double-flaring tool in two stages. (The single flare is not used in brake work.) A special tool is also required to form the ISO flare.

Double-Flaring. The double flare is made with a special tool that has an anvil and a cone. The double flaring process is performed in two steps.

1. The anvil begins to fold over the end of the tubing (Figure 4-11A).
2. The cone is used to finish the flare by folding the tubing back on itself, doubling the thickness and creating two sealing surfaces (Figure 4-11B).

The angle and size of the flare are determined by the tool. Careful use of the double flaring will help to produce strong, leakproof connections. Figure 4-12 shows the importance of the correct flare. Notice that the angle of the flare and the nut are both 45 degrees while the angle of the seat is 42 degrees. When the nut is tightened into the fitting, the difference in angles, called an *interference angle,* causes both the seat and the flared end of the tubing to wedge together. Properly assembled, brake lines connected in this way provide joints that can withstand the high pressures in the hydraulic system.

ISO Flaring. The ISO flare (sometimes called a *bubble flare*) meets the requirements of the International Standards Organization (ISO). It is used on many imports and some domestic front-wheel drive vehicles. The advantages of the ISO flare are:

- Tube seats are no longer needed.
- When tightened, the shoulder of the nut bottoms out in the body of the part, creating a uniform pressure on the tube flare.
- The design is not subject to overtightening. The service brake technician merely tightens the nut firmly on the seat and the correct sealing pressure occurs automatically.

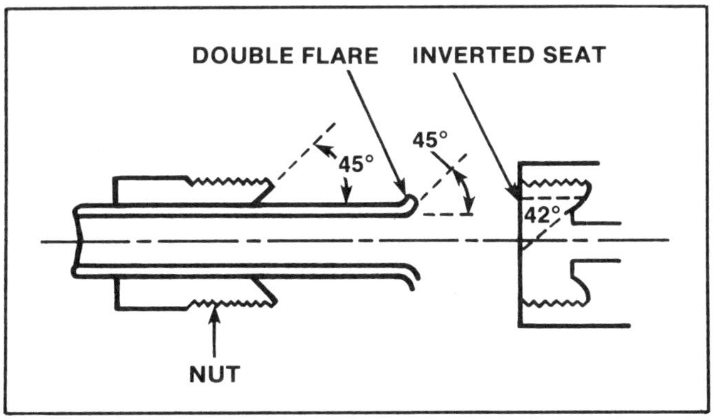

Figure 4-12 Cross-sectional view of brake tubing and fittings

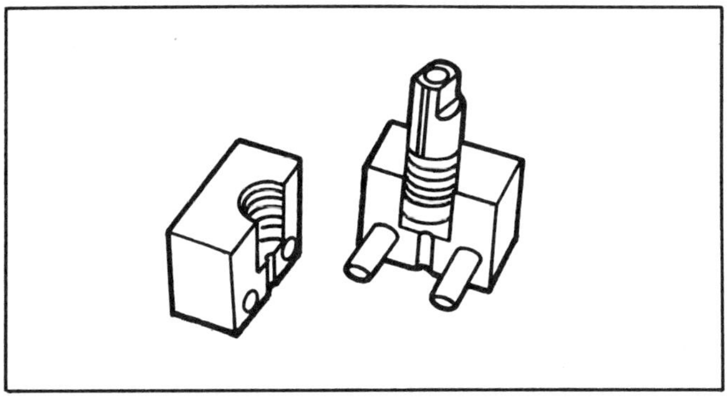

Figure 4-13 ISO flare forming tool

Although there are several tools available to make an ISO flare, the simplest one to use is the forming tool shown in Figure 4-13. With this tool, the ISO flare is made in one step. Once the tubing is cut, deburred, and clamped in a vise, a special anvil is forced down onto the tube. The anvil bends the end of the tube into the bubble-shaped flare.

When making either type of flare, be sure to follow the directions for the flaring tool you are using. Figure 4-14 shows completed flares. After flaring and before installing the tubing, apply air pressure to blow out any metal chips or particles of dirt.

CAUTION: The use of any tubing other than coated steel tubing designed for automotive systems can cause problems. The use of any other attachment methods, such as compression fittings, single flares, brazing, or soldering, will jeopardize the integrity and safety of the entire braking system.

Servicing Brake Line Tubing

The correct size and type of wrench should be used when tightening or loosening the hydraulic

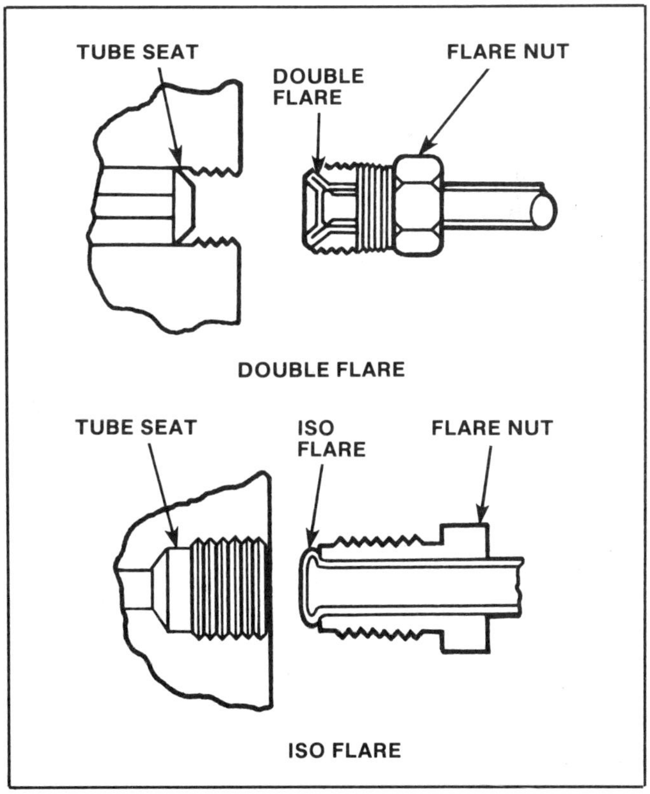

Figure 4-14 Types of flares and their seats

fittings (Figure 4-15). To avoid distortion, a spare tube nut should be screwed into the female fitting to reinforce it. On fittings requiring gaskets, new copper gaskets should always be used since the old ones will have taken a set and might not seal if re-used.

Replacing Brake Hoses

Hoses should be inspected for cracks, cuts, indication of chafing, twists, and loose supports (Figures 4-16 and 4-17). Internal defects, such as those illustrated in Figure 4-18, must be guarded against. To check for internally damaged brake hoses, apply pressure to the brake pedal (using a pumping action) and watch or feel for any bulging or expansion of the hose.

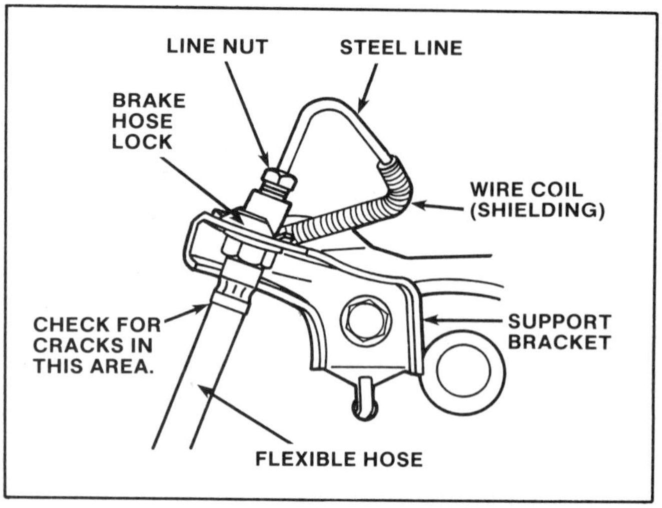

Figure 4-17 Hoses and tubing should be inspected at these points.

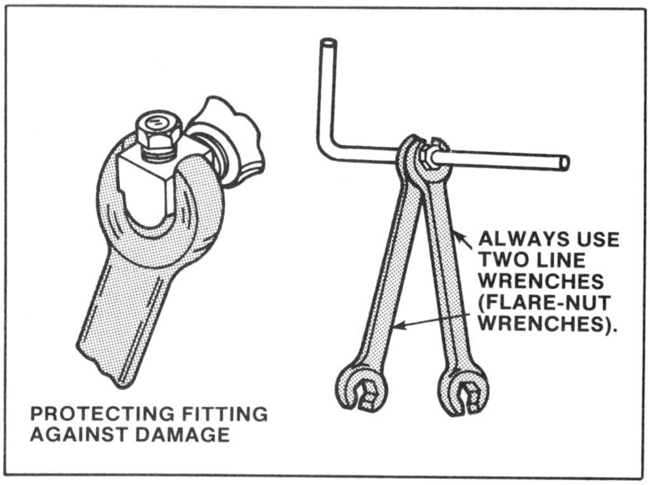

Figure 4-15 Tightening or loosening hydraulic fittings

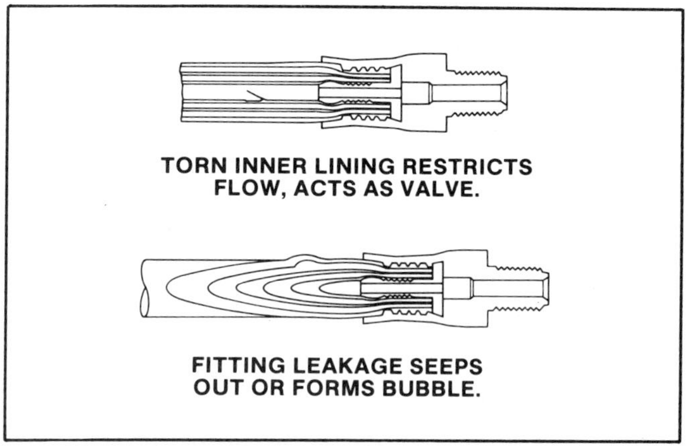

Figure 4-18 Examples of internal defects

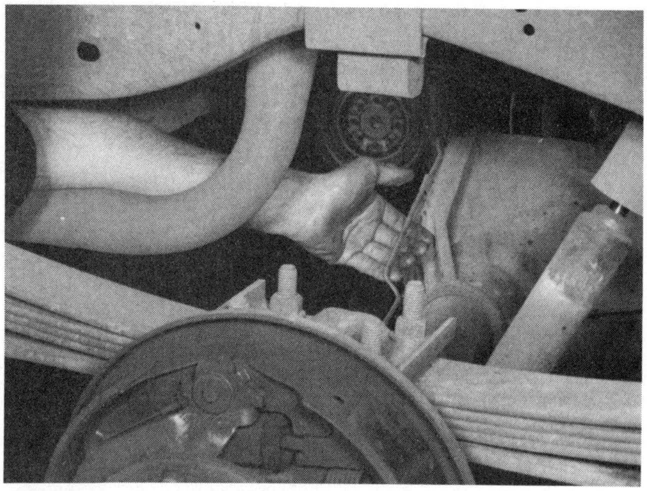

Figure 4-16 Inspecting hose and connections

When replacing brake hoses, use only hoses that meet the OEM safety standards. Before removing the old hose, clean all dirt from the ends of the hose. This reduces the possibility of getting dirt into the system during assembly.

Most hoses are attached to the chassis and secured with a hose lock clip to keep them from rubbing on the frame or suspension members (Figure 4-19). The clip can be removed easily with pliers. To replace a hose, first install and tighten the end that fits into the cylinder. Connect the other end to the brake line that comes from the master cylinder. Use two line wrenches and make sure the hoses are not twisted or kinked. Keep them clear of anything that could wear or puncture them.

Remove the old brake hose by disconnecting the female end first. Remove the clip or the jam nut holding it in place. Then, unscrew the hose with the male fitting (Figure 4-20). Check the area near the threaded hole in the wheel cylinder, the threaded end of the tubing, and the area near the chassis bracket for dirt. Carefully clean away any dirt that you find.

To install a new hose, connect the male end first. If a copper gasket is used, replace it with a new one and throw the old one away. Once the male end is tightened, the female end can be connected.

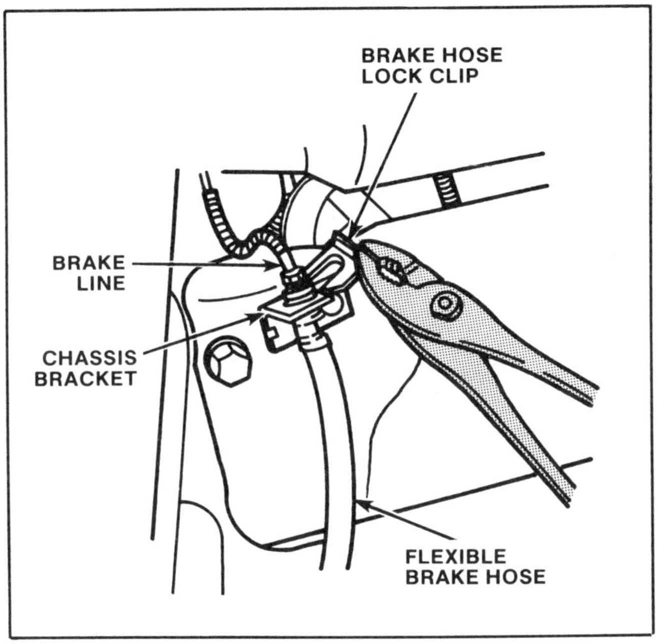

Figure 4-19 Removing lock clip from chassis bracket

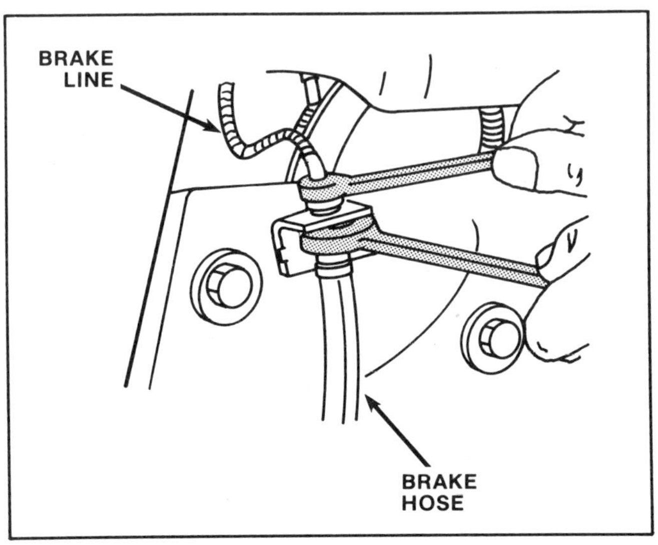

Figure 4-20 Disconnect the female end first

It is important to use the same size hose when replacing hoses. One that is too long will rub on some chassis part, and one that is too short will break when the moveable component reaches the limits of its travel. That is, check clearance during suspension rebound and while turning the wheels. If any contact occurs, reposition the hose, but only adjust the female end. Tampering with the male end once the female end has been tightened could result in twisting the hose. Naturally, the treads and ends must be identical to those used originally.

Though not usually encountered, some caliper components and hoses can be reverse threaded; that is, they turn left to tighten rather than to the right. Sometimes the fasteners are noted with a slash though the flat surfaces of the nut or bolt.

MASTER CYLINDERS

While the brake pedal is the first and strongest indicator of anything wrong in the brake system, there are other components that foretell problems such as the illumination of the brake warning light, and, if so equipped, the brake fluid level in the master cylinder's reservoirs.

Before performing any test procedure on the master cylinder, always check the fluid level. To do this, clean all the dirt from the master cylinder cap. Then, remove the reservoir cover or the cap to check the brake fluid. The fluid level should be within 1/4 inch of the top of the reservoir. The fluid should be clean with no rust or other contamination. The vent holes in the cap should be open and clean. The reservoir diaphragm (if used) should have no holes or other damage.

WARNING: Be careful to avoid spraying brake fluid. To protect the face, never bend directly over the reservoir.

The following conditions are considered normal and are not indications that the master cylinder is in need of service:

- Unequal brake fluid levels in master cylinder reservoirs on front disc/rear drum brake vehicles. This condition is caused by displacement of fluid from the reservoir into the calipers to compensate for normal lining wear. Top off the reservoirs.
- A momentary or slight squirt of brake fluid from one or both master cylinder reservoir chambers upon application of the brake pedal

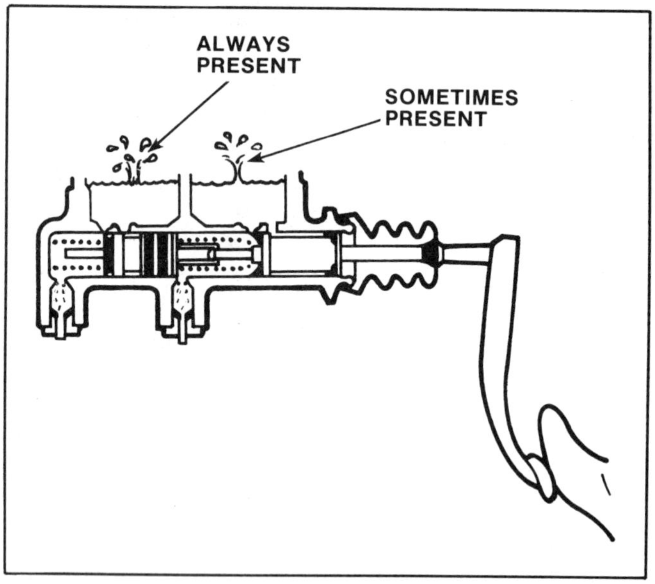

Figure 4-21 Geyser or "squirt" in reservoir

(Figure 4-21). This condition is caused by the fluid displacement through the reservoir replenishing port(s) as the master cylinder pistons move forward in the bore when the brakes are applied.

- A slight turbulence in the reservoir fluid occurring when the brake pedal is released. Turbulence occurs as brake fluid returns to the master cylinder after releasing the brakes.
- A trace of brake fluid exists on the booster shell below the master cylinder mounting flange. This condition results from the lubricating action of the master cylinder wiping seal.

As noted previously, changes in brake pedal feel or travel are indicators that something could be wrong in the brake system. The diagnostic procedures and techniques use brake pedal feel, warning light illumination, and brake fluid level as indicators in diagnosing brake system complaints.

SHOP TALK

Before performing the following diagnosis checks, make certain that the brake system warning light is functional.

To test for air trapped in the brake system, proceed as follows:

1. Fill master cylinder reservoirs if low.
2. Loosely place the gasket and cover assembly on the master cylinder.
3. Have a helper watch the master cylinder as the brakes are pumped rapidly 20 times. Hold the brake pedal down on the last application.
4. Now tell the helper to remove the cap and observe fluid in the reservoirs. Cover the reservoirs with clear plastic wrap or other suitable cover to keep brake fluid off paint. Have water and towels ready, just in case. Release the brake pedal quickly.
5. If air was trapped in the system, it will now be compressed. Upon release of the brakes, air will expand and force the brake fluid to spout as high as 6 inches out of the reservoir.
6. Simply bleed the brake system, which spouts the fluid to rid it of air.

SHOP TALK

Pumping the brake pedal compresses trapped air. When the brake pedal is released, the compressed air expands, forcing brake fluid back into the master cylinder with sufficient force to form a geyser.

To check for external leaks in the master system (one reservoir empty), it is possible that the brake system can have a leak, but no trace of fluid can be found. The car might have been driven for a long period of time with only one system working or rain and snow might have washed off the fluid since it is water soluble.

1. Refill the master cylinder reservoir section that is empty.
2. Apply the brakes several times.
3. Determine whether the fluid level in the master cylinder reservoir is actually dropping by noting the level for 5 or 10 minutes.
4. Examine the master cylinder and all lines, hoses, and fittings to find the source of leakage. Look for puddles on floor under vehicle and at wheels.

To test for a bypass in the master cylinder, proceed as follows:

1. Remove the master cylinder cap and make sure the reservoirs are filled.
2. Observe the fluid levels in the reservoirs while a helper depresses the brake pedal slowly and then releases it quickly.
3. When the level of fluid in one reservoir rises while the level drops in the other as the brake pedal is depressed (This condition reverses when the pedal is released.), but the total (overall) level remains the same, then the master cylinder is bypassing.
4. Replace the master cylinder.

To test for open replenishing ports in the master cylinder, proceed as follows:

1. Remove the cover from the master cylinder and have a helper observe the fluid reservoirs while the brake pedal is pumped.
2. As the brakes are applied, a small ripple or geyser should be seen in the reservoirs.
3. If no turbulence is observed, loosen the master cylinder attaching bolts about 1/4 inch. Pull the master cylinder out and hold it in position. Repeat step 1.
4. If compensation (as evidenced by turbulence) now occurs, adjust the brake pedal pushrod length. If compensation still does not occur, replace the master cylinder.

SHOP TALK

Compensation can only be seen in the front reservoir of the "quick take-up" master cylinders.

To check the master cylinder reserve stroke on vacuum power brake-equipped vehicles where a low pedal or the feel of a bottomed out condition exists, test for brake pedal reserve using the following procedure:

1. Operate the engine at idle with the transmission in either PARK or NEUTRAL position.
2. Apply the brake pedal until it stops moving downward or an increased resistance to the pedal travel occurs.

SHOP TALK

A clunk might be felt as the pedal reaches this point.

3. Hold the pedal in the applied position, and raise the engine speed to approximately 2,000 RPM.
4. Release the accelerator pedal, and observe if the brake pedal moves downward as the engine returns to idle speed.

SHOP TALK

The additional movement of the brake pedal is the result of the increased engine manifold vacuum that exerts more force on the brake booster during engine rundown. This means that additional stroke is available in the master cylinder, and the brake system is not bottoming out as a customer might believe.

If the diagnostic test indicates that the master cylinder is not functioning properly, the unit should be serviced, rebuilt, or replaced with a new one.

MASTER CYLINDER SERVICING

Just like any other brake component, a master cylinder requires periodic service. The usual reason for a master cylinder failure is that the cups lose their ability to seal. Fluid leaks past the cups internally and, sometimes, shows up as an external leak as well. A common symptom is a spongy brake pedal that goes all the way to the floor when all other brake system components are in good shape. The rubber parts wear with usage or can deteriorate with age or fluid contamination. Corrosion or deposits formed in the bore due to moisture or dirt in the hydraulic system might result in wear of the cylinder bore or the parts therein. Also, the fluid level in the reservoirs should be checked periodically. Whenever needed, clean brake fluid should be added to maintain the fluid level 1/4 to 1/2 inch from the top of the reservoir.

The removal, disassembly, cleaning and inspection, reassembly, and installation procedures for the more common master cylinders, whose operation was described and illustrated in Chapter 2, are given here.

Bendix Dual System Master Cylinder

On many vehicles with a dual system master cylinder, it is necessary to disconnect the pushrod from the brake pedal before the master cylinder can be removed. The actual removal and disassembly procedure for a typical dual system master cylinder is as follows:

1. Disconnect hydraulic tubing, remove nuts or bolts that secure master cylinder to firewall or power brake, and remove master cylinder from the vehicle.
2. Remove reservoir cover and drain brake fluid from reservoirs.
3. Remove piston stop bolt from bottom or side of master cylinder (Figure 4–22).

SHOP TALK

Some master cylinders have the stop bolt installed in the bottom of the front reservoir (Figure 4–23).

4. Press in on pushrod or a rod inserted in primary piston, and remove lock ring from cylinder bore or piston retainer from mounting flange, whichever is used.

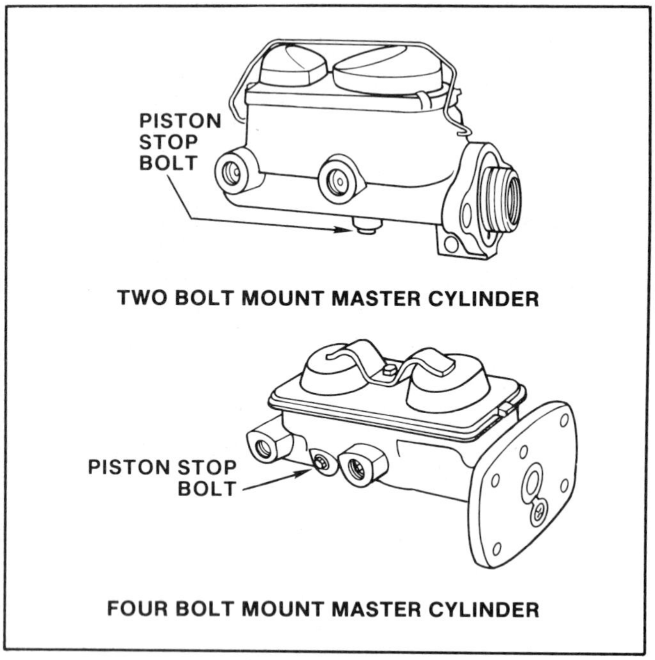

Figure 4-22 Stop bolt locations

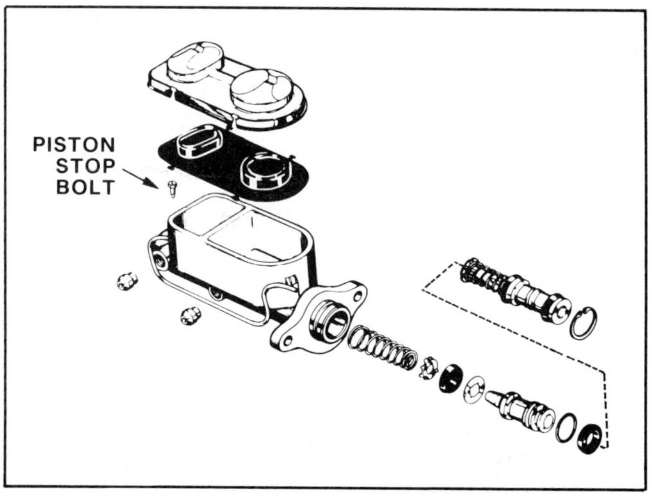

Figure 4-23 Dual system master cylinder—stop bolt in front reservoir

SHOP TALK

If the pushrod is joined to the piston with a plastic or rubber retainer, clamp pushrod in a vise as shown in Figure 4-24. Then use open end wrenches to pry piston from pushrod.

5. Remove parts from cylinder bore, and disassemble secondary piston as shown in Chapter 2.

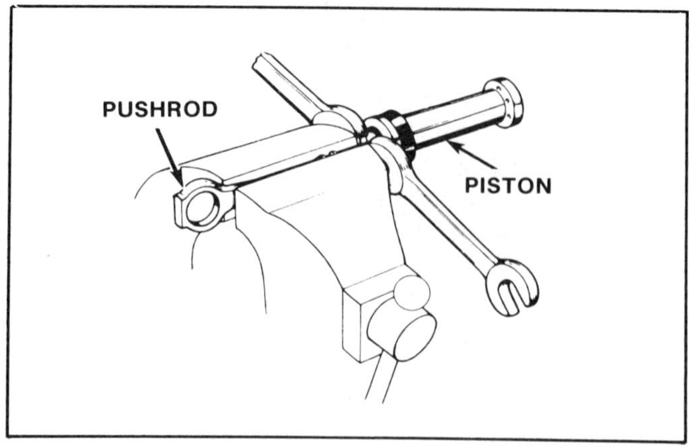

Figure 4-24 Removing pushrod from piston

SHOP TALK

If an outlet port contains a residual pressure check valve, the tube seat insert must be removed so that the check valve can be replaced. If a port does not contain a check valve, do not remove the tube seat unless it is damaged.

6. To remove the tube seat insert and check valve, clamp the master cylinder in a vise with outlet ports facing up. Be careful to avoid damage to casting.
7. Place a flat washer on a 6/32 self-tapping screw, and thread screw into tube seat insert.
8. Use two screwdrivers inserted under the flat washer to pry upward and remove the tube seat (Figure 4-25). Then remove check valve and check valve spring from outlet port.
9. Thoroughly clean master cylinder and any other parts to be reused (except rubber components) in clean denatured alcohol.

CAUTION: Do not clean parts with petroleum products (gasoline, kerosene, etc.).

Check cylinder bore for pits, scoring, or corrosion. Check that fluid inlet and replenishing ports are open. If necessary, a crocus cloth or an approved cylinder hone can be used to clean the cylinder bore. Hone lightly and use brake fluid as a lubricant while honing. If bore does not get clean, master cylinder must be replaced.

Wash master cylinder thoroughly in clean brake fluid after using a crocus cloth or a hone to remove all dust and grit. Wipe cylinder bore with a clean, lint-free cloth. After bore is cleaned, check clearance between the bore wall and a new piston. If a narrow (1/8 to 1/4 inch wide) 0.006-inch feeler gauge can be inserted between the cylinder wall and the new piston, the master cylinder must be replaced.

To reassemble and install the Bendix type dual system master cylinder, proceed in the following manner:

1. Dip secondary piston parts in clean brake fluid and assemble parts as shown in Figure 4-26.
2. Lubricate cylinder bore and all cups with brake fluid.
3. Guide secondary piston assembly, spring end first, into cylinder bore. Use a dull scribe to ease cups into bore and be careful to avoid damage to cups.
4. Guide primary piston into bore in same manner.
5. Press in on primary piston and assemble lock ring in cylinder bore or retainer on mounting flange, whichever is used.
6. Continue to press in on primary piston and assemble piston stop bolt in the position from which it was removed. If tube seat inserts were not removed, omit steps 7 and 8.
7. Insert check valve spring in a check valve and place both parts in vent port, spring end first. Be sure spring is seated in bottom of port.

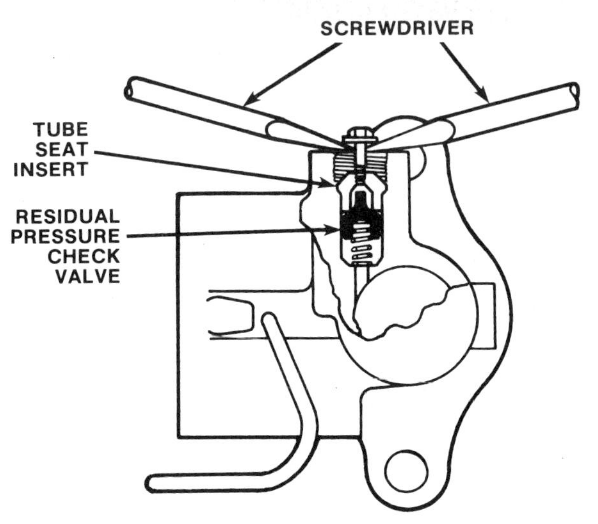

Figure 4-25 Removing tube seat insert

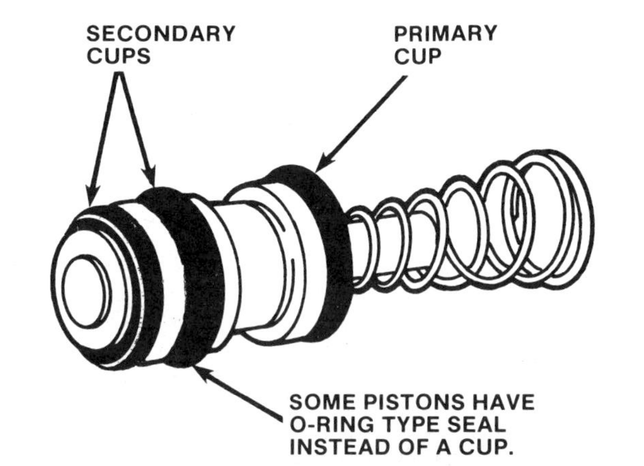

Figure 4-26 Secondary piston assembly

8. Position tube seat insert in port. Press tube seat into place by threading a tube nut into port.
9. Assemble boot, if used, on master cylinder hub.
10. Assemble pushrod retainer, if used, on pushrod. Insert pushrod in primary piston and press in until retainer snaps into place.
11. Install master cylinder on vehicle and connect hydraulic tubing.
12. If disconnected, attach pushrod to brake pedal.

After the master cylinder is installed, the normal bleeding procedure described in Chapter 3 should be followed.

Delco-Moraine Dual System Master Cylinder

The Delco-Moraine dual system master cylinder is removed from the vehicle as follows, remembering that on some cars with manual brakes, the master cylinder pushrod must be disconnected from the brake pedal before the master cylinder can be removed from the car.

1. Disconnect hydraulic tubing, remove nuts or bolts that secure master cylinder to firewall or power brake, and remove master cylinder from car.
2. Remove reservoir cover and drain brake fluid from reservoirs.
3. On manual brake master cylinder, pull back boot to expose pushrod retainer. Bend up the tab that holds the retainer on the master cylinder hub, then remove the pushrod, boot, and retainer.

4. Remove piston stop bolt from bottom of front reservoir and remove lock ring from cylinder bore.
5. Remove parts from cylinder bore.
6. Disassemble secondary piston.

SHOP TALK

If any vent port contains a residual pressure check valve, the tube seat insert must be removed so that the check valve can be replaced. If a port does not contain a check valve, do not remove the tube seat unless it is damaged.

7. To remove tube seat insert and check valve, clamp master cylinder in a vise with vent ports facing up. Be careful to avoid damage to casting.
8. Place a flat washer on an 8/32 inch self-tapping screw and thread screw into tube seat insert.
9. Use two screwdrivers inserted under flat washer to pry upward and remove tube seat. Then remove check valve and check valve spring from vent port.

The following procedure covers fluid reservoir removal and reassembly for cylinders with a plastic reservoir.

1. Separate reservoir from cylinder body. Insert large screwdriver between reservoir and cylinder body.
2. Pry up on screwdriver handle, using cylinder body as pivot point, until reservoir is free.
3. Remove and discard rubber grommets.
4. Lubricate new grommets with clean brake fluid and position into master cylinder.
5. Lubricate inside diameter of grommets with clean brake fluid. Position fluid reservoir, top down, on a hard, flat surface. Start master cylinder onto reservoir at an angle, working lip of reservoir nipples completely through grommets until seated.

To clean this type of dual system master cylinder, follow the procedure given previously for the Bendix unit.

To reassemble and install the master cylinder, follow these steps:

1. Assemble the parts on secondary piston. Be sure that the secondary cups are assembled back to back in grooves near the end of the piston.
2. Dip the secondary and primary piston assemblies in clean brake fluid. Lubricate the cylinder bore with brake fluid.
3. Guide the secondary piston assembly, spring end first, into the cylinder bore. Use a dull scribe to ease cups into the bore and be careful to avoid damage to cups.
4. Insert the primary piston into the bore in the same manner.
5. Press in on the primary piston and assemble the lock ring in the groove in the cylinder bore. Then assemble the piston stop bolt in the threaded hole in the bottom of the front reservoir. If the tube seat inserts were not removed, omit steps 6 and 7.
6. Insert the check valve spring in a check valve and place both parts in the vent port, spring end first. Be sure the spring is seated in the bottom of the port.
7. Position the tube seat insert in the port. Press the tube seat into place by threading a tube nut into the port.
8. On a manual brake master cylinder, assemble the pushrod retainer, pushrod, and boot on the master cylinder.
9. Install the master cylinder on the car and connect the hydraulic tubing.

Once the master cylinder has been re-installed bleed the hydraulic system as described in Chapter 3. Check master cylinder replenishing port clearance by watching for a spurt of brake fluid in front reservoir when brake pedal is depressed slightly.

GM Quick Take-Up Master Cylinder

The quick take-up master cylinder can be removed from the hydraulic system and disassembled in the following manner:

1. Disconnect the four hydraulic lines and electrical lead to the brake warning light switch. Cap open end of lines to prevent fluid loss and system contamination.
2. Unscrew the two mounting nuts attaching the cylinder to the power brake booster. Remove the cylinder.
3. Detach the reservoir cap and discard old brake fluid.
4. Inspect the plastic fluid reservoir for cracks. Replace reservoir as outlined in reservoir removal procedures for Delco-Moraine dual system master cylinder.
5. Remove the proportioning valves and O-rings. Discard the O-rings.

6. Remove the brake warning light switch. Discard the O-ring.
7. Remove lock ring while depressing primary piston.
8. Extract primary and secondary piston assemblies from bore.
9. Remove switch piston assembly and plug.

Clean master cylinder and related hard parts with denatured alcohol or clean brake fluid. Blow-dry with clean, compressed air. Discard old rubber seals, cups, and O-rings. Inspect cylinder bore for pits, scoring, or corrosion. If any of these conditions are evident, replace master cylinder (aluminum units). *Do not attempt to hone cylinder bore.*

To reassemble and install this master cylinder, proceed in the following manner:

1. Install the new primary and secondary cup onto the secondary piston. Install the assembly into the master cylinder bore.
2. Insert the new primary piston assembly into the master cylinder.
3. Depress pistons into the bore while installing the lock ring.
4. Install the brake warning light switch and new O-ring.
5. Install the proportioning valves and new O-rings into the master cylinder. Torque to 18 to 30 foot-pounds.
6. Reassemble the switch piston assembly and plug, using new rubber parts. Reinstall in the cylinder with a plug and O-ring.
7. Mount the master cylinder onto the vehicle. Torque the mounting nuts to 22 to 30 foot-pounds.
8. Reconnect the hydraulic brake lines and electrical lead to the brake warning light switch.

The special bleeding procedures for the quick take-up master cylinder are given in Chapter 3.

Dual Aluminum Master Cylinder

For models equipped with power brakes, disconnect the primary and secondary brake tubing leading from the master cylinder. Plug the master cylinder outlet. Remove the nuts that attach the master cylinder to the power brake unit.

For vehicles equipped with manual brakes, disconnect the primary and secondary tubing from the master cylinder. Plug the outlets. Disconnect the stoplight switch mounting bracket from underneath the instrument panel. Allow it to hang clear. Pull the brake pedal backward to disengage the pushrod from the master cylinder piston.

SHOP TALK

This will require about 50 pounds of pulling force, which will destroy the pushrod retention grommet in Chrysler vehicles. Be sure to remove all traces of the old grommet and always install a new pushrod retainer grommet.

To disassemble an aluminum master cylinder (Figure 4-27), proceed as follows:

1. To disassemble the aluminum master cylinder, first clean the housing and reservoir casing. Remove the reservoir caps and empty the brake fluid from the reservoir. Place the master cylinder in a vise, rock the reservoir from side to side to loosen it, and remove it from the master cylinder housing (Figure 4-28A). Remove the reservoir grommets.
2. With a pair of needle-nose pliers, remove the secondary piston retainer pin from inside the master cylinder housing (Figure 4-28B). This step is usually only necessary with Chrysler cars.
3. Then, use snap-ring pliers to remove the snap ring from the outer end of the cylinder bore (Figure 4-28C). Slide the primary piston out of the cylinder bore (Figure 4-28D).
4. Tap the open end of the master cylinder on a bench to remove the secondary piston (Figure 4-28E). Note the position of the cup lips and remove the rubber cups from the piston. *Do not* remove the primary cup of the primary piston; if it is damaged or worn, install a new primary piston assembly.

Clean master cylinder and related hard parts with denatured alcohol or clean brake fluid. Blow-dry with clean, dry compressed air. Discard all old rubber seals, cups, and O-rings. Inspect cylinder bore for pits, scoring, or corrosion. If any of these conditions are evident, replace master cylinder (aluminum units). *Do not attempt to hone cylinder bore.* If the cylinder bore is scuffed or grooved, replace the complete master cylinder assembly.

Before reassembling the master cylinder, dip all its components in clean brake fluid and place them on a clean shop towel. Then, to reassemble the unit, proceed as follows:

1. Install new cups onto the secondary piston. Insert the assembly into the master cylinder.

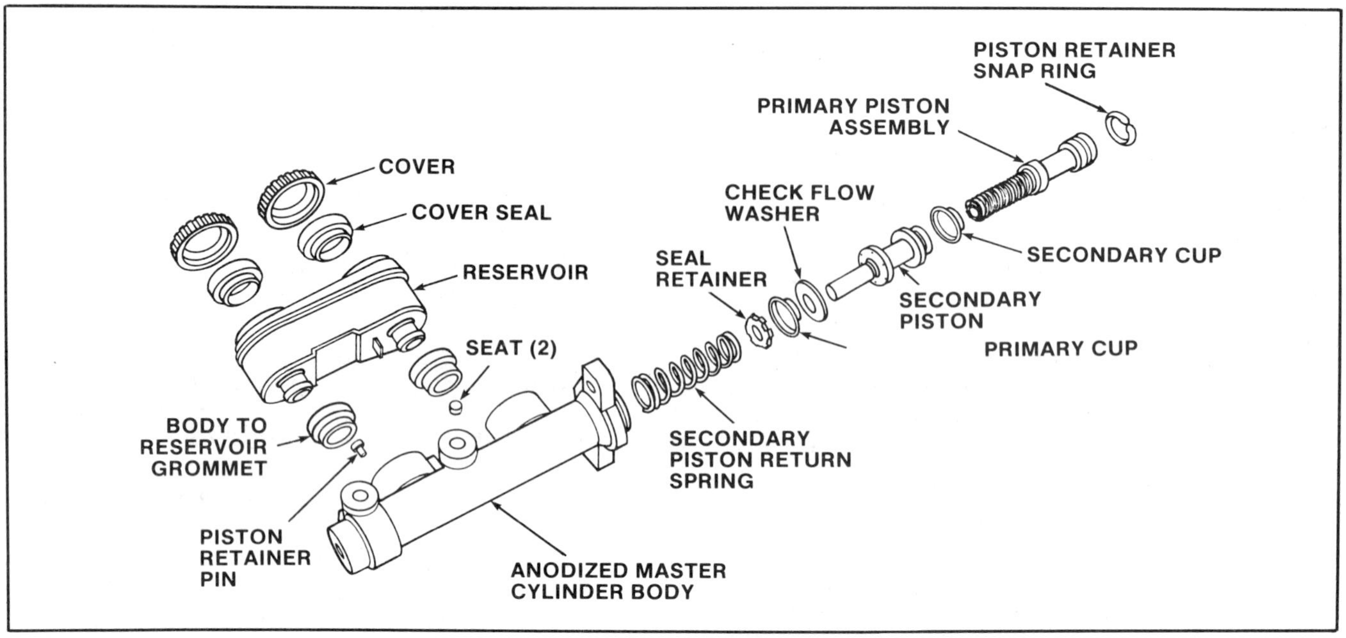

Figure 4-27 Chrysler dual system master cylinder

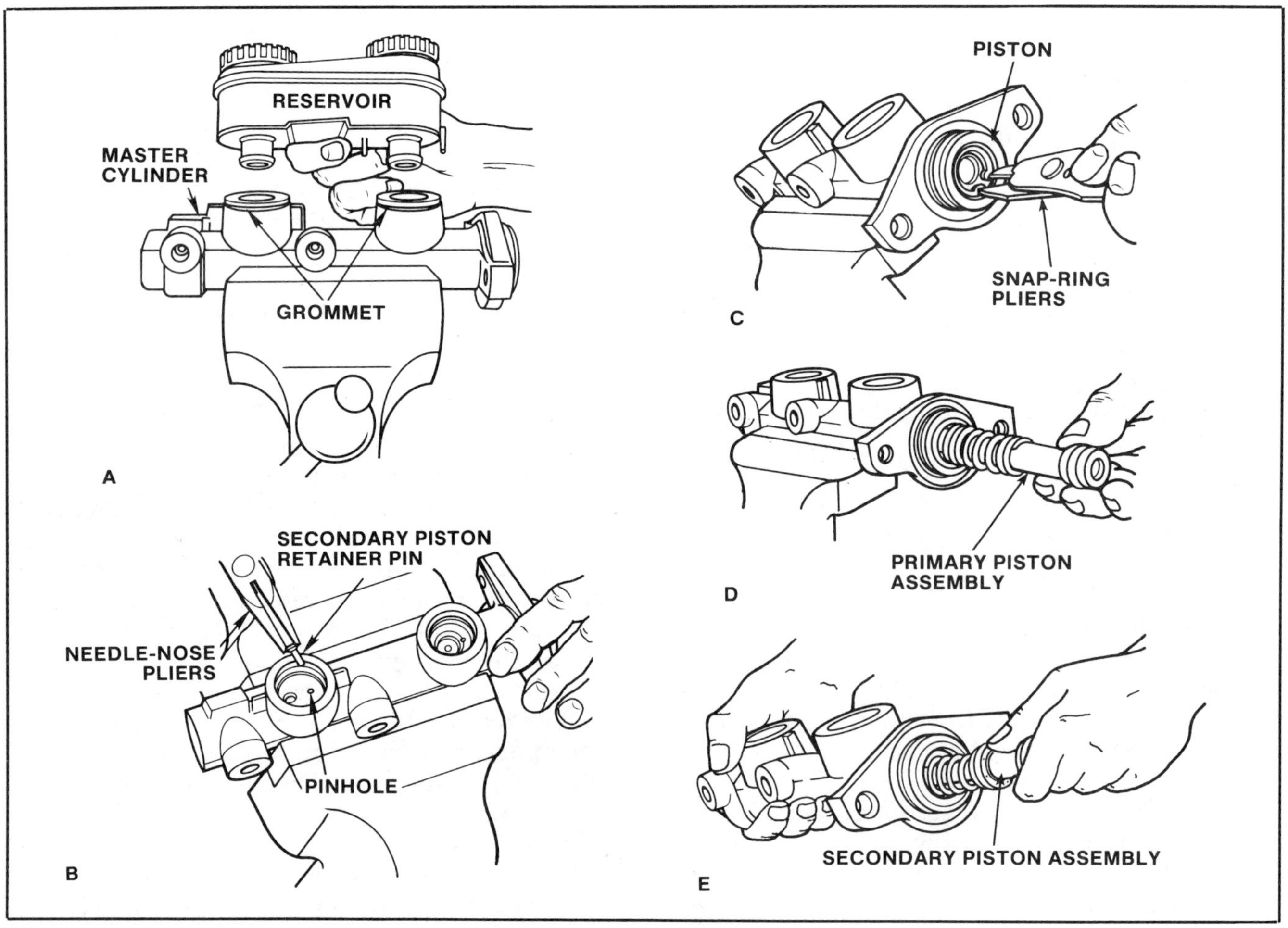

Figure 4-28 Steps to disassemble an aluminum master cylinder

2. Install the new primary piston assembly into the bore.
3. With the pistons depressed, install the lock ring.
4. Insert the secondary piston retainer pin into the housing. Be sure that the pin (if used) is fully seated.
5. Lubricate the new reservoir grommets with clean brake fluid and install them into the master cylinder. Remember that assembling the master cylinder with dry seals can ruin the grommets.
6. With the master cylinder in a vise, install the fluid reservoir using a rocking action. The reservoir bottom should contact the top of the grommets when fully installed.
7. Mount the master cylinder onto the vehicle. Torque the mounting nuts or bolts 16 to 20 foot-pounds.

Follow the procedure given in Chapter 3 when bleeding an aluminum master cylinder.

Bendix Mini-Master Cylinder

There is an early and late model of the mini-master cylinder. The O-ring seal that seats in the recessed area on the bottom of the reservoir is slightly larger in the later model master cylinder. Be sure to determine which O-ring seal is applicable to the master cylinder you are servicing. To disassemble the unit, proceed as follows:

1. Remove the reservoir cover and diaphragm and drain the fluid from the reservoir.
2. Remove the four bolts securing the reservoir to the body using a 5/16-inch six-point socket wrench.
3. Remove the small O-ring and the two replenishing valve seals from the recessed areas on the bottom of the reservoirs.

SHOP TALK

Do not remove the two small filters from the inside of the reservoir unless they are damaged and are to be replaced.

4. Using a suitable tool, depress the primary piston 1.000 inch $\pm$ 0.05 inch. (Spring loads are approximately 90 and 130 pounds. It is very difficult to depress the piston by hand.)
5. With piston depressed, remove the replenishing valve poppets and the replenishing valve springs from the replenishing valve ports and the piston stop pin from the stop pin hole in the master cylinder body.

WARNING: In the following step, if air pressure is used to remove the secondary piston, place the open end of the cylinder bore approximately 1 inch from a padded workbench or other surface to catch the piston when it comes out of the bore. Apply low air pressure very carefully to ease the piston out of the bore. Never point the open end of the bore at anyone when applying air pressure. The piston might come out of the bore with considerable force and cause injury.

6. Using a snap-ring pliers, remove the snap ring at the end of the master cylinder bore. Then release the piston and remove the primary and secondary piston assemblies from the cylinder bore. It might be necessary to plug the front vent port and to apply low air pressure to the front replenishing valve port to remove the secondary piston assembly.
7. Remove tube seats (inserts) only if visual inspection indicates damage to them.

Wash all parts in clean brake fluid. Be sure all foreign material is removed from the filters in the bottom of the reservoir. If filters do not clean up, they must be replaced. Examine the cylinder bore for scoring or corrosion. It is best to replace a corroded cylinder. Corrosion can be identified as pits or excessive roughness. Polish any discolored or stained area with crocus cloth by revolving the cylinder on the cloth supported by a finger. Do not use emery cloth or sandpaper. Clean the cylinder thoroughly after polishing.

CAUTION: Do not hone the master cylinder or wheel cylinder bores. When brake master cylinders or wheel cylinders are overhauled, it is recommended that the cylinder body be replaced rather than "cleaned up" by honing the bore. Master cylinder and wheel cylinder bores have a hard, highly polished "bearingized" surface, which is produced by diamond boring followed by ball or roller burnishing under heavy pressure. Honing will destroy this surface, leaving a softer and rougher surface, which will cause more rapid wear of rubber cups.

Inspect parts for damage and excessive wear and replace as required.

To reassemble the master cylinder, proceed as follows:

1. Lubricate the secondary piston assembly and the master cylinder bore with clean brake fluid.
2. Assemble the secondary spring (the smaller of the two springs) in the open end of the secondary piston actuator and assemble the piston return spring (larger spring) in the open end of the secondary piston.
3. Insert the secondary piston assembly, actuator end first, into the master cylinder bore and press assembly to the bottom of the bore.
4. Lubricate the primary piston assembly with clean brake fluid. Insert the primary piston assembly, actuator end first, into the bore.
5. Place the snap ring over the round-ended rod of a suitable tool and depress the pistons in the bore.
6. Assemble the retaining ring in the groove in the cylinder bore.
7. Assemble the replenishing valve seals and the small O-ring seal in the recesses on the bottom of the reservoir. Be sure that all seals are fully seated.
8. While holding pistons depressed by using a suitable tool, with primary piston stroked 1.000 inch, ± 0.05 inch, assemble the replenishing valve springs and the replenishing valve poppets and piston stop pin in the replenishing valve ports and stop pin hole.
9. Holding the pistons compressed, position the reservoir on the master cylinder body and secure with the four mounting bolts using a special tool or socket wrench. Torque the bolts to 14 foot-pounds.

The mini-master cylinder should be bench bled as follows:

1. Plug the vent ports and fasten the master cylinder in a vise with the front end tilted down slightly.
2. Fill the reservoir with clean brake fluid. Use any tool with a smooth, rounded end to depress and release the primary piston several times.
3. Reposition the master cylinder in the vise with the front end tilted up slightly. Again, depress and release the primary piston several times.
4. Assemble the diaphragm and cover on the master cylinder.

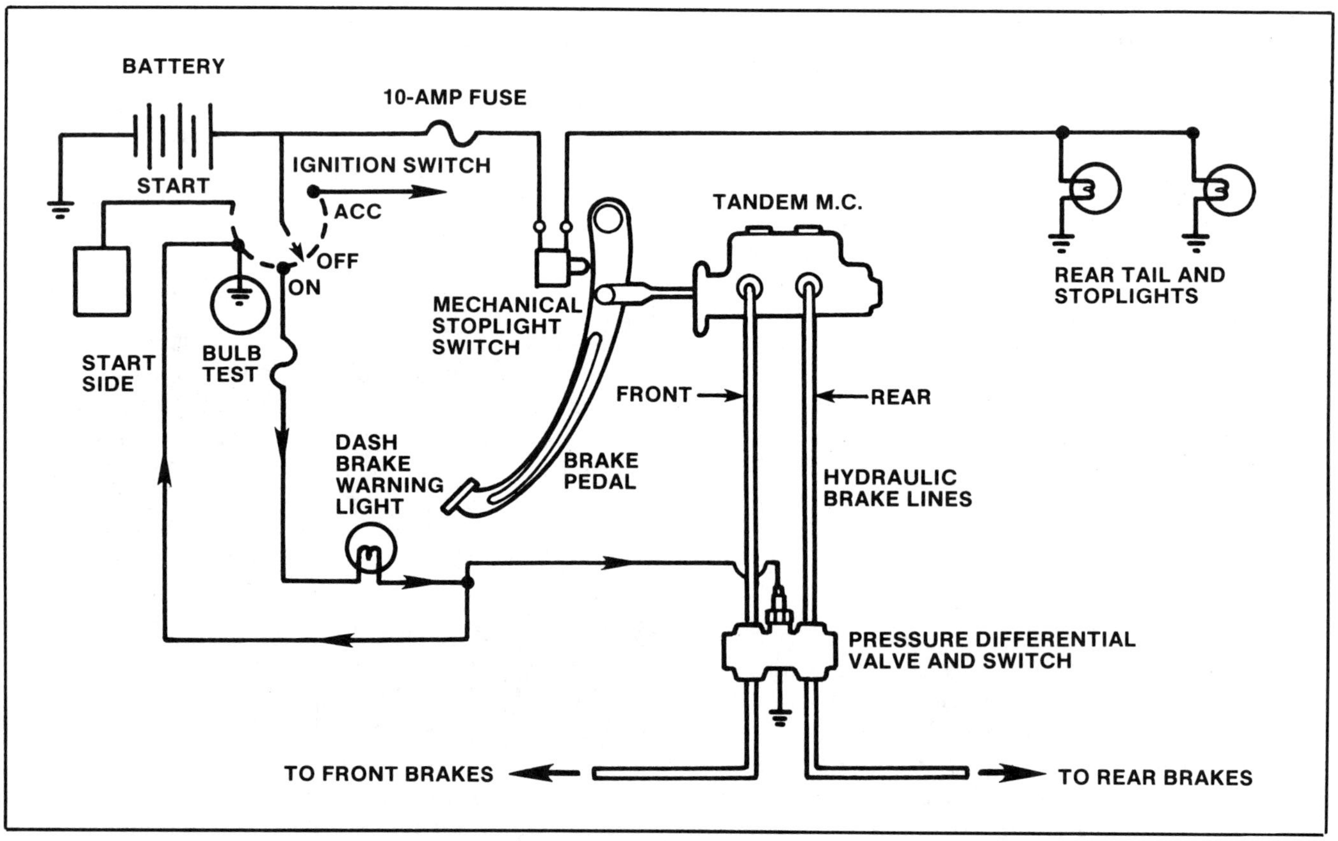

Figure 4-29 Brake warning light—electrical circuit

The master cylinder is ready for installation on the vehicle. Normal bleeding procedures given in Chapter 3 should be followed after the master cylinder is installed.

Most types of master cylinders have repair kits available that contain all the parts ordinarily required when rebuilding or overhauling the unit. Refer to an applicable vehicle parts book for the part number of the repair kit. Complete instructions are furnished with the rebuilding kit.

SAFETY VALVES AND SWITCHES

The various safety valves and switches described in Chapter 2 must be inspected to be sure that they are operating properly.

Testing the Warning Light Switch

If the warning light does not operate, check it out as follows:

Electrical Circuit Check. The brake warning light electrical circuit (Figure 4–29) should be checked prior to testing the switch.

Disconnect the lead wire from the switch terminal. Then, connect a jumper wire from the lead to a ground on the engine or chassis. Turn the ignition key to ON. The brake warning light should come on. If the light does not come on, replace the bulb or repair the electrical circuit, as necessary. After the light comes on with the jumper connected, turn the ignition switch to OFF and reconnect the lead wire to the switch terminal.

Switch Test. Attach a bleeder hose to a rear brake bleeder screw and place the other end of the hose in a container partially filled with brake fluid. Be sure that the master cylinder reservoirs are full. Turn the ignition switch to ON. Open the bleeder screw, while a helper applies heavy pressure to the brake pedal. (Up to 250 PSI is required to operate the switch.) The brake warning light should come on. Tighten the bleeder screw before the helper releases the brake pedal.

If the warning light does not come on during this test but did come on when the jumper wire was connected to the ground, the warning light switch is defective and should be replaced.

After the test is completed, center the piston in the switch bore, if needed.

Servicing Metering Valves

The metering valve should be inspected whenever the brakes are serviced. If there is fluid leakage inside the boot on the end of the valve, the valve is defective and should be replaced.

SHOP TALK

A slight amount of moisture inside the boot does not indicate a defective valve.

Metering valves are non-adjustable and nonrepairable. If a valve is defective, it should be replaced as an assembly. When a metering valve is replaced, be sure to mount the new valve in the same position as the old valve.

Servicing Proportioning Valves

The proportioning valve should be inspected whenever the brakes are serviced. If the valve is leaking, it is defective and should be replaced.

Proportioning valves are non-adjustable and non-repairable. If a valve is defective, it should be replaced as an assembly.

Servicing Combination Valves

The combination valve should be inspected whenever the brakes are serviced. If there is leakage inside the boot or around the large nut on the proportioning end, the valve is defective and should be replaced.

SHOP TALK

A slight amount of moisture inside the boot or a slight dampness around the large nut does not indicate a defective valve.

Combination valves are non-adjustable and nonrepairable. If a valve is defective, it should be replaced as an assembly.

Brake Fluid Level Sensors

Remove the sensor from the master cylinder (Figure 4–30), like the safety valves and switches, according to the instructions in the vehicle's service manual. Be sure the ignition is ON and the wiring is still attached. Hold it in an upright position. Within 30 seconds the instrument panel message "BRAKE FLUID LOW" should appear.

Next invert the sensor. The instrument panel message should turn OFF immediately. This means that the sensor is working properly.

If the message does not appear, remove the wiring from the sensor. Using a jumper wire, connect both sides of the plug. The instrument panel message "BRAKE FLUID LOW" should appear within 30 seconds. If the message does not appear, a

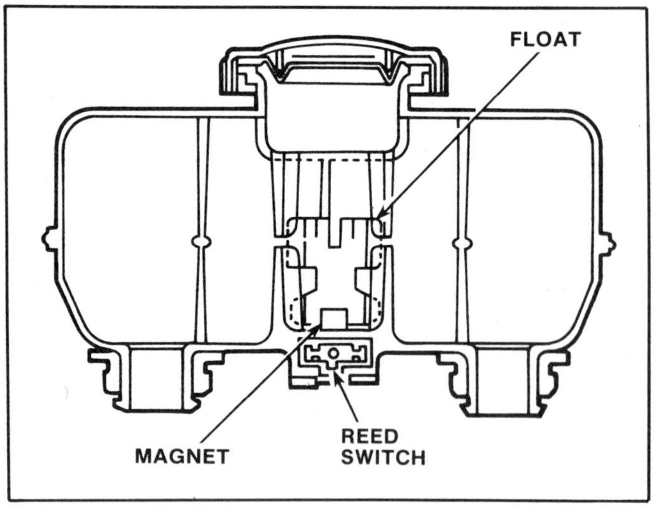

Figure 4-30 Plastic reservoir with fluid level indicator

problem exists with either the wiring or the instrument panel.

If the message does appear, the sensor is faulty and should be replaced.

SHOP TALK

The brake fluid level sensor is not a repairable part.

Servicing Height Sensing Proportioning Valve

The new brake height sensing proportioning valve, as mentioned in Chapter 2, regulates the amount of brake pressure according to vehicle load condition. More cargo weight allows more braking and less cargo weight restricts the brake pressure.

There are several important points to remember about such proportioning valves. The first is the valve must be adjusted correctly if it is to balance rear braking force to the load. The valve linkage is adjusted with the suspension at rest and the normal weight of the vehicle on the wheels (Figure 4-31). The adjustment bracket is then loosened and adjusted to remove all slack from the spring linkage that connects the axle and proportioning valve.

The second point about these valves is that they are calibrated to work with the stock suspension. Any modifications that would improve load-carrying capability (helper springs, overload or air assist shocks, for example) can adversely affect the operation of the valve.

Any modifications that tend to make the suspension stiffer can prevent the suspension from deflecting the normal amount during hard braking or when the vehicle is heavily loaded. As a result, the proportioning valve might not change rear brake effort as much as it should and stopping distance could increase dangerously. Because of

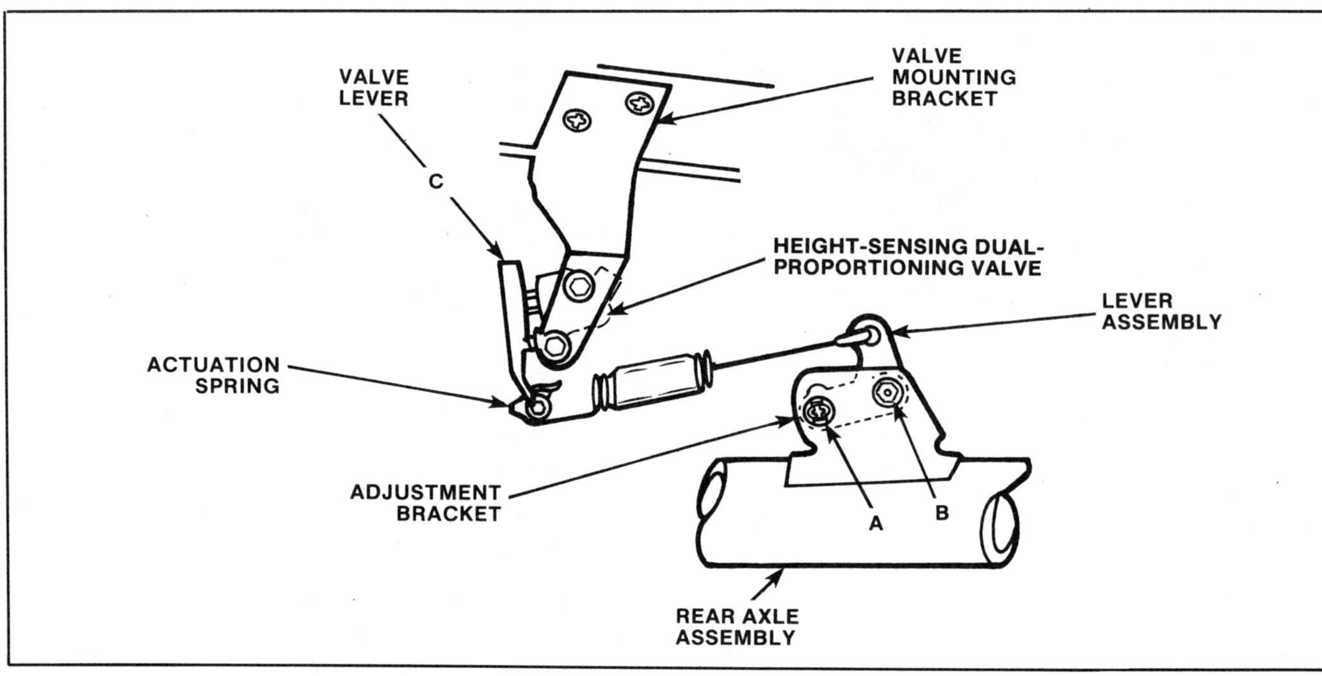

Figure 4-31 Height-sensing dual-proportioning valve adjustment

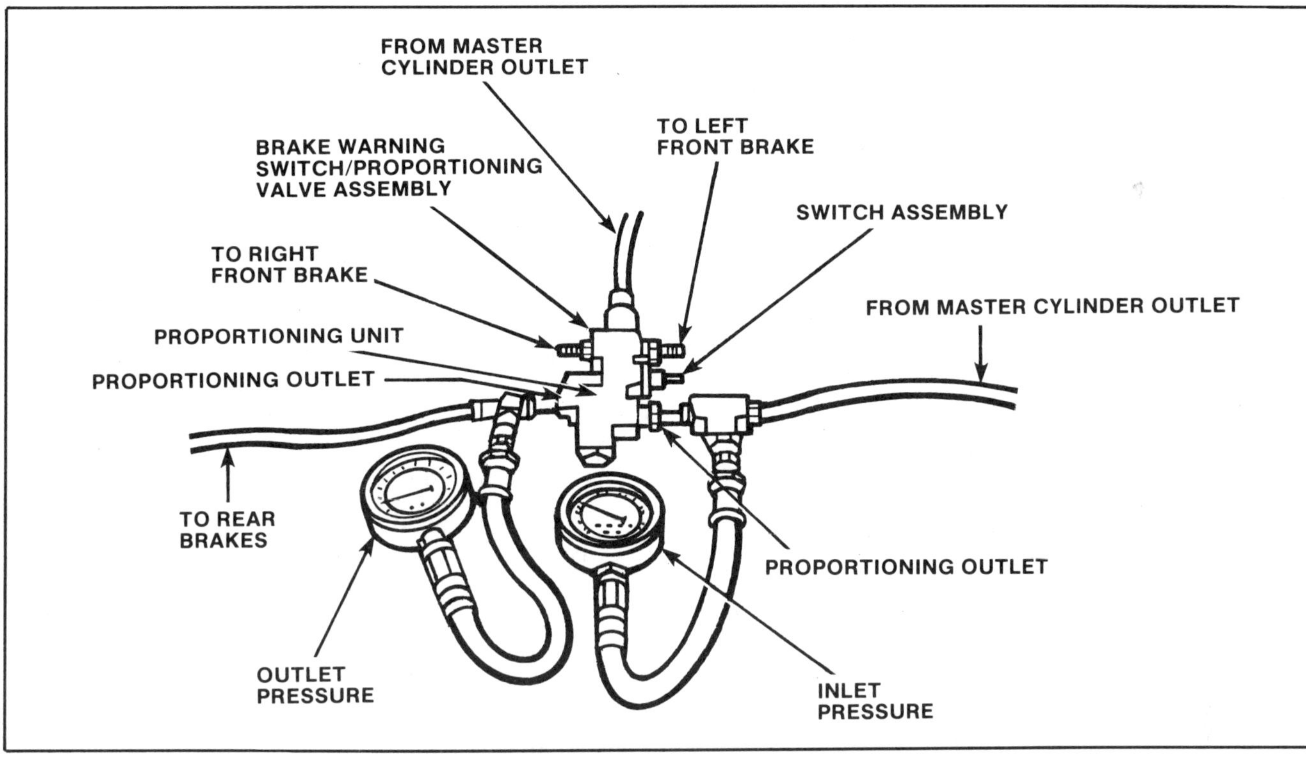

Figure 4-32 Gauges can be used to measure proportioning valve pressures.

this, modifications to the rear suspensions of these vehicles should be avoided.

To check the valve requires a set of 1500 PSI hydraulic gauges and tee fittings. One gauge is installed in the line from the master cylinder to the proportioning valve and the other between the valve and either rear brake (Figure 4-32). Inlet and output pressures as then compared with the valve fully open and closed to the manufacturer's specifications. If the pressures are not within the limits that are specified in the service manual, the valve is defective and should be replaced.

The following procedure can be used for removing and installing a typical valve assembly:

1. Raise the vehicle on a hoist and allow the rear suspension and rear axle to hang in the fully released position. (Follow the procedure in Chapter 5.)
2. Remove both rear wheels and tires to maintain an equally rebounded axle assembly.
3. Remove the linkage assembly from the valve.
4. Remove the brake hose and discard the old washers.
5. Remove the valve from the mounting bracket and replace with a new valve (Figure 4-33).
6. Position and attach the sensing valve to the mounting bracket with two bolts. Torque the two bolts to 14 to 22 foot-pounds.
7. Connect the linkage assembly to the sensing valve shaft. Squeeze the plastic bushing of

Figure 4-33 Heat sensing valve diagram

the link assembly onto the serrated section of the valve shaft, using large groove joint pliers.

8. Install the nut securing link assembly to the valve shaft. Torque the nut to 8 to 10 foot-pounds.
9. Install the rear brake hose assembly to the sensing valve with the existing bolt and two new washers. Torque the bolt to 17 to 25 foot-pounds.
10. Bleed the brake system as per the appropriate service manual and correct any leakage condition.
11. Install the wheels and tires, taking care to torque the lug nuts to the proper torque. (Follow the procedure in Chapter 5.)
12. Lower the vehicle and road test to confirm the repair.

The new brake height sensing proportioning valve will automatically become operational when the vehicle's suspension is in the *curb height* condition. Remember that the addition of extra leaf springs to increase load capacity, spacers to raise the vehicle curb height, and air shocks to allow heavier loads without sagging are not to be used on vehicles with height sensing proportioning valves.

REVIEW QUESTIONS

Review the information in this chapter by answering the following questions.

1. When adjusting brakes, Mechanic A removes all free play at the pedal. When adjusting brakes, Mechanic B leaves a little free play at the pedal. Who is right?
   a. Mechanic A
   b. Mechanic B
   c. both A and B
   d. neither A nor B
2. If the height of a brake pedal is correct, but it still feels spongy, this could indicate
   a. the presence of air in the system.
   b. worn or damaged bushings.
   c. loose attaching bolts.
   d. dirty backing plates.
3. Which of the following statements regarding brake lines is incorrect?
   a. When brakes lines are replaced, fabrication should not be necessary.
   b. All brake lines are steel and should only be replaced with steel.
   c. Rusted brake lines can burst under a panic stop situation.
   d. All of the above.
4. What is the recommended method of attaching brake lines?
   a. brazing
   b. compression fittings
   c. flaring
   d. soldering
5. When checking the fluid level in a master cylinder, the fluid should be within
   a. 1 inch of the top of the reservoir.
   b. 3/4 inch of the top of the reservoir.
   c. 1/2 inch of the top of the reservoir.
   d. 1/4 inch of the top of the reservoir.
6. When checking for external leaks in the master cylinder, no trace of brake fluid can be found. Mechanic A concludes that no leak is present. Mechanic B concludes that there still can be a leak. Who is right?
   a. Mechanic A
   b. Mechanic B
   c. both A and B
   d. neither A nor B
7. When testing for open replenishing ports in the master cylinder, how can you tell if compensation occurs?
   a. turbulence in the fluid reservoirs
   b. dropping fluid level
   c. brake pedal moves downward as the engine returns to idle speed
   d. clunking noise when the brake pedal is applied
8. Which of the following statements regarding master cylinder servicing is incorrect?
   a. Clean the master cylinder and any other parts to be reused in gasoline or kerosene.
   b. The most common reason for master cylinder failure is that the cups lose their sealing ability.
   c. A common symptom of master cylinder failure is a spongy brake pedal.
   d. On vehicles with a dual-system master cylinder, it is often necessary to disconnect the pushrod from the brake pedal before the master cylinder can be removed.
9. When disassembling a typical dual system master clinder, if an outlet port contains a residual pressure check valve, the tube seat insert
   a. should never be removed.
   b. should be removed only if damaged.
   c. must be removed.
   d. takes the place of a check valve.

CHAPTER 5

DRUM BRAKES—OPERATION

OBJECTIVES

Upon completing this chapter on drum brakes, you should be able to

- Explain how a drum brake operates.
- Identify the major components of a typical drum brake and describe the function of each component.
- Describe the different functions of primary and secondary brake shoes.
- Explain the function of brake lining and describe the importance of its proper location on the shoe.
- Describe the installation process for mechanical components such as the shoe return springs, shoe hold-downs, and shoe anchors.
- Explain the differences between duo-servo and non-servo drum brakes.
- Identify the various types of automatically adjusted servo brakes.
- Explain the difference between the Lockheed manual and the adjustable strut type of non-servo adjuster designs.
- Identify the various types of automatically adjusted non-servo brakes.

Drum brakes have been in use since the early days of the automobile. Their operation is fairly simple. The most important fact contributing to the effectiveness of the braking force supplied by the drum brake is the brake shoe pressure or force directed against the drum (Figure 5-1). With the vehicle moving in either the forward or reverse direction with the brakes on, the applied force of the brake shoe, pressing against the brake drum, will increasingly multiply itself. This is because the brake's anchor pin acts as a brake shoe stop and prohibits the brake shoe from its tendency to follow the movement of the rotating drum. The result is a wedging action between the brake shoe and brake drum. The wedging action combined with the applied brake force creates a self-multiplied brake force.

For many years drum brakes were used on all four wheels of nearly all cars and trucks. Recently, design changes in the vehicles have required a change to disc brakes on the front of most cars and small trucks.

The drum brake is still used on the rear wheels of all but a few expensive domestic or import models. With the back wheels doing only about 20 to 40 percent of the braking and the fact that drum

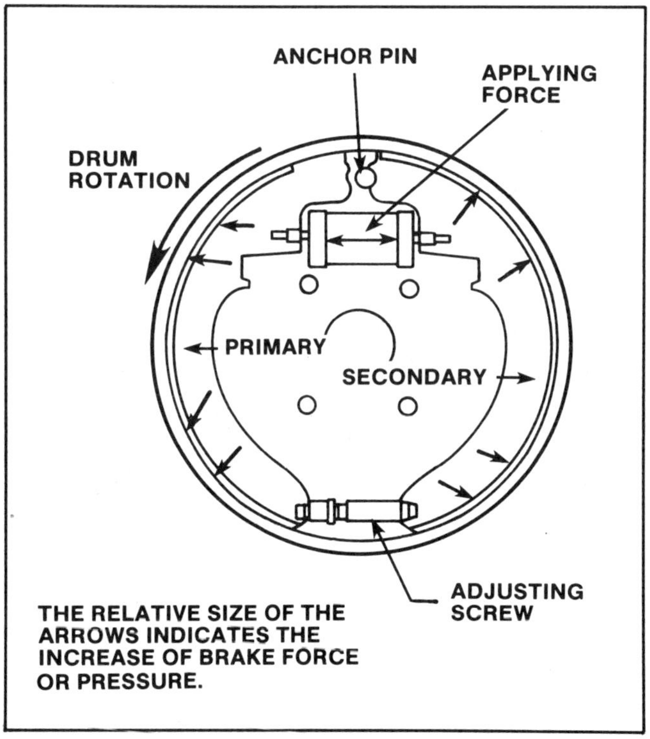

Figure 5-1 Brake shoe pressure against brake drum

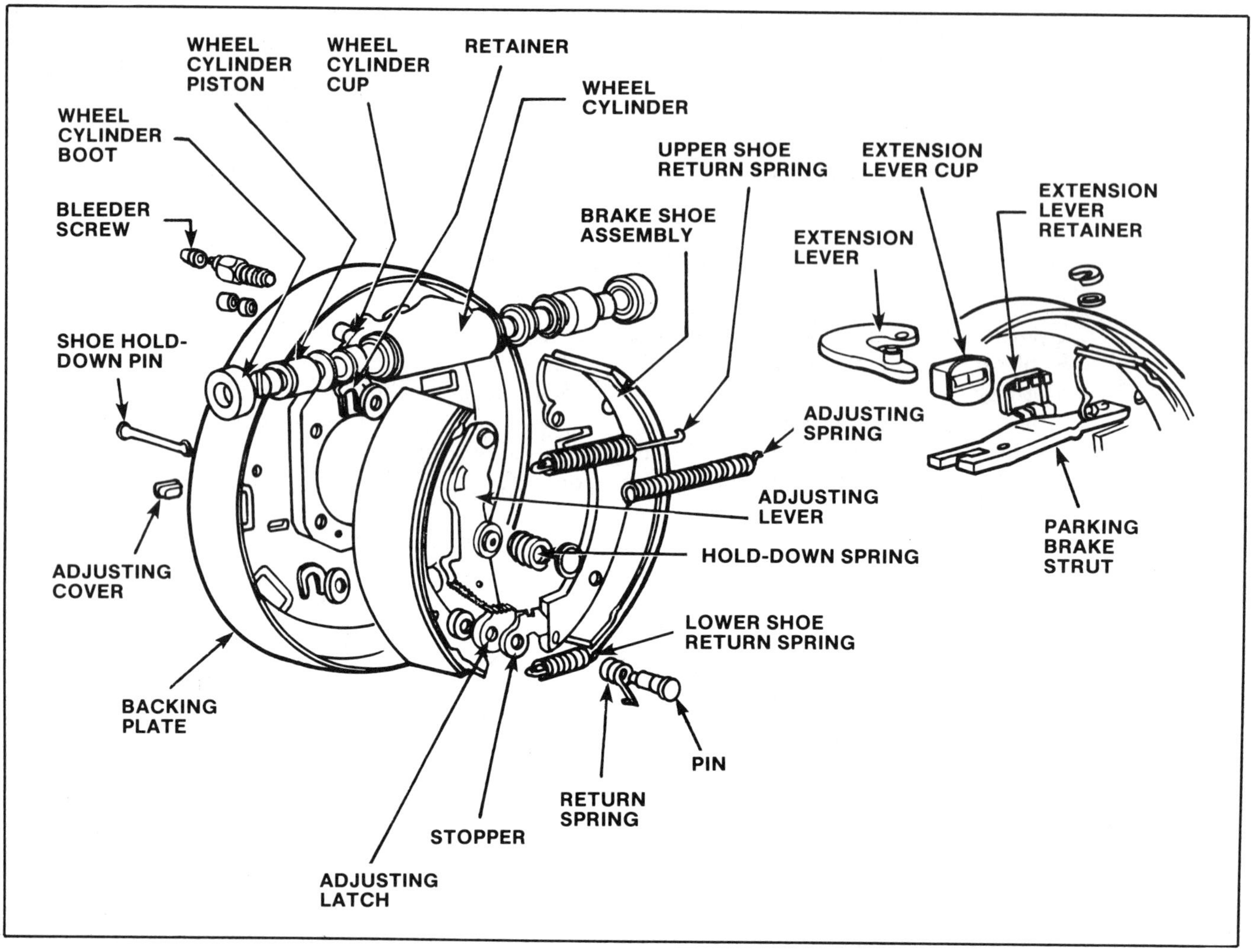

Figure 5-2 Rear drum brake components

brakes can be made with a simple parking brake, means the drum brake will be around for a long time to come.

DRUM BRAKE COMPONENTS

The major components of a typical drum brake like the one shown in Figure 5-2 are:

- *Backing Plate*—The backing plate provides a foundation for the brake shoes and associated hardware. The plate is secured and bolted to the axle flange or spindle.
- *Wheel Cylinder*—The wheel cylinder, under hydraulic pressure, forces the brake's shoes against the drum.
- *Two Lined Brake Shoes Attached to the Backing Plate*—The brake shoe is the backbone of a drum brake. The shoe must support the lining and carry it into the drum so that the pressure is distributed across the lining surface during brake application.
- *Shoe Return Springs and Shoe Hold-down Parts*—These maintain the correct shoe position and clearance (Figure 5-3).

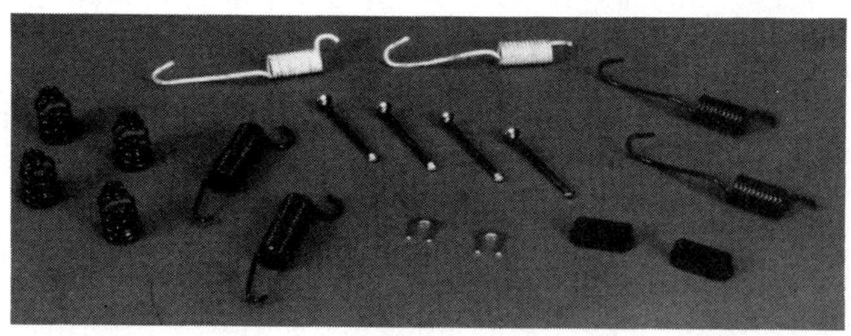

Figure 5-3 Hold-down hardware and return springs

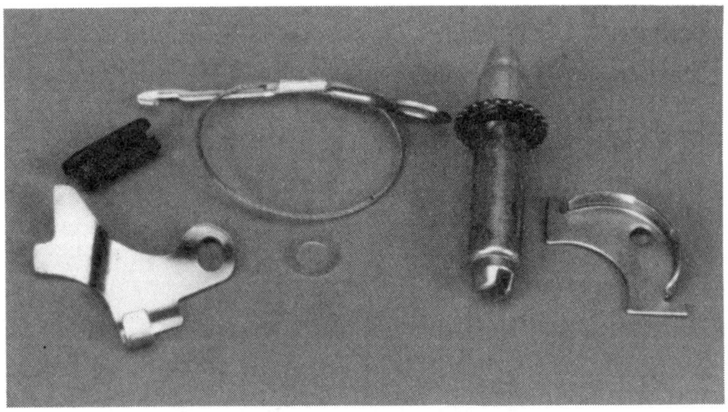

Figure 5-4 Self-adjustment drum brake parts

- *Adjusting Mechanisms*—Some drum brakes are self-adjusting (Figure 5-4); others require manual adjustments.
- *Drum*—Brake drums provide the rubbing surface area for the linings. Drums must withstand high pressures without excessive flexing and also dissipate large quantities of heat generated during brake application.
- *Parking Brake Mechanism*—The rear drum brakes on most vehicles include the parking brakes. (Parking brakes are described fully in Chapter 9.)

WHEEL CYLINDERS

Wheel cylinders convert hydraulic pressure supplied by the master cylinder into a mechanical force at the wheel brakes. The typical wheel cylinders are discussed fully in Chapter 2. The space in the wheel cylinder bore between the cups is filled with fluid. When the brake pedal is depressed, additional brake fluid is forced into the cylinder bore. The additional fluid, which is under pressure, moves the cups and pistons outward in the bore. This piston movement forces the shoe links and brake shoes outward to contact the drum and thus apply the brakes.

BRAKE SHOES AND LININGS

The workhorse of the drum brake is the brake shoe. In the same brake sizes, there can be differences in web thickness, shape of web cutouts, and positions of any reinforcements. Some vehicles require shoes made of higher tensile strength steels. Higher strength shoes usually are coded with a letter symbol stamped on the shoe web. While shoes might vary, there are many points that are common to all shoes.

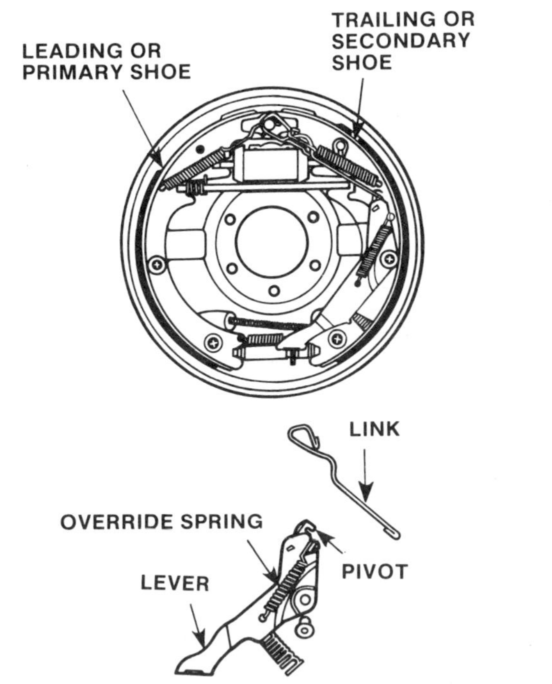

Figure 5-5 Position of leading (primary) and trailing (secondary) shoes

The shoe rim is welded to the web to provide a table surface for the lining. The web thickness might differ to provide the stiffness or flexibility needed for a specific application. Many shoes have nibs or indented places along the edge of the rim. These nibs rest against shoe support ledges on the backing plate and keep the shoe from hanging up.

Each drum in the drum braking system contains a set of shoes (Figure 5-5):

- *Primary Shoe*— The primary (or leading) shoe is the one that is toward the front of the vehicle. The friction between the primary shoe and the brake drum forces the primary shoe to shift slightly in the direction that the drum is turning (an anchor pin permits just limited movement). The piston movement forces the primary shoe against the secondary shoe.
- *Secondary Shoe*—The secondary (or trailing) shoe is the one that is toward the rear of the vehicle. It comes into contact as a result of the movement and pressure from the primary shoe and wheel cylinder piston and increases the braking action.

The brake shoe lining provides friction against the drum to stop the car. It contains heat-resistant fibers. The lining is molded with a high-tempera-

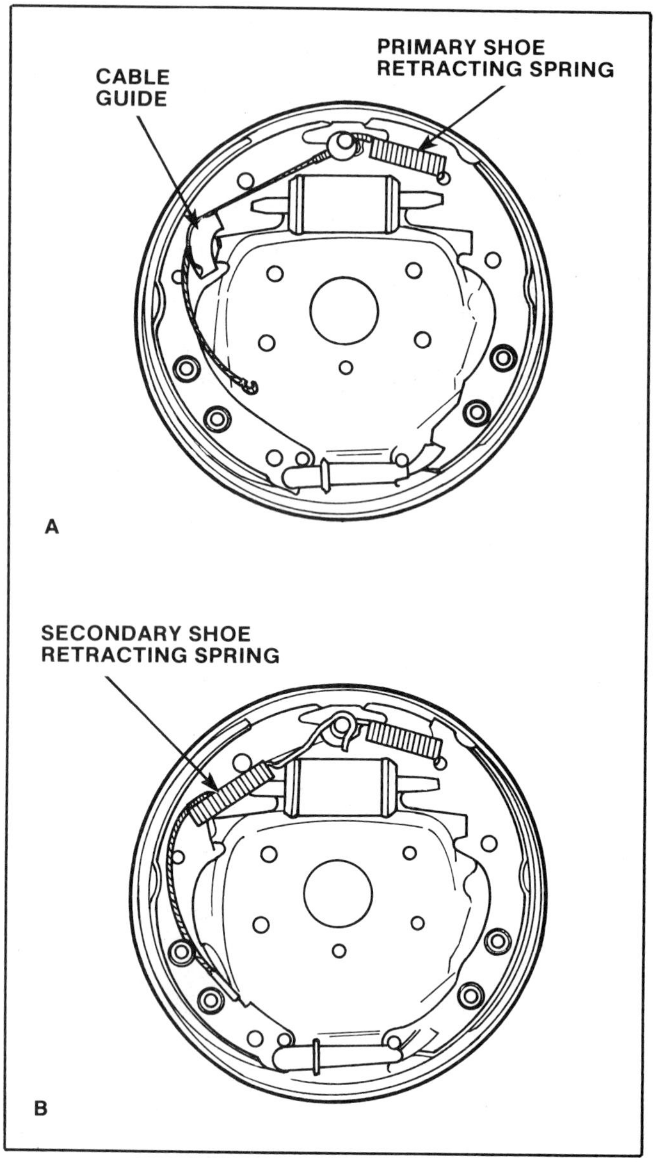

Figure 5-6 (A) Primary shoe retracting spring; (B) secondary shoe retracting spring

ture synthetic bonding agent. Other materials such as zinc, brass, copper, graphite, and ceramic are added to give the lining the desired gripping action (coefficient of friction) and heat-resistance properties. Other new synthetic fibers are also being used. Semi-metallic and synthetic fiber compound linings have two advantages:

1. They have excellent frictional characteristics.
2. They make it possible to avoid using asbestos as a frictional material.

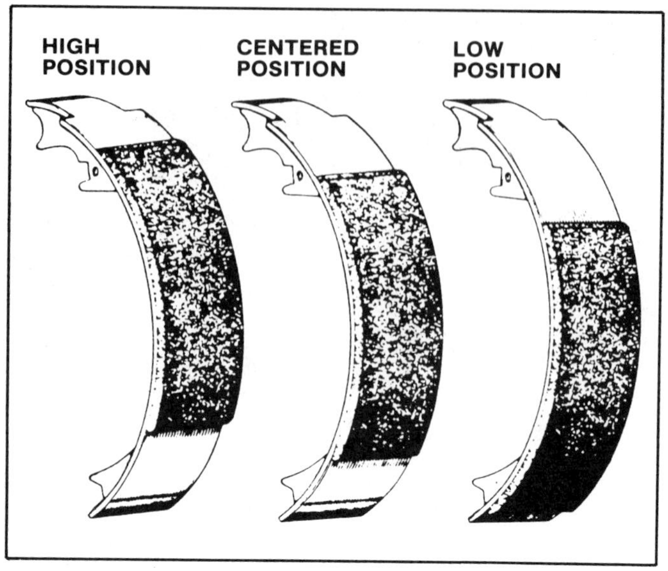

Figure 5-7 Lining position

Semi-metallic and synthetic fiber compounds are fast becoming OEM equipment on many new vehicles.

Most brake linings can withstand temperatures as high as 600 degrees F. But, a stop from 60 MPH will raise the brake temperature to only 450 degrees F. However, wear rate increases as the temperature increases. Passenger cars have about 1 square inch of lining for each 25 pounds of gross vehicle weight.

In a duo-servo brake, the secondary lining does about 75 percent of the braking. So, it is longer and has more surface area than the primary lining. This permits the linings to wear at the same rate.

The two general methods of attaching the lining to the shoe are riveting and bonding (Figure 5-10). Regardless of the method of attachment, brake shoes are usually held in a position by spring tension (Figure 5-6). They are ordinarily held in two ways:

1. Against the anchor by the shoe return springs
2. Against the support plate pads by shoe hold-down springs

The shoe webs are linked together at the end opposite the anchor by an adjuster and a spring. The adjuster holds them apart; the spring holds them against the adjuster ends.

The location of the lining on the shoe also is important. The secondary shoe normally has a full-length lining that is centered on the shoe. The

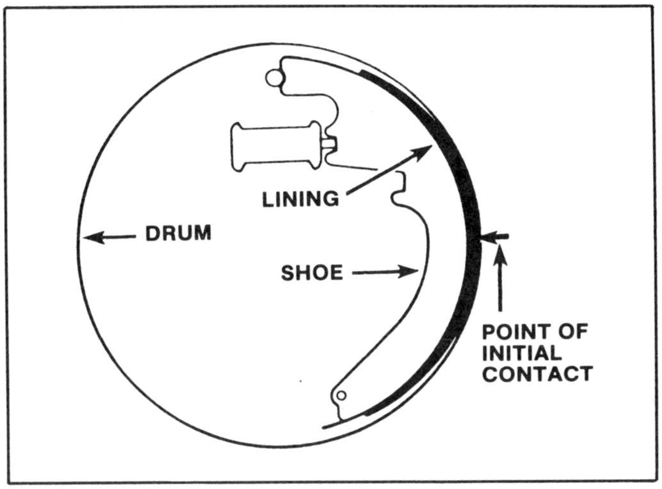

Figure 5-8 Lining-to-drum contact

primary shoe, however, has a shorter length of lining that can be in a high, centered, or low position of the shoe (Figure 5-7). The position of the lining affects the servo action of the brake.

Most brake linings used today are ground so that they are slightly thicker at the center. Thus, when the brakes are applied and the shoes and linings move outward to the drum, the center of the lining contacts the drum first (Figure 5-8). There is a slight clearance between the drum and lining at each end of the lining. As the brake pressure is increased, both the shoe and the drum flex slightly to bring the entire lining surface into contact with the drum (Figure 5-9).

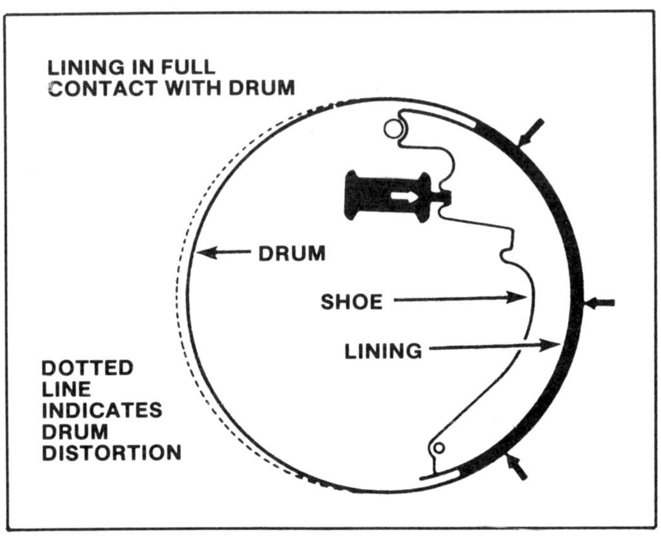

Figure 5-9 Flexing of shoe and drum

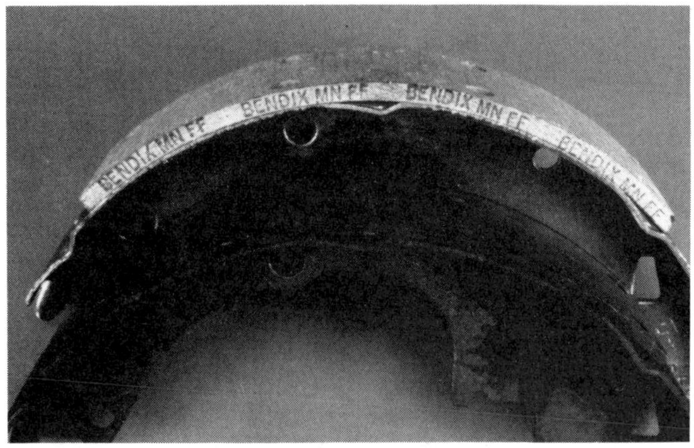

Figure 5-10 New and worn brake shoes—the new shoe is edge branded

SHOP TALK

As a brake technician, you should caution the customer to use care during the break-in period. High-speed brake stops and severe brake applications should be avoided during the first 500 miles. This will help to provide smoother brakes that last longer.

Most friction linings are edge marked or branded. The first marking on the brake lining identifies the manufacturer and is usually a letter code (Figure 5-10). The next identifies the material, and the last is the friction code for the listed materials. That is, the letters "FE," "FF," and so on refer to the friction characteristics of that particular material; they do not indicate quality. Each application has its own requirements and optimum coefficient of friction; hence, the different lining materials and their markings (Figure 5-11). It is important to remember that if the coefficient of friction is too low, excessive pressure on the brake pedal is required to stop the vehicle. If it is too high, the brakes can grab and cause the wheels to slide or skid.

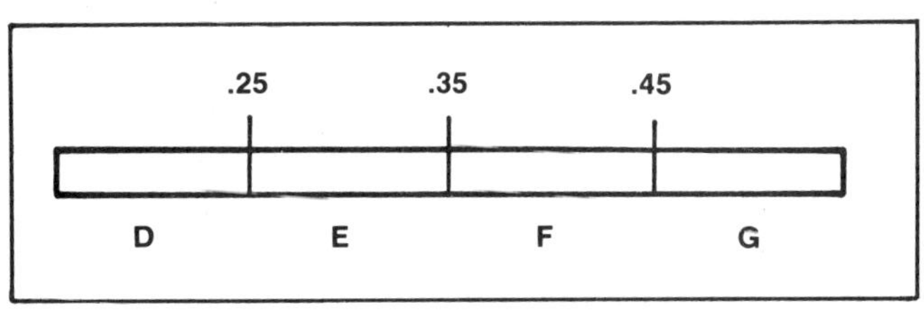

Figure 5-11 Coefficient of friction letter codes

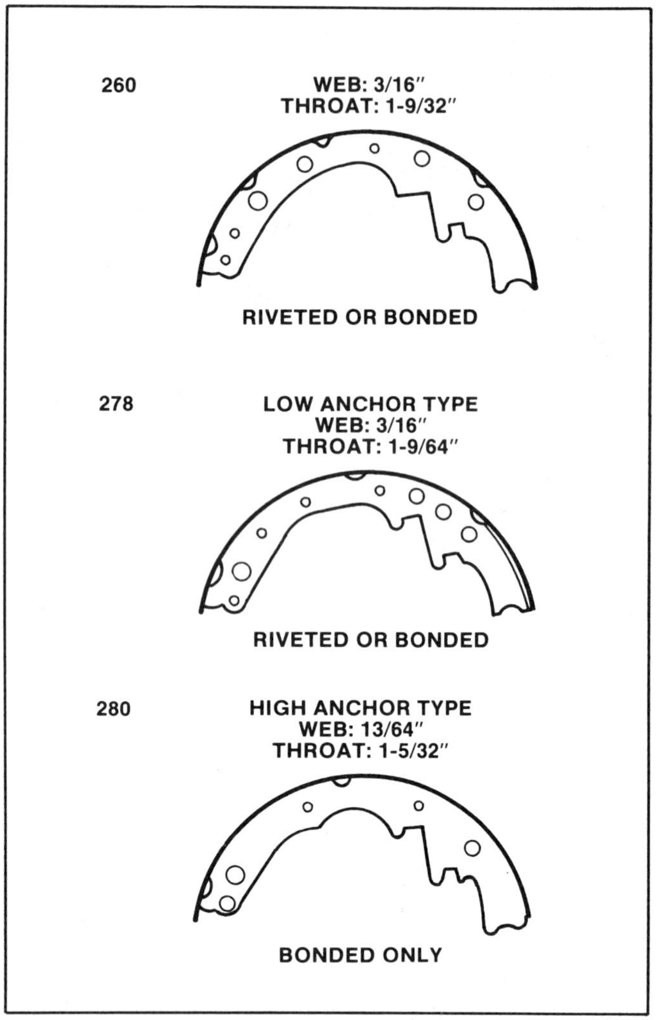

Figure 5-12 Sample brake shoe listings from an FMSI catalog

The correct shoe applications can usually be determined by checking the Friction Material Standards Institute (FMSI) or handbook (Figure 5-12). A copy of this publication can be obtained by writing to the institute at East 210, Route 4, Paramus, New Jersey 07652.

MECHANICAL COMPONENTS

As mentioned earlier, the brake shoes are usually held in position by spring tension. In the unapplied position, the shoes are held against the anchor pin by the return springs. The shoes are held to the backing plate by hold-down springs or spring clips. Opposite the anchor pin, a star wheel adjuster links the shoe webs and provides a threaded adjustment that permits the shoes to be expanded or contracted. The shoes are held against the star wheel by a spring.

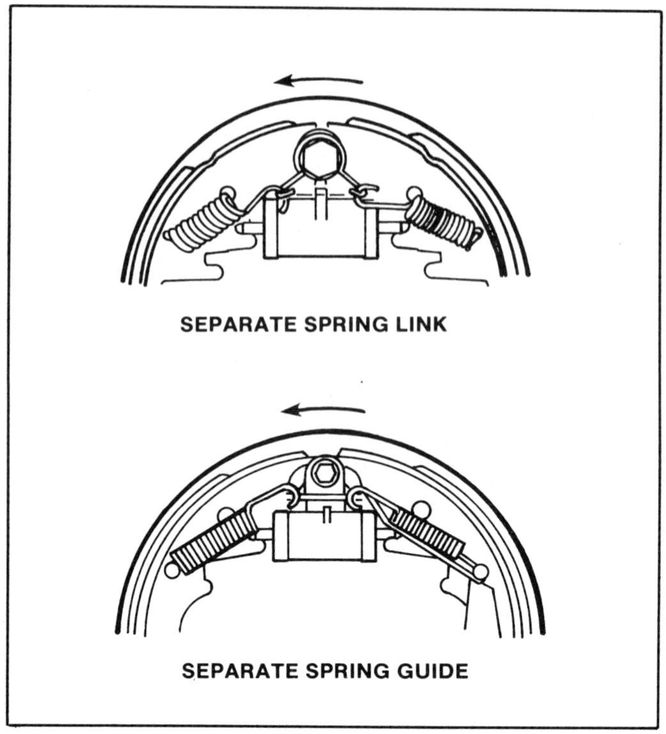

Figure 5-13 Shoe return spring alignments

Shoe Return Springs

Return springs can be separately hooked into a link or a guide (Figure 5-13), or strung between the shoes. Springs are normally installed on the anchor in the order shown under each category listed here (Figure 5-14):

- *Manual Adjusters*—(1) anchor plate, (2) primary spring, (3) secondary spring.
- *GM Lever Adjusters*— (1) anchor plate, (2) adjuster link (the secondary spring hooks into the link loop), (3) primary spring.
- *Cable Adjusters*—(1) anchor plate, (2) cable eye, (3) primary spring, (4) secondary spring.
- *Auxiliary Spring*—When a helper (third) spring is attached between the shoes, it is hooked into the coiled ends of the primary and secondary springs.
- *Crank Adjusters*—Adjuster on primary shoe: (1) anchor plate, (2) primary spring, (3) secondary spring, (4) adjuster link. Adjuster on secondary shoe: (1) anchor plate, (2) secondary spring, (3) primary spring, (4) adjuster link.

While shoe brake springs look the same, they are usually not interchangeable. Sometimes to help distinguish between them, they are color coded. Pay close attention to the colors and the way they are hooked up.

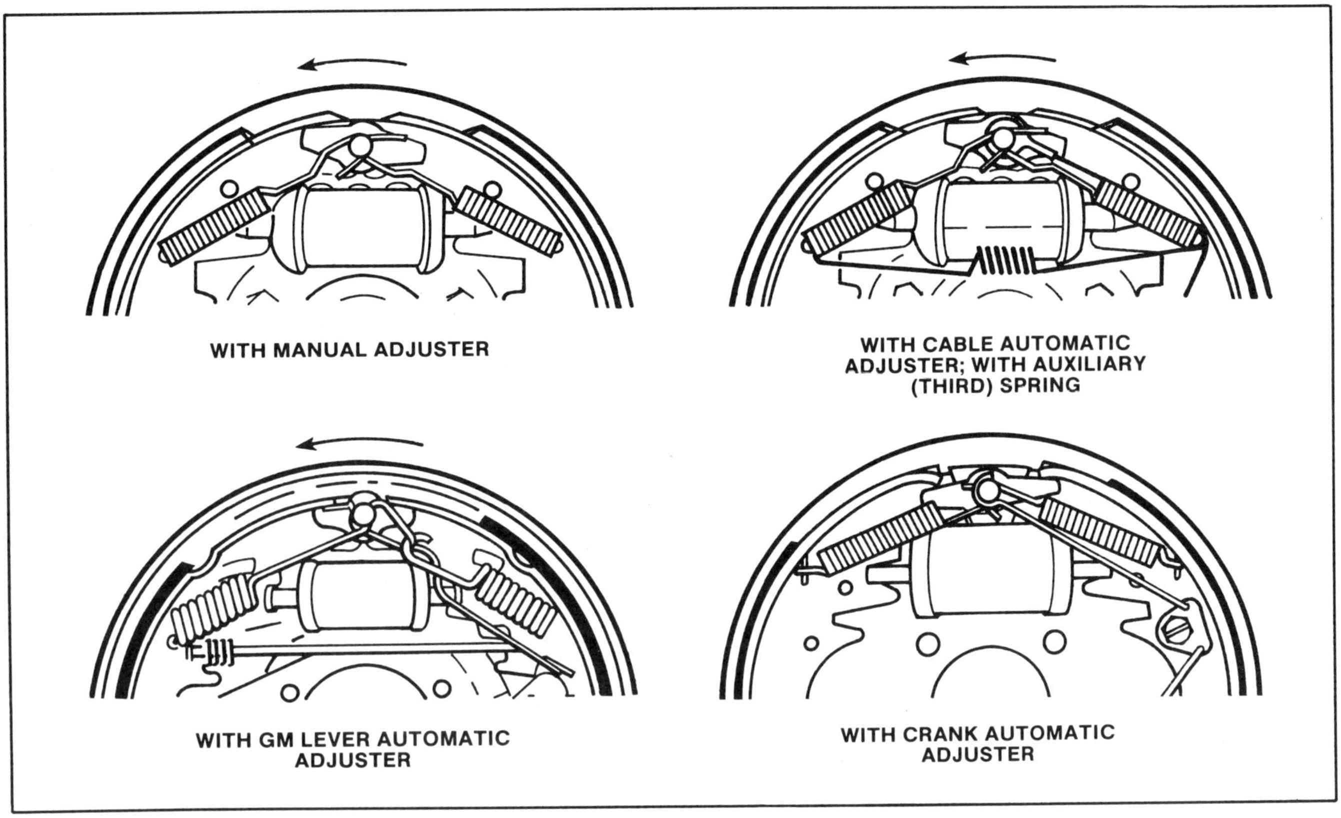

Figure 5-14 **Shoe return spring alignments**

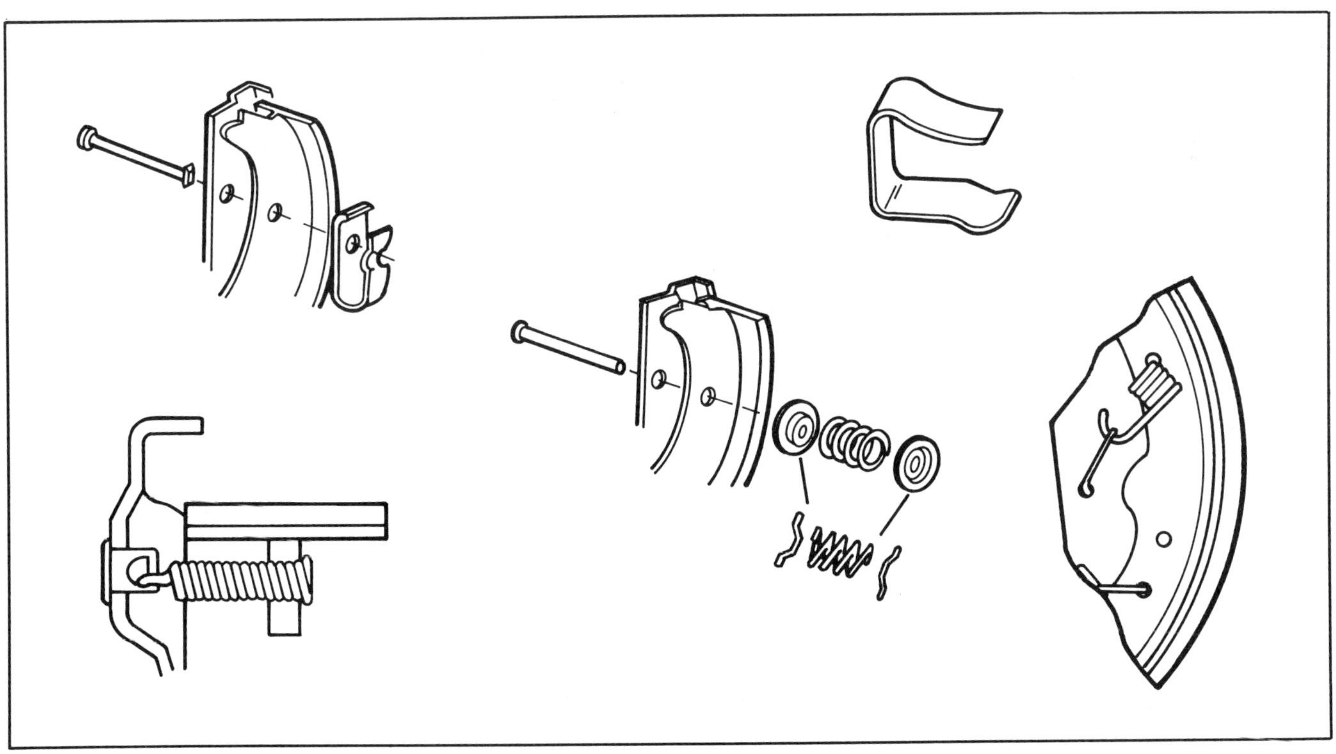

Figure 5-15 **Types of shoe hold-downs**

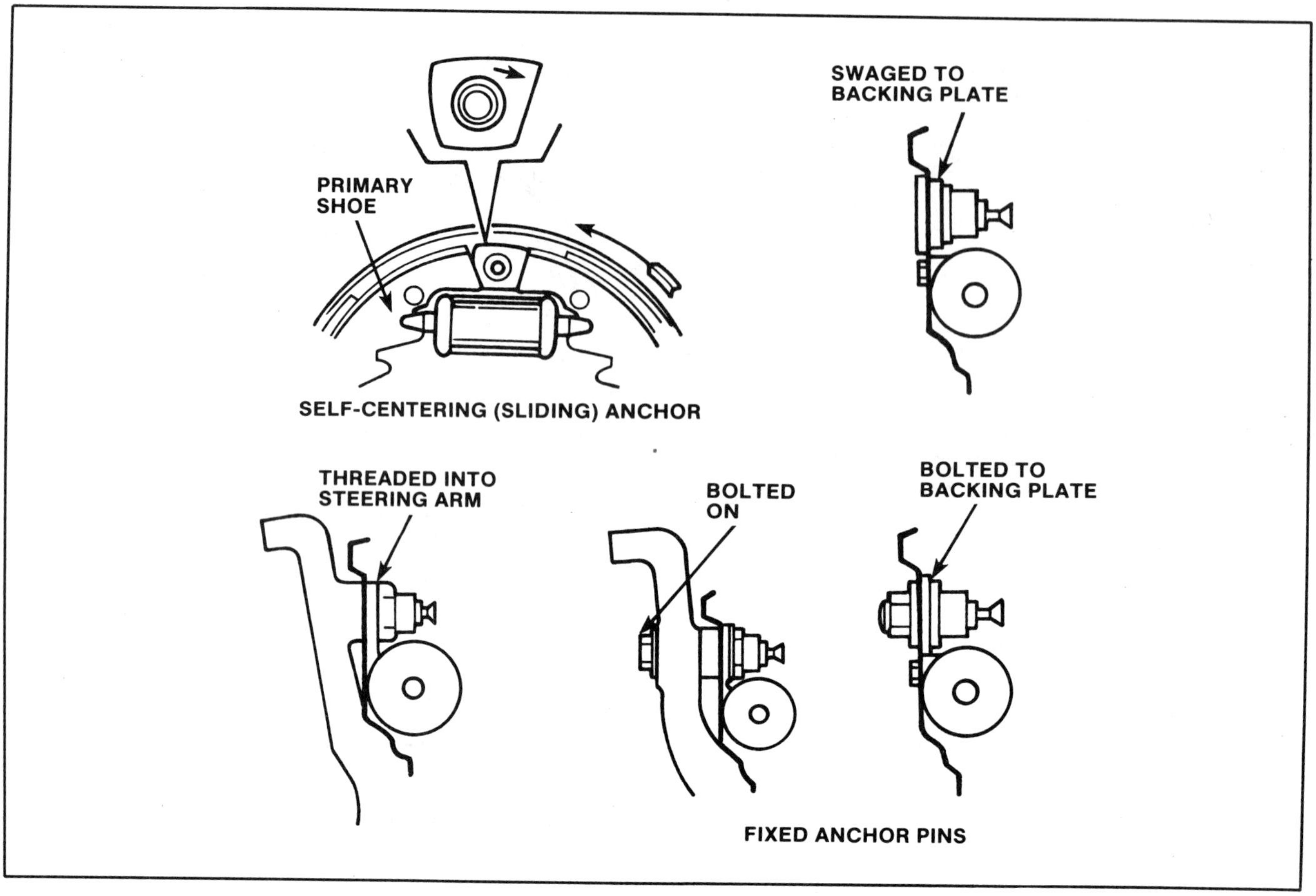

Figure 5-16 Types of shoe anchors

Shoe Hold-downs

Various shoe hold-downs are illustrated in Figure 5-15. To unlock or lock the straight pin hold-downs, depress the locking cup and coil spring or the spring clip, and rotate the pin or lock 90 degrees. On GM lever-type adjusters, the inner (bottom) cup has a sleeve that aligns the adjuster lever.

Shoe Anchors

As shown in Figure 5-16 there are various types of anchors such as the fixed non-adjustable type, self-centering shoe sliding type, or, on some earlier models, adjustable fixed type providing either an eccentric or a slotted adjustment. On some front brakes, fixed anchors are threaded into or are bolted through the steering knuckle and also support the wheel cylinder.

On adjustable anchors, when it is necessary to recenter the shoes in the drum or drum gauge, loosen the locknut enough to permit the anchor to slip but not so much that it can tilt.

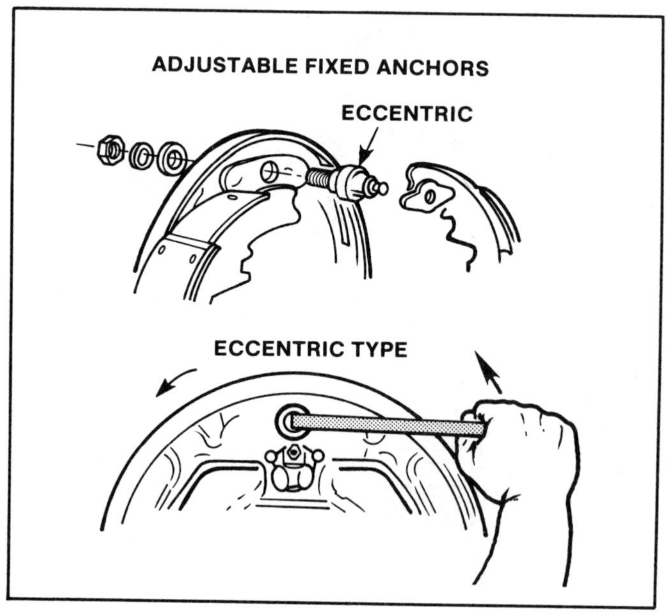

Figure 5-17 Anchor adjustment—fixed and eccentric

On eccentric type anchors, tighten the star wheel to heavy brake drag (Figure 5-17). Rotate the eccentric anchor in the direction that frees the brake until drag is again encountered. Return the eccentric to the position between the two points of drag where the drum rotates freely. Repeat this sequence until drag cannot be relieved. Tighten the anchor nut. Back off the star wheel to a normal manual adjustment.

On the slotted type anchor (Figure 5-18), tighten the star wheel to heavy drag. Tap the support plate until the anchor moves and frees the brake. Repeat this sequence until drag cannot be relieved. Tighten the lock nut to 100 to 200 foot-pounds torque. Back off the star wheel to a normal manual adjustment.

DRUMS

Modern automotive brake drums are made of heavy cast iron (some are aluminum with an iron or steel sleeve or liner) with a machined surface inside against which the linings on the brake shoes generate friction when the brakes are applied. This results in the creation of a great deal of heat, and the inability of drums to dissipate as much heat as disc brakes is one of the main reasons discs have replaced drums at the front of all late-model cars and light trucks, and at the rear of some sports and luxury cars.

Front drums (Figure 5-19) are usually integral with the hub, so they cannot be removed without disassembling the wheel bearing. Rear drums on FWD cars are similar to front drums, but the rear drums of most rear wheel drive cars are held in place by the wheel lugs, so they can be removed without tampering with the wheel bearings.

DRUM BRAKE DESIGNS

As described in Chapter 1, there are two brake designs in common use: duo-servo (or self-energizing) drum brakes and non-servo (or leading-trailing) drum brakes.

Most large American cars use the duo-servo design of brake. However, the non-servo type has become popular as the size of cars has gotten smaller. Because the smaller cars are lighter, this type of brake helps reduce rear brake lockup without reducing breaking ability.

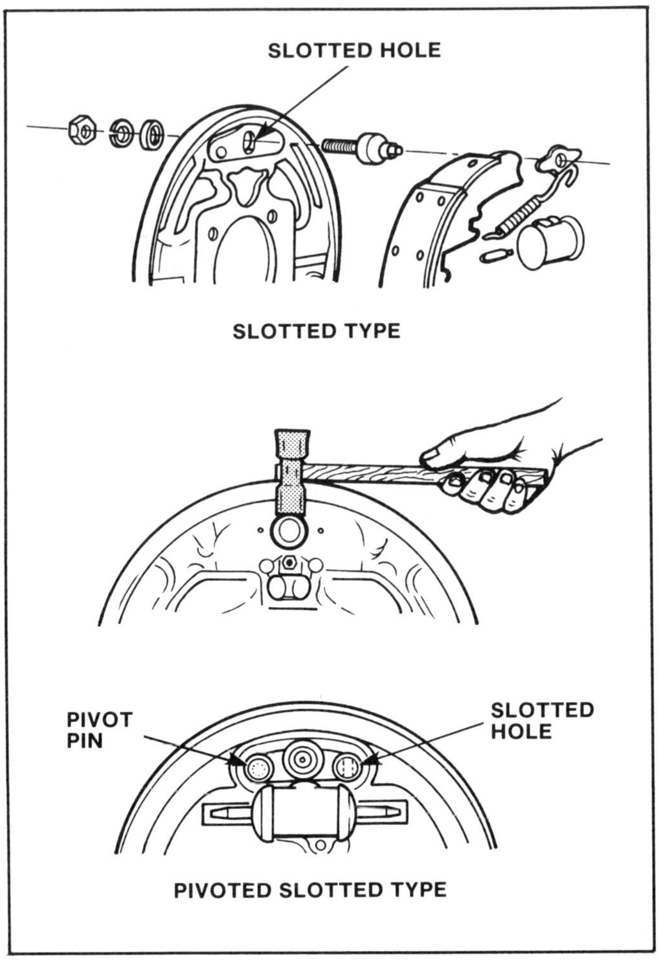

Figure 5-18 Anchor adjustment—slotted type

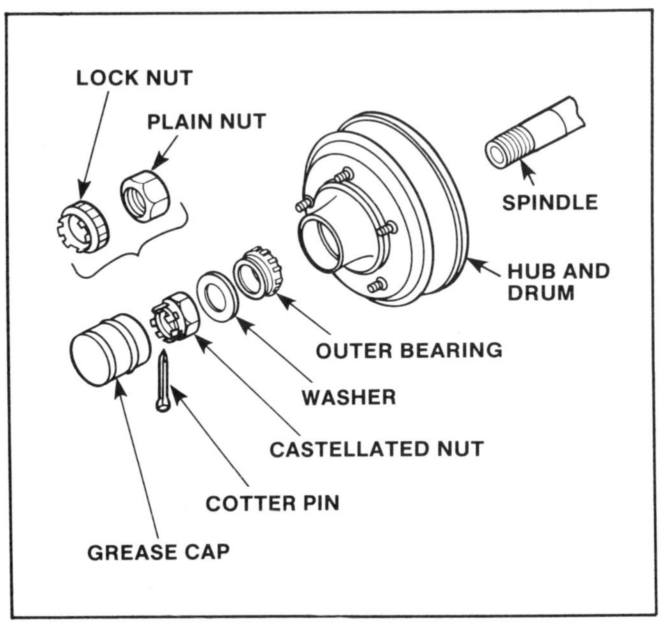

Figure 5-19 Front drum, hub, and bearing

DUO-SERVO DRUM BRAKES

The name of this brake design is derived from the fact that the self-energizing force is transferred from one shoe to the other with the wheel rotating in either direction. This self-energizing force that the wheel cylinder applies to the shoes is supplemented by the tendency of the shoes to wrap into the drum during breaking. Both the primary (front) and secondary (rear) brake shoes are actuated by a double-piston wheel cylinder (Figure 5-20). The upper end of each shoe is held against a single anchor by a heavy coil return spring. An adjusting screw assembly and spring connect the lower ends of the shoes.

The wheel cylinder is mounted on the backing plate at the top of the brake. When the brakes are applied, hydraulic pressure behind the wheel cylinder cups forces both pistons outward causing the brakes to be applied. Some wheel cylinders have extensions on the pistons that contact the brake shoes, while others have separate links (Figure 5-21).

When the brake shoes contact the rotating drum in either direction of rotation, they tend to move with the drum until one shoe contacts the anchor and the other shoe is stopped by the star wheel adjuster link (Figure 5-22). With forward rotation, frictional forces between the lining and the drum of the primary shoe result in a force acting on the adjuster link to apply the secondary shoe. This adjuster link force into the secondary shoe is many times greater than the wheel cylinder input force acting on the primary shoe. The adjuster link force into the secondary shoe is again multiplied by the frictional forces between the secondary lining and rotating drum, and all of the resultant force is taken on the anchor pin. In normal forward braking, the friction developed by the secondary lining is greater than the primary lining; therefore, the secondary brake lining is usually thicker and has more surface area. The roles of the primary and secondary linings are reversed in braking the vehicle when backing up.

Manually Adjusted Servo Drum Brakes

Automatic drum brake adjusters were first introduced in 1957 and have been used since 1963 on all American-made and most import cars.

Most manually adjusted servo brakes have single or dual star wheel adjusters (Figure 5-23). The

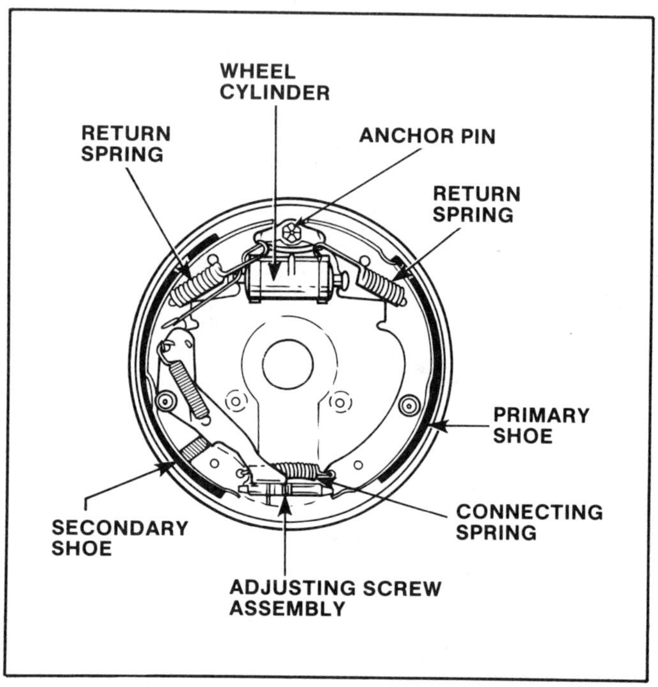

Figure 5-20 Duo-servo brake

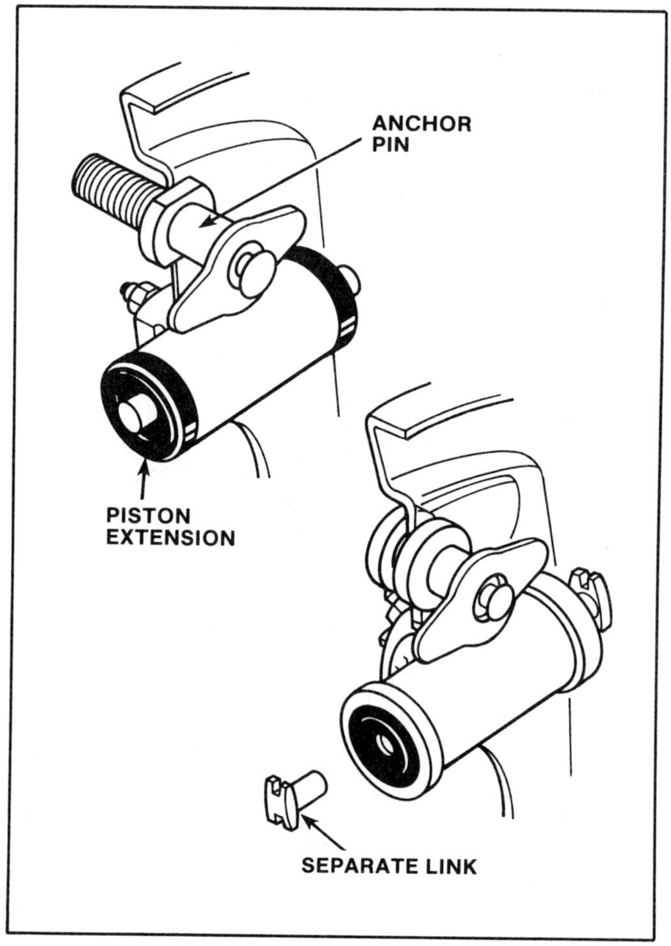

Figure 5-21 Wheel cylinders

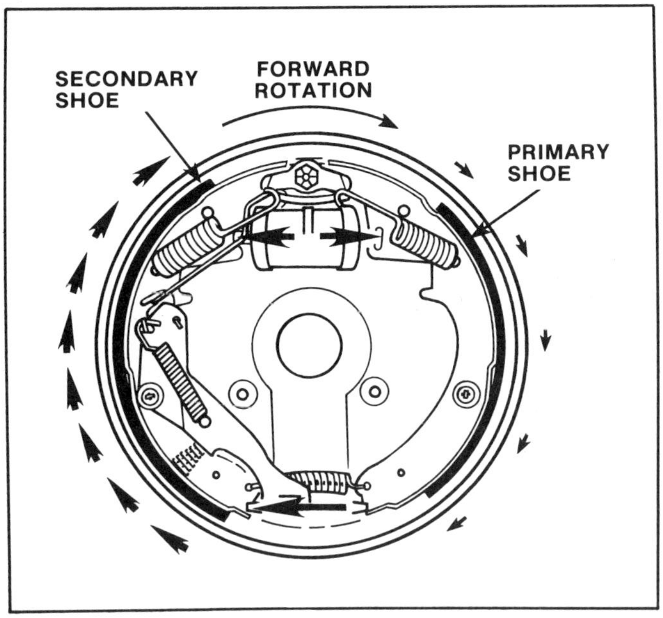

Figure 5-22 Duo-servo brake forces

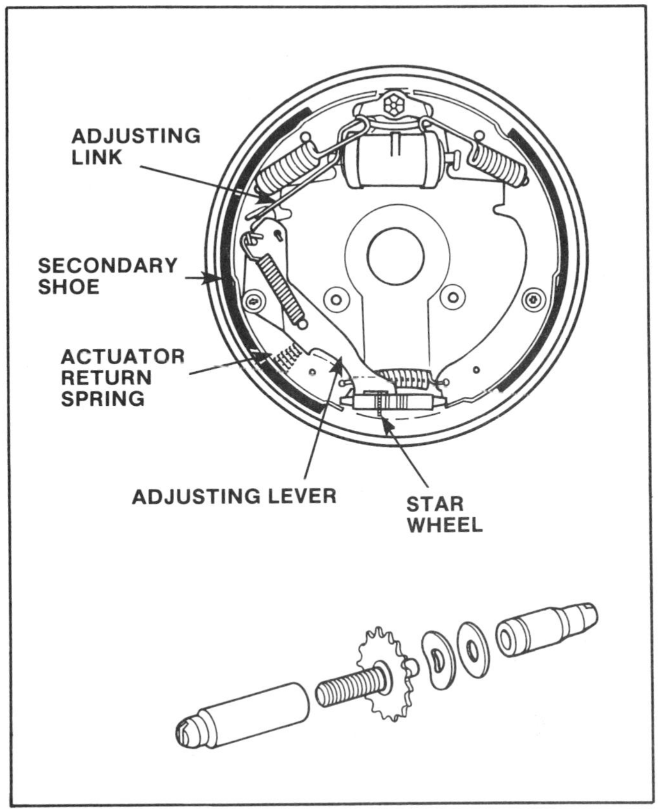

Figure 5-23 Star wheel adjuster link

adjuster spring bears on the star wheel teeth, providing a ratchet lock. A slot in the backing plate provides access for adjusting the star wheel. Star wheel adjusters usually have right-hand threads so that, on opposite brakes, they are rotated in opposite directions to tighten or loosen. On some vehicles, however, the positions of opposite star wheels are reversed end for end on left and right brakes. This is usually indicated by a difference in the location of access slots on opposite sides of the vehicle. Then, in relation to their access slots, the teeth of both star wheels are rotated in an identical manner to make adjustments. In other words, moving the star wheel teeth down, for example, will tighten the adjustment on either side.

Automatically Adjusted Servo Brakes

There are several variations of automatic adjusters used with servo brakes. The more common types available are as follows.

Basic Cable Type. Figure 5-24 shows a typical automatic adjusting system. Adjusters, whether cable, crank, or lever type, are installed on one shoe and operated whenever the shoe moves away from its anchor. The upper link, or cable eye, is attached to the anchor. As the shoe moves, the cable pulls over a guide mounted on the shoe web (the crank or lever types pivot on the shoe web) and operates a lever (pawl), which is attached to the shoe so that it engages a star wheel tooth. The pawl is located on the outer side of the star wheel and, on different styles, slightly above or below the wheel centerline so that it serves as a ratchet lock, which prevents the adjustment from backing off. However, whenever lining wears enough to permit sufficient shoe movement, brake application pulls the pawl high enough to engage the next tooth. As the brake is released, the adjuster spring returns the pawl, thus advancing the star wheel one notch.

On most vehicles the adjuster system is installed on the secondary shoe, and thus operates when the brakes are applied as the vehicle is backing up. On a few models, it is located on the primary shoe and operates when the brakes are applied as the vehicle is moving forward. Left-hand and right-hand threaded star wheels are used on opposite sides of the car, so parts should be kept separated. If the wrong star wheel thread is installed, the system will not adjust at all.

Another system (Figure 5-25) uses a cable with downstroke pawl advance, and the left brake with right-hand threads and the right brake with left-hand threads. The first cable guide is usually retained on the shoe web by the secondary shoe return spring, and the lever-pawl engages a hole in the shoe web. The adjuster operates in either direc-

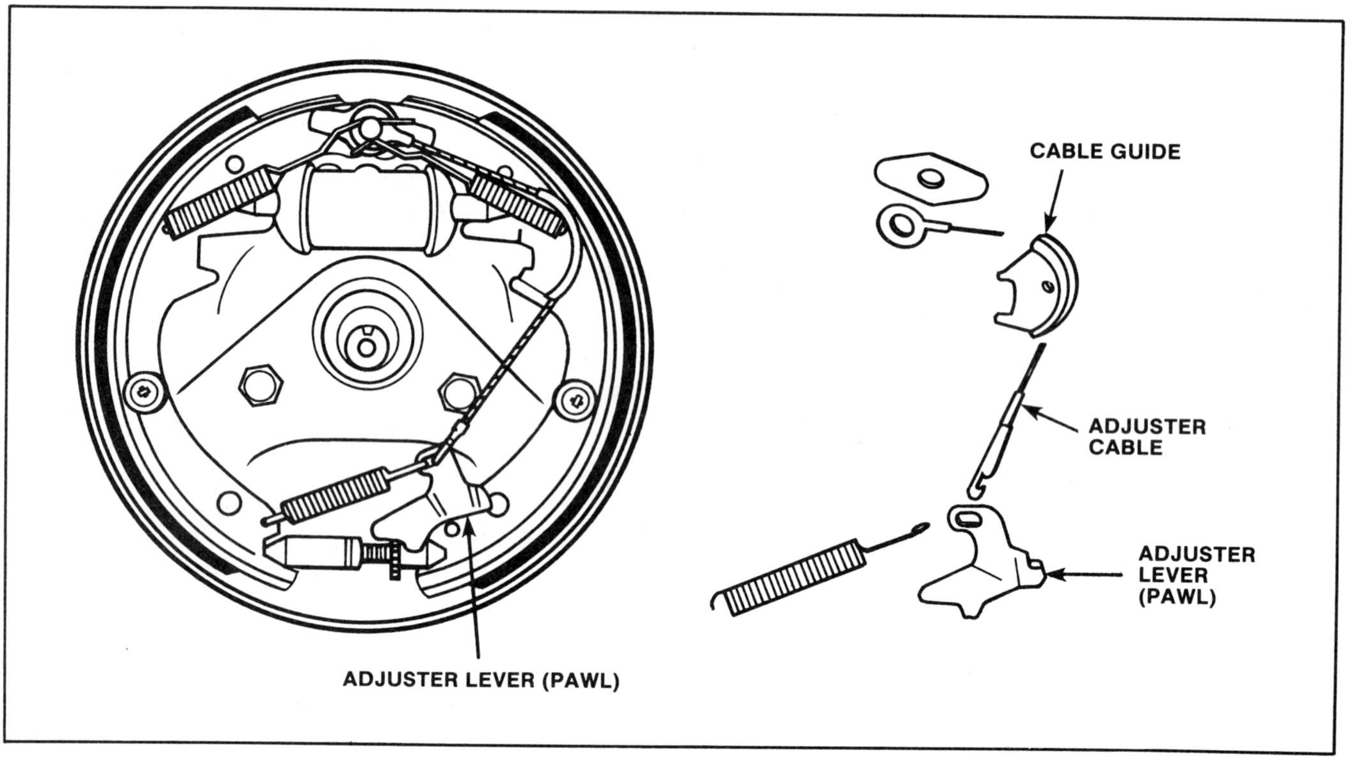

Figure 5-24 Cable type automatic adjustment: Ford, Chrysler, American (some), Rambler 8's—left front shown

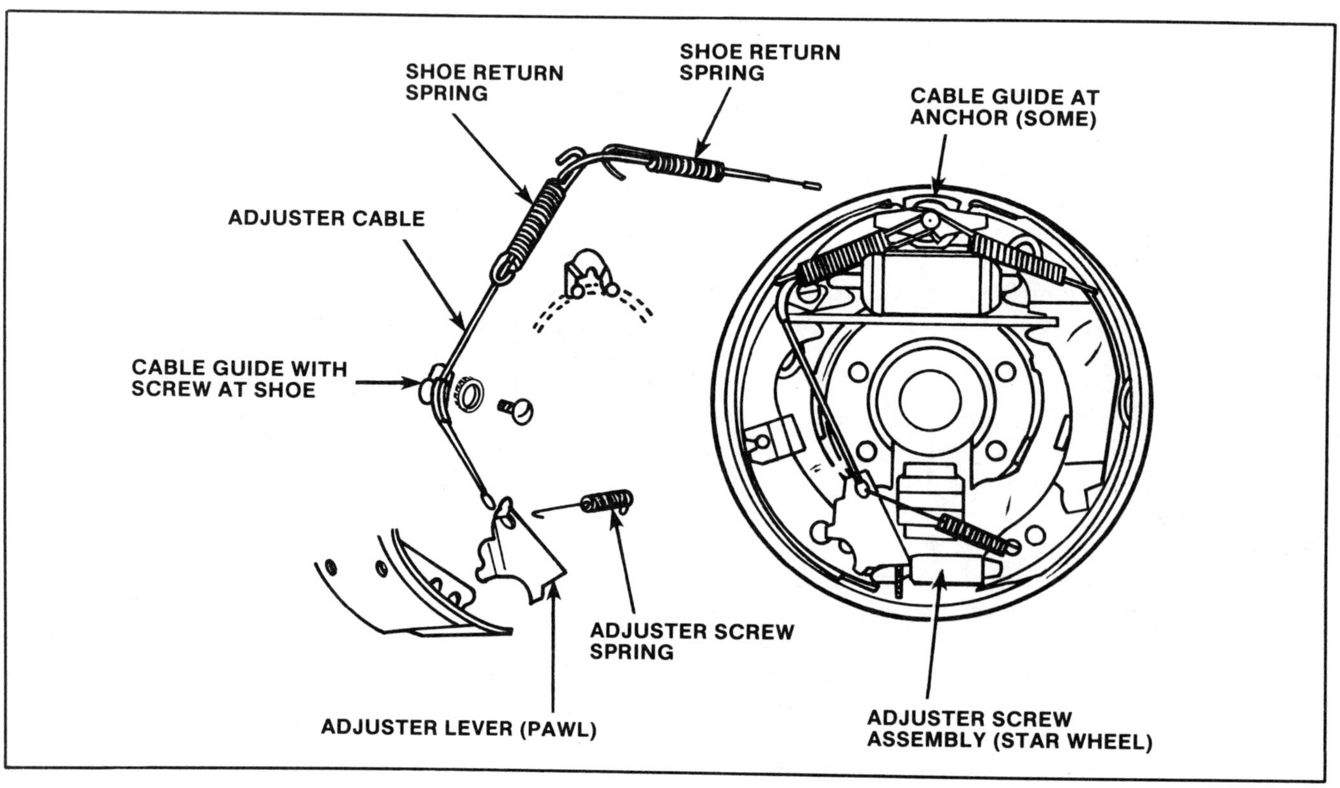

Figure 5-25 Cable type automatic adjustment—right rear shown

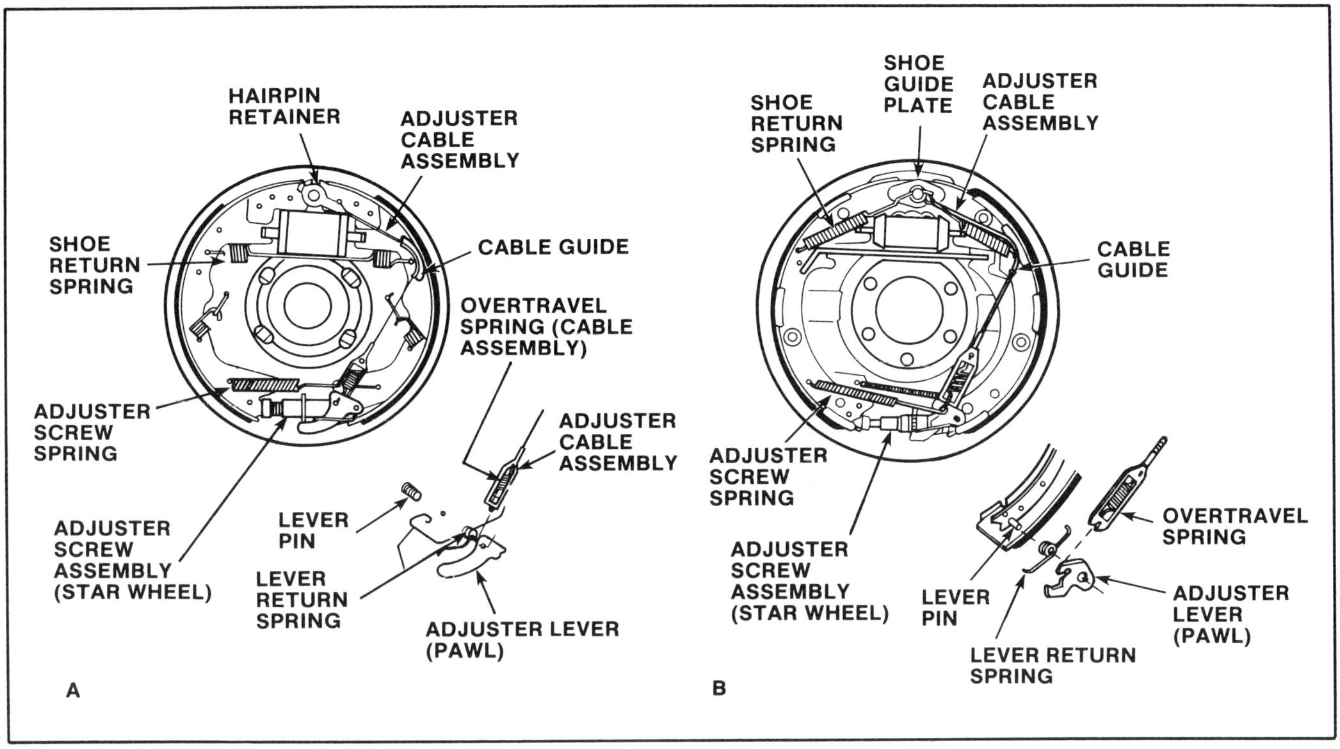

Figure 5-26 Cable type automatic adjustment with overtravel spring

tion of vehicle movement. The cable guide is slotted and attached to the web with a screw. The guide position should be adjusted to align the pawl after the self-centering shoes are located on sliding anchors.

Cable Type with Overtravel Spring. Figure 5-26 shows a system with an upstroke pawl advance; the left brake with left-hand threads, and the right brake with right-hand threads. The lever (pawl) is installed on a web pin with an additional pawl return "mousetrap" spring. The cable is hooked to the lever (pawl) by means of an overtravel spring installed in the cable hook. The overtravel spring dampens movements and prevents unnecessary adjustment should sudden hard braking cause excessive drum deflection and shoe movement.

Lever Type with Override. The system illustrated in Figure 5-27 uses a downstroke pawl advance; the left brake with right-hand threads, and the right brake with left-hand threads.

The lever (pawl) is mounted on a shoe hold-down, pivoting on a cup sleeve, and has a separate lever-pawl return spring located between the lever and the shoe table. A pivot lever and an override spring assembled to the upper end of the main lever dampen movement, preventing unnecessary adjustment in the event of excessive drum deflection.

Lever and Pawl Type. The system illustrated in Figure 5-28 uses a downstroke pawl advance;

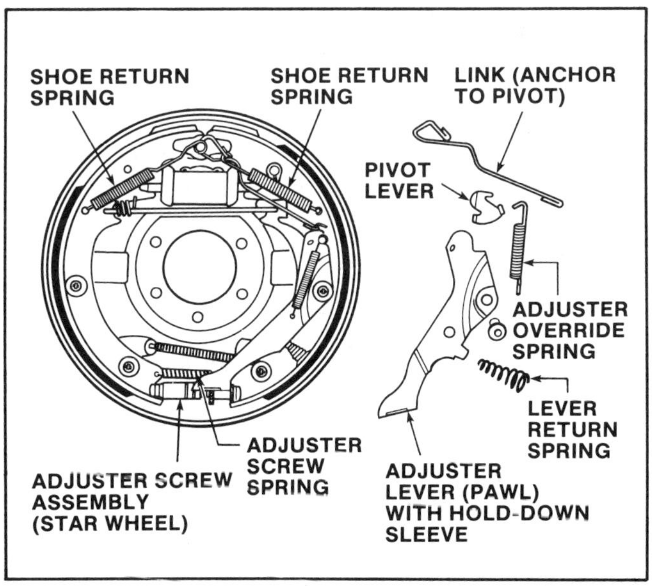

Figure 5-27 Lever with pivot and override spring; General Motors makes left rear shown.

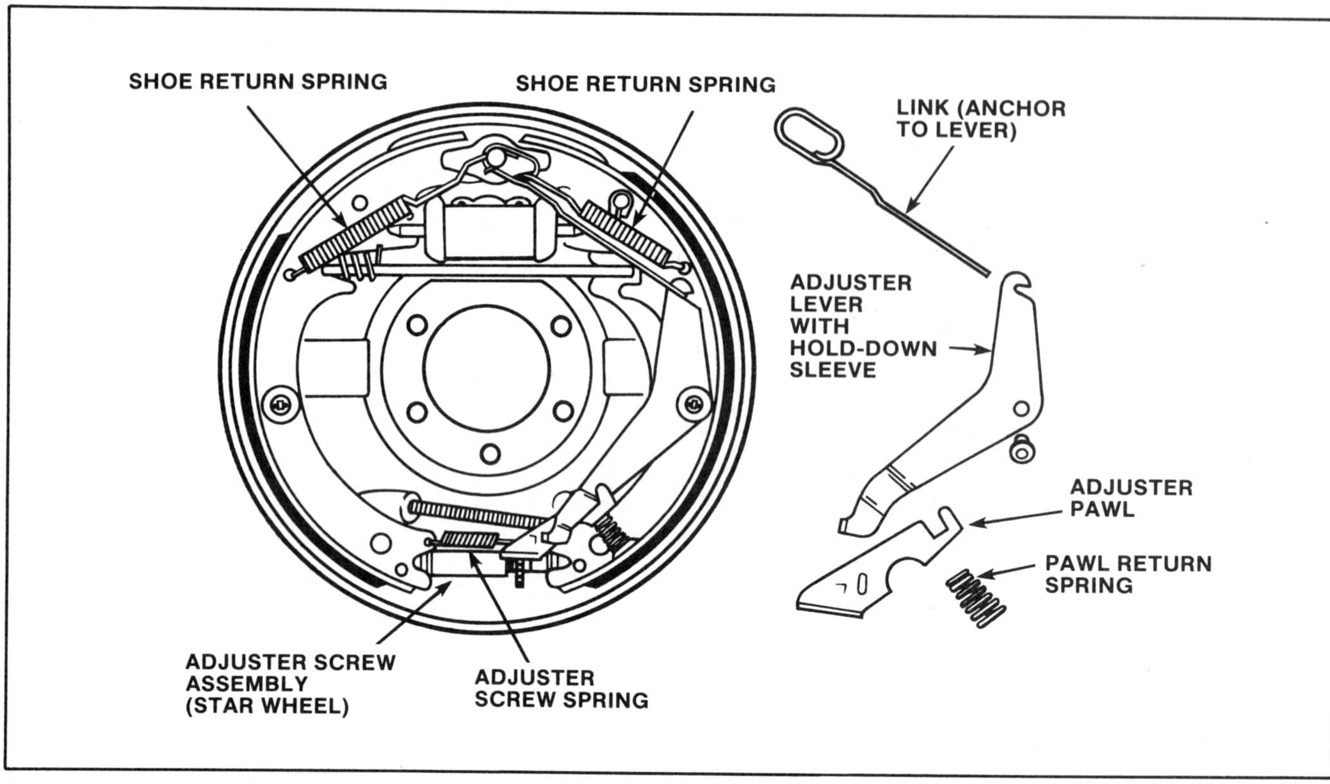

Figure 5-28 **Lever and pawl; General Motors makes left rear shown.**

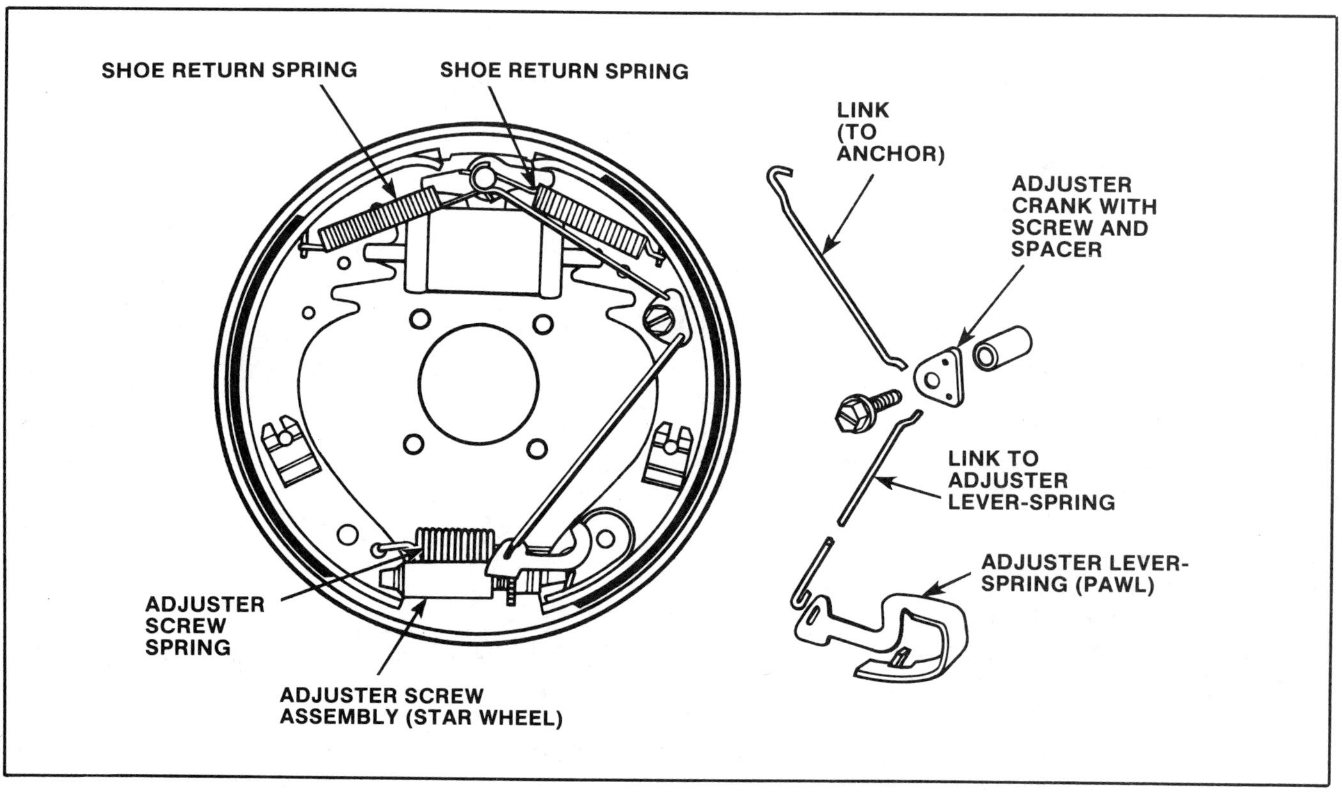

Figure 5-29 **Crank and link type with spring lever pawl**

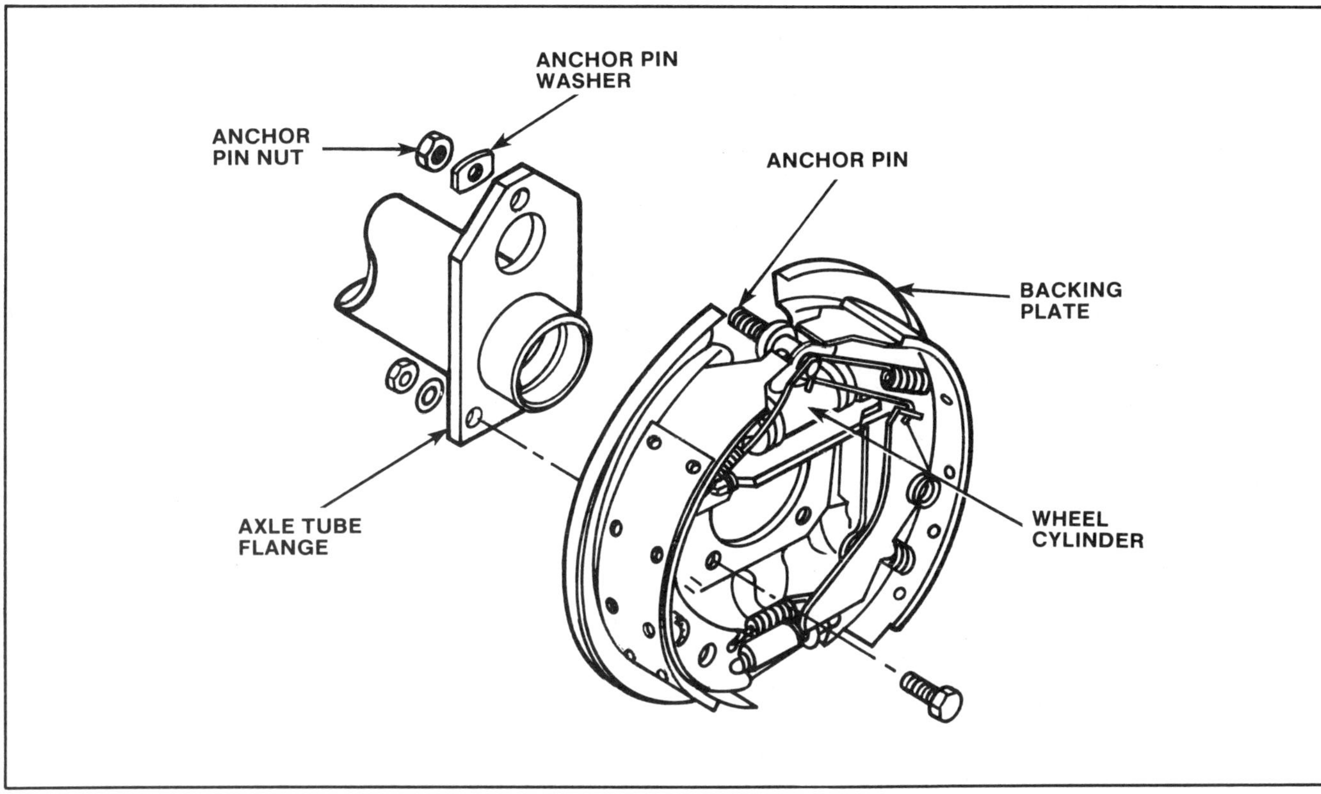

Figure 5-30 Direct torque design

the left brake with right-hand threads and the right break with left-hand threads. The lever is mounted on a shoe hold-down, pivoting on a cup sleeve, and engages the pawl. A separate pawl return spring is located between the pawl and the shoe.

Crank and Link Type with Spring Lever Pawl. The adjustment arrangement illustrated in Figure 5-29 features a pawl and spring that are combined in one component that is clipped onto the shoe.

Direct Torque Design Brake. The duo-servo brakes described in the preceding paragraphs all use a heavy backing plate since the plate must transmit all braking torque absorbed by the anchor pin to the rear axle flange. The new direct torque design drum brake has a much lighter backing plate. This is possible because the anchor pin is mounted directly to the axle tube flange as shown on Figure 5-30.

NON-SERVO DRUM BRAKES

The non-servo (or as it is better known today as the leading-trailing shoe type) drum brake is often used on small cars. The basic difference between this type and the duo-servo brake is that both brake shoes are held against a fixed anchor at the bottom by a retaining spring (Figure 5-31). Non-servo brakes have no servo action.

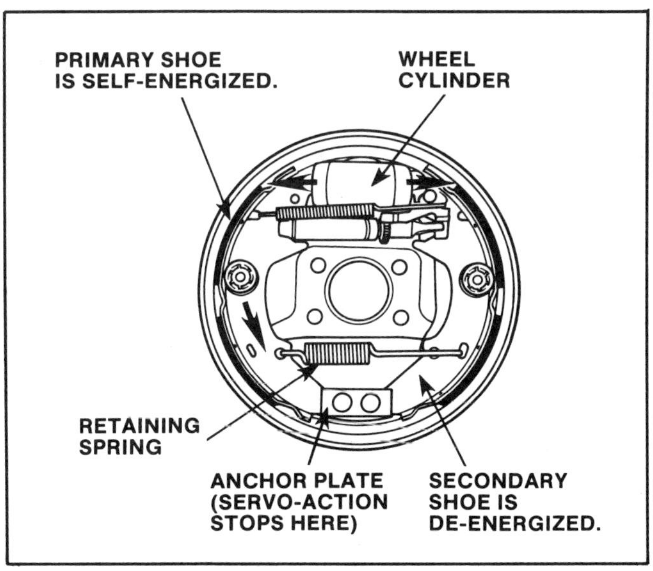

Figure 5-31 Non-servo brake

On a forward brake application, the forward (leading) shoe friction forces are developed by wheel cylinder fluid pressure forcing the lining into contact with the rotating brake drum. The shoe friction forces work against the anchor pin at the bottom of the shoe. The trailing shoe is also actuated by wheel cylinder pressure, but can only support a friction force equal to the wheel cylinder piston forces. The trailing shoe anchor pin supports no friction load. The leading shoe in this brake is energized and does most of the braking in comparison to the non-energized trailing shoe. In reverse braking, the leading and trailing brake shoes switch functions.

SHOP TALK

It is important for the technician to remember that on non-servo drum systems that forward shoe is called the "leading" shoe and the rear one is known as the "trailing" shoe (when the vehicle is moving in the forward direction). On duo-servo designs, the forward shoe is the "primary" and the rear, the "secondary."

Non-servo, or leading-trailing shoe, drum brakes can be either manually adjusted or can have an automatic adjustment feature.

Manually Adjusted Non-Servo Drum Brakes

Some of the manual adjusters vary from common domestic designs used on larger cars. Many have star wheel adjusters, but they can be located on the anchor block, in a strut between the shoes, or on one end of single or double wheel cylinders. Some have conventional adjustment cams that work against the shoe tables. Others might have an adjuster bolt that spreads the anchor to adjust the brakes.

There are several types of non-servo drum brake adjuster designs available, but the two most common are:

Lockheed Manual Adjuster. This is a very simple brake system (Figure 5-32). The shoes are located on the top and bottom rather than the sides. The shoes are held tight against the wheel cylinder and the adjusters by two retracting springs. Each shoe has to be adjusted individually by tightening it against the drum and then backing it off until the drum turns freely.

Depending on what car this type of brake is used on, there will be differences in the brake shoes, adjusters, anchors, wheel cylinders, or parking brake linkage. It is best to work on one wheel at a time and to use the other for reference.

Adjustable Strut Adjuster. On this design (Figure 5-33) the bottoms of the leading and trailing shoes are held against the anchor plate by shoe-to-anchor retaining springs. The top webs engage the wheel cylinder pistons and are straddled by a specially designed adjuster screw assembly. Return springs attached to the adjuster hole in the backing plate hold the shoes in against the adjuster screw assembly. One end engages the trailing shoe web and parking brake lever and the other end engages the leading shoe web.

The upper end of the parking brake lever is pinned to the upper end of the trailing shoe and the lower end is attached to the parking brake cable. When the cable is pulled, force is transmitted through the special adjuster assembly to the leading shoe. Shoe adjustment is done manually by inserting a thin bladed screwdriver through the hole in the backing plate and rotating the adjuster screw to spread the shoes.

Automatically Adjusted Non-Servo Brakes

While some standard automatic adjusters similar to the one already discussed are employed on small cars, some of the automatic adjuster mechanisms are unique and varied, using expanding struts between the shoes, or special ratchet adjusting mechanisms. Among the more common of

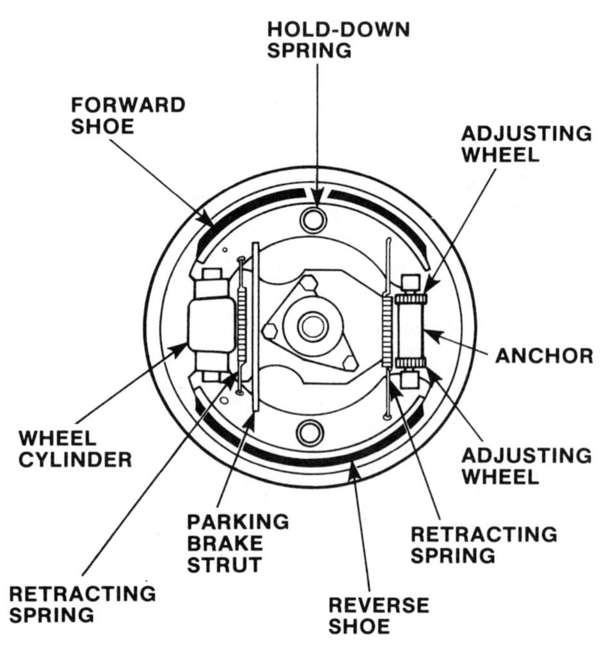

Figure 5-32 Lockheed non-servo (VW)

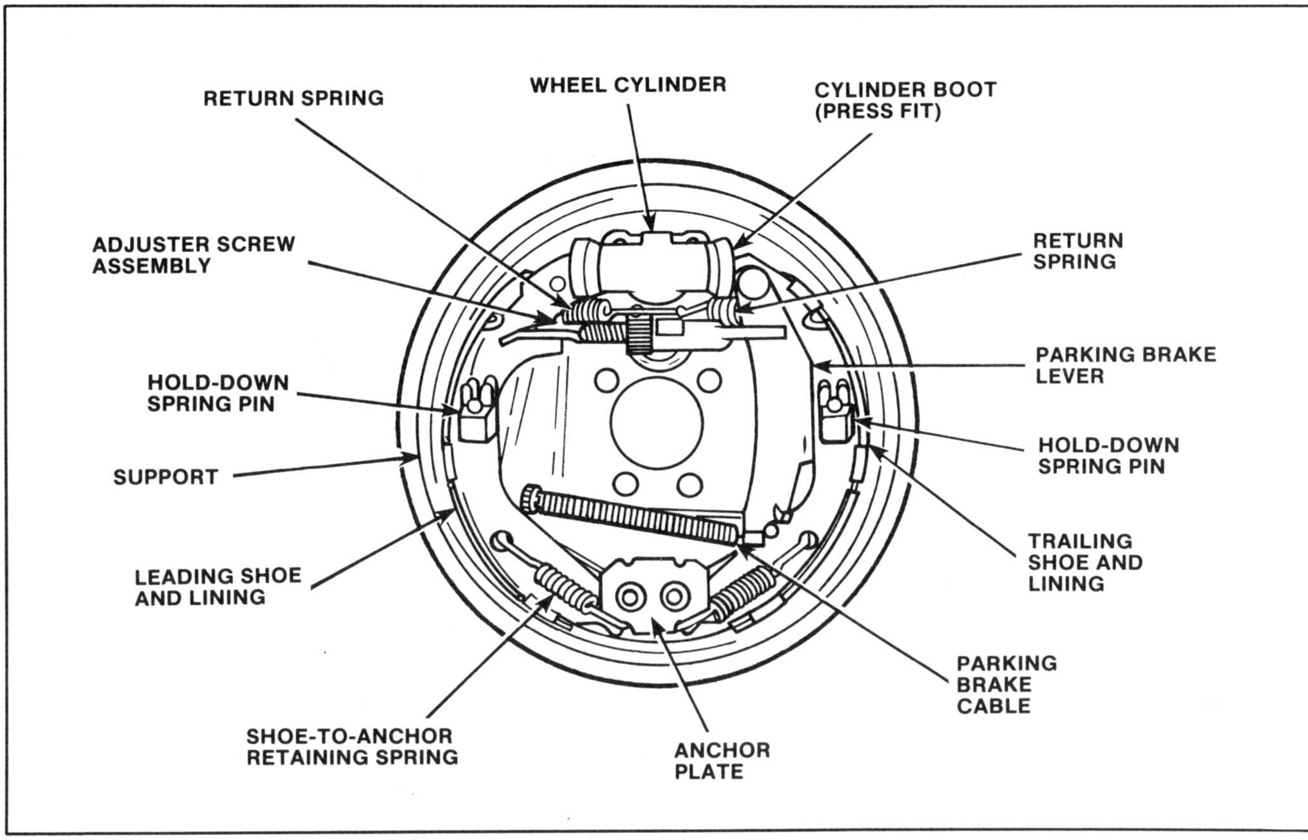

Figure 5-33 Omni/Horizon rear drum non-servo manually adjusted brake

these small car leading-trailing brake shoe drum designs are:

Automatic Cam Adjusters. This rear non-servo drum brake is for use with front disc brakes and has one forward acting (leading and one reverse acting (trailing) shoe. Shoes rest against the wheel cylinder pistons at the top and are held against the anchor plate by a shoe-to-shoe pull-back spring. The anchor plate and retaining plate are riveted to the backing plate. Adjustment of the brake shoes takes place automatically as needed when the brakes are applied. The automatic cam adjusters are attached to each shoe by a pin through a slot in the shoe webbing. As shoes move outward during application, the pin in the slot moves the cam adjuster, rotating it outward. Shoes always return enough to provide proper clearance because the pin diameter is smaller than the width of the slot (Figure 5-34).

Fiesta with Automatic Adjuster. These popular brakes are a leading-trailing shoe design with a ratchet type self-adjusting mechanism. The shoes are held to the backing plate by spring and pin type hold-downs. The shoes are held against the anchors at the top by a shoe-to-shoe spring. At the bottom, the shoe webs are held against the wheel cylinder piston ends by a return spring (Figure 5-35).

The self-adjusting mechanism consists of a spacer strut and a pair of toothed ratchets attached to the secondary brake shoe. The parking brake actuating lever is pivoted on the spacer strut.

The self-adjusting mechanism automatically senses the correct lining-to-drum clearance. As the linings wear, the clearance is adjusted by increasing the effective length of the spacer strut. This strut has projections to engage the inner edge of the secondary shoe via the hand brake lever and the inner edge of the large ratchet on the secondary shoe. As wear on the linings increases, the movement of the shoes to bring them in contact with the drums becomes greater than the gap (Figure 5-36). The spacer strut, bearing on the shoe web, is moved together with the primary shoe to close the gap. Further movement causes the large ratchet, behind the secondary shoe, to rotate inward against the spring-loaded small ratchet, and the

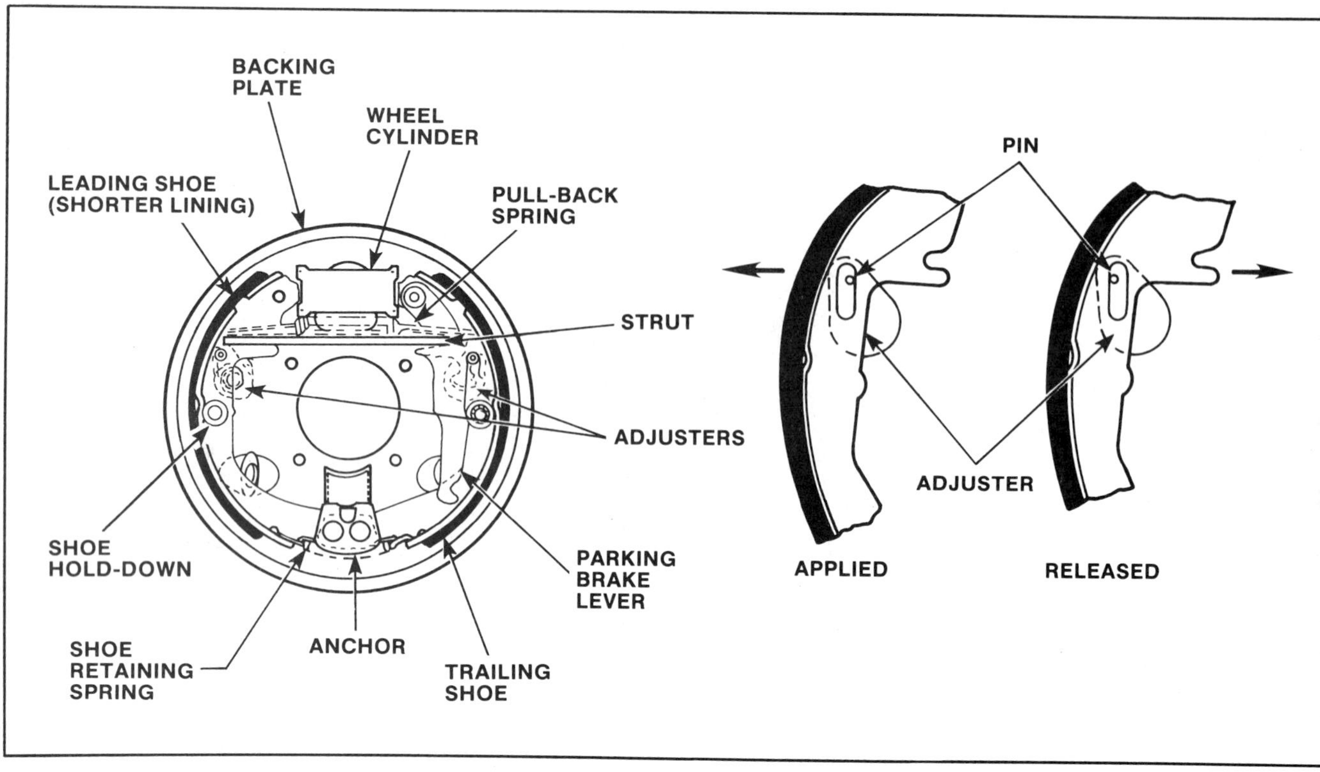

Figure 5-34 Chevette non-servo with automatic cam adjusters

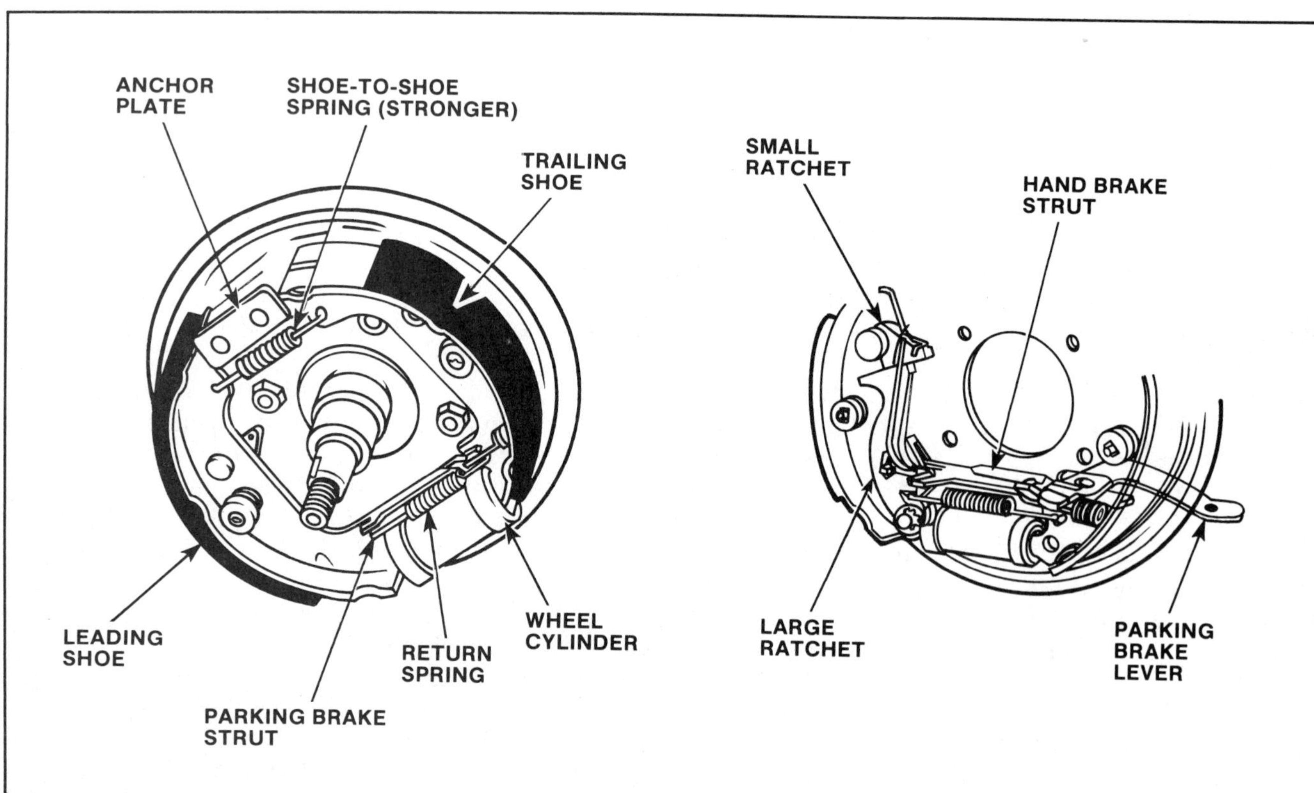

Figure 5-35 Retaining springs hold leading and trailing shoes against anchor plate.

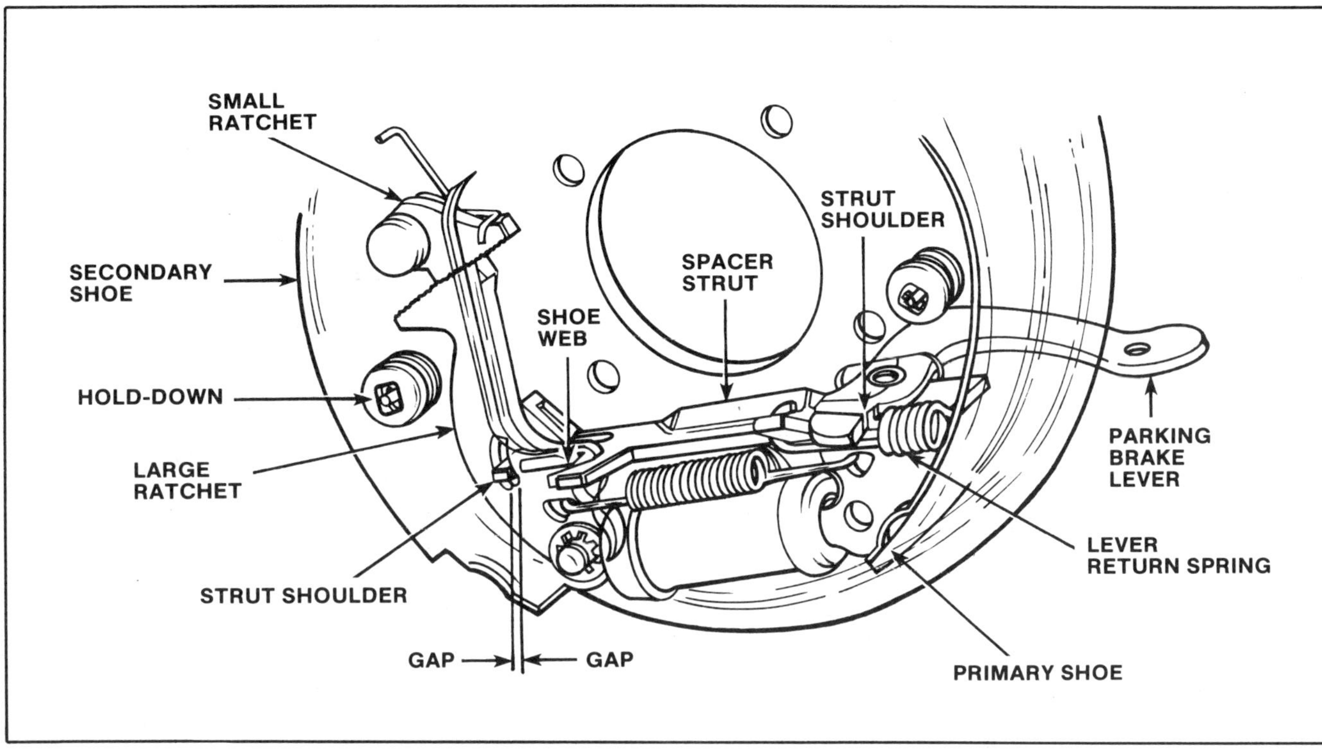

Figure 5-36 Fiesta self-adjusting mechanism

serrations on the mating edges maintain this new setting until further wear on the shoe results in another adjustment. On releasing brake pedal pressure, the return springs cause the shoes to move into contact with the shoulders of the spacer strut/hand brake actuating lever, thus restoring the clearance between the linings and the drum proportionate to the gap shown.

Wagner Link-Crank Type. This linkage is attached to the shoe at an adjustable eccentric mounted near the center of the shoe web. Shoe movement during a reverse stop causes the linkage to rotate the lever (pawl) at the star wheel (Figure 5-37). When this movement is sufficient, the lever (pawl) meshes with the star wheel teeth. This permits the adjuster spring to return the lever (pawl) and advance the adjustment during a forward stop.

Nova with Automatic Adjuster. The brake (Figure 5-38) is similar to other non-servo brakes, having one forward acting and one reverse acting shoe. The lower portion of the brake shoe is held to the anchor plate by an anchor spring. The upper portion of the shoe is held against the wheel cylinder piston by means of the return spring. Brake shoe retaining springs, pins, and washers are used to hold brake shoes to the brake backing plate. The self-adjuster assembly, which is mounted beneath the wheel cylinder, is actuated by an adjuster lever mounted to the parking brake lever. As brake lining wear occurs, proper lining-to-drum clearance is maintained automatically whenever the parking brake is applied.

Vega Non-Servo with Semi-Automatic Adjuster. This brake was designed to be used only as a rear wheel drum for a small car with front disc brakes (Figure 5-39). It has one forward acting and one reverse acting shoe. A hold-down clip has replaced the hold-down spring, cup, and pin. A specially designed parking brake strut and rod assembly straddles the shoe webs just below the wheel cylinder. A shoe-to-shoe spring at the top and another at the bottom under the anchor plate hold the shoes in place.

The strut and rod assembly for the parking brake has the strut part on the forward shoe and the rod part connected to a hole in the reverse shoe. When the parking brake is applied, rear brake shoe adjustment is made. The strut and rod spread the shoes enough to pass the spring locks mounted in the assembly, and the shoes are prevented from returning to their original position. Presure on the brake shoes is released when the rod is relaxed, providing enough clearance for the wheels to turn.

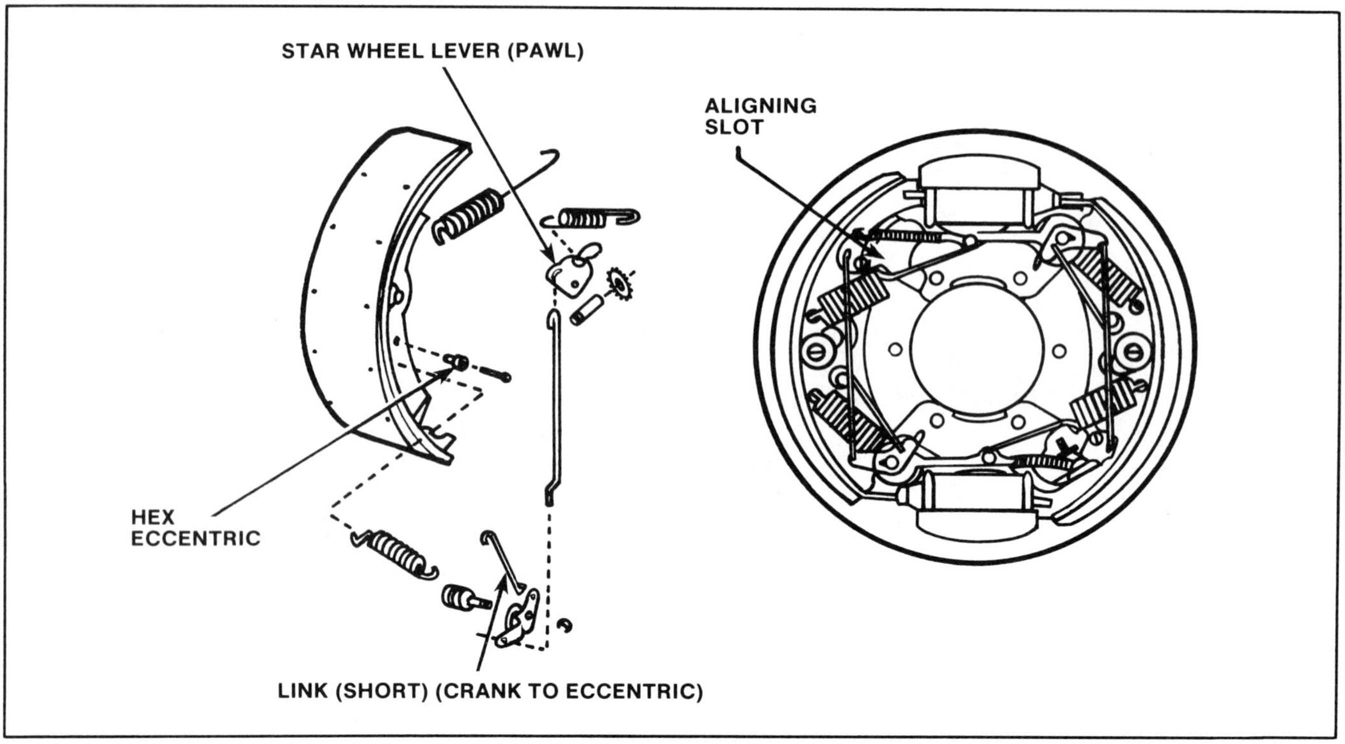

Figure 5-37 Wagner link-crank type brake

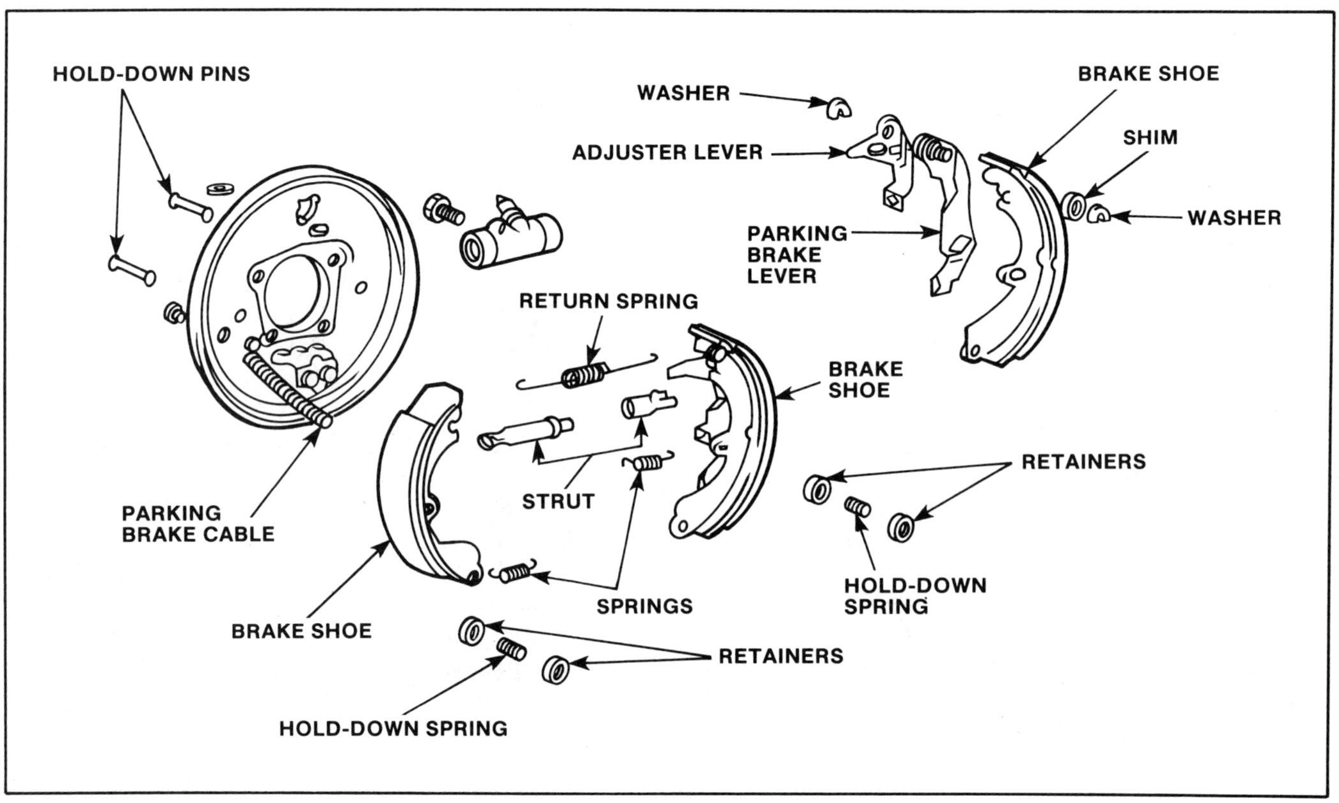

Figure 5-38 Chevrolet Nova non-servo drum brake

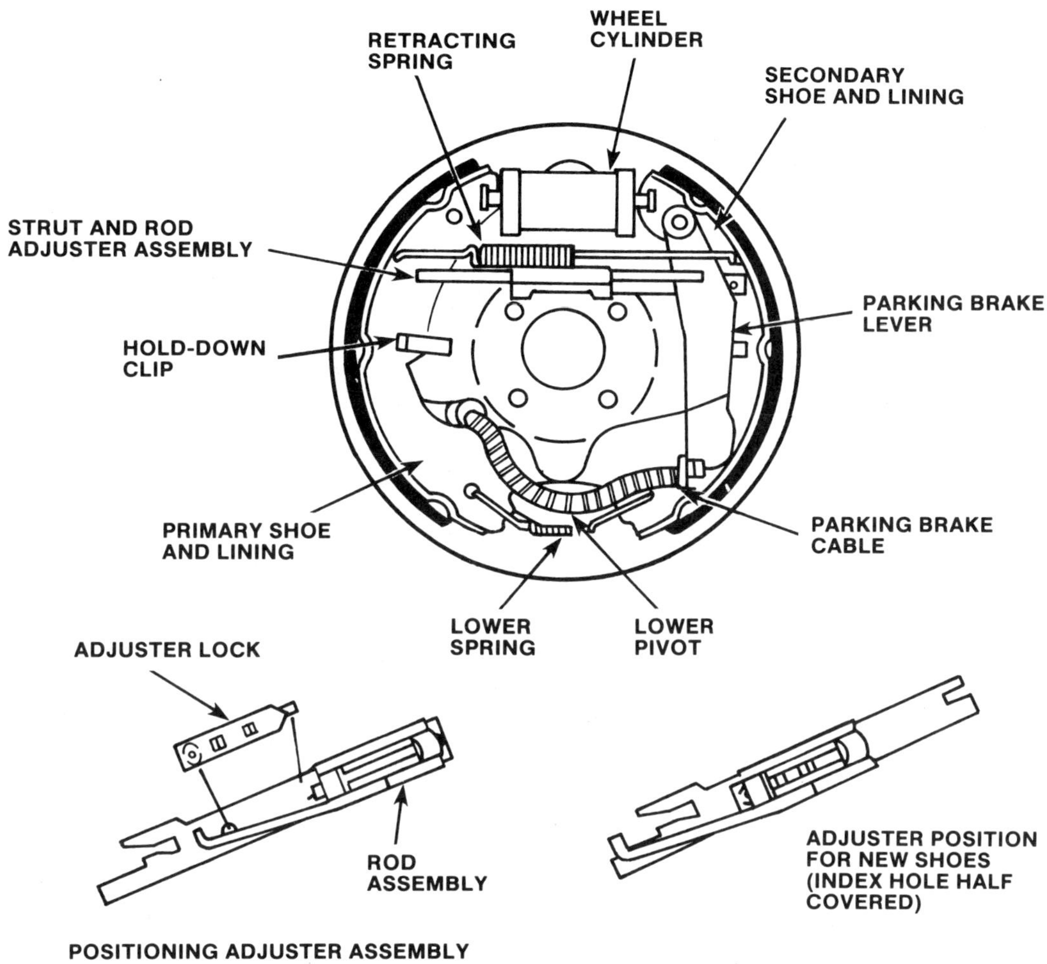

Figure 5-39 Vega non-servo with semi-automatic adjuster

REVIEW QUESTIONS

Review the information in this chapter by answering the following questions.

1. Today, the drum brake is used on
   a. all four wheels of most cars and small trucks.
   b. the front wheels of most cars and small trucks.
   c. the rear wheels of most cars and small trucks.
   d. the rear wheels of only expensive domestic or import models.
2. In a typical drum brake, which component provides a foundation for the brake shoes and associated hardware?
   a. wheel cylinder
   b. drum
   c. backing plate
   d. lining
3. The function of the wheel cylinder is to
   a. provide the rubbing surface area for the linings.
   b. convert hydraulic pressure into a mechanical force at the wheel brakes.
   c. maintain correct shoe position and clearance.
   d. prohibit the shoe from following the movement of the rotating drum.
4. What type of wheel cylinder is most commonly used on today's drum brakes?
   a. double-piston step-bore

b. double-piston straight-bore
c. single-piston step-bore
d. single-piston straight-bore

5. When servicing a single-piston fixed wheel cylinder, Mechanic A lubricates only the cylinder body. When servicing a single-piston fixed wheel cylinder, Mechanic B lubricates both the cylinder body and the slot for the cylinder. Who is right?
   a. Mechanic A
   b. Mechanic B
   c. both A and B
   d. neither A nor B
6. Which of the following statements concerning drum brake shoes is incorrect?
   a. Web thickness never varies.
   b. Many shoes have nibs along the edge of the rim.
   c. The shoe rim is welded to the web.
   d. The primary shoe is the one toward the front of the vehicle.
7. Mechanic A attaches lining to drum brake shoes by bonding. Mechanic B attaches lining to drum brake shoes by riveting. Who is right?
   a. Mechanic A
   b. Mechanic B
   c. both A and B
   d. neither A nor B
8. Which of the following statements concerning drum brake lining is incorrect?
   a. A stop from 60 MPH can raise the brake temperature to 450 degrees F.
   b. Wear rate increases as the temperature of the lining increases.
   c. In a duo-servo brake, the secondary lining does about 25 percent of the braking.
   d. Passenger cars have about 1 square inch of lining for each 25 pounds of gross vehicle weight.
9. When the brakes are applied and the shoes and linings move outward to the drum, which part of the lining should contact the drum first?
   a. center
   b. slightly thinner part
   c. one of the ends
   d. both ends
10. In the unapplied position, drum brake shoes are held against the anchor pin by the
    a. hold-down springs.
    b. star wheel adjuster.
    c. shoe hold-down.
    d. return springs.
11. Duo-servo drum brakes are also known as
    a. leading-trailing brakes.
    b. self-energizing brakes.
    c. non-servo brakes.
    d. none of the above.
12. On most vehicles, the automatic adjuster system is
    a. installed on the secondary shoe.
    b. set up to operate when the brakes are applied as the vehicle moves forward.
    c. installed on the primary shoe.
    d. set up to operate when the brakes are not applied.
13. On a Lockheed manual adjuster, the shoes are located
    a. on the top and bottom.
    b. on the sides.
    c. only on the top.
    d. only on the bottom.
14. Which of the following statements concerning automatic cam adjusters is incorrect?
    a. Adjustment of the shoes takes place automatically as needed when the brakes are applied.
    b. The anchor plate and retaining plate are riveted to the backing plate.
    c. They have one forward acting and one reverse acting shoe.
    d. They are used with rear disc brakes.
15. Some double-piston straight-bore wheel cylinders have a baffle that acts much like a(n)
    a. vacuum.
    b. air shock absorber.
    c. fluid shock absorber.
    d. dust boot.

CHAPTER 6

DRUM BRAKES— INSPECTING AND SERVICING

OBJECTIVES

Upon completing this chapter, you should be able to

- Remove a wheel and tire assembly, including anti-theft wheel lug nuts, in order to inspect and service the drum brake.
- Describe common drum mounting methods.
- Prepare a typical duo-servo drum brake for inspection and servicing.
- Prepare a typical non-servo drum brake for inspection and servicing.
- Perform a cleaning and inspection of a drum brake and its related parts.
- Recognize conditions that adversely affect the structural strength of a drum, including scoring, hard spots, and heat checks, among others.
- Measure a drum using a drum micrometer.
- Identify the general rules of drum machining.
- Clean and mount a newly refaced drum.
- Attach linings to shoes using the riveting method.
- Reassemble a typical duo-servo drum brake.
- Reassemble a typical non-servo drum brake.
- Recondition and replace a wheel cylinder.

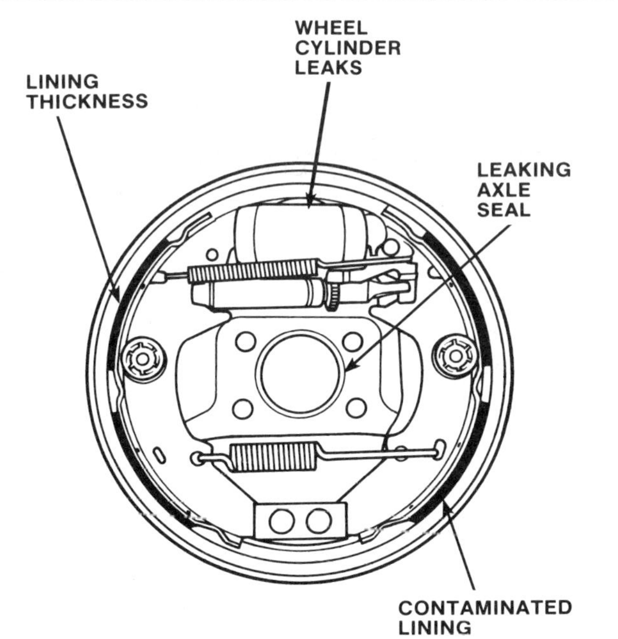

Figure 6-1 Main points of a drum brake inspection

When removing the drum to check the lining, wheel cylinder, and its conditions, you should also evaluate other parts of the braking system. Signs of grease or oil at the center of the brake assembly show that the axle seal needs replacing. If brake fluid drips out of the wheel cylinders when rubber dust boots are pulled away, it indicates that the cylinder needs rebuilding or replacing.

Contaminated linings should be replaced. Dryness or corrosion in the bore means that honing or replacement is called for (Figure 6-1). In the brake line, inspect for cracked or worn hoses and bent or corroded tubes. But, before inspecting drum brakes, it is necessary to release the parking brake and raise the vehicle on jack stands or hoist. If the work is to be done on a hoist, position the car on the hoist and lift at the points recommended by the vehicle manufacturer (Figure 6-2). Mark the wheel-to-drum and the drum-to-axle positions.

CAUTION: Remember that incorrect lift points can damage steering linkage components and front-end suspension struts.

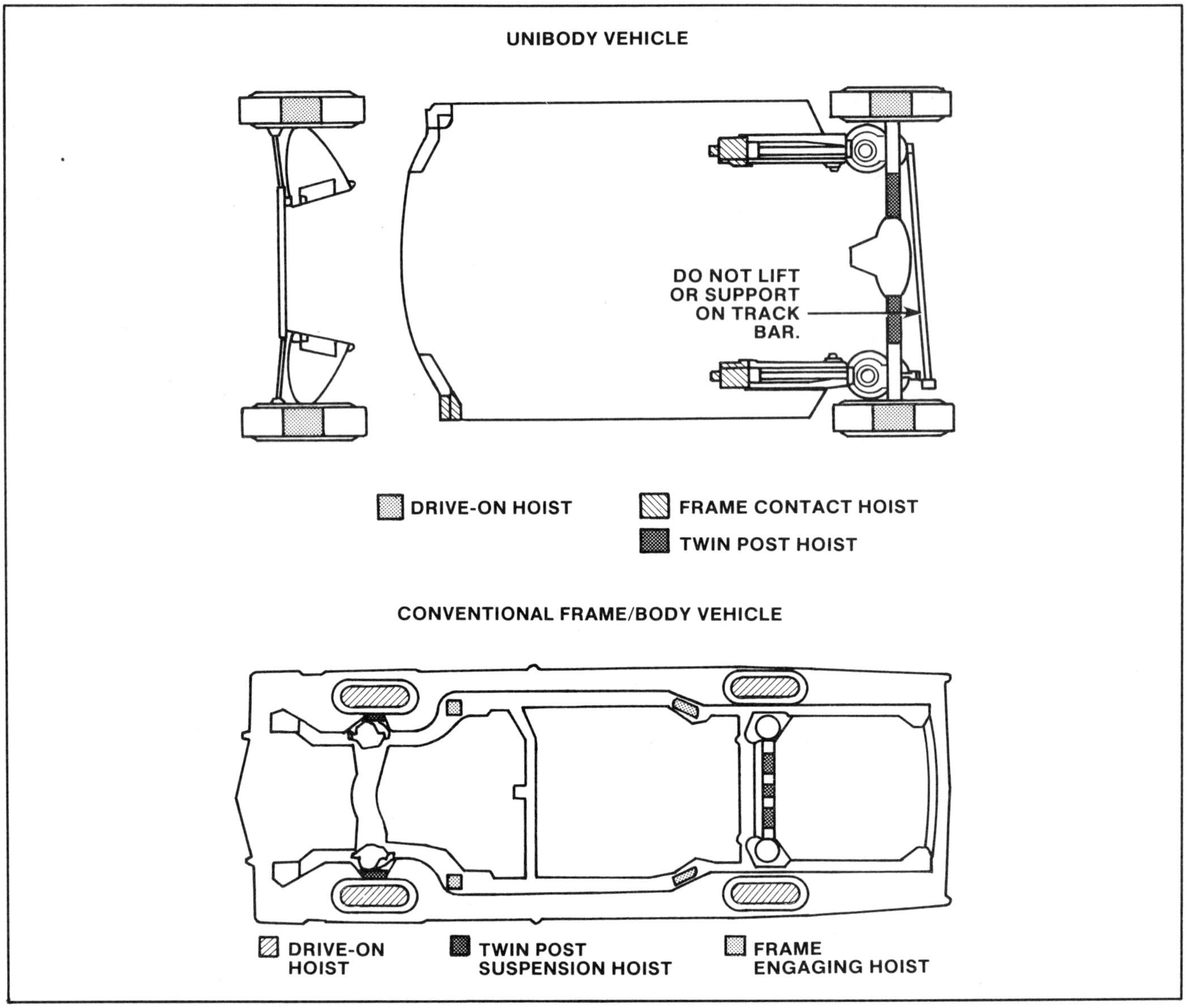

Figure 6-2 Typical lift points for unibody and conventional frame/body vehicles

Place the transmission in the NEUTRAL position. Relieve all tension from the parking brake cables. On most cars this can be done by loosening or removing the adjusting nut at the equalizer. It also might be necessary, as discussed later in this chapter, to back off the brake shoe adjustment before the drums can be removed.

The wheels themselves are made of either stamped steel discs riveted or welded to a circular shape or a die-cast or forged aluminum or magnesium rim or disc. Wheel mounting holes are provided in the rim or disc. The mounting holes are tapered to fit tapered mounting nuts that center the wheel over the hub.

After the vehicle has been securely raised into position, the wheel and tire assembly can be removed as follows:

1. Remove the wheel cover.

SHOP TALK

Some wheel covers are equipped with an anti-theft locking system. The lock bolt for each wheel cover is located behind the hub ornament. A special key wrench is required to pry off the center hub ornament and for removing the lock bolt (Figure 6-3). To allow vehicle service in the event the customer's key has been misplaced, a master key set is available at the vehicle dealer.

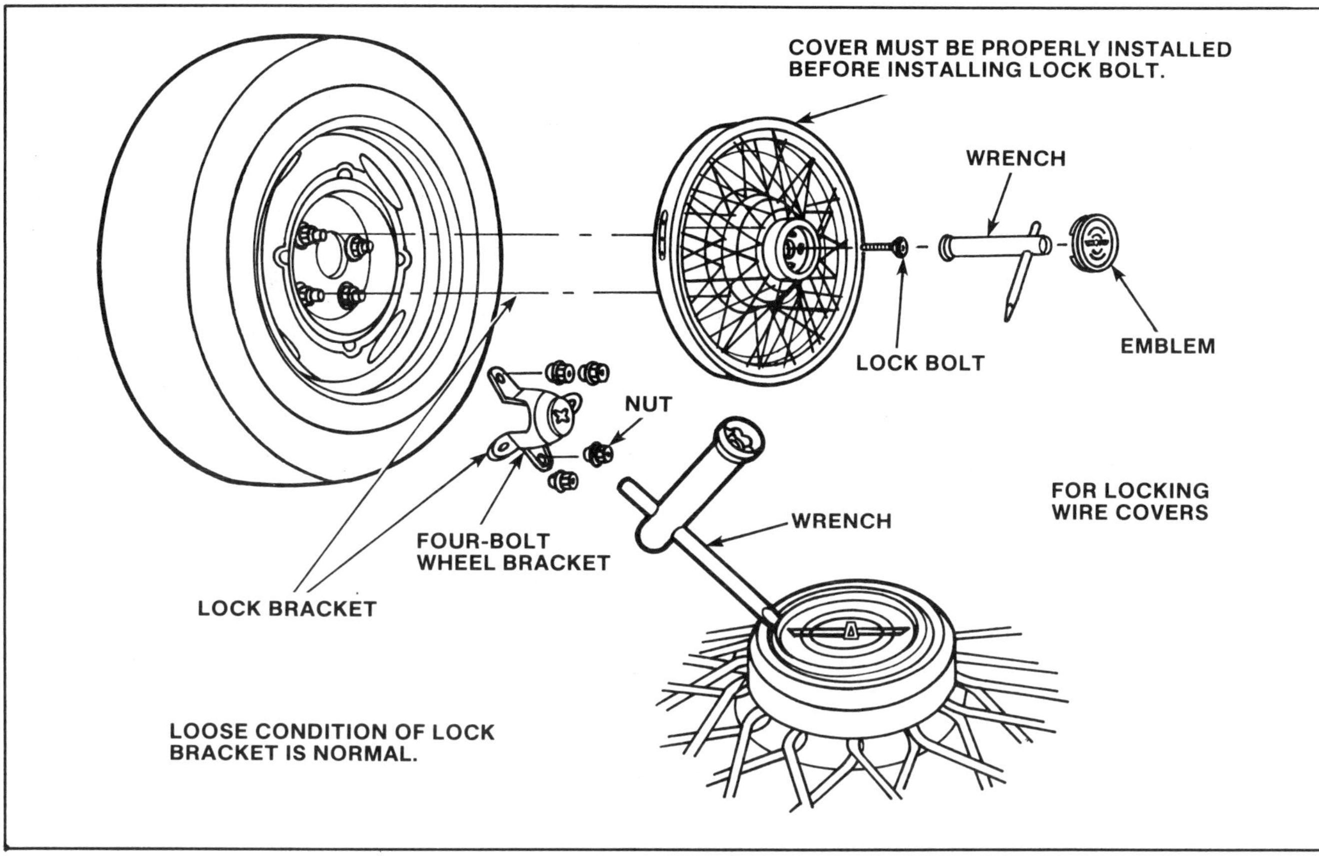

Figure 6-3 Removal of a typical anti-theft wire wheel cover

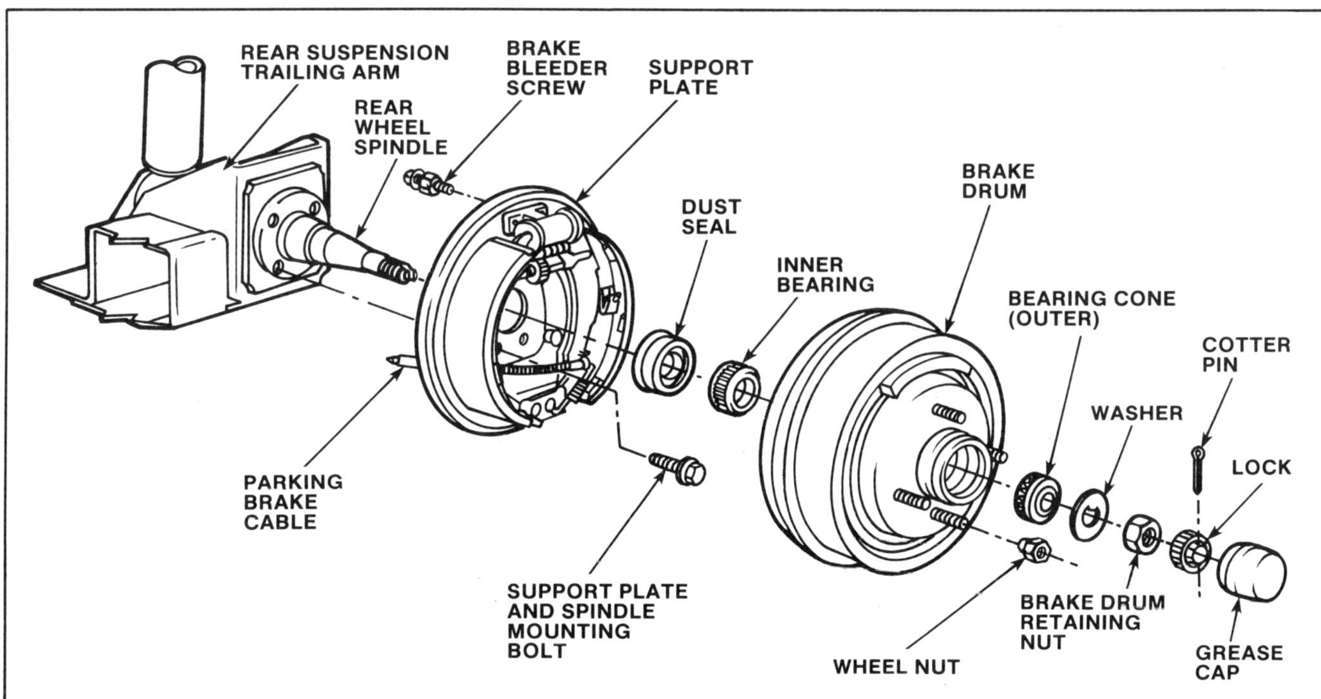

Figure 6-4 Trailing rear axle wheel mounting and bearing arrangement

Some aluminum wheels are also equipped with anti-theft wheel lugs (one per wheel). The key has a circular keyway that is matched to the female slot in the anti-theft wheel lug nut. To remove or install the anti-theft wheel lug nut, insert the key into the slot of the lug nut, place the lug nut wrench on the key, and, while applying pressure on the key, remove or install the lug nut. Again, if the customer does not have the key, a master one can be had from the vehicle's authorized dealership. Without a key the wheels cannot be removed.

2. Remove the hub nuts. Pull the wheel off the hub.
3. If corrosion has formed around the center pilot hole so that the wheel cannot be removed by hand, use a wheel puller to remove the frozen wheel.

DRUM SHOE AND LINING REMOVAL AND INSTALLATION

Certain types of wear or damage typically are found with drums that have seen hard or prolonged service. These may or may not be correctable. To determine whether a drum is usable as is, requires machining, or cannot be machined and must be scrapped—it is important to first find out the type and extent of wear or damage, and if the drum is already close to the minimum safe thickness limit.

The front wheels on most cars have integral drum and hub assemblies. It is necessary to disassemble the front-wheel bearings in order to remove the drum. Remove the dust or grease cap, cotter pin, castellated nut or plain nut and nut lock, thrust washer and outer bearing. Pull the drum and hub assembly from the spindle. Protect the bearings from dirt and be sure to use care not to drop the drum.

SHOP TALK

Late-model Chevrolet cars have demountable front drums. These drums can be removed without removing the hub or disturbing the front-wheel bearing.

Before pulling a rear drum from the axle flange (Figure 6–4), check to see if it has a locating tag. If not, mark one of the studs and the corresponding drum hole to insure that the drum is reinstalled in the same position. On some cars, a couple of speed nuts can retain the drum on the studs. Remove these first. Also be sure to carefully inspect the rear wheel axle gaskets and wheel seals for leaks (Figure 6-5).

Figure 6-5 Checking the rear wheel assembly

SHOP TALK

If difficulty is encountered in removing a rear drum, make sure that the brakes are fully released. Then, try a sharp rap with a plastic mallet to get the drum started.

SHOE AND LINING REMOVAL

In order to hold the drums in their piloted positions while the wheels are dismounted, several assembly methods are used. When the original drum is mounted against a wheel hub flange, it can be fastened with rivets or by swaging the piloting shoulders of the wheel studs, or with speed nut fasteners installed over the threads of the wheel studs (Figure 6–6).

The most common mounting method is using the tire rim and lug nuts as illustrated in Figure 6–7. The drum is a slip fit over the axle flange and studs. Speed nuts are installed at the vehicle manufacturing plant for temporary retention of the drum on the assembly line until the wheel is installed.

A floating drum can be retained by a nut and cotter pin. The drum can also be secured to the axle flange by one or two bolts. Some import applications have two additional holes threaded into the drum face so that bolts can be used to press the drum from the hub or flange.

Efficient methods for replacing riveted or swaged stud types require a shop press (8- to 12-ton capacity) and a press type drum assembly fixture

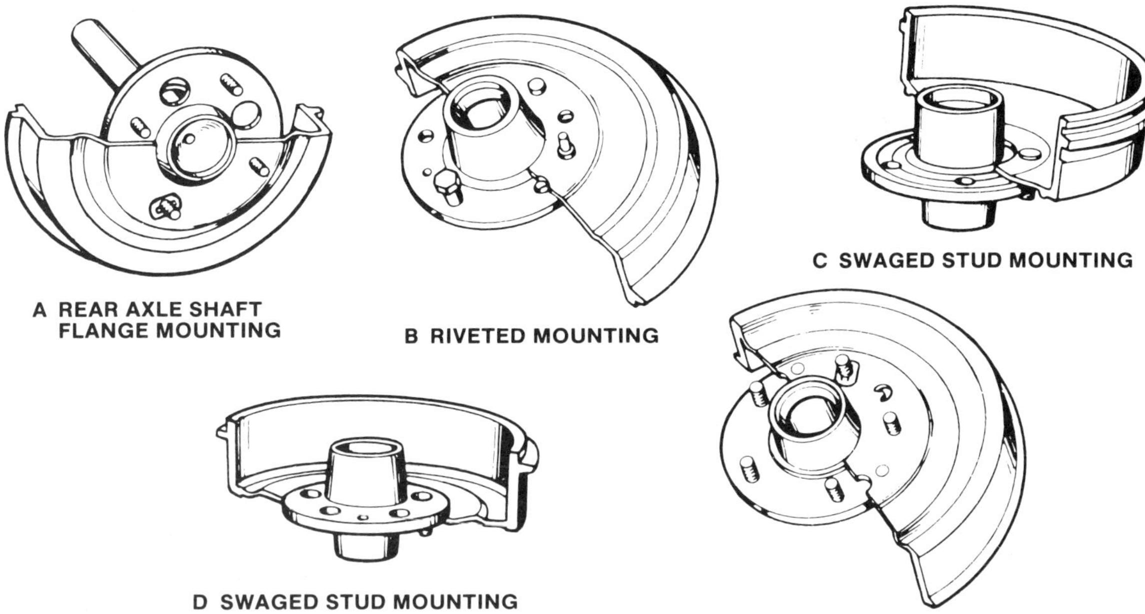

Figure 6-6 Brake drum mountings

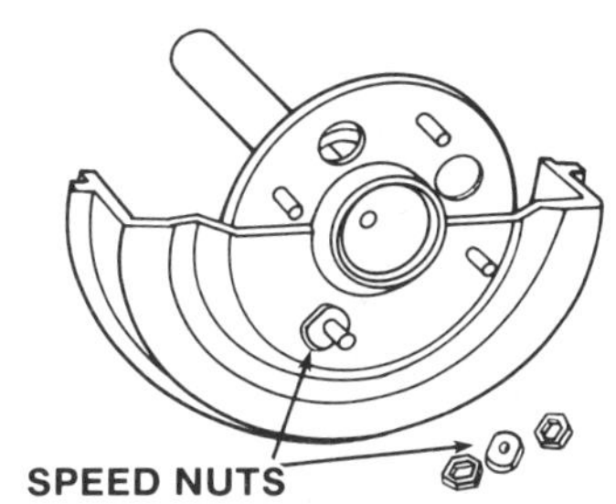

Figure 6-7 Tire rim and lug nut mounting

equipped with several sizes of peen punches and a stud cutter for milling (shaving) down the swaged surface of the stud shoulders (Figure 6-8).

Replacement methods are as follows:

- *Drums with Threaded Fasteners*—Unthread the speed nuts or machine screws and lift the drum off the flange.
- *Drums with Riveted Mounts*—With a chisel, shear off the upset ends of the rivets. Punch or drill out the rivet shanks and separate the hub from the drum.
- *Drums with Swaged Stud Mounts*—Chuck the stud cutter into an electric drill (Figure 6-9).

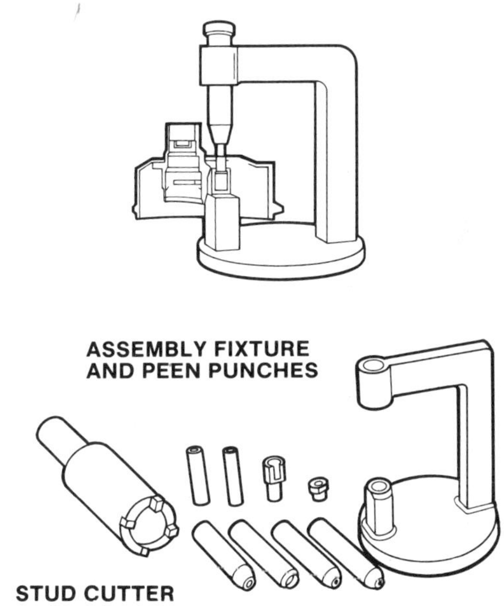

Figure 6-8 Equipment for replacing riveted or swaged stud type drums

Align the cutter over each stud thread in turn, and mill off the swaged material. Do not press off the studs without first removing the swage.

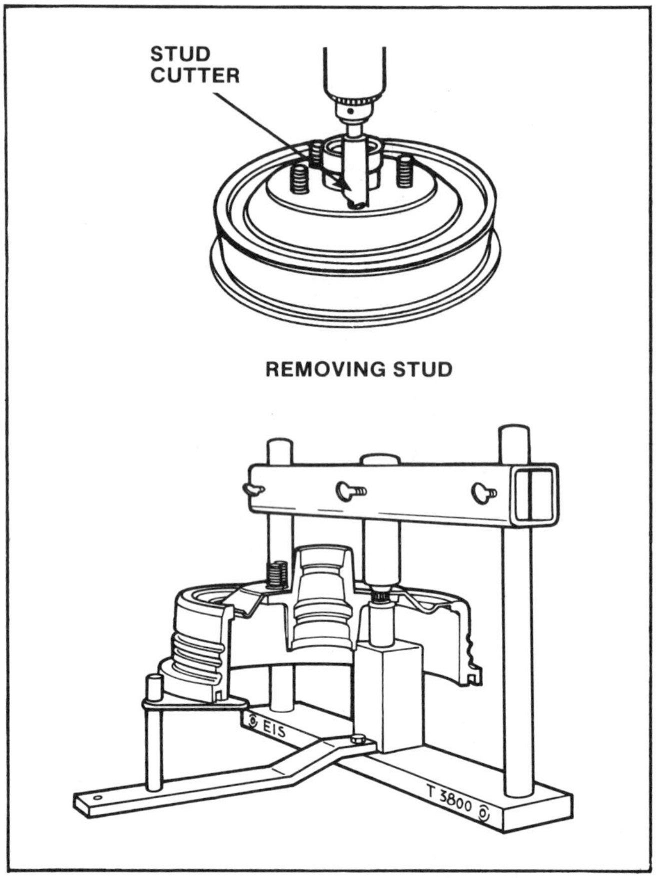

Figure 6-9 Chuck the stud cutter into an electric drill.

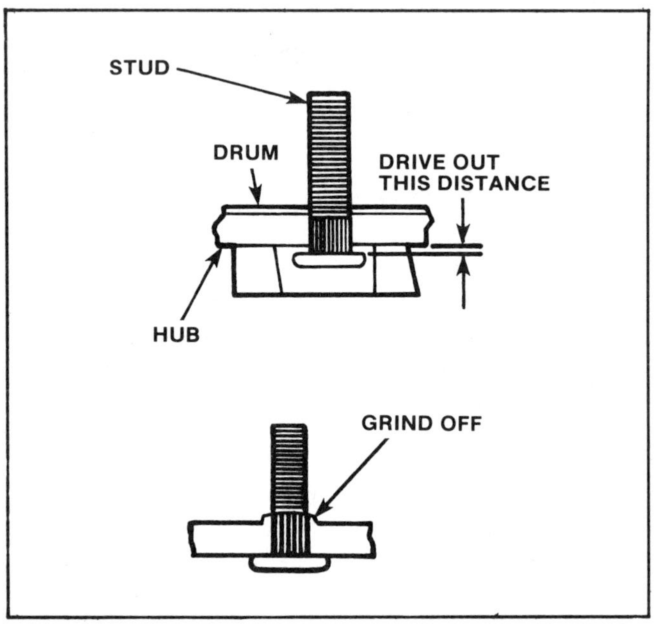

Figure 6-10 Alternate method of removing the drum

Separate the hub from the drum and press the studs from the hub. If the old studs must be re-used, do not shave off more than two-thirds of their pilot shoulders protruding from the hub flange. Otherwise, the studs should be replaced.

Alternately, the studs can be pressed or driven out only enough to bend back the swaging and pop off the drum. Then, reset the studs in the hub and grind or file off the swaged material before forcing the studs through the hub flange (Figure 6-10).

WARNING: When servicing wheel brake parts, do not create dust by cleaning with a dry brush or with compressed air. Asbestos fibers can become airborne if dust is created during servicing. Breathing dust containing asbestos fibers can cause serious bodily harm. To clean away asbestos from brake surfaces, use an OSHA approved washer like the one shown in Figure 6-11. Follow manufacturer's instructions when using the washer.

Figure 6-11 A typical washer used to remove asbestos fibers from brake surfaces (*courtesy of Ammco Tools, Inc.*)

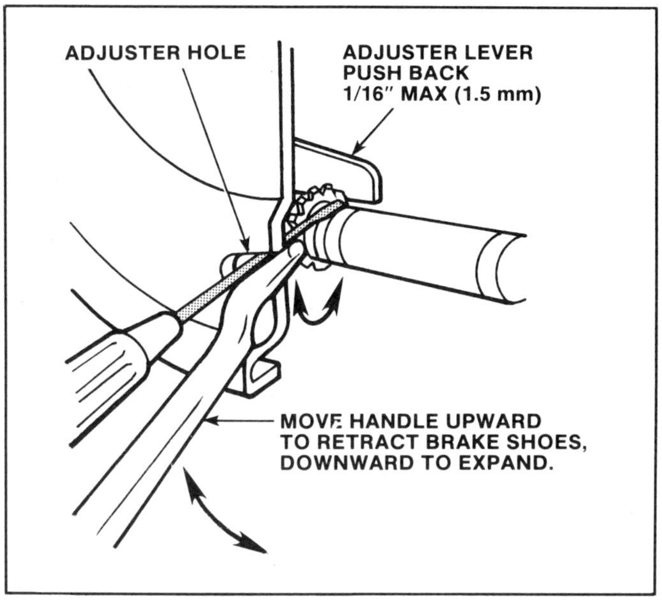

Figure 6-12 Removing drums with self-adjusters

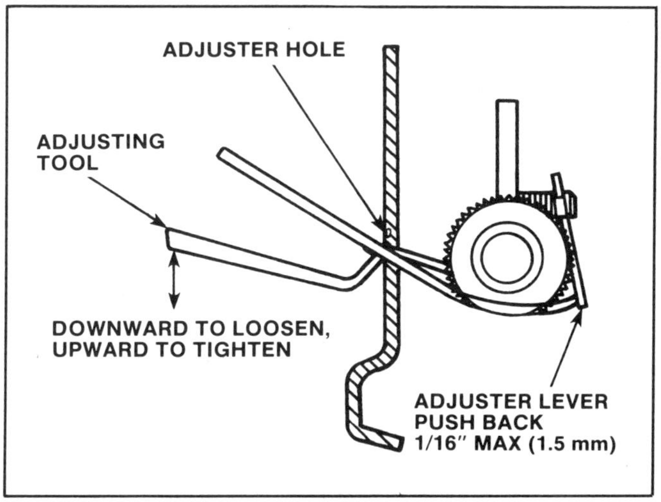

Figure 6-13 Removing drums with self-adjusters on Chrysler models

After removing the retaining devices that hold the drum to the axle flange or hub, the drum can be removed for servicing. If the brake drum is rusted or corroded to the axle flange and cannot be removed, lightly tap the axle flange to the drum mounting surface with a plastic mallet. Remember that if the drum is worn, the brake shoe adjustment will have to be backed off for the drums to clear the brake shoes. Do not force the drum or distort it, and do not allow the drum to drop.

If the brake shoes have expanded too tightly against the drum or have cut into the friction surface of the brake drum, the drums might be too tight for removal. In such a case, the shoes must be adjusted inward before the brake drum is removed. On most cars with self-adjusting mechanisms, reach through the adjusting slot with a thin screwdriver (or similar tool) and carefully push the self-adjusting lever away from the star wheel a maximum of 1/16 inch (Figures 6-12 and 6-13). While holding the lever back, insert a brake adjusting tool into the slot and turn the star wheel in the proper direction until the brake drum can be removed.

On cars that have the adjusting slot in the drum rather than in the backing plate, reach through the slot with a thin wire hook and pull the adjuster lever away from the star wheel while slackening the adjustment. If the slot in the drum is not open, knock out the center of the slot-shaped lanced area with a punch or chisel.

To prepare a typical duo-servo drum brake for inspection and servicing, proceed as follows:

1. Disconnect the cable from the parking brake lever.
2. Back off the adjusting screw or release the shoe adjusting cams to provide ample lining-to-drum clearance. If the brake lining is dragging on the brake drum, back off the brake adjustment by rotating the adjustment screw. Refer to individual car service manuals for the specific disassembly procedures.
3. If required, install wheel cylinder clamps on the wheel cylinders (Figure 6-14) to prevent

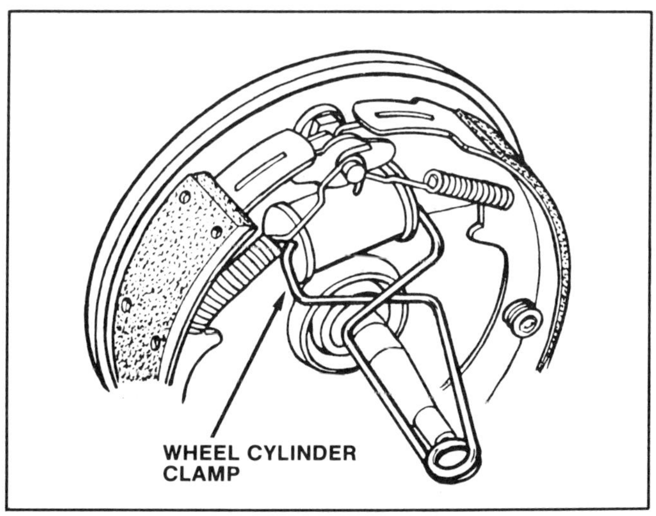

Figure 6-14 Using a wheel cylinder clamp

fluid leakage or air from getting into the system while the shoes are removed. Some brakes have wheel cylinder stops; therefore, wheel cylinder clamps are not required. Regardless of whether or not the clamps are needed, do not press down on the brake pedal after shoe return springs have been removed. To prevent this, block up the brake pedal so it cannot be depressed.

SHOP TALK

Keep the adjusting screws and automatic adjuster parts for left and right brakes separate because these parts usually are different. For example, with Bendix and Delco-Moraine automatic adjusters, the adjusting screws on the right brakes have left-hand threads and the adjusting screws on the left brakes have right-hand threads.

4. Remove the brake shoe return springs (Figure 6-15). Use a brake spring removal and installation tool to unhook the springs from the anchor pin or anchor plate (Figure 6-16). In some cases the return springs might not be attached directly to the anchor pin. One spring might be hooked over the adjusting mechanism link, which is in turn installed over the anchor pin. Or, as in Figure 6-17, both springs can be hooked onto an anchor plate. Regardless of how springs ae attached, carefully note the location, position, and color of all springs before removing any so that they can be reinstalled properly.

CAUTION: Do not use the brake spring tool to remove the adjusting link from the anchor pin or anchor plate.

5. Remove the shoe retaining or hold-down cups and springs. Special tools are available, but the hold-down springs can be removed by using pliers to compress the spring and rotating the cup with relation to the pin.

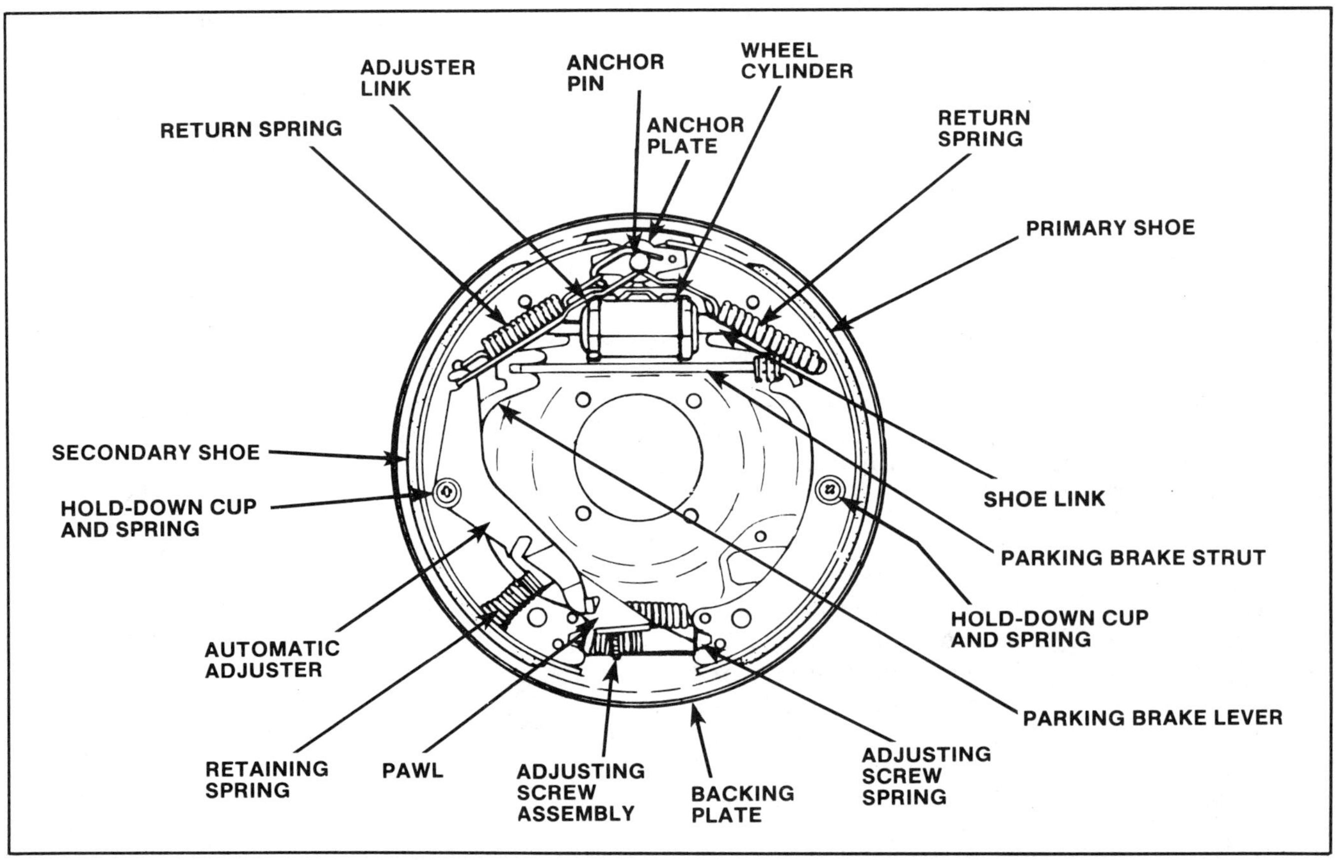

Figure 6-15 Late-model full-size car self-adjusting brake assemblies

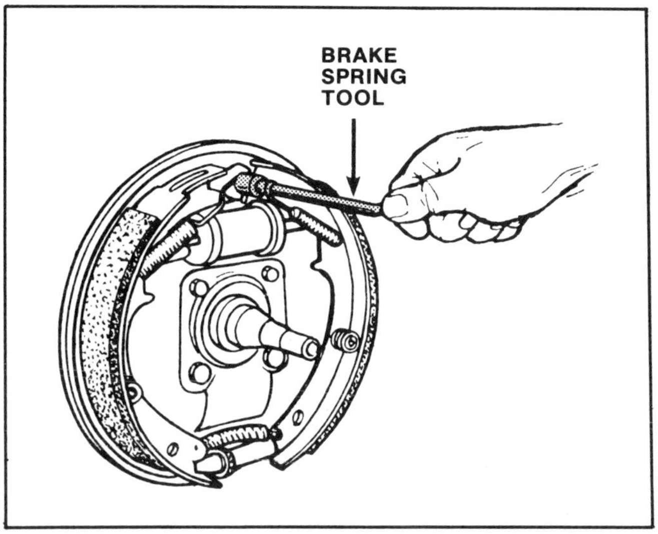

Figure 6-16 Using a brake spring tool

6. Self-adjuster parts can now be removed from brakes that are equipped with them. Lift off the actuating link, lever and pivot assembly, sleeve (through lever), and return spring. No advantage is gained by disassembling the lever and pivot assembly unless one of the parts is damaged.
7. On all rear brakes, spread the shoes slightly to free the parking brake strut and remove the strut with its spring. Disconnect the parking brake lever from the secondary shoe. It can be attached with a retaining clip, bolt, or simply hooked into the shoe.
8. Slip the anchor plate off the pin. No advantage is gained by removing the plate if it is the bolted-on or riveted type. Spread the anchor ends of the shoes and disengage from the wheel cylinder links, if used. Remove the shoes (Figure 6–18) connected at the bottom by the adjusting screw and spring, as an assembly.
9. Overlap the anchor end of the shoes to relieve spring tension, unhook the adjusting screw spring, and remove the adjusting screw assembly.

Although servicing methods for non-servo or leading-trailing brakes vary more than those for servo units, the procedure given for the non-servo unit shown in Figure 6–19 will serve as a guide for

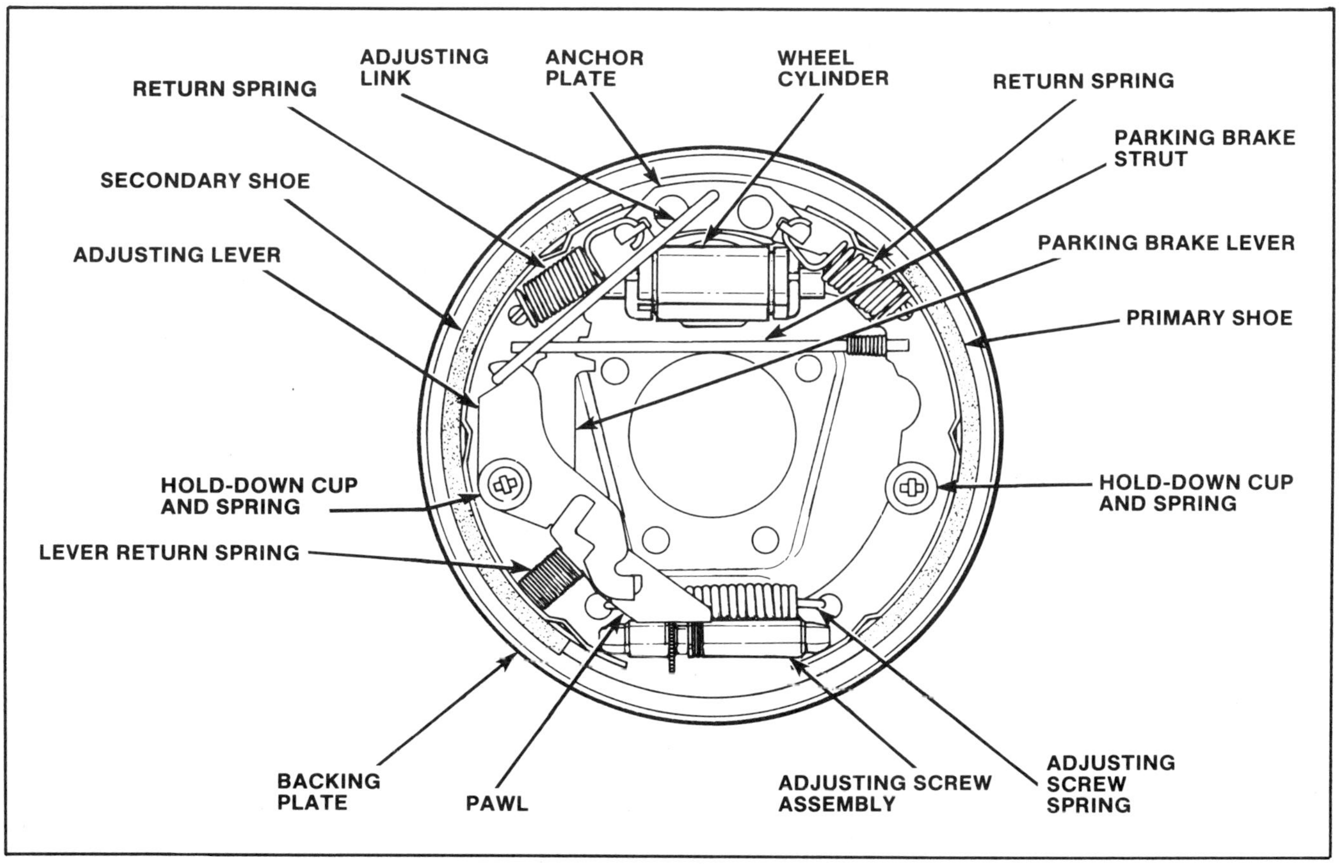

Figure 6-17 Late-model small car self-adjusting brake assemblies

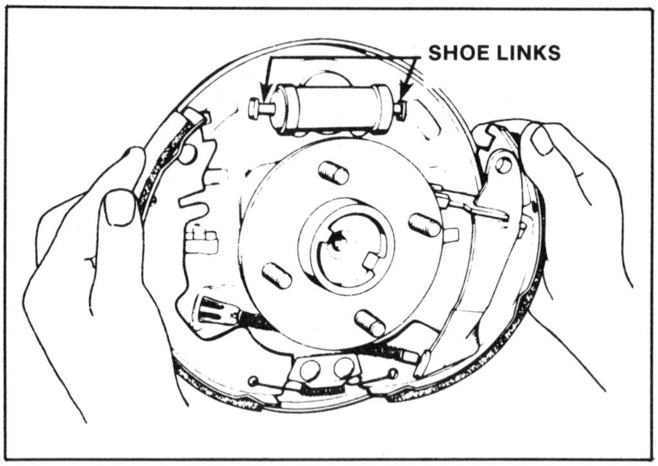

Figure 6-18 Removing shoes

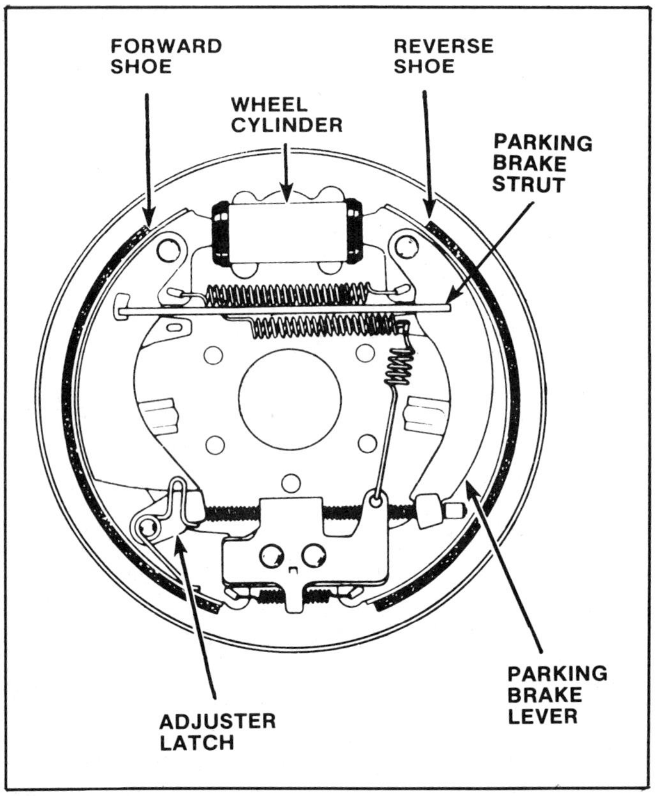

Figure 6-19 Left rear non-servo brake

similar designs. For the exact procedures, check the vehicle's service manual.

With the car on a hoist or jack stands and the wheels and drums removed, proceed as follows:

1. Be sure that the parking brake is released.
2. Remove wheel and drum. If the shoes interfere with drum removal, insert a screwdriver into the access hole (in backing plate) and press down on the adjuster latch to permit the shoes to retract.

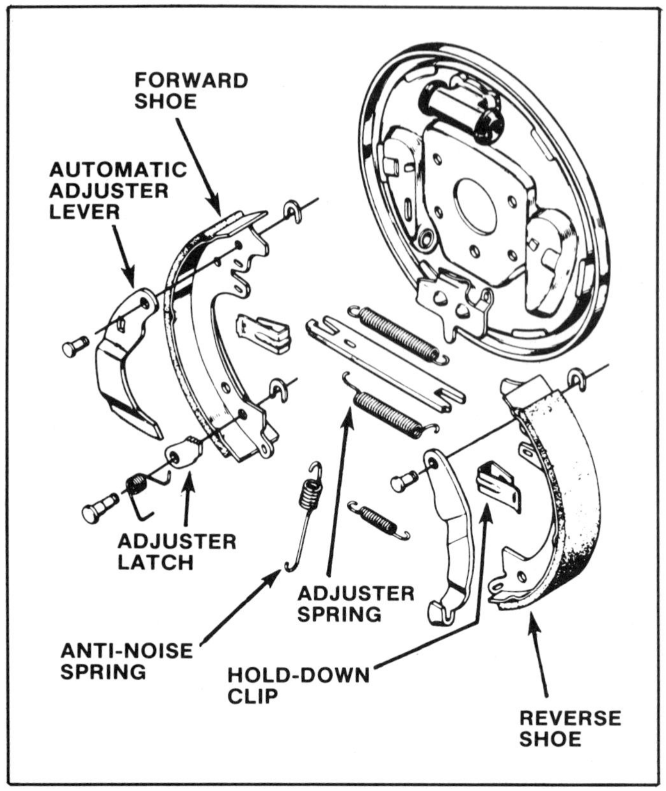

Figure 6-20 Parts of a non-servo brake

3. Install the wheel cylinder clamp. Then unhook the adjuster spring from the parking brake strut and reverse shoe (Figure 6-20).
4. Unhook the upper shoe-to-shoe spring from the shoes and unhook the anti-noise spring from the spring bracket.
5. Remove the parking brake strut and disengage the shoe webs from the flat, clamp type shoe hold-down clips.
6. Unhook the lower shoe-to-shoe spring and remove the forward shoe. Disconnect the parking brake cable, then remove the reverse shoe.
7. Remove the shoe hold-down clips from the backing plate.
8. Press off the C-shaped retainers from the pins and remove the parking brake lever, automatic adjuster lever, and adjuster latch.

SHOP TALK

Mark the shoe positions if shoe and linings are to be reused. When disassembling an unfamiliar brake assembly, work on one wheel at a time and use the other wheel as a reference.

9. Remove the parking brake lever.

CLEANING, INSPECTING, AND LUBRICATING BRAKE PARTS

> **WARNING:** When servicing wheel brake parts, do not create dust by cleaning with a dry brush or compressed air. Asbestos fibers can become airborne if dust is created during servicing. Breathing dust containing asbestos fibers can cause serious bodily harm. See the WARNING on this subject in Chapter 1.

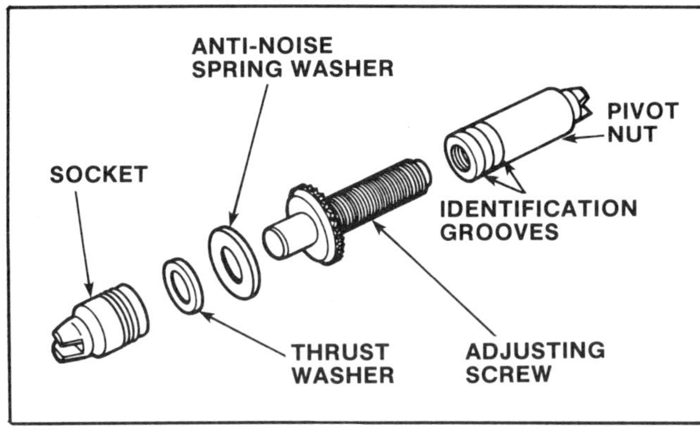

Figure 6-21 Adjusting screw assembly

Regardless of the brake design, it is necessary to perform a cleaning and inspection procedure similar to the following:

1. Clean the backing plates, struts, levers, and other metal parts to be reused using a water-dampened cloth or a water-based solution. Equipment is commercially available to perform washing functions of brake parts. Wet cleaning methods must be used to prevent asbestos fibers from becoming airborne.
2. Carefully examine the raised shoe pads on the backing plate to make sure that they are free from corrosion or other surface defects that might prevent the shoes from sliding freely. Use fine emery cloth to remove surface defects, if necessary; then clean thoroughly.
3. At the rear wheels, look for evidence of oil or grease leakage past the wheel bearing seals. Such leakage could cause brake failure and indicates the need for additional service work.
4. Check that the backing plates are not cracked or bent. If so, they must be replaced. Make sure that backing plate bolts and bolted-on anchor pins are torqued to specifications.
5. If repair of the wheel cylinders is needed, it should be done at this time (the procedure for servicing wheel cylinders is given later in this chapter). To determine wheel cylinder condition, carefully inspect the boots. If they are cut, torn, heat-cracked, or show evidence of excessive leakage, the wheel cylinders should be repaired or replaced. On cylinders with external boots, carefully pull back the lower edge of the boot. If more than a drop of fluid spills out, leakage is excessive and indicates that repair or replacement is necessary. If the wheel cylinders have internal boots, carefully remove one of the wheel cylinder connecting links to check for leakage.

SHOP TALK

A slight amount of fluid behind the boot is normal. This lubricates the piston. However, if enough fluid is present to run or spill out, this indicates excessive leakage.

6. Disassemble the adjusting screw assembly (Figure 6-21) and clean the parts in a suitable solvent. Make sure that the adjusting screw threads into the pivot nut over its complete length without sticking or binding. Check that none of the adjusting screw teeth are damaged, particularly on self-adjusting brakes. Lubricate the adjusting screw threads with brake lubricant, being careful not to get any on the adjusting teeth, and reassemble. Most adjusting screw assemblies use a thrust washer between the adjusting screw and socket. Some also might have the anti-noise spring washer. Thread the adjusting screw in as far as it will go.

> **CAUTION:** Do not use ordinary grease to lubricate drum brake parts. It will not hold up under high temperatures.

7. Apply a thin film of a brake lubricant made for brakes to the raised shoe pads on the backing plate lands. Check that there are no burrs on the edges of the replacement shoes

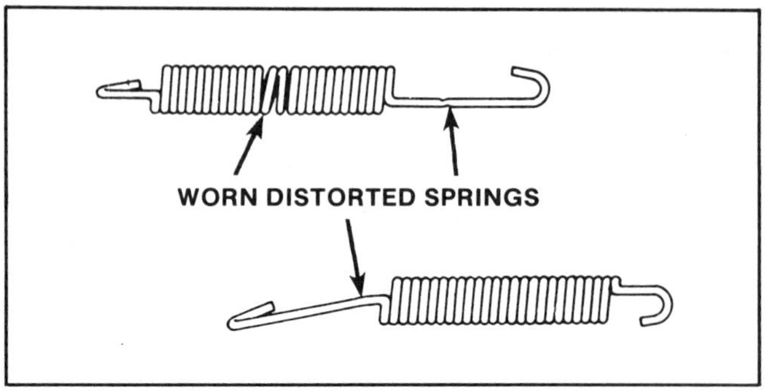

Figure 6-22 Inspect for damaged springs

where they contact the pads. On the rear brakes, lubricate the parking brake lever pivot point.

SHOP TALK

Anti-seize lube works very well as brake lubricant and it will be there on the next brake job.

8. Examine the shoe anchor, support plate, and small parts for signs of looseness, wear, or damage that could cause faulty shoe alignment. Check springs for spread or collapsed coils, twisted or nicked shanks, and severe discoloration (Figure 6-22). Operate star wheel type automatic adjusters by prying the shoe lightly away from its anchor or by pulling the cable to make sure the adjuster advances easily, one notch at a time. Adjuster cables tend to stretch, and star wheels and pawls become blunted after a long period of use. For rear axle parking brakes, pull on the cable and shoe linkage to make sure no binding condition is present that could cause the shoes to drag when the parking brake is released.

SHOP TALK

Some brake technicians check brake spring tension by the drop method. Although not overly scientific and the results are not always correct, it is conducted as follows: drop the spring being tested on a clean concrete floor and if it bounces with a chunky sound, it is good. If, on the overhand, the bounced spring gives off a tinny sound, it is "tired" and should be replaced.

To complete the drum brake inspection, except for the drum itself, examine wheel bearings and hub grease seals for signs of damage. Service or replace, if necessary.

Figure 6-23 Checking the condition of the drum

Drum Inspection

One of the most important safety inspections to be made is that of the brake drum (Figure 6-23). First, visually inspect the brake shoes, as installed on the car. Their condition can, many times, reveal defects in the drums. If the linings on one wheel are worn more than the others, it might indicate a rough drum. Uneven wear from side to side on any one set of shoes can be caused by a tapered drum. If some linings are worn badly at the toe or heel, it might indicate an out-of-round drum.

Thoroughly clean the drums with a water-dampened cloth or a water-based solution. Equipment for washing brake parts is commercially available. Wet cleaning methods must be used to prevent asbestos fibers from becoming airborne. If the drums have been exposed to leaking oil or grease, thoroughly clean them with a non-oil base solvent after washing to remove dust and dirt. It is important to determine the source of the oil or grease leak and correct the problem before reinstalling the drums.

Brake drums act as a heat sink; they absorb heat and dissipate it into the air. As drums wear from normal use or are machined, their cooling surface area is reduced and their operating temperatures increase. Structural strength will also be reduced, leading to overdistortion, a cause for some of the drum conditions covered here.

- *Scored Drum Surface (Figure 6-24A)*—The most common cause of this condition is buildup of brake dust and dirt between the brake lining and drum. A glazed brake lining, hardened by high heat or in some cases by very hard inferior grade brake lining can also groove the drum surface. Excessive lining

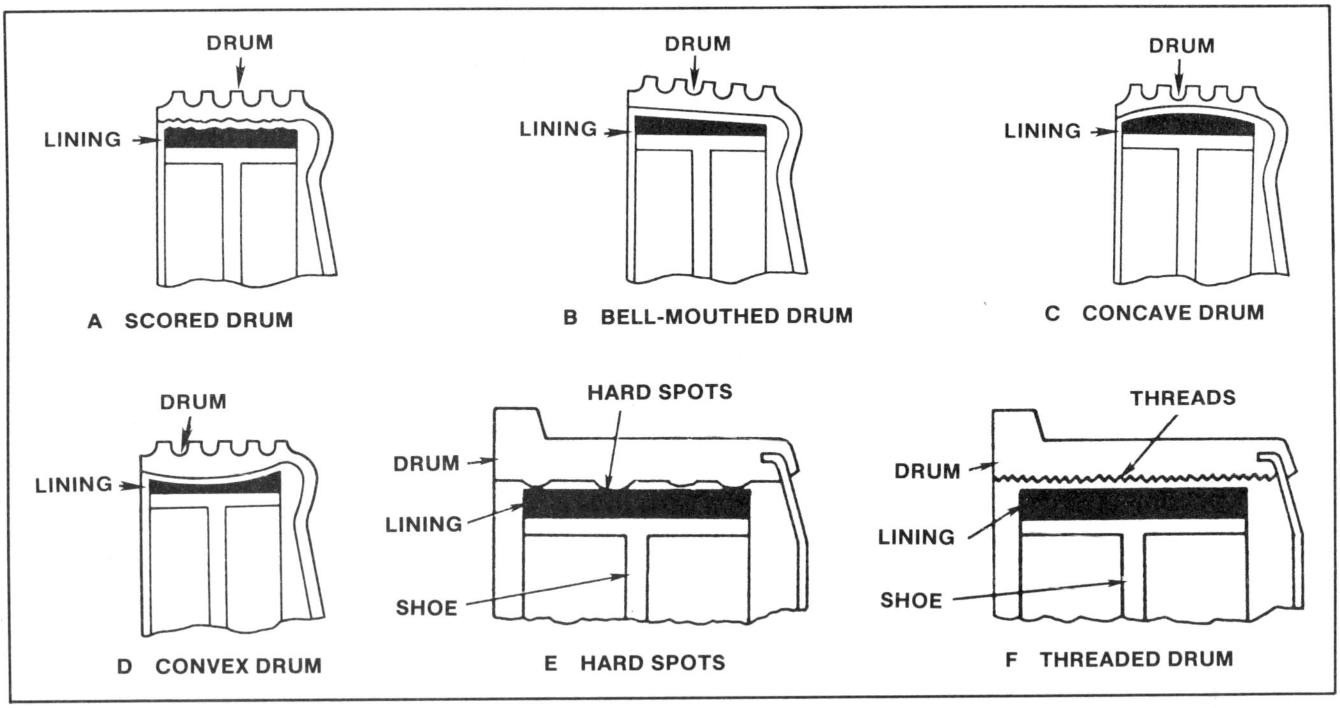

Figure 6-24 Drum conditions

wear that exposes the rivet head or shoe steel will score the drum surface. If the grooves are not too deep, the drum can be turned.

- *Bell-mouthed Drum (Figure 6-24B)*—A distortion due to extreme heat and braking pressure. It occurs mostly on wide drums and is caused by poor support at the outside of the drum. Full drum-to-lining contact cannot be achieved and fading can be expected. Drums must be turned.
- *Concave Drum (Figure 6-24C)*—This is an excessive wear pattern in the center area of the drum brake surface. Extreme braking pressure can distort the shoe platform so that braking pressure is concentrated at the center of the drum.
- *Convex Drum (Figure 6-24D)*—This wear pattern is greater at the closed end of the drum. It is the result of excessive heat or an oversized drum, which allows the open end of the drum to distort.
- *Hard Spots on the Drum (Figure 6-24E)*—This condition, sometimes called *chisel sports* or *island* of steel in the cast-iron surface, results from a change in metallurgy caused by braking heat. Chatter, pulling, rapid wear, hard pedal, and noise will occur. These spots can be removed by grinding. However, only the raised surfaces are removed, and they can reappear when heat is applied. The drum must be replaced.
- *Threaded Drum Surface (Figure 6-24F)*—An extremely sharp or chipped tool bit or a lathe that turns too fast can result in a threaded drum surface. This condition can cause a snapping sound during brake application as the shoes ride outward on the thread, then snap back. To avoid this, recondition drums using a rounded tool and proper lathe speed. Check the edge of the drum surface around the mounting flange side for tool marks indicating a previous rebore. If the drum has been rebored, it might have worn too thin for use. Check the diameter.
- *Heat Checks (Figure 6-25A)*—Heat checks will be visible, unlike hard spots that do not appear until the machining of the drum. Extreme operating temperatures are the major cause. The drum might also show a bluish/gold tint, which is a sign of high temperatures. Hardened carbide lathe bits or special grinding attachments are available through lathe manufacturers to service these conditions. Excessive damage by heat checks or hard spots will require drum replacement.
- *Cracked Drum (Figure 6-25B)*—Cracks in the cast-iron drum are caused by excessive stress.

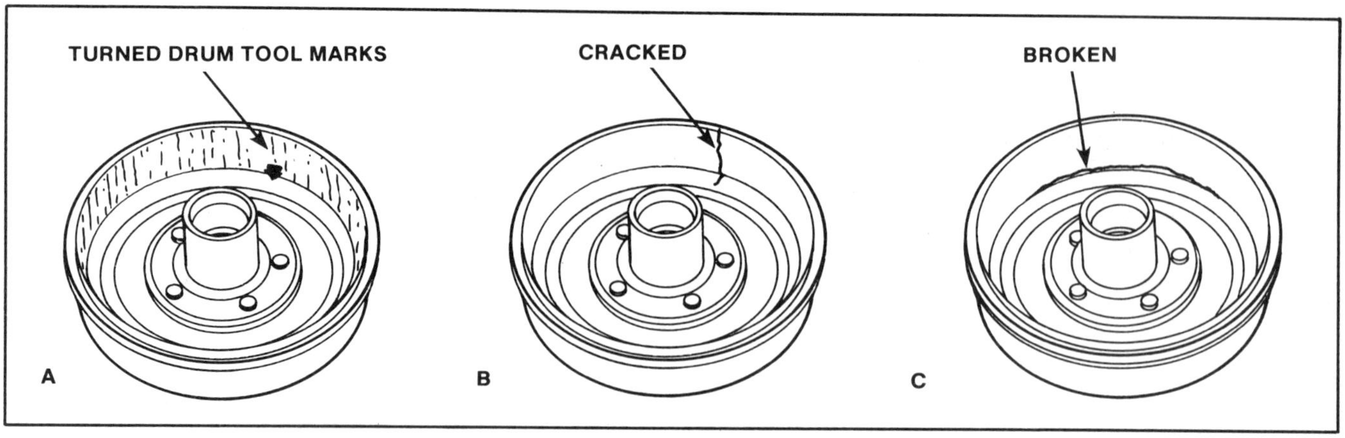

Figure 6-25 Heat checks

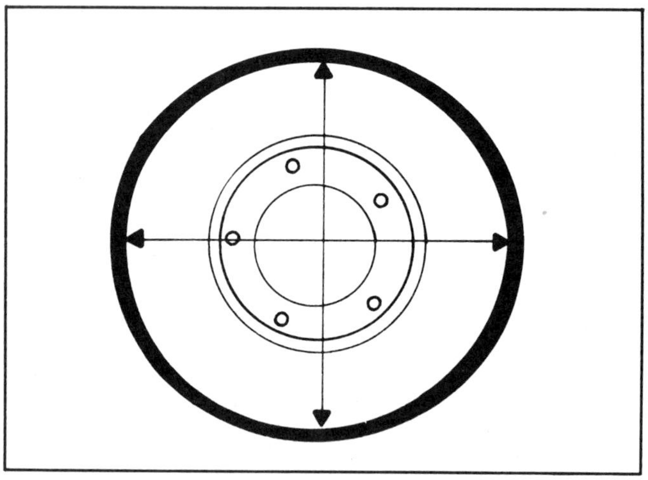

Figure 6-26 Out-of-round drum

They can be anywhere but usually will be in the vicinity of the bolt circle or at the outside of the flange. Fine cracks in the drums are often hard to see and, unfortunately, often do not show up until after machining. Nevertheless, should any cracks appear, no matter how small, the drum must be replaced. Of course, should the drum be broken (Figure 6-25C), it must be discarded.

- *Out-of-Round Drums (Figure 6-26)*—Drums with eccentric distortion might appear good to the eye but will cause pulling, grabbing, and pedal vibration or pulsation. An out-of-round or egg-shaped condition is often caused by heating and cooling during normal brake operation. Out-of-round drums can be detected before the drum is removed by adjusting the brake to a light drag and feeling the rotation of the drum by hand. After removing the drum, gauge it to determine the amount of eccentric distortion. Drums with this defect should be machined or replaced.

Drum Measurements

Measure every drum with a drum micrometer (Figure 6-27) even if the drum passed a visual inspection to make sure that it is within the safe oversize limits. If the drum is within safe limits, even though the surface appears smooth, it should be turned to assure a true drum surface and remove any possible contamination in the surface from previous brake linings, road dust, etc. Remember that if too much metal is removed from a drum, the following unsafe conditions can result:

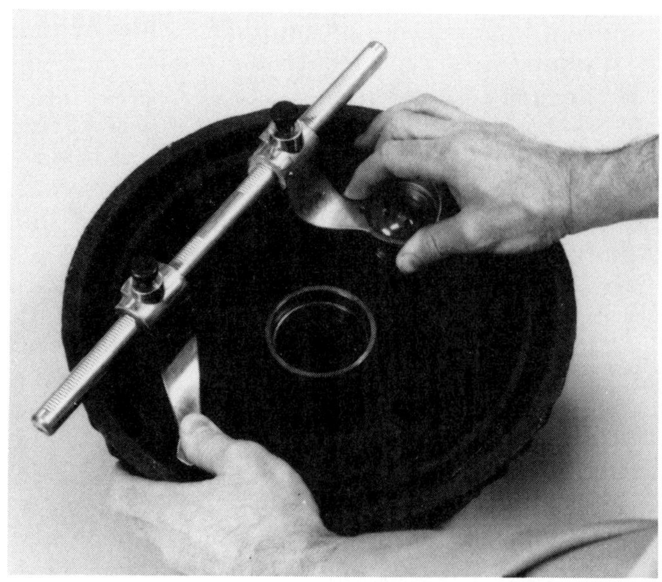

Figure 6-27 Measuring the inside diameter of a brake drum

Figure 6-28 Typical brake lathe (*courtesy of Ammco Tools, Inc.*)

- Brake fade due to the thin drum being unable to absorb the heat generated.
- Poor and erratic braking due to distortion of drum.
- Noise due to possible vibration caused by thin drum.
- A cracked or broken drum on a severe or very hard brake application.

Take measurements at the open and closed edges of the friction surface and at right angles to each other. Drums with taper or out-of-roundness exceeding 0.006 inch are unfit for service and should be turned or replaced. If the maximum diameter reading (measured from the bottom of any grooves that might be present) exceeds the new drum diameter by more than 0.060 inch, the drum cannot be reworked. If the drums are smooth and true but exceed the new diameter by 0.090 inch or more, they must be replaced.

If the drums are true, smooth up any slight scores by polishing with fine emery cloth. If deep scores or grooves are present that cannot be removed by this method the drum must be turned or replaced.

Drum Refinishing

Brake drums can be refinished either by turning (Figure 6-28) or grinding on a brake lathe. Best brake performance is obtained by turning drums with a very fine feed. Regardless of the method used, all work should be done by qualified personnel using reliable machinery.

Only enough metal should be removed to obtain a true, smooth friction surface. The exception is that when one drum must be machined to remove defects, the other drum on the same axle set must also be machined in the same manner and to the same diameter so that braking will be equal.

Brake drums manufactured after December 31, 1970, are stamped with a discard dimension (Figure 6-29). This is the allowable wear dimension and not the allowable machining dimension. There must be 0.030 inch left for wear after turning the drums. Typical drum diameters are shown in Figure 8-7. Some states have laws about measuring the limits of a brake drum.

Machining or grinding brake drums (Figure 6-30) increases the inside diameter of the drum and changes the lining-to-drum fit. When remachining a drum, follow the equipment instructions for the specific tool you are using. However, the following general rules of drum machining should always be kept in mind:

1. Remove the ridge on the inner and outer edge of the drum friction surface before turning to prevent damage to the cutting tool.
2. Always use the correct lathe adapters. These adapters must be in good condition.

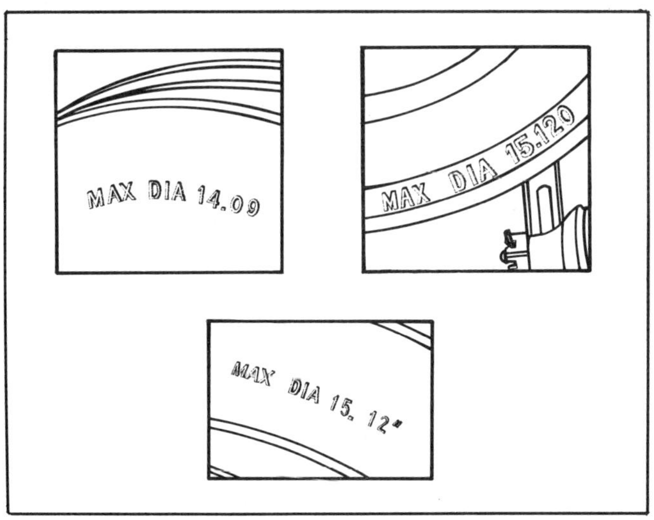

Figure 6-29 Brake drum discard diameters

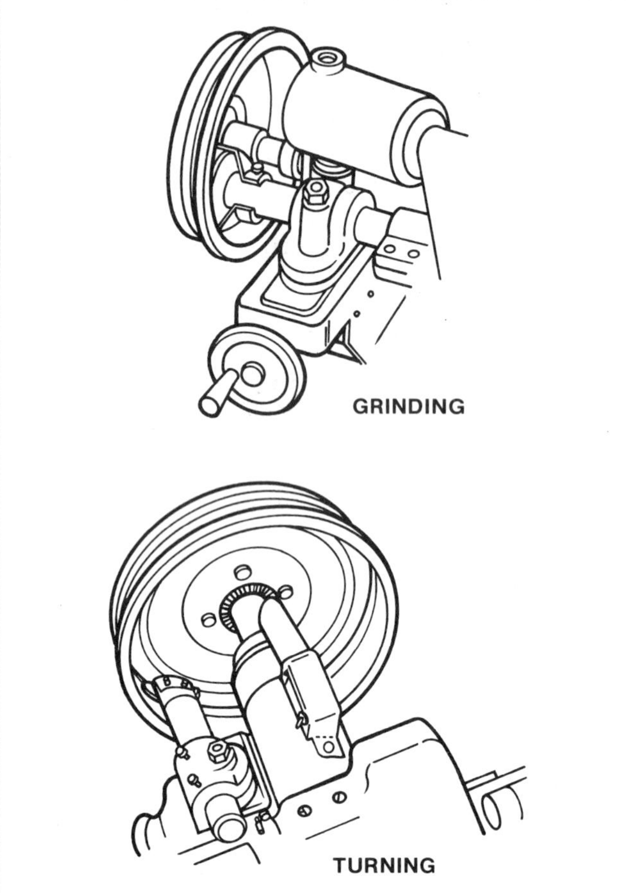

Figure 6-30 Brake drum remachining

3. Never grind a tool bit to a sharp point because this will leave threaded lines in the drum. The tip of the tool must be rounded.
4. With a hub-mounted drum, torque all wheel nuts to the same specifications before remachining (this applies whether the wheel is mounted or dismounted). Torqued nuts reinforce the drum, which reduces the possibility of distortion or chatter.
5. Some front drums can be removed without the hub. These drums must be turned with their individual hubs attached and not with a lathe adapter.
6. Remachine left-hand and right-hand drums in pairs—*do not* machine one side only.
7. Different makes of lathes permit finishing the drum surface by either a fine cut, grinding, or honing to remove spiral tool marks. If spiral marks remain, the brake shoes will be forced to thread out. This can cause faster lining wear, clinking noises as shoes snap back against the support pads and, more importantly, interference with shoe alignment during braking action.
8. If the reboring tool chatters over hard spots in the drum surface, grind the surface or discard the drum.
9. After turning, while the drum is still on the lathe, deburr it with No. 80 grit sandpaper (do not use emery cloth) to remove all minute rough and jagged surfaces.
10. Wash the turned drum with denatured alcohol. Never pick up the drum with your fingers on the newly turned surface.
11. Do not drop the drum because this can cause it to go out-of-round.
12. On drums with integral bearing hubs, check both the inner and outer bearings and races for damage. Replace, if necessary. If the bearings are in good condition, clean and repack them with the recommended grease. Always install new grease seals.
13. Studs on the drums must be in good condition or should be replaced if damaged or missing.
14. If the drum does not clean up when turned to the maximum rebored diameter, it must be replaced. The removal of more metal will affect the ability of the drum to dissipate heat and can cause drum distortion.

Some replacement brake drums are semi-finished and might require additional conditioning or machining to obtain the proper dimensional specifications and surface finish.

Full finished drums do not require additional machining unless it is necessary to match the diameter of an old drum on the same axle set. New drums are protected with a rustproofing coating that must be thoroughly cleaned off the friction surfaces. A volatile, non-oil base solvent, such as carburetor cleaner or lacquer thinner, should be used to remove all traces of oil or grease.

Cleaning Newly Refaced Drums

The friction surface of a newly refaced drum contains millions of tiny metal particles. These particles not only remain free on the surface, they always lodge themselves in the open pores of the newly machined surface. If the metal particles are allowed to remain in the drum, they will become imbedded in the brake lining. Once the brake lining gets contaminated in this manner, it acts as a fine grinding stone and scores the drum.

These metal particles must be removed by washing or cleaning the drum. *Do not* blow out the drum with air pressure. Either of the following methods is recommended to clean a newly refaced brake drum:

- Wash the brake drum thoroughly with hot water and wipe with a lint-free rag. Then, use the air pressure to thoroughly dry it. If the front hub and drums are being cleaned, be very careful to avoid contaminating the wheel bearing grease. Or, completely remove all the old grease, then regrease and repack the wheel bearing after the drum has been cleaned and dried. The wheel bearings and the grease seals *must* be removed from the drum before cleaning.
- Wipe the inside of the brake drum (especially the newly machined surface) with a lint-free white cloth dipped in one of the many available brake cleaning solvents that do not leave a residue. This operation should be repeated until dirt is no longer apparent on the wiping cloth. Allow the drum to dry before reinstalling it on the vehicle.

Both of these procedures are good for disc brake rotors, too (see Chapter 8).

Mounting a Refaced or New Drum

Clean the exposed hub or axle mounting flange using a coarse file. Clean up rivet holes with a rattail file. The wheel studs must have a tight press fit, and if a stud hole in the flange appears to be broached oversize, the flange must be replaced. Place a straightedge across the flat surface of the flange to make sure it is not bent. On the new drum, remove the rust-preventive sealer with a solvent and remove excess paint from the edge of the drum hub hole.

Drums with Threaded Fasteners. Position the drum on the flange and install the speed nuts or machine screws.

Drums with Riveted Fasteners. If the hub design uses fixed wheel studs, reriveting is unnecessary. Instead, position the drum and retain it by installing speed nuts on at least three studs. If the hub is designed for wheel lug bolts instead of studs, then the rivets must be reinstalled. One rivet head should have an extended shank that acts as a pilot to assist in remounting the wheel. Peen the rivets over using the rivet set in the press assembly fixture, or use a punch, anvil, and hammer.

Drums with Swaged Mounts. Some brake mechanics prefer to retain the new drum by installing speed nuts on the studs. Install at least three nuts. If the drum is retained in this manner and an extended stud shoulder prevents tightening a wheel nut against the drum, shave the shoulder or swage it down.

To insert the new studs, locate them in the flange holes. Align each stud with the press assembly fixture so that the stud threads will be protected by the fixture support, and press the stud home. To swage the drum in place, position the drum over the studs and temporarily snug it against the hub flange by placing oversize washers as they are reached.

Alternately, the studs can be pounded and drawn in, using the oversize washers and wheel nuts. Swage the stud shoulders with a punch and hammer.

Front Drum or Wheel Bolt Replacement. The following procedure can be used to replace one or more damaged wheel bolts. Or, if the drum must be replaced and the wheel hub is serviceable, this procedure can be used to replace all wheel bolts and the drum.

1. Secure the hub and drum assembly. Mark the center of the bolt head with a center punch. Drill a 1/8-inch pilot hole in the bolt head. Redrill a bolt head using a 9/16-inch bit. Cut off any remaining part of the bolt head with a chisel and drive out the bolt in the direction of the arrow using a suitable drift (Figure 6-31).
2. Use the procedure described in step 1 to remove any other damaged bolts.
3. If the drum is to be replaced, follow the procedure in step 1 to remove all bolts, and replace the drum.
4. Press the new bolt or bolts into the drum and hub as shown in Figure 6-31.

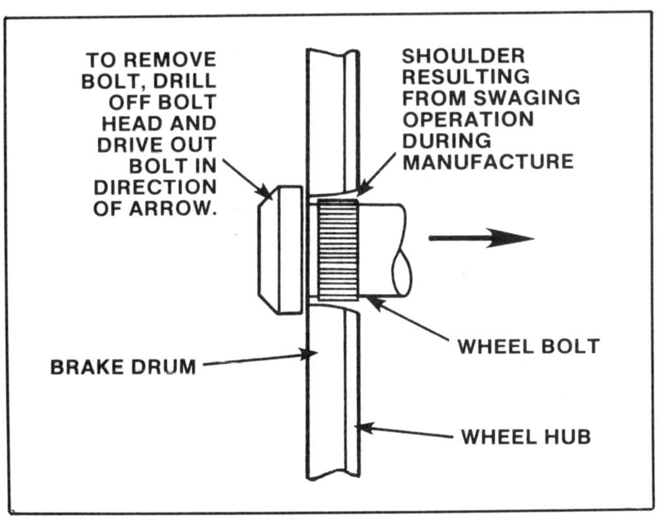

Figure 6-31 Front wheel bolt

BRAKE SHOES AND LININGS

Lining materials influence braking operation. The use of a lining with a friction value that is too high can result in a severe grabbing condition. A friction value that is too low can make stopping difficult because of a hard pedal. Other important lining characteristics are: limited sensitivity to humid or wet conditions, freedom from drum scoring, quiet and odor-free operation, and long-lasting operating stability with suitable wear life in the anticipated range of brake heat.

Overheating a lining accelerates wear and can result in potentially dangerous lining fade—a friction-reducing condition that hardens the pedal and lengthens the stopping distance. Continual overheating will eventually push the lining beyond the point of recovery into a permanent fade condition. In addition to fade, overheating can cause squeal.

Overheating is indicated by a lining that is charred or has a glass-hard glazed surface, or if severe, random cracking of the surface is present. Some lining is grooved above the shoe web (Figure 6-32). This prevents high-stress concentrations in that area to give a more uniform contact pressure.

These grooved linings are made extra wide and should not be field-grooved because the reduction in the contact area would lead to higher braking temperatures. Never burn-in a new lining because the surface must be burnished and seated while the lining is in use on the vehicle. Allow approximately 250 miles for the lining to seat itself effectively. Even a well-fitted lining can have a high spot that can float the shoe away from the drum and cause a rumbling sound. However, this kind of spot will wear off in a short time.

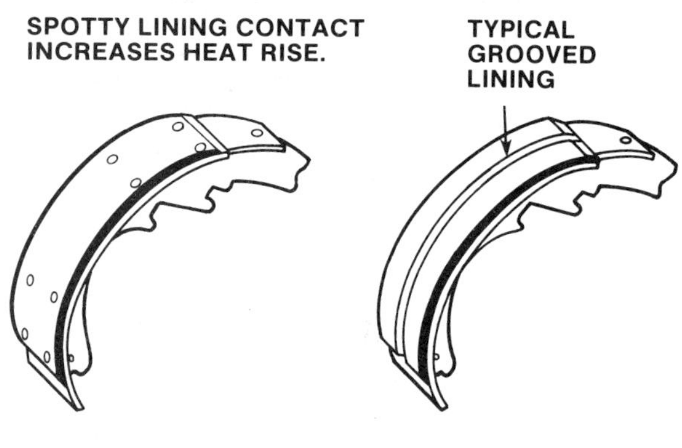

Figure 6-32 Brake linings

Figure 6-33 Checking the depth of the lining with a depth gauge

> **WARNING:** Automotive friction materials often contain substantial amounts of asbestos. Studies indicate that exposure to excessive amounts of asbestos dust can be a potential health hazard. It is important that anyone handling brake linings understands this and takes the necessary precautions to avoid injury. See Chapter 1 for a complete discussion of health protection procedures when working with asbestos.

Inspect the linings for uneven wear, imbedded foreign material, loose rivets, and to see if they are oil soaked. If linings are oil soaked, replace them.

If linings are otherwise serviceable, tighten or replace loose rivets, remove imbedded foreign material, and clean the rivet counterbores.

If linings at any wheel show a spotty wear pattern or an uneven contact with the brake drum, it is an indication that the linings are not centered in the drums. Linings should be circle ground to provide better contact with the drum.

Brake Relining

Brake linings that are worn to within 1/32 inch of a rivet head or have been contaminated with brake fluid, grease, or oil must be replaced (Figure

6-33). Failure to replace worn linings results in a scored drum. When it is necessary to replace linings, they must also be replaced on the wheel on the opposite side of the vehicle. Inspect brake shoes for distortion, cracks, or looseness. If these conditions exist, the shoe must be discarded. Do not attempt to service a damaged brake shoe.

When replacing linings, be sure that the shoes have been thoroughly cleaned and all burrs or rough spots have been removed from the shoe surface. Do not use shoes with elongated rivet holes. Install linings in a manner that will prevent gaps between the lining and shoe.

Do not let brake fluid, oil, or grease touch the brake lining. If a brake lining kit is used to replace the linings, follow the instructions in the kit and install all the parts provided.

As stated in Chapter 5, the two general methods of attaching the linings to the brake shoes are bonding and riveting (Figure 6-34). The bonded linings are fastened with a special adhesive to the shoe, clamped in place, then cured in an oven. Since bonding on linings requires special treatment, most auto shops use the riveting method to reline brakes.

Before riveting new linings and shoes, be sure shoe and lining contact faces are clean. Have tubular rivets of the correct body diameter, head size, length, and material ready for use.

1. Remove old linings by drilling out rivets. Punching out rivets will distort the shoe rim. File off any burrs or high spots.
2. Clamp the lining to the brake shoe with C-clamps so the rivet holes in both pieces are in alignment. Keep hands clean while handling brake linings.

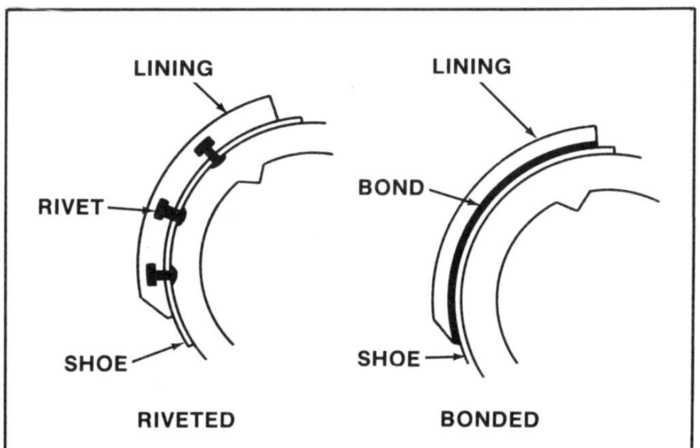

Figure 6-34 Riveted and bonded linings

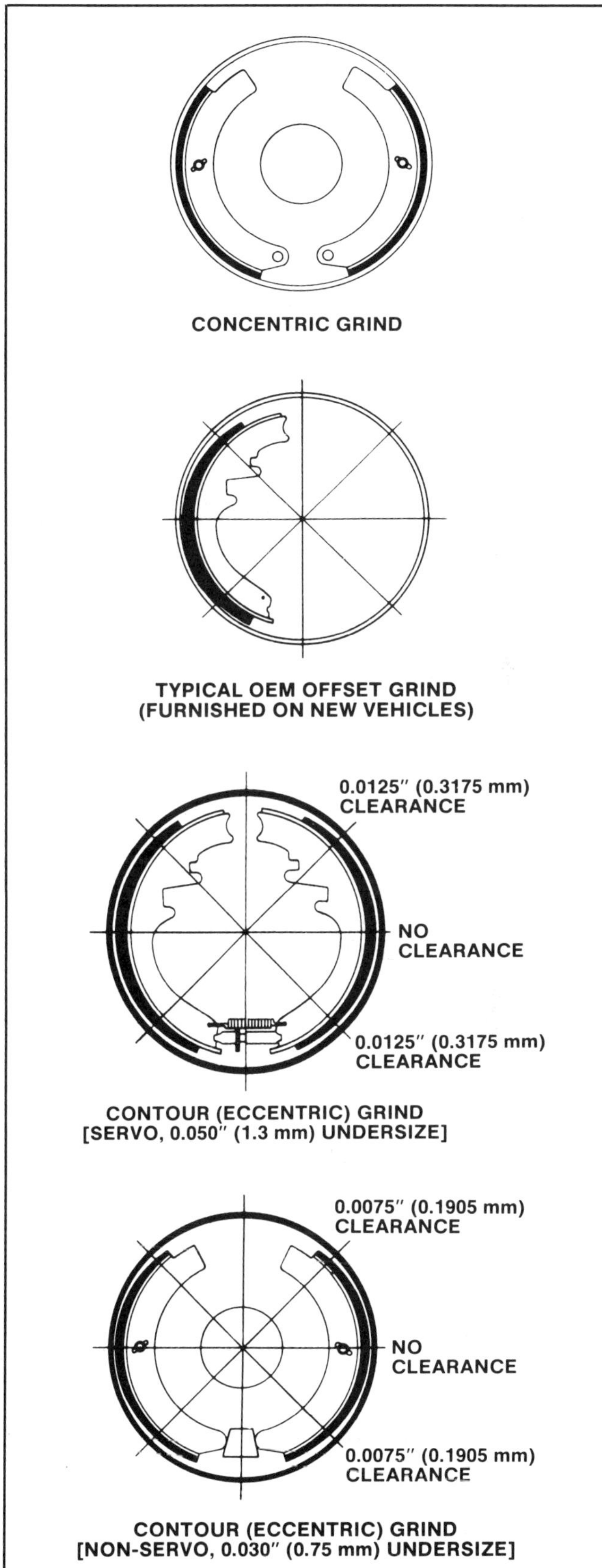

Figure 6-35 Various types of grinds

3. Locate the C-clamps as close as possible to the rivet holes; clamp firmly in position.
4. Start riveting at the center of the shoe and lining and work toward the ends.
5. Drive the rivets squarely into the holes with a flathead drift.
6. Form the rivet heads with the correct tubular rivet set. A roll set is recommended because a star set might split the tubular end and then the rivet will not fill the hole. Rivet holes in the riveted linings are countersunk about two-thirds of the lining thickness.
7. After riveting is completed, check the lining-to-shoe contact with a 0.008-inch feeler gauge. Measure midway between the rivets. It should not be possible to insert a 0.008-inch feeler gauge between the lining and shoe.

Sizing New Linings

Modern brake shoes are usually supplied with what is known as cam, offset, contour, or eccentric shape (Figure 6-35), which is ground in at the factory. That is, the full thickness of the lining is present at the middle of the shoe, but is ground down slightly at the heel and toe. The diameter of the circle the shoes make is slightly smaller than that of the drum. This compensates for the minor tolerance variations of drums, brake mountings, etc., and promotes proper wearing-in of the linings to match the drum. Since most automotive shoes are manufactured this way, the once-common practice of arcing the shoes in the field to match the drums has become unnecessary for the most part.

SHOP TALK

As mentioned in Chapter 5, on duo-servo shoe designs, the forward shoe is the primary and the rear, the secondary (Figure 6-36). Note that on the duo-servo drum brakes, the secondary shoe lining is longer.

Lining Adjustment

New eccentric-ground linings tolerate a closer new lining clearance adjustment than concentric ground linings. With manual adjusters, the shoes should be expanded into the drums until the linings are at the point of drag but not dragging heavily against the drum. With star wheel automatic adjusters, a drum/shoe gauge provides a convenient means of making the preliminary adjustment. This type of gauge, when set at actual drum diameter, automatically provides the working clearance of the shoes. If new linings have been concentrically ground, the initial clearance adjustment must be backed off an amount that provides sufficient working clearance.

Many service personnel take pride in showing the customer a high pedal, but it should be remembered that with new linings an extremely high pedal indicates tight clearances and can cause seating problems. For this reason, customers

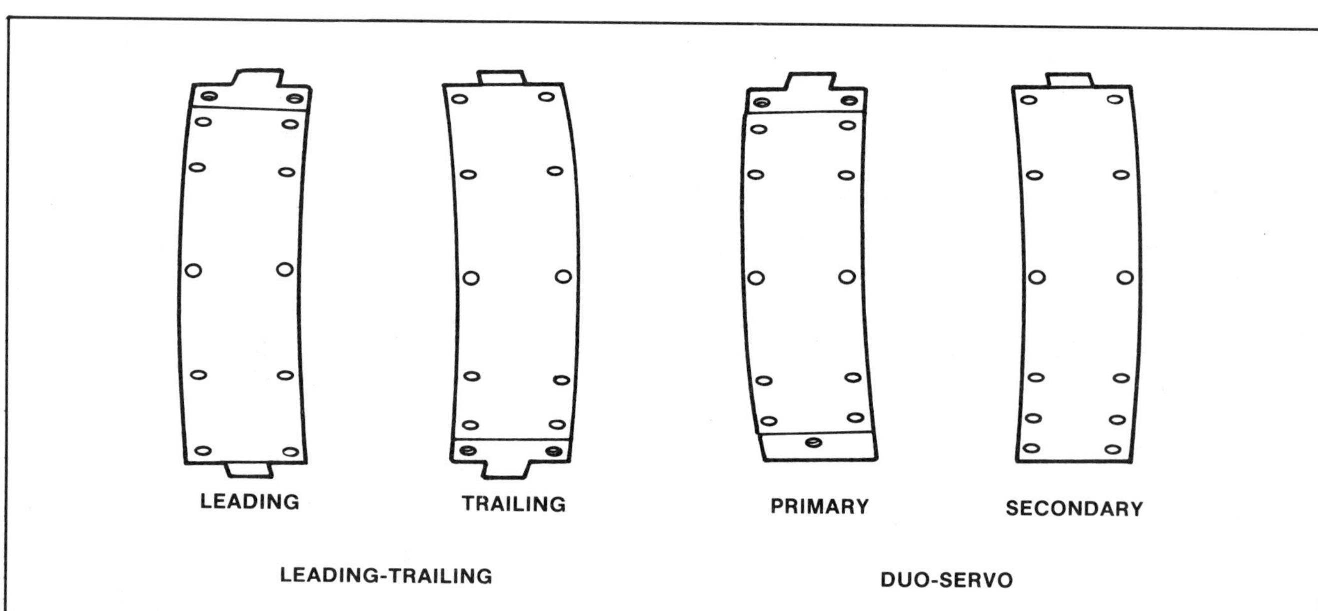

Figure 6-36 Brake shoes

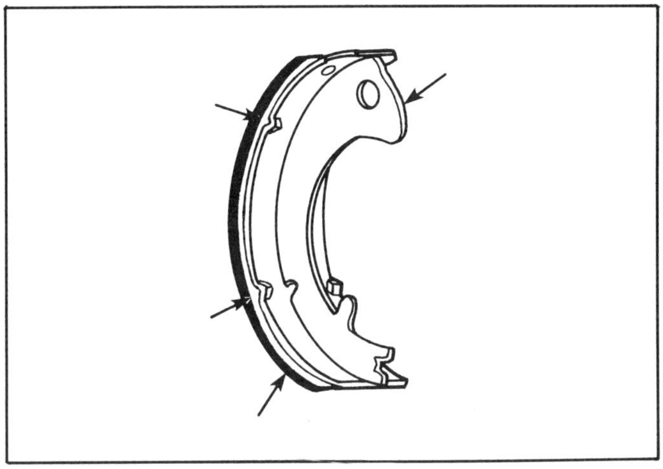

Figure 6-37 Dress down lining overhang, shoe table edge, and shoe web

should be asked to return for a free adjustment after 750 to 1,000 miles of driving.

WARNING: Some brake specialists have made a practice of converting automatic adjusters to manual adjusters. This is especially dangerous on dual (split) brake systems. The design of dual systems requires the smaller clearances provided by automatic adjusters. Larger clearances can result in excessive fluid requirements of the brakes. Periodic brake adjustment is no longer part of normal maintenance, so the chances of brakes becoming grossly out-of-adjustment if the self-adjusters are removed or inoperative are high.

DRUM SHOE AND BRAKE INSTALLATION

Reassemble the brakes in the reverse order of disassembly. Make sure all parts are in their proper locations and that both brake shoes are properly positioned in either end of the adjuster. Also, both brake shoes should correctly engage the wheel cylinder pushrods and parking brake links, and should be centered on the backing plate. Parking brake links and levers should be in place on the rear brakes. With all of the parts in place, try the fit of the brake drum over the new shoes. If not slightly snug, pull it off and turn the star wheel until a slight drag is felt when sliding on the drum. A brake preset gauge will make this job easy and final brake adjustment simple. Then install the brake drum, wheel bearings, spindle nuts, cotter pins, dust caps, and wheel/tire assemblies, and make the final brake adjustments as specified in individual instructions in the vehicle's service manual. Torque spindle and lug nuts to specifications.

Before installing the shoes, be sure that the following steps are performed:

- Sand or stone the inner edge of the shoe to dress down any slight lining or metal nicks and burrs that could interfere with the sliding upon the support pads (Figure 6-37).
- A support (backing) plate must be tight on its mount and not bent. Stone the shoe support pads brightly and dress down any burrs or grooves that could cause the shoes to bind or hang up (Figure 6-24).
- Using an approved lubricant, *lightly* coat the support pads and the threads of servo type star wheel adjusters. On rear axle parking brakes, lubricate any point of potential binding in the linkage and the cable. *Do not* lube non-servo type brake adjusters other than to free a frozen adjuster with penetrating oil (Figure 6-38).

To reassemble the typical duo-servo brake covered previously in the chapter and shown in Figure 6-15, proceed as follows:

1. Lay out the shoes on a workbench in the same relative position as they are to be in-

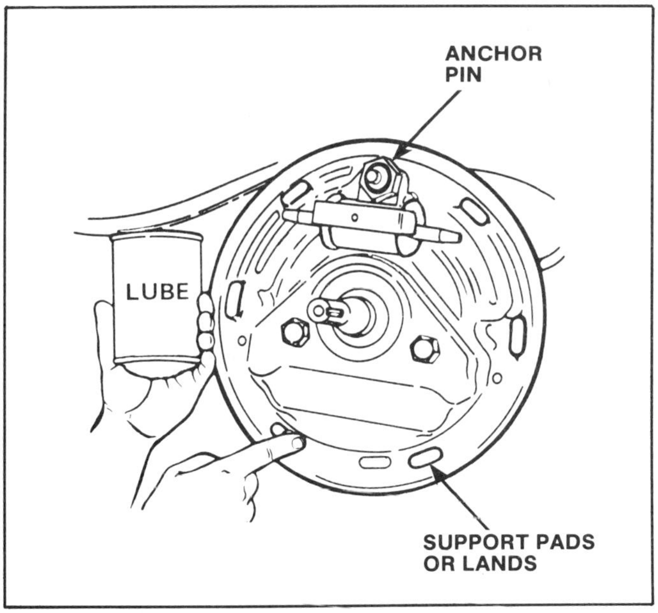

Figure 6-38 Lubricating the support pads or lands

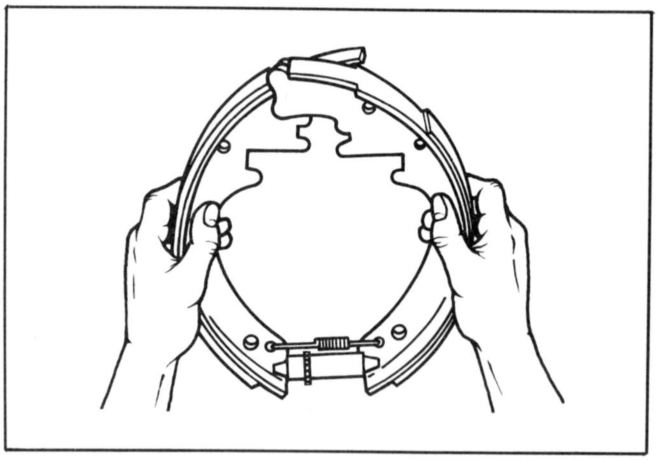

Figure 6-39 Shoe, spring, and adjusting screw assembly

stalled. Hook the adjusting screw spring between the shoes. Overlap the anchor ends of the shoes and engage the adjusting screw assembly between them (Figure 6-39).

CAUTION: Make sure that the adjusting screw spring is installed with its longer free end over the star wheel of the adjusting screw assembly (Figure 6-40). Otherwise, the spring coils will interfere with the operation of the star wheel.

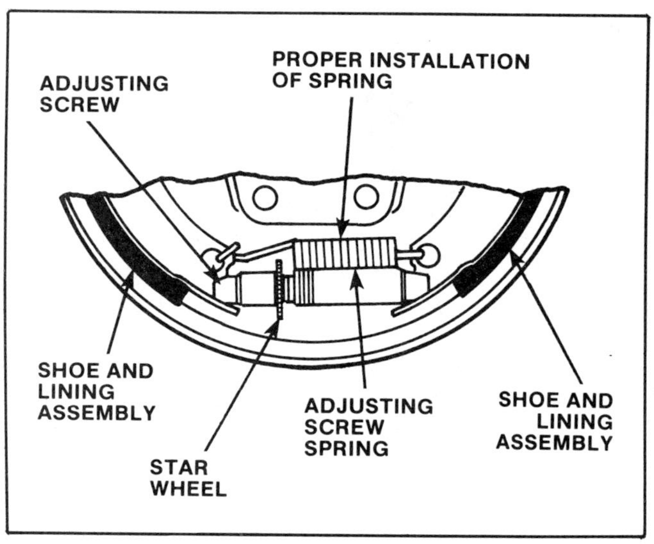

Figure 6-40 Proper installation of adjusting screw spring

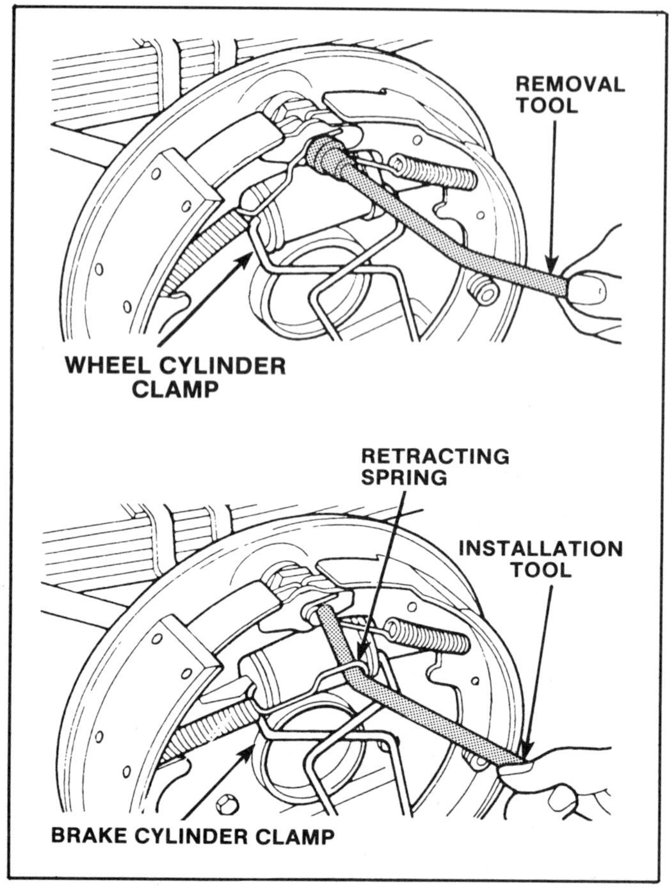

Figure 6-41 Spring replacement

2. Spread the anchor ends of the shoes to retain the adjusting screw in its installed location. Position the shoe assembly on the brake backing plate so that the shoes engage the wheel cylinder connecting links or piston extensions. If removed, install the anchor plate on the anchor pin over the shoe webs.
3. On rear brakes, spread the shoes slightly and install the parking brake strut and spring. Connect the parking brake lever to the secondary shoe.
4. Install the hold-down parts in the primary shoe. Insert the hold-down pin through the backing plate and shoe and hold in position. Grasp the hold-down spring and cup with pliers; position over the pin. Press down and twist one-quarter turn to lock in place.
5. Install the adjusting mechanism on the secondary shoe of self-adjusting brakes. Hook one end of the adjusting link over the anchor pin or anchor plate and the other to the lever and pivot assembly. Position the lever on the secondary shoe with the lever return spring

in position between the lever and the shoe. Install the sleeve. Reinstall the secondary hold-down pin, spring, and cup as described in step 4.

6. Install the shoe return springs (Figure 6–41). Hook the springs into the shoes first. Depending on the attachment method, hook the other end of the return springs as follows:
   - Over the anchor pin, using a brake spring installation tool
   - Onto the bolted-on type anchor plate
   - On some self-adjusting brakes, the secondary return spring into the hook on the adjusting link

CAUTION: Do not use a brake spring tool to remove or install the adjusting link to the anchor pin.

7. Make preliminary manual adjustment of self-adjusting brakes, install drums and wheels, and make final adjustment of both service and parking brakes.

To reassemble the non-servo brake with automatic adjuster illustrated in Figure 6–20, proceed as follows:

1. Apply a light coat of high-temperature lubricant to the shoe support points on the backing plate, to the adjuster lever and latch pins, and to the anchor.
2. Assemble the levers and latch on the shoes with pins and retainers. The reverse shoe, mounted toward the rear of the car, can be identified by its narrower lining.
3. Lubricate the inner surface of the shoe hold-down clips and assemble the clips on the backing plate with the short end of the clips in the slot in the plate.
4. Connect the cable to the parking brake lever and attach the lower shoe-to-shoe spring to the shoes.
5. Position the shoes on the backing plate and slide the shoe webs into the hold-down clips. Be careful not to disengage the clips from the backing plate.
6. Install the parking brake strut with the tab on the strut engaging slot of the adjuster lever.
7. Place the upper end of the shoe webs against the wheel cylinder and install the upper shoe-to-shoe spring. Remove the wheel cylinder clamp.
8. Attach the automatic adjuster spring to the reverse shoe and parking brake strut.
9. Attach the anti-noise spring between the spring bracket and reverse shoe.

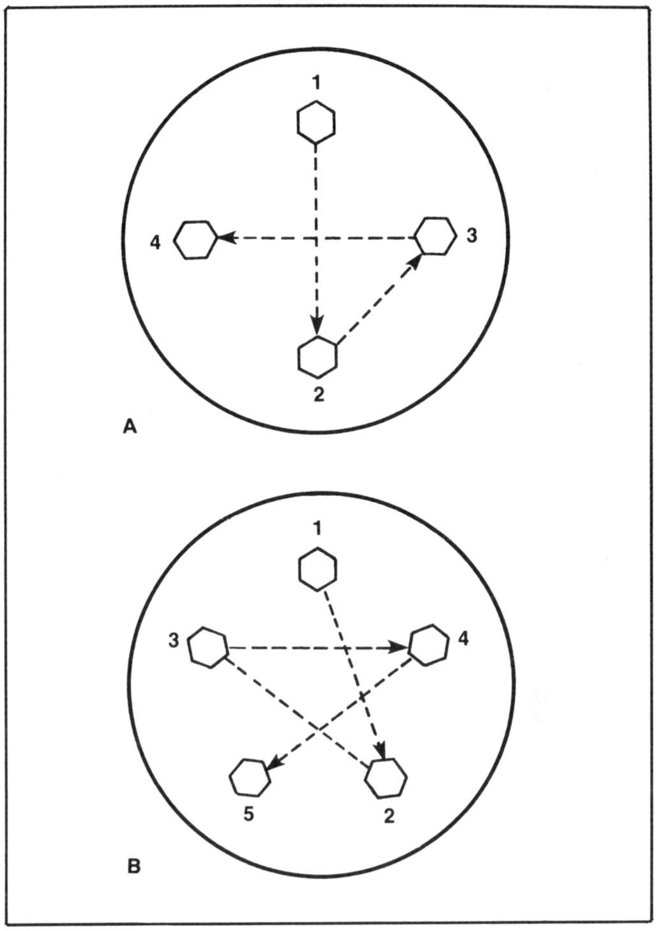

Figure 6-42 (A) Lug nut tightening sequence for a four-bolt wheel. (B) Lug nut tightening sequence for a five-bolt wheel.

Make a number of alternate forward and reverse stops while applying firm pressure to the brake pedal on each stop. This should bring the shoes into proper adjustment.

Once the drum brake is reassembled, replace the wheel/tire assembly on it. To do this, the wheel lugs or nuts should be tightened to a specified torque (see service manual) in the proper sequence (Figure 6–42) to avoid drum distortion. Many brake technicians use an impact wrench to run up the lug nuts snugly. Then, when the car is lowered to the floor, they use a torque wrench for the final tightening.

Once the vehicle is on the ground, check and adjust the air pressure in all the tires. Remember that an improper amount of air in the tires can present braking problems.

Road Test

As a final brake drum check, road test the vehicle as described in Chapter 14. But, before doing this, bleed the brakes as detailed in Chapter 3 to make sure of a high, hard brake pedal. Most self-adjusting mechanisms are activated only during the rearward motion of a car. So, whenever servicing self-adjusting brakes, make sure that the road test includes enough stops traveling in reverse to allow the self-adjusters to perform the proper balancing of the drum brakes. Or, operate the parking brake several times in designs where that activates the automatic adjuster.

WHEEL CYLINDER INSPECTION AND SERVICING

Wheel cylinders might need reconditioning or replacement when the brake shoes are replaced or when they begin to leak. On many designs, the wheel cylinders can be disassembled without removing them from the backing plate. In some types, however, the cylinder is mounted in a recess in the backing plate, or a cylinder piston stop is welded to the backing plate. When servicing this type, the cylinder must be removed from the backing plate before being disassembled.

INSPECTING AND CLEANING WHEEL CYLINDERS

Any hydraulic leak is dangerous. Those at a wheel cylinder reveal themselves in several ways: (1) fluid can be found when the dust boot is peeled back; (2) the cylinder, linings, and backing plate, or the inside of a tire might be wet; or (3) there might be a drop in the level of fluid in the master cylinder reservoir.

Such leaks can cause the brakes to grab or fail and should be immediately corrected. Note the amount of fluid present when the dust boot is pulled back. A small amount of fluid seepage dampening the interior of the boot is normal; a dripping boot is not. Unless other conditions causing a brake to pull, grab, or drag become obvious, the wheel cylinder is suspect and should be included in general reconditioning. If fluid has come in contact with the brake lining, the lining should be replaced on both sides of the car.

CAUTION: Hydraulic system parts should not be allowed to come in contact with oil or grease; neither should they be handled with greasy hands. Even a trace of any petroleum-based product is sufficient to cause damage to the rubber parts.

Cylinder binding can be caused by rust deposits, swollen cups due to fluid contamination, or by a cup wedged into an excessive piston clearance. If the clearance between the pistons and the bore wall exceeds allowable values, a condition called *heel drag* might exist. It can result in rapid cup wear and can cause the piston to retract very slowly when the brakes are released.

Typical examples of a scored, pitted, or corroded cylinder bores is a ring of hard, crystal-like substance is sometimes noticed in the cylinder bore where the piston rests after the brakes are released.

Light roughness or deposits can be removed with crocus cloth or an approved cylinder hone. While honing lightly, brake fluid can be used as a lubricant. If the bore cannot be cleaned up readily, the cylinder must be replaced.

CAUTION: Aluminum wheel cylinders must not be honed. Also, wheel cylinders equipped with baffles between their cups cannot be honed.

At the time of reline, wheel cylinders should be considered a part of that reline. The wheel cylinders should be replaced as a unit or reconditioned with new internal working parts. At first, inspection might not reveal any problems with the wheel cylinders. However, due to the installation of new lined shoes of full thickness, the repositioning of the cups and pistons deep into the cylinder bores might create a hydraulic failure during or soon after the reline. Contamination, cylinder bore pitting, etc., cannot be found without completely disassembling the working parts. This contamination would begin to destroy the lips of the cups once the pistons are repositioned.

Some vehicles are equipped with wheel cylinder piston stops, which protrude from the backing plate and keep the pistons from popping out of the cylinder accidentally during service. It is necessary to remove the cylinder from the backing plate far enough to clear these stops if the cylinder is to be overhauled. Do not try to force the seals, pistons, and boots around the piston stops with the cylinder bolted in place on the backing plate.

Replacing a Wheel Cylinder

It is a common practice to recondition a drum brake wheel cylinder without dismounting it; however, some brakes are equipped with external piston stops that prevent disassembly unless the cylinder is removed. In order to do so, remove the shoe springs and spread the shoes apart, disconnect the brake line, remove the mounting bolts or retaining clips, and pull the cylinder free.

To replace most wheel cylinders, proceed as follows:

1. Since brake hoses are an important link in the hydraulic system, it is recommended that they be replaced when a new cylinder is to be installed or when the old cylinder is to be reconditioned. Remove the brake shoe assemblies from the backing plate before proceeding because the smallest amount of brake fluid will contaminate the friction surface of the brake lining.
2. On front axles, remove the lock clip that holds the brake hose to the chassis bracket. This is a small U-shaped piece found at the end of the brass ferrule of the hose closest to the chassis.
3. Using two appropriate wrenches, disconnect the hydraulic hose from the steel line located on the chassis. On solid rear axles, use the appropriate tubing wrench and disconnect the hydraulic line where it enters the wheel cylinder. Care must be exercised in removing this steel line because it might be bent at this point and difficult to install once new wheel cylinders are mounted to the backing plate.
4. Remove the plates, shims, and bolts that hold the wheel cylinder to the backing plate (Figure 6-43). Some later designed wheel cylinders are held to the backing plate with a retaining ring (Figure 6-44) that can be removed with two small picks.
5. Remove the wheel cylinder from the backing plate and clean the area with a proper cleaning solvent.
6. On front brakes, install a new hose, making sure that the small copper washer is installed between the hose and wheel cylinder casting. Pass the hose through the hole in the backing plate and reinstall the anchor bolts or retaining ring.
7. Connect the hose to the steel tubing by starting the tubing nut into the hose with your fingers to avoid the possibility of stripping the threads. Tighten this connection with sufficient torque to prevent leakage.
8. Replace the small horseshoe lock that clamps this hose to the chassis bracket.
9. Replace the brake shoes and bleed the system as described in Chapter 13.

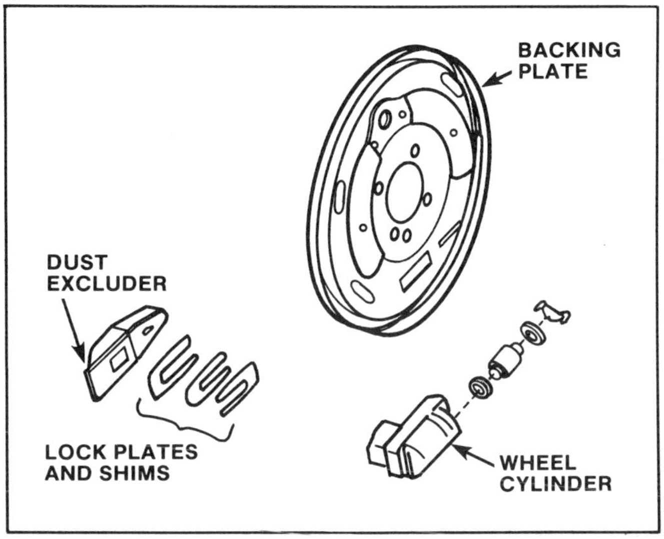

Figure 6-43 Lock plates and shims hold the wheel cylinder to the backing plate

Reconditioning a Wheel Cylinder

It is possible to recondition a wheel cylinder using the following procedure:

1. Remove the brake shoe assemblies from the backing plate.

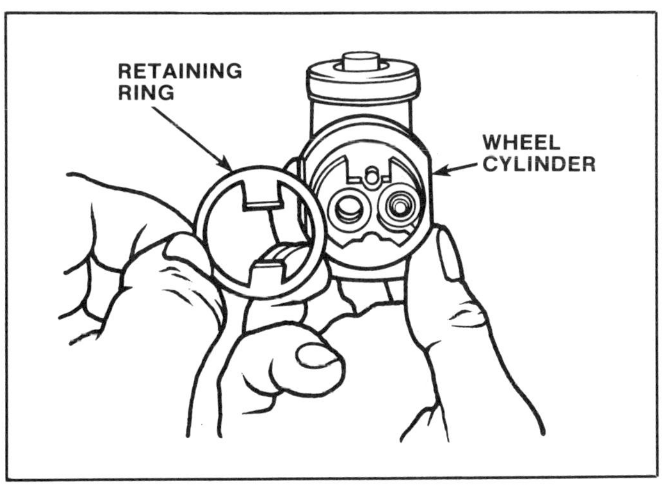

Figure 6-44 Wheel cylinder retaining ring

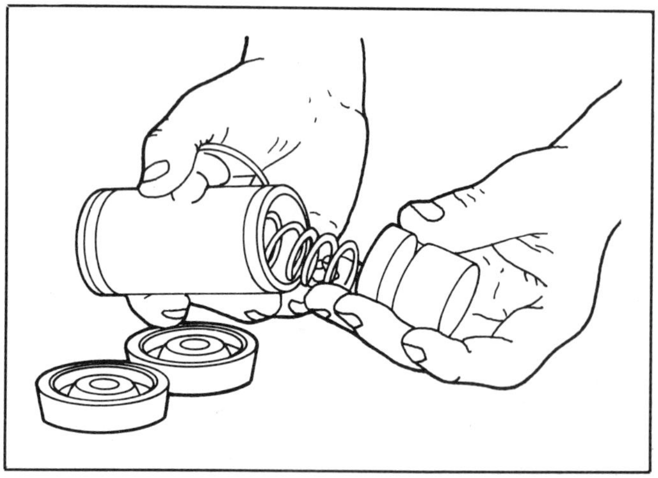

Figure 6-45 Disassembling the wheel cylinder

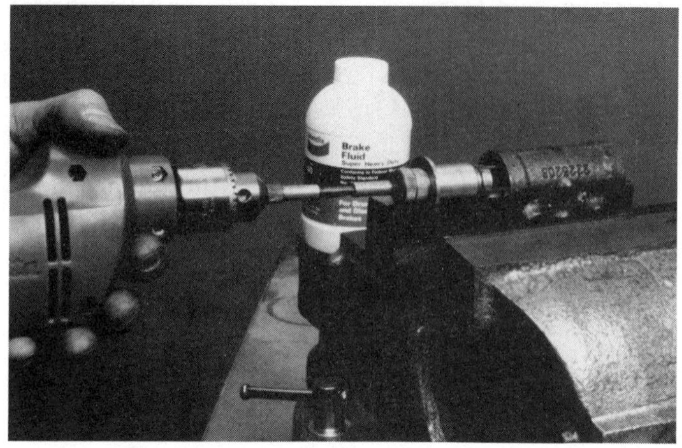

Figure 6-46 Wheel cylinder removed for disassembly and honing

2. Clean the backing plate and wheel cylinder. It is not necessary to disconnect the hydraulic brake hose or tubing. Make sure the bleeder screw can be opened. If it is seized so that it breaks off when turned, the wheel cylinder must be replaced.
3. If it is impossible to remove the internal parts of the wheel cylinder because of piston stops, it will be necessary to unbolt the casting.
4. Remove the dust boots, pistons, cups, and spring (Figure 6-45). If a fluid baffle is present, it can be driven out with a drift so that the cylinder can be honed. But since baffles serve a purpose, it is better to replace cylinders that have them.
5. Clean the entire cylinder and, in particular, cylinder bore. Use only denatured alcohol or fresh brake fluid, never petroleum-based solvents.
6. Hone the cylinder bore at low RPM with a steady back-and-forth motion (Figure 6-46). Dip hone stones in brake fluid before starting the procedure—never hone a dry cylinder.
7. Clean the newly honed cylinder bore with denatured alcohol and dry with a lint-free cloth.
8. Check for proper piston-to-bore clearance. The piston and a 0.006-inch feeler gauge should not be able to slip into the bore together (Figure 6-47). If they do, replace the wheel cylinder.
9. Inspect the cylinder bore for pitting, rusting, cracks, burrs, and any other condition that might be detrimental to its operation. If any of these conditions are found, a new wheel cylinder must be installed.
10. Apply assembly fluid to the cylinder bore walls.
11. Liberally coat the rubber cups with assembly fluid and insert a cup into one end of the cylinder bore. Make sure that the sealing lip is facing the inside of the cylinder bore.
12. Coat the pistons with assembly fluid. Install one piston with its solid end against the solid end of the cup. Note that the other end of the piston is designed to accommodate the brake shoe web or pin.
13. Install the dust boot.

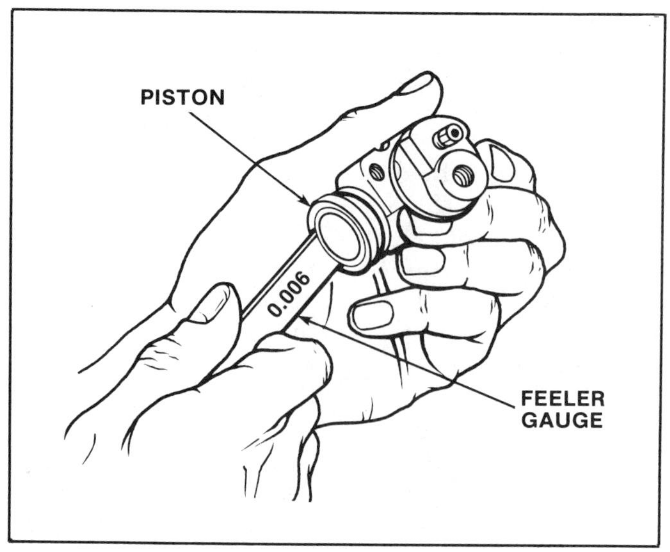

Figure 6-47 Checking the wheel cylinder bore for excessive wear

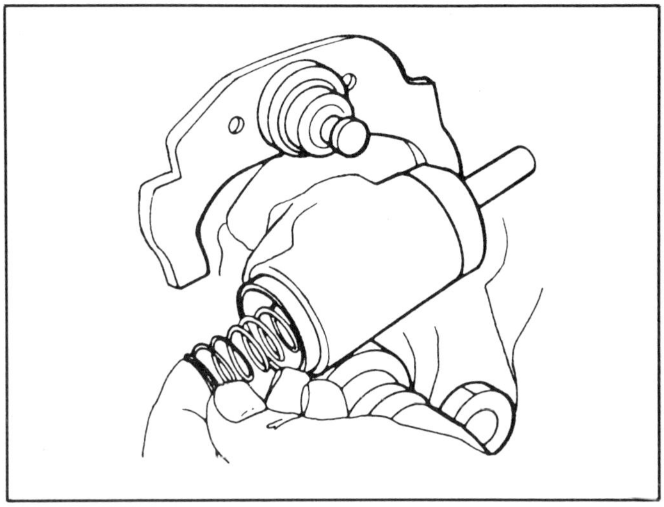

Figure 6-48 Installing the spring

14. Insert the spring (Figure 6-48) and install the remaining cup, piston, and dust boot in the same manner as previously described.
15. Install and tighten any bolts or retainer rings that were removed or loosened prior to assembly with a 1-1/8-inch socket.
16. Bleed the system as described in Chapter 3.

CAUTION: Do not attempt to rebuild aluminum cylinders.

As mentioned earlier, some cars are equipped with wheel cylinder piston stops (Figure 6-49).

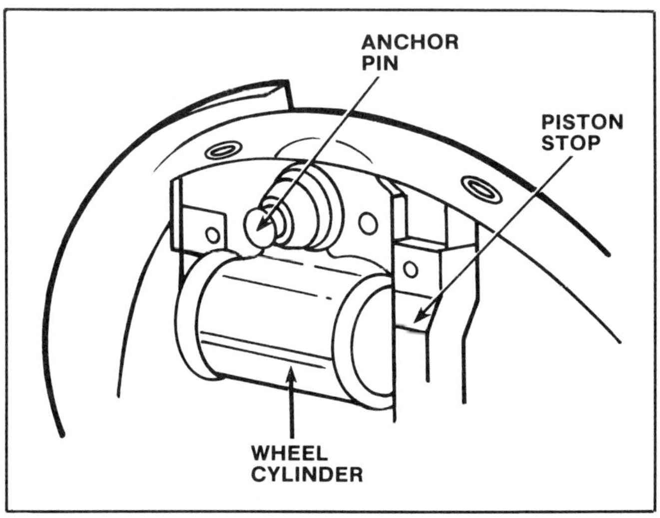

Figure 6-49 Wheel cylinder piston stops

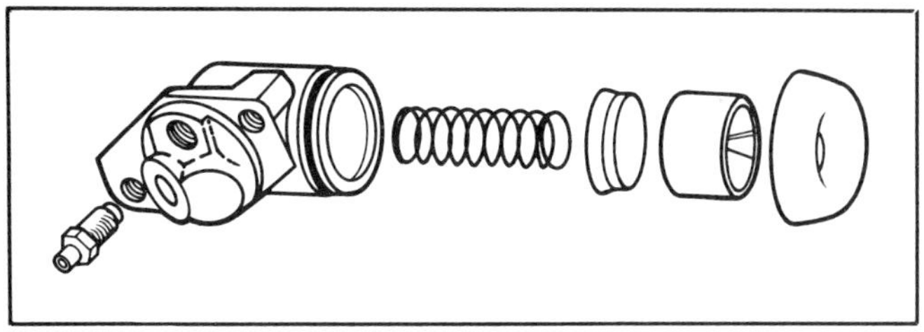

Figure 6-50 Single-end cylinder

These stops prevent the piston from accidentally coming out when the brake drums are removed. Care must be taken when installing new or reconditioned wheel cylinders on cars equipped with wheel cylinder piston stops. The rubber dust boots and the pistons must be squeezed into the cylinder before it is tightened to the backing plate. If this is not done, the pistons will jam against the stops causing hydraulic fluid leaks and erratic brake performance.

Single-Piston Sliding Wheel Cylinder. As mentioned previously in this chapter, some import drum brake systems use a single-piston sliding wheel cylinder (Figure 6-50). The wheel cylinder is secured to the backing plate with shims and spring steel lock plates. When the brakes are applied hydraulic pressure forces the piston outward, applying the forward brake shoe. For the reverse shoe to be applied, the wheel cylinder body slides rearward in its mounting slot.

When servicing this brake design, the wheel cylinder should be removed and the backing plate cleaned. The slot for the wheel cylinder should be lubricated along with the cylinder body. Once installed, the wheel cylinder should move freely, by hand, in the backing plate.

Servo drum systems that use a single-piston wheel cylinder instead of a dual-piston are sometimes called a "uni-servo" brake design. The wheel cylinder is usually mounted so that brake application force from the wheel cylinder acts only on the primary shoe. Other than that, the self-energization and servo action increase breaking force in the same manner as in the duo-servo design.

REVIEW QUESTIONS

Review the information in this chapter by answering the following questions.

1. In the course of removing a brake drum to check the lining, grease is found at the center of the brake assembly. Mechanic A says that this means the wheel cylinder must be rebuilt

or replaced. Mechanic B says that this means the axle seal must be replaced. Who is right?
   a. Mechanic A
   b. Mechanic B
   c. both A and B
   d. neither A nor B

2. What should be used to remove a frozen wheel?
   a. hot-air gun
   b. cutting torch
   c. crowbar
   d. wheel puller
3. Demountable front drums are found on
   a. late-model Chevrolets.
   b. late-model Fords.
   c. Ford Pintos and Bobcats.
   d. most of today's domestic vehicles.
4. When the original drum is mounted against a wheel hub flange, it can be fastened by
   a. swaging the piloting shoulders of the wheel studs.
   b. riveting.
   c. using speed nut fasteners.
   d. all of the above.
5. When adjusting the brake shoes on a vehicle with self-adjusting mechanisms, the self-adjusting lever should be pushed away from the star wheel a
   a. minimum of 1/16 inch.
   b. maximum of 3/16 inch.
   c. maximum of 1/16 inch.
   d. maximum of 3/8 inch.
6. Backing plates, struts, levers, and other metal brake parts should be
   a. wet-cleaned using water or a water-based solution.
   b. wet-cleaned using an alcohol-based solvent.
   c. wet-cleaned using gasoline.
   d. dry-cleaned only.
7. In the course of inspecting a drum brake, a slight amount of fluid is discovered behind the internal boots. Mechanic A says this is normal. Mechanic B says the boots must be replaced. Who is right?
   a. Mechanic A
   b. Mechanic B
   c. both A and B
   d. neither A nor B
8. Except for the drum itself, what is the final step in a drum brake inspection?
   a. Apply a thin film of brake lubricant to the raised shoe pads on the backing plate lands.
   b. Check the springs for spread or collapsed coils, twisted or nicked shanks, and severe discoloration.
   c. Check the wheel bearings and hub grease seals for damage.
   d. None of the above.
9. A buildup of brake dust and dirt between the lining and the drum is the most common cause of
   a. concave/barrel-shaped drum.
   b. convex/tapered drum.
   c. threaded drum surface.
   d. scored drum surface.
10. Heat checks are caused by
    a. excessive stress.
    b. extreme operating temperatures.
    c. heating and cooling during normal brake operation.
    d. a lathe that turns too fast.
11. A drum with a taper or out-of-roundness of 0.006 inch is
    a. unacceptable and should be turned.
    b. unacceptable and must be replaced.
    c. borderline, and probably should be turned.
    d. acceptable.
12. Brake drums can be refinished by
    a. grinding.
    b. turning.
    c. grinding or turning.
    d. none of the above.
13. Which of the following statements concerning brake linings is incorrect?
    a. Lining should never be grooved above the shoe web.
    b. Overheating can be indicated by a charred lining.
    c. The use of a lining with a friction value that is too high can result in a severe grabbing condition.
    d. Never let brake fluid, oil, or grease touch the brake lining.
14. Non-servo type brake adjusters
    a. can be lubed as often as needed.
    b. should be lubed only as a means of freeing a frozen adjuster.
    c. should not be lubed more than once every 5,000 miles.
    d. should never be lubed.

CHAPTER 7

DISC BRAKES— OPERATION

OBJECTIVES

Upon completing this chapter, you should be able to

- Explain the advantages of disc brakes over drum brakes.
- Recognize the difference between a floating and a sliding caliper.
- Explain how a disc brake hub and rotor assembly, caliper assembly, and brake pad assembly operates.
- Describe the basic operation and applications of fixed caliper disc brakes.
- Describe the basic operation and applications of floating and sliding caliper disc brakes.
- Explain how a rear disc/parking brake system operates.

There are many reasons for the popularity of disc brakes. As mentioned in Chapter 1, they can handle much more heat than drum brakes and so are not as apt to fade during heavy braking. They have a large tolerance for water; therefore, they will stop the vehicle even in flood conditions. They stay in adjustment automatically and are generally easy to reline.

When disc brakes first appeared on American cars in 1965, the four-piston caliper (Figure 7-1) was used almost exclusively. Two years later this caliper design began to be replaced by the single-piston type (Figure 7-2). Currently, most domestic vehicles have front discs and rear drums, although many foreign and some special domestic cars have discs on all four wheels. Disc brakes can generate

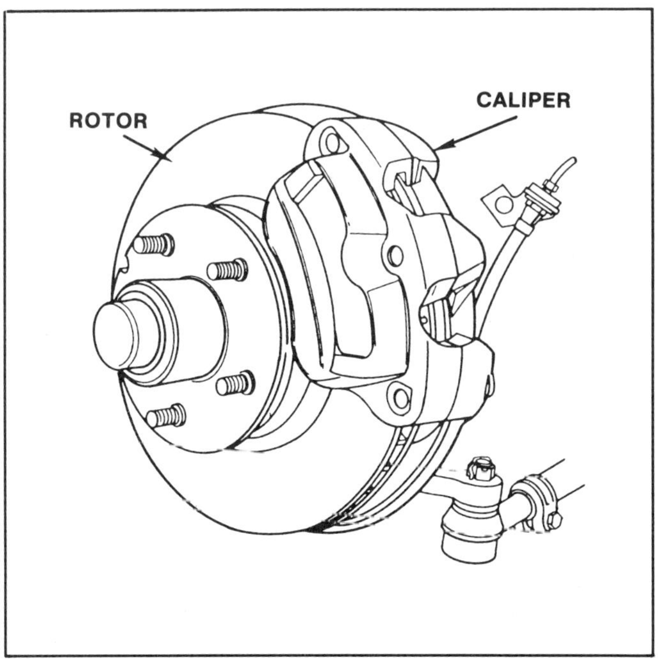

Figure 7-1 Typical four-piston fixed caliper disc brake

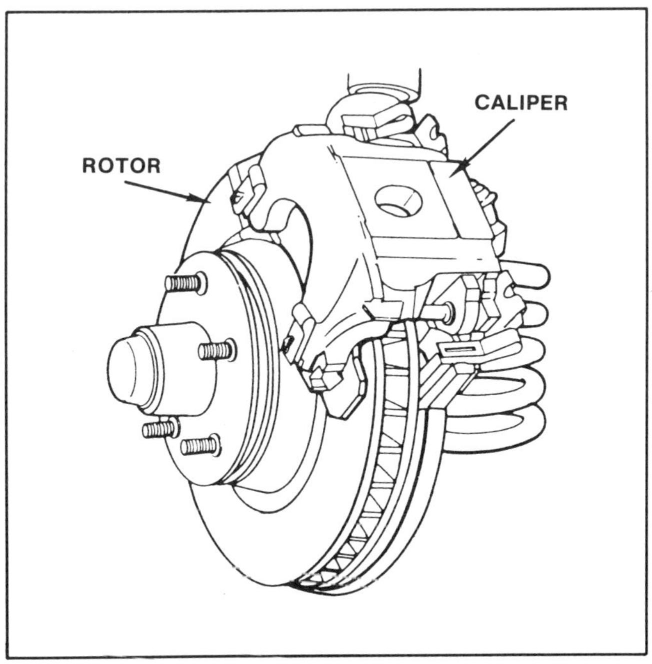

Figure 7-2 Typical one-piston floating caliper disc brake

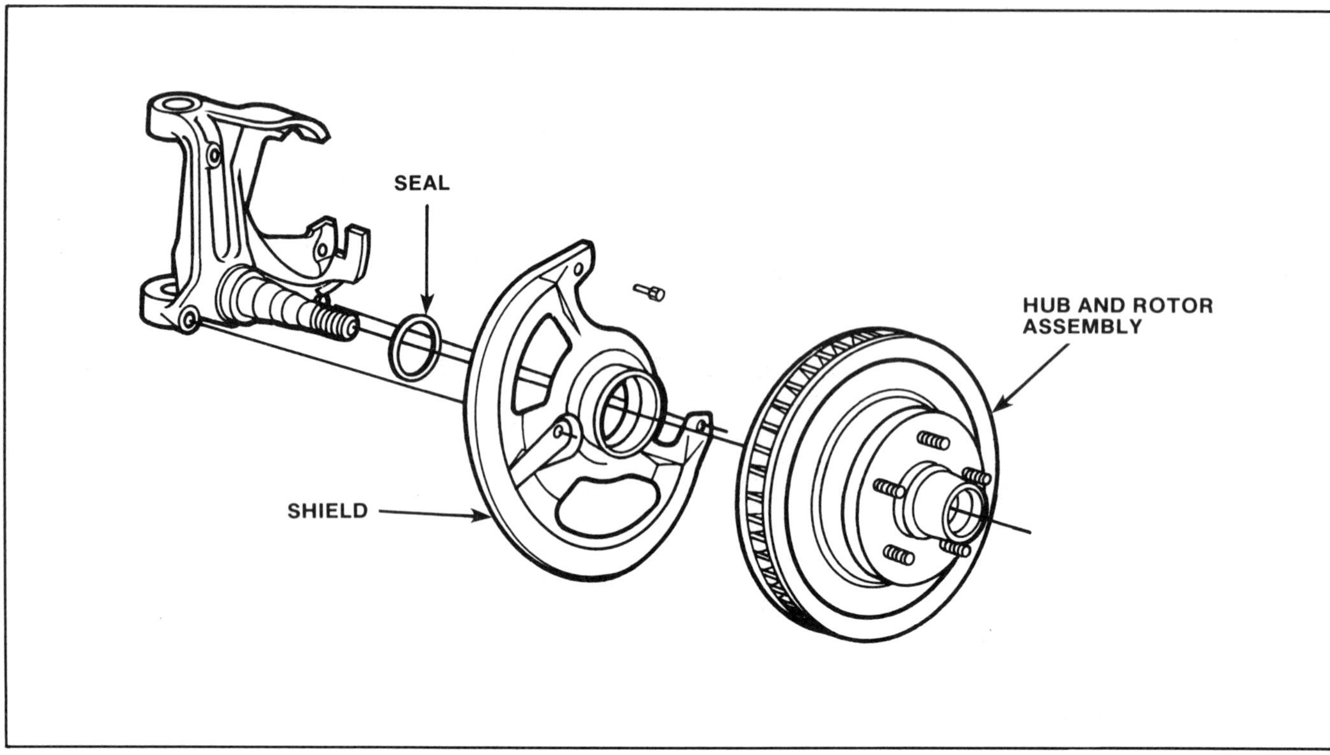

Figure 7-3 Steering knuckle, splash shield, and finned rotor for disc brake.

higher braking forces than drum brakes without distorting the machined parts. This is because the pads bear upon both sides of the disc, canceling out the bending stresses, whereas drum brake shoes only push out from the inside. The same piston forces in a drum brake would distort or even break the drum. Furthermore, disc brakes can endure severe applications more frequently without overheating because the linings cover a smaller portion of the rotor and because the rotor is directly in the stream of fresh air, affording better heat dissipation. Disc brakes run with very low pad-to-rotor clearances. There are several benefits from this: first, disc brakes are naturally self-adjusting; second, they are less susceptible to problems caused by water or dirt, being wiped clean every revolution; third, due to their open construction, foreign matter cannot accumulate. Finally, due to these low clearances, they are rapid acting. There is no travel in springs or linkages that must be taken up before they "go to work." However, since the stopping power is generated by forces perpendicular to the rotor, disc brakes are not self-energizing, and as such require more piston force to achieve the same amount of braking power.

DISC BRAKE COMPONENTS AND THEIR FUNCTIONS

The disc brakes used on American cars are one of two basic designs: fixed or floating caliper. There is also a sliding caliper, but, as stated in Chapter 1, its design is very similar to the floating caliper. The only difference is that sliding calipers slide on surfaces that have been machined smooth for this purpose, and floating calipers slide on special pins or bolts. The disc brake, regardless of its design, consists of a hub and rotor assembly, a caliper assembly, and a brake pad assembly.

HUB AND ROTOR ASSEMBLY

Many of the advantages of disc brakes can be attributed to the disc, or rotor, as it is more commonly called. The rotor is made of cast iron (iron has a high coefficient of friction and withstands wear exceptionally well) and is attached to and rotates with the hub assembly (Figure 7-3). Solid rotors are used on lightweight vehicles, but heavier automobiles, high-performance cars, and trucks generally have ventilated rotors (Figure 7-4).

The ventilated rotor is cast with a web-like construction between two friction surfaces. The webs radiate out from the center of the rotor, much like vanes or fins in a fan. As the rotor turns, air is drawn into the rotor at its center, flows between the friction surfaces, and is discharged along the outer edge. This cools the rotor very effectively, drastically reducing the incidence of brake fade, even during multiple hard stops.

The inner web of some ventilated rotors is shaped like scoops (Figure 7-5). This type of rotor draws air from its outer edge and forces it through the rotor and out through its center. Due to the directional nature of the webs, the rotors are not interchangeable from side to side.

Rotors are also cooled by splash shields or plates. The splash shield protects the rotor and pads from road splashes and dirt. It is also shaped to channel the flow of air over the exposed rotor surfaces. As long as the car is moving, this flow of air helps to cool the rotor. The splash shield cannot be removed unless the rotor and caliper are first removed. Replacement of the splash shield is necessary only when it has been damaged or when the spindle is replaced.

The clamping action of the disc brake also lessens the likelihood of brake fade. As the rotor gets hot, it expands. The expansion decreases the distance between the pads and rotor. Less brake pedal force is needed to engage the brakes. On the drum brake, however, the drum expands as it gets hot, increasing the distance between shoe and drum. To keep the shoes pressed against the drum, the brake pedal must be depressed harder as the drum expands.

CALIPER ASSEMBLY

Caliper action converts hydraulic pressure into mechanical force. That is, the caliper assembly (Figure 7-6) provides a means of applying the brake pads to the rotor. The caliper assembly is connected to the brake system by a flexible brake hose and a hydraulic brake line leading from a metering valve or from the master cylinder.

Caliper Housing

The caliper housing is usually a one-piece construction of cast iron or aluminum and has an inspection hole in the top to allow for lining wear inspection. A cylinder bore is located in that portion of the casting nearest the engine. In the cylin-

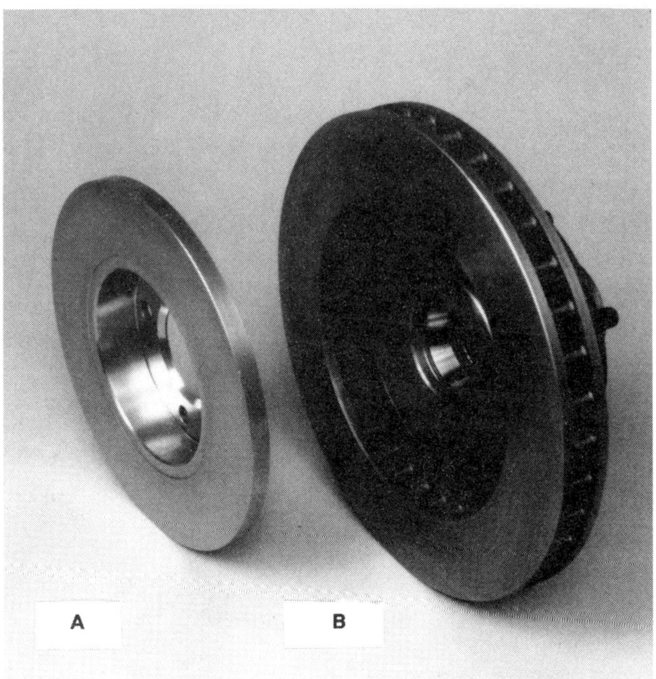

Figure 7-4 (A) Downsized solid rotor; (B) full-size ventilated rotor

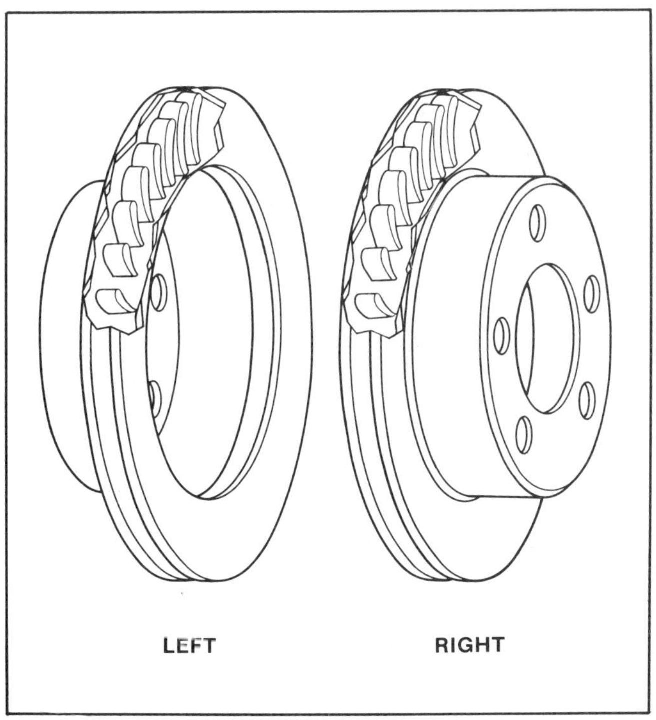

Figure 7-5 Rear rotors

der bore is a groove that seats a square-cut seal. This groove is tapered toward the bottom of the bore to increase the compression on the edge of the seal that is nearest hydraulic pressure. The top of the cylinder bore is also grooved as a seat for the dust boot. A fluid inlet hole is machined into the bottom of the cylinder bore and a bleeder valve outlet hole is located near the top of the casting. The caliper has inner ears with holes for mounting on the anchor plate. Actually, calipers are positioned differently, depending on the vehicle.

Caliper Pistons

A caliper can contain one, two, or four cylinder bores and pistons that provide uniform pressure distribution against the brake's friction pads. Because there is no self-energizing action with disc brakes, the pistons are relatively large in diameter and short in stroke to provide high pressure on the friction pad assemblies with a minimum of fluid displacement.

Basically, the hydraulics of disc brakes are the same as for drum brakes, where the master cylinder piston forces the brake fluid into the wheel cylinders and against the wheel pistons, with the following exceptions:

- *More Brake Fluid Is Required.*—In the disc brake systems, each caliper has greater brake fluid requirements.
- *Higher Force for a Given Pressure.*—The forces are high for a given pressure.

The disc brake piston is made of either steel, aluminum, or fiberglass-reinforced phenolic resin. The steel pistons are usually nickel-chrome plated for improved durability and smoothness. The pistons usually have a larger diameter to provide the necessary braking force with little fluid movement. The top of the pistons is grooved to accept the dust boot.

In many original equipment applications, the phenolic piston has replaced the steel piston. It is lighter than steel pistons and has better resistance to corrosion. The phenolic piston also transfers less heat to the brake fluid, which reduces the amount of fluid loss due to vaporization. The bottom line, however, is safety. The special qualities of phenolic pistons make them considerably more reliable over a longer period of time.

Dust Boot

The dust boot seats in a groove at the top of the cylinder bore and also in a groove in the piston

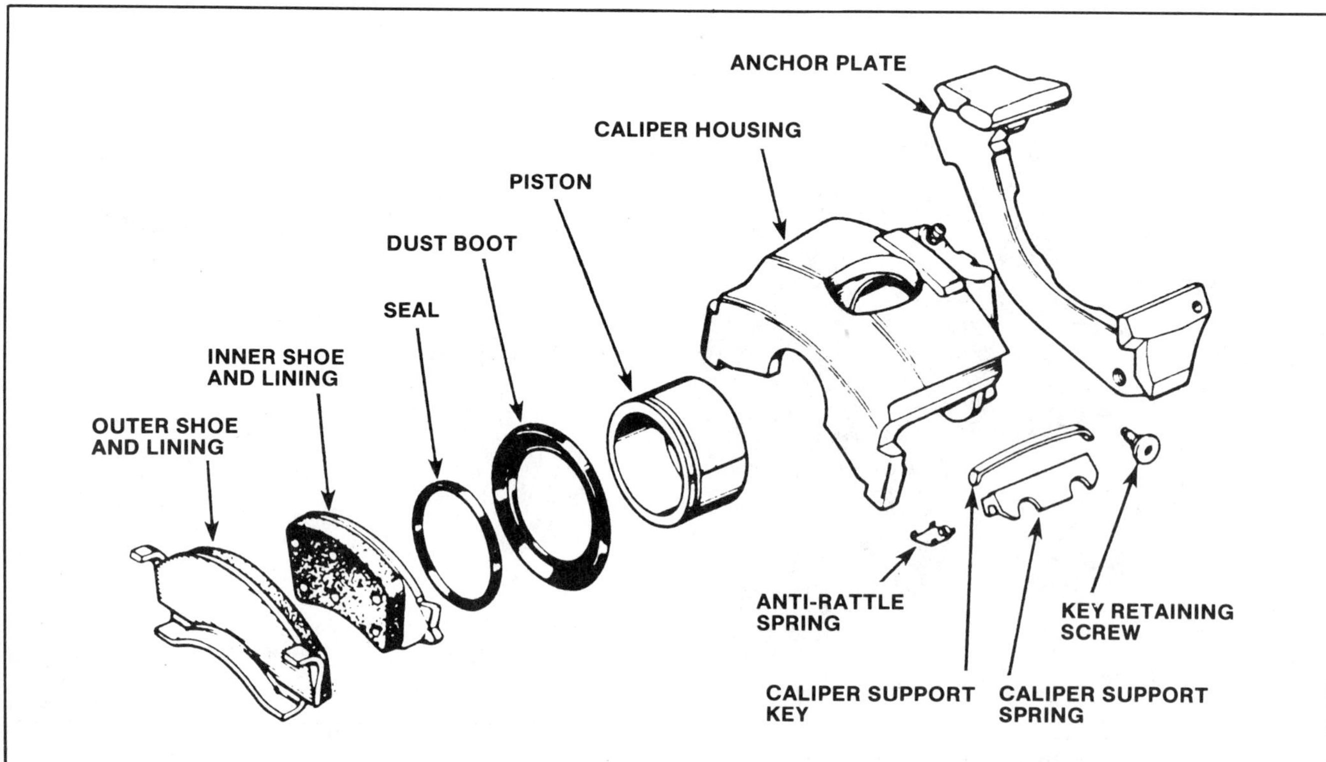

Figure 7-6 Caliper parts

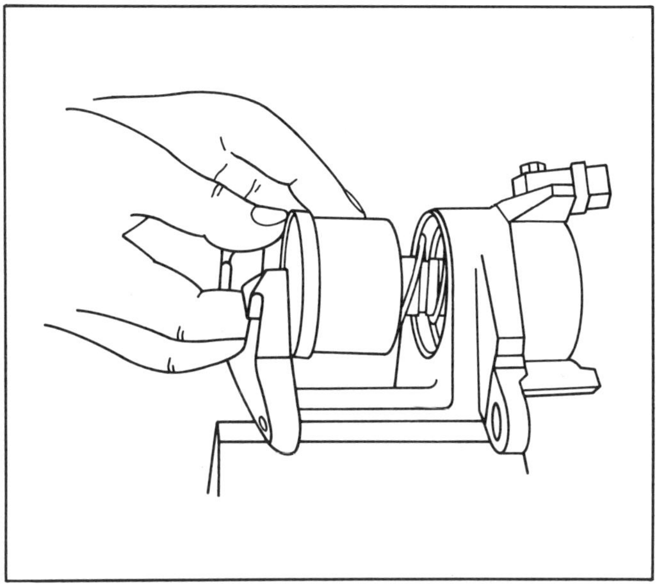

Figure 7-7 Installing a piston

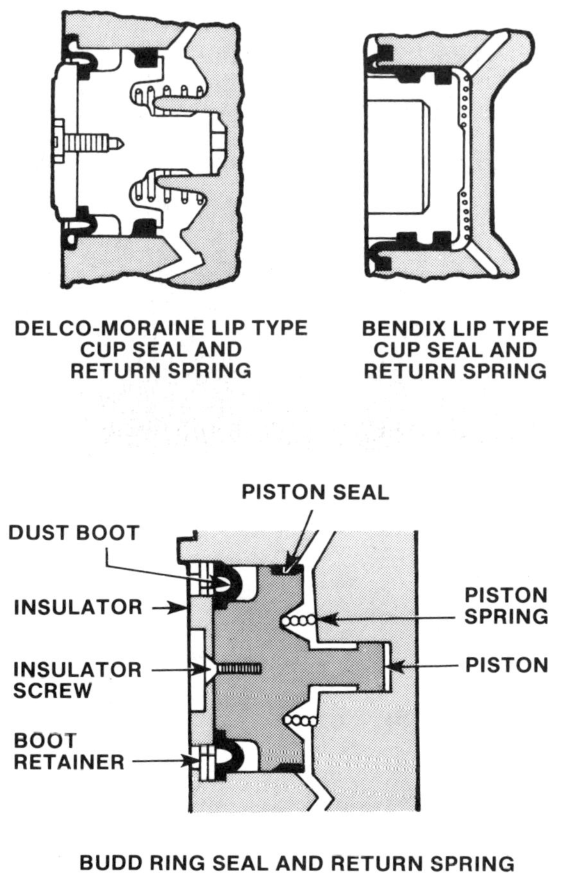

Figure 7-8 Piston hydraulic stroking seal arrangements

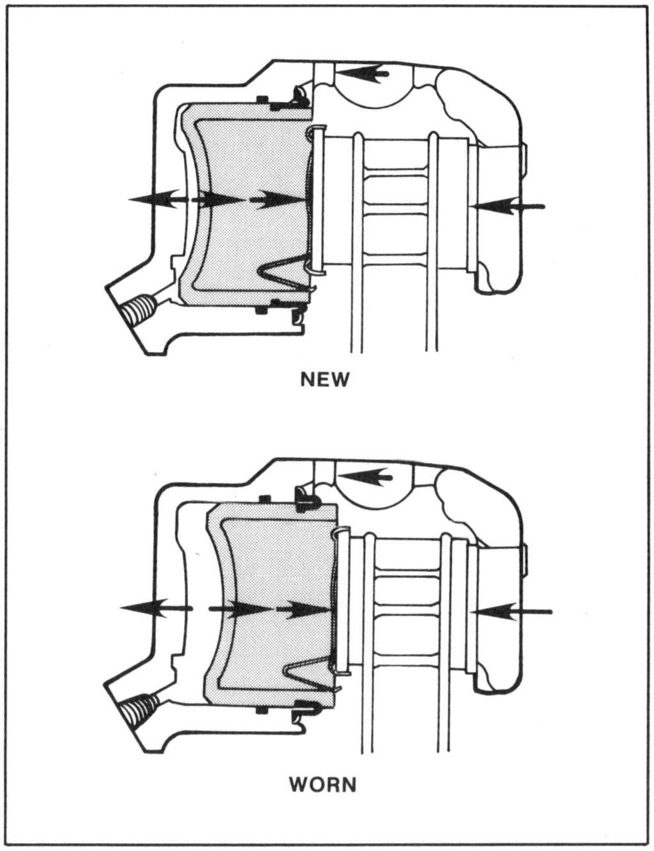

Figure 7-9 Disc brake pad wear

(Figure 7-7). The dust boot prevents moisture and road contamination from entering the bore.

Piston Hydraulic Seals

Seal arrangements at the caliper pistons vary, depending upon the brake manufacturer. Three makes of fixed caliper brakes, Bendix, Budd, and Delco-Moraine, use a ring seal that fits in a circumferential groove on the piston. These arrangements are shown in Figure 7-8.

The piston hydraulic seal prevents fluid leakage between the cylinder bore wall and the piston. This rubber sealing ring also acts as a retracting mechanism for the piston when hydraulic pressure is released, causing the piston to return in its bore. The seal distorts to allow the piston to be pushed out to bring the pad into contact with the rotor. When hydraulic pressure is diminished, the seal functions as a return spring to retract the piston.

In addition, as the disc brake pads wear, the seal allows the piston to move further out to adjust automatically for the wear and to maintain the lining in proper relationship with the rotor (Figure 7-9). Additional brake fluid in the caliper bore compensates for lining wear. In this manner the caliper assembly maintains the inboard and outboard shoe and lining in the proper relationship with the rotor surface throughout the full life of the lining.

Low-Drag Calipers

Beginning in 1980, many brake manufacturers began using drag calipers. This caliper design works the same as a typical caliper, except that it has a lead angle seal groove. A low-drag caliper retracts the piston in the caliper up to 0.030 inch.

Before, during, and after application positions of the piston seal and caliper are shown in Figure 7-10. The relative position of these components does not change where there is no lining wear when hydraulic pressure is applied. As pressure is introduced, the piston moves out and the inside diameter of the seal remains in contact with the piston as the seal flexes. When hydraulic pressure is released, the seal retracts the piston. If, during application, the piston moves out far enough to compensate for lining wear, it will slip on the inside diameter of the seal. But, after hydraulic pressure is released, the seal retracts the piston. This retraction is the same whether or not the piston compensates for lining wear. The design of the seal groove (lead angle) is the primary factor that determines how far the piston can travel before it slides on the inside diameter of the seal. The farther the piston can travel without sliding on the seal inside diameter, the greater the amount of piston retraction that will occur.

If the disc brake seals lose their ability to flex and retract the piston, hydraulic pressure will cause the pads to drag against the rotor.

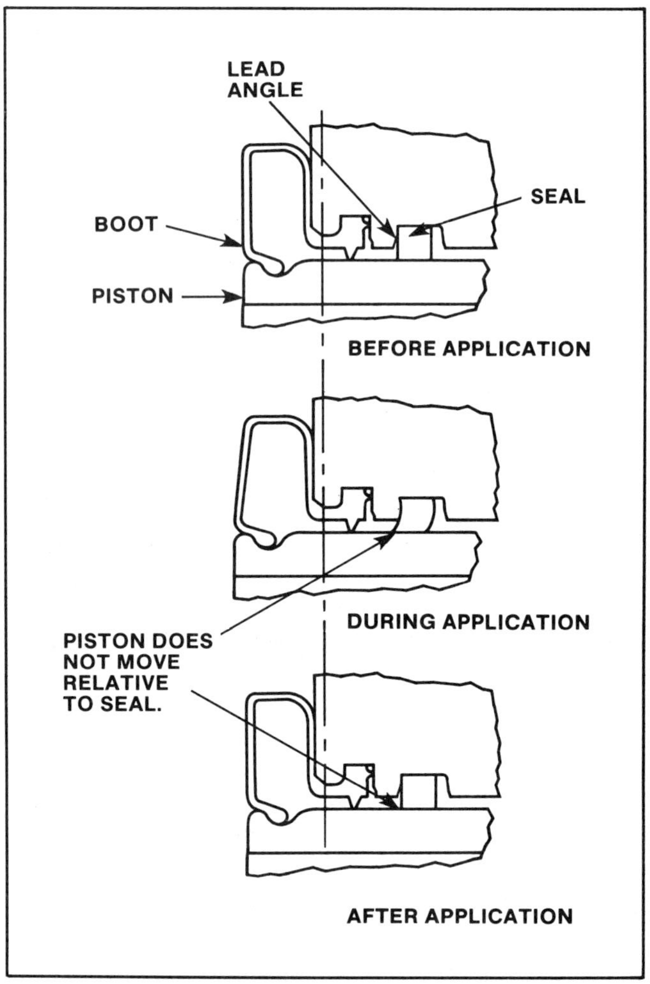

Figure 7-10 Low-drag caliper seal function

Anchor Plate

The anchor plate is bolted to the spindle by two mounting bolts. The upper bolt faces the front of the vehicle and the lower bolt faces toward the engine. This creates a more rigid caliper assembly that is able to withstand high braking torque. A safety wire usually passes through holes in the heads of these bolts and prevents them from turning. Two holes at the top of the anchor plate hold rubber insulator bushings.

Caliper Mounting Hardware

Various forms of caliper mounting hardware are outlined here and illustrated in Figure 7-11.

Caliper Guide Bolts. Floating calipers are supported by guide bolts that fasten to the caliper anchor plate. The caliper moves laterally on these bolts. Always inspect guide bolts for alignment, scoring, and corrosion; service as required. Lubricate with silicone grease or equivalent upon reassembly.

Insulators/Bushings. These are used in conjunction with guide bolts to cushion caliper movement and eliminate brake noise from metal-to-metal contact.

Guide Pins. These locate the disc pads in the caliper housing. Pins are retained in the caliper by pin retainers.

Caliper Support Key. Various sliding caliper designs can be positioned by a support key inserted between the caliper and anchor plate. For example, stabilizer keys are sometimes used to hold the caliper weight off the rotor and allow necessary movement of the caliper relative to the anchor plate. When the car travels over bumpy roads, the stabilizers keep the caliper from bouncing up and down because the caliper is supported only by rubber insulator bushings in this direction.

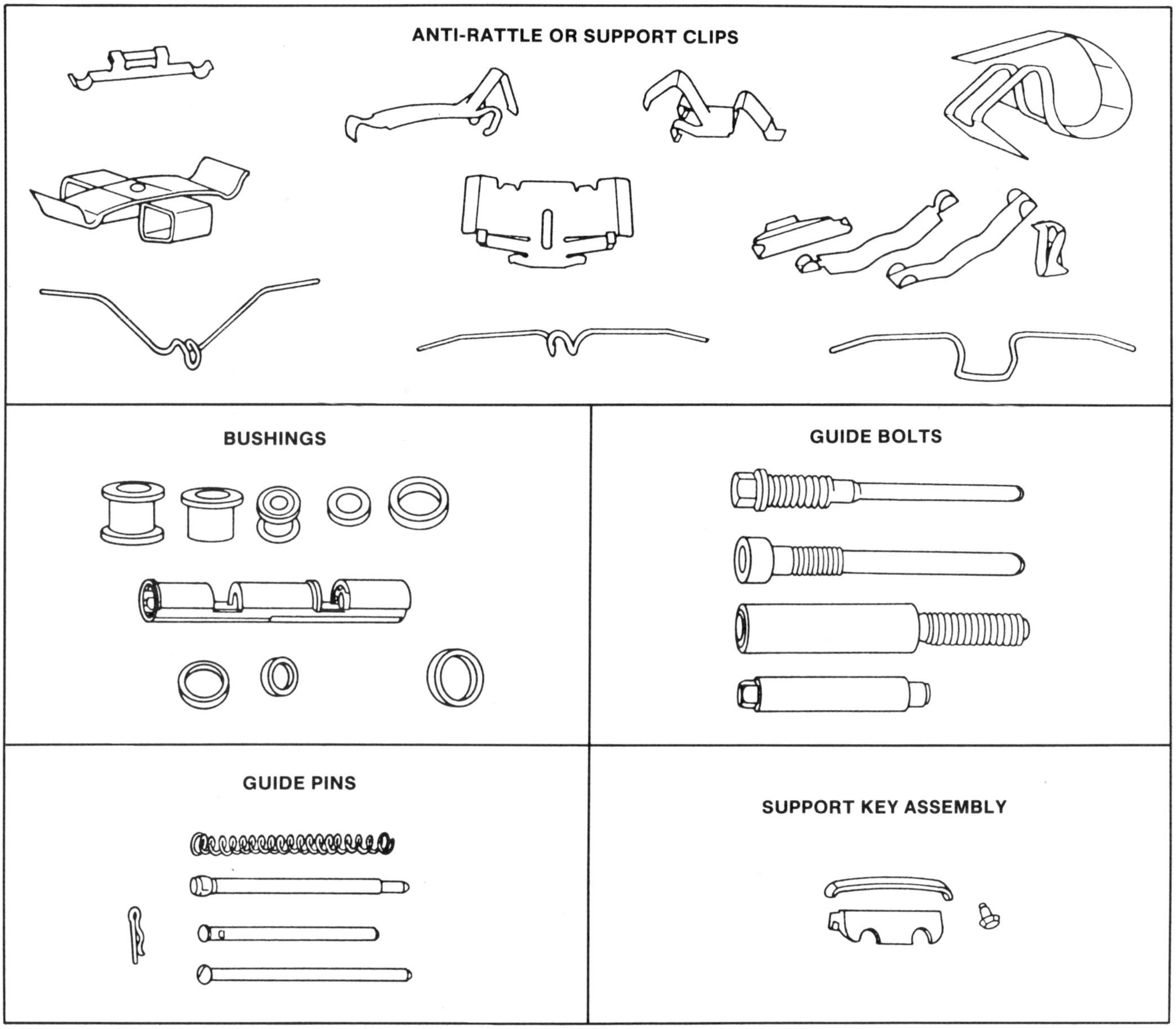

Figure 7-11 Caliper mounting hardware

Anti-Rattle Clips/Springs. These devices secure the brake pads in the caliper housing to prevent vibration. Figure 7-12 shows how two types of anti-rattle or support clips are fastened on the brake pad and lining.

SHOP TALK

Always service caliper hardware to assure maximum life and performance of new disc pads. Hardware life is drastically affected by intense heat, weather conditions, abrasive particles, and normal wear.

BRAKE PAD ASSEMBLY

Brake shoe pads with disc brakes are metal plates with the linings either riveted or bonded to them. Pads are placed one in each side of the caliper and together straddle the rotor. The inner brake pad, which is positioned against the piston, is not interchangeable with the outer brake pad (Figure 7-13).

The inner pad has an ear on each end to support it radially in the caliper. The outer pad might be longer and has two abutments on the back, which mate with the outboard ears of the caliper. A special insulator is sometimes glued to the back of the

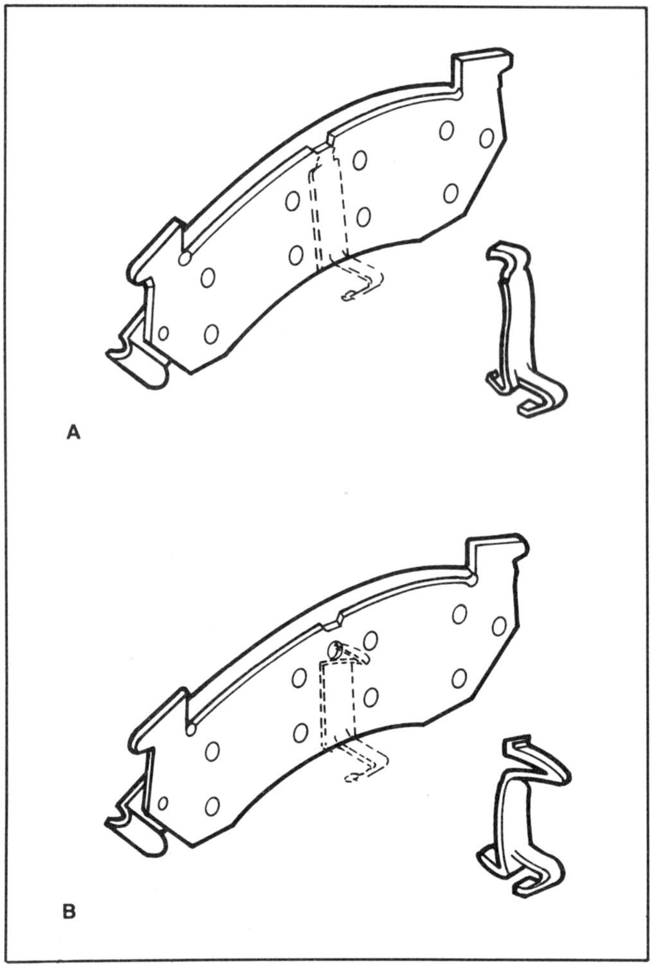

Figure 7-12 (A) Flat notch type and (B) hole type with flat notch

outer pad to minimize noise. Two holes are drilled through the outer pad and lining. The pad guide pins pass through the brake pad and through the caliper outer ears and are held in place by two retainer clips.

Disc brake pads have traditionally used asbestos materials except for certain heavy-duty applications. However, with the downsizing of brake systems and the increased use of front-wheel drive designs (which increases the front braking requirements), a new type of pad material has emerged. It is called semi-metallic lining and is gaining wide application. Instead of the conventional asbestos matrix in a cured cement, a semi-metallic lining has a high metallic content in a resin binder. Although semi-metallic pads have a lower coefficient of friction than asbestos, they are found as OEM in complete sets or in hybrid sets (the inner or outer pad will be semi-metallic and the other pad organic).

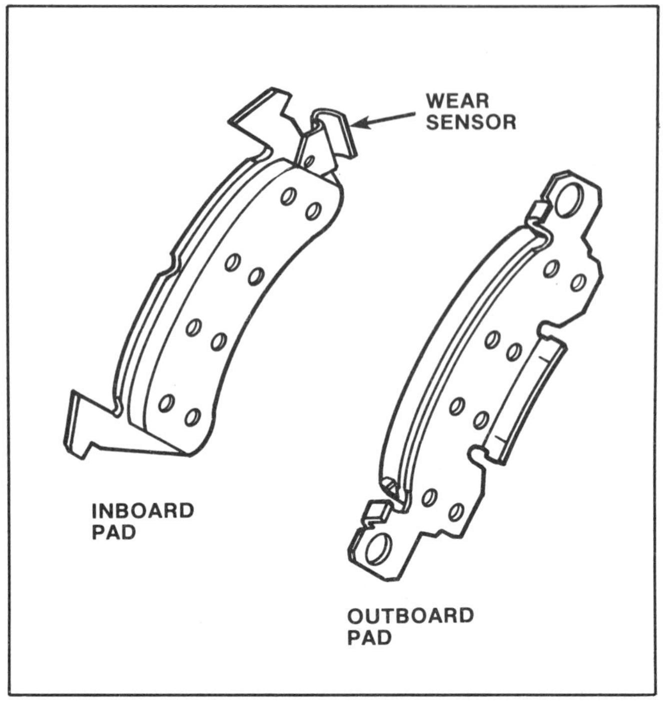

Figure 7-13 Typical pad and linings

Figure 7-14 A new (top) and worn brake pad. Note edge code markings on new pad.

Like brake drum linings, pad linings are edge branded using the Automotive Friction Material Edge Codes. As a general rule, most asbestos pads now available are marked "FF" or "FG," while semi-metallics are edge branded "EE" (Figure 7-14). Disc brake linings are usualy smaller in area than drum brake linings.

When replacing disc brake pads, it is always wise to use the same material as previously used (Figure 7-15). That is, replace semi-metallic pads with semi-metallics. The use of asbestos pads in semi-metallic applications can result in short pad life. A good guide to the proper selection of pad material is an updated brake catalog (Figure 7-16) or the vehicle's service manual.

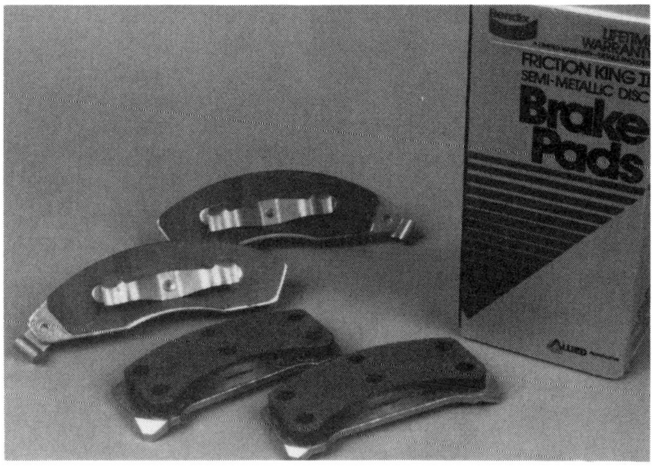

Figure 7-15 Typical semi-metallic pads

Disc Pad Wear Sensors

Some brake shoe pads have wear sensing indicators. These indicators are electrical rivets that complete the circuit to the navigator when the pad wears down to the point where the rivet makes contact with the rotor. The navigator then alerts the driver that the brake pads have worn to the minimum allowable thickness and should be replaced. The three most common design wear sensors are audible, visual, and tactile.

Audible Sensors. These are thin, spring steel tabs that are riveted to the edge of the pad backing plate and are bent so that they will contact the rotor when the lining wears down to a point that replacement is necessary. At that point, the sensor will cause a high-pitched squeal when the brakes are applied, warning the driver that brake service is needed and perhaps saving the rotor from destruction (Figure 7-17). The tab is generally installed toward the rear of the vehicle.

Visual Sensors. This method of informing the driver of the need for new linings employs electrical contacts recessed in the pads that touch the rotor when the linings are worn out. This completes a circuit and turns on a dashboard warning light (Figure 7-18). This system is found mostly on imports such as Toyota and Mercedes-Benz.

Tactile Sensor. These sensors create pedal pulsation as the sensor on the rotor face contacts the sensor attached to the lower portion of the disc pad (Figure 7-19).

> **CAUTION:** As disc brake pad linings wear thin, more brake fluid is needed in the system.

FIXED CALIPER DISC BRAKES

Fixed caliper disc brakes include a caliper assembly that is bolted in a fixed position and does not move when the brakes are applied. The pistons in both sides come inward to force the pads against the rotor (Figure 7-20). Import cars use the fixed type caliper more widely than domestics.

OPERATION

A cross-sectional view of a fixed caliper disc brake in the released position is shown in Figure 7-21. The pistons are near the bottom of the cylinder bores. Normally, the pads would be touching the rotor or the clearance between the pads and the rotor would be very small. The clearance is exaggerated in the illustration to show how the brake operates.

When the brake pedal is depressed, hydraulic brake fluid is forced into the caliper cylinder bores. This forces the pistons out of the bores and against the pad and plate assemblies as shown in Figure 7-22.

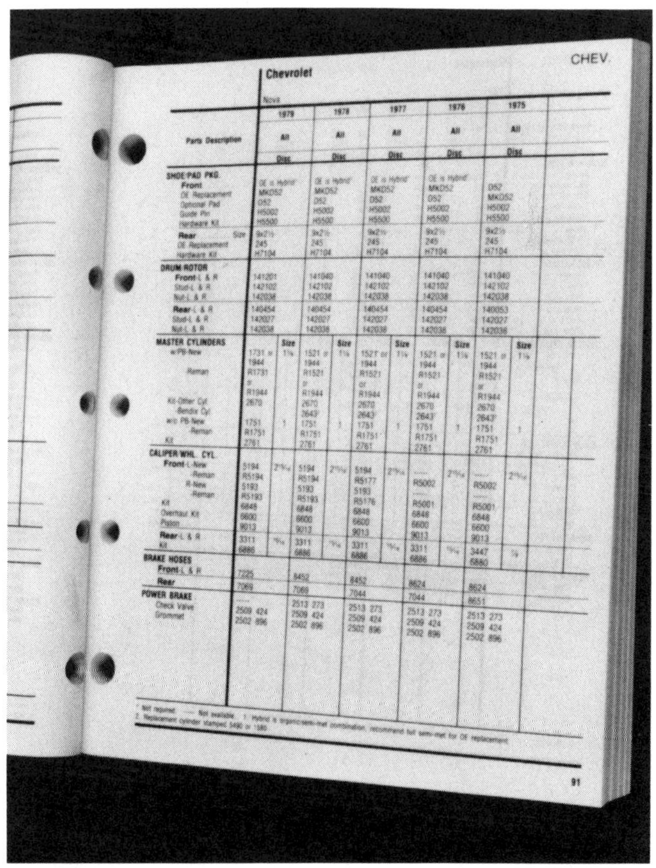

Figure 7-16 Brake catalog applications page

Figure 7-17 Audible disc brake wear indicator

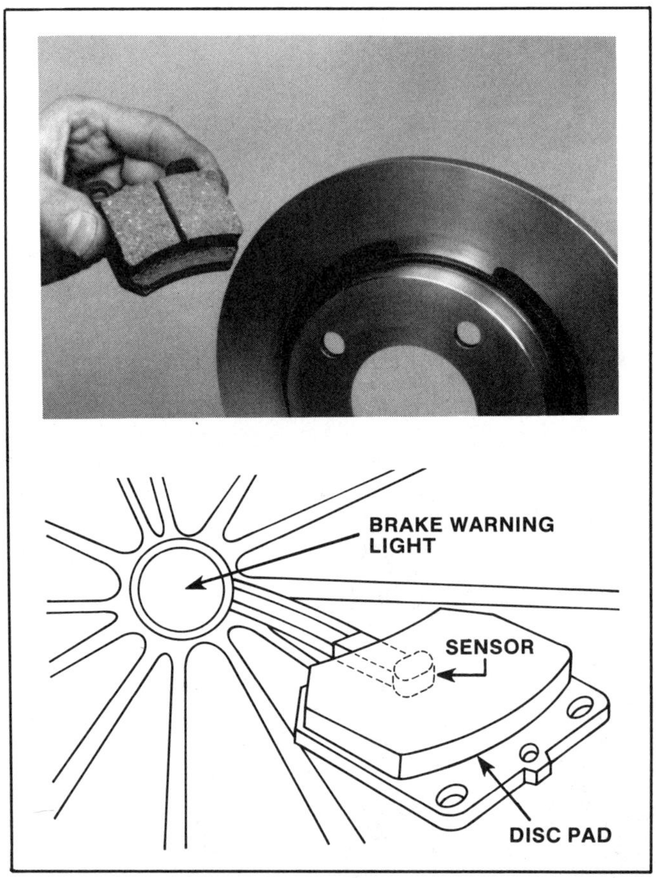

Figure 7-18 Visual type sensor

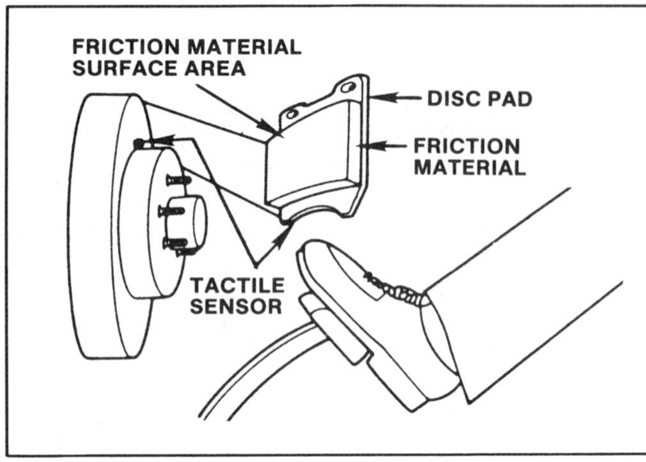

Figure 7-19 Tactile type sensor

With a pad pressing on each side of the rotor, a clamping action takes place. This slows or stops the rotation of the rotor and the wheel. When the brake pedal is released, a small amount of fluid returns to the master cylinder. This reduces the pressure in the caliper cylinders, and the pistons retract very slightly into the cylinder bores. The pressure of the pads against the rotor is removed, and the clamping action is released.

FOUR-PISTON FIXED CALIPER BRAKES

The fixed caliper can have two, three, or four pistons. The four-piston was the first type of disc

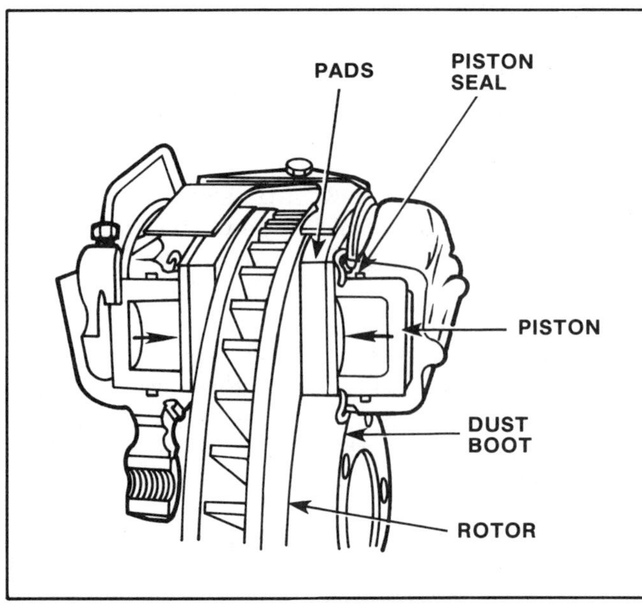

Figure 7-20 Cross section of a fixed caliper

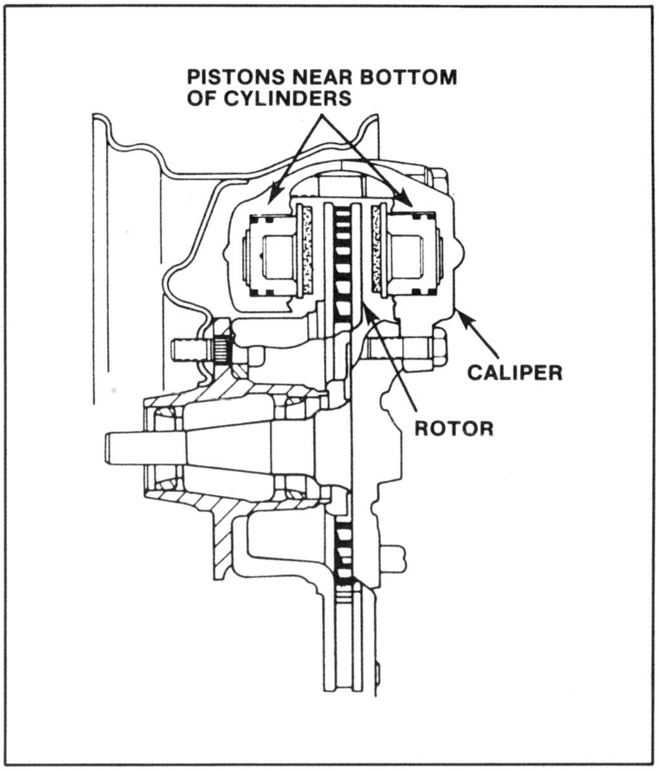

Figure 7-21 Disc brake in the release position

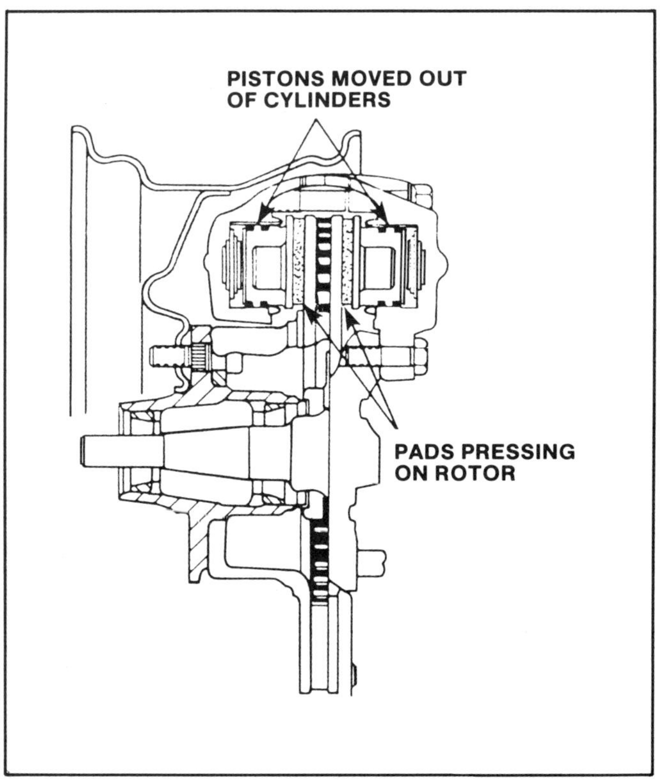

Figure 7-22 Disc brake in the applied position

brake to appear in significant numbers on American cars. Introduced in 1965, this design was superceded by the single-piston type beginning in the late 1960s, and was completely phased out by 1972, except on the Corvette and certain Camaros and Firebirds, which used it up to 1984.

There are four major types of four-piston fixed caliper: Bendix, Budd, Delco-Moraine, and Kelsey-Hayes. All have a two-piece caliper with two pistons in each side. The caliper is rigidly mounted on the steering knuckle, making this design less forgiving of rotor defects, such as runout and non-parallelism, than a floating or sliding caliper. Figure 7-23 shows the various parts that go into the assembly of the Bendix four-piston disc brake caliper—considered typical for this design of disc brake.

TWO- OR THREE-PISTON FIXED CALIPER BRAKES

Two-piston calipers have one piston on each side of the rotor (Figure 7-24); three-piston calipers have one piston on one side of the rotor and two pistons on the other side (Figure 7-25). Two- or three-piston fixed caliper brakes are used primarily on domestic trucks and import vehicles (see Chapters 12 and 13).

FLOATING AND SLIDING CALIPER DISC BRAKES

Floating and sliding caliper designs usually have only one piston. The only basic difference between these two disc brakes is

- Floating calipers have pins that the calipers slide on
- Sliding calipers have a support key (or anchor) that they slide on

TYPICAL FLOATING CALIPER DISC BRAKES

The Ford floating caliper disc brake shown in Figure 7-26 is a typical example of a floating caliper. The caliper on this brake is a one-piece casting that has one hydraulic cylinder and a single piston.

The single hydraulic cylinder has a groove in the bore wall that contains a square sectioned O-ring seal. The piston is inserted through the seal as shown in Figure 7-27. A boot is seated in a groove in the piston and on the adjoining caliper surface

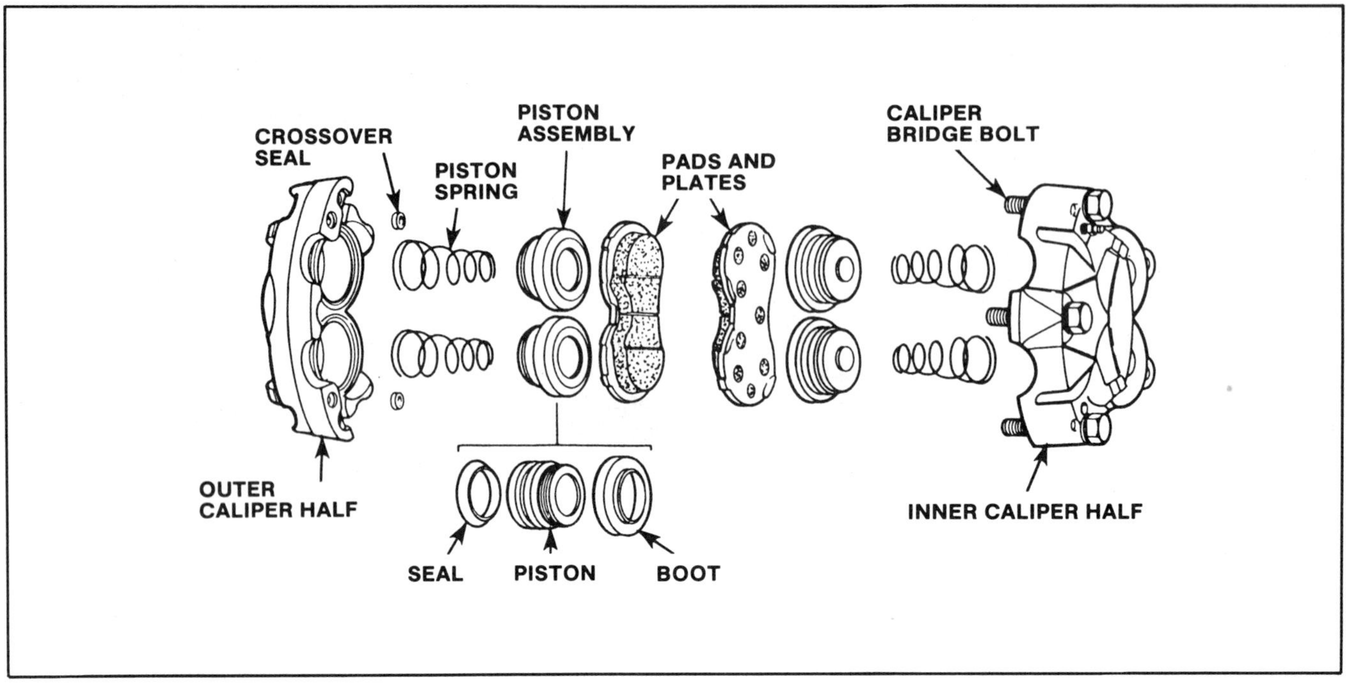

Figure 7-23 Bendix four-piston disc brake

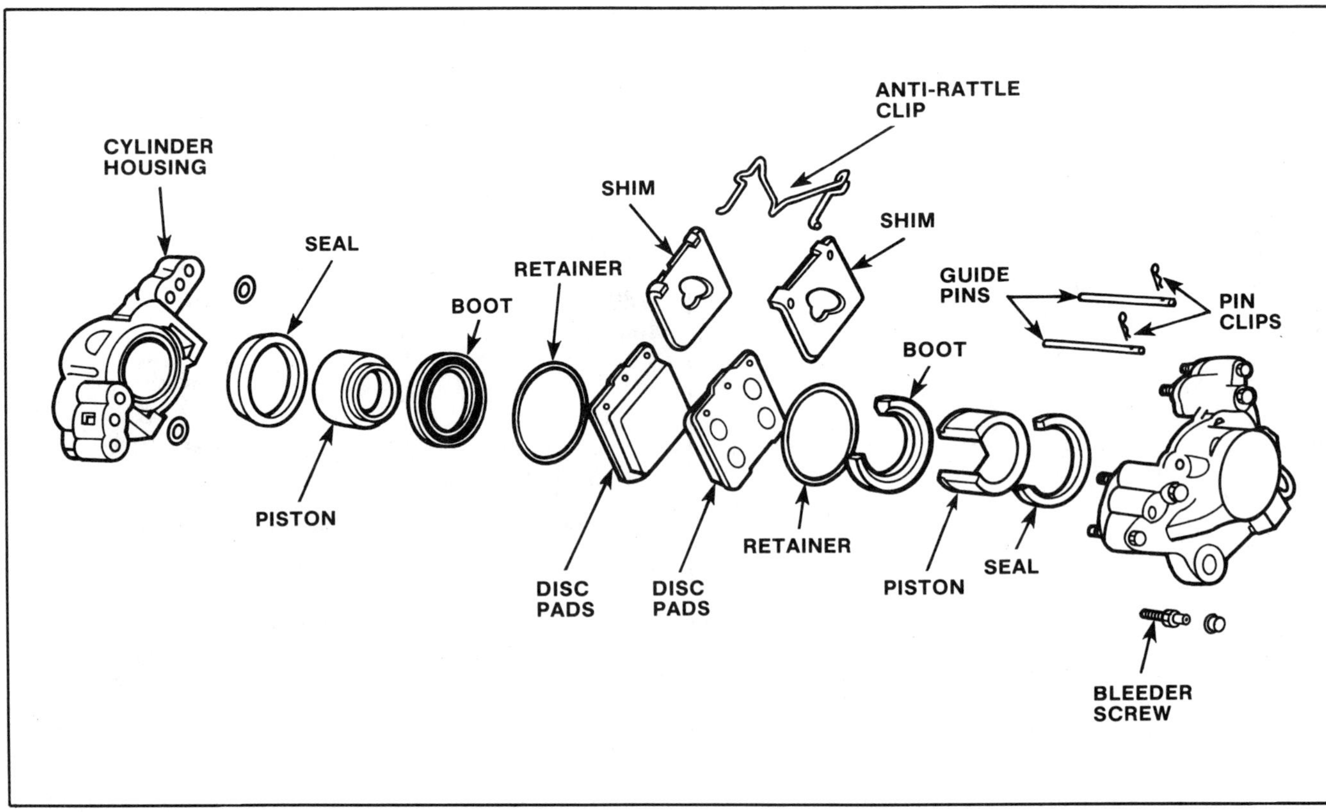

Figure 7-24 Two-piston fixed caliper

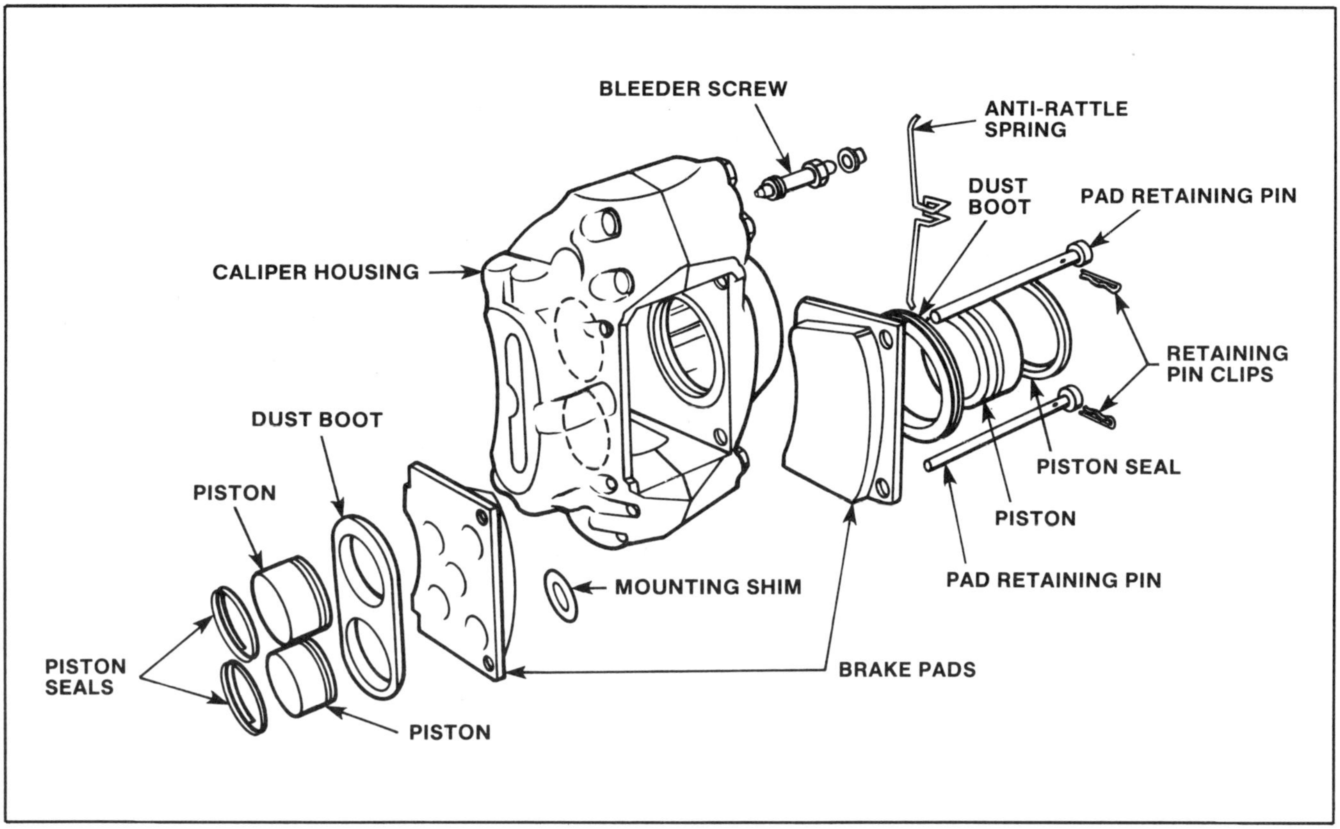

Figure 7-25 Three-piston fixed caliper

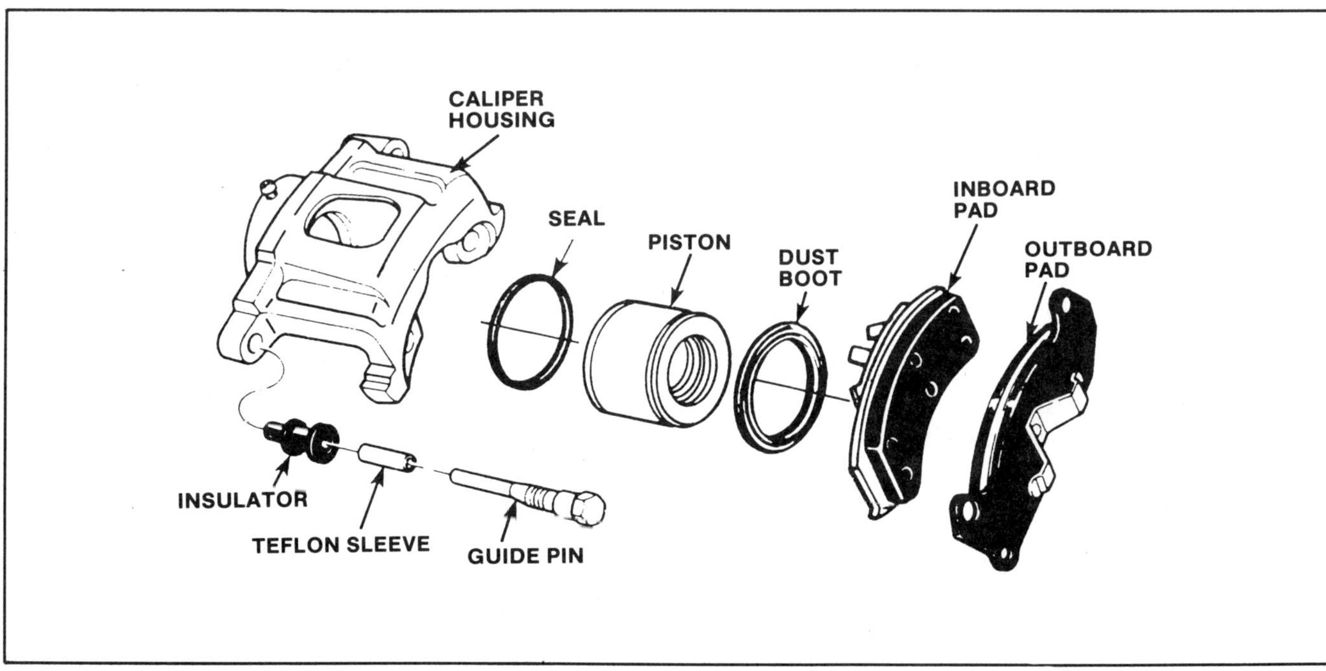

Figure 7-26 Ford/Kelsey floating caliper disc brake

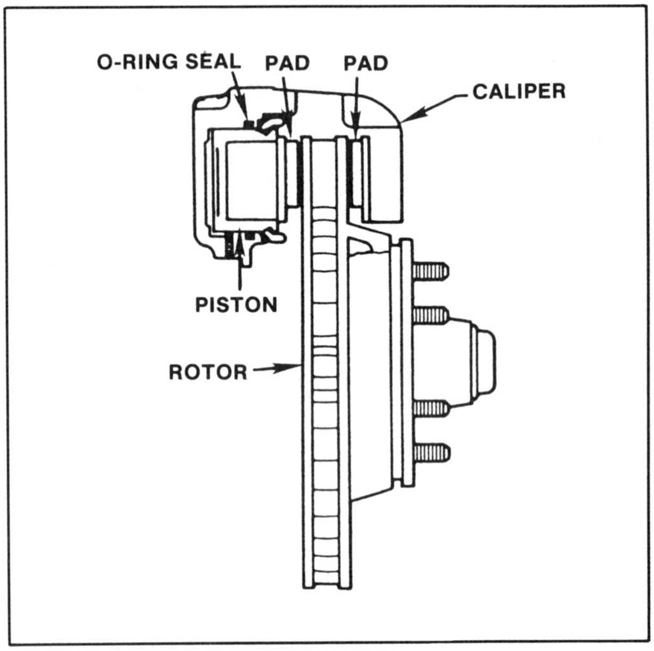

Figure 7-27 Cross section of a floating caliper and rotor

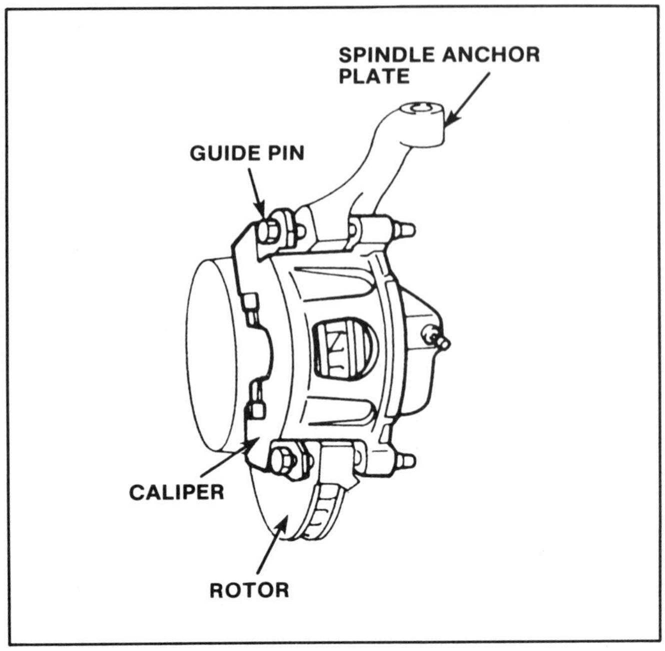

Figure 7-28 Ford caliper mounting

to protect the piston and cylinder bore from contaminants.

Two threaded locating pins are used to secure the caliper assembly to the spindle anchor plate (Figure 7-28).

Operation

As noted in Figure 7-27, a single piston is inserted through a square sectioned O-ring seal that is recessed in a machined groove of the caliper housing. The piston seal exerts drag on the movement of the piston, thus serving as a self-adjuster.

Figure 7-29 shows the floating caliper in a released position with the pads seated slightly away from the rotor face and the seal in a relaxed position. Upon brake application, hydraulic pressure builds in the caliper cavity behind the piston and seal. Because hydraulic pressure exerts equal force in all directions, pressure against the piston is equal to that exerted against the caliper housing. The piston seal offers the point of least resistance, so initial movement is of the piston outward in the bore.

The piston presses the inboard pad against the rotor. As the pad contacts the revolving rotor, greater resistance to outward movement is increased, forcing pressure to push the caliper away from the piston. This action results in outboard pad application (Figure 7-30). However, both pads are applied with equal pressure.

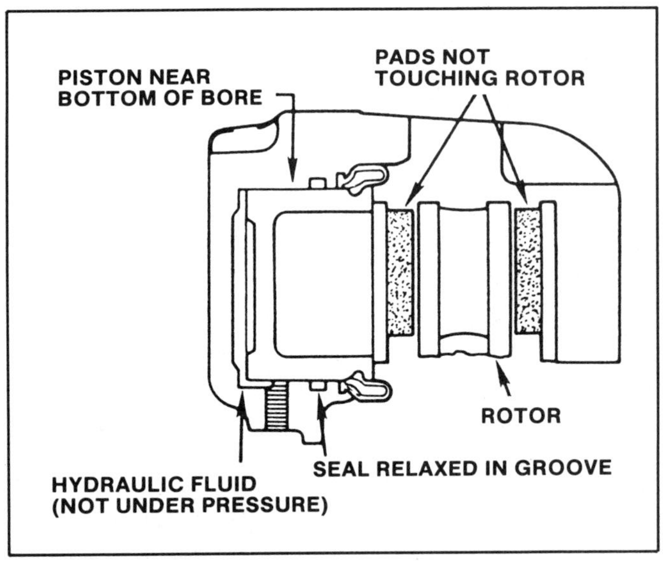

Figure 7-29 Cross section of a caliper in the released position

DELCO-MORAINE LOW-DRAG CALIPER

The Delco-Moraine low-drag caliper shown in Figure 7-31 is a single-piston, floating caliper introduced on the General Motors front-wheel drive X-car series, redesigned to reduce pad drag, which benefits pad life and vehicle gas mileage. Caliper bore is 2-1/4 inches. The two guide bolts secure the caliper to the anchor. The explanation of how the low-drag caliper functions was mentioned previously in this chapter.

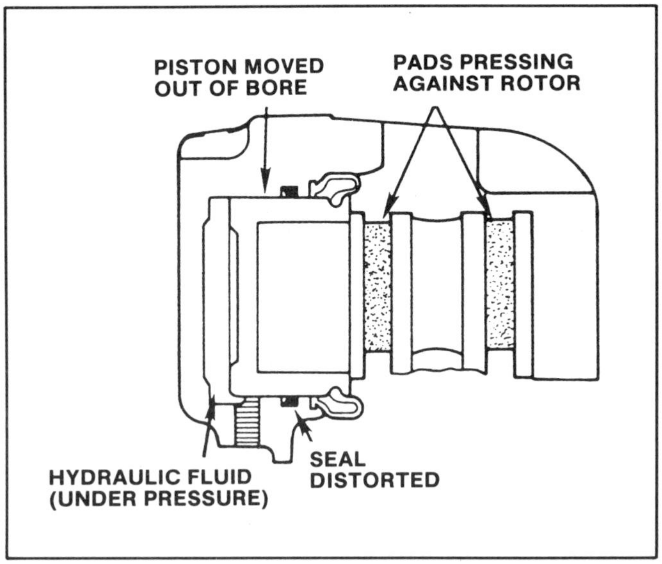

Figure 7-30 Cross section of a caliper in the applied position

TYPICAL SLIDING CALIPER DISC BRAKES

Figure 7-32 is an example of a so-called typical sliding caliper disc brake. The caliper on this brake is a one-piece casting that has one hydraulic cylinder and a single piston. Machined surfaces on the caliper are positioned against mating machined surfaces on the anchor.

The caliper anchor can be one of two designs: a one-piece casting that includes both the anchor and spindle or separate castings for the anchor and spindle. When the castings are separate, the anchor is secured to the spindle arm by two attaching bolts. The upper bolt is parallel to the centerline of the vehicle, and the lower bolt is at a right angle to the centerline. This arrangement resists the twisting motion of the brakes.

The inner pad and plate assembly is installed with the ends of the plate confined within the anchor. An anti-rattle clip is positioned between the leading edge of the plate and the anchor. The outer pad and plate assembly is installed with the plate flange bearing against the locating and torque abutment surfaces on the caliper.

A caliper support key is installed between the leading edge of the caliper and the anchor, with a support spring installed between the key and cali-

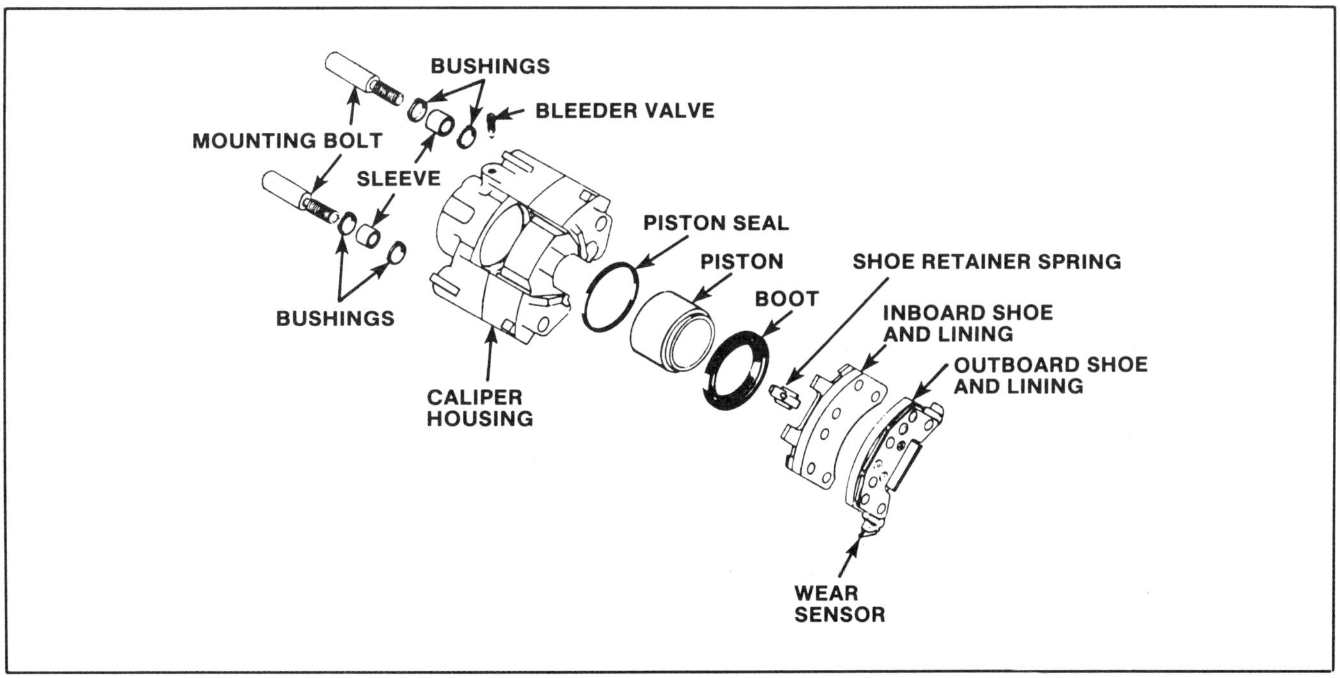

Figure 7-31 Delco-Moraine low-drag caliper

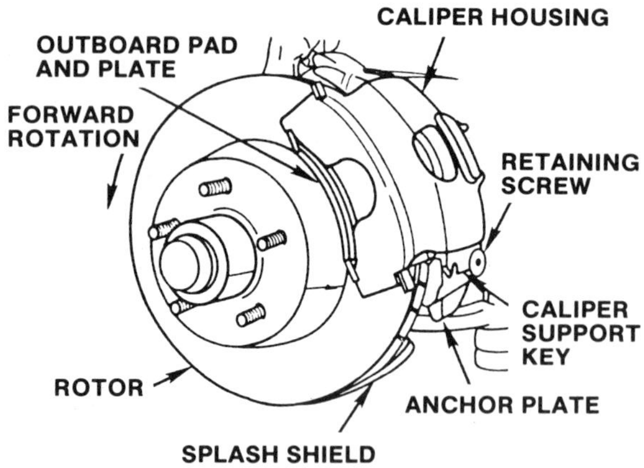

Figure 7-32 Sliding caliper disc brake

per. A key retaining screw holds the key in the caliper. A square section seal is fitted into an annular groove in the caliper cylinder bore, and a rubber boot seals the piston and bore from road splash and contamination.

A cast-iron, ventilated rotor is attached to the wheel hub. A splash shield protects the inboard surface of the rotor.

Operation

Sliding caliper disc brakes get their name from the fact that the caliper "slides" or moves sideways when the brakes are applied. As mentioned previously, in operation these brakes are almost identical. Hydraulic pressure forces a piston out of its cylinder to press the inboard pad against the rotor. The caliper itself moves in the opposite direction to force the outboard pad against the opposite side of the rotor.

But unlike the floating caliper, the sliding caliper does not float on pins or bolts attached to the anchor plate. The sliding caliper has angular machined surfaces at each end that slide in mating machined surfaces on the anchor plate. Except for this basic difference, the operation of the sliding caliper disc brake is the same as the Ford floating caliper disc brake covered earlier in this chapter.

PIN-SLIDER DISC BRAKES

The pin-slider disc brake is a lightweight version of the sliding caliper design. As shown in Figure 7-33, the pin slider consists of an aluminum alloy piston, piston seal, and dust boot. It is held in position on the suspension knuckle with two caliper guide pins and one hold-down clip. The two holes that accommodate the caliper guide pins are equipped with rubber insulator bushings and plastic sleeves.

As the brakes are applied and lining wear takes place, the guide pins allow the caliper to slide inward along the machined abutments of the suspension knuckle. The inboard pad is held in the cavity of the piston by a retainer clip. The outboard pad is retained by two clips snapped into the outboard section of the caliper. A caliper hold-down clip positioned between the suspension knuckle and the caliper stabilizes the outboard section of the caliper.

The Chrysler FWD model's pin slider (Figure 7-34) is a one-piece casting with the inboard side containing a single-piston cylinder bore. It is connected to the hydraulic system by a flexible brake hose and is held in position on the anchor plate with two caliper guide pins. The two holes that accommodate the caliper guide pins are equipped with rubber insulator bushings.

As the brakes are applied and lining wear takes place, the guide pins allow the caliper to slide inward along the machined abutments of the anchor plate. During braking, the caliper reacts against the anchor plate and not against the caliper guide pins.

The anchor plate is constructed to withstand high braking torque. It is an integral part of the steering knuckle and does not have to be removed for brake service. The brake pads are positioned on each side of the anchor plate facing the rotor. An anti-rattle spring is positioned at the top of the anchor plate and is fitted under the ears of the brake shoes. The inner and outer brake shoes are interchangeable.

The only major difference between the pin- and rail-slider designs is that the pin-slider type uses pins and rail sliders use rails. There is no functional difference, however.

Floating and sliding caliper designs are normally single piston, but some import and domestic truck models have two pistons. Other imports are designed so that the cylinder assembly can be removed from the frame. The most common import models are Girling and ATE. Other import car disc brake calipers are very similar. Import floating, sliding, yoke, and pivoting caliper disc brakes are covered in Chapter 13; truck disc types are covered in Chapter 12.

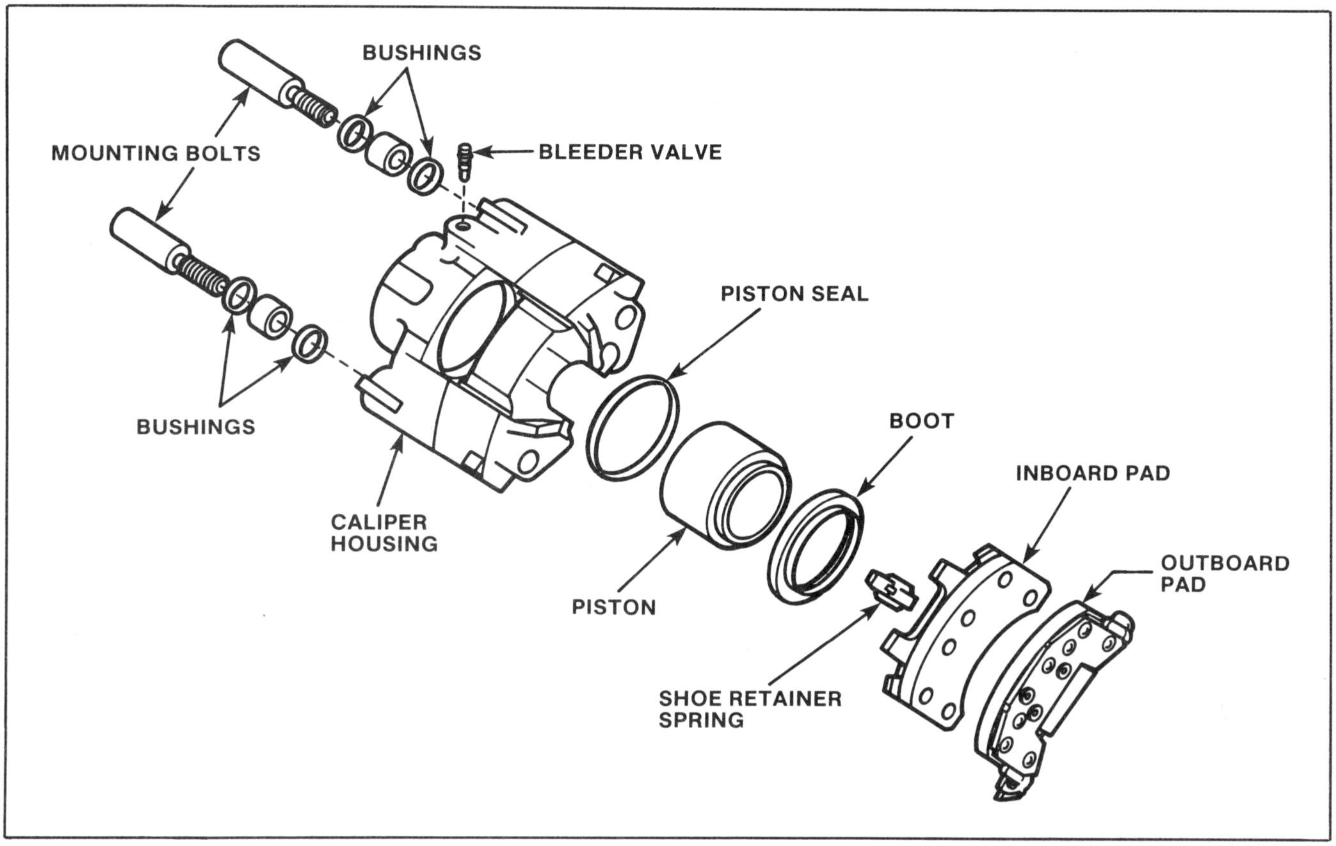

Figure 7-33 Ford front-wheel drive pin-slider disc brake

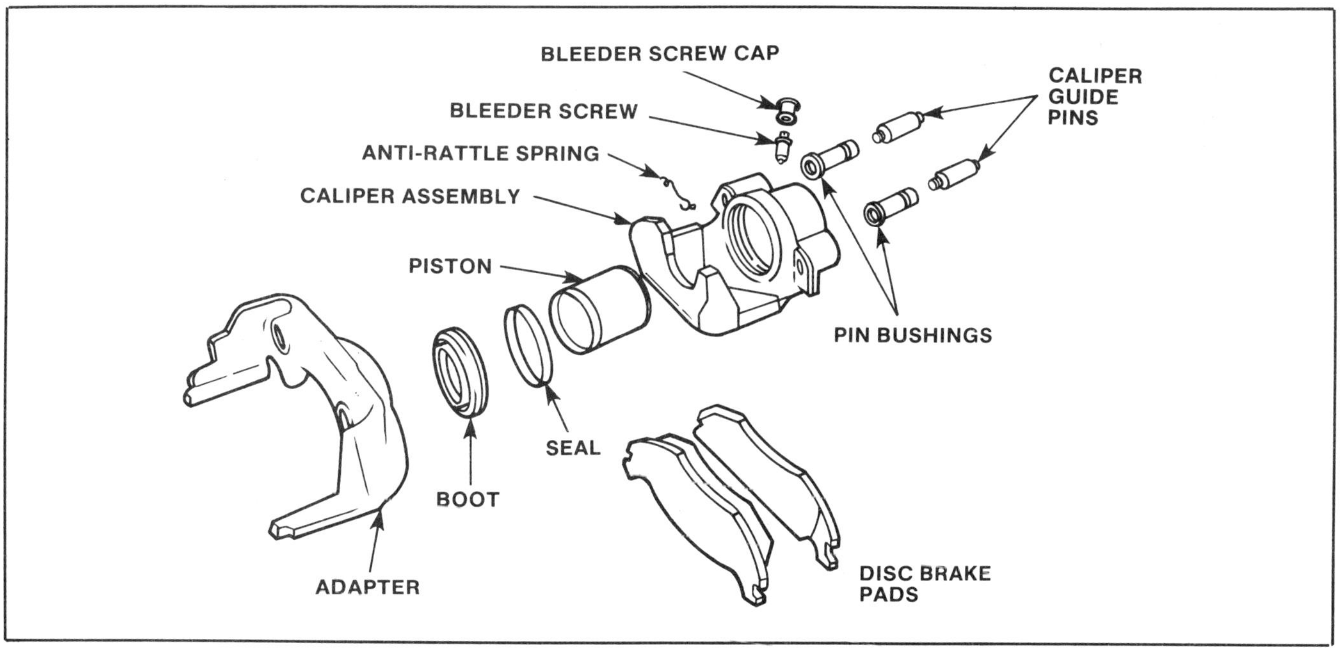

Figure 7-34 Chrysler pin-slider disc brake

REAR DISC BRAKES

Disc brakes are used on some vehicles on all four wheels because of the advantage disc brakes have over drum brakes. Rear-wheel disc brakes are similar in design to front-wheel disc brakes, and they operate in essentially the same manner. The same type of components—calipers, pads, rotors—are used, and the hydraulic system is the same. The major difference is that a parking brake is required at the rear wheels.

There are several methods of adopting the front brake floating caliper with a parking brake mechanism; for example, the Delco-Moraine single-piston rear disc brake illustrated in Figure 7-35, which mechanically operates a pre-assembled parking brake mechanism through the back of the caliper. The parking mechanism consists of the parking brake lever, the return spring, actuator screw, balance spring, and piston.

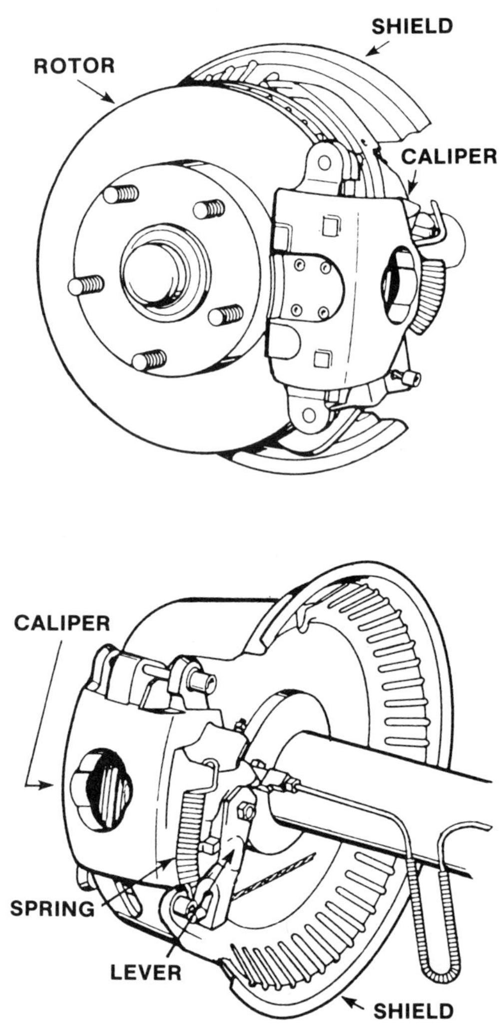

Figure 7-35 Delco-Moraine single-piston rear caliper

The nut in the piston assembly (Figure 7-36) is spring loaded, which allows the piston to move outward when hydraulic pressure is applied. As lining wear takes place and the piston moves out farther in the bore, a 50-pound adjusting spring will mechanically cause the nut to rotate along the high lead screw and reposition itself.

When the parking brake pedal is applied, the brake cable pulls the lever at the back of the caliper housing. The parking brake lever is bolted to the actuator screw, so as cable tension pulls on the lever, the actuator screw rotates. The teeth of the actuator are helical and thread into the back of the caliper piston. Thus, rotation of the screw pushes out the piston to apply the brake mechanically. When the parking brake is released, opposite rotation of the actuator allows the piston to retract.

With the Ford/Keysey-Hayes rear disc brake system shown in Figure 7-37, the parking brake mechanism is actuated by the parking brake cable. Whenever the parking brake is applied, the cable operating lever rotates the operating shaft located in the piston assembly. There are three steel balls located in pockets between the operating shaft and the thrust screw. When the operating shaft is rotated, the steel balls are forced up ramps.

As the steel balls are forced up against the ramps, the thrust screw is forced away from the operating shaft, which drives the piston and disc brake pad against the rotor. An automatic adjuster (Figure 7-38) in the piston assembly moves on the thrust screw to compensate for lining wear and to maintain proper clearance in the parking brake mechanism.

One design of single-piston sliding caliper rear disc brake that is used on some Chrysler products incorporates a small drum type parking brake (Figure 7-39). The brake drums are located within the hub section of the rotor and are thus an integral part of the brake disc. The conventional type of brake control mechanism is employed to operate the parking brake. The drum brake unit consists of an anchor plate on which the two shoes are mounted in a similar manner like the duo-servo drum brake arrangement. A star wheel type adjuster is used to provide proper shoe-to-drum clearance.

Further information on rear disc/parking brake systems can be found in Chapter 9.

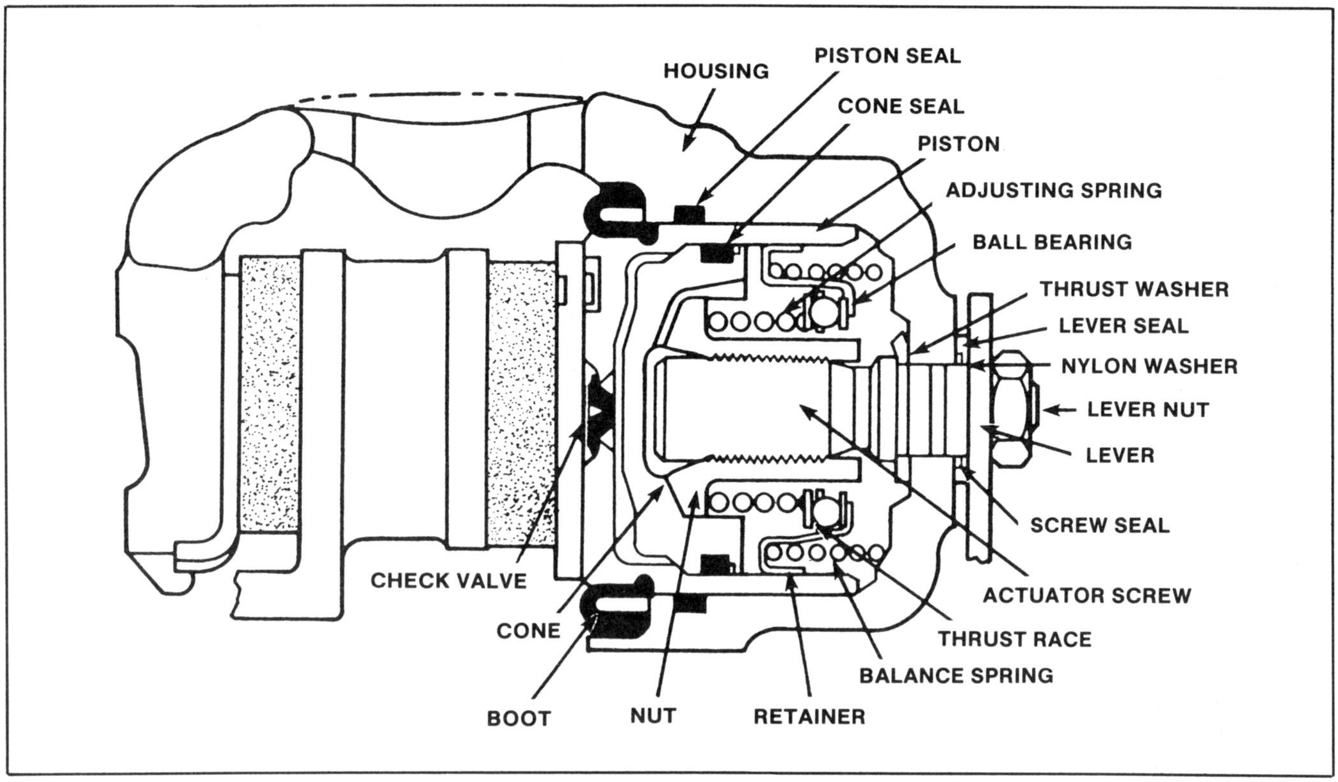

Figure 7-36 Cross section of a Delco-Moraine rear brake caliper

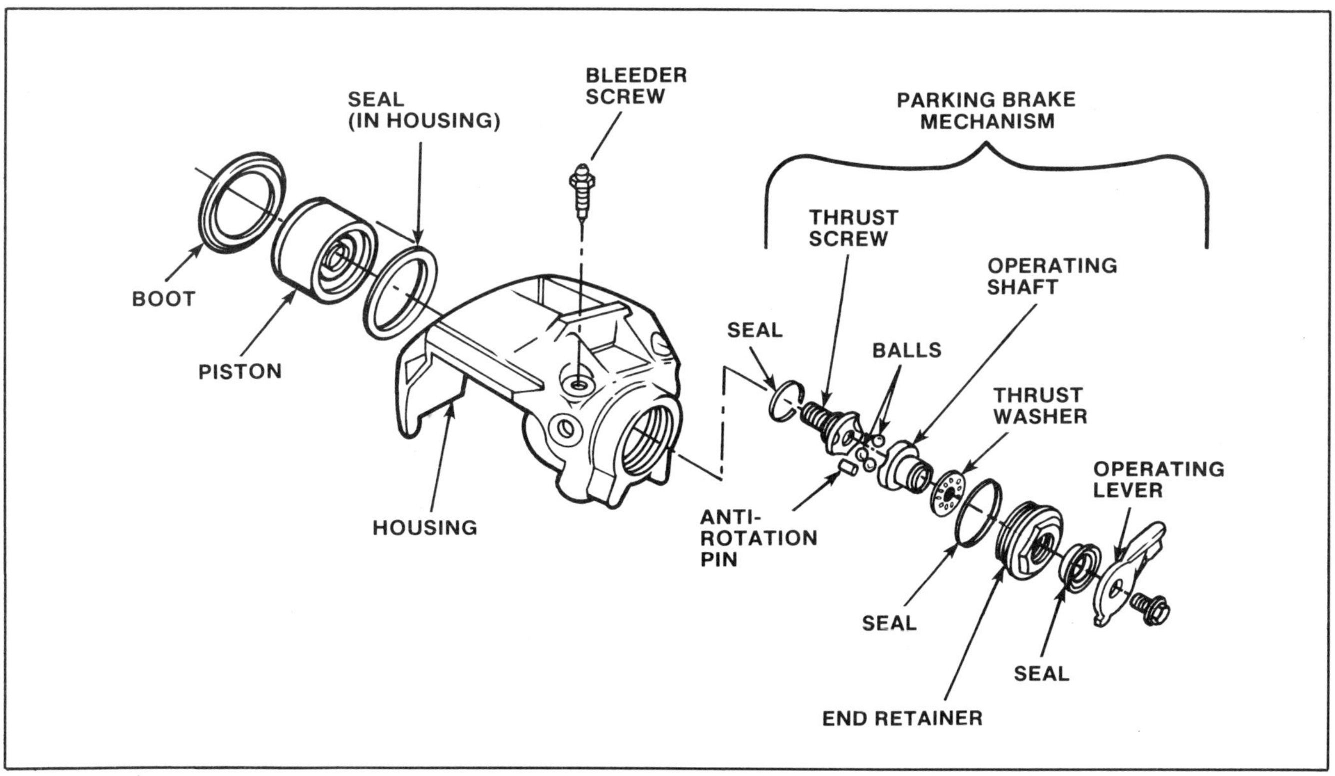

Figure 7-37 Rear disc parking brake mechanism

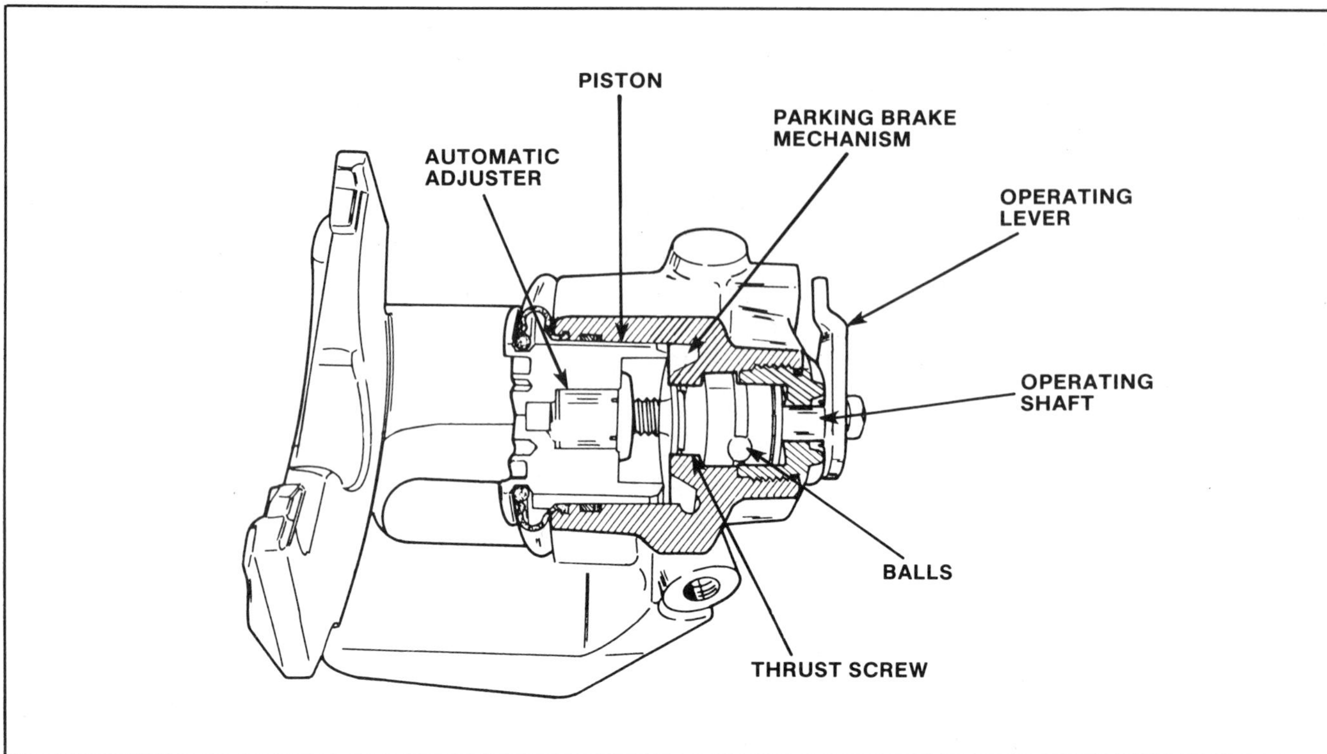

Figure 7-38 Automatic adjustment

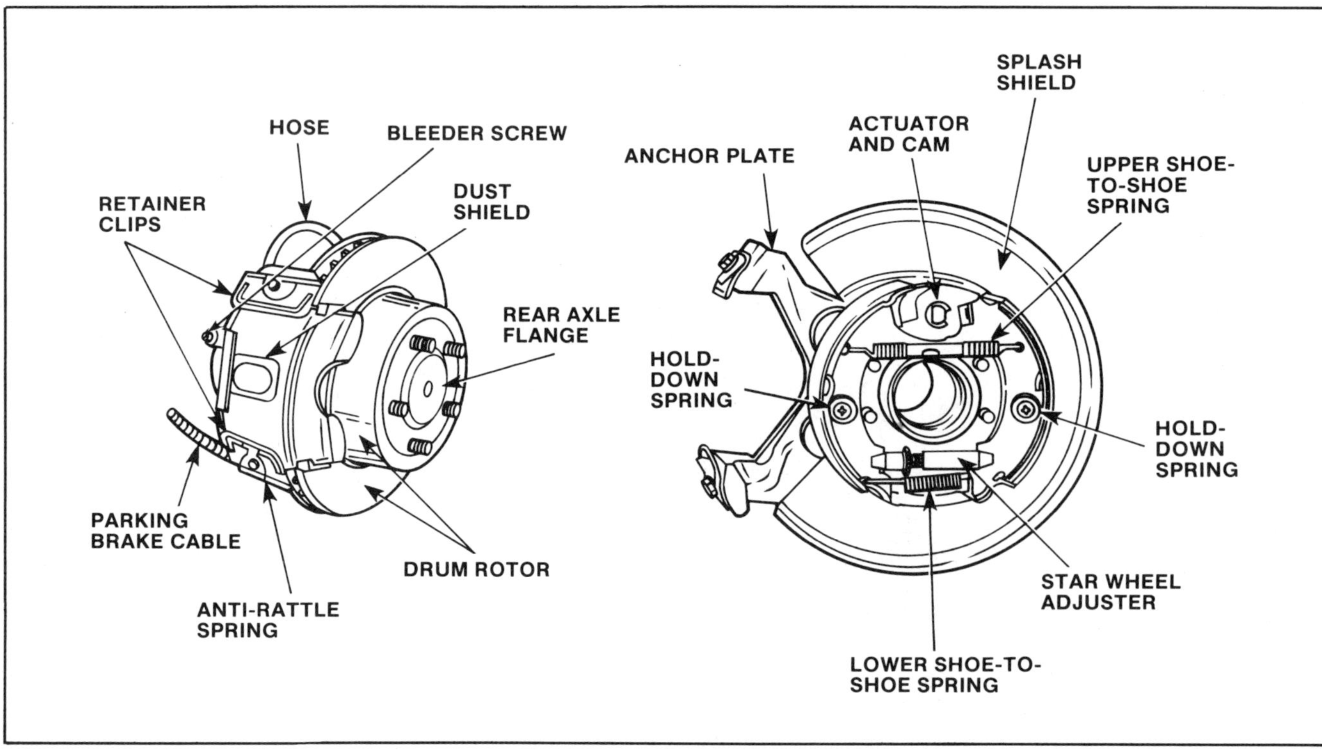

Figure 7-39 Chrysler rear disc brake

REVIEW QUESTIONS

Review the information in this chapter by answering the following questions.

1. Which of the following statements concerning disc brakes is incorrect?
   a. Disc brakes can handle more heat than drum brakes.
   b. Disc brakes are not as likely to fade during heavy braking as are drum brakes.
   c. Disc brakes stay in adjustment automatically.
   d. Disc brakes do not tolerate water well.
2. Disc brakes are not
   a. self-adjusting.
   b. rapid acting.
   c. easy to reline.
   d. self-energizing.
3. Solid rotors are used on
   a. lightweight vehicles.
   b. heavy vehicles.
   c. high-performance vehicles.
   d. none of the above.
4. What channels the flow of air over the exposed rotor surfaces?
   a. webs
   b. hub assembly
   c. splash shield
   d. caliper assembly
5. Which of the following statements concerning the caliper assembly is incorrect?
   a. The caliper housing is usually a two-piece construction of aluminum and cast iron.
   b. Caliper action converts hydraulic pressure into mechanical force.
   c. A caliper can contain as many as four cylinder bores and pistons.
   d. The caliper assembly provides a means of applying brake pads to the rotor.
6. What prevents moisture from entering the cylinder bore?
   a. phenolic piston
   b. drag caliper
   c. splash shield
   d. dust boot
7. To compensate for lining wear, Mechanic A removes some brake fluid from the caliper bore. To compensate for lining wear, Mechanic B adds extra brake fluid to the caliper bore. Who is right?
   a. Mechanic A
   b. Mechanic B
   c. both A and B
   d. neither A nor B
8. If the disc brake seals lose their ability to flex and retract the piston, hydraulic pressure will cause
   a. the caliper to bounce up and down.
   b. the pads to drag against the rotor.
   c. a larger clearance between the pads and rotor.
   d. moisture and/or dirt to enter the cylinder bore.
9. In addition to cushioning caliper movements, what other function does the insulator/bushings perform?
   a. secure the pads in the caliper housing to prevent vibration
   b. eliminate brake noise from metal-to-metal contact
   c. prevent the caliper from bouncing up and down
   d. hold the caliper weight off the rotor
10. Semi-metallic brake pads are edge branded
    a. EE.
    b. FF.
    c. GG.
    d. FG.
11. Which type of disc brake was the first to appear in significant numbers on American cars?
    a. five-piston
    b. four-piston
    c. three-piston
    d. two-piston
12. How many pistons do floating and sliding caliper designs usually have?
    a. four
    b. three
    c. two
    d. one
13. For better corrosion resistance, Mechanic A replaces a phenolic piston with a steel piston. To reduce the amount of fluid loss due to vaporization, Mechanic B replaces a phenolic piston with a steel piston. Who is right?
    a. Mechanic A
    b. Mechanic B
    c. both A and B
    d. neither A nor B

14. Which of the following statements concerning rear disc brakes is incorrect?
    a. Rear-wheel disc brakes are similar in design to front-wheel disc brakes.
    b. On the Ford/Keysey-Hayes rear disc brake system, the parking brake mechanism is actuated by the parking brake cable.
    c. Rear-wheel disc brakes have the same hydraulic system as front-wheel disc brakes.
    d. The Delco-Moraine single-piston rear disc brake incorporates a small drum type parking brake.

15. A pin-slider disc brake is a lightweight version of the
    a. sliding caliper design.
    b. two-piston fixed caliper design.
    c. four-piston fixed caliper design.
    d. none of the above.

CHAPTER 8

DISC BRAKES—INSPECTING AND SERVICING

OBJECTIVES

Upon completing this chapter, you should be able to

- Identify general service precautions to be followed when working on disc brake systems.
- Inspect and service many types of calipers.
- Remove and install brake pads.
- Identify and rectify malfunctions in rotors using inspecting, replacing, and machining procedures.
- Clean and repack wheel bearings.
- Service front and rear disc brakes on many popular domestic cars.

Procedures outlined in the beginning of this chapter are of a general nature and apply to virtually any make of car. These steps should be followed as a matter of course whenever servicing a disc brake system. Following this general discussion is a series of service procedures for specific brake system designs used on popular domestic cars. (Specific truck and import model brake information is given in Chapters 12 and 13.) Use these steps in conjunction with the general procedures below.

Disc brake problems are usually the result of improperly fitting parts or pads or distortions in rotor surfaces. Disc brake specifications do not conform to basic standards as drum brakes do. Inspect each disc brake using the correct manufacturer's shop manuals or charts.

SERVICE PRECAUTIONS

The following general service precautions apply to all disc brake systems and should be reviewed before studying the specific servicing techniques in this chapter.

1. Road test the vehicle to determine the condition of the brakes.
2. Be sure the vehicle is properly centered and secured on stands or a hoist. (See Chapter 6 for details.)
3. Disconnect the battery ground cable.
4. Before any service is performed, inspect the wheel and brake assembly for obvious damage that could affect braking, and so on. Check the following:
   - Tires for excessive wear or improper inflation
   - Wheels for bent or warped rims
   - Wheel bearings for looseness or wear
   - Suspension components to see if worn or broken
   - Brake fluid level
   - Master cylinder, brake lines or hoses, and each wheel for leaks
5. During servicing, grease, oil, brake fluid, or any other foreign material must be kept off the brake linings, caliper, surfaces of the disc, and external surfaces of the hub. Handle the brake disc and caliper in such a way to avoid deformation of the disc and nicking or scratching brake linings.
6. Work on one wheel at a time to avoid "popping" pistons out of the other caliper and to allow the other caliper assembly to be used as a guide.
7. When removing wheels, be sure to avoid damage to the rotor, external lines, bleeder screws, and splash shield.
8. When the hydraulic hose is disconnected, plug it to prevent any foreign material from entering.

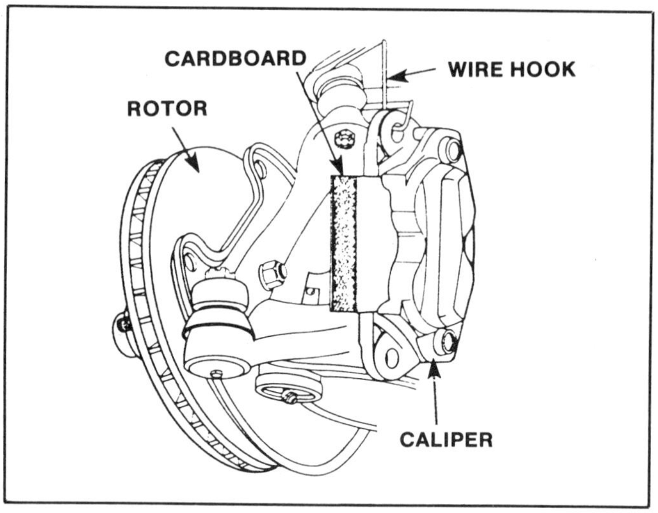

Figure 8-1 Caliper dismounted

9. Do not attempt to remove the hub with the wheel and tire still mounted. The wheel and tire must be dismounted and the caliper removed before the hub and rotor can be dismounted.
10. Never permit the caliper assembly to hang with the weight on the brake hose. Support it on the suspension or hang it by a piece of wire (Figure 8-1).
11. Inspect the caliper for leaks. If leakage is present, the caliper must be overhauled.
12. Use crocus cloth to remove rust, corrosion, pitting, and scratches from the piston bore. If the bore cannot be cleaned with crocus cloth, light honing is permitted. Do not hone a plated bore.
13. When using compressed air to remove caliper piston(s), avoid high pressures. A safe pressure to use is 30 PSI.
14. Clean the brake components in either denatured alcohol or clean brake fluid. Do not use mineral-based cleaning solvents such as gasoline, kerosene, carbon tetrachloride, acetone, or paint thinner to clean the caliper. It will cause rubber parts to become soft and swollen in an extremely short time.
15. Lubricate any moving member such as the caliper housing or mounting bracket to assure a free-moving action. Use only recommended lubricant.
16. Before the brake pads are installed, apply a disc brake noise suppressor to the back of the pads to prevent brake squeal. For best results, follow the directions on the container.
17. The front-wheel bearing should be adjusted to the manufacturer's specifications.
18. Check the master cylinder fluid level and be sure the reservoirs are filled when the brake job is completed.
19. Obtain a firm brake pedal after servicing the brakes and before moving the vehicle. Be sure to road test the vehicle (see Chapter 14).
20. Always torque the lug nuts when installing a wheel on a vehicle with disc brakes. Never use an impact gun to tighten the lug nuts. Warpage of the rotor could result if an impact gun is used.

Before beginning brake work, remove about two-thirds of the brake fluid from the front or disc brake reservoir on a front/rear split system (Figure 8-2). On a diagonally split system, remove fluid from both reservoirs. If this is not done, the fluid could overflow and spill when the piston(s) is forced back into the caliper bore, possibly damaging the painted surfaces. Replace the cover. Discard old brake fluid.

Another common procedure is to open the caliper bleeder screw and run a hose down to a container to catch the fluid that is expelled when the piston is forced back into its bore. This also makes it easier to move the piston; sometimes just moving the caliper by hand or prying with a screwdriver will do it rather than having to use a C-clamp. It is still a good precaution, however, to remove fluid from the reservoir.

If the bleeder screws are frozen tight with corrosion, it is sometimes possible to free them using a propane torch and penetrating oil. Of course, the caliper will have to be removed from the car, taken to a bench, and stripped. If the bleeder screws cannot be loosened, they can be drilled out and the

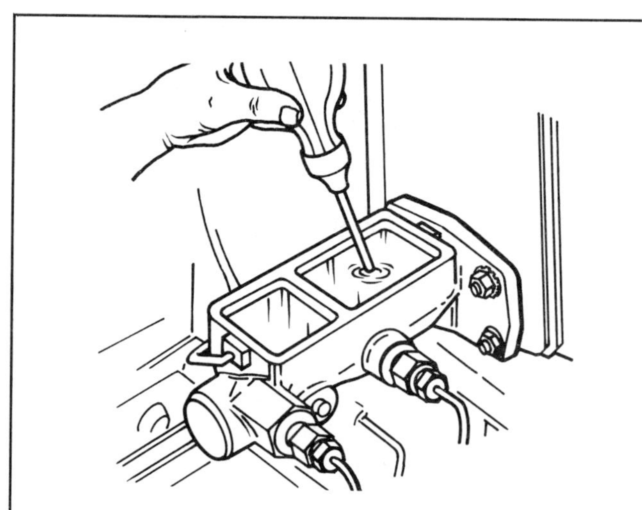

Figure 8-2 Removing brake fluid from master cylinder reservoir

caliper retapped for an insert, or the caliper can be replaced with a new or rebuilt unit. The bleeder screw should be removed when doing an overhaul.

CAUTION: When using the propane torch to loosen a bleeder screw, use it with extreme care.

GENERAL CALIPER INSPECTING AND SERVICING

The general procedures involved in a complete caliper service or overhaul include tasks such as:

- Caliper removal
- Brake pad removal
- Caliper disassembly
- Caliper assembly
- Brake pad installation
- Caliper installation

Not all caliper jobs include every one of these tasks. Frequently, the caliper service job involves only the removal and installation of the brake pads. However, since the new pads will be thicker than the worn-out set they replace, they will locate the piston farther back in the bore where dirt and corrosion might cause the seals to leak. For this reason, it is often a good procedure to rebuild the calipers whenever installing new pads. Of course, sometimes it might be necessary to add the task of rotor machining and service to do a thorough job.

When bench working a caliper assembly, use a vise that is equipped with protector jaws. Excessive vise pressure causes bore distortion and bending of the piston.

SHOP TALK

Many smaller brake operations do not rebuild calipers; they farm them out to specialty shops. However, the brake technician must at least know how to rebuild calipers.

CALIPER REMOVAL

The first step in proper caliper service is to remove the caliper assembly from the vehicle. Depending upon the caliper design, this operation is accomplished in various ways. Specific caliper removal procedures are given later in this chapter. General details of the task are as follows:

1. Raise the vehicle and remove the wheel and tire assembly as described in Chapter 6. Use care to avoid damage to or interference with bleeder screw fitting during removal.

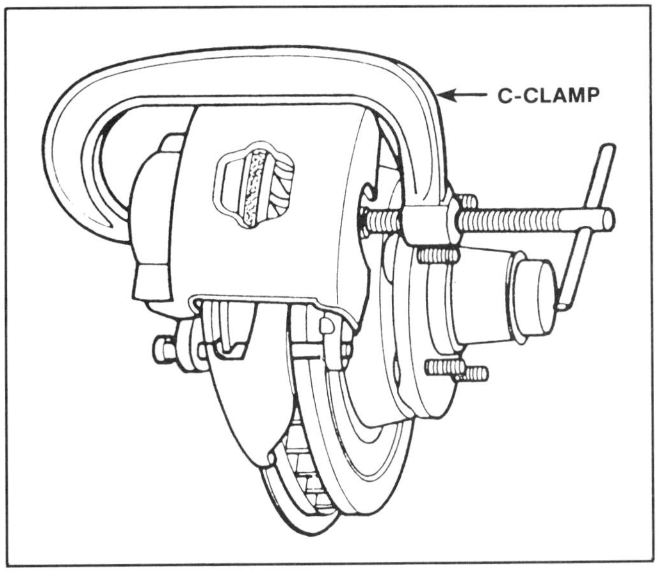

Figure 8-3 Bottoming piston in bore

2. Mark the right-hand and left-hand caliper assemblies with chalk prior to removal from the vehicle, so they can be positioned correctly during installation.
3. On a sliding or floating caliper, install a C-clamp on the caliper with the solid end of the clamp on the caliper housing and the screw end on the metal portion of the outboard brake pad. Tighten the clamp until the piston bottoms in the caliper bore (Figure 8-3), then remove the clamp. Bottoming the piston allows room for the brake pad to slide over the ridge of rust that accumulates on the edge of the rotor.
4. Disconnect the brake hose from the caliper and remove the copper gasket or washer and the cap end of the brake end. Typical examples of front-wheel brake line mounting methods are shown in Figure 8-4. If only the brake pads are to be replaced, do not disconnect the brake hose.
5. Remove the two mounting brackets to the steering knuckle bolts. Support the caliper when removing the second bolt to prevent the caliper from falling.
6. On a sliding caliper, remove the top bolts, retainer clip, and anti-rattle springs (Figure 8-5). On a floating caliper, remove the two special pins that hold the caliper to the anchor plate (Figure 8-6). On the older type fixed caliper, remove the bolts holding it to the steering knuckle. On all three types, get

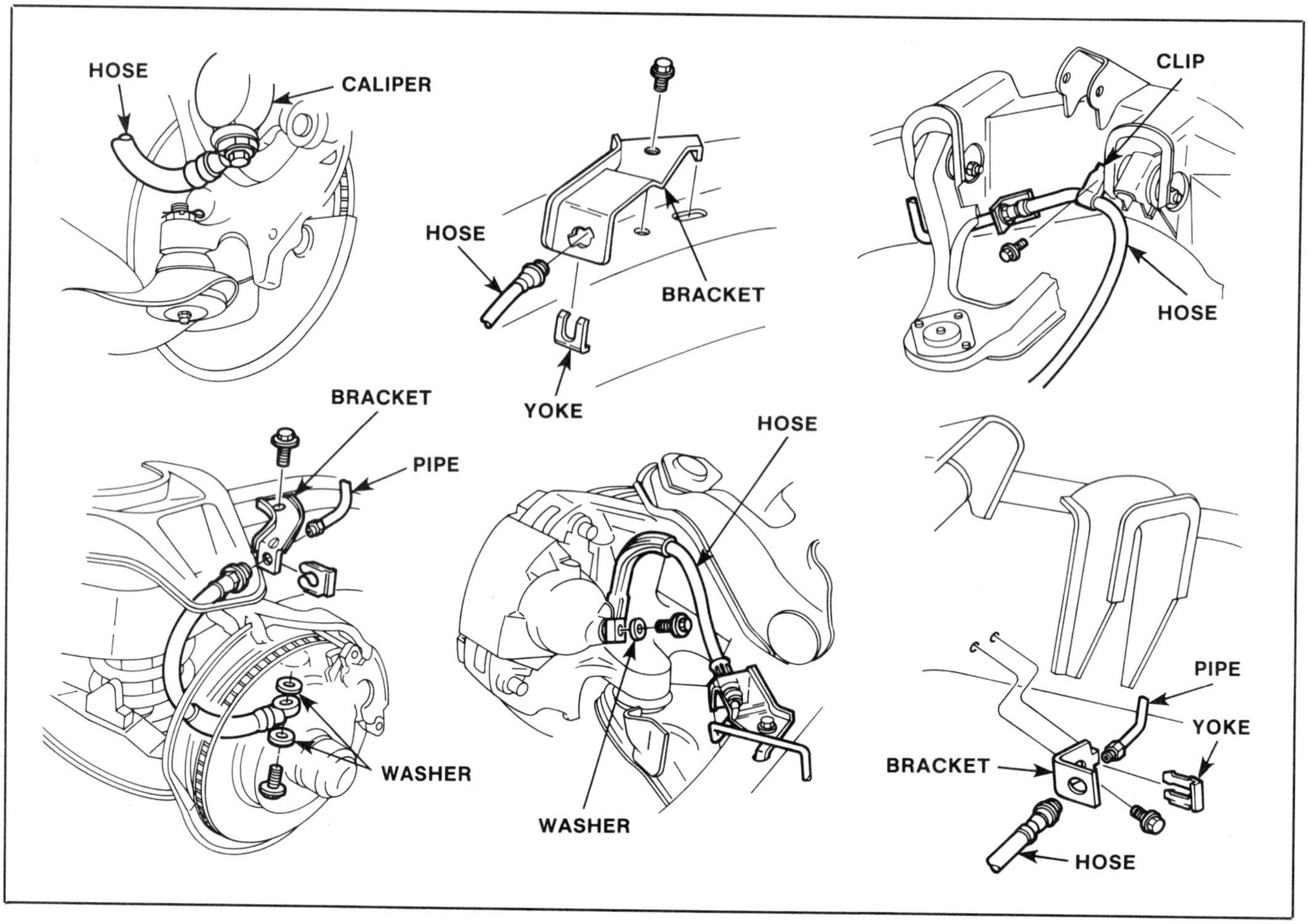

Figure 8-4 Typical examples of front-wheel brake line mounting methods

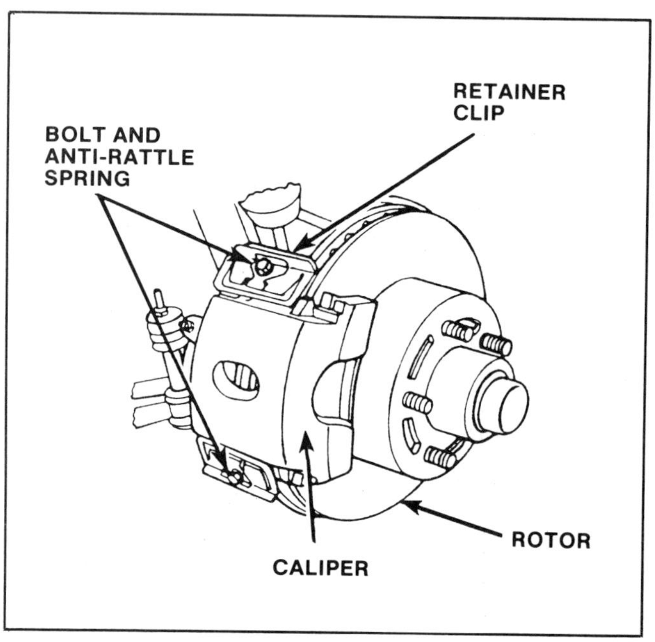

Figure 8-5 Sliding caliper removal

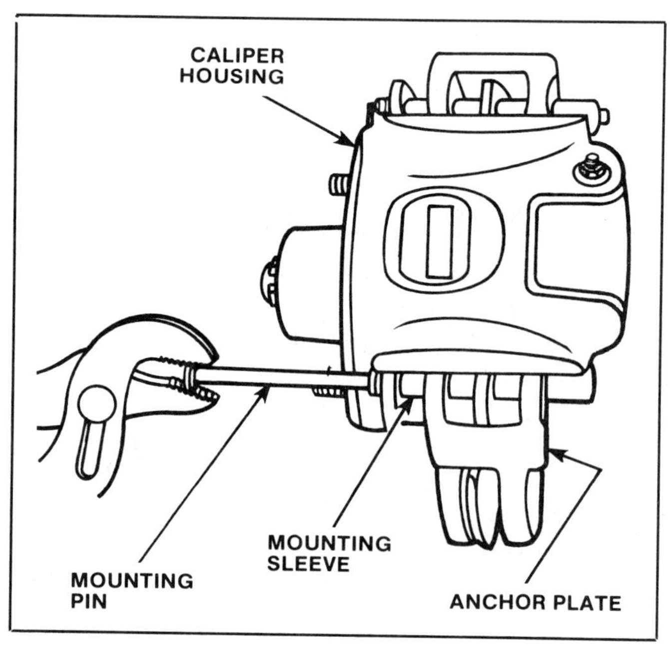

Figure 8-6 Floating caliper removal

MANUFACTURER'S SPECIFIED BRAKE SPECIFICATIONS

DOMESTIC PASSENGER CARS

NOTICE: Should state inspection regulations exceed manufacturer's specifications for lining or rotor/drum reserve, the state inspection regulation specification **must** be used.

VEHICLE	BRAKE SHOE	BRAKE DRUM		BRAKE PAD	BRAKE ROTOR				CALIPER			WHEEL	WHEEL BEARING SETTING		
	*	DIAMETER			MIN. THICKNESS										
YEAR, MAKE, MODEL	Minimum Lining Thickness	Standard Size	Machine To	Minimum Lining Thickness	Machine To	Discard At	Variation From Parallelism	Runout T.I.R.	DESIGN	Mounting Bolts Torque (ft-lbs)	Bridge, Pin or Key Bolts Torque (ft-lbs)	Lugs or Nuts Torque (ft-lbs)	STEP 1 Tighten Spindle Nut (ft-lbs)	STEP 2 Back Off Retorque (in-lbs)	STEP 3 Lock, or Back Off and Lock
AMERICAN MOTORS—ALLIANCE, ENCORE, GTA (SEE RENAULT)															
87-82 Eagle	.030	10.000	10.060	.030*	.815	.810	.0005	.004	1	100	30	75	25	6	Step 2
83-82 Concord exc. Wagon; Spirit	.030	9.000	9.060	.030*	.815	.810	.0005	.003	1	85	30	75	25	6	Step 2
83-82 Concord Wagon	.030	10.000	10.060	.030*	.815	.810	.0005	.003	1	85	30	75	25	6	Step 2
81 All exc. 6 cyl. Concord Wagon and Eagle	.030	9.000	9.060	.030*	.815	.810	.0005	.003	1	80	15	75	25	6	Step 2
81 6 cyl. Concord Wagon and Eagle	.030	10.000	10.060	.030*	.815	.810	.0005	.003■	1	80	15	75	25	6	Step 2
80-78 Spirit exc. 4 cyl., Concord, Gremlin, exc. 4 cyl; Pacer, Eagle, AMX	.030	10.000	10.060	.062	.815	.810	.0005	.003■	1	80	15	75	25	6	Step 2
80-78 Spirit & Gremlin w/4 cyl.	.030	9.000	9.060	.062	.815	.810	.0005	.003	1	80	15	75	25	6	Step 2
78 Matador	.030	10.000	10.060	.062	—	1.120	.0005	.003	1	80	15	75	25	6	Step 2
77 Gremlin, Hornet, Pacer	.030	10.000	10.060	.062	.815	.810	.0005	.003	1	80-90	15-18	75-90	25	6	Step 2
77 Matador	.030	10.000	10.060	.062	—	1.120	.0005	.003	1	80-90	15-18	75-90	25	6	Step 2
76-75 Pacer w/drum brakes, Front	.030	10.000	10.060	—	—	—	—	—	—	—	—	60-90	22	6	Step 2
Rear	.030	9.000	9.060	—	—	—	—	—	—	—	—	75	—	—	—
76-75 Pacer w/disc brakes	.030	9.000	9.060	.062	—	1.120	.0005	.003	1	80	15	60-90	22	6	Step 2
76-75 Hornet & Gremlin w/6 cyl	.030	9.000	9.060	.062	1.130	1.120	.0005	.003	1	80	15	75	22	6	Step 2
76-75 Hornet & Gremlin w/V8 and All Matador	.030	10.000	10.060	.062	1.130	1.120	.0005	.003	1	80	15	75	22	6	Step 2
74-72 Hornet, Gremlin & Javelin 6 cyl. w/drum brakes	.030	9.000	9.060	—	—	—	—	—	—	—	—	75	20	6	Step 2
74-72 Ambassador, Matador, Hornet V8, Gremlin V8 & Javelin V8 w/drum brakes	.030	10.000	10.060	—	—	—	—	—	—	—	—	75	20	6□	Step 2
74-73 Hornet, Gremlin & Javelin w/6 cyl. w/disc brakes	.030	9.000	9.060	.062	—	.940	.0005	.005	2	80	—	75	20	6	Step 2
74-72 Hornet V8, Gremlin V8 Javelin V8, All Matador & Ambassador w/disc brakes	.030	10.000	10.060	.062	—	.940	.0005	.005	2	80	—	75	20	6□	Step 2
72 Hornet, Gremlin & Javelin w/6 cyl	.030	10.000	10.060	.062	—	.940	.0005	.005	2	80	—	75	20	12	Step 2

■Eagle-.004 □1972-12 in/lbs. *.030″ over rivet head; if bonded lining, use .062″

Design 1 is sliding caliper type (Bendix, Kelsey-Hayes).
Design 2 is floating caliper type (Kelsey-Hayes).

Figure 8-7 Typical disc brake torque and specification chart

the caliper off by prying it straight up and lifting it clear of the rotor.

BRAKE PAD REMOVAL

Disc brake linings should be checked periodically or whenever the wheels are removed. Some calipers have inspection holes in the caliper body; if not, the brake pads can be visually inspected from the outer ends of the caliper.

If the friction pads appear worn and in need of replacement, measure them at the thinnest part of the pad. If it is 1/32-inch thick or less, replace the pads. Refer to the disc brake torque and specification chart for minimum brake pad lining thickness (Figure 8-7).

Uneven pad wear on a sliding caliper often means that the caliper is sticking and not giving equal pressure to both pads. On a sliding caliper, the problem could be that the caliper "ways" are not allowing a smooth sliding movement. Check these machined ways for proper clearance. A slightly tapered wear pattern on the pads of certain models is caused by caliper twist during braking. It is normal if it does not exceed 1/8-inch taper from one end of the pad to the other (Figure 8-8).

Sliding or floating calipers must always be lifted off the rotor for pad replacement. Fixed calipers might have pads that can be replaced by removing the retaining pins or clips instead of having to lift off the entire caliper. They can be held in position by retaining pins (Figure 8-9), guide pins (Figure

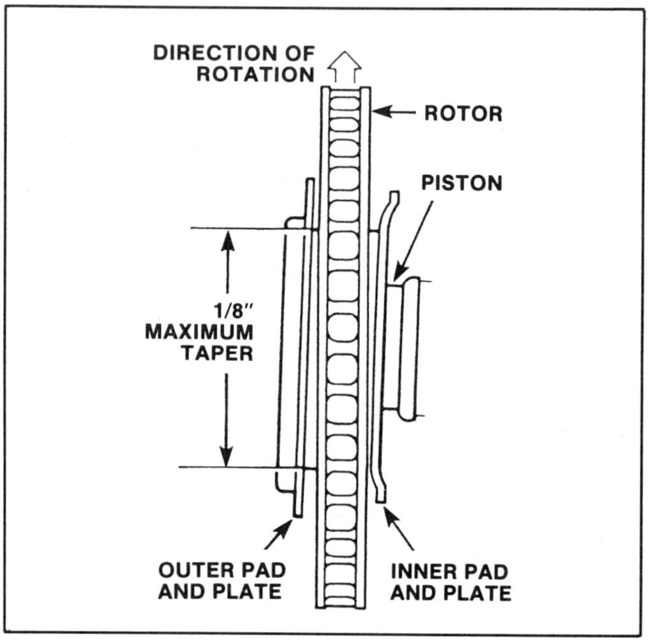

Figure 8-8 Pad wear pattern

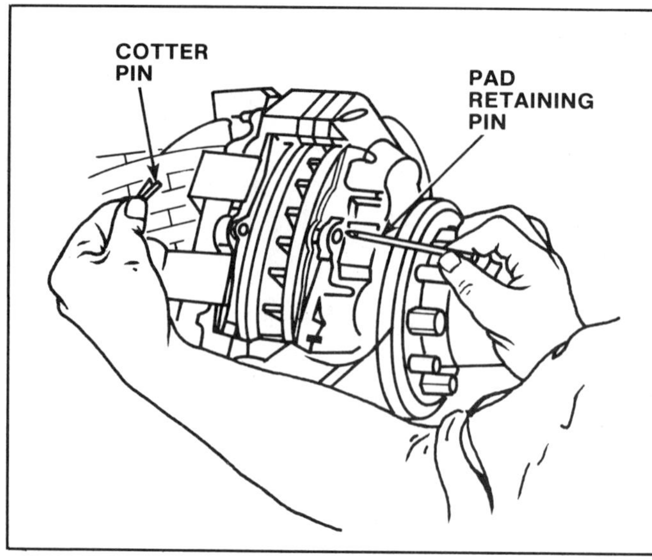

Figure 8-9 Removing brake pad retaining pins

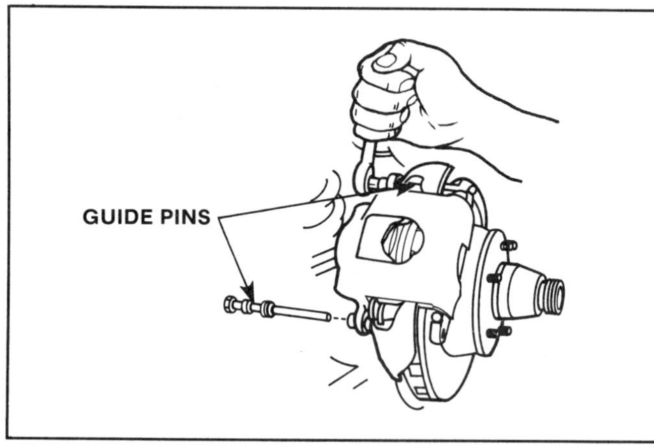

Figure 8-10 Removing guide pins

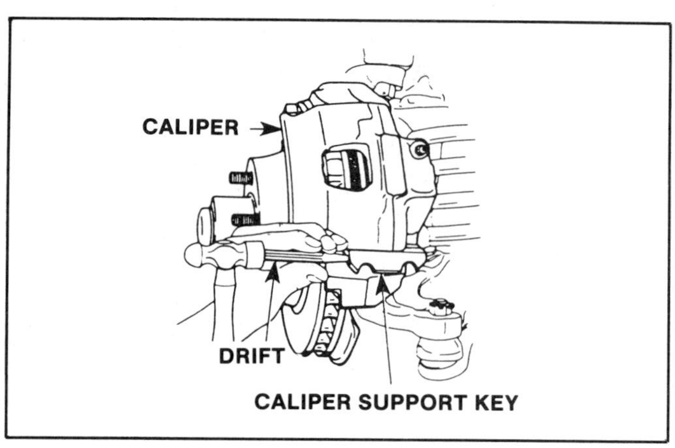

Figure 8-11 Removing caliper support key

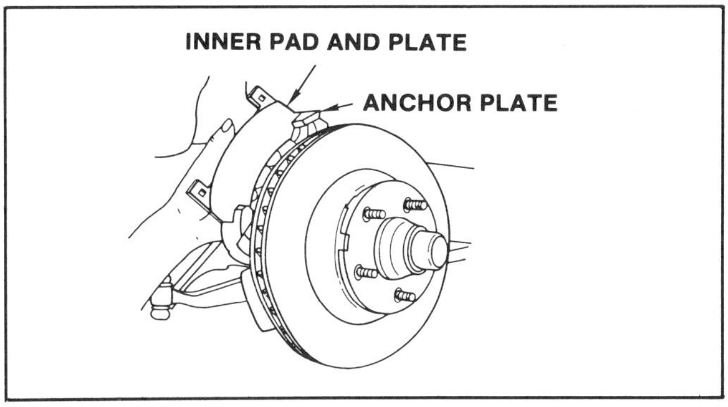

Figure 8-12 Removing inner pad and plate

8–10), or a support key (Figure 8–11). Note the position of the shims, anti-rattle clips, keys, bushings, or pins during disassembly.

If only the pads have to be replaced, lift the caliper off the rotor and hang it up by a wire. Remove the outer pad and inner pad (Figure 8–12). Remove the old sleeves and bushings and install new ones. Replace rusty pins on a floating caliper to provide for free movement. Transfer shoe retainers, which can be clips or springs, onto the new pads.

SHOP TALK

If new pads and plates are not to be installed immediately, insert several thicknesses of clean cardboard or plastic in the caliper. This keeps the pistons from coming out of caliper cylinders.

CALIPER DISASSEMBLY

If the caliper must be rebuilt or completely overhauled, it should be taken to the workbench for servicing. Drain any brake fluid from the caliper by way of bleeder screws. Remove the bleeder valve protector, if so equipped.

On a floating caliper, examine the mounting pins for rust that could limit travel. Most manufacturers recommend that these pins (Figure 8–13) and their bushings be replaced each time the caliper is removed. This is a good idea because the pins are inexpensive and a good insurance against costly comebacks. On a fixed caliper, check the pistons for sticking and rebuild the caliper if this problem is found.

To disassemble the caliper, the piston and dust boot must first be removed. Use compressed air to pop the piston out of the bore. To do this, proceed as follows:

1. Position the caliper face down on a bench (Figure 8–14).
2. Insert the used outer pad into the caliper. Place a folded shop towel on the face of the lining to cushion the piston.
3. Apply low air pressure (never more than 30 PSI) to the fluid inlet port of the caliper to force the piston from the caliper housing.

WARNING: Be careful to apply air pressure very gradually, and be sure there are enough cloths to catch the piston when it comes out of the bore. Never place your fingers in front of the piston for any reason when applying compressed air. Personal injury could occur if the piston is "popped" out of the bore.

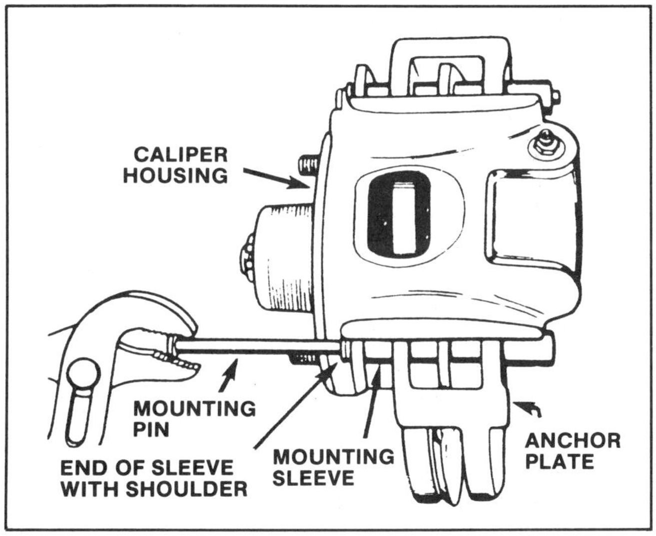

Figure 8-13 Removing mounting pins

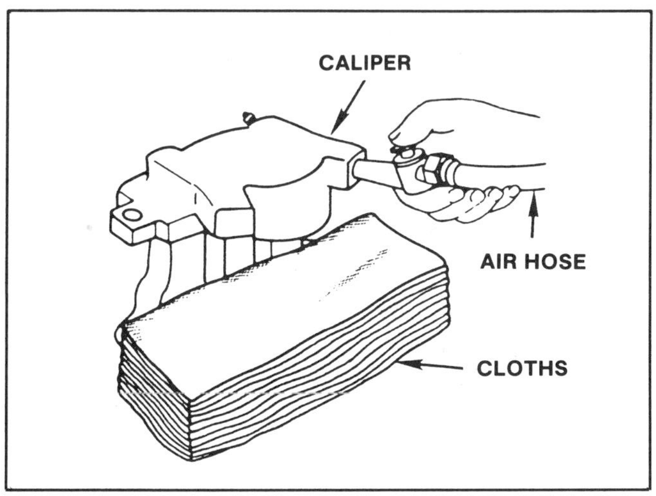

Figure 8-14 Removing piston

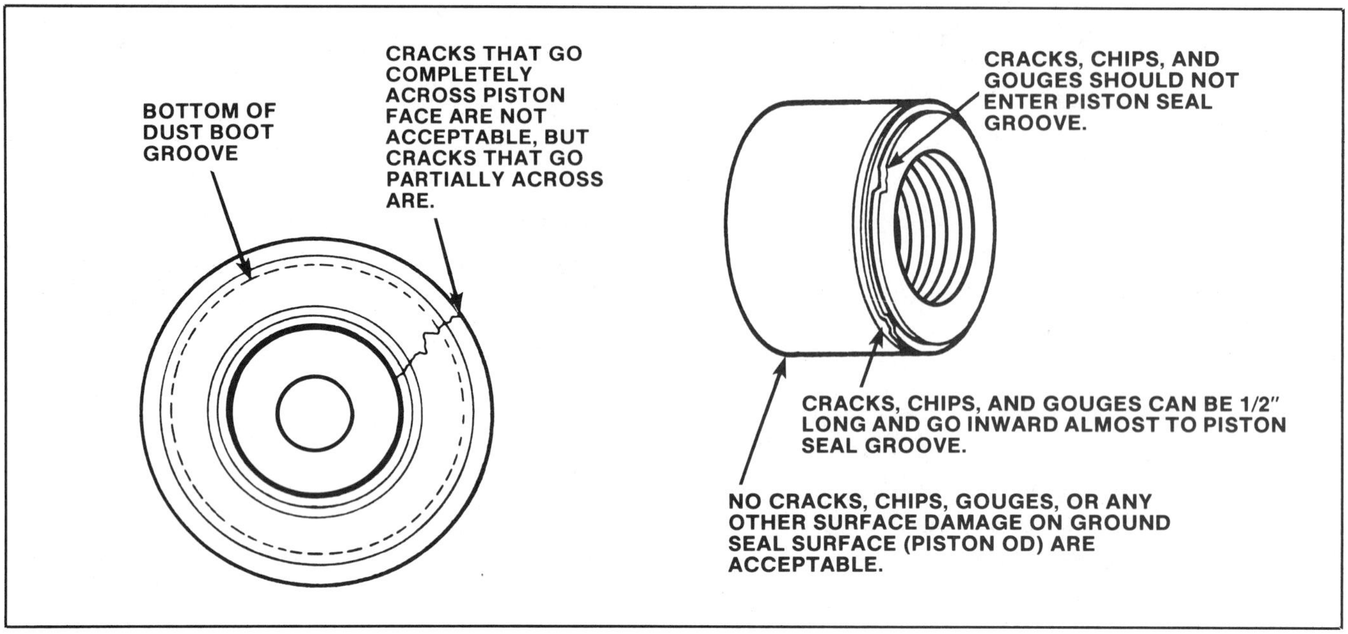

Figure 8-15 Inspect phenolic piston for surface irregularities

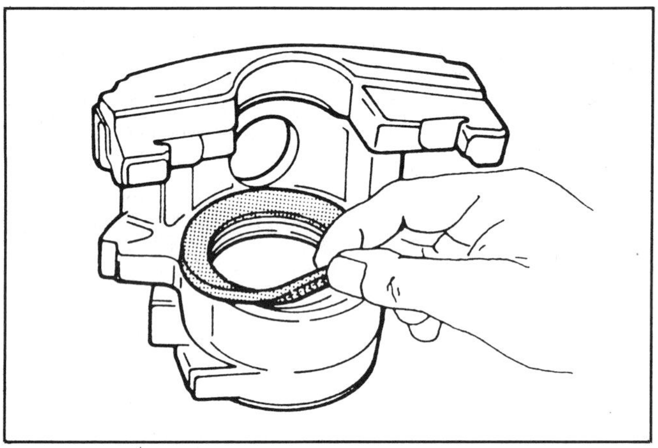

Figure 8-16 Peeling off dust boot

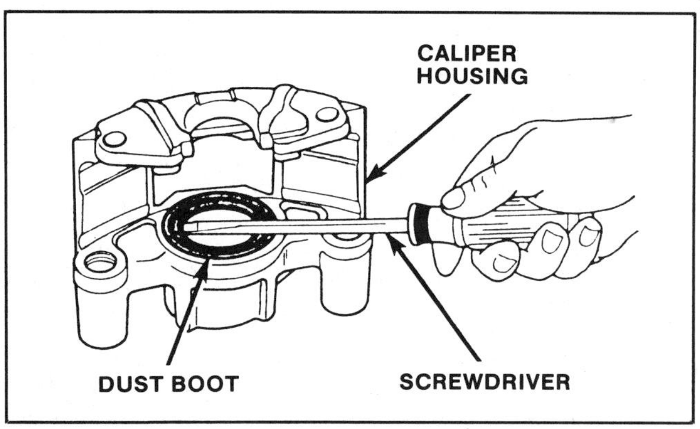

Figure 8-17 Prying out dust boot with a screwdriver

4. If a piston is frozen, release air pressure and tap the end of the piston with a soft-headed hammer or mallet. Reapply air pressure.
5. Frozen phenolic (plastic) pistons can be broken into pieces with chisel and hammer.

CAUTION: Protect the eyes with safety glasses. Avoid damaging the caliper bore and seal groove with the chisel.

6. Internal expanding pliers are sometimes used to remove pistons from caliper bores.

Inspect phenolic pistons for cracks, chips, or gouges (Figure 8-15). Replace the piston if any of these conditions are evident. If the plated surface of a steel piston is worn, pitted, scored, or corroded, it also should be replaced.

Dust boots vary in design depending on the type of piston and seal, but they all fit into one groove in the piston and another groove in the cylinder. One type comes out with the piston and peels off. Another type stays in place and the piston comes out through the boot, and then is removed from the cylinder (Figure 8-16). In either case, peel the boot from its groove. In some cases it might be necessary to pry out the dust boot with a screwdriver (Figure 8-17). When prying the dust boot, be careful not to scratch the cylinder bore. The old boot can be discarded since it must be replaced along with the seal.

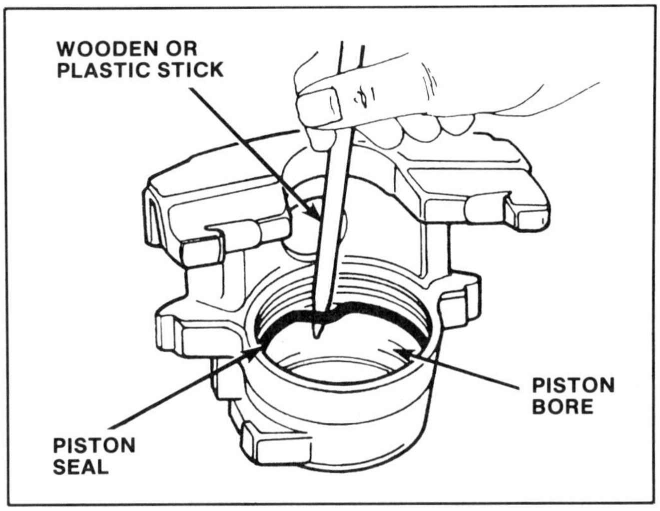

Figure 8-18 Removing piston seal with a wooden or plastic stick

Remove a stroking seal by prying it out of the piston with a wooden or plastic tool. Pry a fixed seal out of the cylinder with the same type of tool (Figure 8-18). Do not use a screwdriver or other metal tool. Any of these could nick the metal in the caliper bore and cause a leak. Inspect the bore for pitting or scoring. A bore that shows light scratches or corrosion can usually be cleaned with crocus cloth. However, a bore that has deep scratches or scoring can be honed, provided the diameter of the bore is not increased more than 0.002 inch. If the bore does not clean up within this specification, a new caliper housing should be installed (black stains on the bore walls are caused by piston seals and will do no harm).

When using a hone (Figure 8-19), be sure to install the hone baffle before honing the bore. The baffle protects the hone stones from damage. Use extreme care in cleaning the caliper after honing.

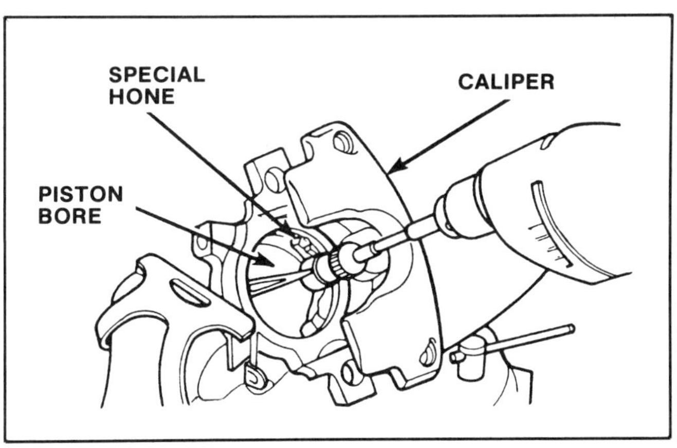

Figure 8-19 Honing caliper bore

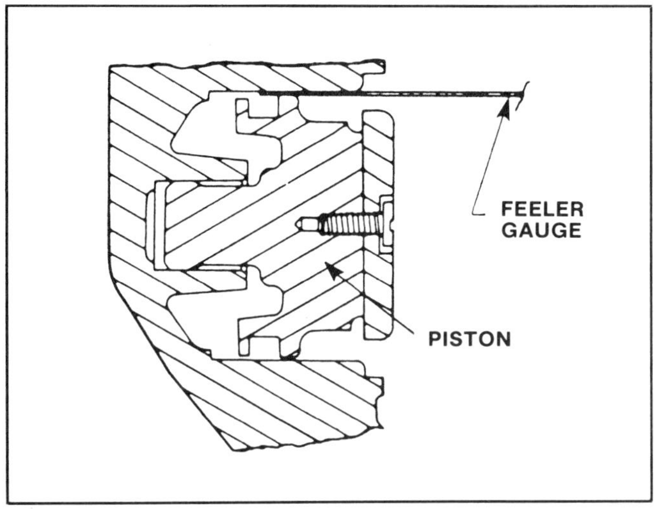

Figure 8-20 Checking piston clearance with feeler gauge

Remove all dust and grit by flushing the caliper with alcohol. Wipe it dry with a clean lint-free cloth and then clean the caliper a second time in the same manner.

CALIPER ASSEMBLY

Before assembling the caliper, clean all metal parts that will be reused and the phenolic piston (if so equipped) in clean denatured alcohol or brake fluid. Then, clean out and dry the grooves and passageways with compressed air. Make sure that the caliper bore and component parts are thoroughly clean.

Some designs of disc brakes require that the piston-to-bore clearance be checked with a feeler gauge (Figure 8-20). If clearances exceed the limits listed in the service manual, the piston must be replaced.

To replace a typical piston seal, dust boot, and piston, proceed as follows:

1. Lubricate the new piston seal with clean brake fluid or assembly lubricant (usually supplied with the caliper rebuild kit). Make sure the seal is not distorted. Insert it into the groove in the cylinder bore so it does not become twisted or rolled.
2. Install a new dust boot by setting the flange squarely in the outer groove of the caliper bore.
3. Coat the piston with brake fluid or assembly lubricant and install it in the cylinder bore (Figure 8-21). Be sure to use a wood block or other flat stock when installing the piston back into the piston bore. Never apply a C-clamp directly to a phenolic piston, and be

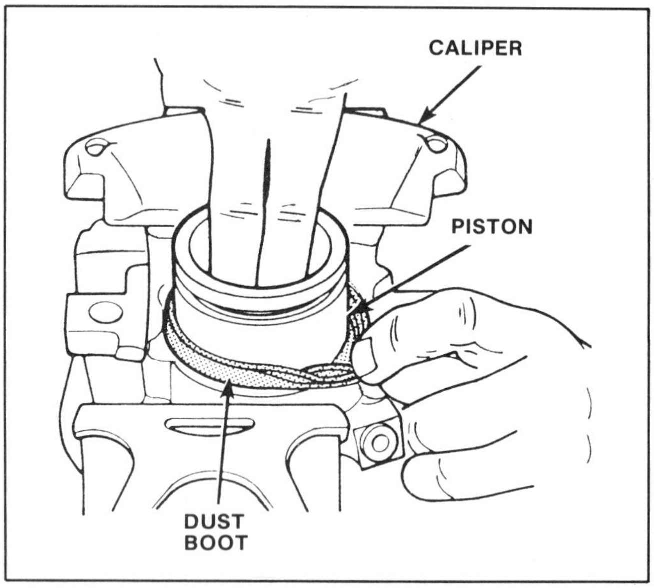

Figure 8-21 Installing piston

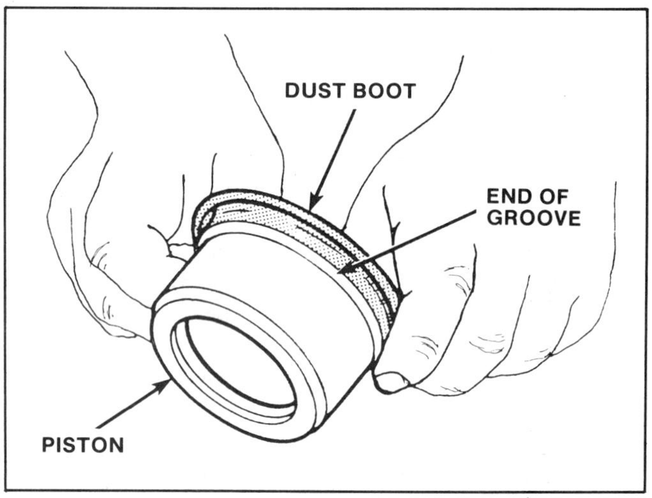

Figure 8-22 Some installation procedures require dust boot to be pulled over the end of piston.

sure the pistons are not cocked. Spread the dust boot over the piston as it is installed. Seat the dust boot in the piston groove.

With some types of boot/piston arrangements, the procedure of installation is slightly different from that already described. That is, the new dust boot is pulled over the end of the piston (Figure 8-22). Lubricate the piston with brake fluid before installing it in the caliper. Then, by hand, slip the piston carefully into the cylinder bore, pushing it straight, so the piston seal is not damaged during

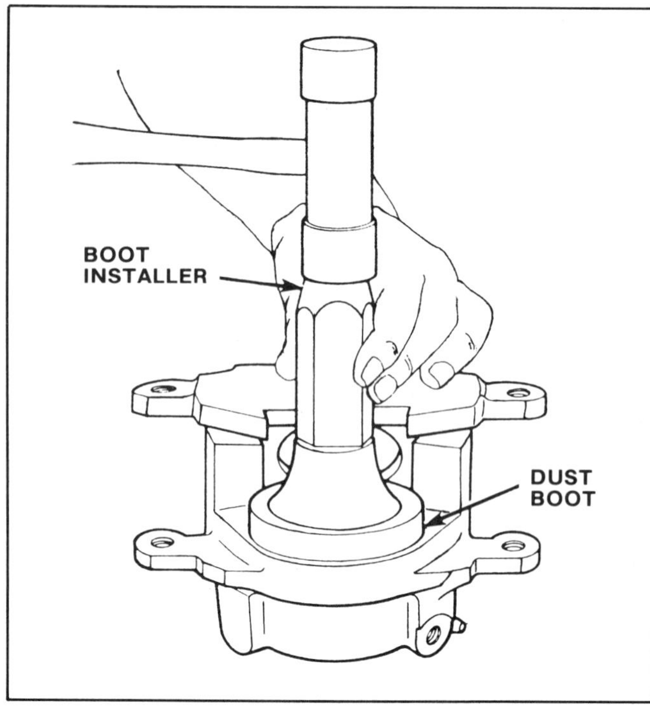

Figure 8-23 Seating dust boot with a boot installer

installation. Use an installation tool or wooden block to seat the new dust boot (Figure 8-23).

The installation of some dust boots requires its groove to be filled with a bead of silicone sealer (Figure 8-24). After lubricating the dust boot, insert the piston. Bottom the piston in the cylinder bore and seat the dust boot in the recess next to the

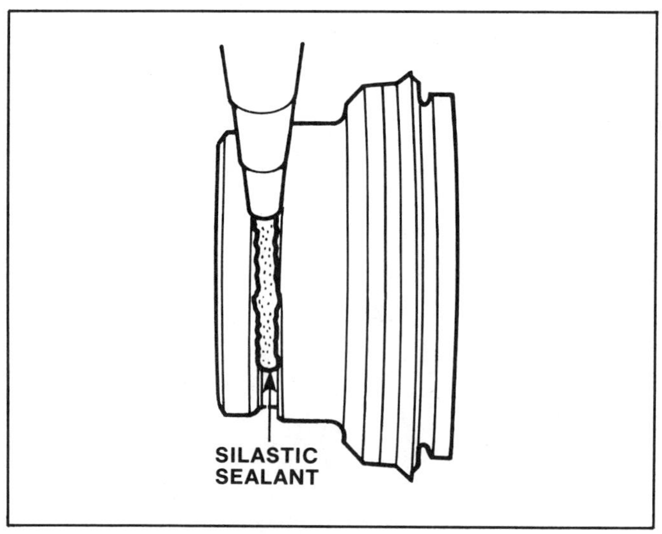

Figure 8-24 Applying sealer to dust boot groove

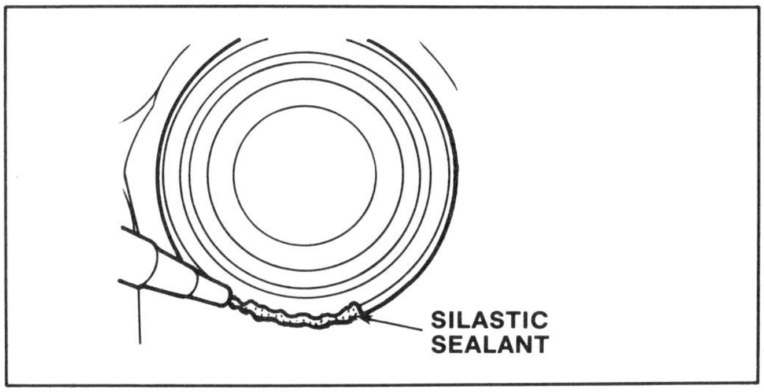

Figure 8-25 Applying sealer between dust boot ring and caliper housing

bore. Seal the area between the dust boot ring and caliper housing with a bead of silicone sealer (Figure 8-25).

CAUTION: Be sure no sealant gets between the piston and cylinder since it retards its action.

Another point to keep in mind is that some caliper designs have a slot cut in the face of the pistons that must align with an anti-squeal shim. Make sure that the piston and shim align. It might be necessary to turn the piston to achieve proper alignment.

To complete the caliper assembly job, install the bleeder screw.

CAUTION: On fixed calipers, bridge bolts are used to hold the two caliper halves together. These are high-tensile bolts ordered only by specific part number. They require accurate torque tightness to prevent leakage. Do not attempt to use standard bolts in place of bridge bolts.

BRAKE PAD INSTALLATION

It is a good practice to replace disc brake hardware when replacing disc brake pads. Hardware life is drastically affected by intense heat, weather conditions, abrasive particles, and normal wear. Replacement of the disc brake hardware will assure proper caliper movement and brake pad retention. Hardware replacement will aid in preventing brake noise and uneven brake pad wear.

Fixed Caliper Brake Pads

Although the designs of fixed caliper disc brakes vary slightly, the general procedure for replacing the pad is as follows:

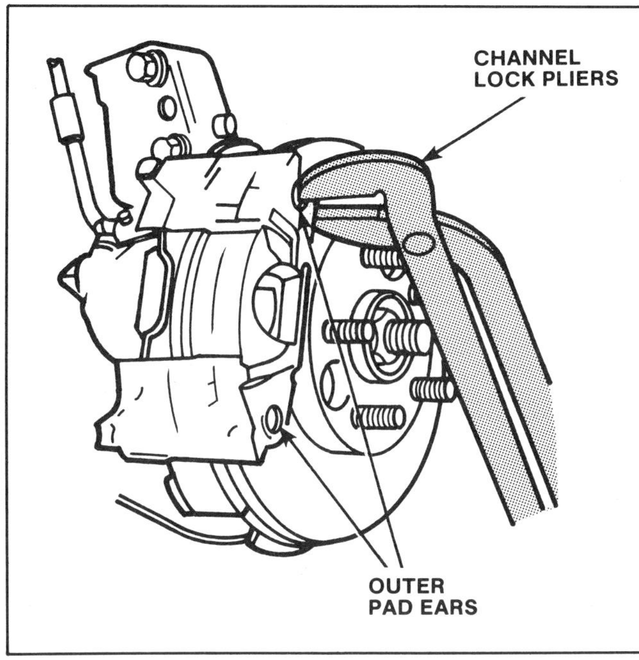

Figure 8-26 Clinching outer pad ears with a channel lock pliers

1. Insert new pads and plates in the caliper with the metal plates against the end of the pistons. Be sure that the plates are properly seated in the caliper.
2. Spread the pads apart and slide the caliper into position on the rotor. With some pads, mounting bolts are used to hold them in place. These bolts are usually tightened 80 to 90 foot-pounds. On some fixed disc brakes, the pads are held in place by retaining clips or, as with some Delco-Moraine designs, both retaining pins and clips are used. Figure 8-26 shows a channel lock pliers being used to clinch down the ear clips of the outer pads.
3. Reinstall the anti-rattle spring/clips and other hardware (if so equipped).

Sliding Caliper Brake Pads

Push the piston carefully back into the bore until it bottoms. Slide a new outer pad and lining assembly into the recess of the caliper (Figure 8-27). No free play between the brake pad flanges and caliper fingers should exist. If free play is found, remove the pad from the caliper and bend the flanges to eliminate all vertical free play (Figure 8-28). Install the pad.

Place the inner pad into position on the adapter with the pad flange in the adapter's machined

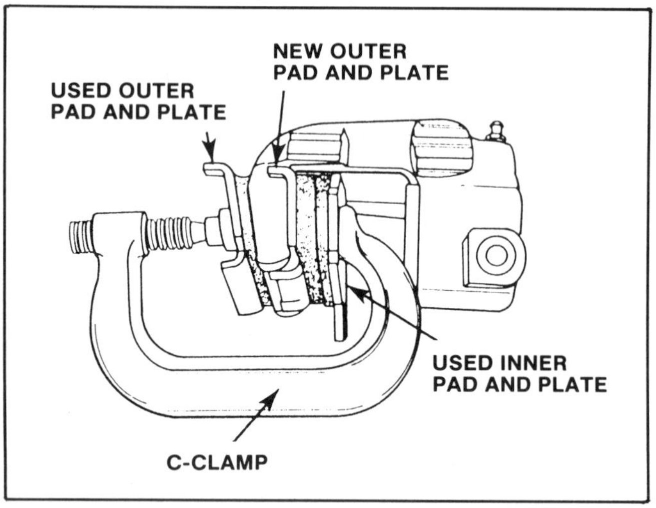

Figure 8-27 Installing outboard pad on the caliper

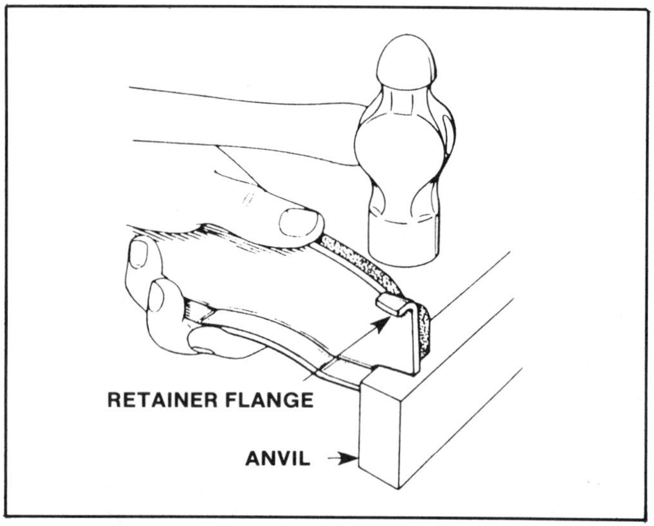

Figure 8-28 Bending retaining flange

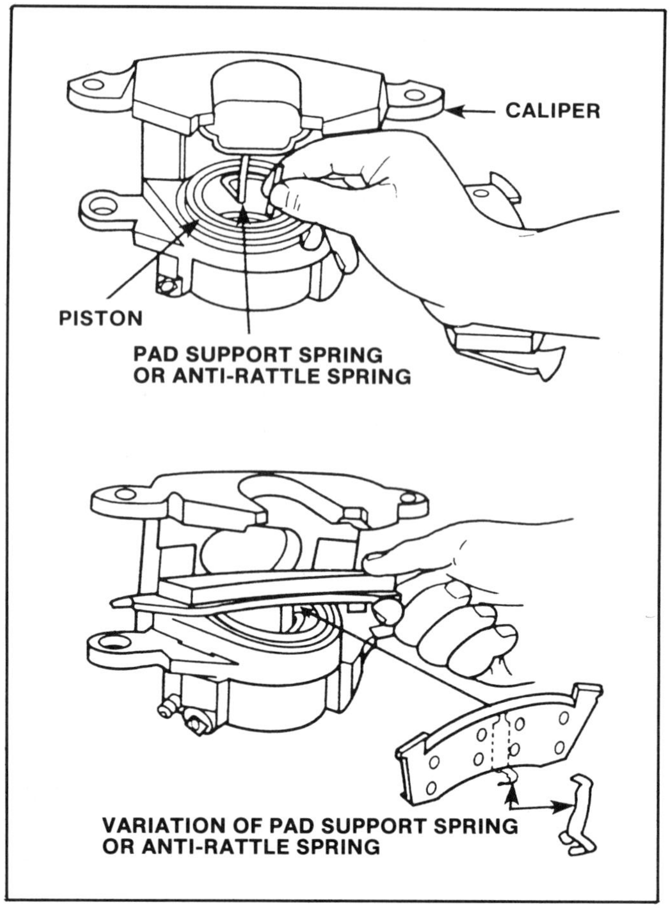

Figure 8-29 Removing and installing pad support spring

ways. Slide the adapter assembly into position in the adapter and over the disc. Align the caliper on the adapter's ways. Do not pull the dust boot from its groove when the piston and boot slide over the inboard pad. Install the anti-rattle springs (if so equipped) on top of the retainer plate and tighten the retaining screws to specification.

Floating Caliper Brake Pads

For floating or pin type caliper disc brakes, compress the flanges of the outer bushing in the caliper fingers and work them into position in the hole from the outer side of the caliper. Compress the flanges of the inner guide pin bushings and install them.

Slide the new pad and lining assemblies into position in the adapter and caliper. Be sure that the metal portion of the pad is fully recessed in the caliper and adapter and that the proper pad is on the outer side of the caliper (Figure 8-29).

Hold the outer pad and carefully slide the caliper into position in the adapter and over the disc. Align the guide pin holes of the adapter with those of the inner and outer pads. Install the guide pins through the bushings, caliper, adapter, and inner and outer pads into the outer bushings in the caliper and anti-rattle spring.

CALIPER INSTALLATION

To reinstall the caliper on the vehicle, proceed as follows:

1. Install the caliper assembly over the rotor with the outer brake pad against the rotor's

braking surface. This prevents pinching the piston boot between the inner brake pad and the piston. Make sure the correct caliper is installed on the correct anchor plate according to the way they were marked during disassembly.

2. Lubricate the rubber insulators (if so equipped) with silicone dielectric compound.
3. Install the caliper assembly back on its mounting brackets.
4. Connect the brake hose to the caliper. If copper washers or gaskets are used, be sure to use new ones—the old ones might have taken a set and might not form a tight seal if reused.
5. Fill the master cylinder reservoirs and bleed the hydraulic system.
6. Check for fluid leaks under maximum pedal pressure.
7. Lower the vehicle and road test it.

ROTOR INSPECTING AND SERVICING

The cause for braking complaints on cars equipped with drum brakes can usually be found by visually examining the brake drum. But in cases where the same complaints involve disc brakes, visual inspection of the rotor will not give the answer. Tolerances on rotor thickness, parallelism, runout, flatness, and depth of scoring are very critical and must be measured with exacting gauges and micrometers. To perform a proper servicing job on rotors, accurate measuring tools and up-to-date rotor resurfacing equipment are required.

> **CAUTION:** Never turn the rotor on one side of the vehicle and not the other.

The rotors should be inspected during every brake reline and when the wheels are removed for other types of service. The following are typical disc brake rotor conditions that need careful inspection.

Lateral Runout

Excessive lateral runout is a wobbling of the rotor from side to side when it rotates. The excessive wobble knocks the pads farther back than normal, causing the pedal to pulse and vibrate during braking; chatter can also result. It also causes excessive pedal travel because the pistons have farther to travel to reach the rotor. A warped rotor or loose wheel bearings might be the cause of the condition. If this condition seems most pronounced after a sharp turn, it could be caused by excessive play in the suspension or steering systems.

Runout specifications vary from one manufacturer to another. If in doubt, a good standard figure to use is 0.0035. The single-piston disc brake with floating caliper is capable of following minor irregularities without problems, while the rigidly mounted two and four piston types are more affected by runout.

To check the runout of the rotor, first make sure that the wheel bearings are correctly adjusted to the manufacturer's specifications. Mount a dial indicator on the spindle (Figure 8-30), caliper, or knuckle so that the point of the stylus makes contact with the rotor surface about 1 inch from the rotor edge. Set the dial at zero, and slowly turn the rotor one complete turn while checking the reading. If the total indicator reading (TIR) exceeds specifications, the rotor must be turned or replaced.

Figure 8-30 Checking rotor runout

SHOP TALK

Overtorquing or unevenly torquing the lug nuts on the wheel with an impact wrench is one of the primary causes of warpage. Uneven torque loads can distort the rotor. Never use an impact wrench for the final tightening of the lug nuts.

Overheating is the other major cause. Front-wheel drive is especially hard on brake rotors because it puts about 80 percent of the brake load on the front wheels. Combined with the higher operating temperatures common to FWD brake systems, overheating can lead to rotor distortion.

Lack of Parallelism

Parallelism refers to variations in thickness of the rotor. If the rotor is out of parallel, it can cause excessive pedal travel, front end vibration, pedal pulsation, chatter, and on occasion, grabbing of the brakes.

Parallelism can be checked with a micrometer at twelve spots around the rotor (Figure 8-31). Variations between the maximum and minimum rotor thickness is the lack of parallelism. A perfectly parallel rotor would have no thickness variation. If measurements do not meet the manufacturer's specifications, the rotor must be resurfaced or replaced.

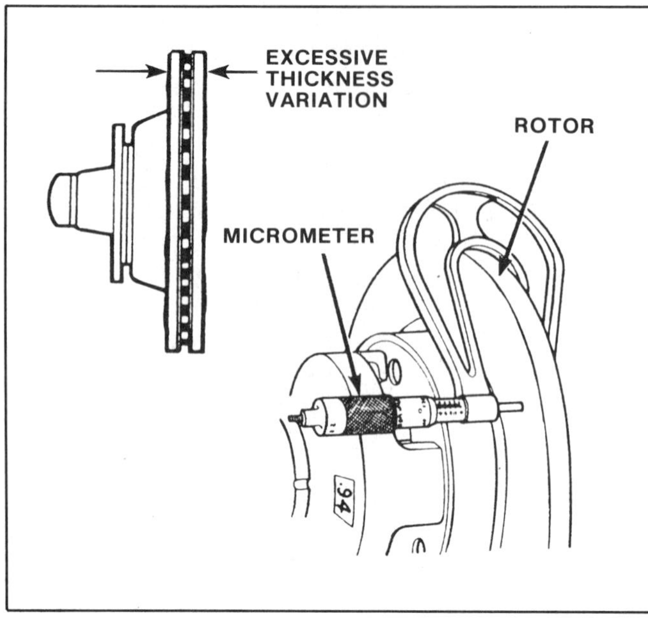

Figure 8-31 Checking for rotor parallelism

Scoring

Rotor wear or scoring can be caused by linings that are worn through to the rivets or backing plate or by friction material that is harsh or unkind to the mating surface. Rust, road dirt, and other contamination could also cause rotor scoring.

Light scoring (less than a depth of 0.015 inch) of the disc braking surface can occur during normal brake use. This will not affect brake operation; however, this can result in a higher-than-average brake shoe lining wear rate. But, any rotor having score marks more than 0.15 inch should be refinished or replaced.

Figure 8-32 Inspect the brake rotor for scoring, heat cracks, blue spots, and out-of-roundness.

SHOP TALK

Some rotors, such as those used on many GM models, are manufactured with one deep groove in the middle of the friction surface. This is present to keep the pads from moving radially outward and to reduce noise.

If cracks are evident, the hub and rotor assembly must be replaced.

Bluing or Heat Checking

Inspect the rotor for signs of overheating (Figure 8-32). The surface of the pads will provide a clue. In some instances the pads can be removed easily. Or it might be possible to move them back far enough to inspect the friction surface. (Remember to remove about half of the fluid from the master cylinder reservoir before doing this.) If the lining surface is charred, blued, or hard-ended with a heavy glaze, or if the rotor is severely heat checked, machine the rotor or replace it.

Hard or chill spots of steel in a rotor cast-iron surface usually result from a change in the metallurgy caused by braking heat. Pulling, rapid wear, hard pedal, and noise will occur. These spots can be removed by grinding. However, only the raised surfaces are removed, and they could reappear when heat is again encountered. The rotor must be replaced.

Rotor Finish with Semi-Metallic Lining

A proper rotor finish is important with semi-metallic brake linings to prevent noise and achieve good break-in and performance. The directional finish left by a lathe should not be used. Instead, the rotor should be ground using the right equipment to provide a non-directional, swirl-like finish (Figure 8-33). The correct grit to use is #120, which is finer than the #80 normally recommended with asbestos lining.

If grinding equipment is not available, an acceptable finish can generally be achieved by sanding the rotor in a non-directional pattern with #120 grit sandpaper.

Rusty Rotor

If the vehicle has not been driven for a period of time, the discs will rust in the area not covered by the lining and cause noise and chatter. Excessive wear and scoring of the discs and lining will result. Wear ridges on the discs can cause temporary improper pad lining contact if ridges are not removed before installation of new lining.

Figure 8-33 A non-directional swirl-like rotor finish

Lining deposits on the disc can cause erratic friction characteristics if a new lining is installed without resurfacing or cleaning the disc. The simplest method of removing rust and pad lining deposits is to use a sanding disc held in an electric drill. Remove the sanding disc against the rotor in a non-directional pattern as it rotates on the lathe. This action will remove the surface contamination without removing much material. It will usually follow the variations in thickness in the rotor as well as give the desirable swirl finish.

ROTOR SERVICING

Here are some important points to keep in mind when performing any rotor servicing:

- If a rotor does not need resurfacing, sand it with #80 grit sandpaper (#120 grit for semi-metallic linings) to remove previous lining contamination.
- Do not drop a rotor because this could throw it out of specification.
- Most rotors cannot be removed from the hub and must be replaced as an assembly.
- After resurfacing a rotor, sand the surface with #80 grit sandpaper (#120 grit for semi-metallic linings) and wash it with denatured alcohol.
- Make sure the lathe tool bit has a rounded cutting point and is not chipped.
- Always use the correct adapters to mount the rotor on the lathe.
- Do not remove or rechuck a rotor until both sides have been resurfaced.
- Make sure that the studs on the flange of the hub are not loose or missing.
- After the rotor has been resurfaced and washed, do not touch the machined surface with your hands.
- It is very important to follow the manufacturer's instructions when torquing the wheel bearing adjusting nut.

GENERAL ROTOR REPLACEMENT PROCEDURES

It is necessary in any replacement of the rotor to first remove the hub and rotor assembly. To do this, proceed as follows:

1. Remove the wheel and tire assembly as described in Chapter 6. The disc brake rotor

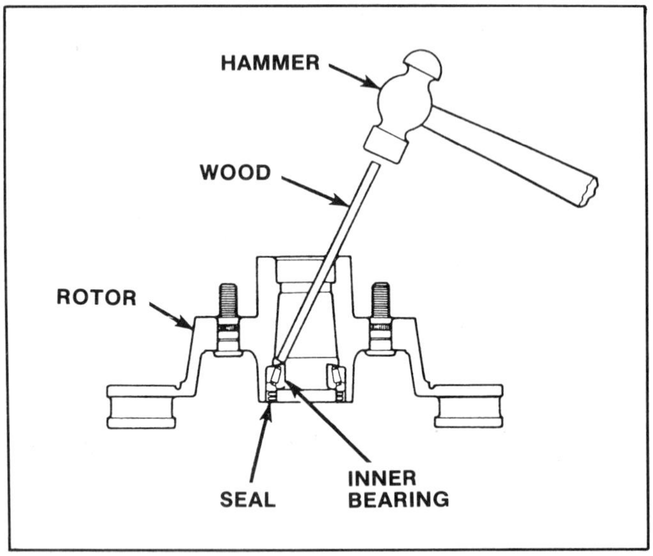

Figure 8-34 Removing inner bearing and grease seal

cannot be removed with the wheel as the caliper straddles the rotor.

2. Remove the caliper assembly, but do not disconnect the brake hose unless the caliper is to be removed for service. If the caliper does not require servicing, suspend it from a wire hook or loop on the front suspension to avoid strain on the hose or damage to the crossover line or other hydraulic connections.
3. If the rotor is integral with the wheel bearing hub, remove the grease cap from the hub (Figure 8-34) and the cotter pin, adjusting nut, and thrust washer from the spindle.
4. Remove the outer wheel bearing cone and roller assembly from the hub.
5. Remove the hub and rotor assembly from the spindle.
6. If the splash shield has been damaged, it can be replaced by removing the bolts that attach it to the knuckle.

SHOP TALK

On most FWD cars, the rotor is easily removed without tampering with the hub and knuckle bearing. Often, the rotor is attached to the hub with small screws. However, numerous Japanese FWD cars have rotors that cannot be removed without using special pullers or an arbor press to remove and replace the hub. In some cases, this operation destroys the wheel bearing in the knuckle.

When reinstalling the hub and rotor assembly, proceed as follows:

1. Before installing the hub and rotor, make sure that the inner and outer bearings are not damaged and that they are properly greased. Install a new grease seal.
2. Place the hub and rotor assembly over the spindle and push it in as far as it will go.
3. Install the outer bearing over the spindle followed by the thrust washer and nut.
4. Tighten the bearing adjusting nut while rotating the hub, then back it off and adjust it to the proper torque as given in the vehicle service manual.
5. Install the lock washer, cotter pin, and grease cup.
6. Install the caliper over the rotor as described for the type of disc brake being serviced.
7. Install the wheel and tire; road test the vehicle.

MACHINING ROTORS

For best performance and to provide a safe wear allowance for reuse with new replacement disc pads, hold rotor remachining to within the minimum allowable thickness, parallelism, surface finish, and lateral runout tolerances specified for each brake type and/or vehicle manufacturer. To maintain the necessary parallelism, machine both sides of the rotor without rechucking. On rotors used with fixed calipers, machine equal amounts from both sides. In certain states, reuse of a rotor

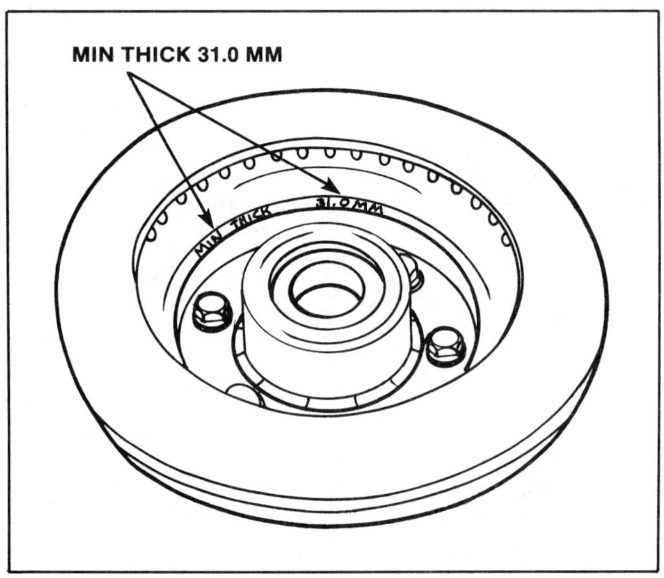

Figure 8-35 Minimum thickness dimension cast into a disc brake

Figure 8-36 Resurfacing rotor

thinner than minimum allowable thickness is a violation of safety laws.

Starting in 1971, all disc brake rotors, like drum brake drums, have a minimum thickness dimension cast into them on an unmachined surface of the disc (Figure 8-35). This dimension is the minimum wear thickness (sometimes called the discard thickness) and not the refinishing dimension. The following example illustrates the typical dimension that applies to rotor thickness:

New rotor thickness	1.290 inches
Minimum thickness after refinishing	1.230 inches
Minimum wear or discard dimension	1.215 inches

CAUTION: Never refinish a rotor to the minimum wear or discard dimension.

The surface finish is most important when refinishing a rotor and is the smoothness of the rotor friction surface, which is measured in microinches. In a service situation, an actual microinch measurement is impractical and unnecessary. It serves only to determine the grit of a grinding wheel, which might or might not be used. But, the finish left by a properly shaped tool bit and correct lathe speed and procedure will be adequate in almost every case. In some case a rotor can be swirl resurfaced with a special swirl grinder (Figure 8-36).

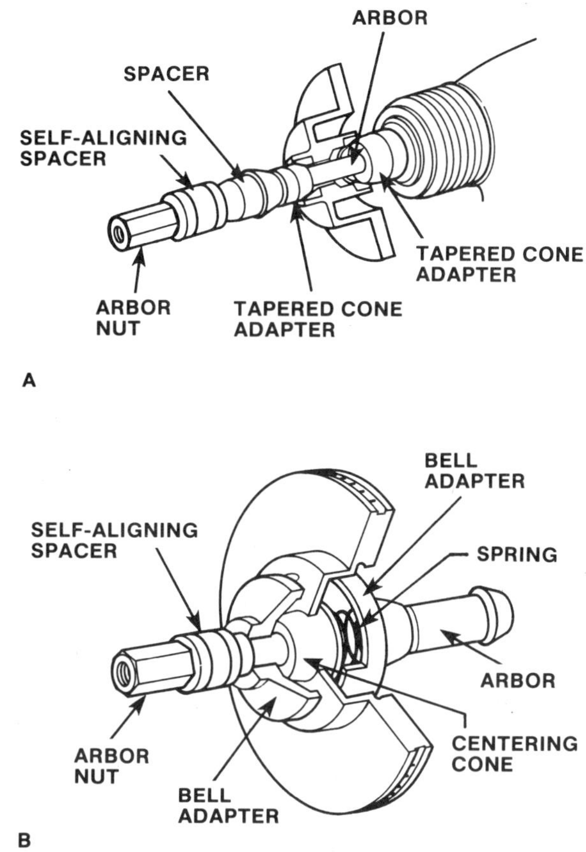

Figure 8-37 (A) Hubbed rotor mounting setup; (B) hubless rotor mounting setup

Lathe Setup and Operation

Before mounting the rotor on the lathe, knock out the inner bearing and grease seal as described earlier in the chapter so that the lathe indexes on the outer races of both the inner and outer wheel bearings. Use the appropriate cones and spacers to lock the rotor firmly to the shaft. As shown in Figure 8-37, index the rotor on the wheel bearing races to assure that the machining job is true. Remove all grease and dirt from the bearing races before mounting the rotor. Sometimes it is necessary to steam clean the grease out of the hubs. Also, to keep metal chips from getting into the hub, cut the bottom out of a paper cut and place it over the hub when mounting the rotor.

Installing a vibration dampener on the rotor will keep the cutting blades from chattering during the refinishing procedure, and it also helps reduce unwanted noise. Determine the amount of metal to be removed by bringing the cutting tool up against

Figure 8-38 Making the test scratches on the rotor. Note vibration damper on the rotor.

the rotating disc until signs of a slight scratch are noticed (Figure 8-38). Then turn the depth of cut, dial to zero, and advance the tool bit into the deepest groove visible on the face of the rotor. If the deepest marks are more than 0.015 inch deep, take two or more cuts rather than one very deep cut for best results.

> **WARNING:** When operating a lathe, be sure to wear the proper eye protection.

When using a lathe with only one cutting bit, never cut one side of the rotor and then take the rotor off at the lathe and reinstall it before cutting the other side. Even a 0.001- to 0.002-inch variation in positioning the rotor on the lathe can produce a thickness variation of the rotor. This can cause tremendous pedal pulsation and loss of braking efficiency. With single- or double-bit lathes, always cut both sides of the rotor without removing it from the lathe to insure that both sides of the rotor will be parallel after refinishing. Check the accuracy of the work by using a dial indicator and an outside micrometer.

When making the test scratch, if it does not make a complete circle on the rotor but appears as a crescent, the rotor is wobbling more than 0.004 inch. Do not proceed with the resurfacing of the rotor because a wobble could be machined into the rotor. This could cause the brake pedal to pulsate when the brakes are applied. Three conditions could cause the rotor to wobble excessively on the lathe's arbor:

1. A bent arbor
2. Distorted mounting adapters
3. Excessive rotor runout

A dial indicator mounted on the lathe can be used to determine if the arbor is bent. Release the pulley belt tension by moving its controlling lever and rotate the arbor slowly by turning the drive pulleys. Observe the movement of the dial indicator needle. Movement of the needle in excess of one division or 0.001 inch would indicate a bent arbor. Contact the lathe manufacturer's representative to check.

A distorted mounting adapter can be corrected by installing it on the lathe's arbor and machining it.

The rotor might have excessive wobble or runout because it was previously machined incorrectly or the rotor was distorted when the wheel lug nuts were overtightened. The rotor can be resurfaced on the lathe to correct these conditions if the amount and position of the rotor's runout on the car is determined and noted on the rotor. This information is used to mount the rotor accurately on the lathe. However, to correctly mount the rotor on the lathe's arbor, a few steps should be followed:

1. The rotor should be removed from the hub and inside surface of the rotor, and the axle hub face should be thoroughly cleaned of rust and grease.
2. Install the rotor on the hub and torque the wheel stud nuts backward on the studs.
3. Rotate the rotor by hand and notice the maximum movement of the indicator's needle. This movement is the amount the rotor wobbles in thousandths of an inch. The minimum and maximum runout can be found in the brake's specifications. The maximum travel of the needle in a clockwise direction indicates the place where the rotor has its maximum outward deflection. Mark this spot on the rotor hub with a plus (+) mark.

Install the rotor on the lathe with the proper mounting adapters. Install the dial indicator on the lathe and adjust the indicator's point to the outside rotor surface. Slowly turn the rotor clockwise and observe whether the runout amount and location are the same as they were on the car. If the readings are not the same as they were on the car, loosen the arbor nut and reposition the rotor and the adapters on the arbor. Repeat this procedure until the rotor runout matches the runout measured on the car as previously described.

If the rotor's runout on the lathe cannot be made to match the runout found on the car, the rotor hub is distorted and should be replaced. To determine the runout of the rotor hub, mount the dial indicator point against the flange. Rotate the axle and observe the movement of the gauge needle.

Another solution to the rotor hub runout problem on front-wheel drive cars is to recondition the rotor on the car with an on-the-car rotor lathe (Figure 8-39). To use this lathe, remove the caliper from its mounting bracket and hang it out of the way with a wire hook. Then torque the wheel mounting nuts equally.

The lathe's operation manual explains how the lathe is attached to the car using the hardware that comes with the lathe. These knobs are used to adjust the tool bits. The dial marks determine the depth of cut each tool is taking from the rotor surface. The engine is used to rotate the rotor, but a problem exists because the differential gearing in the transaxle transmits the power to the opposite front axle not to the rotor to be resurfaced. To keep that axle from rotating, the opposite wheel can be lowered to the floor or lift. However, this can make it difficult to operate the lathe. A vice grip pliers can be clamped to the axle, but this can damage the axle, so it is not usually recommended.

Another method would be to block off the brake hose on the caliper that is removed from its anchor.

Figure 8-39 On-the-car rotor lathe

Apply the brakes and clamp the other front brake hose to keep the caliper locked on. Start the engine; shift the transmission into first gear. Idle the engine as slowly as possible. The rotor must turn as slowly as possible to provide a smooth surface on the rotor. (Tool bits will not overheat, and they last longer.) Turn each feed knob inward until the tool bit just touches the rotor face. Rotate the hand wheel to position the tool bit.

Reconditioning a rotor with an on-the-car lathe is recommended because it saves time when the rotor hub must be disassembled from the car's spindle to mount it on a lathe's arbor. New rotors should not be machined before they are installed on the vehicle. If a new rotor has excessive runout after it has been installed on a front-wheel drive vehicle, the on-the-car lathe should be used. The new rotor must be machined because the axle hub is distorted or the new rotor was not machined accurately by the manufacturer.

Some 4 x 4 vehicle front disc brake rotors can be resurfaced on the car. This will save time if the front wheel bearings have been recently serviced.

Rotor Touch-up and Cleaning

Frequently, after the rotor has been machined, both outside edges could be razor sharp. These sharp edges could make the chipping of the brake pads difficult and could cause damage to the dust boots when installing the caliper. To reduce these sharp edges, cut a 1/32-inch chamfer on them with a mill-smooth metal file. When performing this operation, the rotor should be revolving in the lathe and care must be taken to prevent the file from catching on the rotor surface.

The friction surface of the newly refaced rotor is covered with metal particles—on the surface and lodged in the open pores of the newly machined metal. If these particles are allowed to remain on the rotor, they will imbed themselves in the brake lining. Once the brake lining gets contaminated this way, it scores the rotor with every brake application. The metal particles must, therefore, be removed.

Wipe the surface of the rotor, especially the newly machined surface, with a clean lint-free cloth saturated in brake cleaner or denatured alcohol. This will remove the fine bits or chips of metal left from machining.

CAUTION: Do not use compressed air to remove the particles from the rotor.

WHEEL BEARING REPACKING AND ASSEMBLY

Clean and repack the wheel bearings before reinstalling the rotor. To do this, proceed as follows:

1. Clean the inner and outer bearings and cups with solvent. Inspect the bearings and cups for scratches, pits, excessive wear, and other damage. If the cups are worn or damaged, replace parts as necessary, following manufacturer's directions.
2. Wipe all old lubricant from the spindle and the inside of the hub. Cover the spindle with a clean cloth and brush all loose dust and dirt from the dust shield. Carefully remove the cloth from the spindle to prevent dirt from falling on it.
3. If the inner or outer bearing cups were removed, install replacement cups using the proper tools shown in Figure 8-40. Be sure to seat the cups properly in the hub.

CAUTION: When installing bearing cups, support the rotor on the hub barrel with a wood block to avoid damage to the wheel studs.

4. Thoroughly clean old grease from the surrounding surfaces.
5. Using a bearing packer, pack the bearing cone and roller assemblies with the recommended grade of wheel bearing grease or multipurpose grease. If a packer is not available, work as much grease as possible between the rollers and cages. Grease the cup surfaces.
6. Place the inner bearing cone and roller assembly in the inner cup. Apply a light film of multipurpose lubricant to the lips of a new grease retainer and install the retainer (Figure 8-41). Be sure the retainer is properly seated.

CAUTION: When installing the grease retainer, support the rotor on the hub barrel with a wood block to avoid damage to the wheel studs.

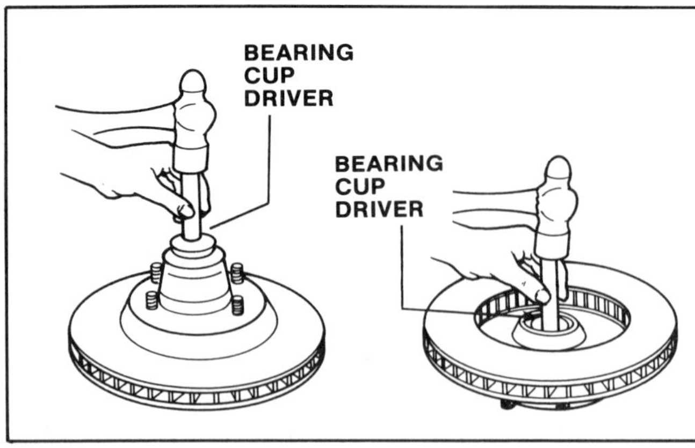

Figure 8-40 Installing bearing cups

Figure 8-41 Installing grease seal

7. Install the hub and rotor assembly on the spindle. Keep the hub centered on the spindle to prevent damage to the retainer and the spindle threads.
8. Install the outer bearing cone and roller assembly and flatwasher on the spindle. Install the adjusting nut fingertight. Then using an inch-pound torque wrench, adjust the wheel bearings.

SHOP TALK

The front wheel bearings of all rear-wheel drive cars and light trucks are adjusted by turning the nut that holds the hub to the spindle. However, this adjustment is more critical on disc brake-equipped vehicles than on those with drum brakes at the front. This is because too much lateral runout or sideplay of the rotor tends to knock the pads farther away from the rotor surface, and this results in excessive brake pedal travel or even pumping being necessary to bring the pads up against the rotor when stopping. This must be corrected immediately.

9. If a retainer or ring is used, install it over the nut so that its slots line up with the hole in the spindle. If the slots in the retainer or those between the castellations on the nut do not line up with the hole in the spindle, back off the adjustment slightly until they do. In some cases, both a maximum and minimum back-off specification is given. Insert a new cotter pin to lock the adjustment. (Never reuse an old cotter pin.) Bend and trim the ends of the pin so that they do not contact the dust cap.

SHOP TALK

The basic idea of wheel bearing adjustment is to achieve a setting that does not preload the bearing or allow excessive clearance. Preloading causes accelerated bearing wear, and too much clearance can result in braking problems, premature wear, and wheel shimmy.

10. Install the grease cap taking care not to damage it. Be sure the cap is fully seated.
11. Install the rotor and caliper to the spindle as described earlier in this chapter. Be sure that brake fluid hoses are correctly connected (Figure 8-42).
12. Place the wheel and tire on the hub and rotor. Install the wheel attaching nuts and tighten them alternately to draw the wheel evenly against the hub and rotor.
13. Lower the vehicle and tighten the lug nuts with a foot-pound torque wrench to the manufacturer's recommended specification. Install the wheel cover.
14. Before driving the vehicle, pump the brake pedal several times to restore normal brake pedal travel.

Cars or trucks with front- or four-wheel drive have a bearing and hub setup that is different from the normal dead spindle system used on the front wheels of rear-wheel drive cars. Additional components are needed to drive the spindle, and many four-wheel drive vehicles incorporate locking hubs into the assembly (see Chapter 12).

SERVICE PROCEDURES FOR POPULAR DOMESTIC CARS

Before beginning any repair work on a specific model car, be sure you are familiar with the procedures previously given in this chapter. These procedures are intended to be used in conjunction with the data found in this section and explain many operations in greater detail than given here. Also check the vehicle's service manual for more information.

CHRYSLER SINGLE-PISTON SLIDING CALIPER (FIGURE 8-43)

The caliper housing is a one-piece construction with an inspection hole on top to check lining

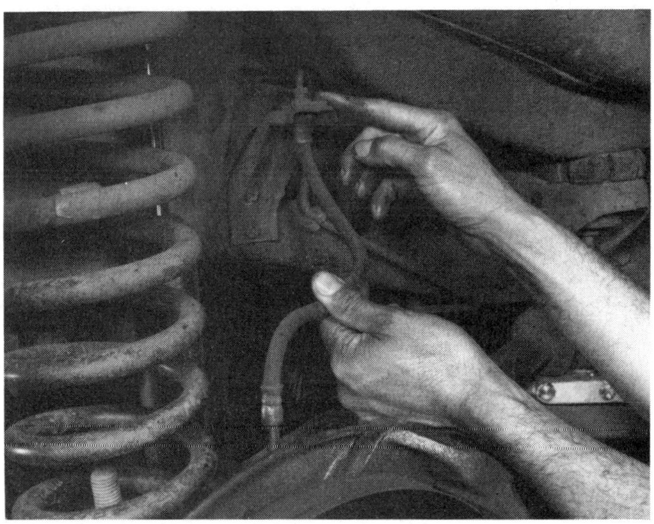

Figure 8-42 When installing the rotor and calipers, be sure the hoses are properly connected.

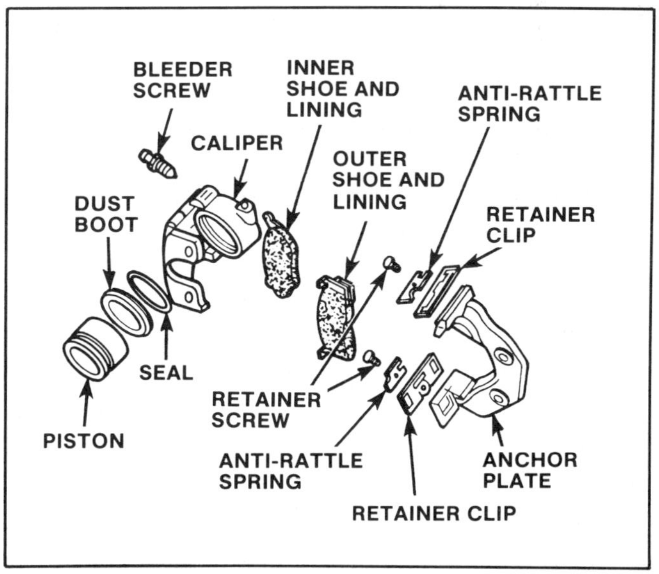

Figure 8-43 Chrysler single-piston sliding caliper

wear. The cylinder bore is grooved to fit a square-cut seal. A taper toward the bottom of the bore increases compression on the seal edge nearest the hydraulic pressure. The seal also acts to retract the piston into the bore. There is a groove at the top of the bore to seat the dust boot. Holes are machined through the caliper into the cylinder bore for fluid inlet and bleeder valve outlet. The piston is nickel-chrome plated and cannot be cleaned with any type of abrasive. It is hollow for heat insulation. Any black stain on the piston from the seal does not affect its ability to seal.

The caliper is mounted on an anchor plate that is bolted to the steering knuckle. The caliper slides on machined guides and is held in position by retaining clips. The shoes are different in design—the inner shoe has an ear on each end and is mounted on the machined abutments of the anchor plate. The outer shoe is flanged and seated in the caliper housing.

When servicing a Chrysler single-piston sliding caliper, proceed as follows:

1. Take out the screw that holds each of the anti-rattle springs and caliper retaining clips.
2. Rotate and lift the caliper from the anchor plate.
3. Remove the outer shoe from the caliper housing and the inner shoe from the anchor plate.
4. Check for leaks.
5. Overhaul the caliper.
6. Brush the machined guides clean with an abrasive or a wire brush.
7. Seat the piston with a C-clamp and a block of wood.
8. Position a new inner pad on the anchor plate with the ears of the shoe on the machined abutments.
9. Install the new outer pad with the metal flanges snugly fitting the caliper.
10. Bend the flanges, if necessary, and snap into place with fingers or use a C-clamp.

CHRYSLER SINGLE-PISTON FLOATING CALIPER (FIGURE 8-44)

The rotor has cooling fins to cool the assembly; this improves lining wear. Weights are placed between the two surfaces to balance the rotor. A splash shield, bolted to the steering knuckle, protects the inner rotor surface from water and mud.

The caliper is a one-piece design with an inspection hole for checking pad wear. The cylinder bore has a square cut seal groove. At the bottom is a fluid inlet hole and at the top is a bleeder outlet hole. The square-cut seal keeps the system from losing fluid, but also retracts the piston and serves as a self-adjuster for the pads as they wear. The dust boot fits in a groove in the top of the cylinder bore. It prevents moisture and dirt from contaminating the bore.

When servicing a Chrysler single-piston floating caliper, proceed as follows:

1. Remove the anti-rattle spring.
2. Remove the guide pins.
3. Remove the caliper and outboard shoe.
4. Remove the inboard shoe.
5. With the brake hose attached, step on the brake pedal to force the piston out of the caliper bore.
6. Disconnect the brake hose.
7. Rebuild the caliper.
8. Install the caliper using new inner and outer bushings.
9. Install the brake hose using a new copper gasket.
10. Install the caliper and torque the guide pins 25 to 40 foot-pounds

FORD FLOATING CALIPER (FIGURE 8-45)

This caliper is used on several Ford models, beginning with 1978. Its anchor plate is part of the spindle casting.

When servicing a Ford floating caliper, proceed as follows:

1. If the caliper needs rebuilding, disconnect and remove the brake hose.
2. Mark the left and right calipers with chalk.
3. Remove the caliper guide pins.
4. Lift off the caliper.
5. Rebuild the caliper if necessary.
6. Install the brake pads in the caliper, making sure the two round torque buttons on the outer shoe are firmly located in the two holes of the outer caliper leg. The pad must be held tightly in place by its spring clip.
7. Install the caliper assembly over the rotor.
8. Install the caliper guide pins by hand.
9. Torque the guide pins 20 to 30 foot-pounds.
10. Install the brake hose.

SHOP TALK

This is a special self-sealing fitting that does not use a gasket.

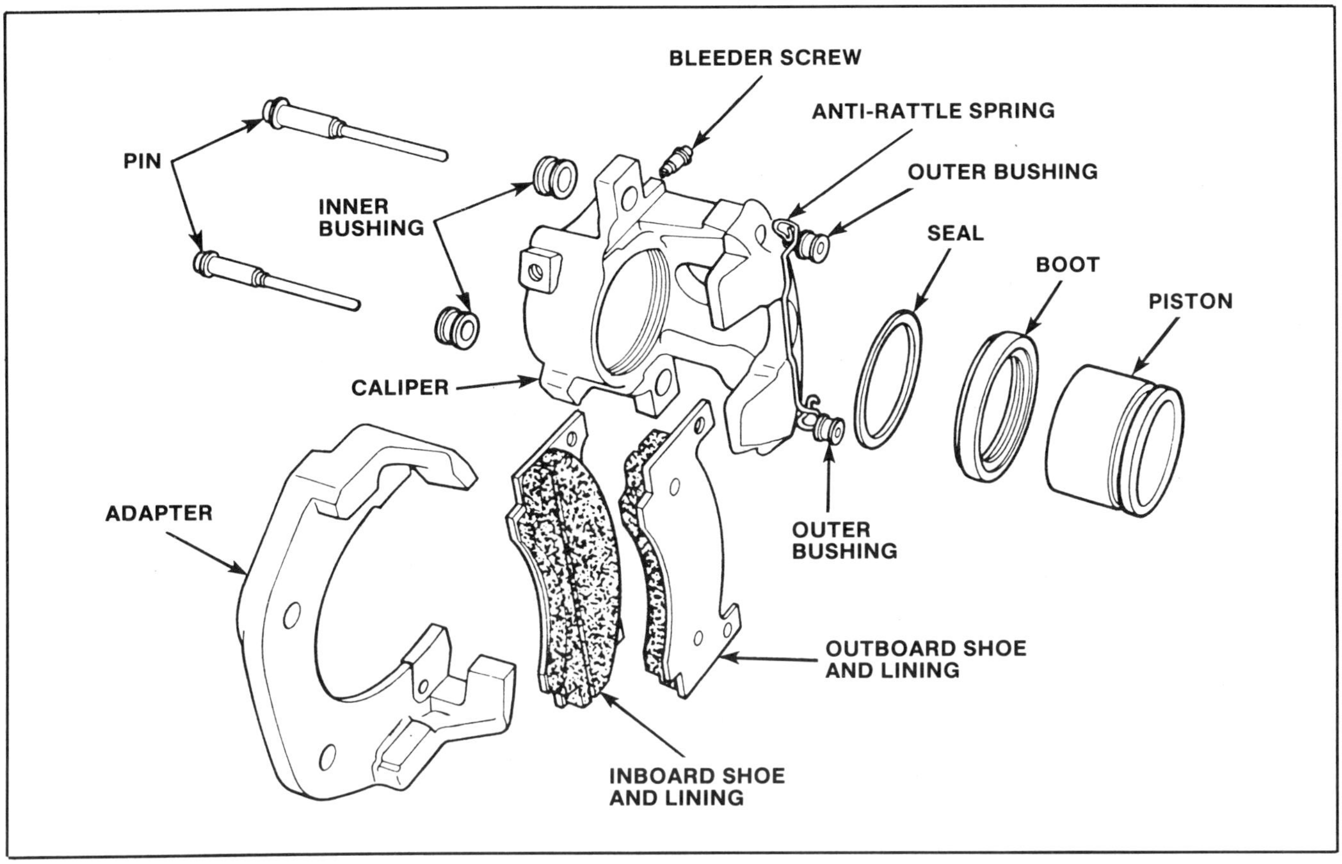

Figure 8-44 Chrysler single-piston floating caliper

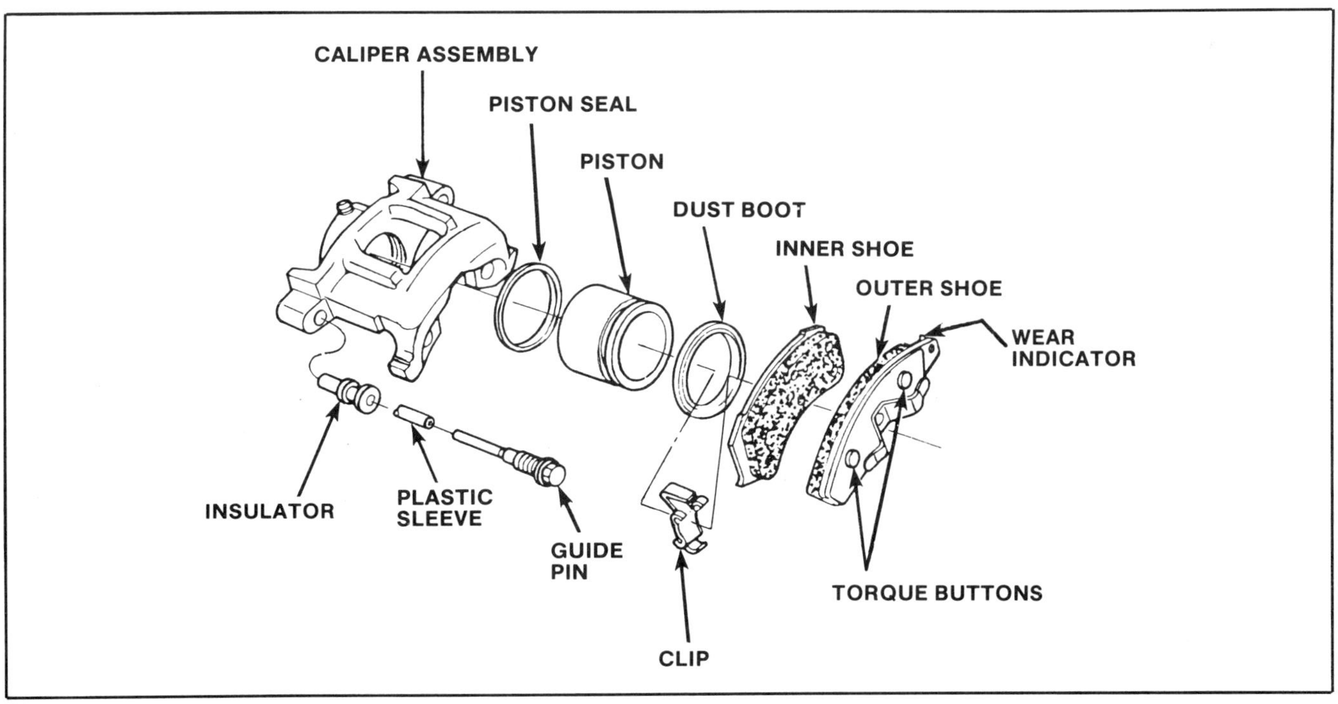

Figure 8-45 Ford floating caliper

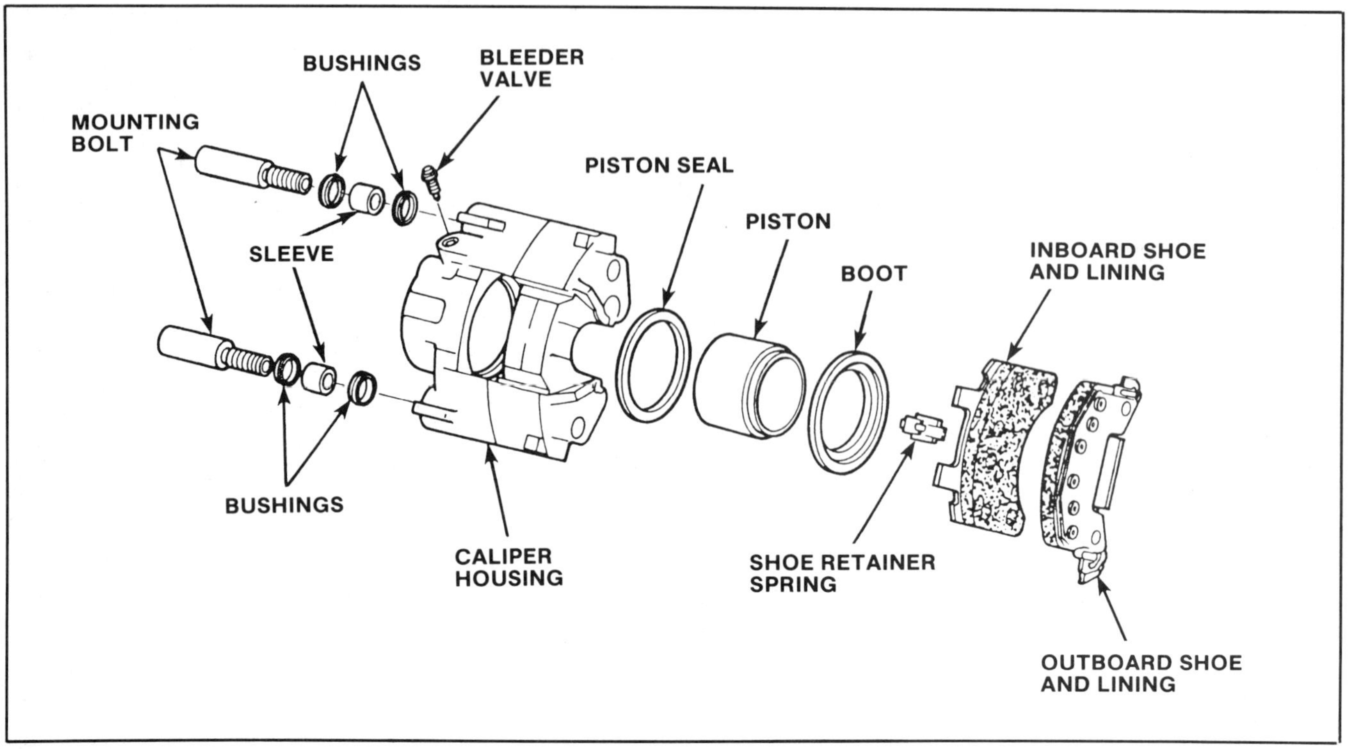

Figure 8-46 GM X- and J-car single-piston floating caliper

11. Bleed the brake system.
12. Road test the vehicle.

GM X- and J-CAR SINGLE-PISTON FLOATING CALIPER (FIGURE 8-46)

This floating caliper was introduced on GM X-cars in 1980 and is now being used on all X- and J-cars. It is a one-piece caliper that mounts on two bolts with sleeves and bushings.

When servicing a GM X- and J-car single-piston floating caliper, proceed as follows:

1. Remove two-thirds of the fluid from the master cylinder.
2. Use a C-clamp to bottom the caliper piston.
3. If the caliper must be rebuilt, remove the bolt holding the inlet fitting.
4. Remove the allen head caliper mounting bolts.
5. Inspect the mounting bolts for rust and corrosion. If they are corroded, discard them.
6. Remove the caliper.
7. Remove the sleeves and bushings from the mounting bolt holes.
8. Remove the shoes.
9. Rebuild the caliper if necessary.
10. Replace the shoes, being sure to install the retainer spring on the inboard shoe.
11. Install the new sleeves and new bushings.
12. Position the caliper and install the mounting bolts. Torque the mounting bolts 21 to 35 foot-pounds.
13. Use a large pair of channel lock pliers to clinch the outboard shoe ears to the caliper.
14. Install the brake hose and torque the inlet-fitting bolt 18 to 30 foot-pounds.
15. Bleed the brakes.
16. Test the brake pedal, then road test the vehicle.

DELCO-MORAINE FOUR-PISTON FIXED CALIPER (FIGURE 8-47)

This caliper design was introduced in the mid 1960s by GM. It is now being used only on the Corvette. The caliper is fixed; that is, it remains stationary and does not float or slide. The caliper has four pistons, two on each side of the rotor. The caliper is made in two halves, which are separated when the caliper is rebuilt. The shoes are held in position by a retaining pin.

When servicing a Delco-Moraine four-piston fixed caliper, proceed as follows:

SHOE REPLACEMENT

1. Remove two-thirds of the fluid from the disc brake reservoir.
2. Remove the cotter pin from the shoe retaining pin and remove the retaining pin.
3. Remove the inboard shoe by pulling up on it.
4. Use two screwdrivers to push the pistons back far enough to install the new shoe.
5. Install the new inboard shoe.
6. Replace the outboard shoe the same way.
7. Install the shoe retaining pin from the outboard side. Secure the retaining pin with a new 3/32-inch by 5/8-inch plated cotter pin.

When servicing a Delco-Moraine four-piston fixed caliper, proceed as follows:

CALIPER OVERHAUL

1. Remove the brake hose from a front caliper or the brake line from a rear caliper.
2. Remove the two caliper mounting bolts and remove the caliper from the mounting bracket.
3. Remove the shoe retaining pin and the brake shoes.
4. Remove the two caliper bolts and separate the caliper halves.
5. Remove the O-rings from around the fluid transfer holes.
6. Remove the piston boots by pushing the pistons down in the bore and using them as a fulcrum as you pry out the dust boots with a screwdriver.
7. Remove the pistons and piston springs.
8. Remove the boots and seals from the pistons.
9. Service the caliper bores as you would any caliper.
10. When all components have been serviced and cleaned, install the seal in the groove closest to the flat end of the piston. The lip on the seal must face the large end of the piston.
11. Place the piston spring in the bottom of the bore.
12. Lubricate the piston seal with brake fluid and install the piston into its bore using a piston ring compressor to protect the piston seal as it goes in.
13. After the piston is in its bore, insert a bead of silastic sealant into the boot groove and assemble the boot into the groove.
14. Check the piston movement in the bore. It must slide smoothly until the end of the piston is flush with the end of the bore. If it does not, check the location of the piston seal and piston spring.
15. Use a boot seal installer to seat the steel boot retaining ring in its counterbore. The boot

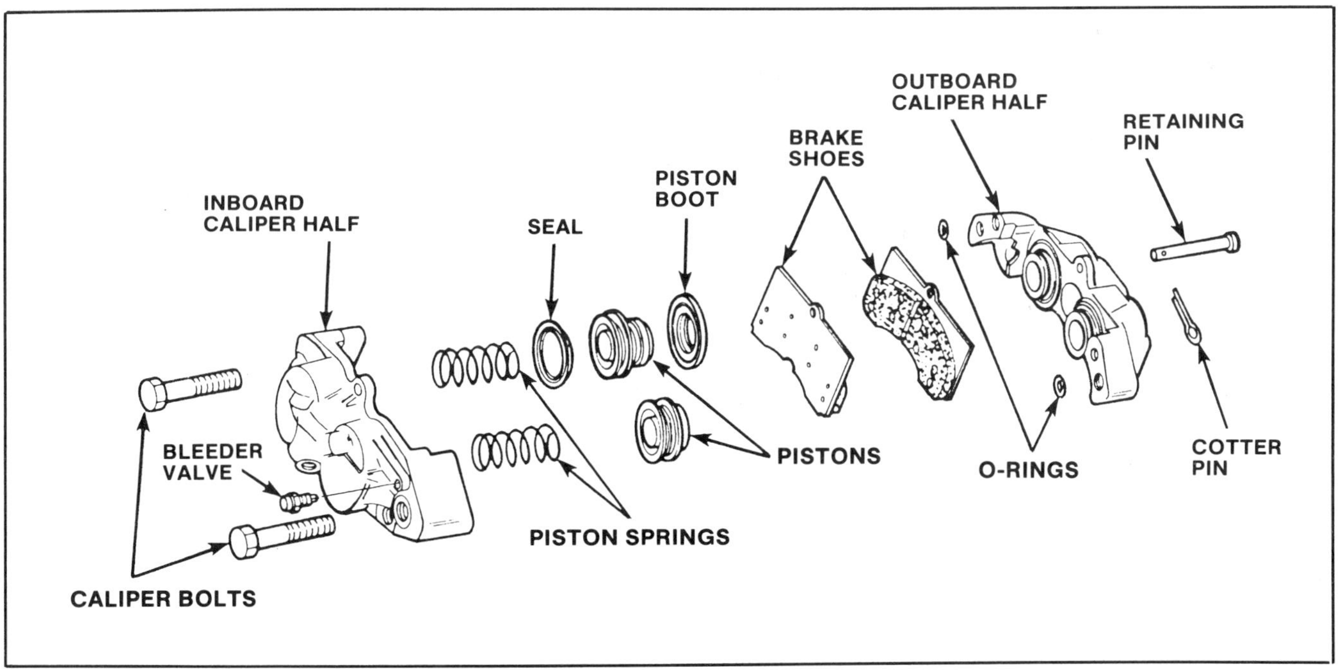

Figure 8-47 Delco-Moraine four-piston fixed caliper

retaining ring must be flush or below the machined surface of the caliper when you are through.

16. Depress the piston and place a bead of silastic sealant between the boot retainer ring and the housing.
17. Perform the previous procedure for each piston in both caliper halves.
18. When the caliper halves are ready to be assembled, place new O-rings in the grooves around the fluid transfer holes in the outboard caliper.
19. Dip the caliper bolts in clean brake fluid. Fit the caliper halves together. Install the caliper bolts and torque them to 130 foot-pounds for front disc calipers and 60 foot-pounds for rear disc calipers.
20. Carefully position the caliper over the rotor. You will probably have to use two screwdrivers to depress the pistons to do this.
21. Install the two caliper mounting bolts and torque them to 70 foot-pounds.
22. Install the shoes.
23. Install the brake hose (use a new copper gasket) or brake line.
24. Bleed the brakes (Figure 8-48).
25. Test the brake pedal and road test the vehicle.

REAR DISC BRAKES

Rear disc brakes are being used on some models of domestic and import cars. In most cases, they are identical to the equivalent front disc calipers on the same cars. However, most of the rear brakes employ some type of parking brake mechanism.

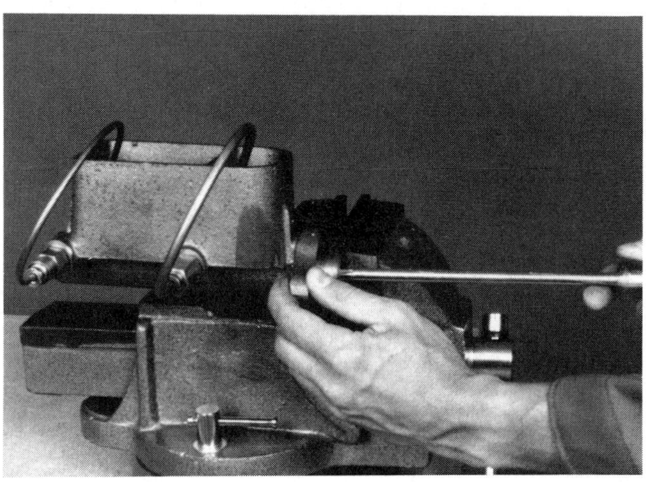

Figure 8-48 Bleeding master cylinder

CHRYSLER SINGLE-PISTON REAR DISC CALIPER (FIGURE 8-49)

The difference between the rear disc and front disc calipers on this brake is a set of parking brake shoes and drum inside the hub of the rear rotor. The rear caliper assembly is serviced the same as the Chrysler single-piston front disc caliper. To remove and replace the drum brake shoes, the caliper and rotor have to be removed.

When servicing a Chrysler single-piston rear disc caliper, proceed as follows:

1. Remove the caliper retaining clips and anti-rattle springs.
2. Rotate and lift the caliper from the anchor plate. Hang it by a wire.
3. Remove the inner pad from the anchor plate.
4. Remove the rotor. (If drum shoes prevent removal, insert an adjusting spoon through the access hole in the splash shield and back off the star wheel adjuster to release the shoes. Then pull off the rotor.)
5. Remove the lower shoe-to-shoe spring.
6. Spread the shoes apart and remove the star wheel adjuster.
7. Remove the upper shoe-to-shoe spring.
8. Move the shoes out of the support platform, and mark them for identification.
9. Remove the hold-down springs and shoes.
10. Inspect the brake lining and wheel cylinder; replace and rebuild as necessary.
11. Reposition the brake shoes and install the hold-down springs.
12. Attach the upper shoe-to-shoe return spring.
13. Install the star wheel adjuster mechanism. (On the left side, the star wheel is forward. On the right side, it is rearward.)
14. Attach the lower shoe-to-shoe spring.
15. Replace the rotor.
16. Seat the parking brake shoes against the drum by rotating the star wheel adjuster. Use an adjusting spoon through the splash shield access hole.
17. Back off the star wheel twelve clicks to provide enough running clearance.
18. Install the inboard shoe onto the adapter ways of the anchor plate.
19. Slide the caliper assembly slowly over the rotor until it contacts the machined abutments on the anchor plate.
20. Install the retaining clips and torque the screws to 15 foot-pounds.
21. Install the anti-rattle springs on top.

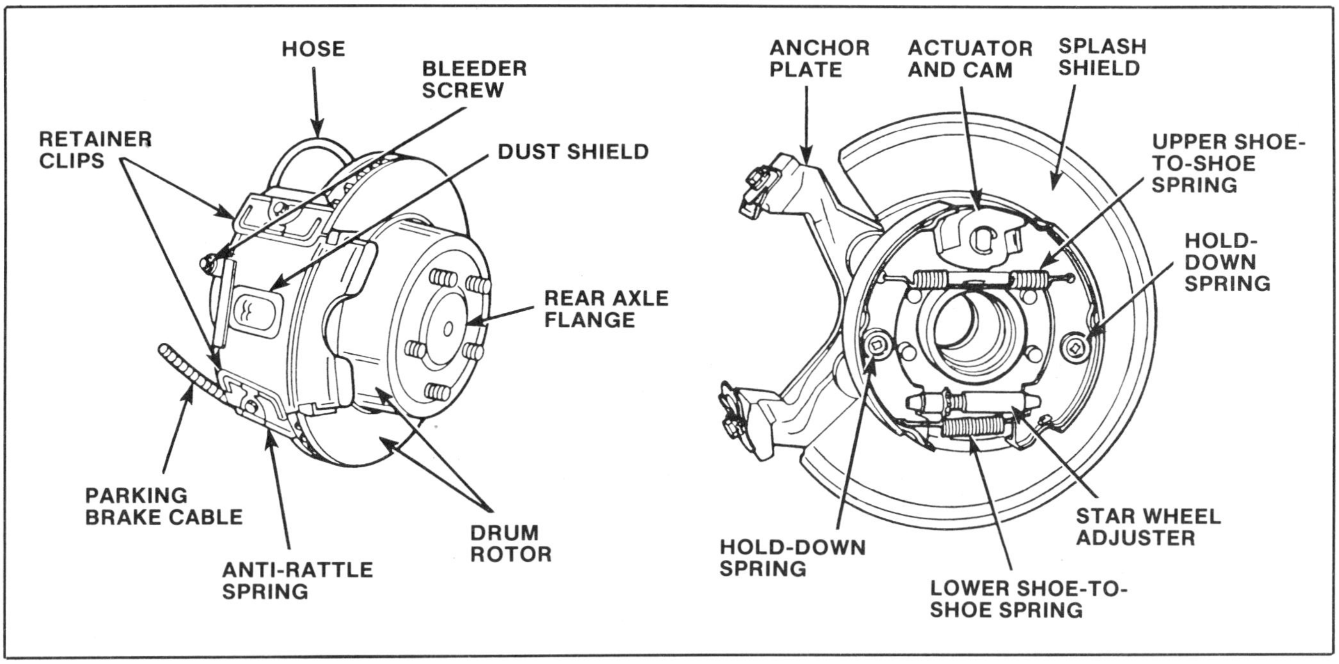

Figure 8-49 Chrysler single-piston rear disc

22. Bleed the brakes.
23. Test the pedal and road test the car.

KELSEY-HAYES (FORD) SLIDING CALIPER (FIGURE 8-50)

This rear sliding disc uses a ball and ramp mechanism to activate the single piston for use as a parking brake. The removal and replacement of pads is the same as for front brakes, but the parking brake has to be disconnected from the actuating lever. Also, the piston has to be screwed back into the caliper bore. When the parking brake cable is pulled, the actuating lever is rotated, which operates the shaft in the piston assembly. Three steel balls are in pockets between the operating shaft and thrust screw. When the shaft rotates, the balls are forced up the ramps formed by these pockets, forcing the thrust screw away from the operating shaft. This, in turn, drives the piston and disc pad out to contact the rotor. To compensate for lining wear and to keep proper clearance, an automatic adjuster in the piston assembly moves on the thrust screw.

When servicing a Kelsey-Hayes (Ford) sliding caliper, proceed as follows:

1. Disconnect the parking brake cable from the actuating lever.
2. Unscrew the allen head locking screw on the retaining key.
3. Tap out the caliper retaining key and caliper support spring.
4. Rotate and lift the caliper from the anchor plate.
5. If a damaged rotor prevents the caliper from being removed, remove the parking brake lever and loosen the end retainer no more than half a turn. The piston can then be pushed into the bore far enough to remove the caliper.
6. If the end retainer is loosened more than half a turn, the seal can break between the thrust screw and housing and brake fluid could leak into the parking brake mechanism. If this happens, it will have to be taken apart and cleaned.
7. Hang the caliper by a wire.
8. Remove the inner pad and anti-rattle clip.
9. Remove the outer pad. If necessary, free the pad by tapping lightly with a hammer.
10. Check the caliper and overhaul if necessary, removing from the brake hose.
11. Remove the end retainer.
12. Lift out the operating shaft and thrust bearing.
13. Remove the steel balls and anti-rotation pin using a magnet.
14. Remove the thrust screw with an allen wrench rotated counterclockwise.

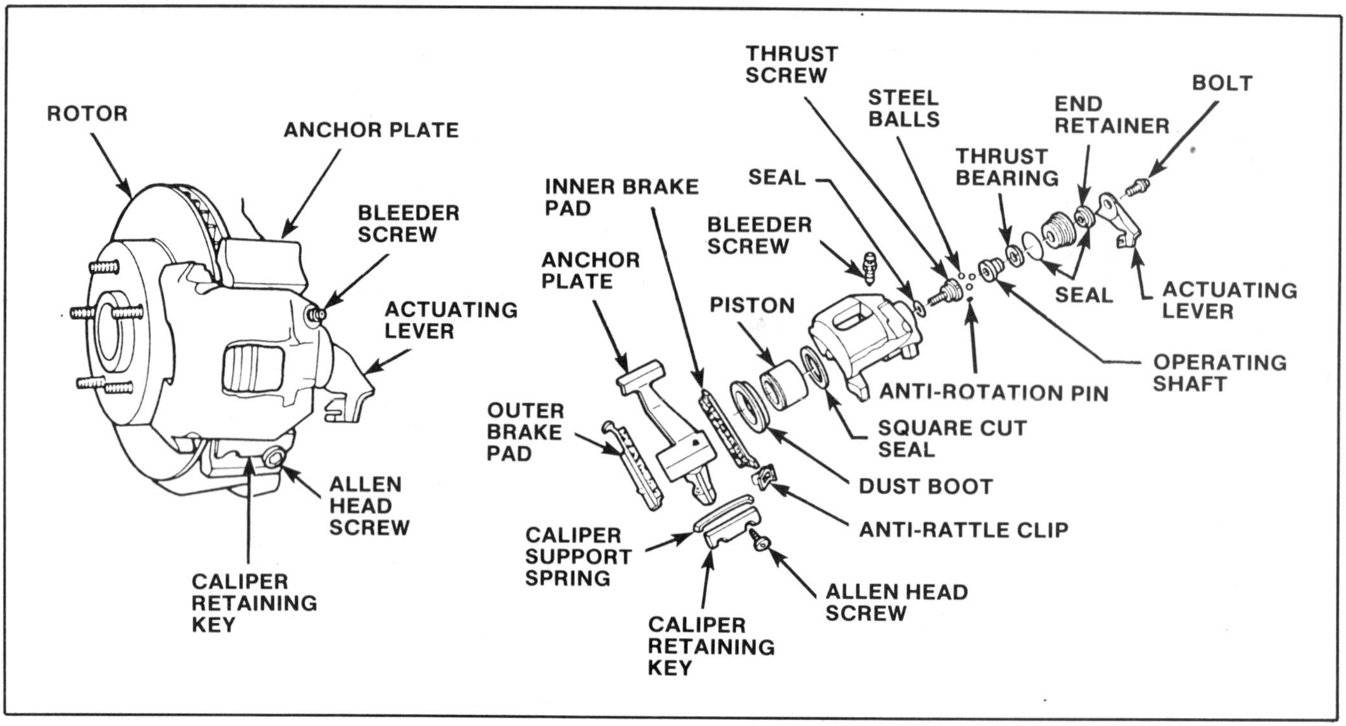

Figure 8-50 Major components of Kelsey-Hayes (Ford) sliding caliper

15. Remove the piston assembly with a special tool through the back of the caliper housing to push out the piston.
16. Overhaul the caliper.
17. Mount the caliper in a vise and fill the piston adjuster assembly with clean brake fluid to the bottom edge of the thrust screw bore.
18. Coat the new O-ring with brake fluid and install it on the thrust screw.
19. Install the thrust screw into the piston adjuster. Turn clockwise with an allen head wrench until the top surface is flush with the bottom of the threaded bore.
20. Align the notches on the screw with those on the caliper.
21. Install the anti-rotation pin.
22. Place one of the steel balls in each of the three pockets of the thrust screw.
23. Use silicone grease on all parts of the mechanism.
24. Install the operating shaft on the steel balls. The thrust screw and operating shaft are not interchangeable from one side to the other.
25. Coat the thrust bearing with silicone grease and install it on the operating shaft.
26. Install a new O-ring and lip seal on the end retainer with a coating of silicone grease.
27. Install the end retainer in the caliper while holding the operating shaft firmly seated against the steel balls. Torque 75 to 95 foot-pounds.
28. Install the parking brake actuating lever onto its keyed spline. The lever arm must point down and rearward for the cable to pass freely under the axle. Torque the lever retainer screw 16 to 22 foot-pounds.
29. Bottom the piston in the caliper using a special tool rotated counterclockwise while holding the shaft of the tool. When firmly seated against the piston, back off the tool handle one-quarter of a turn. Rotate the shaft clockwise until there is no further movement of the piston. The piston will keep turning after bottoming.
30. Attach the brake hose.
31. Position the anti-rattle spring onto the lower edge of the inner abutment of the anchor plate with the loop of the clip away from the rotor.
32. Install the inner pad in the anchor plate.
33. Install the outer pad in the caliper housing so that the metal flanges fit over the caliper legs.
34. Install the caliper. Pull the caliper outward to seat the inner pad on the rotor.
35. There should be a maximum 1/16-inch clearance between the outer pad and the rotor. If clearance is greater, the piston self-

adjuster can be pulled out of position each time the parking brake is used. Adjust the piston by rotating the tool shaft counterclockwise.

36. If the clearance is 1/16 inch or less, hold the caliper against the anchor plate with a screwdriver. Slide the key and caliper support spring into the slot so the notch in the key lines up with the screw hole in the anchor plate.
37. Install the lock screw with an allen head wrench until the screw head seats on the anchor plate.
38. Connect the parking brake cable to the actuating lever.
39. Bleed the brakes.
40. Road test the vehicle.

DELCO-MORAINE FLOATING REAR CALIPER (FIGURE 8-51)

This caliper is similar to the front caliper, but it has a parking brake mechanism to apply piston force against the rotor. It operates the piston assembly through the back of the caliper. An actuating lever is attached to a high lead screw. When the parking brake cable is pulled, the actuating lever rotates the high lead screw through a nut in the preassembled piston. The piston moves mechanically on the high lead screw and forces the pads against the rotor. When the pads wear, the piston moves further out in the bore.

When servicing a Delco-Moraine floating rear caliper, proceed as follows:

1. Remove the wheel. Screw one wheel lug back on to keep the rotor in place.
2. Loosen the parking brake cable at the equalizer, and remove the cable from the actuating lever.
3. Remove the actuating lever return spring.
4. Hold the actuating lever in place and remove the actuating lever lock nut.
5. Remove the actuating lever seal and antifriction washer.
6. Bottom the pistons with a C-clamp. Do not position the C-clamp on the actuator screw.
7. Disconnect the brake line.
8. Remove the caliper guide pins.
9. Lift and drain the caliper.
10. Push the two metal sleeves out of the inner caliper ears.

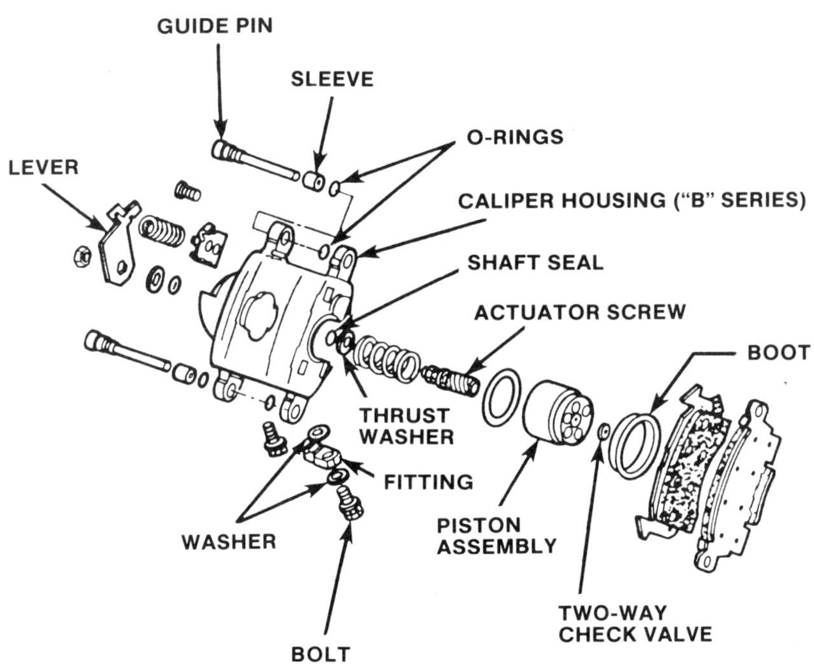

Figure 8-51 Major components of Delco-Moraine floating rear caliper

11. Remove the O-rings from the inner and outer ears.
12. Replace the lever and rotate it back and forth until the piston comes out of the bore.
13. Remove the piston assembly and balance spring, lock nut, lever seal, and anti-friction washer.
14. Push the high lead actuating screw out of the caliper.
15. Clean and overhaul the caliper and all parts. Left- and right-hand parts are not interchangeable.
16. Assemble the new thrust screw with a new washer and seal into the piston.
17. Install the spring and piston assembly into the caliper bore.
18. Bottom the piston with a special tool. Lubricate and install the anti-friction washer, new lever seal, lever, and lock nut.
19. Install the lever away from the stop, rotate it forward, and hold in place. Torque to 25 foot-pounds, and remove the special tool.
20. Rotate the lever back to stop on the caliper. Install the return spring. The red is for the right side; the black is for the left side.
21. Use a special tool to seat the dust boot ring.
22. Lubricate and install the O-rings into the inner and outer ears.
23. Lubricate the outer surface of the sleeves and install into the inner ear with the special tool.
24. Install the inner pad with the D-shaped tab turning piston, if necessary.
25. Install the outer piston with the ears of the shoe over the two outer ears of the caliper and the bottom flange fitting the cutout section of the caliper.
26. Clean and coat the guide pins with assembly fluid.
27. Slide the caliper into place so that the caliper holes and the holes on the support bracket correspond.
28. Push the pins under the inner pad ears, into the outer pad ears, and thread them into the support bracket threads. Torque the pins properly.
29. Connect the brake hose.
30. Bleed the system and pump the pedal to seat the pads.
31. Crimp the shoe ears with pliers.
32. Connect the parking brake cable to the actuating lever.
33. Operate the parking brake. If the actuating lever does not return to stop, loosen the cable at the equalizer. If it still fails to return, you might have to replace the return spring or piston and screw assembly.

REVIEW QUESTIONS

Review the information in this chapter by answering the following questions.

1. Disc brake problems are usually the result of
   a. improperly fitting parts of pads.
   b. distortions in rotor surfaces.
   c. both a and b
   d. neither a nor b
2. When servicing disc brakes on a car, Mechanic A works on one wheel before beginning work on another. Mechanic B performs the same procedure on each front wheel at the same time. Who is right?
   a. Mechanic A
   b. Mechanic B
   c. both A and B
   d. neither A nor B
3. What should be used to remove rust, corrosion, pitting, and scratches from the piston bore?
   a. an oily rag
   b. crocus cloth
   c. abrasive paper
   d. all of the above
4. Brake components should be cleaned in
   a. denatured alcohol or clean brake fluid.
   b. oil brake fluid.
   c. paint thinner.
   d. gasoline or kerosene.
5. The first step in proper caliper service is
   a. to disassemble the calipers.
   b. to remove the brake pads from the vehicle.
   c. to remove the rotor from the vehicle.
   d. to remove the caliper assembly from the vehicle.
6. When replacing fixed calipers, Mechanic A uses standard bolts in place of the bridge bolts. Mechanic B uses bridge bolts. Who is right?
   a. Mechanic A
   b. Mechanic B
   c. both A and B
   d. neither A nor B
7. When examining disc brakes, Mechanic A visually inspects the rotor for wear. Mechanic B uses exacting gauges and micrometers to examine the rotor. Who is right?
   a. Mechanic A

b. Mechanic B
c. both A and B
d. neither A nor B

8. Which term refers to variations in thickness of the rotor?
a. torque
b. lateral runout
c. parallelism
d. pedal pulsation

9. Which term refers to wobbling of the rotor from side to side when it rotates?
a. parallelism
b. torque
c. chatter
d. lateral runout

10. Rotor wear or scoring can be caused by
a. worn linings.
b. rust or road dirt.
c. harsh friction material.
d. all of the above.

11. The simplest method for removing rust and pad lining deposits on a rotor is
a. to use a sanding disc held in an electric drill.
b. to use a lathe.
c. both a and b
d. neither a nor b

12. What is the minimum wear thickness of a rotor?
a. the rotor's dimensions after refinishing
b. the discard thickness
c. the rotor's dimension when new
d. none of the above

13. What causes a rotor to wobble excessively on a lathe's arbor?
a. bent arbor
b. distorted mounting adapters
c. excessive rotor runout
d. all of the above

14. The anchor plate is part of the spindle casting on the floating caliper in what car model?
a. Chrysler
b. GM
c. Ford
d. all of the above

15. The Delco-Moraine four-piston fixed caliper was introduced in the mid-1960's by GM and is now being used only on what car?
a. Corvette
b. Ford
c. Buick
d. none of the above

16. Unlike drum brakes, disc brake specifications
a. always conform to basic standards.
b. never conform to basic standards.
c. sometimes conform to basic standards.
d. are seldom needed.

17. Which of the following should be done before any service is performed on a disc brake system?
a. Check the wheels for bent or warped rims.
b. Check the brake fluid level.
c. Check the suspension components.
d. All of the above

18. If the bleeder screws are frozen tight with corrosion,
a. they can only be freed using a drill.
b. it is virtually impossible to free them.
c. never use a propane torch to free them.
d. they can be freed using a propane torch and penetrating oil.

19. Which of the following statements concerning caliper removal is incorrect?
a. If only the brake pads are being replaced, do not disconnect the brake hose.
b. Support the caliper when removing the second bolt to prevent the caliper from falling.
c. On the older type fixed caliper, remove the bolts holding it to the steering knuckle.
d. On the sliding caliper, remove the special pins that hold the caliper to the anchor plate.

20. To disassemble a caliper, what should be removed first?
a. piston
b. dust boot
c. both a and b
d. neither a nor b

21. To force the piston from the caliper housing, Mechanic A uses a minimum of 50 PSI. Mechanic B uses a maximum of 30 PSI. Who is right?
a. Mechanic A
b. Mechanic B
c. both A and B
d. neither A nor B

22. The installation of some dust boots requires the groove to be filled with
a. a bead of silicone sealer.
b. several drops of silicone sealer.
c. a bead of brake fluid.
d. none of the above

23. Which of the following statements concerning rotors is incorrect?
    a. It is proper to turn the rotor on one side of the vehicle and not the other.
    b. Rotors should be inspected during every brake reline.
    c. A proper rotor finish is important with semi-metallic brake linings to prevent noise.
    d. It is necessary in any replacement of the rotor to first remove the hub and rotor assembly.
24. On most FWD cars, the rotor
    a. can only be removed using special pullers.
    b. can only be removed using an arbor press.
    c. is easily removed.
    d. both a and b.
25. Preloading a wheel bearing causes
    a. accelerated wear.
    b. braking problems.
    c. wheel shimmy.
    d. all of the above
26. The floating caliper is a
    a. four-piece caliper.
    b. three-piece caliper.
    c. two-piece caliper.
    d. one-piece caliper.

CHAPTER 9

PARKING BRAKES—OPERATION AND SERVICING

OBJECTIVES

Upon completing this chapter, you should be able to

- Explain the function of parking brakes.
- Identify the types of parking brake systems.
- Explain the methods for adjusting and servicing parking brakes.
- Describe the procedures for replacing a rear and a front cable.
- Recognize and correct problems in a brake control assembly.
- Explain the procedure for testing parking brakes.

The parking brake (or emergency brake) keeps a vehicle from rolling while it is parked. It is important to remember that the parking brake in most vehicles is not part of the car's hydraulic brake system. It works mechanically, using a ratio level pedal assembly connected through a cable system to the rear drum service brakes. It is not designed to overcome the kinetic energy of a moving car. The parking brake is meant only to keep an already stopped vehicle from moving.

WARNING: Do not use the automatic gearshift selector in place of the parking brake. Always set the parking brake fully and make sure the gearshift selector is securely latched in P (PARK) position.

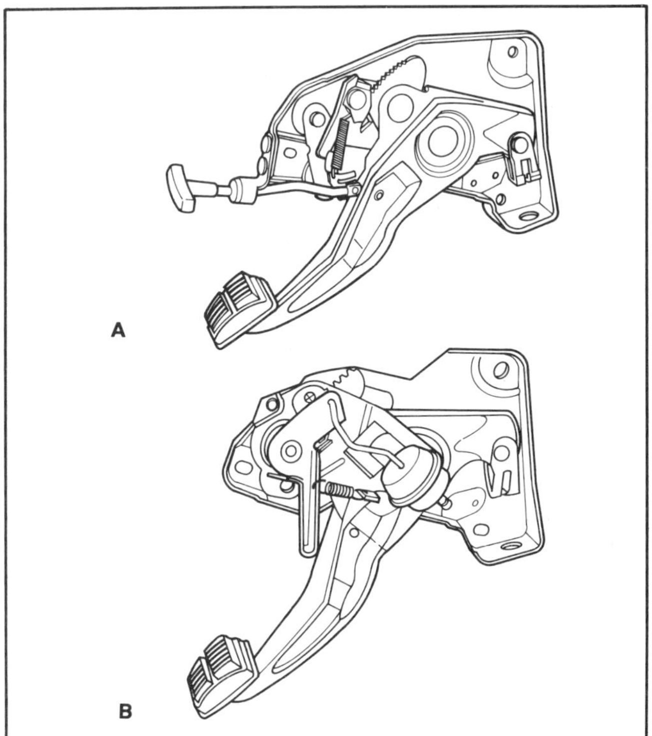

Figure 9-1 Parking brake control assembly—(A) manual release and (B) automatic vacuum release

TYPES OF PARKING BRAKE SYSTEMS

For many years the parking brake was hand operated. With the advent of the automatic transmission, the hand lever was replaced on most automobiles by a foot-operated pedal. While some downsized cars and light-duty trucks have reverted to the hand-operated, self-adjusting parking brake, the majority are set by depressing the parking brake foot pedal. The pedal or manual control assembly (Figure 9-1A) is designed to latch into an applied position and is released by pulling the brake release handle or lever. On some newer cars a vacuum power unit (Figure 9-1B) is connected by a rod to the upper end of the release lever.

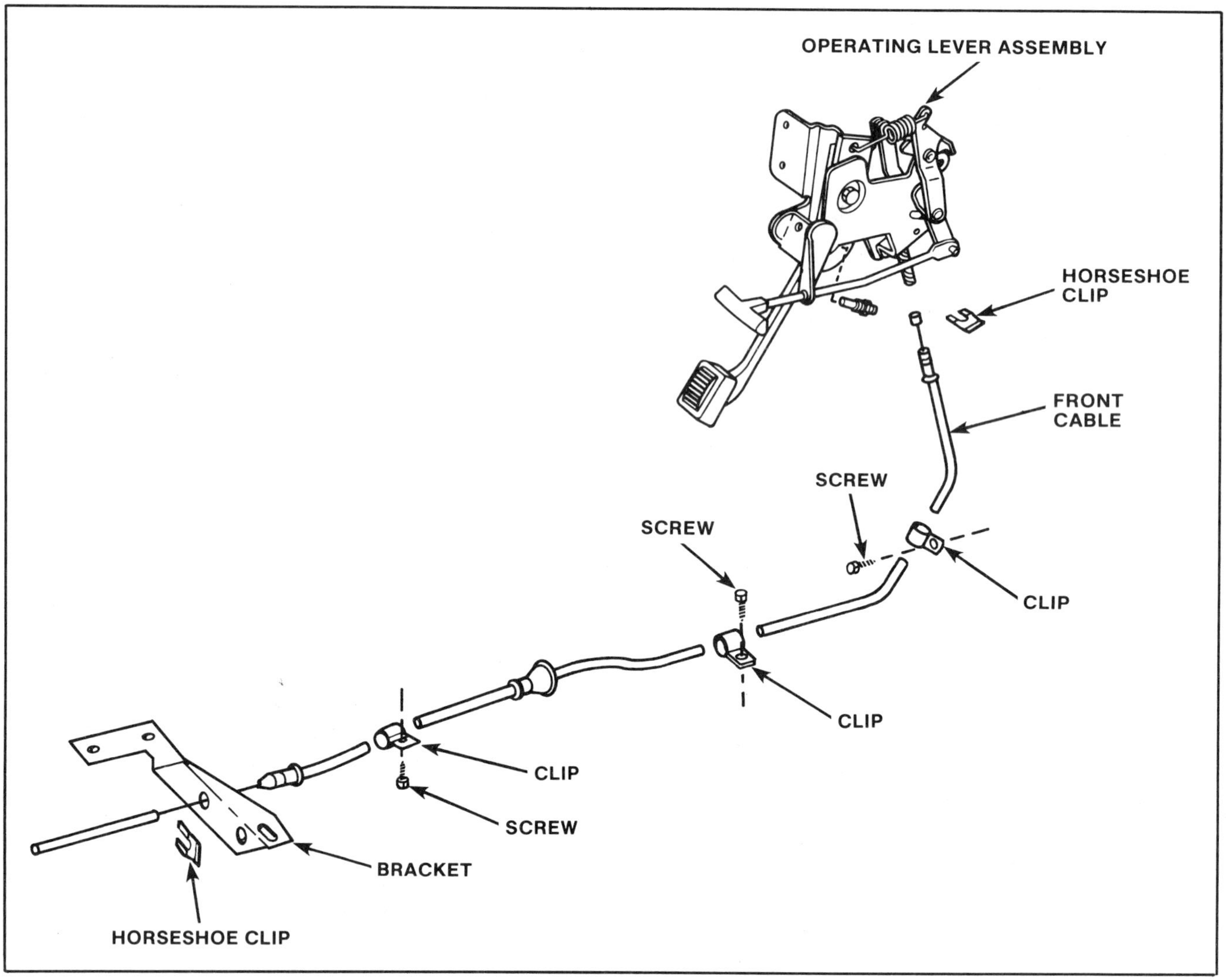

Figure 9-2 Typical parking brake cable and lever system

The vacuum motor is actuated to release the parking brake whenever the engine is running and the transmission is in forward driving gear. The lower end of the release lever extends down for alternate manual release in the event of vacuum power failure or for optional manual release at any time. Hoses connect the power unit and the engine manifold to a vacuum release valve on the steering column.

The starting point of a typical parking brake cable and lever system (Figure 9-2) is the foot pedal or control assembly. This assembly is a variable ratio lever mechanism that converts input effort of the operator and pedal travel into output force with less travel. Thus, as the pedal is being depressed, the attached front cable assembly tightens increasingly as it moves a short distance (by comparison to the pedal's travel distance). Tensile force from the front cable is transmitted through the car's brake cable system to the rear brakes. This tension pulls the flexible steel cables attached to each of the rear brakes. It serves to operate the internal lever and strut mechanism of each rear brake, expanding the brake shoes against the drum. Springs return the shoes to the unapplied position when the parking brake pedal is released and tensile forces in the cable system are relaxed.

An electronic switch, triggered when the brake pedal is applied, lights the brake indicator in the instrument panel when the ignition is turned on. The light goes out when either the pedal or control is released or the ignition is turned off.

As shown in Figure 9-3, the cable/lever routing system in a typical parking brake arrangement employs a three lever setup to multiply the physical effort of the operator. First, of course, is the

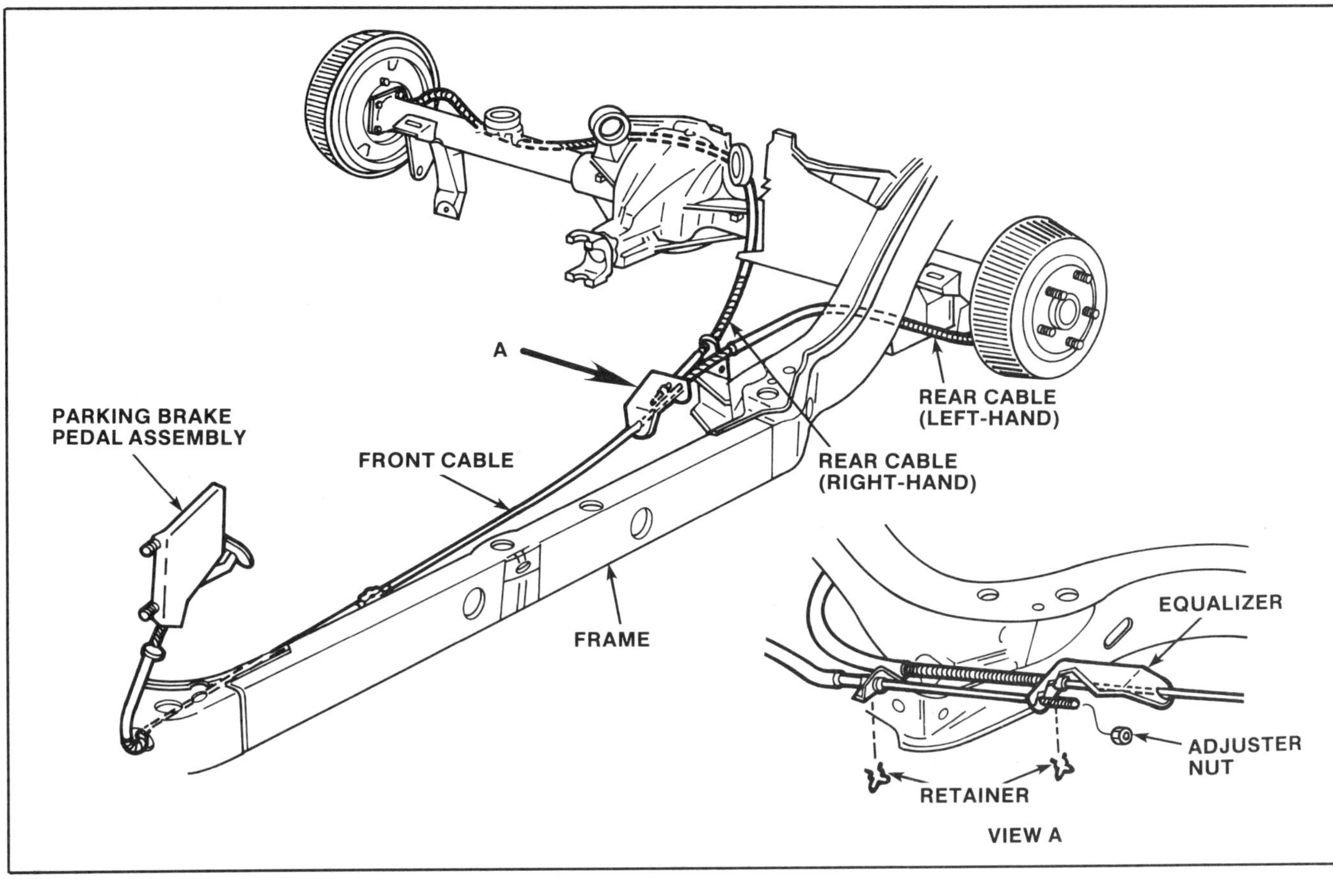

Figure 9-3 Typical parking brake routing system

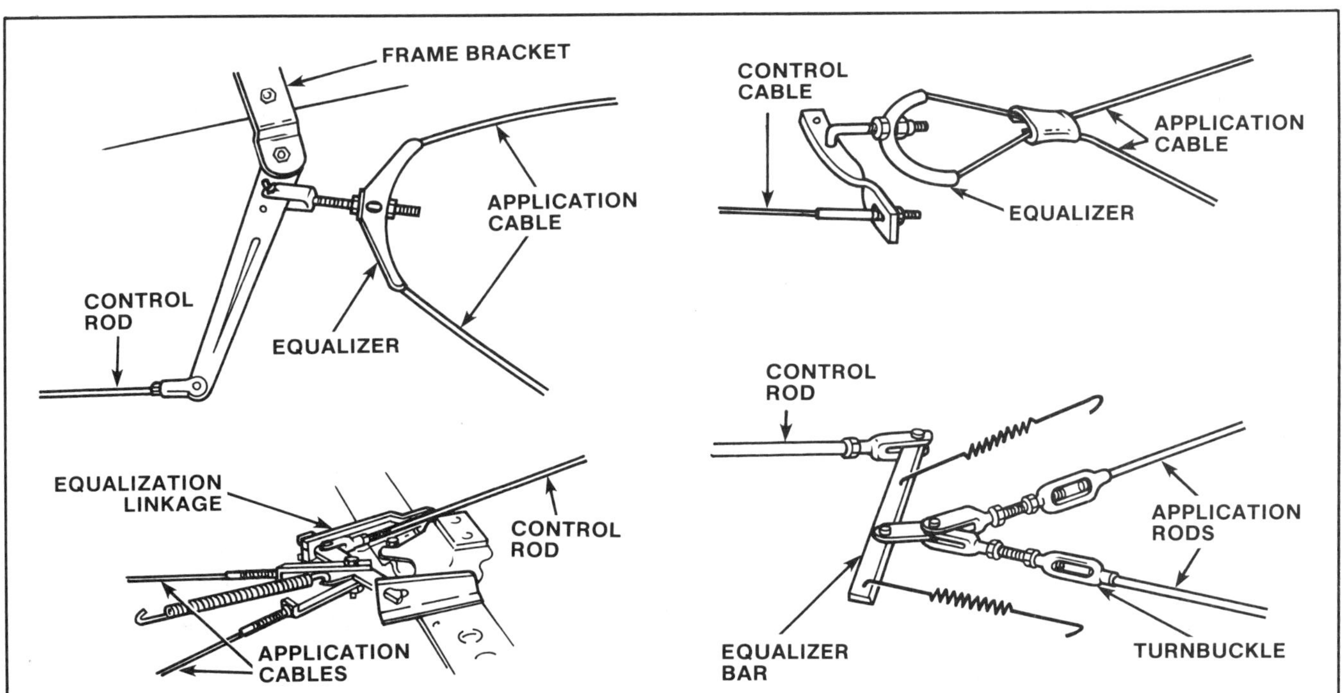

Figure 9-4 Typical equalizers

pedal assembly (or hand grip), which when moved multiplies the operator's effect and pulls the front cable. The front cable, in turn, pulls the equalizer lever.

The equalizer lever multiplies the effort of the pedal assembly, or hand grip, and pulls the rear cables. This pulling effort passes through an equalizer, which ensures equal pull on both rear cables. The equalizer functions by allowing the rear brake cables to slip slightly so as to balance out small differences in cable length or adjustment. Typical equalizer arrangements are shown in Figure 9-4.

The rear cables pull the parking brake levers (Figure 9-5). As for the parking brakes themselves, there are several types in use. They include the following:

- *Integral or Conventional Rear Drum Parking Brake*—This is part of the vehicle's ordinary rear drum system and utilizes the same friction elements as the hydraulic brakes.
- *Rear Disc/Drum Parking Brake (also known as Auxiliary Drum Parking Brake)*—This system has a separate drum brake that is used in conjunction with the rear discs.
- *Rear Disc Parking Brake*—Here the disc calipers themselves, by way of a mechanical actuating mechanism, are used to hold the car while parked. It is interesting to note that while the majority of parking brake systems usually act upon the rear wheels, some import models employ arrangements that hold the front wheels.
- *Transmission or Drive Shaft Parking Brake*—This system operates completely independent of the rear wheel brakes. These parking brakes are known in the trade as *drive shaft brakes.*

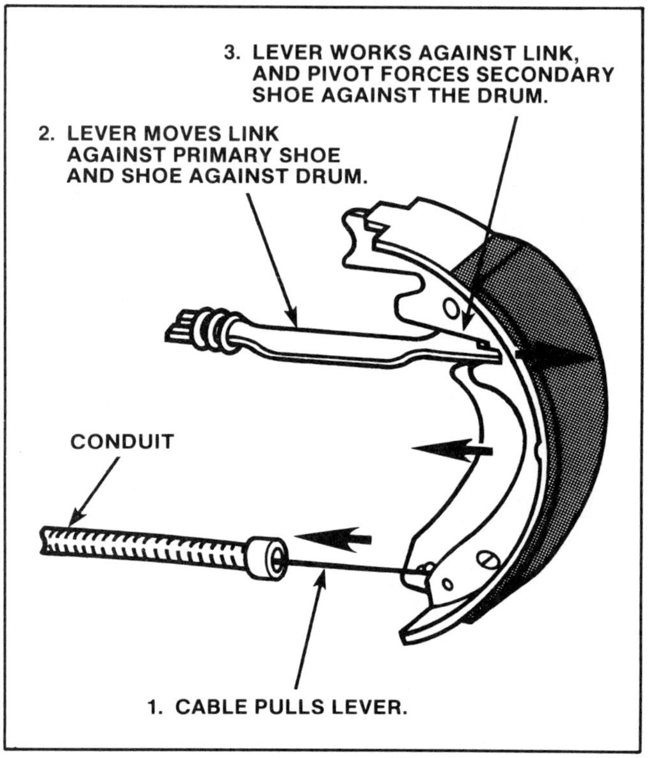

Figure 9-5 Parking brake operation

INTEGRAL PARKING BRAKES

Integral parking brakes are for vehicles with rear wheel drum brakes. Figure 9-6 shows a typical integral parking brake. When the parking brake pedal is applied, the cables and equalizer exert a balanced pull on the parking brake levers of both rear brakes. The levers and the parking brake struts move the shoes outward against the brake drums. The shoes are held in this position until the parking brake pedal is released. The portion of the parking brake from the equalizer on back to the rear wheel drum brake is shown in Figure 9-7. Also, the parking brake components are shown removed from the wheel brake for identification.

The rear cable enters each rear brake through a conduit (Figure 9-8). The cable end engages the lower end of the parking brake lever. This lever is hinged to the web of the secondary shoe and linked with the primary shoe by means of a strut. Thus the lever and strut expand both shoes away from the anchor and wheel cylinder and into contact with the drum as the cable and lever are drawn forward. The shoe return springs reposition the shoes when the cable is slacked.

To remove and replace the rear brake shoes, it might be necessary to relieve the parking brake cable tension by backing off the adjusting check nut(s) at the equalizer. Count the turns backed off in order to restore the nuts to their original position.

ADJUSTING AND REPLACING PARKING BRAKES

Regular wheel brake service should be completed before adjusting the parking brake. Then, check the parking brake for:

- Free movement of the parking brake cables in the conduits. If necessary, apply a lubricant to free the cables.
- Worn equalizer and linkage parts. Replace any defective parts.

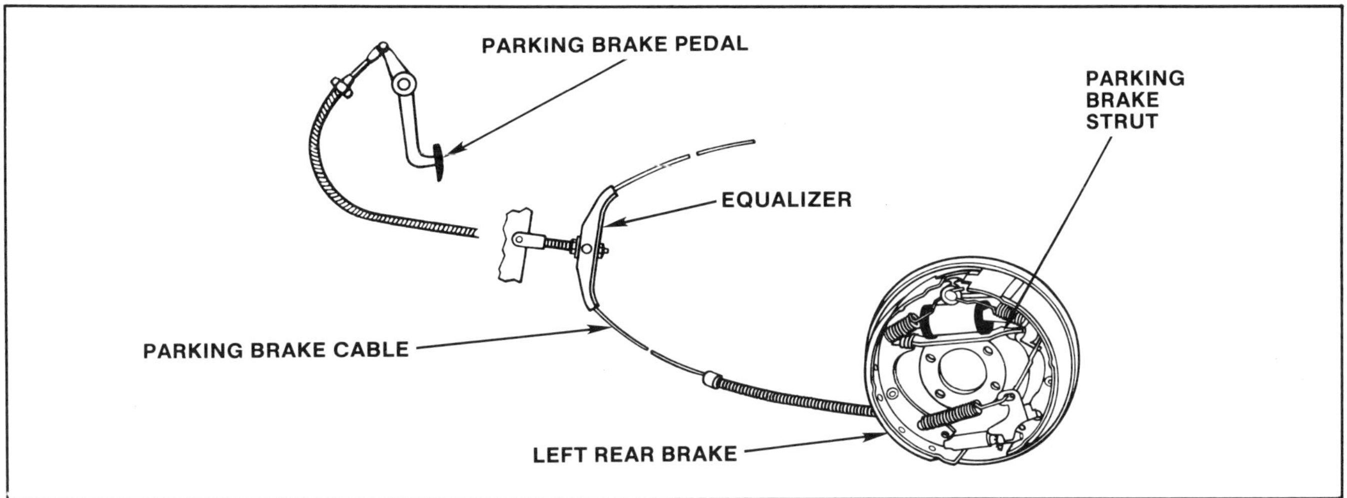

Figure 9-6 Typical integral parking brake

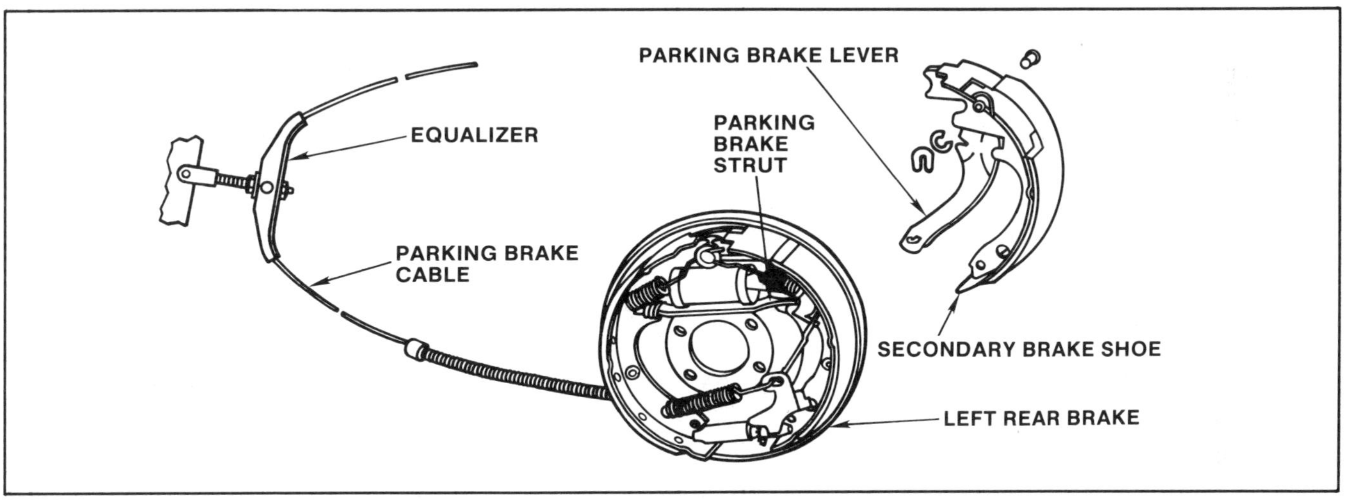

Figure 9-7 Integral parking brake components

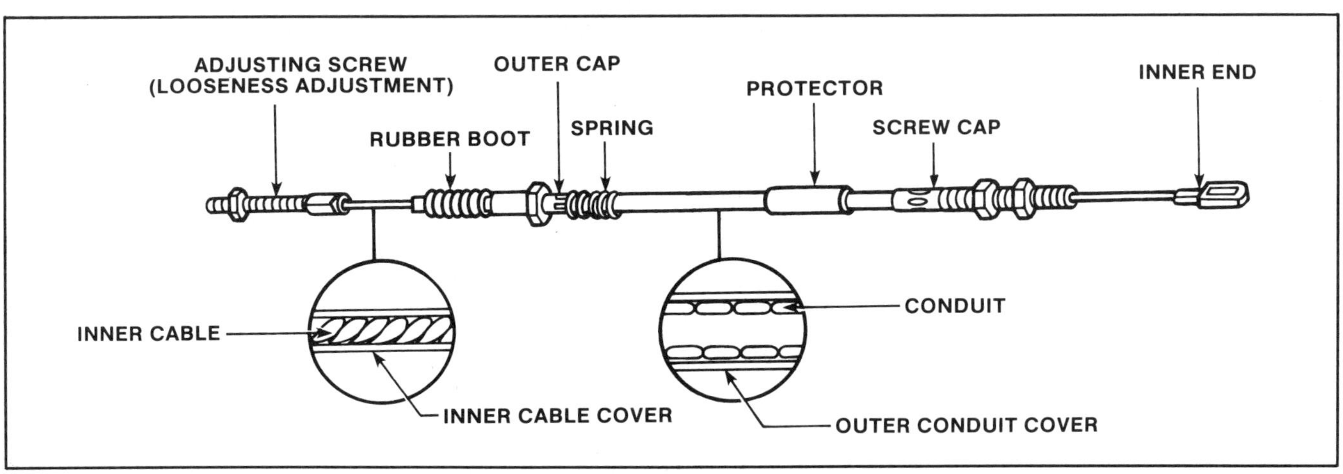

Figure 9-8 Rear cable enters each rear brake through a conduit.

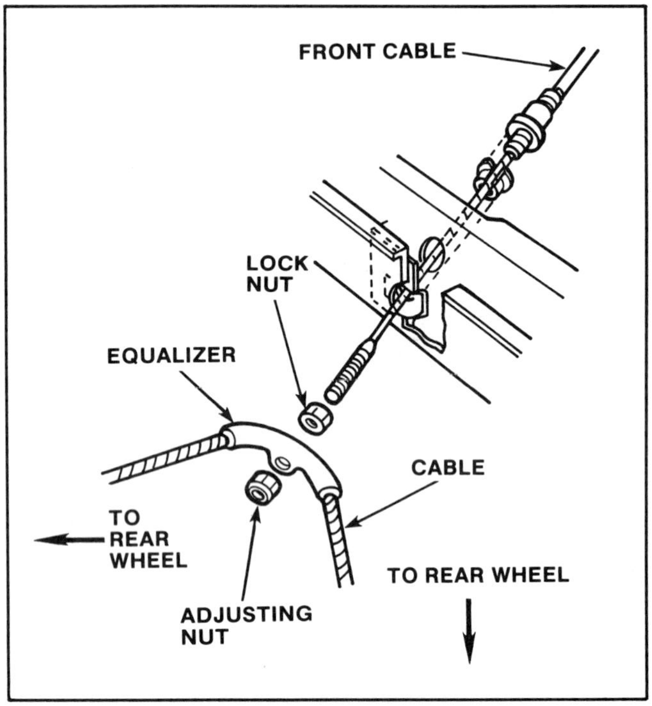

Figure 9-9 Parking brake adjustment

- Broken strands in the cables. Replace any cable that has broken strands or shows signs of wear.

SHOP TALK

While it is always wise to check the cable adjustment and replacement procedures in the appropriate vehicle manual, the following steps are considered typical.

Adjusting a Parking Brake

The following procedure is typical for adjusting parking brakes in a rear drum system:

1. Raise the vehicle and support it with jack stands placed under the suspension.
2. Set the transmission shift lever in the NEUTRAL (N) position.
3. Release the parking brake operating lever and loosen the cable adjusting nut to make sure that the cable is slack (Figure 9-9).
4. Clean the cable threads with a wire brush, and lubricate to prevent rusting.
5. Tighten the cable adjusting nut at the equalizer until a slight drag is felt while rotating the rear wheels.
6. Loosen the cable adjusting nut until both rear wheels can be rotated freely.
7. Back off the cable adjusting nut two full turns. There should be no drag at the rear wheels.

SHOP TALK

If a front cable must be adjusted, do this first. Then, engage the operating lever at least one notch before adjusting the rear cables. This helps avoid overtightening, which would expand the brake shoes into contact with the drum.

8. Make a final operational check of the parking brake and then lower the vehicle.

On some import vehicles, the parking brake adjustment can be made inside the car (Figure 9-10).

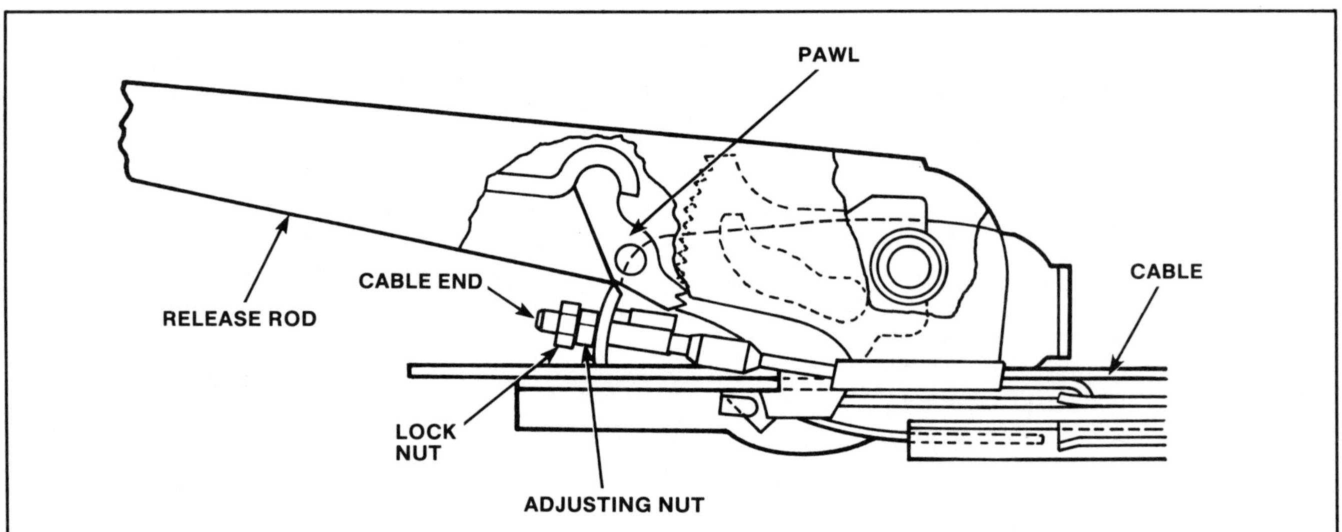

Figure 9-10 Parking brake adjustment components inside the car

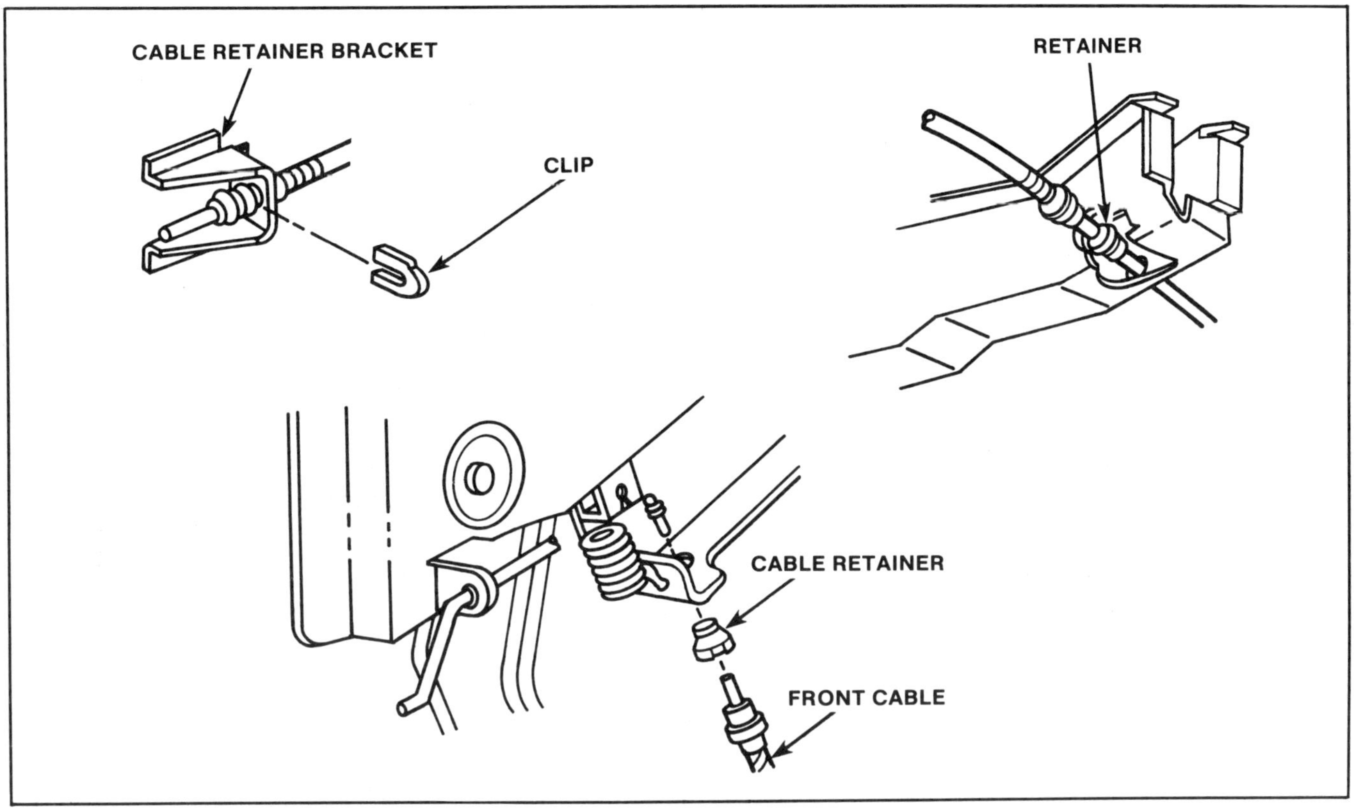

Figure 9-11 Brake cable bracket with retaining clip

REPLACING A REAR CABLE

When working on the parking brake cables, it is usually necessary to raise the vehicle and support it on jack stands or a lift as previously described. Then, to remove the rear cable, proceed as follows:

1. Remove the rear wheels and disconnect the brake cable.
2. Detach the retaining clip from the brake cable retainer bracket (Figure 9-11). Remove the brake drum (from the rear axle), the brake shoe return springs, and the brake shoe retaining springs.
3. Remove the brake shoe strut and spring from the brake support plate.
4. Disconnect the brake cable from the operating lever.
5. Compress the retainers on the end of the brake cable housing and remove the cable from the support plate.

To install a rear cable, proceed as follows:

1. Insert the brake cable and housing into the brake support plate, making sure that the housing retainers lock the housing firmly into place.
2. Hold the brake shoes in place on the support plate.
3. Engage the brake cable into the brake shoe operating lever.
4. Install the parking brake strut and spring.
5. Install the brake shoe retaining springs, brake shoe return springs, and brake drum and wheel.
6. Insert the brake cable and housing into the bracket and install the retaining clip.
7. Insert the brake cable into the equalizer.
8. Adjust the rear service brakes and parking brake cable.

REPLACING A FRONT CABLE

To remove the front cable, it is necessary to disengage the front parking brake cable from its connector and then proceed as follows:

1. Using a screwdriver, force the cable housing and attaching clip forward out of the body crossmember.
2. Fold back the left front edge of the floor covering to get at and pry out either the rubber grommet or rubber cable cover (Figure 9-12)

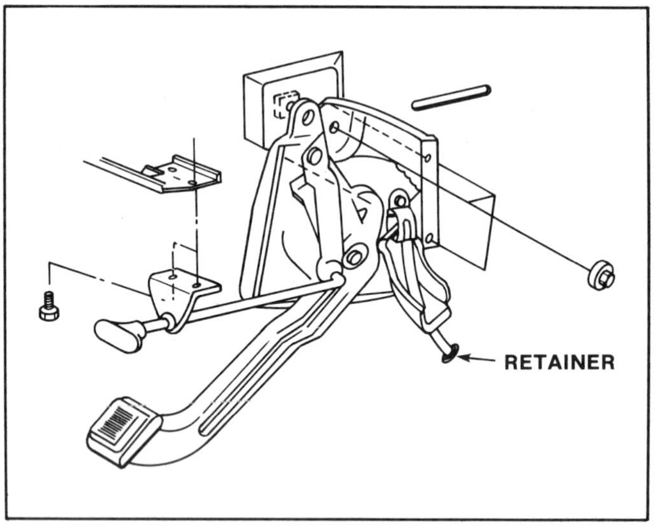

Figure 9-12 Parking brake pedal assembly bracket

from the floor pan (depending on car model). Remove the floor pan clip.

3. Engage the parking brake and work the brake cable up and out of the clevis linkage or guide bracket.
4. Using a screwdriver, force the cable assembly and clip down and out of the pedal assembly bracket. If necessary, attach a substantial wire about 50 inches long to the lower end of the cable housing (to be used as a guide for cable installation.)
5. Work the cable and housing assembly up through the floor pan. If used, disconnect the guide wire, but leave it in place for new front cable installation.

To install a front cable, proceed as follows:

1. Pull the guide wire (if used) that is connected to a new cable through the floor pan. Otherwise, insert new brake cable through the floor pan (Figure 9-13).
2. Insert the retainer into the hole in the bottom of the parking brake pedal assembly bracket.
3. Insert the cable through the hole and the cable end fitting into the clevis linkage or guide bracket.
4. Force the upper cable housing into the retainer until it is firmly seated against the pedal assembly bracket.
5. Insert the retainer into the hole in the crossmember or bracket.
6. Insert cable through the hole in the crossmember and force the cable housing into the retainer until it is firmly seated.

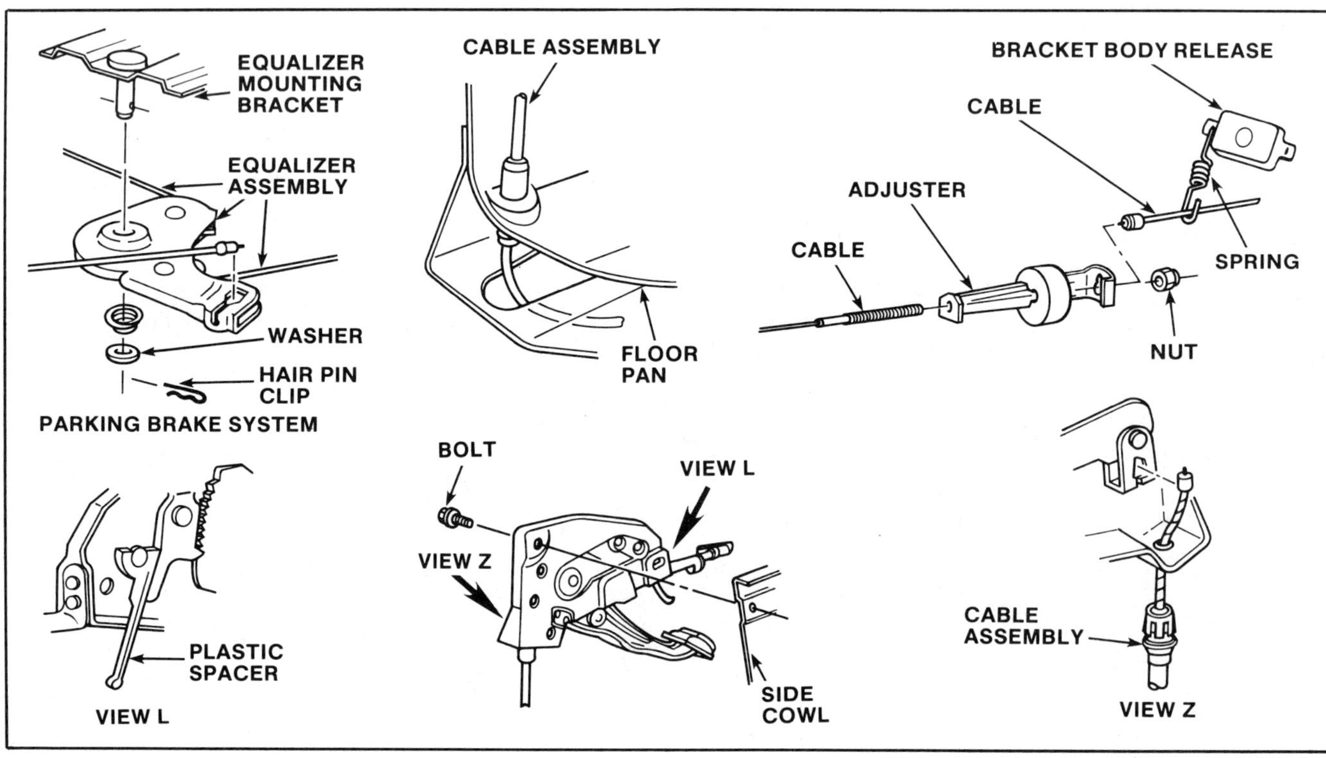

Figure 9-13 Typical fastening of parking brake system *(courtesy of Ford Motor Company)*

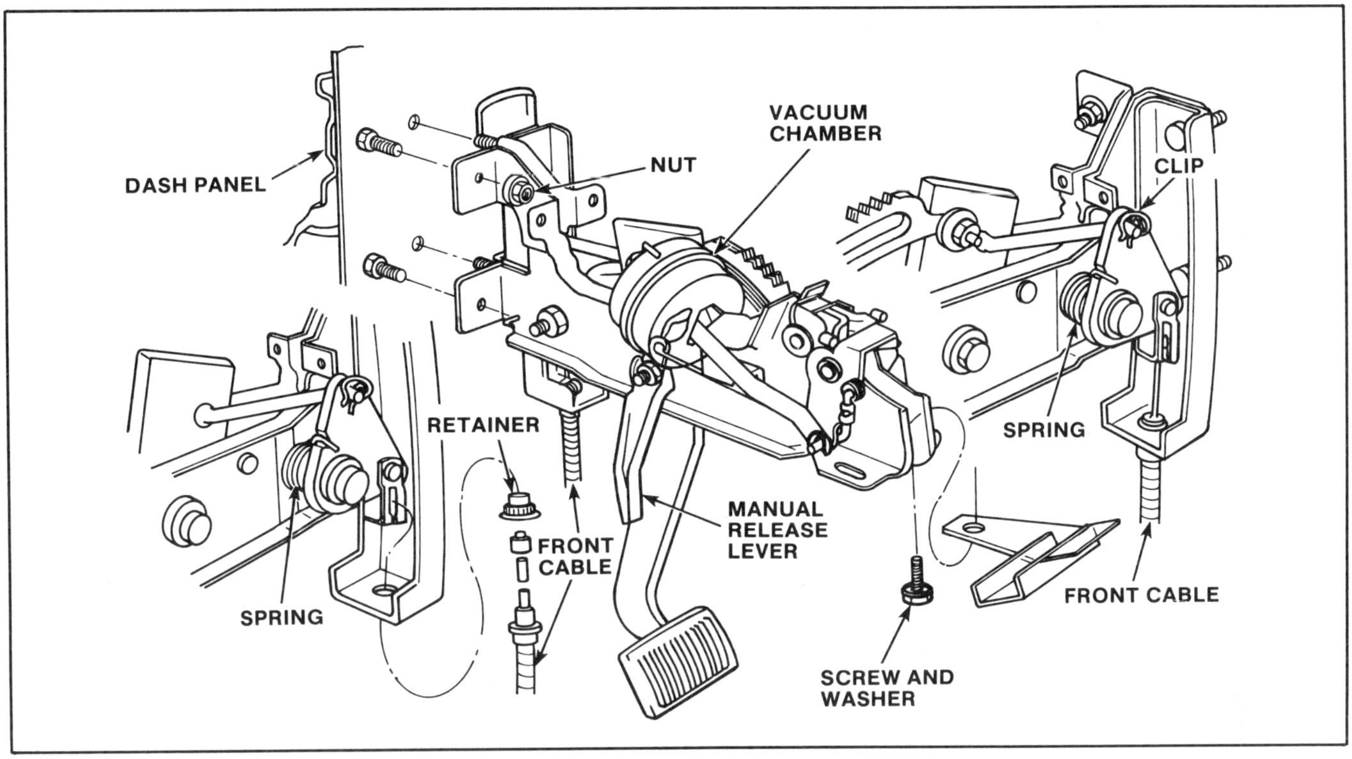

Figure 9-14 Parking brake vacuum release system *(courtesy of Ford Motor Company)*

7. Where applicable, push the cable seal into the hole in the dash panel with enough force to seat the locking tabs. Otherwise, reinstall the rubber cable cover and floor pan clip.
8. Attach the cable (through bracket if necessary) to the connector.
9. Adjust the rear service brakes and the parking brake.
10. Apply the parking brake several times and test for free wheel rotation each time you release it.

SERVICING BRAKE CONTROL ASSEMBLY

Look closely at the operation of the brake linkage as the brake pedal is depressed. Then, check the operation of the brake linkage when the manual release lever is activated. These checks will indicate whether the manual parking brake control linkage is operating properly, requires service, or adjustment due to inability of the parking brake to hold against moderate vehicle movement. Perform tests of the parking brake system and controls after making certain the linkage and manual controls operate properly.

When testing a parking brake vacuum release system, a minimum pressure (specified in the service manual) must be available at all points where vacuum is applied. This can be checked with a vacuum tester gauge. Failure to maintain pressure during vacuum system tests could be caused by a loose hose connection, resulting in a vacuum leak. When checking for vacuum between two points, trace the hose along its entire routing to be sure it is not crossed with another hose or connected to the wrong connection.

All of the vacuum parking brake control checks are to be performed with the engine running at idle speed. To detect any leaks in the parking brake hoses or to find disconnected or improperly connected hoses, listen for a hissing sound along the hose routing.

CAUTION: Do not apply air pressure to the vacuum system under any circumstances, because the actuator diaphragm in the parking brake vacuum motor could be damaged.

To check the operation of the parking brake vacuum release system (Figure 9-14), proceed as follows:

1. Start the engine and run it at idle speed.

2. With the transmission shift control in NEUTRAL, depress the parking brake pedal to apply the parking brake.
3. Move the transmission shift control to DRIVE and observe the parking brake sector to see that the sector returns to its zero travel position when the parking brake releases.
4. If the parking brake releases, the parking vacuum control is working properly. It is important to remember that the parking brake vacuum release does not operate with the transmission in REVERSE.
5. If the parking brake does not release, test for vacuum at the vacuum line connected to the parking brake release vacuum motor.
6. This can be accomplished by removing the hose from each component and attaching it to the vacuum gauge.
7. Connect two distributor tester vacuum hose adapters together with a coupling as a connector attaching the gauge.
8. A minimum as per specifications is required to actuate the parking brake vacuum motor. If a minimum reading is not present when checking each of the aforementioned components, they must be replaced.

If it is necessary to remove a vacuum valve:

1. Place the transmission shift selector lever in DRIVE with the engine off.
2. Remove the vacuum hoses from the engine manifold and vacuum cylinder connections of the unit.
3. Take out the screws that attach the vacuum unit to the steering column jacket and remove the vacuum valve.

To install the vacuum valve unit, move the actuating arm on the valve against the spring to the extreme position, or until the locating holes line up. Insert a No. 42 drill in the hole to correctly position the valve for installation (Figure 9-15). Move the transmission shift selector into PARK. Place the valve on the steering column jacket and install the attaching screws. Do not tighten them. Rotate the valve clockwise until the actuating arm contacts the tabs inside the steering column jacket. Tighten the screws securely.

Remove the drill from the locating hole and install the vacuum hoses. Be sure the hose from the engine manifold is attached to the center fitting on the valve. Start the engine and check to see that the parking brake can be set in NEUTRAL and PARK positions and will release in REVERSE and DRIVE positions.

REAR DISC/DRUM (AUXILIARY DRUM) PARKING BRAKE

The rear disc/drum or auxiliary drum parking brake arrangement is found on some GM and Chrysler vehicles that use discs on the rear wheels. On these brakes, the inside of each rear wheel hub and rotor assembly is used as the parking brake drum. Figure 9-16 shows a typical rear disc parking brake.

If it is necessary to service the disc brake, do this before adjusting the parking brake. Lining wear under normal use is negligible; however, should it be necessary to adjust for this, it should be done before adjusting for cable wear. Shoe adjustment should be done with the parking brake lever or pedal released and the cable slack. Access to the adjuster hole, which is in the hub, is gained by removing the wheel. Shoe adjustment is done with a star wheel and performed in the same manner as a conventional drum brake adjustment (Figure 9-17).

To adjust for cable wear, engage the foot pedal or hand lever one notch. Then remove all slack by tightening the cable at the equalizer. Make sure that the wheels rotate freely when the brake is released. Whenever you adjust the cable, examine it for binding in the conduits and broken strands. Lubricate or replace as necessary. Inspect also for wear in the control assembly, equalizer, and linkage parts and replace as necessary as described earlier in this chapter.

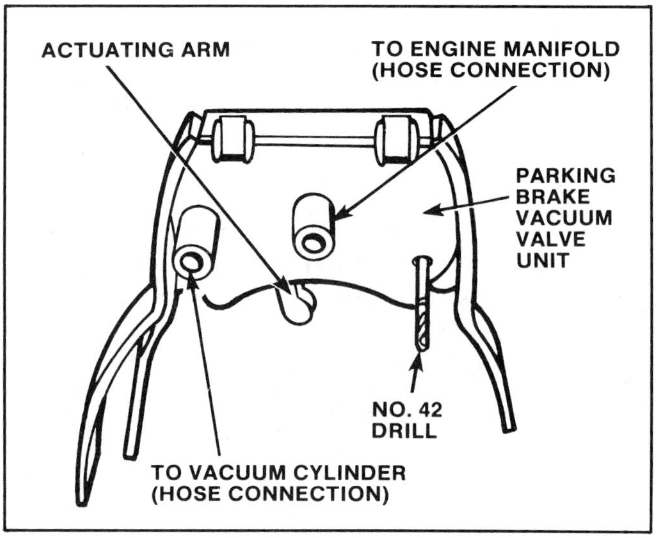

Figure 9-15 No. 42 drill inserted to correctly position the valve for installation

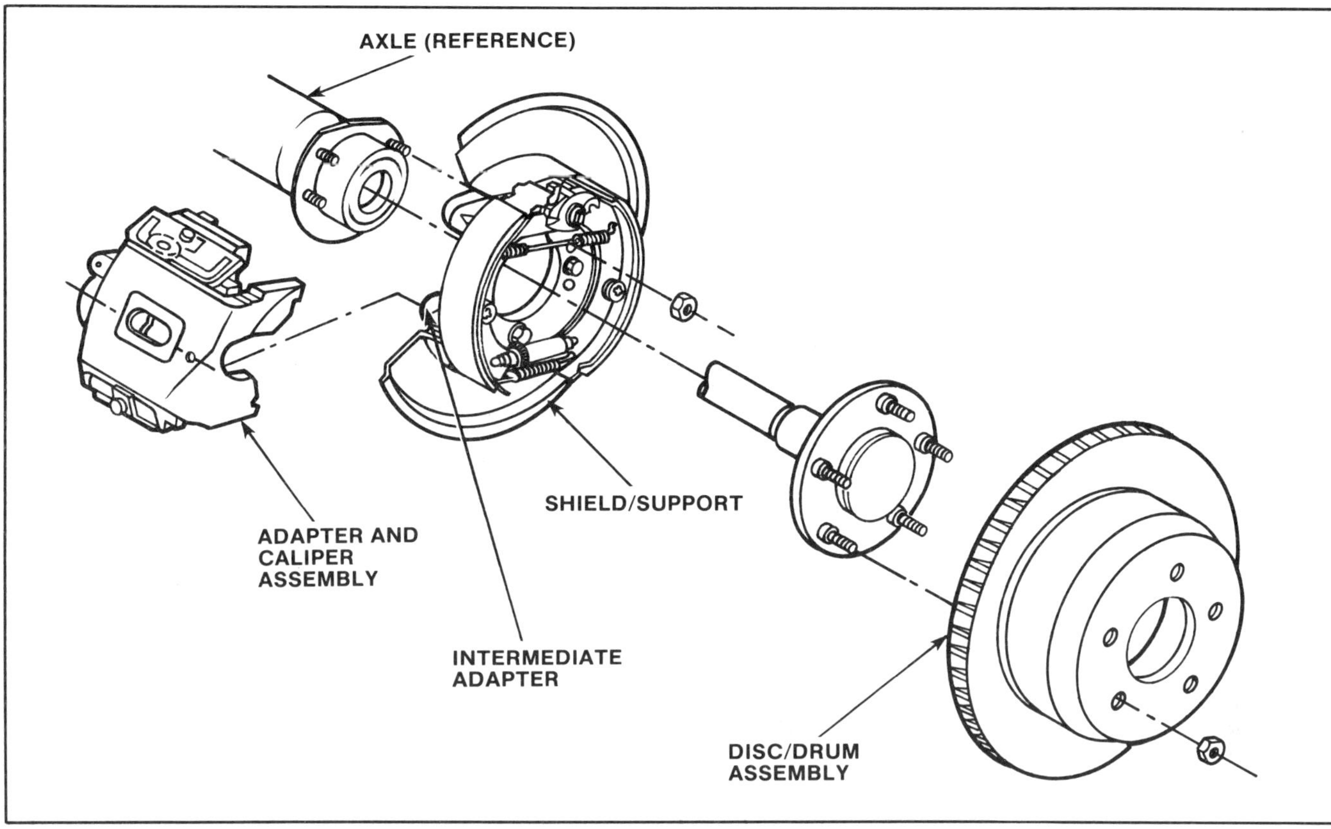

Figure 9-16 A typical parking brake that uses a drum built into the center of a rotor

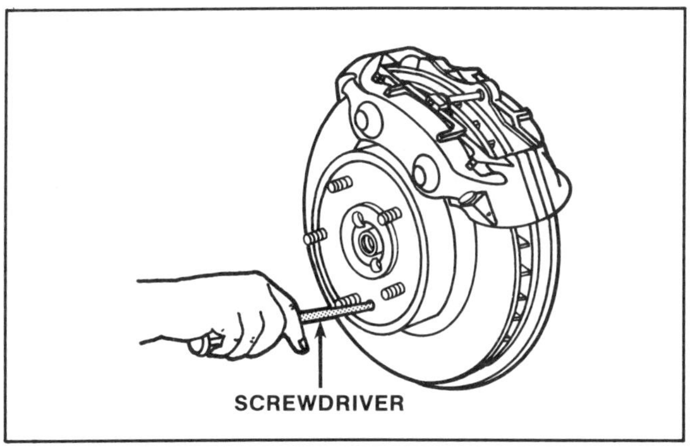

Figure 9-17 Making auxiliary parking brake adjustment

GIRLING-LOCKHEED DISC/DRUM PARKING BRAKE

This disc/drum parking brake system (Figure 9-18) introduced in 1984 is used on several GM four-wheel disc brake system vehicles. While parking brake operation is basically the same as on older models, it is important for the brake technician to understand the following overhaul instructions:

REMOVAL

To remove and inspect the Girling-Lockheed disc/drum parking brake, proceed as follows:

1. Release the parking brake and remove the wheel and tire assembly.
2. Remove the disc/drum combination. If necessary, back off the adjustment of the shoes by turning the star wheel through the slot in the disc/drum hub.
3. Clean away dust, dirt, and rust with solvent and a brush or an approved filtered vacuum cleaner. *Do not* blow the dust off the brake with compressed air.
4. Spread the shoes and remove the adjuster assembly.

CAUTION: Be careful not to damage the wear shims on the mounting plate during any step of this operation. If the wear shims are damaged, the entire mounting plate assembly will have to be replaced.

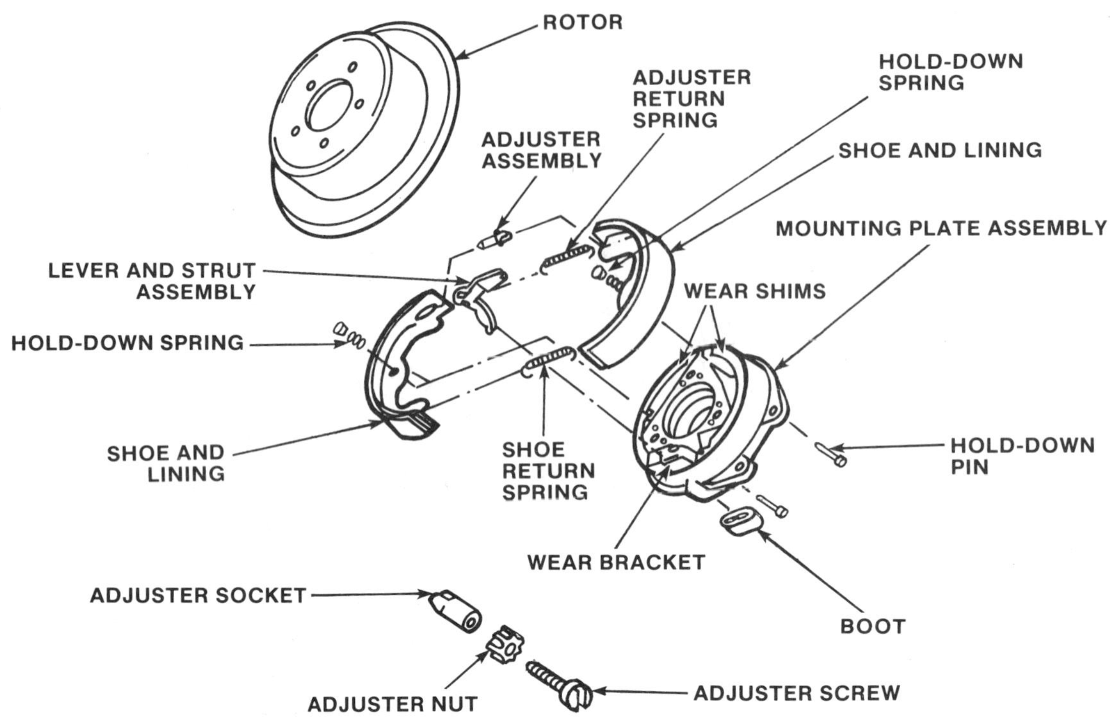

Figure 9-18 Girling-Lockheed disc/drum parking brake

5. Remove the adjuster spring.
6. Remove the hold-down springs by pushing down on the spring and sliding the pin over until it can be removed through the hole at the end of the slot.
7. Remove the shoe return springs.
8. Remove both shoes.
9. Clean and inspect the adjuster assembly. Lubricate the threads.
10. Clean and inspect the wear bracket and wear shims. If they are unserviceable, the mounting plate assembly must be replaced.
11. Clean and inspect the rubber boot. If it is cracked or torn, it must be replaced.
12. Clean and inspect the hardware.
13. Clean all the parts and lubricate the wear bracket and wear shims.

INSTALLATION

When installing the parking brake, proceed as follows:

1. Install the shoes. Make sure the operating lever and strut assembly is positioned on the wear shims, not down in the hole.
2. Install the shoe return spring.
3. Install the hold-down springs and pins.
4. Install the adjuster return spring.
5. Install the adjuster assembly.
6. Adjust the shoe-to-drum clearance and cable after installing the disc/drum. Hold it on with two lugs.
7. Install the wheel and tire assembly.

REAR DISC PARKING BRAKES

Instead of using an auxliary drum and shoes to hold to the vehicle when parked, these brakes have a mechanism that forces the pads against the rotor mechanically. One method for doing this is the ball-and-ramp arrangement. Another method, used in the GM rear calipers for E-body cars, uses a high lead thrust screw to hold the piston out and maintain contact with the pads and rotors. Other types of actuating systems for rear disc brakes include a variety of cams. The two major types in use, as discussed in Chapter 7, are the Kelsey-Hayes and the Delco-Moraine design.

KEYSEY-HAYES REAR DISC PARKING BRAKE

A sectional view of the Kelsey-Hayes sliding caliper rear disc brake as used on Ford Motor Company vehicles is shown in Figure 9-19. The parking brake is a part of the regular brake. A mechanical lever works the ball and ramp device within the parking brake.

Check adjuster operation by first assembling the thrust screw into the piston assembly. As shown in Figure 9-20, pull the two pieces apart by hand approximately 1/4 inch, then release them.

When pulling the pieces apart, the brass drive ring must remain stationary, causing the nut to rotate. When releasing the two parts, the nut must remain stationary and the drive ring must rotate. The piston and adjuster assembly should be replaced if the action of the components does not follow this pattern. Then, inspect the ball pockets, threads, grooves, balls, anti-rotation pin, lever, and other components for wear or damage. Replace worn or damaged components.

When reassembling the unit, remember that the thrust screws and operating shaft are not interchangeable from side to side due to the ramp direction in the ball pockets. The pocket surface of the operating shaft and the thrust screw are stamped with the letter R or L, indicating the side on which each is to be used.

The balls should be placed in each of the three pockets of the thrust screw. Liberal amounts of silicone grease should be applied to all components of the parking brake mechanism. The operating shaft goes on the balls and the thrust bearing on

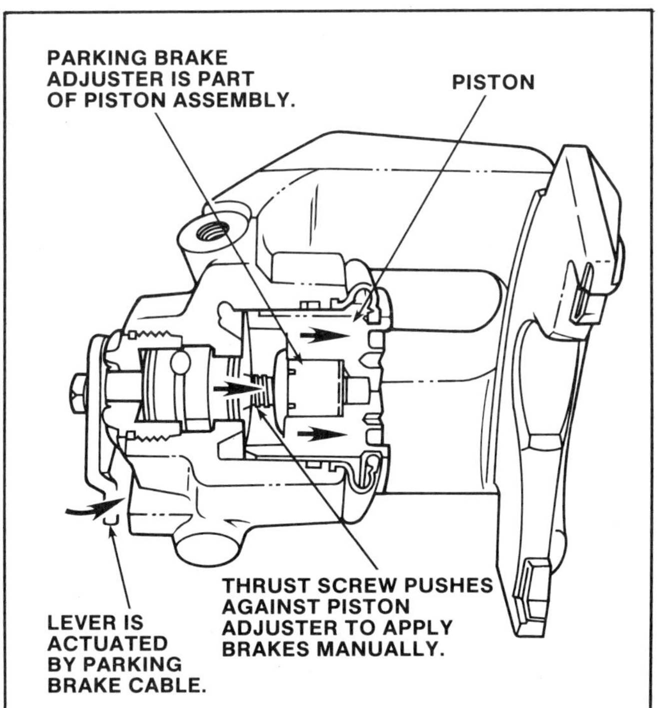

Figure 9-19 Kelsey-Hayes rear disc parking brake

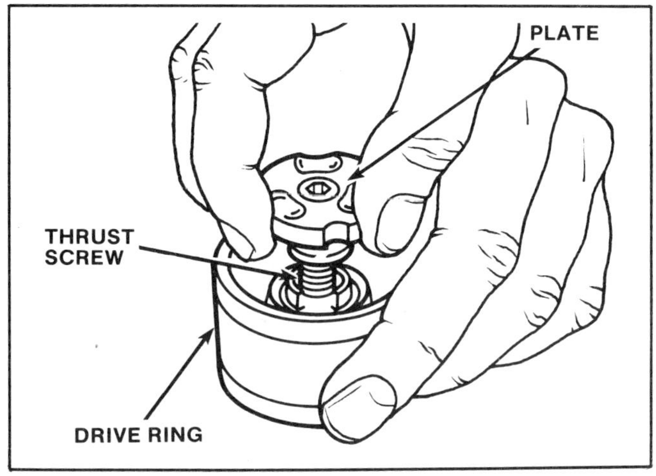

Figure 9-20 Checking the Kelsey-Hayes adjuster

the operating shaft. The parking brake lever must rotate freely after torquing the end retainer to 75 to 95 foot-pounds. While installing the parking brake lever on its keyed spline, the lever arm must point down and rearward. The lever retainer screw should be torqued 16 to 22 foot-pounds. Further disassembly and assembly information is given in Chapters 7 and 8 under Rear Disc Brakes.

DELCO-MORAINE REAR DISC PARKING BRAKE

In this design (Figure 9-21), the actuating lever is connected to a high lead screw that passes through a nut in the piston. Rotation of the lever moves the piston along the screw, thus engaging the brake.

The nut in the piston is spring-loaded, permitting piston travel during hydraulic braking. As lining wear allows the piston to move further out of the bore, the nut is rotated along the lead screw by means of an adjustment spring. No manual adjustment is possible on this brake. When the caliper is overhauled, the thrust washer, seals, and any damaged parts of the parking brake mechanism should be replaced. Lead screws are marked right and left.

Normally, there is no reason to disassemble the parking brake mechanism except during pad replacement or servicing of the regular disc brake. However, it is necessary to remove the actuating lever to get the caliper off the car.

REMOVAL

To do this, follow the normal procedure for caliper removal, up to and including removing the wheel. Then, proceed as follows:

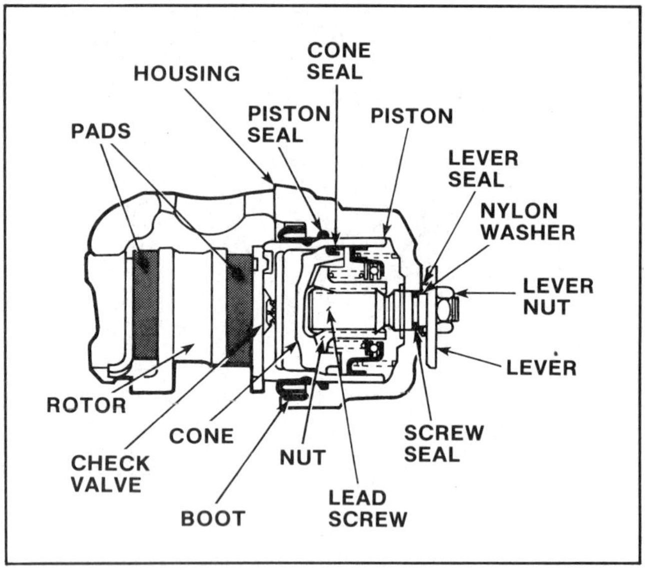

Figure 9-21 Delco-Moraine parking brake

1. Loosen the tension on the parking brake cable at the equalizer.
2. Remove the cable from the lever (Figure 9-22).
3. Remove the lever spring.
4. Holding the lever in place, remove the lock nut, lever, seal, and washer.

Proceed with normal caliper removal.

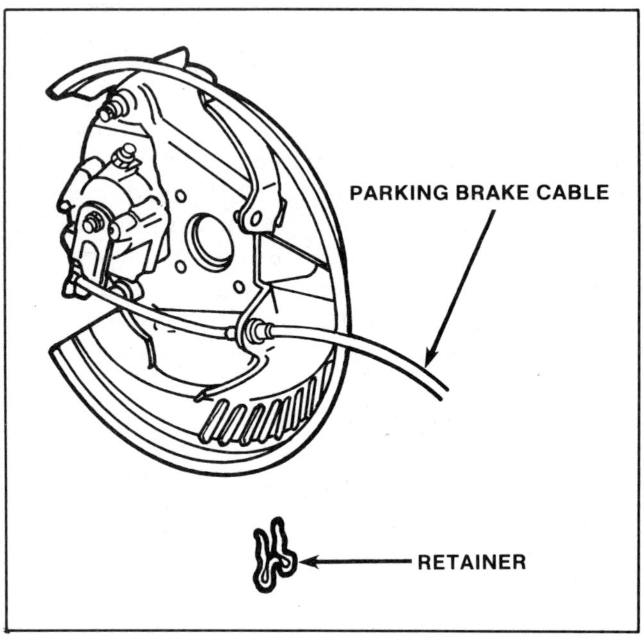

Figure 9-22 Rear brake cable

ASSEMBLY

When it comes to reassembling the brake unit, proceed as follows:

1. Ensure that the parking brake pedal is completely released and that the lever is against its stop.
2. Work the parking brake pedal back and ck to forth. Chesee if the lever returns all the way to the stop. If not, more cable slack is needed—adjust it at the equalizer nut.
3. If it still does not return, even with plenty of cable slack, check the return spring and piston and screw assembly and replace as necessary. Continue with normal caliper reinstallation.

Further assembly and disassembly information is given in Chapter 8 under Rear Disc Brakes.

TRANSMISSION OR DRIVE SHAFT PARKING BRAKE

There are two different types of transmission or drive shaft parking brakes, both of which are completely separate from the basic brake system: the external band type and the internal shoe type.

EXTERNAL BAND BRAKE

As shown in Figure 9-23, this brake consists of a drum, which is normally attached to the drive

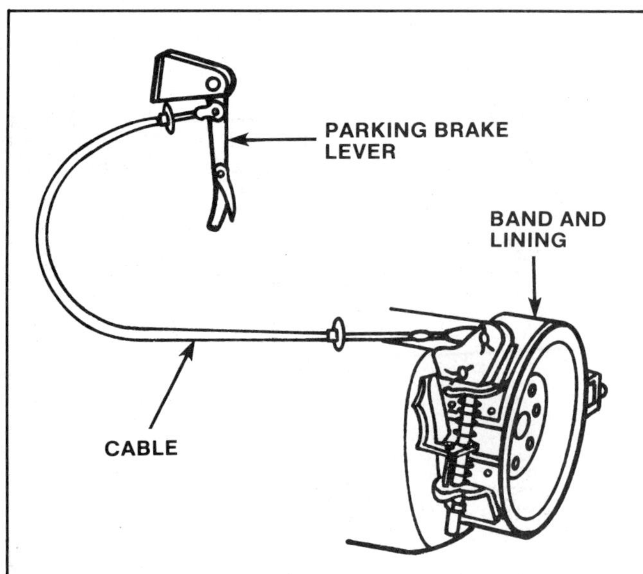

Figure 9-23 External band parking brake

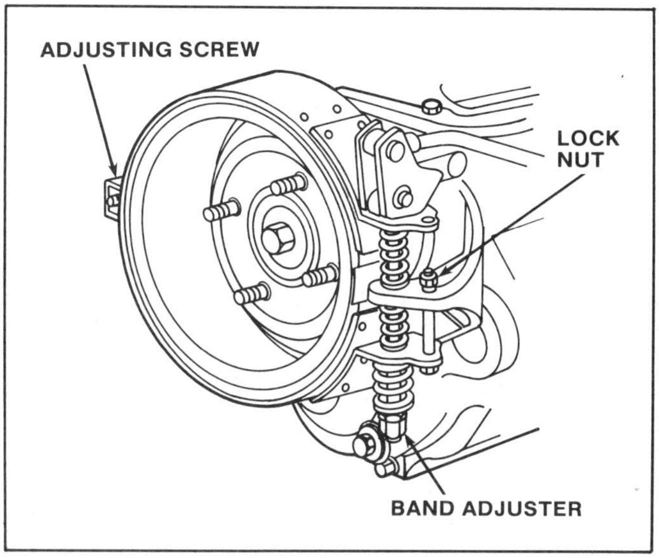

Figure 9-24 Adjustment of the external band parking brake

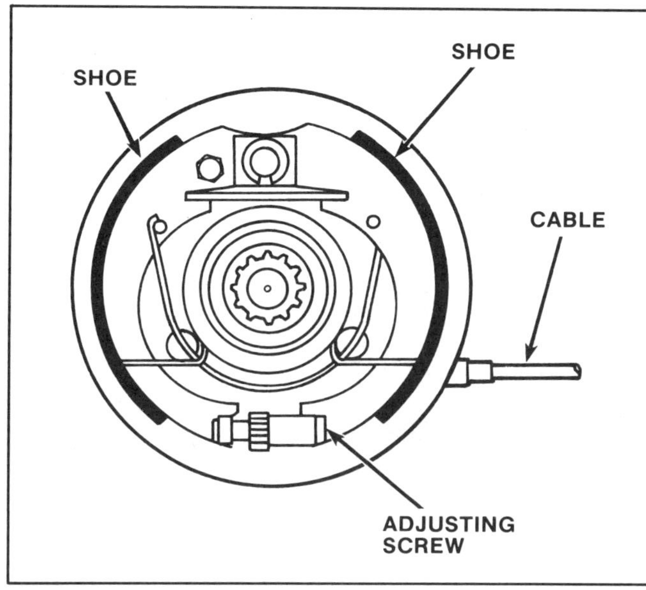

Figure 9-25 Internal shoe parking brake

shaft, and an external band. The lining is attached to the inside of the band. The lever arrangement acts to squeeze the band on the brake drum and keep the drum and drive shaft from turning. This keeps the rear wheels from turning and holds the car in place.

The adjustment procedure (Figure 9-24) is as follows:

1. Cut the locking wire and adjust the adjusting screw to provide 0.020 to 0.030 inch of clearance between the lining and the drum.
2. Loosen the lock nut on the adjusting nut and rotate the lock nut until the lower portion of band clears the drum by 0.020 to 0.030 inch. Tighten the lock nut.
3. Rotate the band adjuster nut until the clearance at the top of the band is 0.020 to 0.030 inch.
4. Recheck at the three positions mentioned and readjust if necessary.
5. Install the lock wire in the adjusting screw.

To adjust the cable, loosen the clevis lock nut, remove the pin, and adjust the yoke on the threads to remove all slack until the clevis can just be installed. Replace the pin and tighten the lock nut.

CAUTION: Always check for proper band adjustment before adjusting the cable.

INTERNAL SHOE

As shown in Figure 9-25, this brake consists of a drum that is attached to the drive shaft and a shoe assembly at the rear of the transmission. The brake shoes are mechanically adjusted by a lever or a cam inside the assembly.

To adjust the internal shoe type transmission parking brake (Figure 9-26), proceed as follows:

1. Place the transmission in NEUTRAL and release the parking brake.
2. Raise and support one of the rear wheels so the drum can be rotated by hand.
3. Remove the adjusting screw cover at the lower edge of the brake shield. This cover has one slotted end, which permits it to be rotated to expose an access opening when the two screws are loosened.
4. Loosen the clamp bolt to free the cable adjusting nut.
5. Back out the cable adjusting nut or slide the cable sleeve in the clamp.
6. Expand the shoes by rotating the shoe adjusting nut until the lining drags on the drum.
7. Back off the nut to provide 0.010 inch clearance between the linings and the drum (at least one notch of the nut.)
8. Make sure the shoulders on the nut are seated in the grooves in the sleeve.
9. On vehicles equipped with a hand control lever, rotate the cable adjusting nut against

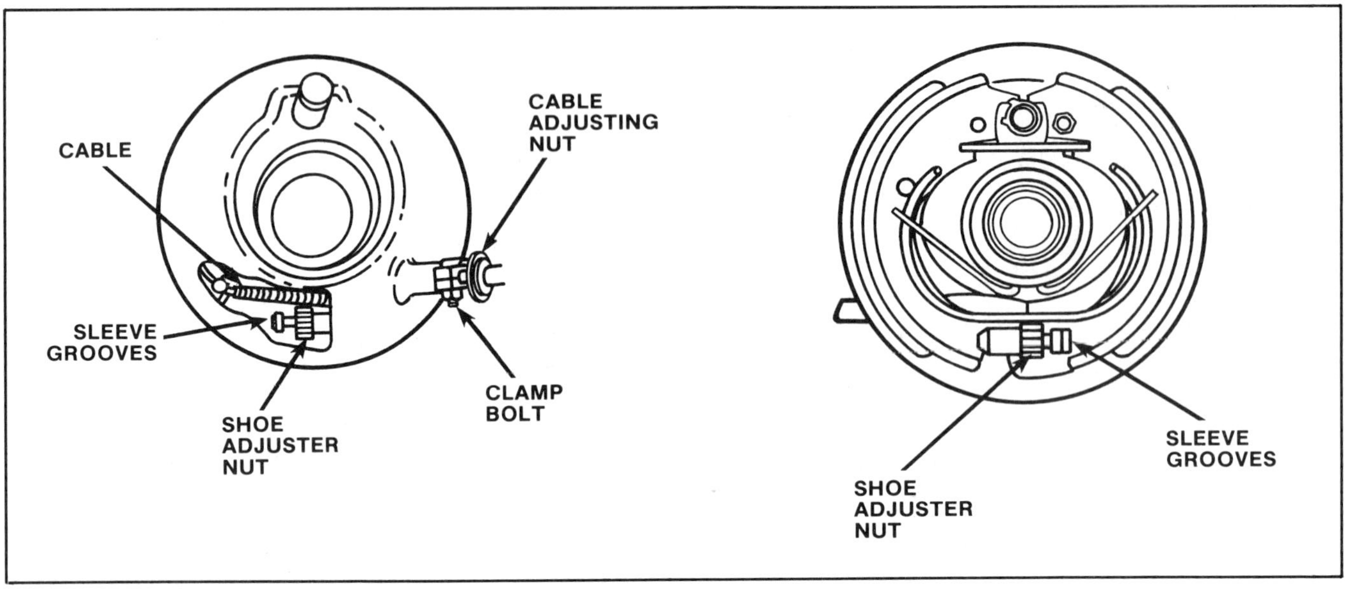

Figure 9-26 Adjustment of the internal shoe parking brake

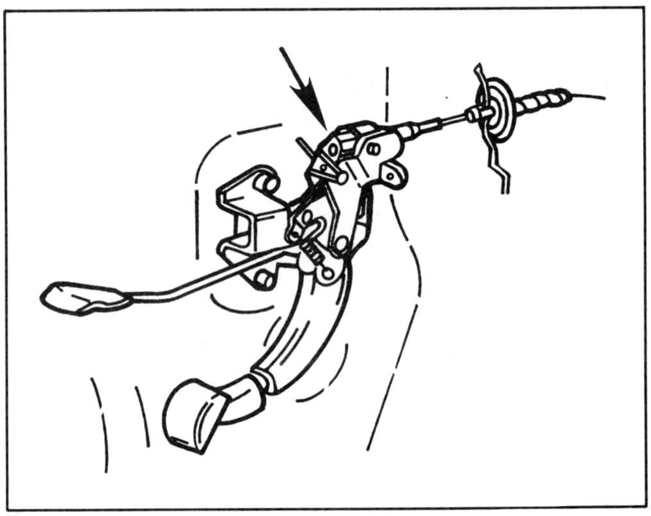

Figure 9-27 Internal shoe parking brake foot pedal control

the cable clamp until 0.010 inch of clearance exists between the brake shoe cable and the operating lever at point. On some models, this nut is not used and the adjustment is made by sliding the cable sleeve in or out of the clamp. On vehicles equipped with a foot pedal control, the cable adjusting nut is located under the dash on the pedal control linkage (Figure 9-27).

10. Tighten the cable clamp bolt.
11. Pull up on the brake control lever (or depress the foot lever), which should advance four to six notches to firmly apply the brake. This must not be more than half of the available travel.

TESTING PARKING BRAKES

Test the parking brake by parking the vehicle facing up on an incline of 30 degrees or less. Depress the parking brake pedal fully to set the brakes. Release the pedal and park the car facing down the incline. Again, apply the pedal fully. The car must hold its position in both cases on such a grade. If the car creeps or rolls under these conditions, adjust the parking brake as described in Chapter 6.

Parking brake troubleshooting problems and solutions can be found in Chapter 14.

REVIEW QUESTIONS

Review the information in this chapter by answering the following questions.

1. The starting point of a typical parking brake system is
   a. the front cable.
   b. the rear cable.
   c. the control assembly.
   d. tensile force.
2. The majority of parking brake systems are set by

a. putting the gearshift selector in the PARK position.
b. depressing the parking brake foot pedal.
c. pulling the parking brake hand lever.
d. pushing the wheel brake pedal.

3. The effort of the pedal assembly or hand grip is multiplied and passed on to the rear cables by the
   a. equalizer lever.
   b. front cable.
   c. rear disc.
   d. drive shaft.

4. A unit in some newer vehicles that releases the parking brake whenever the engine is running and the transmission is in forward driving gear is called a(n)
   a. control assembly.
   b. equalizer lever.
   c. brake lever.
   d. vacuum power unit.

5. When checking a vacuum system, which of the following should never be done?
   a. Depress the brake pedal.
   b. Apply air pressure to the vacuum system.
   c. Remove the hoses.
   d. Connect two tester vacuum hose adapters together.

6. When servicing a parking brake system, it is important to check for
   a. broken strands in the cables.
   b. free movement of the cables in the conduits.
   c. worn equalizer and linkage parts.
   d. all of the above.

7. A parking brake system needs to be adjusted. Mechanic A says to adjust the front cable first. Mechanic B says to adjust the rear cable first. Who is right?
   a. Mechanic A
   b. Mechanic B
   c. both A and B
   d. neither A nor B

8. The parking brake system on an import vehicle needs to be adjusted. Mechanic A says that the adjustment can possibly be made inside the car. Mechanic B says that the adjustment must always be made from underneath the car. Who is right?
   a. Mechanic A
   b. Mechanic B
   c. both A and B
   d. neither A nor B

9. In a rear disc parking brake system, which of the following mechanisms forces the pads against the rotor?
   a. ball-and-ramp arrangement
   b. high lead screw
   c. a variety of cams
   d. all of the above

10. Which of the following is a major type of rear disc parking brake system?
    a. Kelsey-Hayes
    b. Delco-Moraine
    c. both a and b
    d. none of the above

11. All of the vacuum parking brake control checks are to be performed with the engine
    a. running at idle speed.
    b. turned off.
    c. moving the vehicle forward slowly.
    d. moving the vehicle backward slowly.

12. A car needs to have both the regular wheel brakes and the parking brakes serviced. Mechanic A says to service the parking brake system first. Mechanic B says to service the regular wheel brakes first. Who is right?
    a. Mechanic A
    b. Mechanic B
    c. both A and B
    d. neither A nor B

13. The type of parking brake system that is completely separate from the basic brake system is the
    a. transmission parking brake.
    b. drive shaft parking brake.
    c. both a and b.
    d. integral parking brake.

14. Which of the following is a type of drive shaft parking brake?
    a. external band
    b. internal shoe
    c. both a and b
    d. rear disc

15. The parking brake system must be tested while the vehicle is positioned
    a. on an incline of 30 degrees or less.
    b. on an incline of more than 30 degrees.
    c. both a and b
    d. on the level.

16. It is important to remember that the parking brake in most vehicles
    a. is not part of the car's hydraulic brake system.
    b. works mechanically.
    c. is operated by pushing a foot pedal or pulling a hand lever.
    d. all of the above.
17. Tensile force from the front of a vehicle when braking is transmitted to the rear brakes through the vehicle's
    a. discs.
    b. cable system.
    c. control assembly.
    d. vacuum power unit.
18. To remove and replace rear brake shoes, it might be necessary to relieve the parking brake cable tension how?
    a. by backing off the adjusting check nut(s) at the equalizer
    b. by replacing the cable
    c. by replacing the equalizer
    d. by backing off the adjusting check nut(s) at the brake pedal
19. What should be done first when removing a vacuum valve?
    a. Take out the screws that attach the vacuum unit to the steering column jacket.
    b. Place the transmission shift selector lever in DRIVE with the engine off.
    c. Remove the vacuum hoses from the engine manifold.
    d. Place the transmission shift selector lever in DRIVE with the engine on.
20. What is another name for a rear disc/drum parking brake?
    a. drive shaft
    b. integral
    c. auxiliary drum
    d. transmission

CHAPTER 10

POWER BRAKES—OPERATION AND SERVICING

OBJECTIVES

Upon completing this chapter, you should be able to

- Describe the two basic types of power assist mechanisms: vacuum assist diaphragm assemblies and hydraulic pressure assist mechanisms or hydro-boost.
- Identify and rectify problems in any power brake system.
- Explain the four basic phases of braking action involved in both types of power assist mechanisms.
- Describe the procedures for bleeding a power unit.
- Explain the function and operation of a vacuum pump.
- Service a vacuum pump system and components, including vacuum pump, controller, and piston assembly.
- Describe the operation of the Powermaster hydraulic booster, as well as other power booster units such as Hydrovac, Air-Pak, and Master Vac.

In order to properly service and repair available brake systems, a thorough understanding of the various power assist systems is necessary. Two basic types of power assist mechanisms are used (Figure 10-1): vacuum assist diaphragm assemblies, which use engine vacuum and, in some cases, vacuum pressure developed by an external vacuum pump, and hydraulic pressure assist mechanisms, which use pressure developed by an external pump (usually the power steering pump). Both types of systems act to multiply the force exerted on the master cylinder piston by the operator in order to increase the hydraulic pressure delivered to the wheel cylinders, while decreasing the effort necessary to obtain acceptable stopping performance.

Vacuum assist units are similar in operation and get their energy by opposing engine vacuum to atmospheric pressure. A piston and cylinder, flexible diaphragm, or bellows utilize this energy to provide brake assistance. The fundamental difference between the different types of vacuum assist

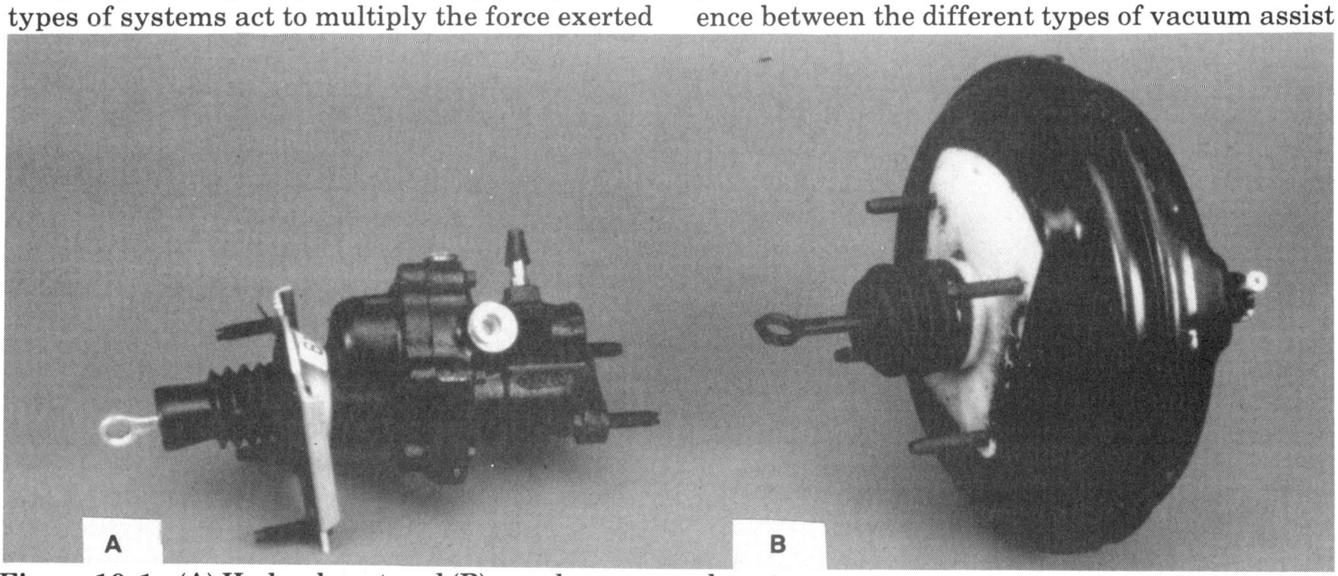

Figure 10-1 (A) Hydro-boost and (B) regular vacuum boost

systems lies simply in how the diaphragm within the power unit is suspended when the brakes are not applied.

In order to properly diagnose vacuum assist system malfunctions, it is important to know whether the diaphragm within a power unit is air suspended or vacuum suspended. Air-suspended units are under atmospheric pressure until the brakes are applied, then engine vacuum is admitted, causing the piston or diaphragm to move or the bellows to collapse. Vacuum-suspended types are balanced with engine vacuum until the brake pedal is depressed, allowing atmospheric pressure to unbalance the unit and apply force to the brake system.

CAUTION: Regardless of whether the brakes are vacuum or hydraulically assisted, certain general service procedures apply. Only top quality, clean brake fluid should be used in brake systems and, on hydro-boost systems use of the specified hydraulic fluid in the boost circuit is essential to proper system operation. On hydro-boost systems care must be taken not to mix the fluids of the two separate operating circuits since the use of improper fluids will do damage to the various seals and valves. For the same reason, be sure all dirt is kept out of the system.

The fact that brakes will operate even if the power unit fails gives us a clue to successful power brake service. This means the conventional brake system is left intact and a power unit is simply added to the existing system. Troubleshooting is then exactly the same until we get to the power unit. As with conventional hydraulic brakes, a spongy pedal with power brakes still means air is in the system and grease on the linings will still make the brakes grab. Keep in mind, however, that power brakes give a higher line pressure, thus making leaks more critical.

Power units do not require adjustment in normal service. Either they work or they do not. If they do not, the various valves and connections are simply replaced. The only exception is that the power units themselves sometimes have an adjustable connection to the brake pedal or master cylinder.

Another thing that is helpful in power brake overhaul is the manner in which the units wear. Repairs are infrequent until the unit has been in service for a long time. Then, when a malfunction does occur, it means the whole unit should be overhauled. Complete overhaul kits are available and all the parts in the kit should be used in the overhaul.

GENERAL POWER BRAKE SERVICE PROCEDURES

When working on power brakes, it is important that the following general precautions be kept in mind:

1. After disassembling a power brake unit, soak all metal parts in a suitable solvent.
2. Use only alcohol on rubber parts or parts containing rubber. After the parts have been thoroughly cleaned and rinsed in solvent, the metal parts that come in contact with hydraulic brake fluid or rubber parts should be rewashed in clean denatured alcohol before assembly.
3. Use an air hose to blow dirt and cleaning fluid from the recesses and internal passages.
4. Always use all the parts furnished in the repair kit. Discard all old rubber parts.
5. Use extreme caution when disassembling power assist mechanisms, because releasing internal spring tension suddenly could cause damage or personal injury.
6. On models with hydro-boost assist systems, do not disassemble the accumulator. If the accumulator is damaged or diagnosed as defective, it must be replaced as an assembly.

When servicing power brakes the overall operation should be handled as if two separate brake systems exist. That is, check for faults in the basic hydraulic system first. If this system is operating properly, check the power brake circuit. For a quick determination of proper power unit operation, press the brake pedal firmly and then start the engine. The pedal should fall away slightly and less pressure should be needed to maintain the pedal in any position. On vacuum-suspended power units, air will rush into the air intake when the brakes are applied.

Another check begins with installation of a suitable pressure gauge in the brake hydraulic system. Take a reading with the engine off and the power unit not operating. Maintaining the same pedal height, start the engine and take another reading. There should be a substantial pressure increase in the second reading.

Pedal free travel and total travel are critical on cars equipped with power brakes. As a general rule, brakes should be adjusted or relined if the pedal is closer than 1 inch from the floor with the brakes applied.

On models with vacuum boost brakes, take a manifold vacuum reading or check the operation of the external vacuum pump if the power unit is not giving enough assistance. Remember that newer emission-controlled engines produce manifold vacuum readings that can be less than specifications at idle. If manifold vacuum is abnormally low, tune the engine and then try the power brakes again. Loose vacuum lines and clogged air intake filters will cut down brake efficiency. Most units have a check valve that retains some vacuum in the system when the engine is off. A vacuum gauge check of this valve will tell you when it is restricted or stuck open or closed.

On models equipped with hydro-boost brakes, check the operation of the power steering pump and check for improperly adjusted drive belts, defective pressure relief valve, leaking hoses and fittings, or a defective accumulator. Most causes of failure on these systems is not within the power assist unit itself. However, if all other components are satisfactory but a malfunction persists, the power unit should be disassembled and inspected for damage, wear, and contamination.

Failure of the brakes to release in most instances is caused by a tight or misaligned connection between the power unit and the brake linkage. If this connection is free, look for a broken piston, diaphragm, or bellows return spring on models with vacuum boost brakes. On models with hydro-boost brakes, check for contamination of the fluid that results in restrictions within the spool valve, sticking assist piston within the power unit, or a defective accumulator. The power unit will have to be disassembled for this check and, if the spring is in good condition, continue going through the power unit, checking the fluid, air, and vacuum valves.

A simple check of the hydraulic system should be made before proceeding. Loosen the connection between the master cylinder and the brake booster. If the brakes release, the trouble is in the power unit. If the brakes still will not release, look for a restricted brake line or similar difficulties in the regular hydraulic circuit.

Power brakes that have a hard pedal usually lead to complete power unit failure. Collapsed or leaking vacuum lines or insufficient manifold vacuum, as well as punctured diaphragms or bellows and leaky piston seals, all lead to weak power unit operation. A steady hiss when the brake is held down means a vacuum leak that will cause poor power unit operation.

Do not immediately condemn the power unit if the brakes grab. First, look for all the usual causes, such as greasy linings or scored drums. Then, investigate the power unit. When the trouble has been traced to the power unit, check for a damaged reaction control. The reaction control is usually made up of a diaphragm, spring, and valving that tends to resist pedal action. This feature is designed into the system to give the brake pedal "feel" to the operator.

VACUUM POWER BRAKES

Vacuum assist power brakes, as previously stated, take advantage of the difference in engine manifold vacuum and external atmospheric pressure. The pressure differential acts on both sides of a single diaphragm power piston, which divides the power brake into two chambers. (Smaller sized tandem-type boosters installed where space is limited, use two diaphragms in tandem. Both single and tandem-type units [Figures 10-2 and 10-3] operate the same.)

The power piston or vacuum control valve suspended in the diaphragm between the chambers performs two operations:

1. It controls the vacuum and atmospheric valves as the booster pushrod moves.
2. It transmits additional force to the pushrod of the master cylinder.

Vacuum from the intake manifold keeps both front and rear chambers at relatively low pressures when the engine is running and the brakes are not applied.

When the brakes are released, air circulates freely on both sides of the diaphragm. Depressing the brake pedal allows atmospheric pressure to enter the rear chamber of the booster. With the engine running and evacuating the front chamber, the pressure differential (low pressure in the front chamber and higher pressure in the rear chamber) acts across the diaphragm and the booster's power piston pushes forward to contact the master cylinder's pushrod. This action, in turn, actuates the master cylinder's pistons to provide conventional output hydraulic pressure to the brakes.

A check valve in the engine intake manifold inlet of the booster retains sufficient vacuum in the booster for additional braking when the engine stops. The booster can then deliver enough output for several moderate brake applications. Beyond this point, braking effort becomes very hard.

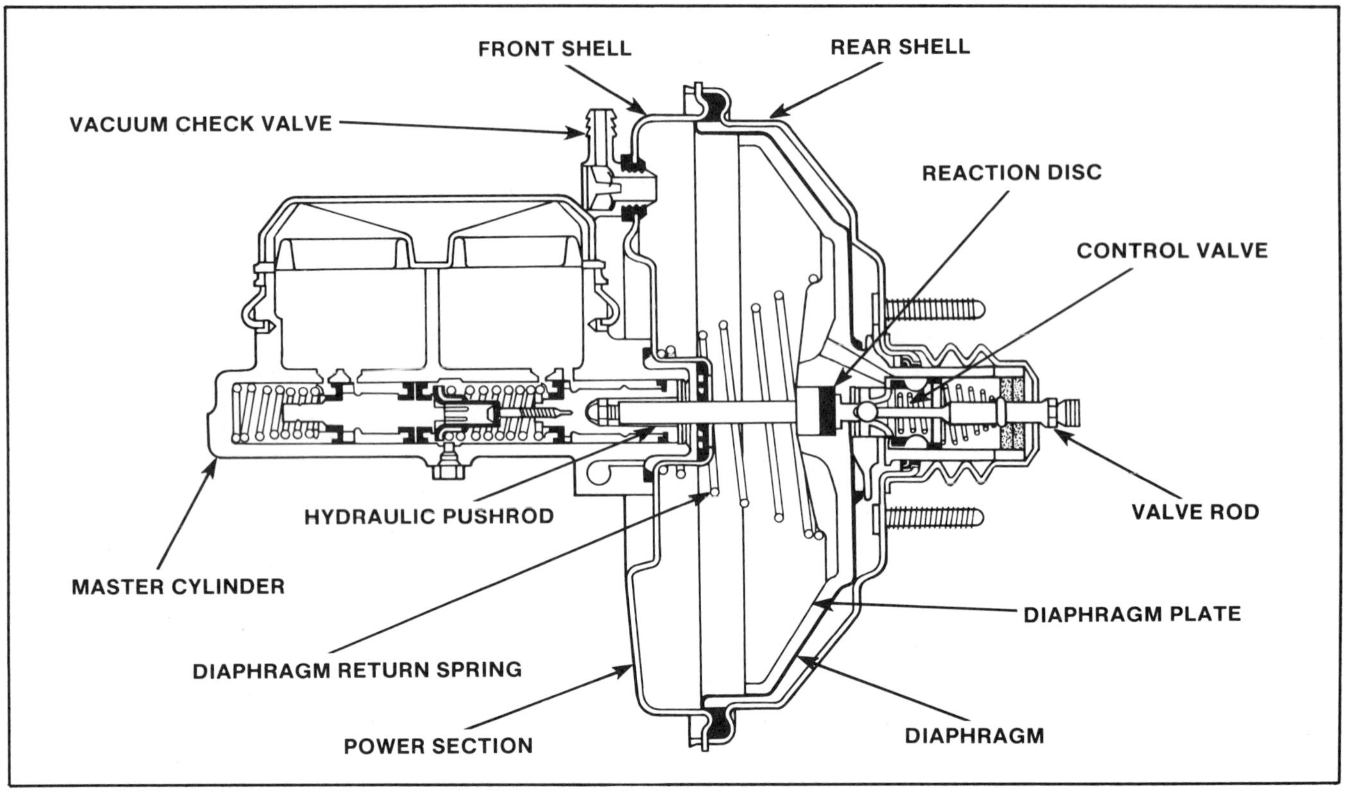

Figure 10-2 Cross section of Bendix single diaphragm master vac

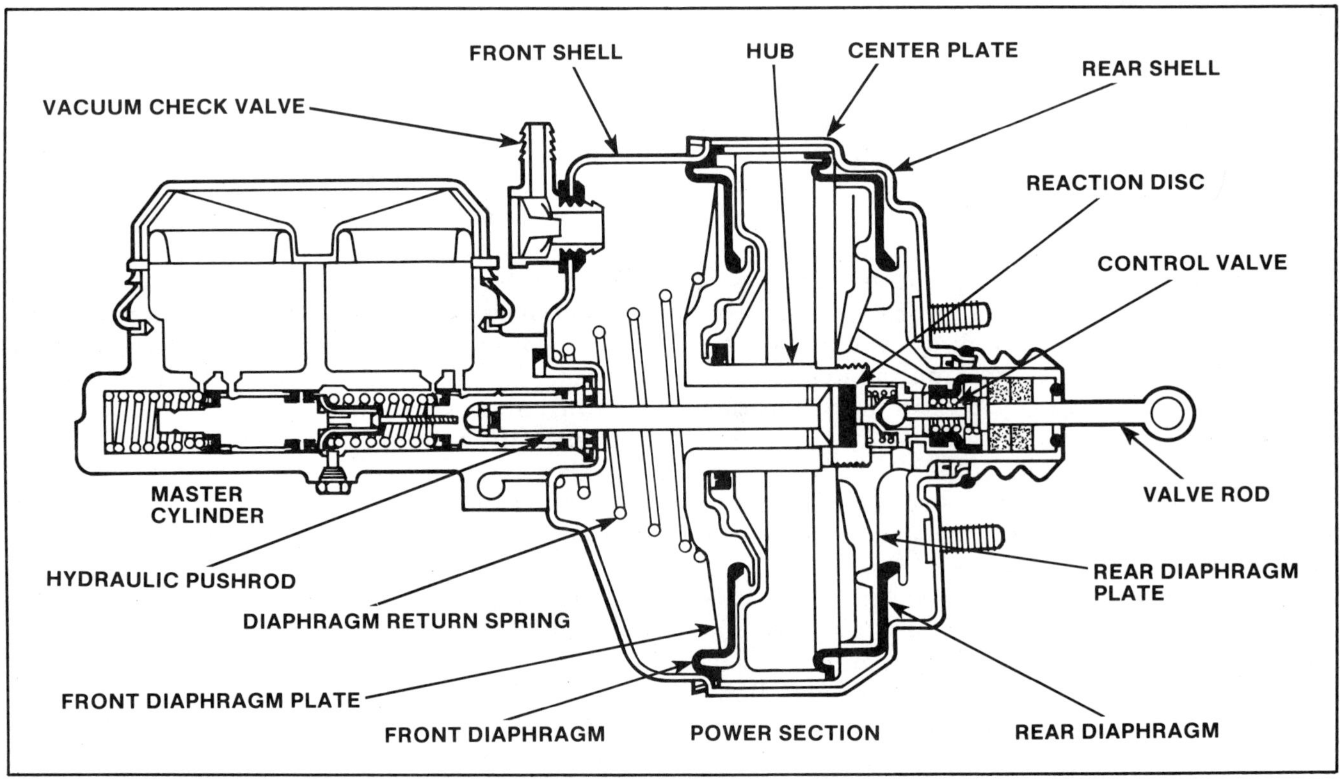

Figure 10-3 Cross section of tandem diaphragm master vac

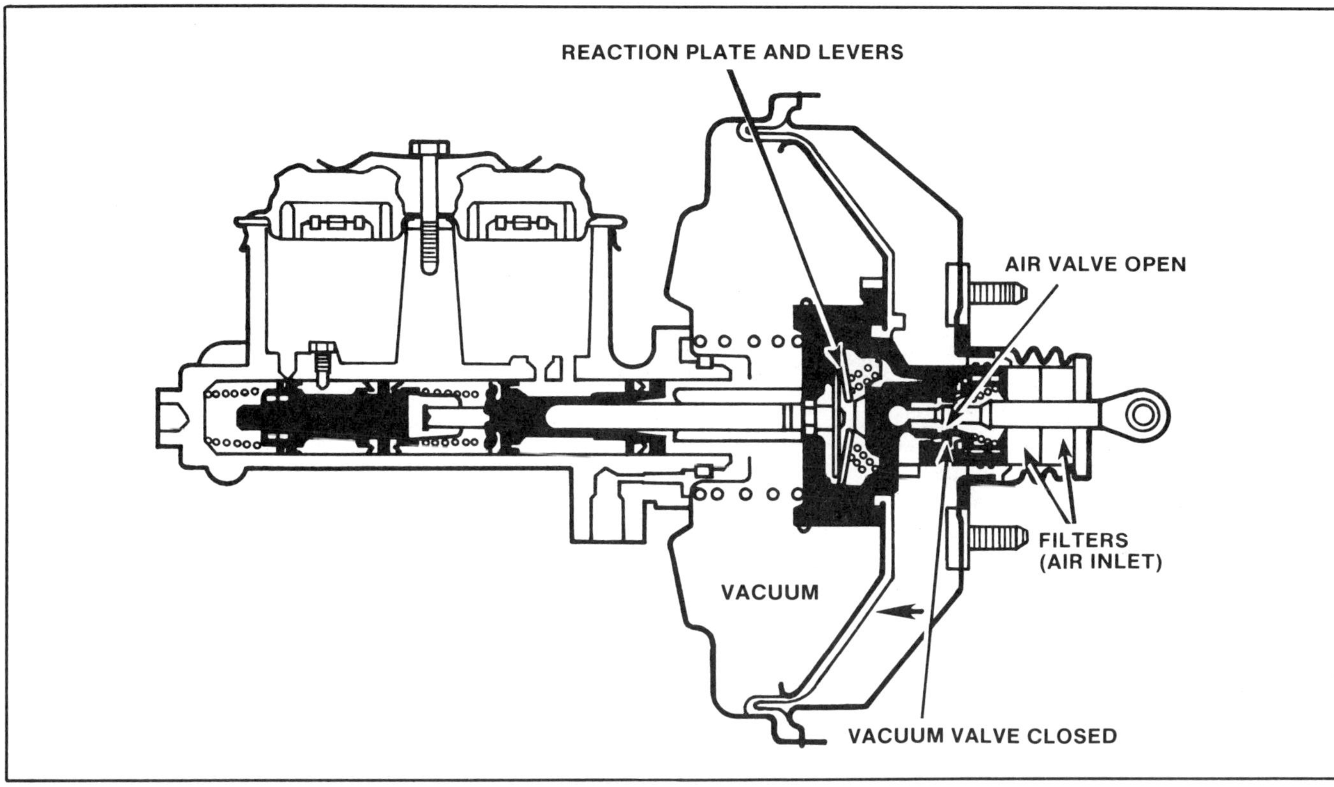

Figure 10-4 Delco-Moraine (GM) vacuum suspended lever reaction—applying position

Two types of vacuum boosters most commonly used in modern vehicles are: the lever booster, which reacts force back to the pedal through a bridge/lever mechanism; and the disc booster, which reacts force back to the pedal through a rubber reaction disc.

In the reaction *lever booster* (Figure 10-4), force from the master cylinder reacts back along the booster output pushrod to a set of levers through a bridging mechanism. During initial braking, fixed ends of each lever are in contact with the power piston. The other ends are spring loaded and free to move.

Initial reaction back to the driver's foot at the brake pedal is kept low until the springs are deflected enough to allow contact between the moveable ends of the levers and the vacuum and atmospheric control valve mechanism. From the time of contact, the reaction force, divided proportionally between the input pushrod of the master cylinder and the power piston of the booster, reacts back to the pedal.

The reaction *disc booster* (Figure 10-5) contains a rubber disc, which absorbs reaction force from the master cylinder when the brakes are applied. Then force, transferred back through the output pushrod from the master cylinder to the brake pedal, is eliminated in a two-step process. First, the pushrod chamber lengthens until the pushrod, in effect, stops short of contact with the reaction disc and the full reaction load is accepted by the power piston.

Second, when a specified load is reached, the force from the master cylinder pushrod becomes great enough that the rubber disc pushes outward to the point where it contacts the input pushrod head. At this point of contact, the force divides between the booster power piston and the input pushrod, and, thus, results in a share of the force returning to the brake pedal.

There are two general methods of mounting vacuum power brakes. One method is to use mounting brackets (Figure 10-6A) between the power section and the fire wall. This mounting uses a lever arrangement and a pedal link that is attached to the brake pedal inside the car. The other method is to attach the power section directly to the fire wall as shown in Figure 10-6B. When this flush-mounted method is used, the power brake valve rod is connected directly to the brake pedal.

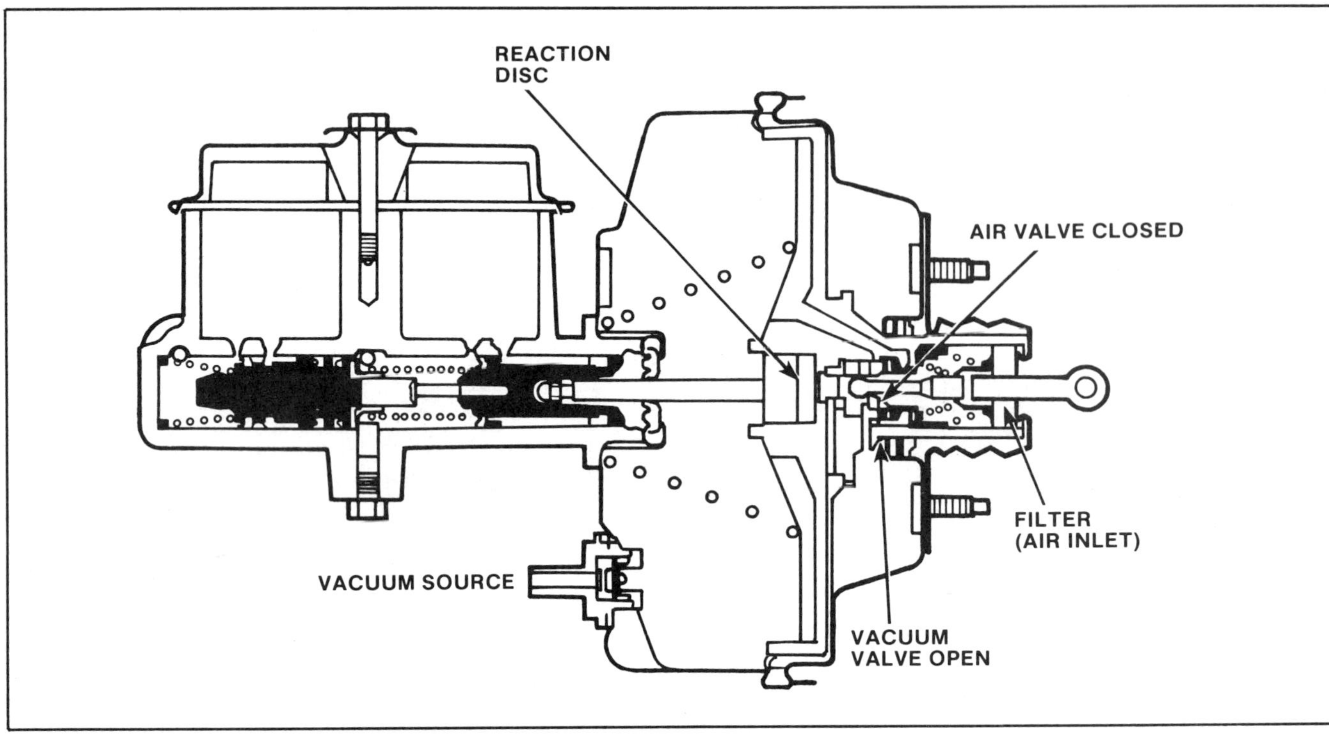

Figure 10-5 Bendix vacuum suspended disc reaction—released position

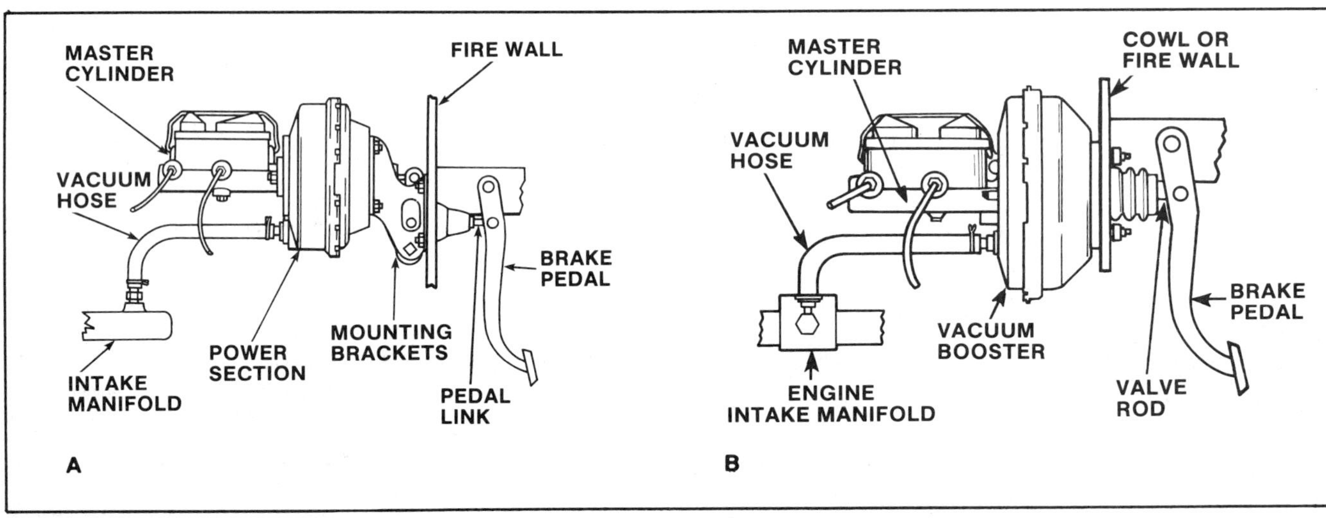

Figure 10-6 (A) Bracket-mounted master vac and (B) vacuum booster

VACUUM BRAKES OPERATION

Most vacuum boosters use a single diaphragm. However, their operation, as well as the tandem or dual diaphragm design, has four basic phases:

1. **Brakes Released (Figure 10-7A).** With the brakes released, there is no pedal force applied to the valve operating rod. The valve rod return spring holds the rod and attached valve plunger to the right, seating the plunger on the poppet assembly. The vacuum control port is open and the atmosphere control port is closed.

When the engine is running, atmosphere is evacuated from the vacuum line, check valve, and power section. This creates a partial vacuum in the unit. The unit, now suspended, has equal vacuum levels on both sides of the return spring, so that no force is applied to the master cylinder. Its components are also in released position.

2. **Brakes Partially Applied (Figure 10-7B).** The car's brake fluid pressure applies a reaction force to the master cylinder pushrod and rubber reaction disc. A portion of the resultant reaction force acts directly on the power piston. The remainder forces the reaction disc to extrude into the space at the left of the valve plunger and is transmitted to the valve plunger. As the plunger is forced to the right and reseats on the poppet assembly, the atmosphere port closes and prevents the entrance of atmosphere. This is the "lap" position in which the brakes are partially applied and both atmosphere and vacuum ports of the booster are closed.

   The portion of the reaction force transmitted by the reaction disc to the valve plunger, is also transmitted by the valve operating rod and brake pedal to the driver's foot. This driver a "feel" of the braking effort. The reaction force against the power piston gives the and the force of the compressed power piston return spring are balanced by the force due to the pressure differential across the power piston. In all cases, the total force on the master cylinder pushrod is equal to the sum of the valve rod and power piston forces. This means that part of the braking effort is supplied by the driver, while the remainder (and larger portion) is supplied by the booster. If the driver feels more braking is needed, he/she applies more force on the brake pedal. The booster functional cycle repeats and the increased pressure is applied to the hydraulic fluid for greater braking effort until the vacuum runout point is reached.

3. **Brakes Applied (Figure 10-7C).** Brake pedal application forces the valve operating rod and plunger to the left against the opposing force of the rod return spring. This allows the poppet assembly to also move left and seal the vacuum passages to the left side of the power piston. Further movement of the valve rod allows the valve plunger to unseat from the poppet assembly. This permits atmosphere to enter and fill the right side of the power piston.

   The resulting pressure differential across the power piston forces it further to the left, compressing the power piston return spring. In moving, the power piston transmits force to the master cylinder pushrod, which, in turn, forces the master cylinder piston to the left against the return springs and pressure is applied to the brake fluid for actuating the brakes.

   When the booster is in the fully applied position, movement of the valve plunger away from the poppet assembly allows more atmosphere to enter. Full atmospheric pressure can thus be brought into the right side of the power piston. Maximum pressure differential across the power piston is obtained. This is referred to as the "vacuum runout" point of the booster and is usually felt as a definite "hardening" of the pedal.

   Any additional hydraulic output pressure must be obtained from increased effort by the driver on the brake pedal without power assist from the booster. When runout is reached, the valve plunger seats in the power piston and directly transmits the valve operating rod force from the brake pedal, to the power piston, and, in turn, to the master cylinder pushrod.

4. **Brakes Released (Figure 10-7D).** To decrease the amount of braking effort, the driver reduces the force applied to the brake pedal. This action decreases the force on the valve operating rod and valve plunger and allows the plunger to move to the right, closing the atmosphere port (if the brakes were in the fully applied position) and opening the vacuum port. A portion of the atmosphere to the right of the power piston is removed and the pressure differential across the power piston is reduced. The power piston return spring forces the power piston to the right until the return spring and reaction forces are balanced by the force due to the pressure differential and the valve operating rod's input force from the brake pedal.

   The booster is now in the partially applied brake lap position and both the atmosphere and vacuum ports are closed. The reduced force on the master cylinder pushrod allows the hydraulic fluid pressure and the master cylinder return springs to force the pistons back in the master cylinder bore. Pressure to the wheel cylinders is thus relieved and the brake shoes or pads start to retract.

   The driver removes all force on the brake pedal to completely release the brakes. The valve plunger moves to the right, as previously described, and the vacuum port op-

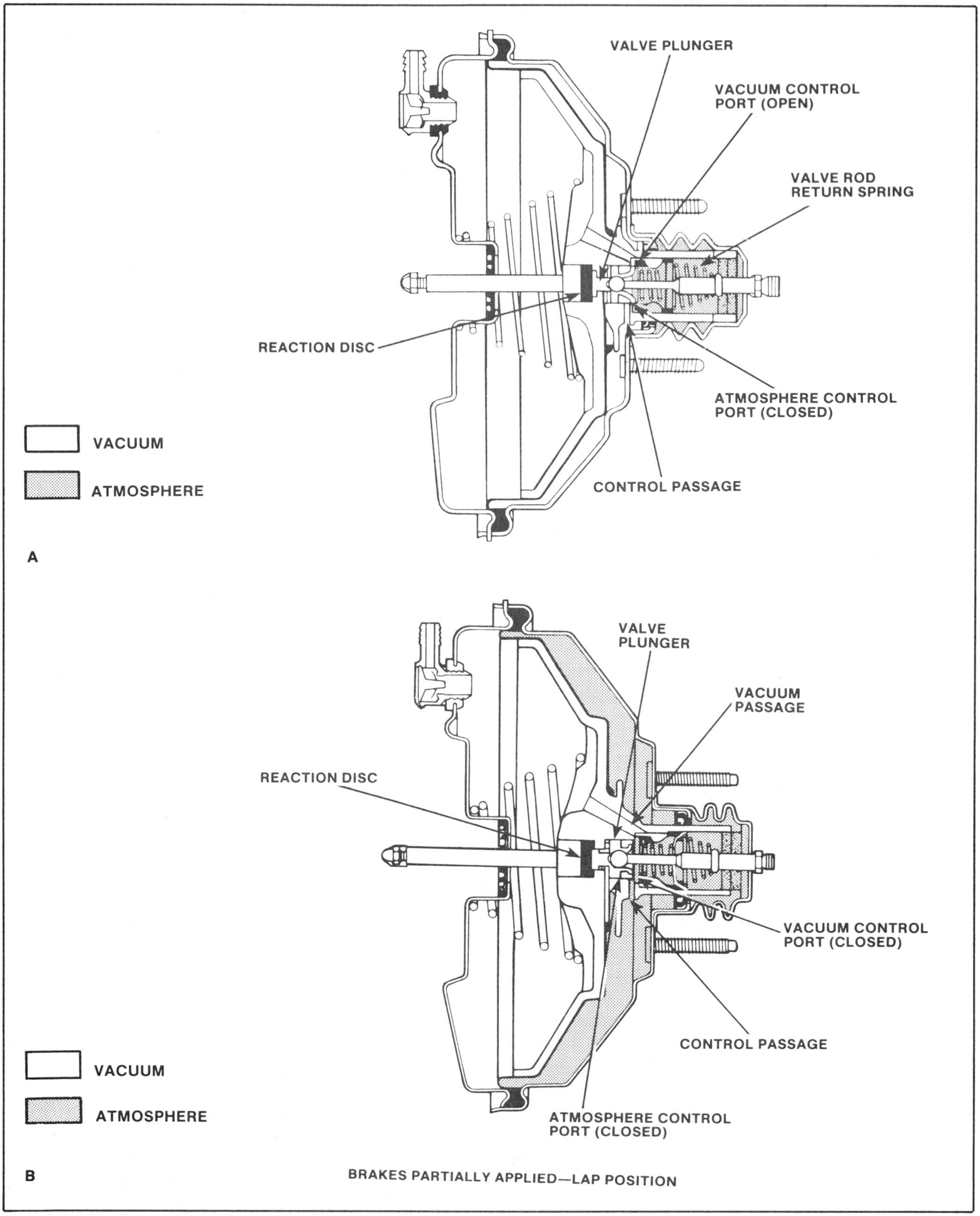

Figure 10-7 Brake system components in the (A) unapplied, (B) partially applied, (C) applied, and (D) released positions

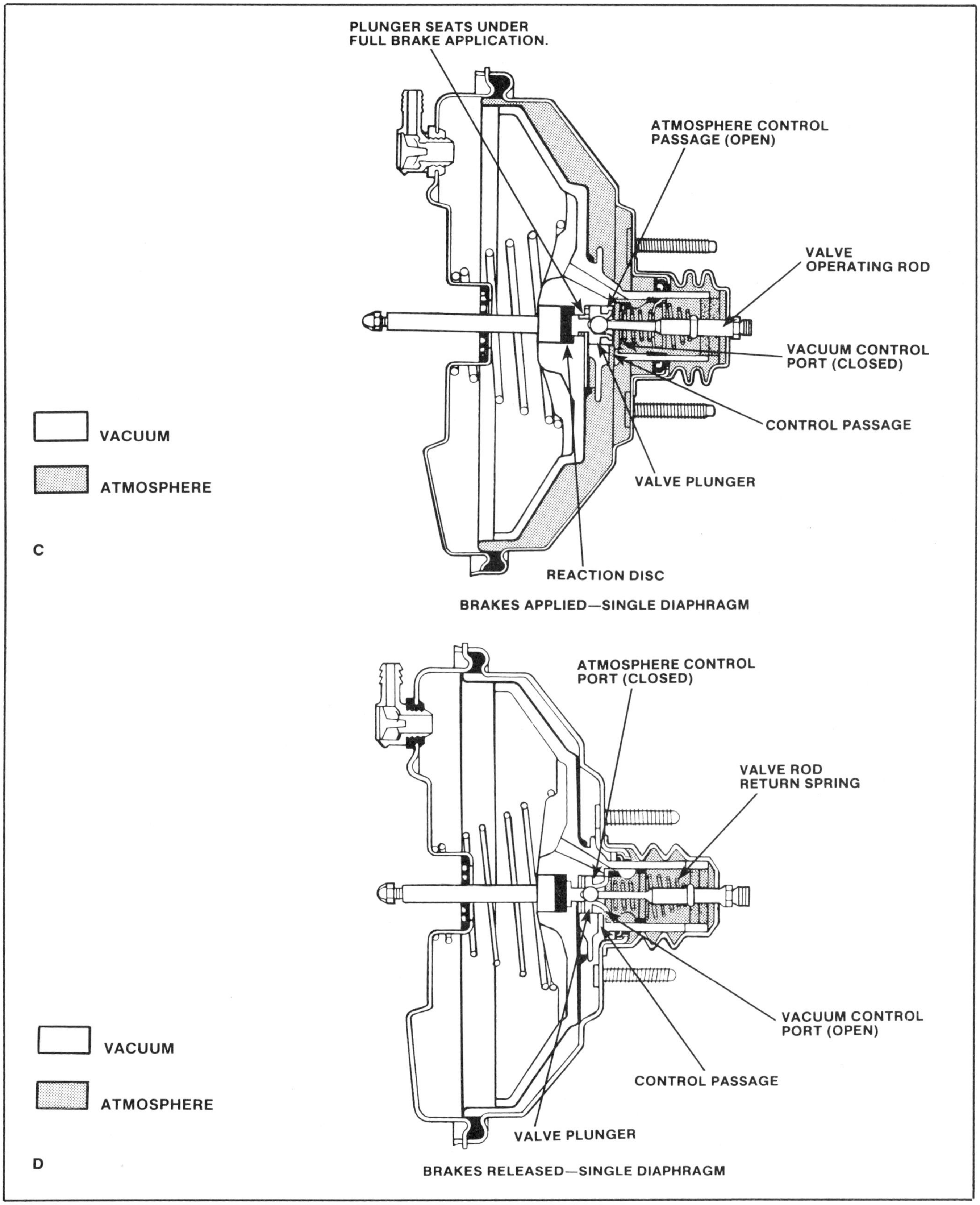

Figure 10-7 Continued

ens. Atmosphere is evacuated from the right side of the power section. This allows the power piston, master cylinder pushrod, and master cylinder pistons to return to their released positions. The replenishing ports of the master cylinder are opened to the bore and pressure is removed from the hydraulic fluid allowing the car's brakes to return to their released position.

With the engine not operating, there is no source of vacuum for the booster. A reserve of vacuum is held in the booster by the check valve and is sufficient for one or more brake applications. When this vacuum reserve is exhausted, the brakes must be applied manually without power assist. Hard application of the brake pedal moves the valve operating rod and valve plunger until the plunger seats in the power piston. The power piston then forces the master cylinder pushrod to the left to pressurize the hydraulic fluid and apply the brakes.

BLEEDING THE POWER UNIT

On power booster-equipped vehicles, the engine should be turned off and the power system purged of vacuum or compressed air by depressing the brake pedal several times. After bleeding the master cylinder, bleed the power brake unit if it is equipped with a bleeder screw (Figure 10-8).

Pressure-multiplying type power units often have bleeder screws to remove the air trapped within the unit. If there is more than one bleeder screw, bleed the one at the pressure (main) cylinder first; the control valve second. When bleeding, manually close the bleeder screw before the pedal is allowed to backstroke each time.

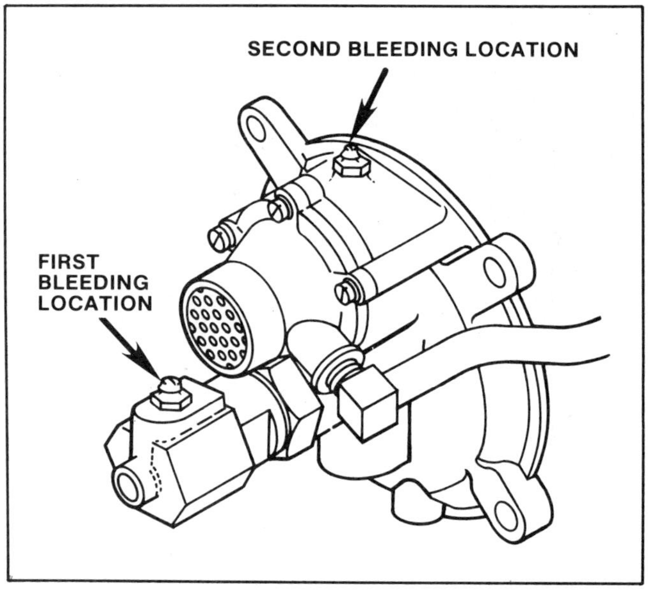

Figure 10-8 Bleeding the power brake unit

VACUUM BRAKE TESTING

It is important to note that any condition that reduces the amount of vacuum the engine generates will affect power brake performance. These conditions include vacuum leaks, faulty valves, improper valve timing, and the installation of a high-performance camshaft. Therefore, engine vacuum should be checked with a vacuum gauge before a complaint of poor stopping action is blamed on the power system. As a general rule, a minimum of 14 inches Hg of vacuum is required to provide proper power brake operation.

SHOP TALK

Vacuum is measured on a gauge in inches of mercury (Hg).

Also, always make sure the vacuum tube or hose that runs from the engine intake manifold to the booster unit is securely connected and not collapsed, clogged, punctured, or torn.

Some vehicles such as diesels or newer FWD models use an engine-driven or electric vacuum pump to supply the power unit with vacuum. Test it as the manufacturer recommends.

Basic Operational Test

The following basic operational test steps should be made:

1. With the engine off, pump the brake pedal numerous times to be sure any residual vacuum is exhausted from the booster unit.
2. Hold firm pressure on the brake pedal and start the engine.
3. If the system is operating properly, you should feel the pedal move downward a short distance, then stop. Only a small amount of pressure should be needed to hold the pedal down.
4. If you do not get these results, proceed to the other tests.

Vacuum Supply Test

To check the vacuum system proceed as follows:

1. With the engine idling, attach a vacuum gauge to an intake manifold port (Figure

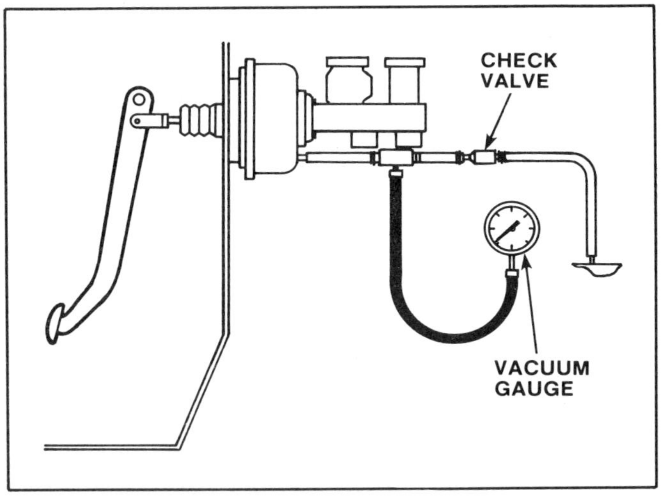

Figure 10-9 A quick method for detecting system leaks

10-9). A low reading indicates an engine problem.

2. Disconnect the vacuum tube or hose that runs from the intake manifold to the booster unit, and place a thumb over it (do this quickly or the engine might stall).
3. If you do not feel a strong vacuum at the end of the tube or hose, shut off the engine, remove the tube or hose and see if it is collapsed, crimped, or clogged. Replace it if necessary.

SHOP TALK

Do not mistake normal booster breathing for a vacuum leak. When the pedal is moved, air rushing through the filter installed in the rubber boot of the booster's input pushrod causes a slight breathing sound, which is normal.

Vacuum Check Valve Test

The operation of the vacuum check valve can be tested as follows:

1. Shut the engine off and wait 5 minutes.
2. Apply the brakes. There should be power assist on at least one pedal stroke.
3. If no power assist is felt, remove the check valve from the booster unit.
4. Test the check valve by blowing into the intake manifold end of the valve. There should be a complete blockage of airflow. Alternately, apply vacuum to the booster unit end of the valve. Vacuum should be blocked.
5. If the above does not result, replace the check valve.

Brake Drag Test

The power brake drag test is made in the following manner:

1. With the wheels of the vehicle raised off the floor, pump the brake pedal to exhaust residual vacuum from the booster.
2. Turn the front wheels by hand and note the amount of drag that is present.
3. Start the engine and allow it to run for 1 minute, then shut it off.
4. Turn the front wheels by hand again. If drag has increased, the booster control valve is faulty.

Fluid Loss Test

If the master cylinder reservoir level has fallen noticeably, but there is no sign of an external brake fluid leak, remove the vacuum tube or hose that runs between the intake manifold and the booster unit, and inspect it carefully for evidence of brake fluid. If any fluid is found, the master cylinder secondary seal is leaking.

PUSHROD ADJUSTMENT

Proper adjustment of the master cylinder pushrod is necessary to ensure proper operation of the power brake system. A pushrod that is too long will cause the master cylinder piston to close off the replenishing port, preventing hydraulic pressure from being released and resulting in brake drag. A pushrod that is too short will cause excessive brake pedal travel and cause groaning noises to come from the booster when the brakes are applied. A properly adjusted pushrod that remains assembled to the booster with which it was matched during production should not require service adjustment. However, if the booster, master cylinder, or pushrod are serviced, the pushrod might require adjustment.

There are two methods that can be used to check for proper pushrod length and installation.

1. Gauge Method
2. Air Method

Gauge Method

Usually, if the power unit pushrod requires an adjustment, the power unit repair kit for the unit being serviced includes a gauge. The gauge measures from the end of the pushrod to the power unit shell (Figure 10-10).

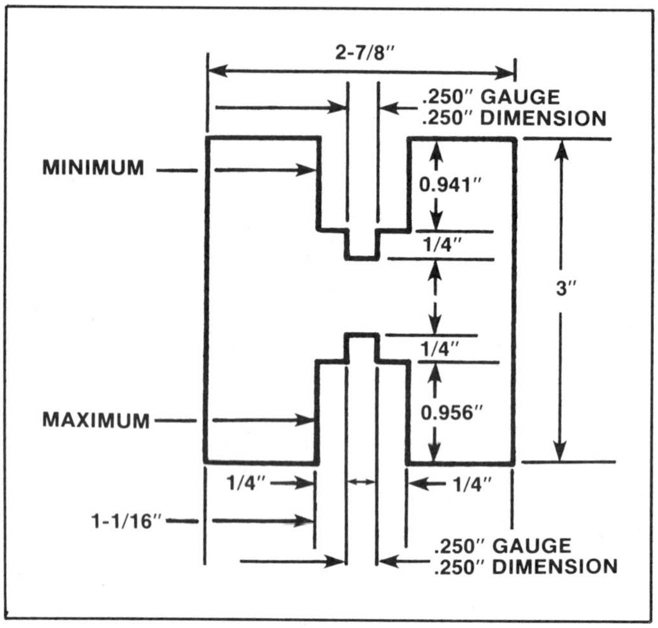

Figure 10-10 Pushrod gauge dimensions

To adjust Bendix type vacuum power brakes (Figure 10-11), proceed as follows:

1. Disconnect the master cylinder from the booster leaving the brake lines connected. Secure the cylinder to prevent the lines from being damaged.
2. Start the engine and allow it to run at normal curb idle speed.
3. With the engine running, position the gauge over the pushrod. The gauge should bottom against the booster housing with a force of approximately 5 pounds applied to the pushrod.
4. If the force required to seat the gauge exceeds 5 pounds, shorten the length of the pushrod. If the force required to seat the gauge is less than 5 pounds, lengthen the pushrod. Ensure that the pushrod is properly seated in the booster when performing the gauge check.
5. Reinstall the master cylinder and remove the reservoir cover.
6. Have a helper pump the brake pedal. Observe the fluid surface when the brakes are applied and released rapidly. If no change of the fluid level is observed, the pushrod is too long.

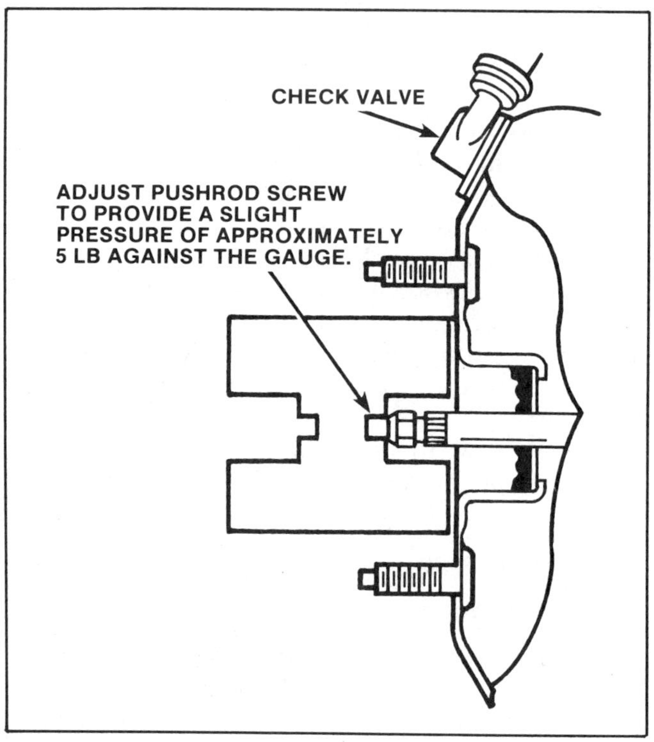

Figure 10-11 Bendix pushrod adjustment

On Delco-Moraine vacuum power brake models (Figure 10-12), the master cylinder pushrod length is fixed, and the length is usually only checked after the unit has been overhauled or replaced (Figure 10-13). Note that the construction and operation of the Delco-Moraine passenger car power brake is similar to the Bendix unit shown in Figure 10-2.

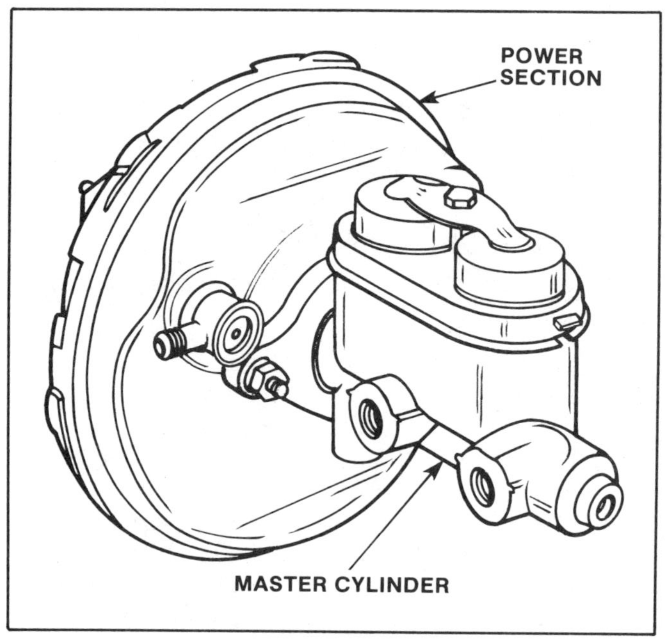

Figure 10-12 Delco-Moraine power brake

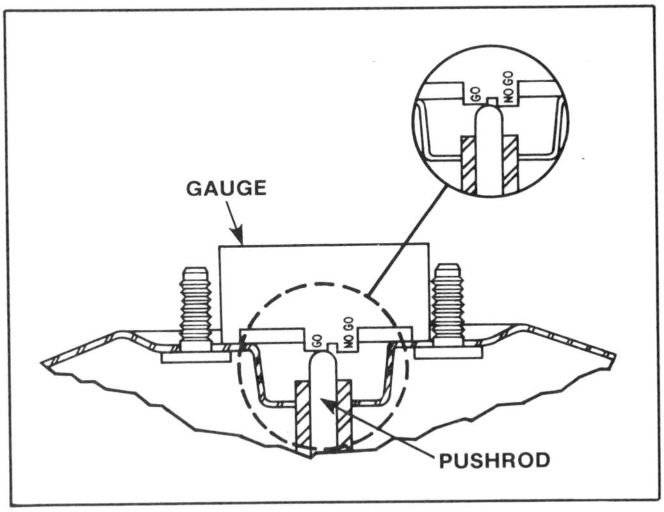

Figure 10-13 Delco-Moraine pushrod adjustment

1. Assemble the booster unit and install the pushrod. Be sure that the pushrod is fully seated.
2. Position the go/no go gauge over the pushrod. (The gauge is furnished in the repair kit.)
3. If the pushrod height is not within the limits of the gauge, install the service adjustable pushrod and adjust the rod to obtain the correct height.
4. Install the power unit and check the adjustment, ensuring that the master cylinder replenishing port is open with the engine running and the brake pedal released.

Air Method

The air testing method is as follows:

1. Be sure the master cylinder attaching nuts are tight.
2. Remove the master cylinder filler cap.
3. With the brakes released, force compressed air into the hydraulic outlet of the master cylinder. Regulate air pressure to a value of approximately 5 PSI to prevent spraying brake fluid from the master cylinder.

> **WARNING:** Care must be taken not to allow brake fluid to contact skin or eyes or any painted surface of the vehicle. Personal injury or vehicle damage will result.

4. If air passed through the replenishing port, which is the smaller of the two holes in the bottom of the master cylinder reservoir, the adjustment is satisfactory.
5. If air does not flow through the replenishing port, adjust the pushrod as required, either by means of the adjustment screw (if provided) or by adding shims between the master cylinder and power unit shell until the air flows freely.
6. Reconnect the brake lines and bleed the system.

VACUUM BRAKE SERVICING

REMOVAL

The removal procedures for both the single diaphragm and tandem diaphragm type boosters are the same:

1. Remove the nuts attaching the master cylinder to the power brake unit.
2. Carefully slide the master cylinder off the mounting studs. Rest it against the fender shield.
3. Disconnect the vacuum hose from the power brake unit. For hose identification and replacement, refer to the manufacturer's service manual.

INSTALLATION

When installing a booster:

1. Rotate the linkage as necessary and position the power brake booster into the dash panel.

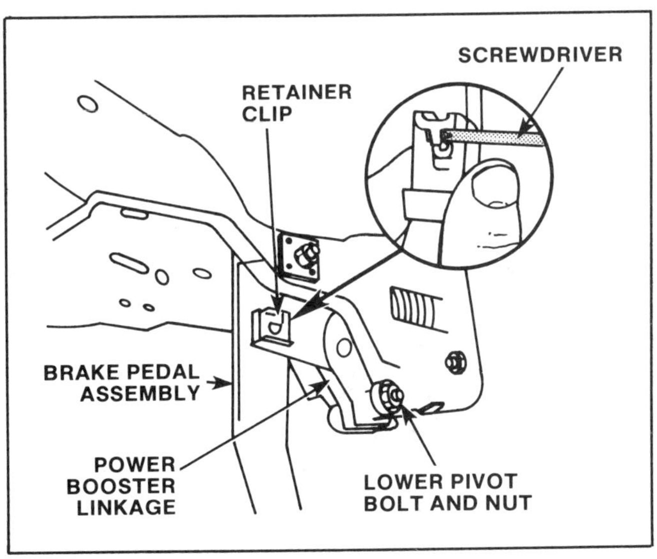

Figure 10-14 Removing retainer clip

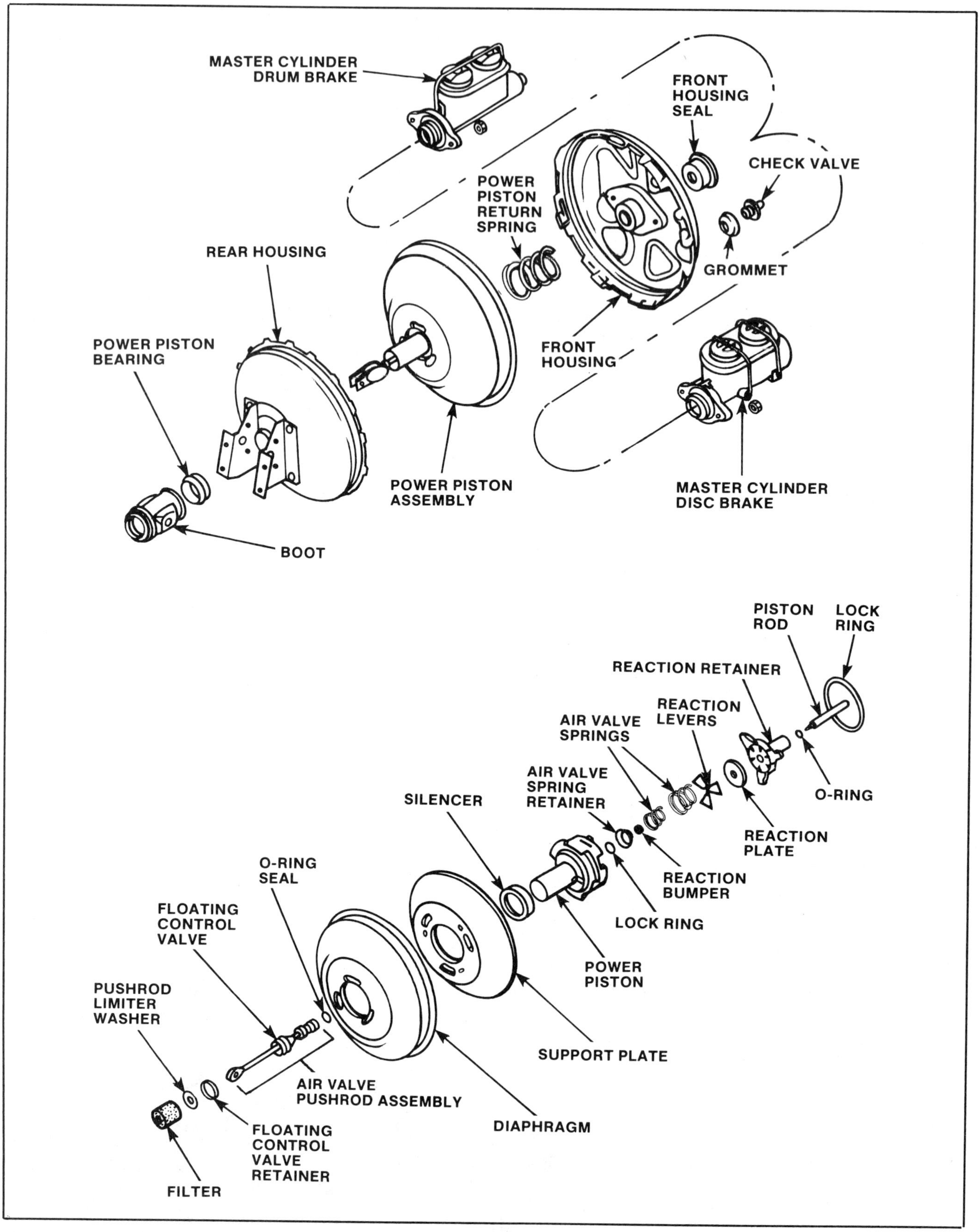

Figure 10-15 Delco-Moraine single diaphragm booster

2. Install the mounting nuts and washers. Tighten to specification.
3. Carefully position the master cylinder on the power brake unit. Install the mounting nuts and tighten them to specification.
4. Connect the vacuum hose.
5. Coat the bearing surface of the pedal pin with a suitable lubricant.
6. Connect the linkage to the pedal pin and install a new retainer clip (Figure 10-14).
7. Coat the bushings and sleeve with lubricant and install them into the linkage.
8. Position the linkage and install the lower pivot bolt. Tighten as specified in the service manual.

Overhauling a Single Diaphragm Booster

With a proper overhaul kit, a typical single diaphragm booster (Figure 10-15) can be disassembled for overhauling as follows:

1. Remove the boot and silencer from the rear of the housing.
2. Remove the check valve and grommet, front seal, and vacuum switch and grommet, if equipped, from the front housing.
3. Scribe matching marks between the front and rear housings to aid assembly, then mount the booster assembly in the suitable holding fixture with the rear housing facing up (Figure 10-16).
4. Apply pressure to hold the housing together, then rotate the front housing counterclockwise to unlock the housing.
5. Slowly release pressure on the booster assembly, then remove the rear housing assembly, power piston, and diaphragm assembly and diaphragm spring.

> **WARNING:** When separating the booster housing be careful not to allow sudden release of spring tension, as damage or personal injury might result.

6. Remove the power piston bearing from the rear housing.
7. Remove the master cylinder piston rod and reaction retainer from the power piston.
8. Pry out the filter using the suitable pointed tool.
9. Grasp the power piston at the outer edges of the diaphragm and support plate, then separate the power piston by rapping the pedal pushrod against the work surface.
10. Clean all components.

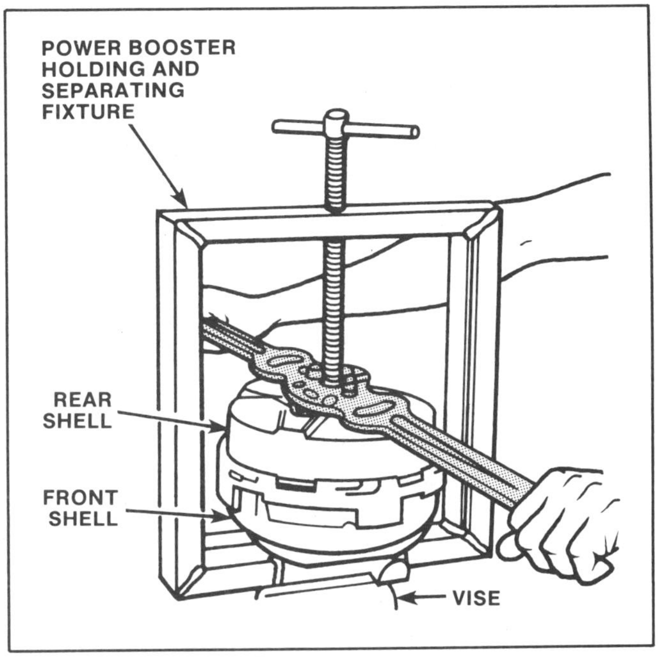

Figure 10-16 Typical brake booster holding and support fixture installation

> **CAUTION:** The power piston and pushrod assembly should not be disassembled. If replacement is required, these components are serviced as an assembly.

To assemble the vacuum power booster, proceed as follows:

1. Lubricate the inner lip of the diaphragm and fit the diaphragm into the support plate.
2. Place the power piston on the work surface with the pedal pushrod facing up, then mount the diaphragm and support plate assembly over the piston.
3. Install the new diaphragm retainer, then seat the retainer and diaphragm assembly using a recommended driver and a suitable hammer.
4. Seat a new filter into the rear of the power piston assembly, then install the reaction retainer and master cylinder piston pushrod.
5. Mount the front housing in the holding fixture, then position the diaphragm return

spring in the housing, with the white coil flush against the housing.

6. Install the power piston bearing and piston assembly in the rear housing, then position the rear housing assembly over the front housing ensuring that matching marks are aligned.
7. Ensure that all components are properly aligned. Apply pressure to the rear housing with the holding fixture forcing screw, then rotate the rear housing clockwise to the lock position.

SHOP TALK

Assembly of the housing can be aided by applying vacuum to the port in the front housing.

8. Stake housing tabs in two locations, 180 degrees apart to secure the assembly, but do not reuse tabs that have already been staked.
9. Lubricate and install the front housing grommets, then press the check valve and vacuum switch, if equipped, into the grommets.
10. Install the front housing seal, silencer, and boot.

The basic procedure for overhauling a typical tandem diaphragm is the same. However, always follow the specific instructions for either type given with the overhaul kit.

AUXILIARY VACUUM PUMPS

Because of a deficiency in manifold vacuum available for power brake operation, auto manufacturers, especially General Motors, have equipped some of their gasoline- and diesel-powered vehicles with auxiliary pumps to provide the vacuum necessary for power brake booster operation. On such cars, without enough vacuum for booster operation, the boosters cannot provide enough assist in braking action. This requires drivers to use excessive pedal effort in order to stop their cars.

In addition to supplementary increased vacuum power to power the brake system and the engine, it also helps to provide the needed vacuum necessary in the operation of the electronic ignition systems, heat and air conditioning system, EGR systems, and cruise control. Of the several auxiliary vacuum pumps available, the most common for passenger cars are those driven by:

- Belt (Figure 10-17)
- Gear or cam/lever from the engine (Figure 10-18)
- Electric motor (Figure 10-19)

Some vehicle models have a vacuum pump that is mounted to the rear of the alternator and driven by the alternator rotor shaft. Vacuum pumps used on cars and light-duty trucks are usually of the

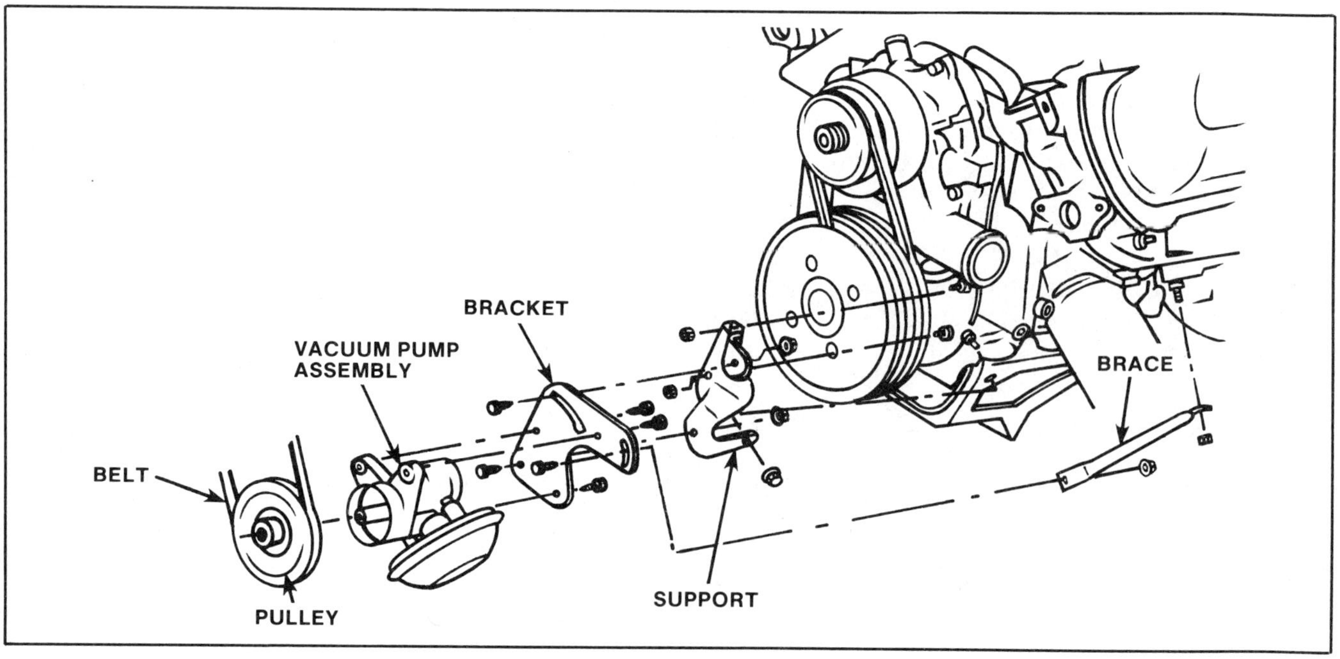

Figure 10-17 Typical installation of belt-driven vacuum pump

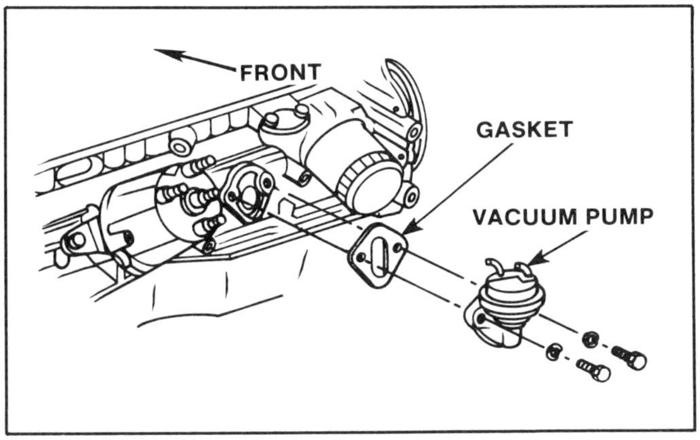

Figure 10-18 Typical installation of cam lever-actuated vacuum pump

diaphragm or vane type. The electrically driven vacuum pump used on some GM models consists of an electric motor controller assembly, which contains an integral on-off switch, timer/relay, and a piston and umbrella valve assembly. The vacuum pump inlet hose is connected to the vacuum-operated components by means of vacuum hoses. The vacuum pump outlet hose, if equipped, is usually connected to the intake manifold.

The electrically operated vacuum pump used on some passenger models is supplied with current whenever the ignition switch is in the RUN position. Under low vacuum conditions, the internal vacuum switch closes, sending current to the timer/relay, which activates the vacuum pump motor to maintain 13.9 inches Hg of vacuum in the system. When normal vacuum conditions are reached, the internal vacuum switch opens, and the timer/relay will continue to supply current to the pump motor for an additional 5 to 10 seconds. The timer/relay contacts will open after the 5- to 10-second period, stopping current to the pump motor. Some pumps incorporate an external pressure switch that will operate the pump, bypassing the internal vacuum switch. A low vacuum warning switch located in the inlet hose line will illuminate a warning lamp on the instrument panel when a low vacuum condition exists. A charcoal filter located in the outlet line prevents harmful vapors from entering the vacuum pump.

When excessive brake pedal effort problems are encountered, check the brake hydraulic system and brake booster before checking the vacuum pump system. Check the vacuum system for loose connections and leaks at the vacuum hoses and components before checking the vacuum pump.

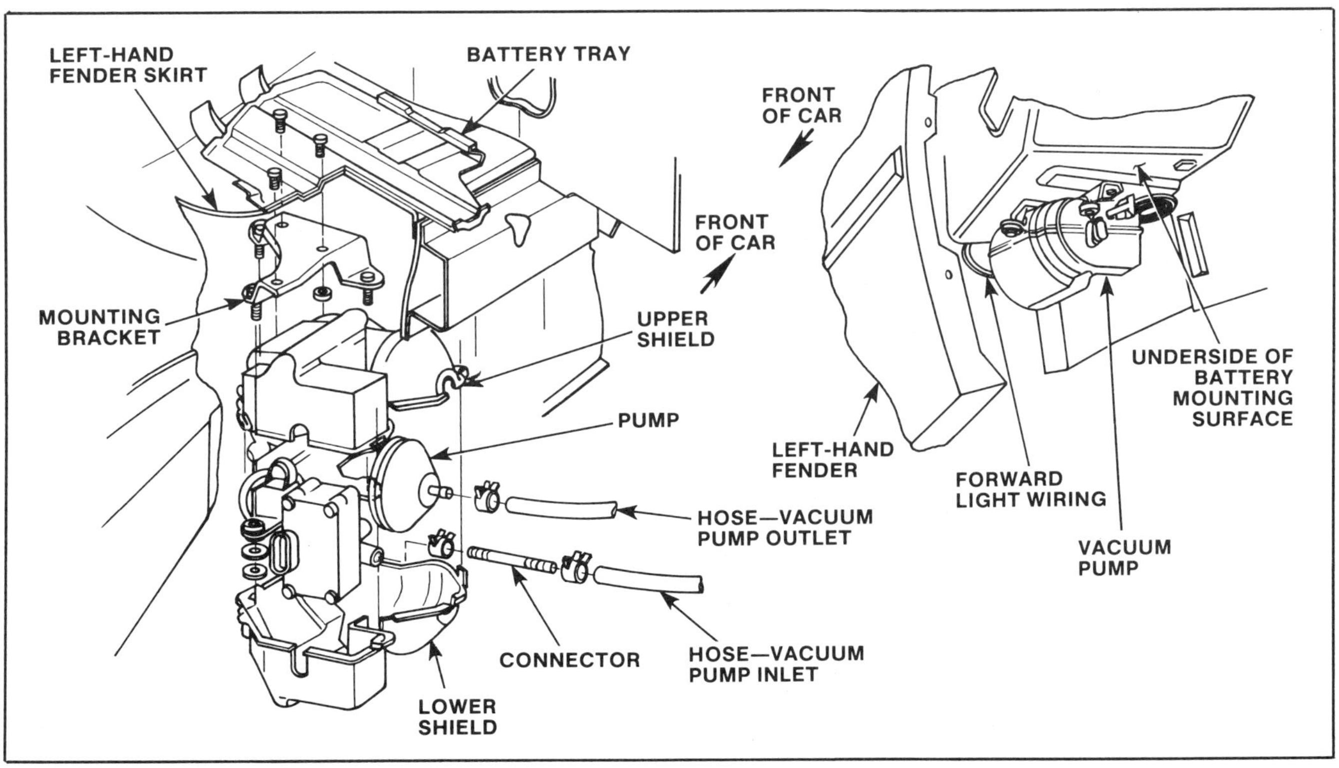

Figure 10-19 Electrical vacuum pump

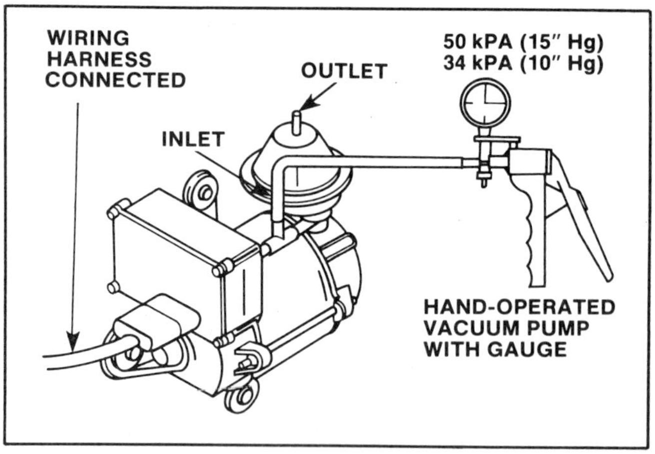

Figure 10-20 Vacuum pump test connection

Also check for loose mounting bolts, improperly tensioned drive belt, or improper engine idle speed on mechanically operated vacuum pumps. On inoperative electrically operated vacuum pumps, check for current at the pump motor electrical connector before considering a pump motor to be defective.

For most diagnostic procedures given in the manufacturer's service manual and in this chapter, a hand-operated vacuum pump is used (Figure 10-20). The pump utilizes a vacuum gauge, usually calibrated in inches of mercury (Hg). For these diagnostic procedures, kilopascals (kPa) are specified and should be noted that 47 plus or minus 6 kPa equals 13 plus or minus 1.8 inches Hg.

CAUTION: At no time during the testing of the vacuum pump should the outlet nozzle be blocked. Permanent damage to the diaphragm can result from a blocked outlet nozzle.

ELECTRICAL VACUUM PUMP SYSTEM

Diagnosis

Before attempting to diagnose problems in vacuum pump systems, be sure that the brake hydraulic system and the brake power booster are working properly. The following are various problems that can occur in an electrical vacuum pump system and steps that should be taken for diagnosis and correction.

EXCESSIVE BRAKE PEDAL EFFORT, VACUUM PUMP INOPERATIVE, AND BRAKE WARNING LIGHT ON

1. With the ignition ON, disconnect the electrical connector from the vacuum pump and

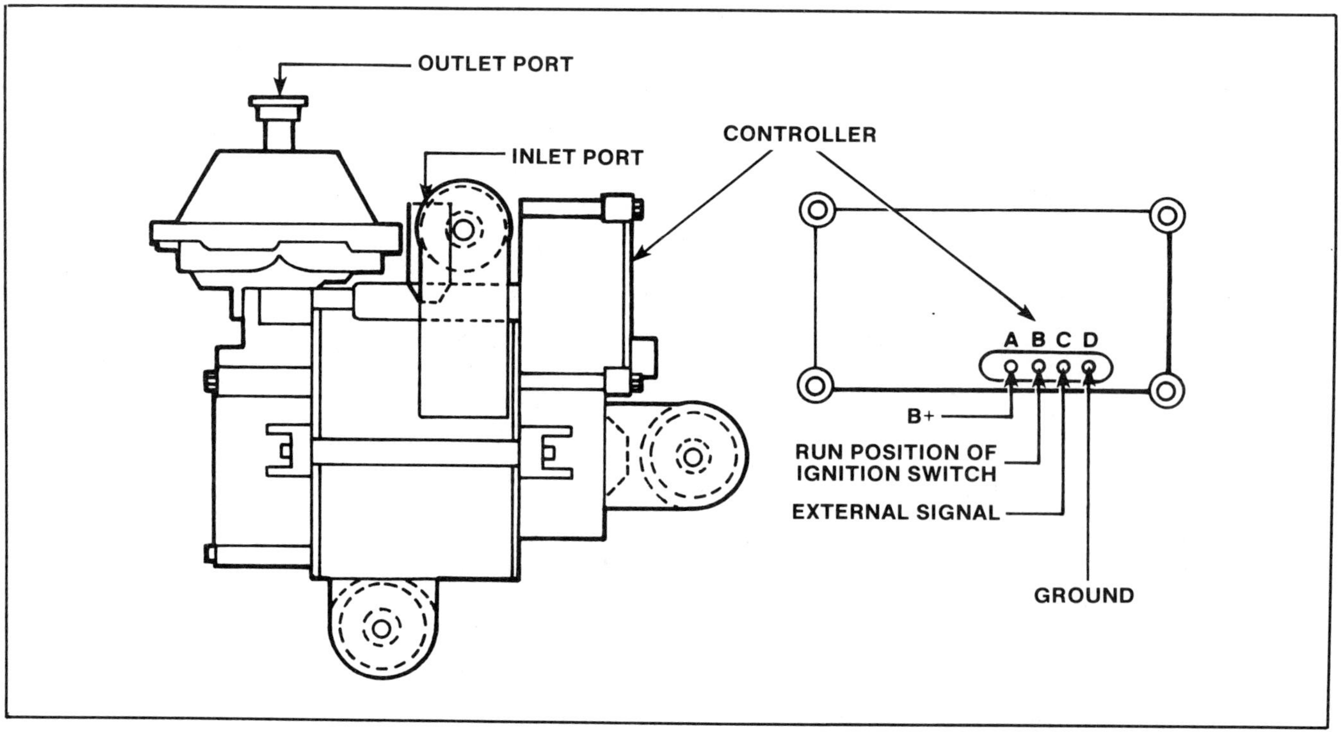

Figure 10-21 Vacuum electrical connections

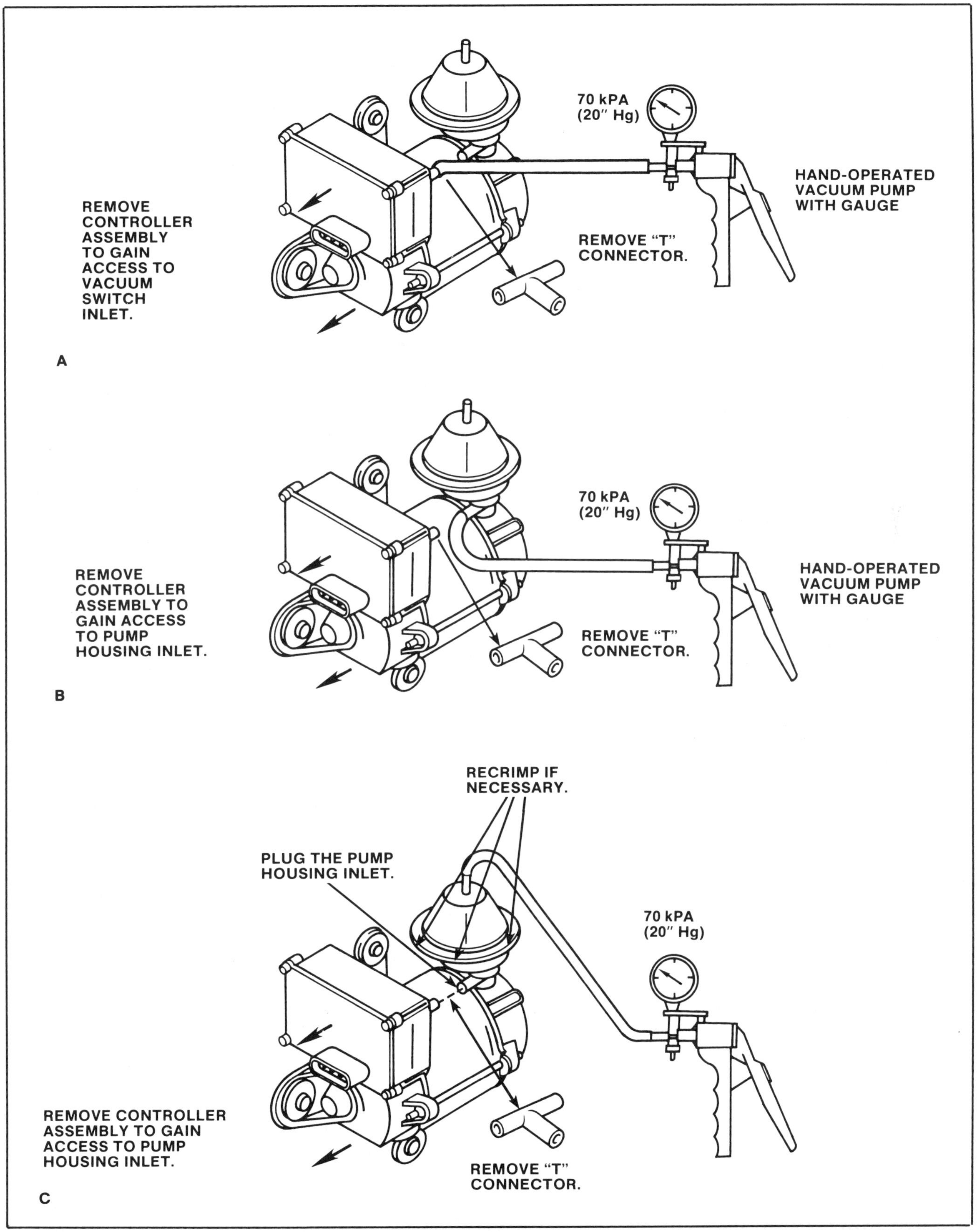

Figure 10-22 Vacuum pump test connections

check for battery voltage at terminals A and B and for proper ground at terminal D (Figure 10-21). If battery voltage is indicated, proceed to step 2. If not, check the fusible link, fuse, and wiring circuits and repair as necessary.
2. Connect a 12-volt power source to terminals A and B and ground terminal D. If the pump is still inoperative, go on to step 3. If the pump operates, refer to the "Excessive Brake Pedal Effort, Vacuum Pump Operative, and Brake Warning Light ON" diagnosis.
3. Remove the vacuum pump and check for sticking or shorted brushes. Check the controller for damaged wires and repair as necessary. If no defects are found, replace the controller.

EXCESSIVE BRAKE PEDAL EFFORT, VACUUM PUMP OPERATIVE, AND BRAKE WARNING LIGHT ON

1. Remove the pump inlet hose, connect the hand-operated vacuum pump to the inlet hose, and turn the ignition ON.
2. If the vacuum pump draws vacuum of 10 to 15 inches Hg and continues to run for 5 to 10 seconds, check for vacuum leaks in other parts of the system.
3. If the pump runs intermittently and will not maintain a steady vacuum, proceed to step 4.
4. Remove the vacuum pump, controller, and "T" vacuum connector (Figure 10-22A). Connect the hand-operated vacuum pump to the vacuum switch inlet and apply approximately 20 inches Hg. If the switch leaks down more than 2 inches Hg per minute, replace the controller. If the switch maintains a vacuum, proceed to step 5.
5. Connect the hand-operated vacuum pump to the pump housing inlet and apply 20 inches Hg of vacuum (Figure 10-22B). If the pump leaks down more than 2 inches Hg per minute, replace the umbrella valve. If the pump maintains a vacuum, proceed to step 6.
6. Connect the hand-operated vacuum pump to the pump outlet, plug the pump housing inlet, and apply 20 inches Hg of vacuum (Figure 10-22C). If the pump leaks down more than 2 inches Hg per minute, check for a loose bonnet. Repeat the test. If the pump still leaks, replace the piston. If the pump maintains a vacuum, the system is operating satisfactorily.

VACUUM PUMP RUNS CONTINUOUSLY WITH NORMAL OR EXCESSIVE BRAKE PEDAL EFFORT AND BRAKE WARNING LIGHT ON OR OFF

1. Check all vacuum hoses for leaks and repair as necessary. If no leaks are found, proceed to step 2.
2. With ignition OFF, disconnect the inlet hose from the vacuum "T" connection, attach the hand-operated vacuum pump, and turn the ignition ON (Figure 10-23). If the pump runs continuously drawing a vacuum of 10 inches Hg or more, replace the controller. If the pump runs then stops when the vacuum reaches 10 to 15 inches Hg, the system is normal; bleed off the vacuum to below 10

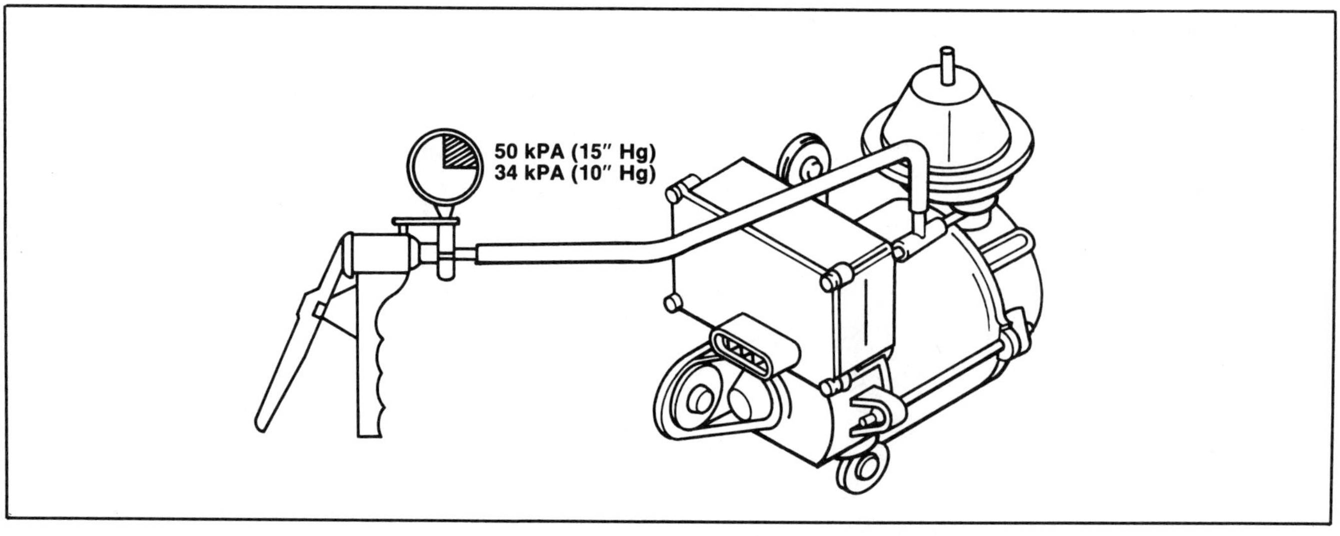

Figure 10-23 Vacuum pump test connection

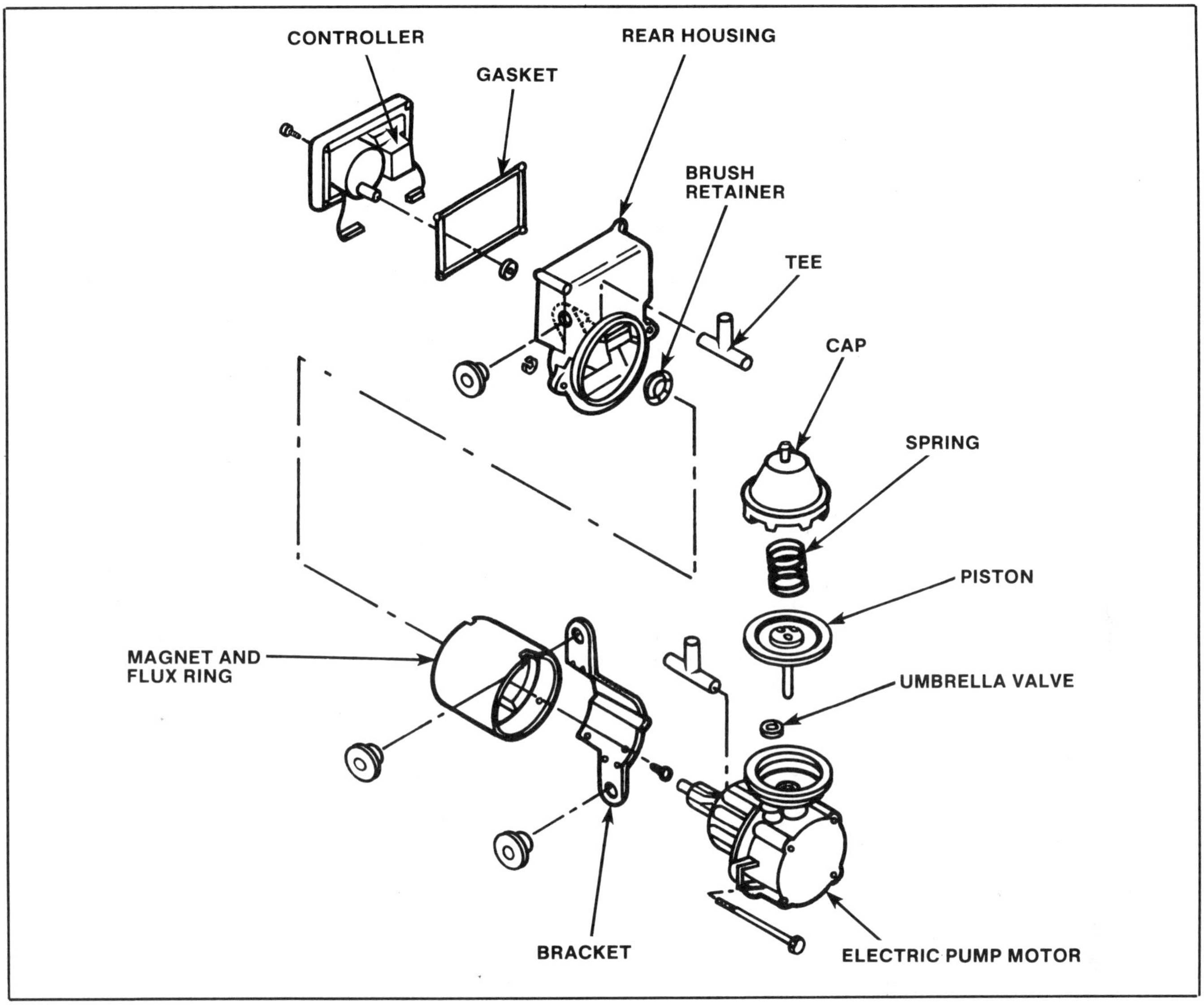

Figure 10-24 Electrically driven vacuum pump

inches Hg. The pump should start automatically and run until the vacuum exceeds 10 to 15 inches Hg, then automatically shut off.

Service

The following are the replacement and installation steps for the various components of the system (Figure 10-24):

VACUUM PUMP

1. Raise the vehicle and support it properly.
2. Remove the left-hand side splash shield.
3. Carefully remove the locking type electrical connectors and vacuum hoses from the pump.
4. Remove the three vacuum pump retaining nuts and the pump.
5. Reverse the procedure to install.

CONTROLLER

1. Remove the vacuum pump as previously described.
2. Remove the upper and lower shields.
3. Remove the two tie bolts.
4. Carefully separate the rear housing from the pump housing.
5. Remove the four screws securing the controller to the rear housing. Remove the brushes from the brush holders, then remove the controller, gasket, and washer.

6. Install the controller using a new gasket and washer. Place both brushes in the rear brush holder cavity.
7. Pre-load brush springs by locking them in the slots provided above the spring access slots. Position the brushes in channels, checking to ensure shunts are properly positioned.
8. Install the brush retainer in the proper position and return springs to the load position.
9. Reverse steps 1 through 4 to complete the installation.

PISTON ASSEMBLY

1. Remove the vacuum pump as previously described.
2. Remove the upper and lower shields.
3. Carefully release the bonnet tabs, then remove the bonnet and spring.

CAUTION: The bonnet is spring loaded. The bonnet and pump housing must be held together during disassembly.

4. Remove the piston from the housing.
5. Remove the umbrella valve.
6. Replace the umbrella valve, piston assembly, and spring.
7. Place the bonnet over the spring and compress the spring and bonnet onto the pump housing. Bend the bonnet tabs around the pump housing, checking to ensure that the bonnet is secure.
8. Reinstall the shields and install the pump.

BELT-DRIVEN VACUUM PUMP SYSTEM

Diagnosis

The vacuum pump is operated by a belt-driven pulley, which turns the pump shaft. The shaft runs a cam, which operates a spring-loaded diaphragm in the pump. Resolve the problem described by following the step-by-step procedure:

EXCESSIVE BRAKE PEDAL EFFORT AND BRAKE WARNING LIGHT ON

1. Block the wheels, apply the parking brake, and shift the transmission to the PARK or NEUTRAL position.
2. Connect a suitable vacuum gauge to the vacuum pump inlet.

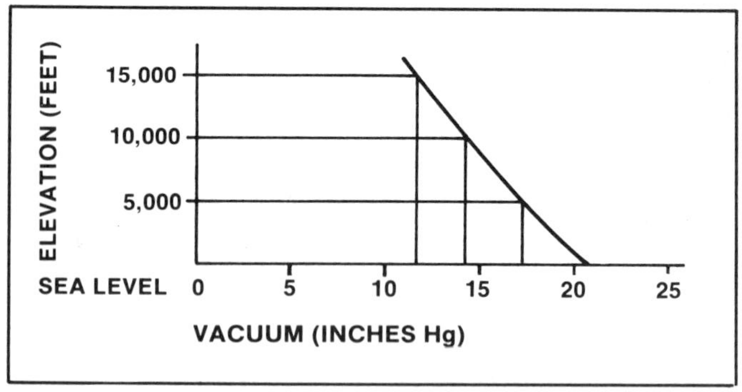

Figure 10-25 Belt-driven vacuum pump test specification chart

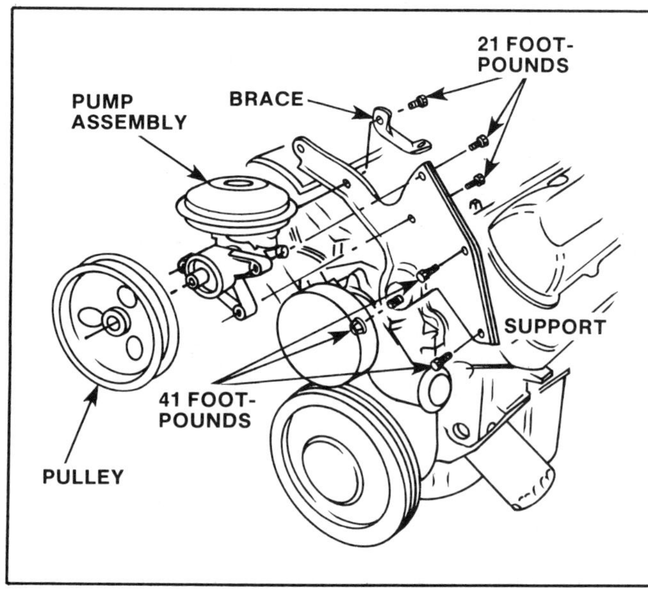

Figure 10-26 Vacuum pump mounting 4.3 liter diesel without air-conditioning

3. On models equipped with a gasoline engine, disconnect the outlet hose from the vacuum pump and plug the hose.
4. On all models, run the engine at idle and observe the vacuum gauge. The vacuum level should reach a specified level (Figure 10-25) within 30 seconds of starting the engine.
5. If the vacuum meets specifications, check the system (other than the vacuum pump) for vacuum leaks and repair as necessary.
6. If the vacuum is low or the gauge reading fluctuates, inspect the gauge and connections for leaks. Then check the belt, pulley, and idle speed; make any necessary adjustments or repairs.

7. If the vacuum is still low, replace the vacuum pump and recheck. If the vacuum now meets specifications, proceed to step 8.
8. On models equipped with a gasoline engine, remove the plug from the outlet hose and reconnect the hose to the vacuum pump.
9. On all models, reconnect the vacuum inlet hose with a vacuum gauge "teed" into the line near the pump.
10. Run the engine at idle and observe the vacuum gauge. After 1 minute, vacuum should not drop more than 3 inches Hg from the reading in step 4.
11. If the vacuum does not drop more than 3 inches Hg, the system is functioning properly.
12. If the vacuum does drop more than 3 inches Hg, check all attaching hoses for leaks and repair as necessary. If the vacuum is still low, check all vacuum accessories for leaks and repair or replace as necessary.

Service

The belt-driven vacuum pump is serviced as an assembly only (Figure 10–26).

GEAR-DRIVEN VACUUM PUMP SYSTEM

Diagnosis

The vacuum pump is a diaphragm pump that requires no periodic maintenance. This pump is driven by a cam inside the drive assembly on which the pump is mounted. The drive housing assembly includes a gear on the lower end that meshes with the camshaft gear in the engine. This gear causes the cam in the drive housing to turn. The gear also operates the engine oil lubricating pump. The diagnostic procedure for this pump is as follows:

1. Start the engine, then attach the hose and vacuum gauge to the pump. The gauge should read at least 20 inches Hg.
2. Shut off the engine. The vacuum level should remain above 19 inches Hg for at least 1.5 seconds.

Service

The removal and installation procedure of the vacuum pump is as follows:

REMOVAL

1. Disconnect the vacuum hose from the vacuum pump inlet.
2. Remove the bolt and clamp the holding drive assembly to the engine block.
3. Carefully lift the pump and drive assembly out of the engine block.
4. Cover the hole in the engine block with a cloth to prevent foreign matter from entering the engine block.

INSTALLATION

1. Remove the protective cloth from the engine.
2. Install the pump and drive assembly in the engine, ensuring that the gears on the drive assembly mesh with the gears on the engine camshaft.
3. Rotate the pump into position so the bracket and bolt can be installed.
4. Install the clamp and bolt, torquing the bolt to 20 foot-pounds.
5. Connect the vacuum hose to the pump.

ALTERNATOR-DRIVEN PUMP SYSTEM

Service

Some models equipped with diesel engines use an alternator-driven vacuum pump to aid the engine in maintaining a proper vacuum level for the power brake system. The system problem diagnosis is as follows:

1. Connect a suitable vacuum gauge to the vacuum pump inlet.
2. With the parking brake fully applied and the transmission in PARK, start the engine and operate at 600 to 5,000 RPM, while noting the vacuum gauge.
3. The vacuum gauge should maintain a reading of 25.6 inches Hg. If the vacuum does not meet specifications, the pump should be disassembled and inspected and serviced as necessary.

Service

The removal, inspection, and installation procedures of the vacuum pump are as follows:

REMOVAL

1. Disconnect the oil lines and vacuum hose from the vacuum pump.
2. Remove the vacuum pump attaching bolts.
3. Remove the vacuum pump from the rotor shaft, in the direction of the shaft, while holding the center plate.
4. Carefully remove the vanes from the pump housing.

INSPECTION

1. Measure the length of the vanes. If the vanes do not measure 0.511 to 0.531 inch, they must be replaced.
2. Measure the inside diameter of the housing. If the housing does not measure 2.2440 to 2.2441 inches, replace the vacuum pump.
3. Apply light pressure to the check valve with a screwdriver. If the valve does not operate smoothly, it should be replaced.
4. Inspect the inner face of the rear housing cover for oil leakage and the inner face of the oil seal for wear or damage. If the oil leakage is evident, or the seal is worn, it should be replaced.

INSTALLATION

1. Position the rotor on the center plate housing, with the serrated boss on the rotor turned up.
2. Align the holes in the center plate and the rotor.
3. Install the vanes in the rotor, ensuring that the chamfered side of the vanes are turned outward.
4. Install the housing, ensuring that the O-ring does not protrude past the slot in the center. Use care to avoid scratching the vane surfaces.

CAUTION: If the holes in the center and housing are not properly aligned, adjust by slowly turning the housing.

5. Install the vacuum pump to the alternator and secure it with mounting bolts.
6. Connect the oil lines and vacuum hose to the pump.
7. Add approximately 5 cc of clean engine oil to the filler port on the pump; then, ensure that the alternator pulley turns freely by hand.

HYDRAULICALLY OPERATED POWER BRAKES

Hydro-boost is a hydraulically operated power assist mechanism that is used to increase fluid pressure in the brake system while at the same time decreasing pedal effort. This unit performs the same job as a vacuum booster and is connected to the hydraulic system and the brake pedal in much the same way. But instead of using manifold vacuum and atmospheric pressure, the hydro-boost uses pressurized fluid from the power steering pump to obtain its assist power.

One reason for using hydro-boost instead of vacuum boost is that federal regulations require that newer vehicles stop in a shorter distance with less pedal effort. This could be accomplished with a larger vacuum unit, but its size would be a problem. Also, there is a limit to how much vacuum can be drawn off for braking assist before other vacuum-related functions are affected, like the exhaust pollution device, which itself greatly reduces available engine vacuum.

Another reason for using the hydro-boost unit is that it is much smaller than a vacuum booster, especially since the exhaust pollution devices reduce available engine space. It is easier to fit it into today's smaller engine compartments and has eliminated the need for a remote or frame-mounted booster previously used on some truck applications.

Hydro-boost is also suitable for use in diesel-powered vehicles. Since diesel engines are unthrottled, they produce no useable manifold vacuum, and a vacuum pump would have to be added to operate vacuum-powered accessories.

HYDRO-BOOST OPERATION

The hydro-boost brake system consists of a booster assembly and accumulator; the power steering fluid circuit, including the gear, pump and reservoir; hydraulic lines; master cylinder; and the rest of the brake system (Figure 10–27).

Hydro-boost I (Figure 10-28A) includes a spring or gas-operated accumulator mounted in a "can" attached to the main housing. The accumulator provides reserve stopping capacity if power steering pump flow to the hydro-boost unit is interrupted. Hydro-boost II (Figure 10-28B), first used in 1980 models, has fewer parts and the accumulator is contained within the power piston.

The booster assembly itself consists of an open center spool valve and sleeve assembly, a lever assembly, an input rod assembly, a power piston, an output pushrod, and the accumulator. The booster assembly is mounted on the vehicle in much the same manner as a vacuum booster. The pedal rod is connected at the booster input rod end.

A dual master cylinder is bolted to the brake booster and is operated by a pushrod projecting from the booster cylinder bore.

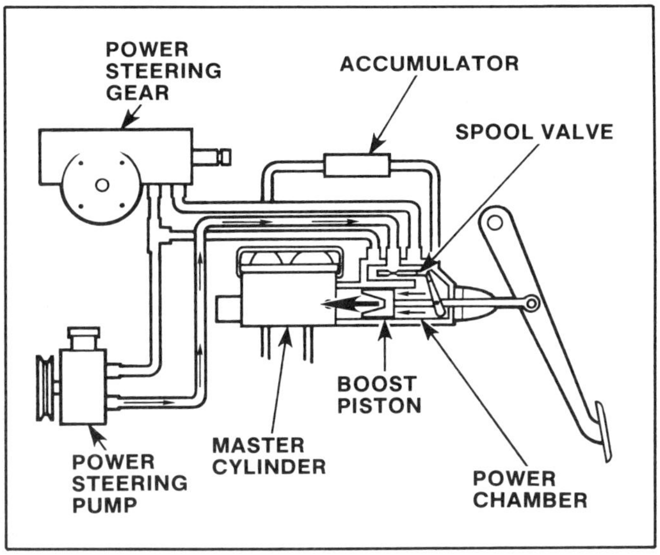

Figure 10-27 Hydro-boost components

Figure 10-28 Hydro-boost power brake boosters

The power steering pump, which is unmodified, except for a larger reservoir and an additional return port for the hydro-boost return line, supplies hydraulic pressure to the hydro-boost unit through external lines. When the engine is running, the fluid enters the unit through the pump port from which most of the fluid is directed to the gear port and then to the steering gear by an external line.

The small quantity of power steering fluid that does not go to the power steering gear circulates inside the hydro-boost unit and then returns to the steering pump reservoir by an external line.

Power steering fluid flow in the hydro-boost unit is controlled by a hollow center spool valve. The spool valve has lands, annular grooves, and drilled passages. These mate with grooves and lands in

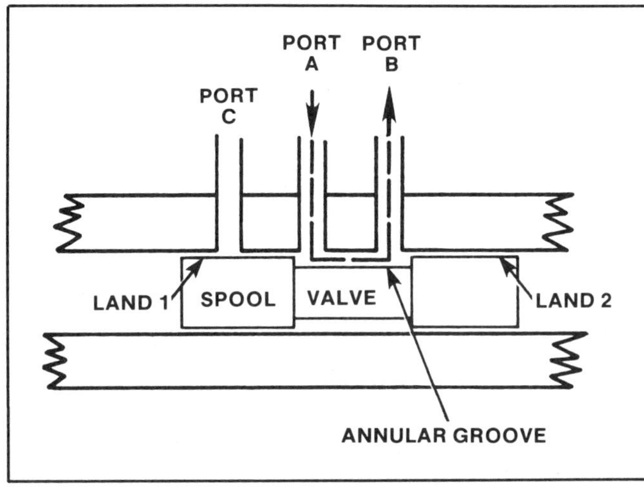

Figure 10-29 Spool valve operation

the valve bore. The flow pattern of the fluid depends upon the alignment of the valve in the bore.

Shown in Figure 10-29 is a typical spool valve in a valve bore that has three ports. In this position, the valves, lands, and annular groove mate with the ports in such a manner that fluid entering the valve bore through port "A" will be directed to port "B." Land 1 will stop the fluid from entering port "C." If the valve were moved to the left, it would close off port "B" and open port "C."

Both the valve bore and spool valve are highly finished parts. The spool valve depends on this finish for proper sealing and smooth movement of the valve in the bore. Any leakage around the spool valve is vented back to the reservoir. When the brake pedal is depressed, the power piston is moved. At the same time, the spool valve is moved by lever action. Spool valve positions for each of the following cycles are shown in Figure 10-30.

1. **Brakes Unapplied.** Pressurized fluid from the power steering pump at 100 to 150 PSI enters the hydro-boost assembly through the pump port in the housing and is directed by an internal passage to the spool valve. The spool valve in its unapplied position directs most of this fluid to the gear port where an external line carries it to the steering gear to perform its steering function (Figure 10-30A). In this position the spool valve also vents its center and the power cavity to the reservoir.
2. **Brakes Applied.** The movement of the spool valve (Figure 10-30B) first closes the reservoir port so fluid cannot escape from the power section, then opens the pump port

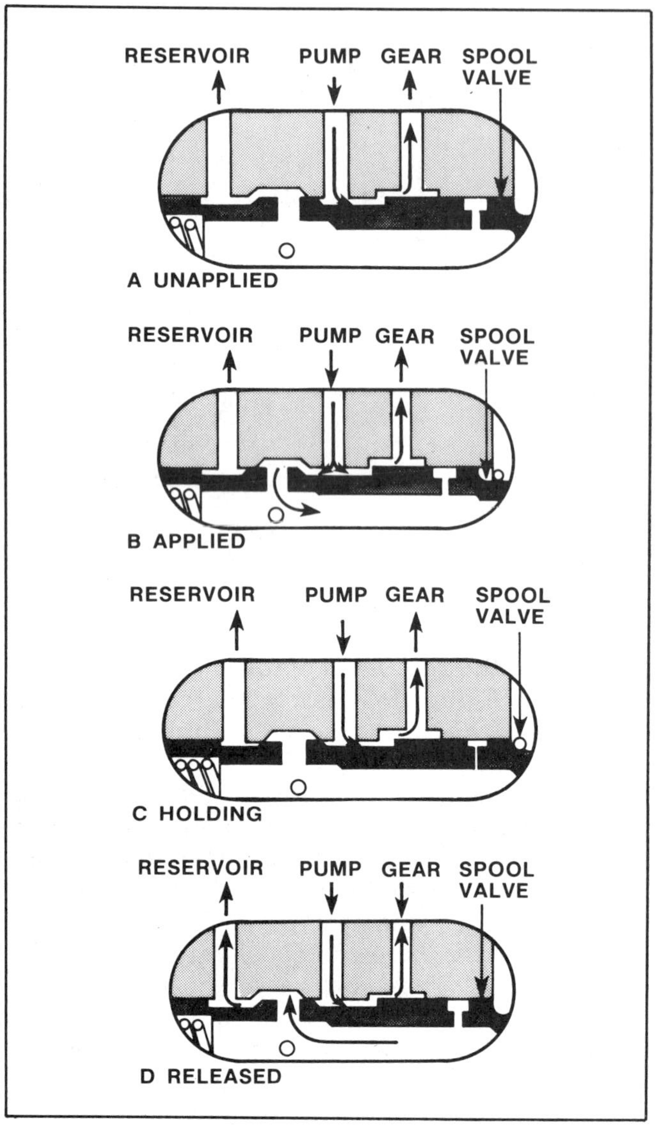

Figure 10-30 Spool valve positions

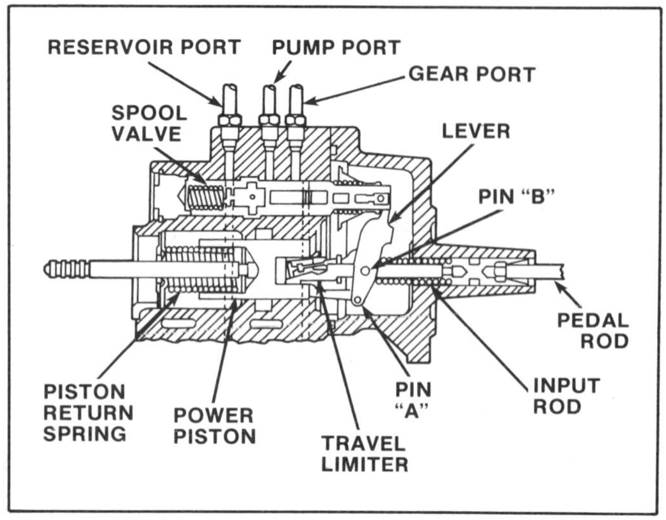

Figure 10-31 Hydro-boost—brakes applied

to allow pressurized fluid into the cavity behind the power section. This pressurizes the power section, forcing the power piston forward, applying the brakes through the output pushrod.

The movement of the spool valve during application also reduces the opening of the gear port in the valve bore. This restricts the flow of fluid to the gear, but at no time is the flow completely stopped.

By restricting the flow of fluid, the power steering pump increases the pressure in the hydraulic lines. The pressure is increased because the pump is of the constant displacement type, which means that it displaces the same amount of fluid every revolution. When there is no restriction in the system, the pump will maintain a pressure in the lines of 100 to 150 PSI. As the fluid flow is restricted, the pressure in the system will begin to rise. If the restriction is great enough, the pressure will rise until the pressure relief valve opens. This relief valve limits pump output pressure to 1,450 PSI (1,300 PSI on light trucks).

With this high pressure applied to the power piston, greater hydraulic pressure is generated in the brake system than that produced by a normal vacuum booster.

3. **Brakes Holding.** When the brake pedal is depressed, pin "B" (Figure 10-31) moves forward, positioning the spool valve so that high-pressure fluid is diverted to the power cavity. This causes the power piston to move forward, developing pressure in the brake hydraulic system, and pulling pin "B" forward. The lever pivots clockwise, pulling the spool valve rearward (Figure 10-30C), blocking flow to the power cavity and closing off the fluid in the power cavity, which creates a holding condition. Once the additional fluid flow stops, the forces between the boost unit and brake system balance. The brakes will remain applied at this rate until a change in pedal pressure is made.

   If more pressure is applied to the pedal, the travel limiter valve again moves forward. This will move the spool valve forward and open the pump port, increasing the pressure in the power section. When pedal pressure is removed, the boost unit goes into release.

4. **Brakes Release.** When the pedal pressure is removed, fluid pressure in the unit acts on the travel limiter and input rod and moves the input rod assembly toward the pedal. The power piston remains stationary due to the balance of forces between the booster unit and the brake system. The lever rotates, pulling the spool valve out of the bore and opening the reservoir port, depressurizing the power cavity.

   Once the pressure is removed from the power cavity, the piston return spring pushes the power piston rearward and releases the brakes (Figure 10-30D). The unit is now in the unapplied position, ready for another application.

ACCUMULATOR OPERATION

The hydro-boost unit includes a reserve brake system that provides one to three power-assisted brake applications if the flow of fluid from the power steering pump is interrupted. This system uses a spring or gas-charged accumulator to store fluid under pressure. The external spring type accumulator provides three reserve stops. The external gas-charged type gives two, and the integral power piston type provides one.

During normal operation, the accumulator is charged by fluid taken from the pump port before it gets to the spool valve. The charging fluid flow is controlled by an accumulator valve assembly, which consists of a check valve, check ball, spacer, plunger seat, and plunger.

The check valve allows fluid under greater pressure than that in the accumulator to go through the valve and into the accumulator. Once past the check valve, the fluid travels by internal passage or external line to the accumulator. In the accumulator, fluid pressure exerts force on the accumulator piston, compressing the accumulator spring. This fluid is then stored under pressure until normal fluid pressure is not available. The spring type accumulator has a pressure relief port that allows pressurizing fluid to return to the reservoir when the piston is compressed to a predetermined position. Points to check for fluid leaks are shown in Figure 10-32.

CAUTION: Power steering fluid and brake fluid cannot be mixed. If brake seals contact steering fluid or steering seals contact brake fluid, seal damage will result.

TESTING THE HYDRO-BOOST

Any investigation of a hydro-boost complaint should begin with an inspection of the power steering pump belt, fluid level, and hose condition and connections. Hydro-boost will not work properly if

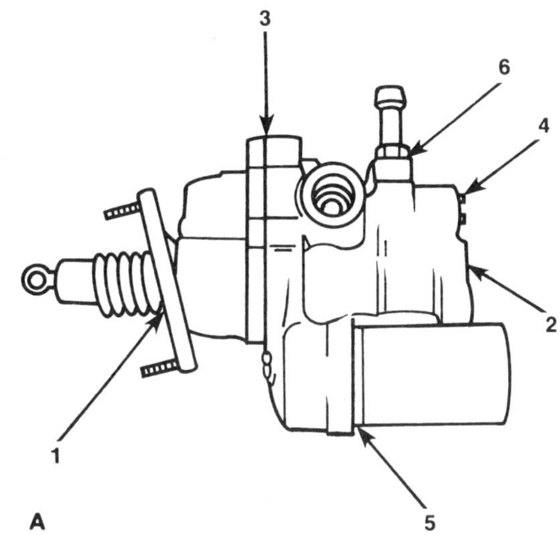

1. **INPUT SEAL LEAK—Fluid leakage from housing cover end of booster near reaction bore. Replace booster.**
2. **PISTON SEAL LEAK—Fluid leakage from vent at front of unit near master cylinder. Replace seal.**
3. **HOUSING—Fluid leakage between the housing and housing cover. Replace seal.**
4. **SPOOL VALVE SEAL—Fluid leakage near plug area. Replace seal.**
5. **ACCUMULATOR CAP SEAL—Fluid leakage near plug area. Replace seal.**
6. **RETURN PORT FITTING SEAL—Replace seal.**

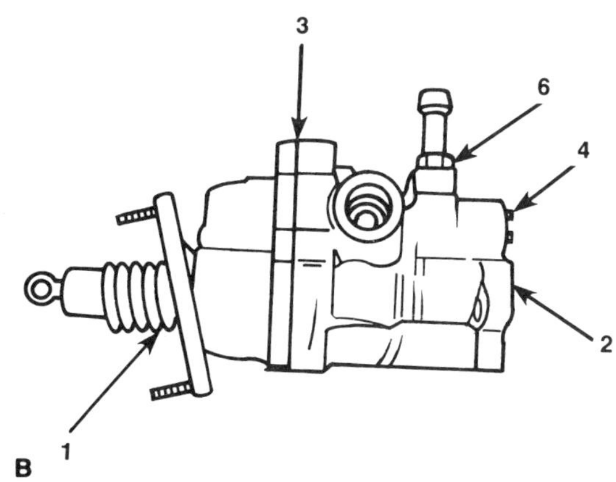

Figure 10-32 (A) Hydro-boost I seal leaks and (B) hydro-boost II seal leaks

it is not supplied with a continuous supply of clean, bubble-free power steering fluid at the proper pressure.

Basic Operational Test

The basic operational test of hydro-boost is as follows:

1. With the engine off, pump the brake pedal numerous times to bleed off the residual hydraulic pressure that is stored in the accumulator.
2. Hold firm pressure on the brake pedal and start the engine.
3. The brake pedal should move downward, then push up against the foot.

Accumulator Test

To be sure the accumulator is performing properly, proceed as follows:

1. With the engine running, rotate the steering wheel until it stops and hold it in that position for no more than 5 seconds.
2. Return the steering wheel to the center position and shut off the engine.
3. Pump the brake pedal. You should feel two to three power-assisted strokes.
4. Now repeat steps 1 and 2. That will pressurize the accumulator.
5. Wait 1 hour, then pump the brake pedal. There will be two or three power-assisted strokes.
6. If the system does not perform as just described, the accumulator is leaking and should be replaced.

Noise Troubleshooting

Hydro-boost differs from vacuum brake boosters not only in the source of power (hydraulic versus vacuum), but in the fact that it is also a part of another major subsystem of the vehicle—the power steering system. Therefore, problems or malfunctions in the steering system can affect the operation of the booster, just as a problem in the booster can affect the steering system.

The following noises often occur with the hydro-boost system and might or might not be cause for customer complaint. Some are normal and for the most part temporary in nature. Others might be a sign of excessive wear or the presence of air in either the booster or the steering system.

1. Moan or low-frequency hum usually accompanied by a vibration in the pedal and/or steering column might be encountered during parking or other very low-speed maneuvers. This can be caused by a low fluid level in the power steering pump, or by air in the power steering fluid due to holding the pump at relief pressure (steering wheel held all the way in one direction) for an excessive amount of time (more than 5 seconds). Check the fluid level and add fluid if necessary. The system must sit for 1 hour with the cap removed to eliminate the air. If the condition persists, it might be a sign of excessive pump wear. Check the pump according to the vehicle manufacturer's recommended procedure.
2. At or near power runout (brake pedal near fully depressed position), a high-speed fluid noise (like a faucet can make) might occur. This is a normal condition and will not be heard, except in emergency braking conditions.
3. Whenever the accumulator pressure is used, a slight hiss is noticed. It is the sound of the hydraulic fluid escaping through the accumulator valve and is completely normal.
4. After the accumulator has been emptied and the engine is started again, another hissing sound might be heard during the first brake application or the first steering maneuver. This is caused by the fluid rushing through the accumulator charging orifice. It is normal and will only be heard once after the accumulator is emptied. However, if this sound continues even though no apparent accumulator pressure assist was made, it could indicate that the accumulator is not holding pressure. Check for this using the Accumulator Test procedures discussed previously.
5. After bleeding, a "gulping" sound might be present during brake applications, as noted in the bleeding instructions. This is normal and should disappear with normal driving and braking.

SERVICING THE HYDRO-BOOST

Before removing a hydro-boost unit from a vehicle, do the following:

1. With the engine off, depress and release the brake pedal several times to deplete accumulator pressure prior to disconnecting hoses or lines.
2. Disconnect the master cylinder from hydro-boost, leaving hydraulic lines connected. Carefully lay the master cylinder aside, being careful not to kink or bend the steel lines.

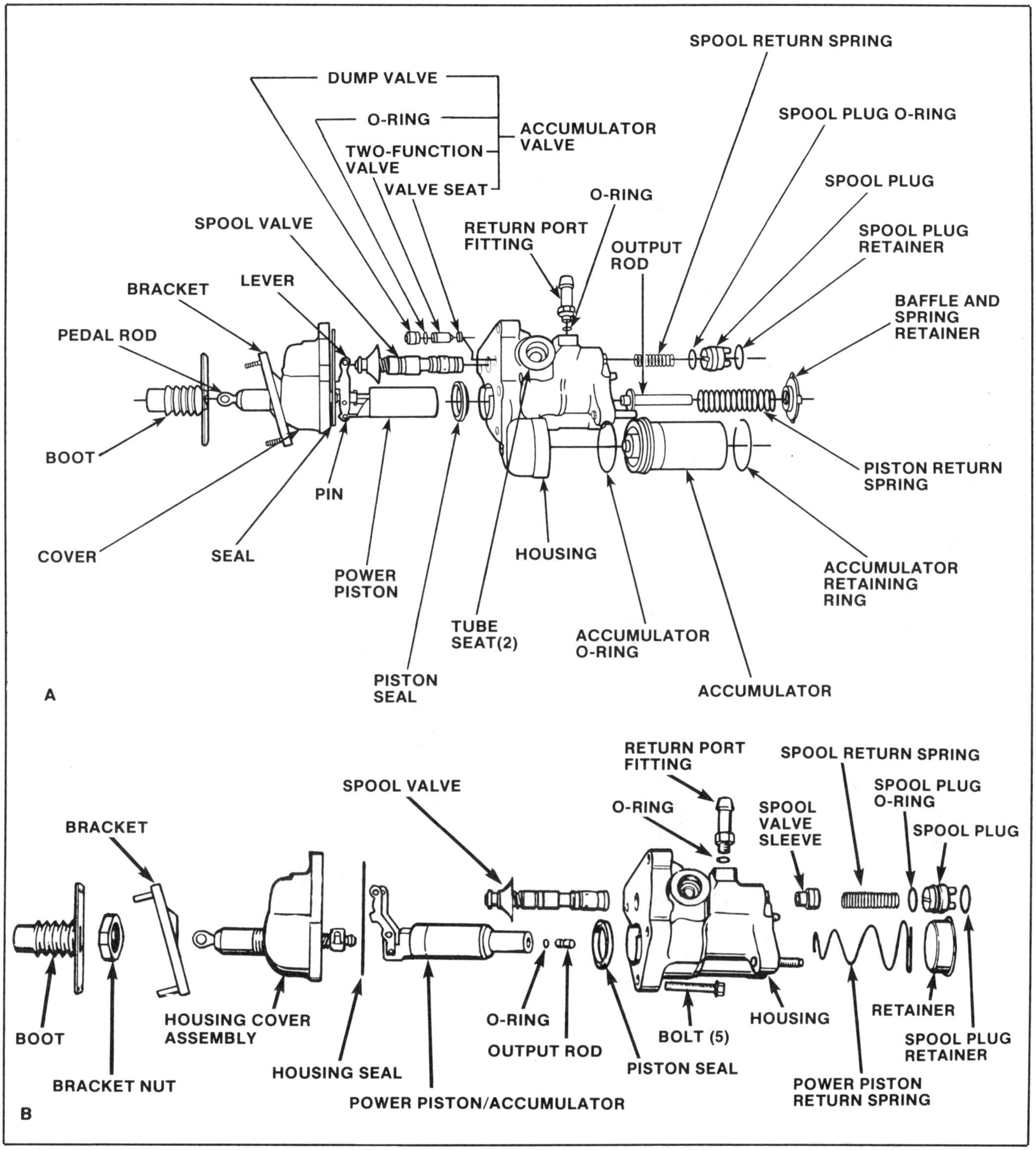

Figure 10-33 (A) Hydro-boost I components and (B) hydro-boost II components

3. Disconnect the hydraulic hoses from the booster ports. Plug all tubes and the booster ports to prevent fluid loss and system contamination.
4. Detach the pedal rod from the brake pedal.
5. Remove the nuts and bolts from the support bracket and remove the booster from the vehicle.

The basic components of both hydro-boost I and II are illustrated in Figure 10-33. Although Figure 10-35 shows the disassembly procedure for hydro-boost I, the basic method is the same for both units. The exception, of course, is the need to remove the external accumulator before starting to disassemble hydro-boost I.

REMOVAL

To remove the external accumulator, proceed as follows:

1. Mount the housing and accumulator in a vise and compress about 1/8 of an inch.
2. Insert a punch into the hole in the unit housing (Figure 10-34A) and dislodge the C-ring.
3. Using a small screwdriver, remove the C-ring (Figure 10-34B).
4. Slowly release the vise and remove the accumulator.

A

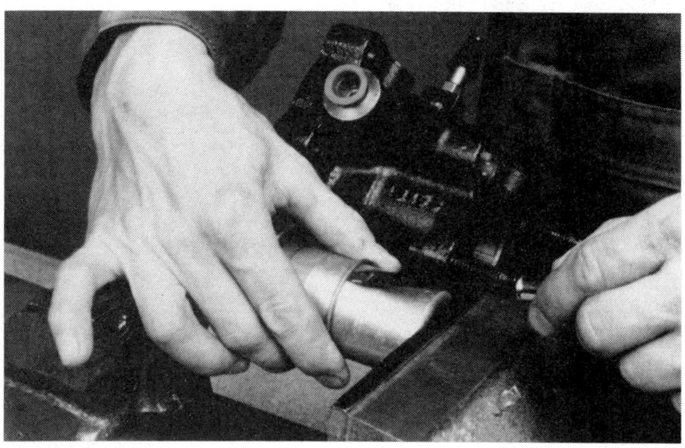

B

Figure 10-34 (A) Preparing to remove retainer and (B) capturing accumulator

WARNING: The accumulator is still charged if it does not compress 1/8 inch. Do not continue the removal procedure until all pressure is relieved.

DISASSEMBLY

To disassemble a typical hydro-boost—either I or II—proceed as follows:

1. Secure the hydro-boost in a vise.
2. Depress the spool valve plug; remove and discard the retainer. Some units have an access hole in the spool valve housing. Use a small punch to unseat the retainer.
3. Remove the spool valve plug and O-ring. Discard the O-ring.
4. Split the housing (Figure 10-35A) leaving the spool valve and power piston assembly on the housing (hydro-boost I only).
5. Remove the spool valve sleeve (if present) and the spool valve return spring (Figure 10-35B).
6. Remove the spring retainer from the power piston bore (if present) and discard. Extract the return spring from the power piston bore.
7. Unscrew the return line fitting and discard O-ring.
8. Remove the housing-to-cover bolts. Holding the unit over a drip pan, separate the cover from the housing and allow the unit to drain. Carefully separate the units and be sure the spool valve is not dropped (hydro-boost II only).
9. With hydro-boost I only, remove the pin holding power piston to the ratio lever (Figure 10-35C).
10. With hydro-boost I only, next remove the power piston (Figure 10-35D).
11. Remove the housing-to-cover seal (Figure 10-35E) and discard.
12. Detach the spool valve from the ratio lever (hydro-boost II only).
13. Remove the input rod and seals from the cover and discard seals. Most hydro-boosts have staked pedal rods, and the pedal rod should not be removed. To service seals or other parts of the cover assembly, purchase a new cover assembly that includes the required rod.
14. Remove the power piston/accumulator assembly (hydro-boost II only).

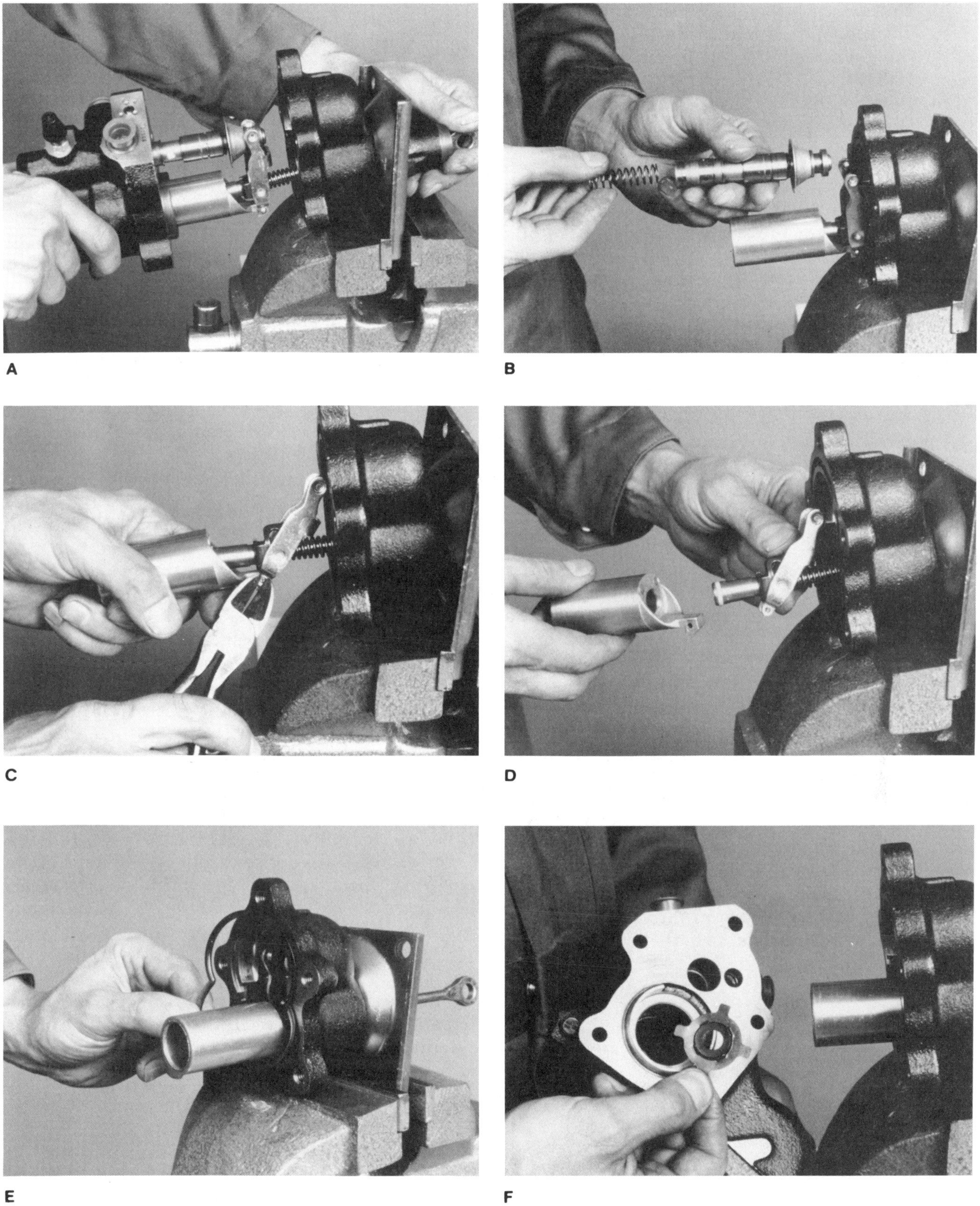

Figure 10-35 **(A) Disassembly of hydro-boost; (B) removal of hydro-boost spool valve; (C) removing pin; (D) removing power piston; (E) removing seal; and (F) removing pushrod baffle and retainer**

15. Extract the power piston seal from bore and discard. With a hydro-boost I unit, remove the pushrod and baffle retainer (Figure 10-35F).
16. Remove the tube seats from the hose ports. Discard seats.

Clean all parts in denatured alcohol or clean power steering fluid. (Do not use transmission fluid.) If the parts are left exposed for 8 hours or more, they must be rewashed at time of assembly.

Inspect spool valve and valve bore for scratches or signs of wear. If scratches can be detected by rubbing a finger over the surface, the complete hydro-boost unit must be replaced. Due to critical clearance tolerances, the spool valve or housing cannot be substituted.

REASSEMBLY

Upon reassembly, lubricate all parts and bores with clean power steering fluid and proceed as follows:

1. If the units used tube seats, set the new seats in position in the ports. Use a spare tube nut to reseat the new tube seat. Do not tighten the tube nut because the seat might be damaged. Remove the tube nut and inspect the seat for metal flakes or foreign material. Plug the ports until it is time to connect the lines. If the unit does not use tube seats, O-rings on the hose lines should be replaced.
2. Place the new O-ring on the return hose fitting and install it into the housing. Torque to 7 foot-pounds.
3. Using the special tool, install the new seal or seals onto the input rod (Figure 10-36). Insert the rod and seals into the cover.
4. Install the new power piston seal.
5. Install the new housing-to-cover seal.
6. Align the power piston into the bore, then attach the spool valve to the ratio lever and install the power piston/spool valve into place (Figure 10-37).
7. Install the housing-to-cover bolts; torque to 20 foot-pounds.
8. Install the spool valve sleeve (where used) and spool valve return spring.
9. Place the new O-ring on the spool plug and install. Depress the plug and insert the new retainer.
10. Insert the power piston return spring. Be sure the spring is seated on the ledge or bore.
11. Mount the hydro-boost on the vehicle.
12. Connect the hydraulic hoses; torque to 20 foot-pounds.
13. Install the master cylinder.

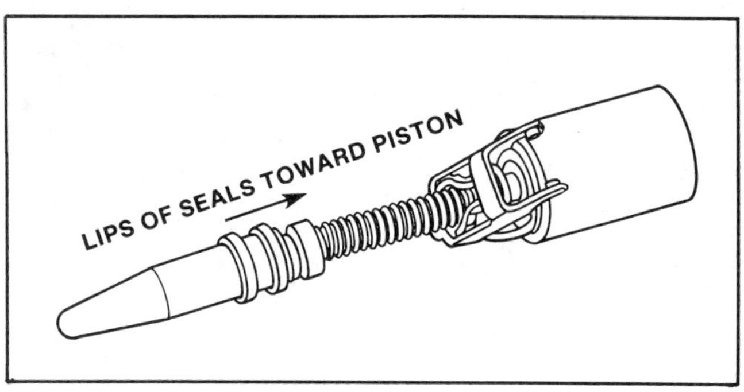

Figure 10-36 Seal installation—input rod

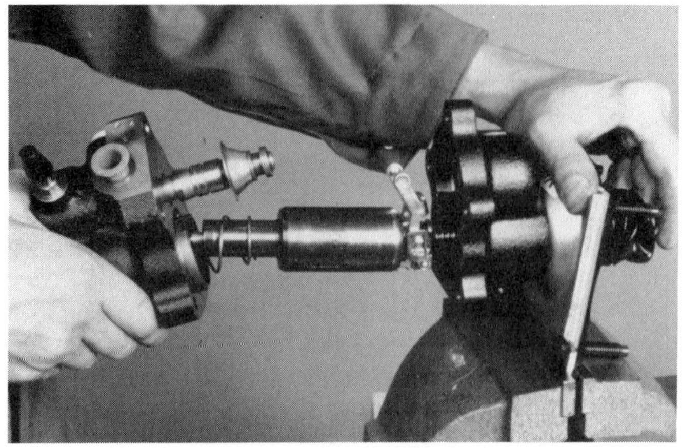

Figure 10-37 Reassembly of hydro-boost

The preceeding steps are basically for the reassembly of the hydro-boost II type unit. To reassemble a hydro-boost I unit (Figure 10-37), follow a reverse procedure of the disassembly steps given earlier in this chapter. This also includes the reassembly of the external accumulator to the outside of the housing.

Bleeding Procedure

The bleeding procedure for both types of hydro-boosts can be considered the same as is done in the following manner:

1. Fill the power steering pump reservoir.
2. Crank the engine for several revolutions. (Do not start the engine.)
3. Check the reservoir fluid level; add fluid, if necessary.
4. Start the engine; turn the steering wheel from stop to stop twice.
5. Stop the engine; depress the brake pedal four or five times to relieve accumulator pressure.

6. Check the fluid level; add fluid, if necessary.
7. Start the engine; turn the steering wheel from stop to stop. Stop the engine; check fluid.
8. If foaming occurs, wait for 1 hour. Check the fluid level; add fluid, if necessary.
9. Road test the vehicle.

SHOP TALK

The accidental mixing of power steering and brake fluid occurs mainly because master cylinders and hydro-boosts are similar in color and appear to be one unit. Do not put automatic transmission fluid in the power steering unit. The fluids are of slightly different viscosities, and automatic transmission fluid contains detergent additives that can damage both the hydro-boost and power steering unit. Dot 3 brake fluid is preferred with most hydro-boost systems.

POWERMASTER HYDRAULIC BOOSTER

In 1985, General Motors introduced its "Powermaster" power brake system. Used primarily on full-sized passenger cars, the system incorporates a self-contained hydraulic booster that is built directly onto the master cylinder. Instead of relying on the power steering pump for hydraulic pressure as is done in the hydro-boost system, the Powermaster has its own vane-type pump and electric motor to provide the hydraulic pressure required for booster operation.

POWERMASTER COMPONENTS

As illustrated in Figure 10–38, the major components of the Powermaster are the pump, accumulator, master cylinder, dual-pressure switch, reservoir, and booster.

Pump

The vane-type pump is driven by an electric motor which is mounted below the master cylinder. Brake fluid is drawn through a low-pressure hose from the reservoir to the pump; there it is pressurized and exits through a high-pressure hose to the accumulator. Pump motor operation is controlled through a dual-pressure switch and relay. Electrical protection for the motor and electrical circuit is provided by a 30 ampere fuse.

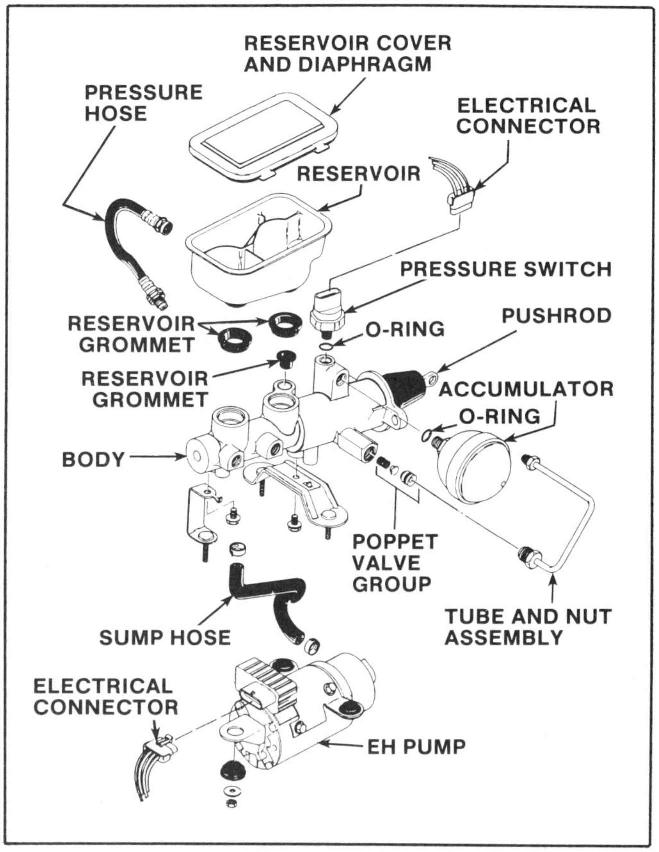

Figure 10-38 Powermaster components

Accumulator

The accumulator stores brake fluid under pressure at approximately 675 PSI (maximum). It has a heavy, flexible diaphragm that separates brake fluid on one side from high-pressure nitrogen gas on the other.

Master Cylinder

In the Powermaster system, the master cylinder is similar to a conventional dual or tandem master cylinder in both design and operation.

Dual-Pressure Switch

This switch controls the pump operation in conjunction with a relay attached to the pump motor. Three lead wires are attached to the switch; one leads to the pump motor relay, one leads to the brake warning light (dash mounted), and the third leads to ground. When accumulator pressure drops below 510 PSI, one contact set in the switch closes, completing the ground for the relay. At this point, the relay is able to supply battery voltage to the pump motor. With the pump operating, accumulator pressure increases to approximately 675 PSI.

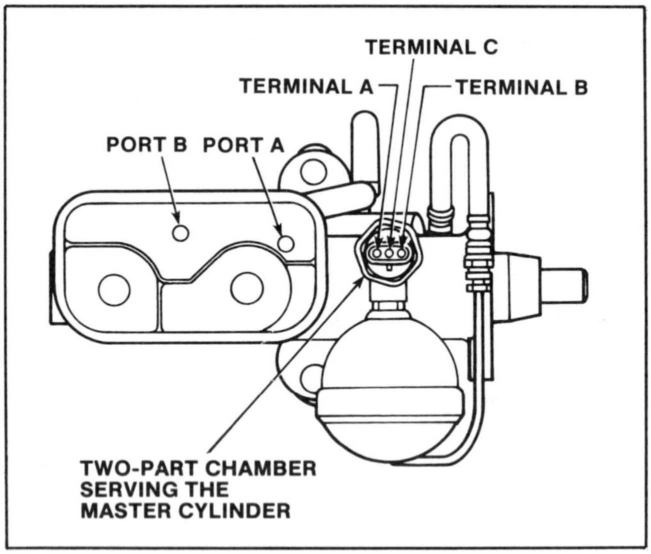

Figure 10-39 Pressure switch terminals and reservoir

At that point the contact set opens, breaking the ground to the relay and shutting off the pump. If accumulator pressure should drop below 400 PSI, the second contact closes. This illuminates the dash-mounted brake warning light.

Reservoir

The reservoir is the removable, plastic type internally divided into three chambers. Illustrated in Figure 10-39, two chambers (primary and secondary) supply the master cylinder with brake fluid as in a conventional master cylinder; both chambers are sealed to the atmosphere. The third chamber supplies the booster pump with brake fluid and is vented to the atmosphere because of the large amount of fluid movement in this portion of the reservoir. A stepped fluid level gauge is incorporated in one corner of the booster reservoir. Before checking fluid level, turn off the ignition switch and pump the brake pedal at least 10 times to bleed off accumulator pressure and return the fluid to the reservoir. Next, remove the master cylinder reservoir cover and add brake fluid (use DOT 3 only) to the first step on the fluid level gauge. To facilitate proper sealing (and venting) of the respective reservoir chambers, the reservoir cover is designed to install in the correct position only.

Booster

The booster assembly consists of a power piston, inner and outer control valves, and the reaction components. The power piston uses the high-pressure fluid from the accumulator against its large diameter to produce the force needed to apply (push) the master cylinder pistons. The reaction components fit between the power piston and master cylinder primary piston. These components consist of a reaction body group, reaction disc, and reaction piston. In operation, the reaction components provide pedal "feel." The apply (outer control) valve and discharge (inner control) valve regulate fluid movement through the booster according to brake pedal movement and brake pedal pressure (force). Both valves, illustrated in Figure 10-40, are located in the power piston.

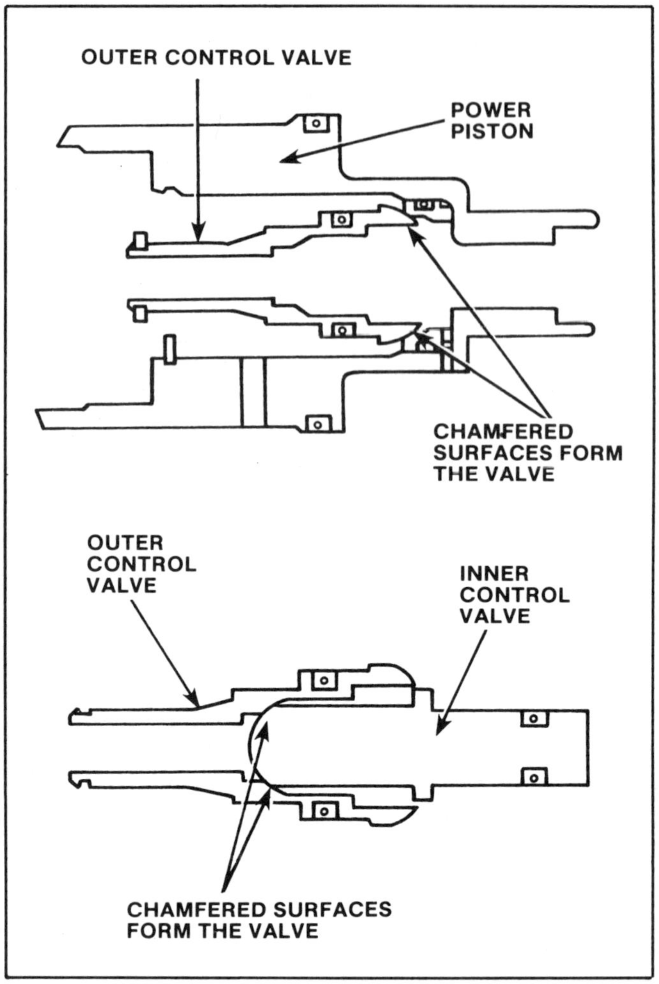

Figure 10-40 Apply valve in power piston.

POWERMASTER OPERATION

With the brakes released, the apply (outer control) valve is held in the closed position by spring pressure on the valve. This prevents high-pressure fluid from passing through the valve and pressurizing the booster cavity (Figure 10-41).

In the event of leakage past the closed apply valve when the brakes are released, the discharge (inner control) valve is held open by its spring. The result is that any seepage through the apply valve will not be able to build up pressure in the booster cavity. Instead, it simply passes through the open discharge valve and returns to the reservoir.

On brake application, the brake pedal pushrod force against the discharge valve pushes that valve closed against spring force, thereby preventing fluid return to the reservoir (Figure 10-42).

With the discharge valve closed, input (pedal rod) force next overcomes the spring force of the apply valve and moves that valve off its seat. With the apply valve open, high-pressure fluid from the accumulator is now able to pass through the apply valve and pressurize the booster cavity. This high-pressure fluid, in turn, pushes against the booster (power) piston, causing the piston to move to the left.

The booster piston, working through the reaction components, forces the master cylinder primary piston to move, resulting in brake application. Assist is accomplished by simply using the high-pressure fluid to bear (push) against the large diameter power piston to increase the physical force on the master cylinder pistons.

The amount of force the driver initially applies to the brake pedal will determine how far the apply valve opens, thereby controlling the rate of hydraulic (boost) pressure buildup in the booster cavity. A hard initial application of the brake pedal, as in a panic stop condition, will therefore result in a fast, high-pressure application of the brakes.

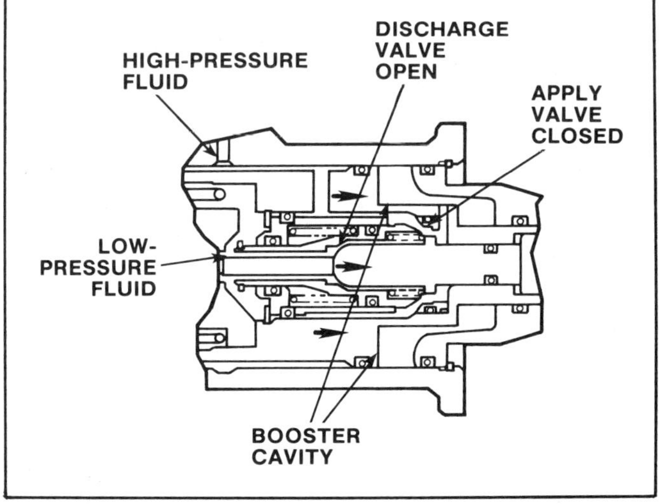

Figure 10-41 Fluid flow-through valves—released

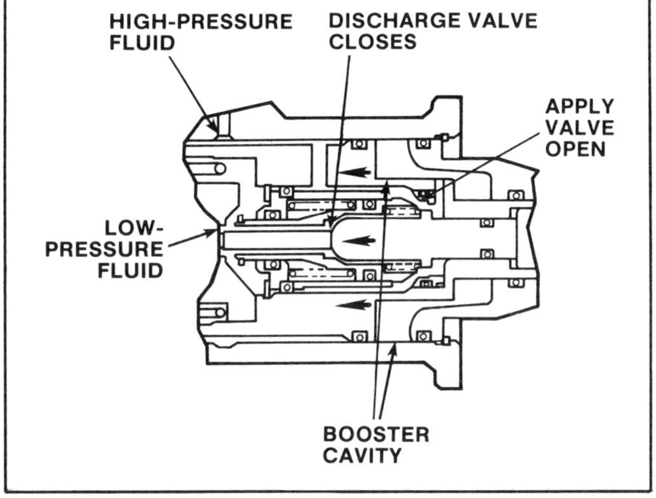

Figure 10-42 Fluid flow-through valves—applied

When the pedal pressure is reduced, the braking pressure in the master cylinder acts against the master cylinder pistons, exerting a higher force back against the brake pedal pushrod and inner control valve. This causes the discharge valve to open partially and thereby reduce the pressure in the booster cavity. This lower pressure reduces the assist force delivered by the booster piston, which reduces the pressure delivered to the master cylinder pistons in proportion to the reduction in force on the brake pedal. This reaction permits the driver to vary the pressure delivered to the brakes by merely increasing or decreasing the force applied to the brake pedal.

With pushrod force no longer present to hold the apply valve open, the apply valve spring pushes that valve closed, preventing additional pressure from entering the booster cavity (Figure 10-43).

Next, the discharge valve spring opens that valve, allowing booster cavity pressure to bleed off quickly from the booster cavity. With no pedal force nor booster pressure present, the master cylinder piston springs are now able to return the master cylinder pistons to their "at rest" position.

POWERMASTER SERVICE

Brake service on a Powermaster-equipped vehicle is similar to brake service on a conventional vehicle with the following minor exceptions:

1. Before starting servicing, depressurize the accumulator. Because of the high pressure in the accumulator, depressurizing of the unit is required to prevent possible injury.

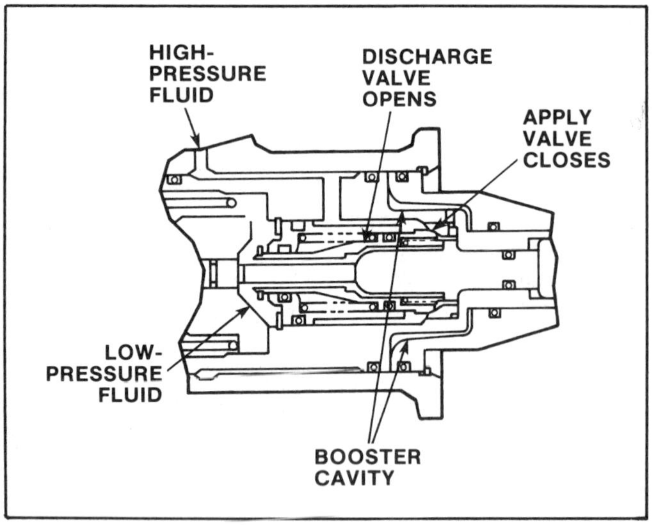

Figure 10-43 Fluid flow-through valves—discharged

This is done by pumping the brake pedal at least 10 times with the ignition switch off.

2. If internal service is performed on the master cylinder or if the master cylinder portion of the Powermaster is changed, bench bleeding is required. This is accomplished as on any master cylinder. The booster portion of the Powermaster will bleed when installed on the vehicle and the pump is actuated by turning on the ignition switch.
3. When adding fluid to the booster reservoir, follow the instructions mentioned earlier. Do not overfill and use DOT 3 fluid only.
4. In order to perform a functional check on a Powermaster unit, a special test gauge adapter or aftermarket equivalent must be employed. The pressure switch is removed, the adapter threaded pressure switch is removed, the adapter threaded into its hole, then the switch screwed into the adapter (reattach the switch electrical connector), then proceed following the instruction that comes with the test gauge.

Actually, all diagnosis and service of the Powermaster system can be accomplished by following instructions given in the appropriate service manuals available from General Motors.

OTHER POWER BOOSTER UNITS

There are several power booster units other than those already described. Most of these are used in vehicles such as trucks and heavy-duty vans. The most common of these boosters units are the hydrovac, air-pak, and master vac.

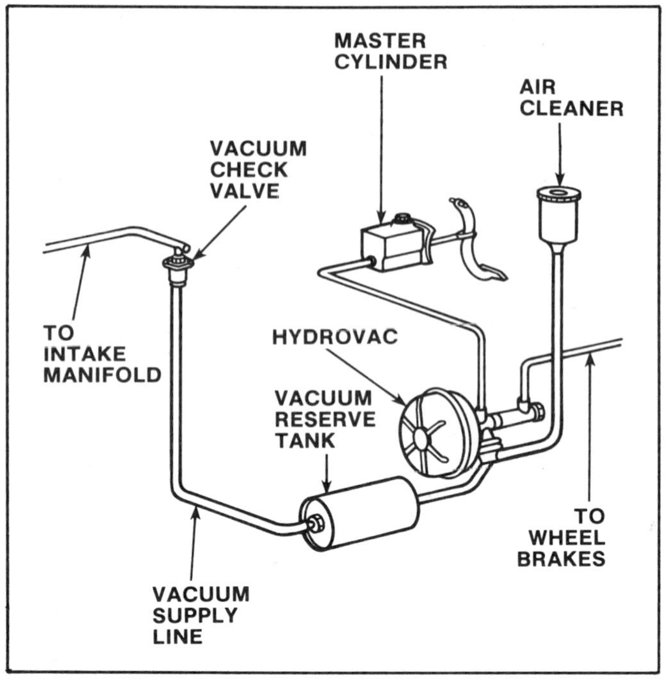

Figure 10-44 Typical Bendix hydrovac installation

HYDROVAC

The vacuum and the hydraulic sections of this power brake system are combined into a single package, called the "Hydrovac." Bendix developed the system in 1948, and the name is a registered Bendix trademark. It is designed for the full weight range of medium and medium-heavy vehicles. It is compact, with no external moving parts, and can be mounted in any convenient location on the vehicle chassis. Bendix Hydrovac units are equipped with a new "self-mount," factory-welded bracket as an integral unit for each of installation. A typical Hydrovac installation is shown in Figure 10-44 and a disassembled view of the Bendix Hydrovac is shown in Figure 10-45.

AIR-PAK

Later, Bendix combined the air and hydraulic sections into "Air-Pak," another registered Bendix trademark. This is used on medium and medium-heavy vehicles with air supplies and a hydraulic braking system displacement requirement of 2.3 to 6.5 cubic inches. It is a compact and durable power braking system.

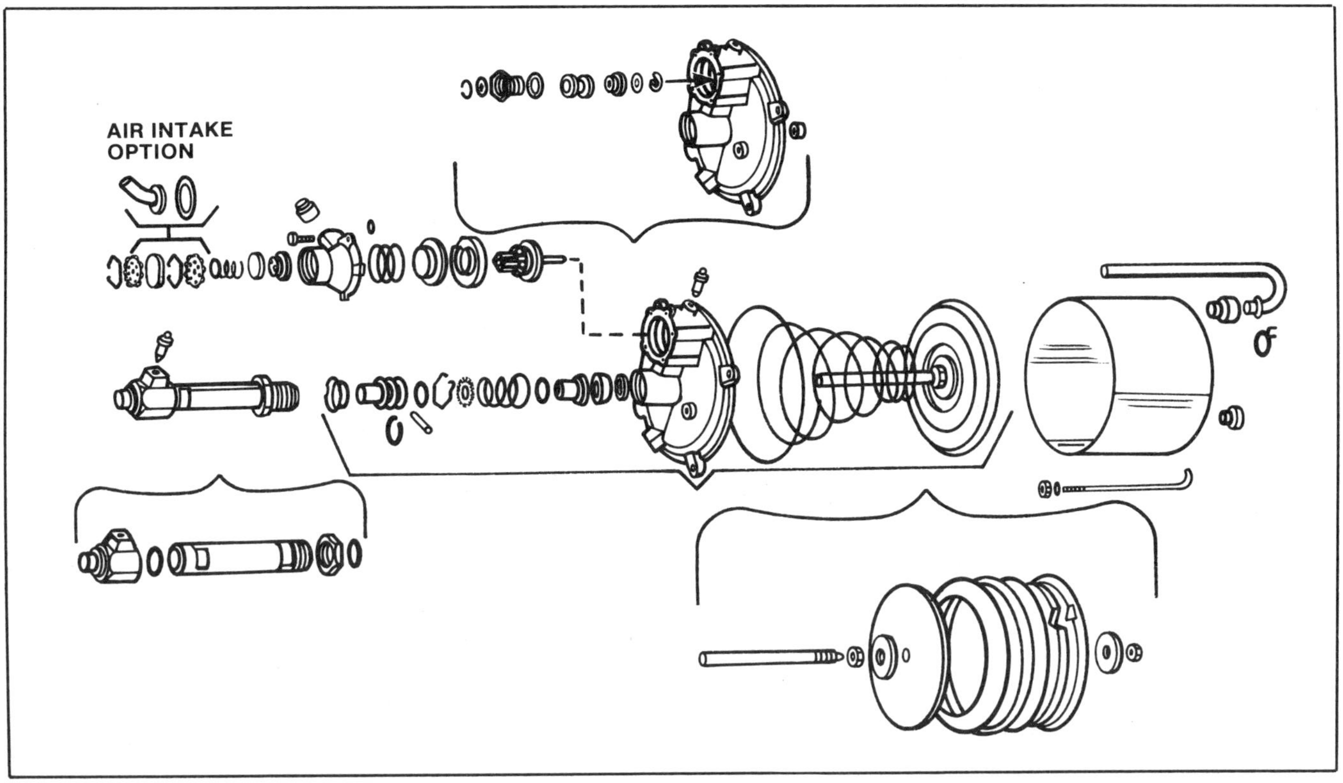

Figure 10-45 Bendix hydrovac

MASTER VAC

Master Vac was developed for light trucks with dual hydraulic systems, and hydraulic fluid displacement requirements of 1.5 cubic inch or less. Mounted on the fire wall, it is linked directly to the conventional brake pedal. Master Vac and a tandem master cylinder provide the added safety of a dual hydraulic system. Bendix Master Vac can be either a single diaphragm or a tandem diaphragm type. The tandem diaphragm Master Vac is used on many domestic automobiles where limited mounting space prevents the use of a large diameter single diaphragm unit, but relatively high hydraulic pressures are required. It provides the same output as two single diaphragm power units with the same size diaphragm. A typical unit is shown previously in Figure 6-4.

REVIEW QUESTIONS

Review the information in this chapter by answering the following questions.

1. Which type of power assist mechanism uses vacuum pressure developed by an external pump, usually the power steering pump?
   a. hydraulic pressure assist mechanism
   b. hydro-boost
   c. both a and b
   d. vacuum assist diaphragm assembly
2. The fundamental difference between the different types of vacuum assist systems lies in
   a. how the diaphragm within the power unit is suspended when the brakes are not applied.
   b. the shape of the flexible diaphragm.
   c. how the brakes are applied.
   d. the location of the bellows.
3. As a general rule, brakes should be adjusted or relined if the pedal is closer than
   a. 1/2 inch from the floor with the brakes applied.
   b. 1 inch.
   c. 2 inches.
   d. 4 inches.
4. Which of the following leads to weak power unit operation?
   a. collapsed or leaking vacuum lines
   b. punctured diaphragms or bellows
   c. leaky piston seals
   d. all of the above

5. The power brakes in a vehicle are grabbing. Mechanic A immediately investigates the power unit. Mechanic B first checks for other causes, such as greasy linings or scored drums. Who is right?
   a. Mechanic A
   b. Mechanic B
   c. both A and B
   d. neither A nor B
6. Which of the following is a type of vacuum booster commonly used in modern vehicles?
   a. lever booster
   b. disc booster
   c. both a and b
   d. neither a nor b
7. What is the minimum gauge measurement of vacuum pressure necessary to provide proper power brake operation?
   a. 14 inches of water
   b. 4 inches of water
   c. 14 inches of mercury
   d. 4 inches of mercury
8. When the brake pedal in a vehicle is moved, a slight breathing sound can be heard. Mechanic A says the sound is normal. Mechanic B says the sound indicates a vacuum leak. Who is right?
   a. Mechanic A
   b. Mechanic B
   c. both A and B
   d. neither A nor B
9. Which of the following should not be disassembled when servicing a power brake system with vacuum assist assembly?
   a. master cylinder power piston and pushrod
   b. single diaphragm booster
   c. hydro-boost
   d. tandem diaphragm booster
10. Hydro-boost differs from vacuum brake boosters in which of the following ways?
    a. The power source differs.
    b. Hydro-boost is also a part of the power steering system.
    c. Both a and b
    d. There is no difference.
11. A leak in a line is more critical with power brakes than with the regular wheel brakes because
    a. the vehicle will have no braking ability.
    b. the power brakes are linked with the power steering system.
    c. leaks in a power brake system are more difficult to repair.
    d. power brakes have a higher line pressure.
12. A spongy brake pedal in a power brake system is a sign of a problem. Mechanic A checks the power unit itself. Mechanic B checks for air in the system. Who is right?
    a. Mechanic A
    b. Mechanic B
    c. both A and B
    d. neither A nor B
13. Why should extreme caution be used when disassembling power assist mechanisms?
    a. because internal spring tension is suddenly released
    b. to avoid contact with brake fluid
    c. to avoid contaminating the system
    d. because several parts are sharp
14. Which of the following is *not* a method of mounting vacuum power brakes?
    a. Use mounting brackets between the power section and the firewall.
    b. Attach the power section directly to the firewall.
    c. Attach the power section under the seat.
    d. All of the above
15. A master cylinder reservoir has fallen noticeably, but there is no sign of an external brake fluid leak. Both Mechanic A and Mechanic B remove the vacuum tube or hose that runs between the intake manifold and the booster unit and find evidence of brake fluid. Mechanic A decides the master cylinder secondary seal is leaking. Mechanic B decides the vacuum hose is leaking. Who is right?
    a. Mechanic A
    b. Mechanic B
    c. both A and B
    d. neither A nor B
16. Which of the following should never be done when working on power brakes?
    a. Soak all metal parts in a suitable solvent.
    b. Use only alcohol on rubber parts.
    c. Use an air hose to blow dirt and cleaning fluid from the recesses and internal passages.
    d. Reuse old rubber parts
17. Which of the following is a function of the power piston or vacuum control valve suspended in the diaphragm?
    a. controlling the vacuum and atmospheric valves as the booster pushrod moves
    b. transmitting additional force to the pushrod of the master cylinder
    c. both a and b
    d. neither a nor b

CHAPTER 11

ANTI-LOCK BRAKE SYSTEMS—OPERATION AND SERVICING

OBJECTIVES

Upon completing this chapter, you should be able to

- Identify the advantages of the principal anti-lock brake systems.
- Recognize and explain the functions of the major components of typical anti-lock systems.
- Describe a control cycle in the operation of the anti-lock brake.
- Explain the procedures for brake system bleeding.
- Explain the three basic steps in diagnosing anti-lock brake system malfunctions.
- Service and repair anti-skid brake systems.

Although anti-skid control systems have been optional on certain luxury cars beginning with the 1969 models, only recently have the systems been in demand. And for good reason: With an anti-lock or anti-skid brake system (ABS), the vehicle's wheels are prevented from locking during panic braking stops.

Modern anti-skid brake systems can be thought of as electronic/hydraulic "pumping" of the brakes for straightline stopping under panic conditions (Figure 11-1). Good drivers have always pumped the brake pedal during panic stops to avoid wheel lockup and loss of steering control. Anti-skid brake systems simply get the pumping job done much faster and in a much more precise manner than the fastest human foot. Not only that, during a panic stop the average driver is only thinking of stopping the car, not pumping the brake pedal. Conventional brake systems cannot prevent unintentional wheel lockup. It must be remembered that a tire on the verge of slipping produces more friction with respect to the road

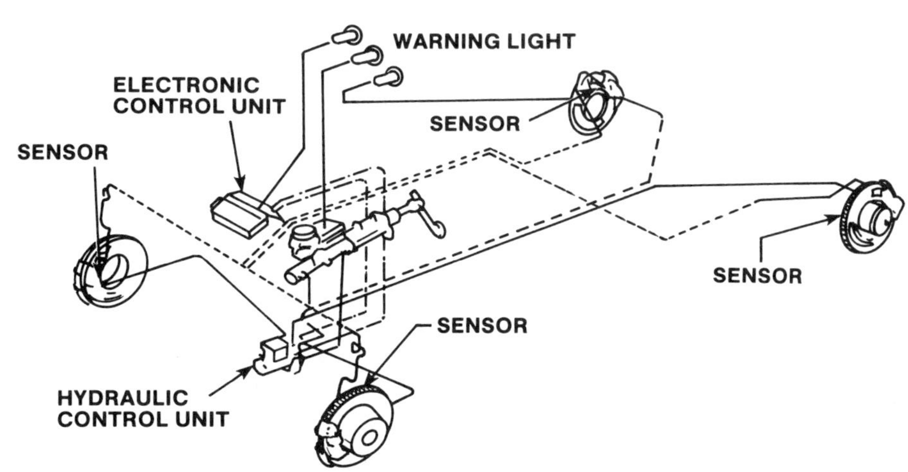

Figure 11-1 Basic elements of ABS four-channel or four-sensor system

than one which is locked and skidding. Once a tire loses its grip, friction is reduced and the vehicle takes longer to stop.

The only exception to this rule is when a tire is on loose snow. A locked tire allows a small wedge of snow to build up ahead of it, which allows it to stop in a shorter distance than a rolling tire (that is why some car manufacturers offer an on/off switch on their ABS-equipped vehicles for deactivating the system when driving on snow).

The ABS system allows the shortest possible braking distance. When wheels lock-up (skid), their frictional forces are lost to the skid, whereas a rolling wheel can transfer higher frictional forces, thus obtaining a shorter stopping distance.

Another condition to consider, which has nothing to do with stopping distances, is a skidding tire that has no directional stabilty. As long as a tire does not slip, it will only go in the direction it turns. But once it skids, it has little or no directional stability. One of the big advantages of ABS, therefore, is the ability to keep control of the vehicle under all conditions. The driver can slam on the brakes and still steer the car—and he/she does not have to worry about the rear end braking loose and spinning around. The ABS maintains vehicle stability during such stops and all other stops over approximately 8 MPH. In other words, it keeps the car from swerving one way or the other.

Obviously, the driver must have common sense. If he/she exceeds cornering speeds, "rides bumpers," or otherwise drives foolishly, no brake system in the world will save the vehicle, the driver's neck, or the lives and property of others. To summarize, the anti-skid brake system has the following advantages:

- Driving stability is maintained.
- The vehicle can be steered.
- Trailers do not jackknife.
- The braking distance is reduced to an absolute minimum.
- Accidents are avoided.

ANTI-LOCK BRAKE DESIGNS

The ABS concept has been around for quite some time. The anti-lock brakes were developed for the aircraft industry to help jet aircrafts to stop straight and quickly on slick runways. The first known use of ABS was on B-47 bombers way back in 1947. Today, you'll find it on everything from 747 jumbo jets to F-15 fighter planes.

Currently there are three basic passenger vehicle designs of anti-lock brake systems: Robert Bosch and Alfred Teves of Germany and Lucas-Girling of England. Originally, the Teves system was called **T**raction **C**ontrol **S**ystem (TCS), while the Bosch system was known as **A**cceleration **S**kid **C**ontrol (ASC). Today, both electronic-controlled arrangements are considered anti-lock brake systems (ABS).

Mercedes-Benz and BMW have both used the Bosch/Daimier system in Europe since 1978. In 1985 Mercedes, BMW, and Audi introduced the Bosch system in America. At about the same time, the domestic manufacturers also recognized the value of ABS, and Ford was the first US manufacturer to offer ABS, choosing a system made by Teves. Actually, this was Ford's second venture the ABS. Ford introduced a two-wheel, vacuum-actuated anti-lock brake system way back in 1968 to improve the car's straight line stopping ability. The system worked, but it cycled so slowly that it actually increased stopping distance on dry pavement.

The ABS system designed by Lucas-Girling is a lower cost mechanical system that may make anti-lock brakes affordable for all cars, not just the high-priced luxury sedans and sports models. Lucas-Girling originally developed their anti-lock brake system for motorcycles. Ford noticed the system and set up a joint venture with Lucas-Girling to develop a low-cost stop control system (SCS) for front-wheel-drive applications. Ford has since adopted the system as an option for its European Escort and Orion.

The Bosch, Teves, and Lucas-Girling anti-lock braking systems control tire slip by monitoring the relative deceleration rates of the wheels during braking. If one wheel starts to slow at a faster rate than the others, it is an indication that the wheel is starting to slip and is in danger of locking up. The system responds by momentarily reducing brake pressure to the affected wheel. The hydraulic control valve pulsates the pressure, which can be felt with the foot in the brake pedal during hard braking. It is the same as pumping the brakes, except that the system does it for each individual wheel or pair of wheels. The Bosch system can produce up to 10 pulsations per second, the Teves system up to 15, and the Lucas-Girling about three per second. Pulsating the braking pressure reduces the strain on the affected wheel and allows it to regain its grip. Then, as the rate of deceleration for that wheel comes back in line with the rest, normal braking pressure is resumed.

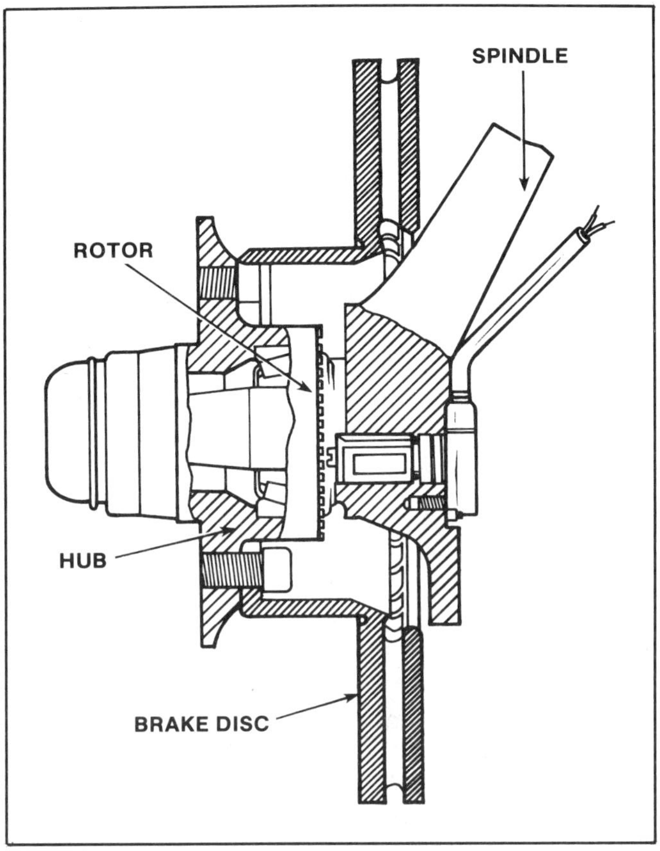

Figure 11-2 The toothed ring on the back of the hub generates pulses in a magnetic pickup.

The Bosch and Teves systems keep track of wheel deceleration rates with magnetic sensors (Figure 11-2). One some applications (Mercedes, for example), three sensors are used: one for each front wheel as well as one for the rear axle differential to monitor the rear brakes as a pair. Using a single sensor for the rear brakes simplifies the control system by dividing it into three circuits instead of four. On other applications (General Motors and Ford, for example), each wheel gets its own sensor, and the system is divided into four control circuits.

The Bosch and Teves speed sensors use a magnetic pickup and a toothed sensor ring. As the wheel turns, the teeth on the ring induce a voltage signal in the pickup. The computer counts these pulses and uses them to calculate the relative speed of each wheel.

If the computer detects a difference in the deceleration rate of one or more wheels during braking, or if the overall rate of deceleration is too fast and exceeds the limits programmed into the computer, the computer pulsates brake pressure in the affected circuit or circuits until its sensors tell it things are back under control. The modulator valve body (which is part of the master cylinder in the Teves systems but separate in the Bosch) has solenoid valves for each brake circuit while Teves uses a pair. By opening or closing the modulator valves, brake pressure within any given circuit can be increased, decreased, or held constant.

When the ABS function comes into play (which is only when the computer determines it is necessary to prevent wheel slip), the solenod valves can be actuated in a few thousandths of a second. Thus, the systems can respond extremely fast to changing conditions that require rapid pressure changes.

Teves, Bosch, and Lucas-Girling are the three passenger car ABS systems that brake technicians will see come into the shop. Other anti-lock brake systems will come into the marketplace in the future, but these are still some time away. Regardless of the system, here are some general facts that the brake technician should know about ABS:

- ABS does not affect normal servicing of brake lining, calipers, wheel cylinders, hoses, or parking brake cables.
- Master cylinders and boosters may change with the addition of ABS.
- Wheel speed sensors should not be disturbed or removed from the vehicle when servicing ABS brakes. The air gap between the speed sensor and toothed signal wheel is very critical, as is the routing of the speed sensor cables.
- ABS systems may control rear brakes individually or together. Front brakes are always controlled individually.
- Rear brakes can either be disc or drum with ABS.
- Hydraulic systems may be "split" in many ways, for example, front/rear, diagonal, etc.
- At this time, the ABS components are not serviceable in the field. They may be replaced as a complete unit.
- Bleeding air from the hydraulic system may be more difficult.
- If a drum or rotor includes the toothed signal wheel, replacement parts must also include the toothed signal wheel.
- If the ABS actuation system has a malfunction, it will shut itself off and signal the driver by means of a light on the instrument panel; however, power-boosted brakes will function normally. In other words, ABS is a "fail safe"

system. The anti-lock malfunction should be diagnosed and corrected by the vehicle dealer or authorized ABS service representative.

- The brake pedal force required to actuate the anti-lock function depends on the road surface condition.
- Some pedal pulsations will be felt when in the ABS mode, but not during normal braking.

TEVES ANTI-SKID BRAKE SYSTEMS

Anti-skid brake systems, developed by the Alfred Teves Corporation, are types found on most American-built production vehicles. Because of this, it is the system that will be detailed in this chapter. Of course, understanding the basic Teves system, the brake technician is well on his/her way to being schooled in modern anti-skid brake systems.

The Teves brake system (Figure 11-3) is divided into three system circuits with individual front brake circuits and a combined rear brake circuit. Major component parts of a typical system include the following:

- *Master Cylinder and Hydraulic Booster*—The master cylinder and brake booster are arranged in the conventional positions, as shown in Figure 11-4, with the booster behind the master cylinder. The booster control valve is above the master cylinder and is operated by a lever mechanism connected to the brake pedal.
- *Electric Pump and Accumulator*—The anti-lock brake system has a dedicated electric pump, shown in Figure 11-5 (the pump does not supply the power steering system). This pump runs at frequent intervals for short periods. It charges the hydraulic accumulator that supplies the service brake system. When pressure in the accumulator drops below 2,230 PSI, the pump turns on; it remains on until pressure in the accumulator reaches 2,610 PSI, at which point it shuts off.

  The accumulator is a gas-filled pressure chamber that is part of the pump and motor assembly. The electric motor, pump, and accumulator assembly is shock mounted to the master cylinder/booster assembly.
- *Solenoid Valve Body Assembly*—The valve body contains three pairs of solenoid valves (Figure 11-6), one pair for each front wheel and

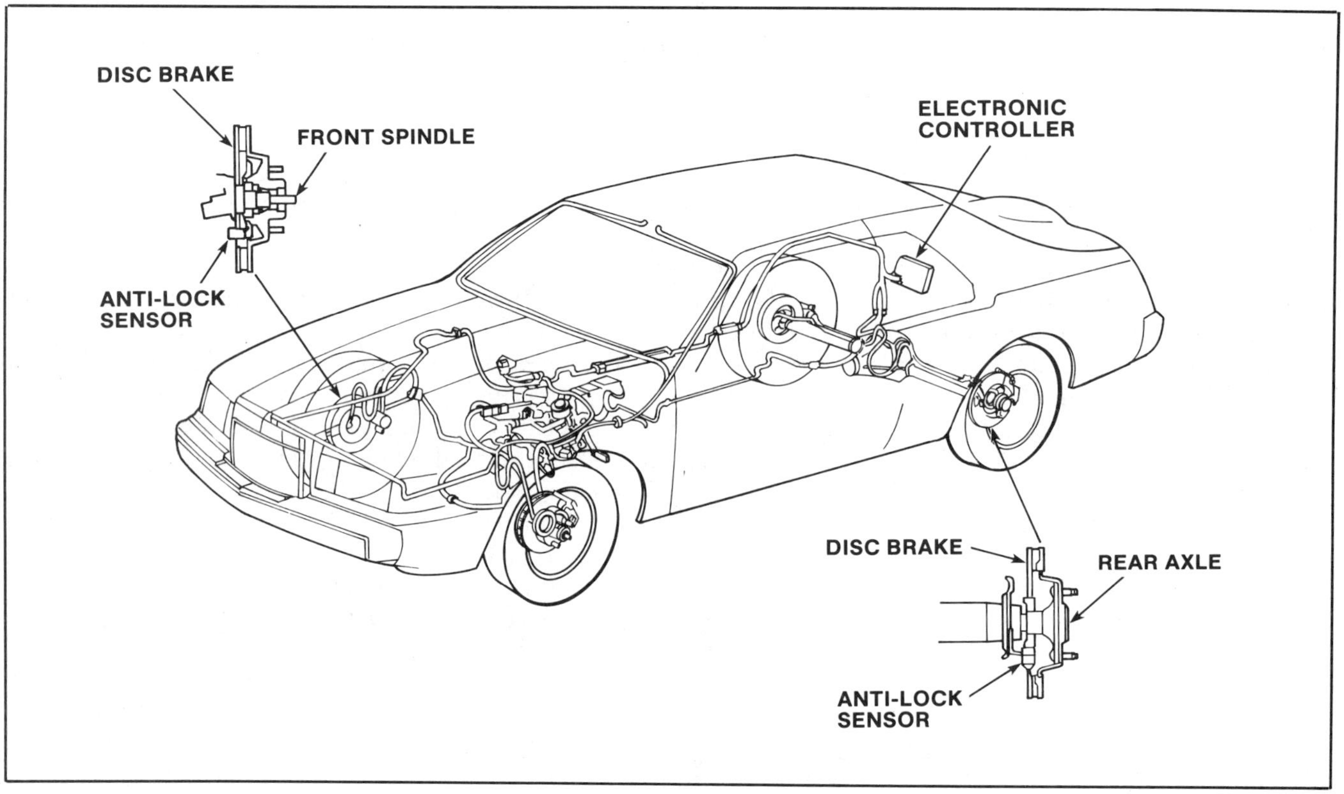

Figure 11-3 The Teves anti-lock brake system

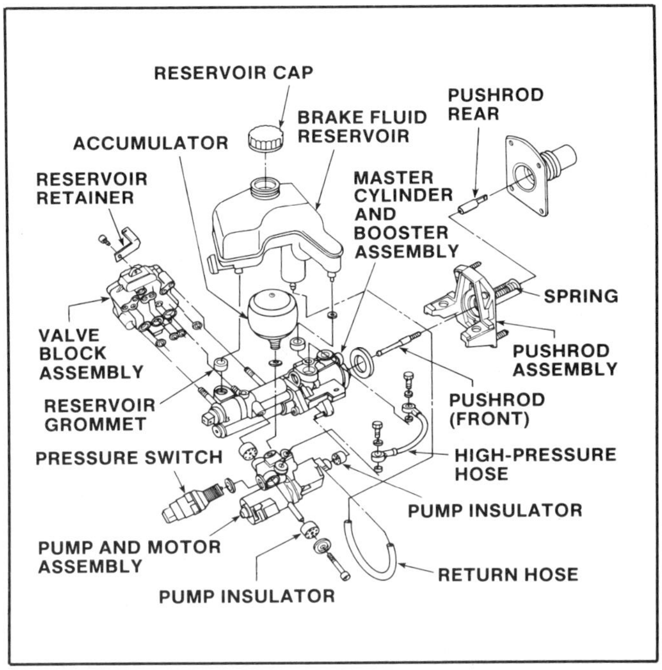

Figure 11-4 Master brake cylinder in Teves system uses an electric pump for power boost.

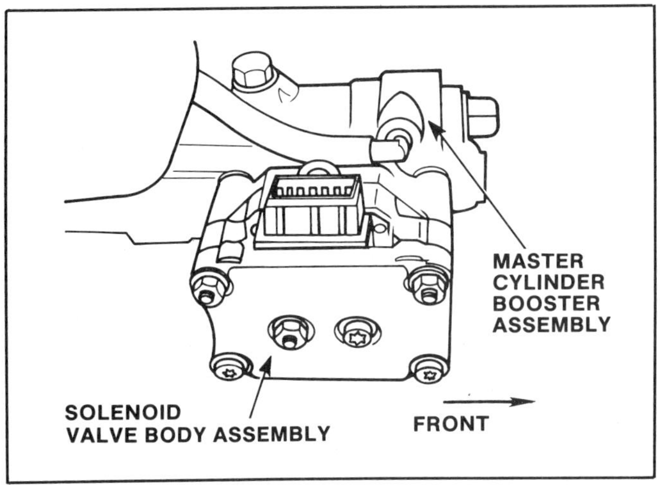

Figure 11-6 Solenoid valve body assembly

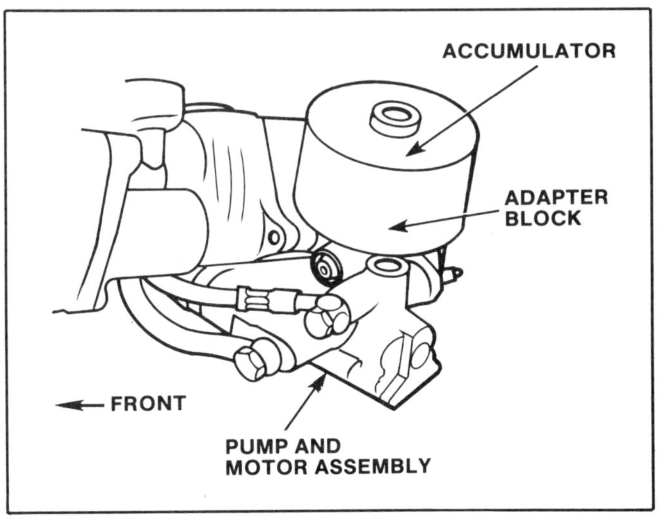

Figure 11-5 Dedicated electric pump

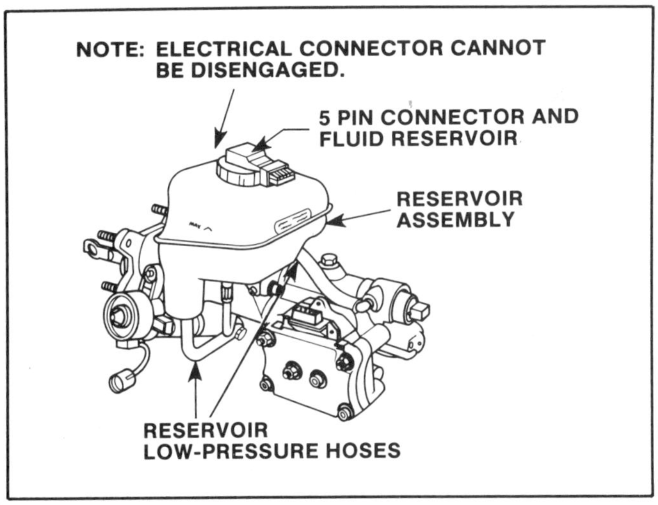

Figure 11-7 Reservoir with fluid level warning switches

one pair for both the rear wheels combined. The paired solenoid valves are inlet/outlet valves with the inlet valve normally open and the outlet valve normally closed. The valve body is bolted to the inboard side of the master cylinder/booster assembly.

- *Reservoir with Fluid Level Warning Switches*—The reservoir assembly, as shown in Figure 11-7, is a translucent plastic container with two main chambers. Integral fluid level switches are part of the reservoir cap assembly with one electrical connector pointing forward for wire harness connections. Two low-pressure hoses lead from the reservoir. One hose attaches to the hydraulic pump assembly and the other to the master cylinder housing.

  The reservoir is mounted to the hydraulic unit with a screw and bracket and a push-in tube outlet that seats in a grommet located in the brake booster housing (Figure 11-8).
- *Wheel Sensors*—There are four variable reluctance electronic sensor assemblies, each with a 104-tooth ring in the anti-lock system. Each sensor is connected to the electronic controller

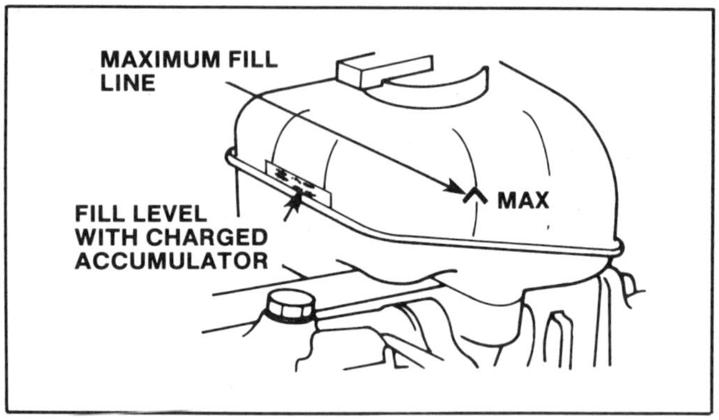

Figure 11-8 The reservoir mounts to the hydraulic unit with a bracket and screw and push-in tube outlet.

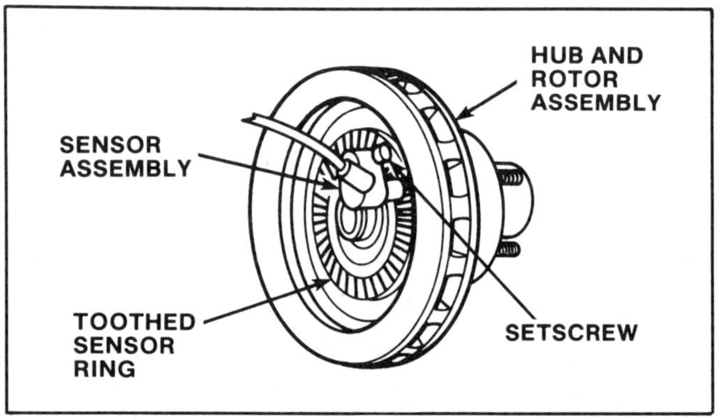

Figure 11-9 Front sensors and sensor rings

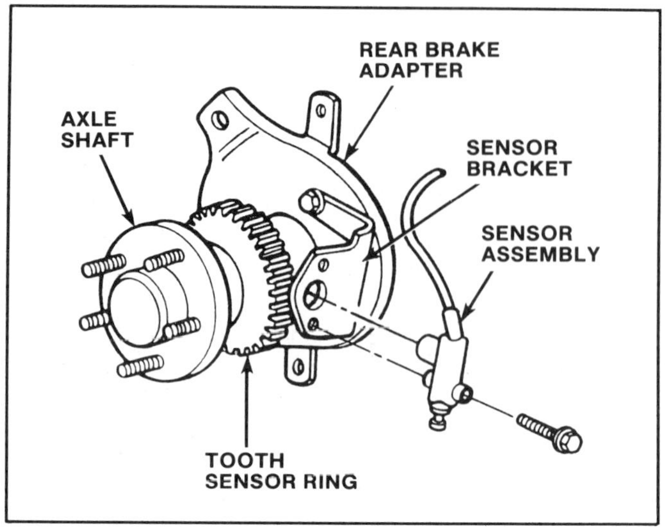

Figure 11-10 Rear sensors and sensor rings

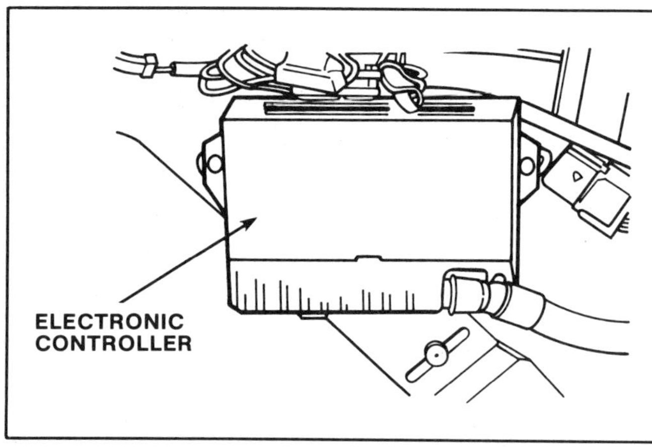

Figure 11-11 Electronic controller

through a wiring harness. The front sensors are bolted to brackets that are bolted to the front spindles (Figure 11-9). The front toothed sensor rings are pressed onto the inside of the front rotors. Each sensor has an adjustable air gap between the sensor head and tooth sensoring ring. The front sensor air gap is 0.043 inch and is set using a paper spacer.

The rear sensors are bolted to brackets that are bolted to the rear disc brake axle adapters (Figure 11-10). The toothed rear sensor rings are pressed on the axle shafts, inboard of the axle shaft flange. The rear sensor air gap is 0.026 inch and can be set using either a paper spacer or a feeler gauge.

- *Electronic Controller*—The electronic controller (Figure 11-11) is a self-contained, nonrepairable unit consisting of two microprocessors and the necessary circuitry for their operation. These microprocessors are programmed identically and operate on the principle of two-channel redundant data processing and plausibility criteria monitoring. The controller monitors system operation during normal driving as well as during anti-lock braking. Under normal driving conditions, the microprocessors produce, to the solenoid valves, short test pulses that check the electrical system without any mechanical reaction. Wheel lock conditions trigger signals from the controller that open and close the appropriate solenoid valves and open the main valve that supplies pressure to a reaction sleeve. This results in moderate pulsations in the brake pedal accompanied by a change in pedal height. During normal braking, the brake pedal feel will be identical to a standard brake system.
- *Anti-Lock Warning Light*—Vehicles equipped with the anti-lock system will have an amber

anti-lock warning light in the instrument panel. The ANTI-LOCK warning light will illuminate if a malfunction in the system is detected by the electronic controller. In case of an electronic malfunction, the controller will turn on the ANTI-LOCK warning light and disable anti-lock braking function. If the ANTI-LOCK and BRAKE warning lights come on at one time, there might be something wrong with the hydraulic brake system.

The ANTI-LOCK light will come on when the engine is first started and will usually stay on for approximately 3 seconds after the ignition switch is returned to the RUN position. In some cases, the ANTI-LOCK light will stay on for as long as 30 seconds. This might be normal operation. If the light stays on longer than 30 seconds after the engine is started or comes on during vehicle operation, the brake system should be inspected for a malfunction.

On some vehicles, the pump circuit includes a timer flasher module, located in the instrument panel behind the driver's side close-out panel, which monitors pump running time. If the pump runs for more than 3 minutes, the timer flasher module will cause the BRAKE warning light to flash.

- *Brake Warning Light*—Most anti-lock brake systems use a three circuit design so that some braking capacity is still available if hydraulic pressure is lost in one circuit. A BRAKE warning light is located at the left-hand side of the instrument cluster and is designed to alert the driver of conditions that would result in unsatisfactory braking ability.

The BRAKE warning light should turn on during engine starting and should remain lit whenever the parking brake is not fully released. If the BRAKE warning light stays on any longer than 30 seconds after starting the engine or comes on during the vehicle operation, there could be a malfunction in the brake hydraulic system.

ANTI-LOCK BRAKE OPERATION

When the brakes of a system regulated by ABS are applied in an emergency, the braking pressure is automatically adjusted to prevent wheel lockup while the brake pedal force remains constant. The following basic processes are involved: Keep pressure constant, reduce pressure, or increase pressure. The Teves ABS hydraulic schematic is shown in Figure 11-12.

If a wheel speed sensor signals a high rate of wheel deceleration to the electronic controller, that is, the wheels are likely to lock, the braking pressure at the wheel concerned is initially kept constant and not increased (Figure 11-13A). If the wheel still continues to decelerate, the pressure in the wheel brake cylinder is reduced so that this wheel is not braked as much (Figure 11-13B).

The wheel accelerates again as a result of the reduced braking pressure. Once a specific limit has been reached, the controller registers the fact that the wheel is not being sufficiently braked. The wheel is decelerated again by increasing the pressure, which was initially reduced (Figure 11-13C). The control cycle starts again. Approximately 4 to 10 control cycles are executed every second depending upon the road conditions.

In operation, when the brakes are released, the piston in the master cylinder is retracted. The booster chamber is vented to the reservoir, and the fluid in the chamber is at the same low pressure as the reservoir, as shown in Figure 11-14.

When the brakes are applied, under normal conditions, the brake pedal actuates a pushrod, as shown in Figure 11-15. This operates a scissors lever, which moves a spool valve. When the spool valve moves, it closes the port from the booster chamber to the reservoir and partially opens the port from the accumulator in proportion to the pressure on the brake pedal. This allows hydraulic fluid under pressure from the accumulator to enter the booster chamber. As hydraulic pressure enters, it pushes the booster piston forward, providing

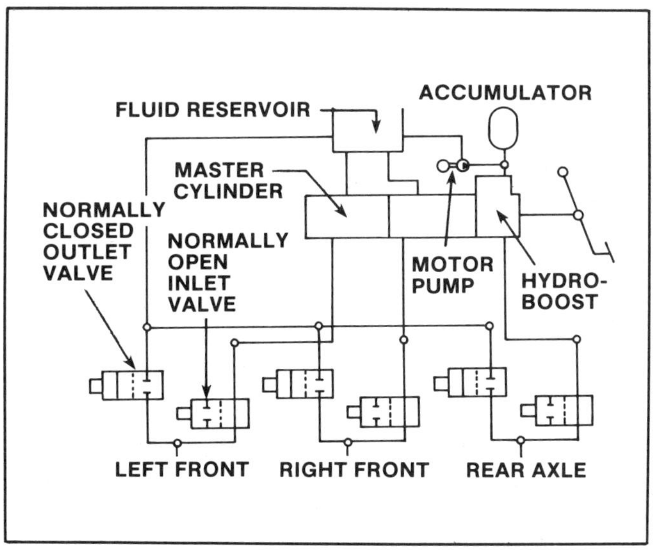

Figure 11-12 Teves ABS system hydraulic schematic

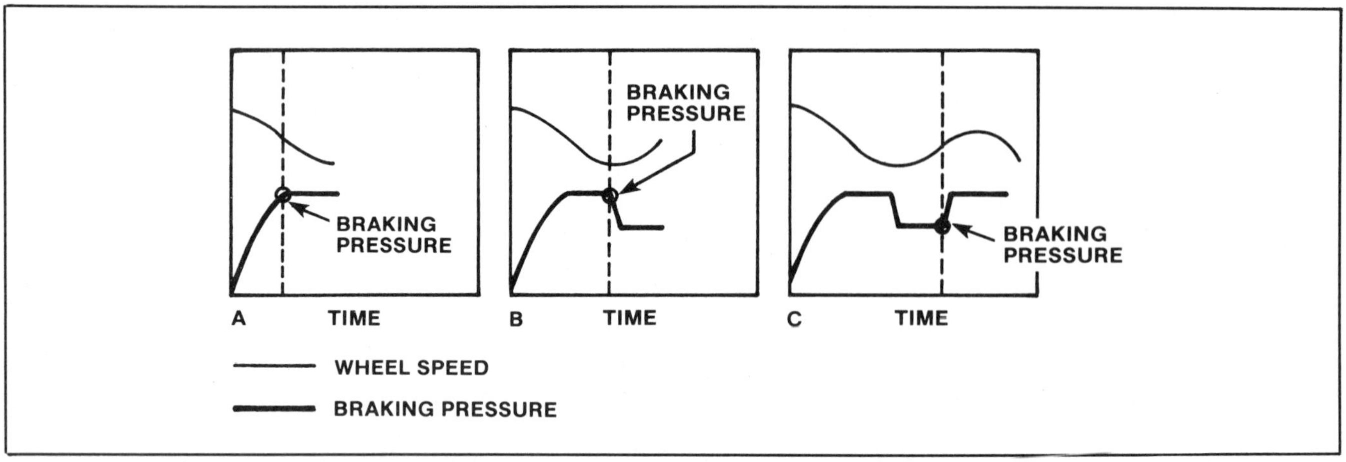

Figure 11-13 Wheel speed and braking pressure during ABS-controlled braking

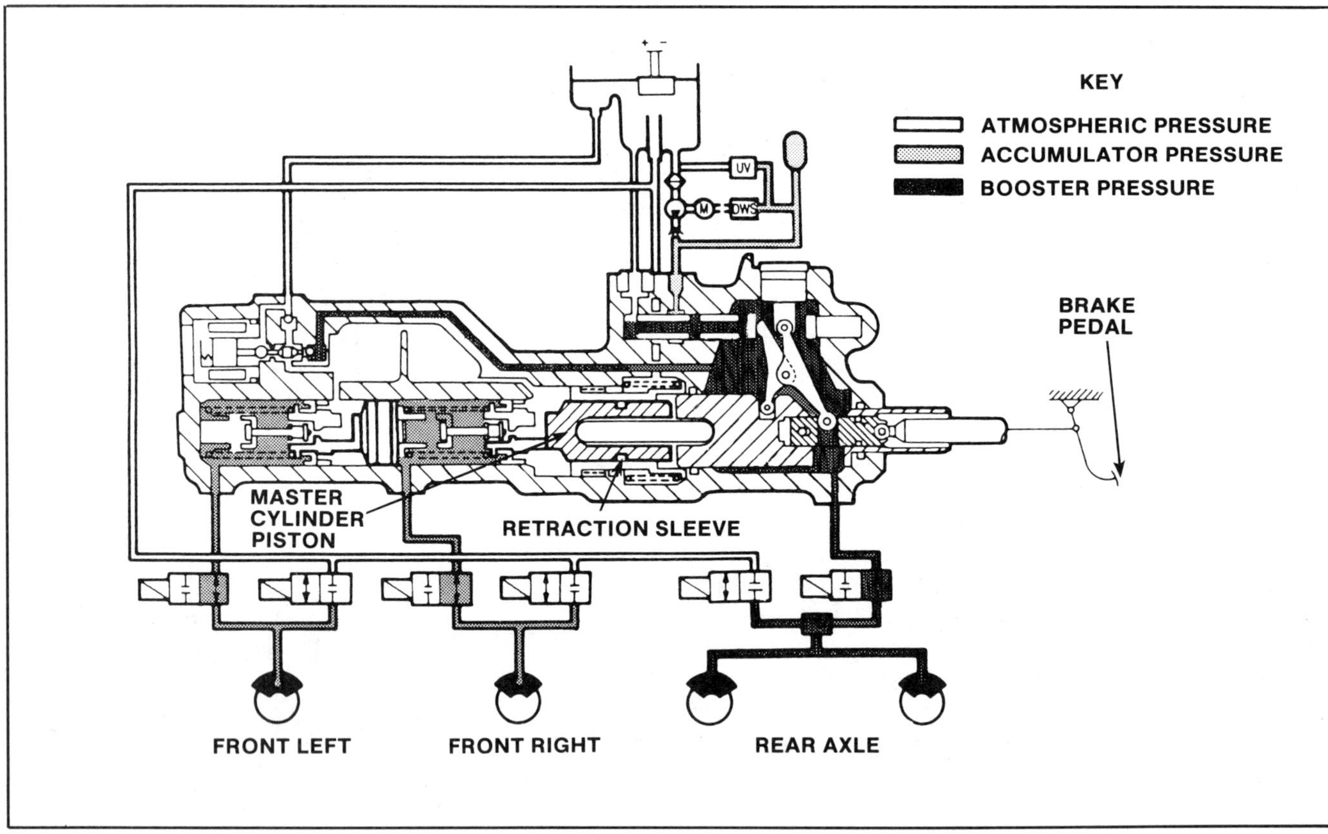

Figure 11-14 Anti-lock—no braking

hydraulic assist to the mechanical thrust from the pushrod.

When the controller determines that the wheels are locking up, it opens a valve that supplies one chamber between the two master cylinder pistons and another chamber between the retraction sleeve and the first master cylinder piston (Figure 11-16). The hydraulic pressure on the retraction sleeve retracts the pushrod, pushing back the brake pedal. In effect, the hydraulic pressure to the front wheels is now supplied by the accumulator, not by the brake pedal action. The controller also opens and closes the solenoid valves to cycle the brakes on the wheels that have been locking up.

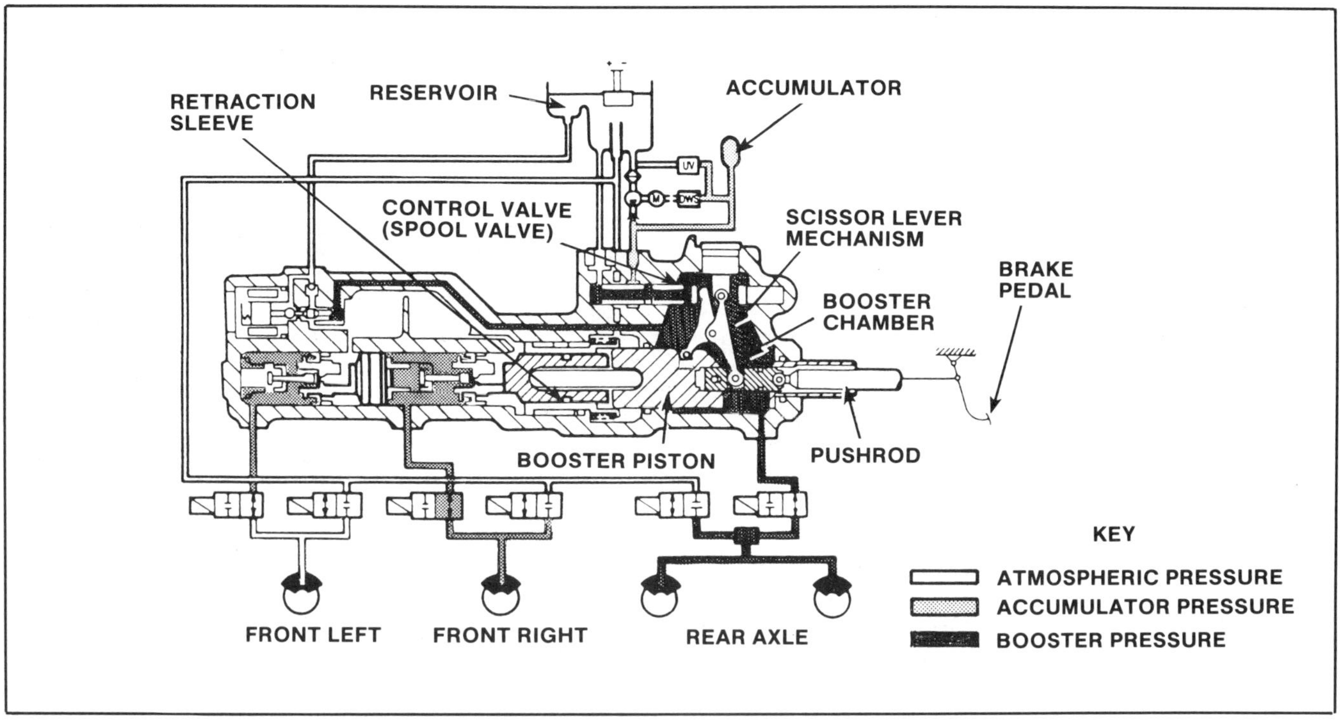

Figure 11-15 Normal braking with anti-lock system

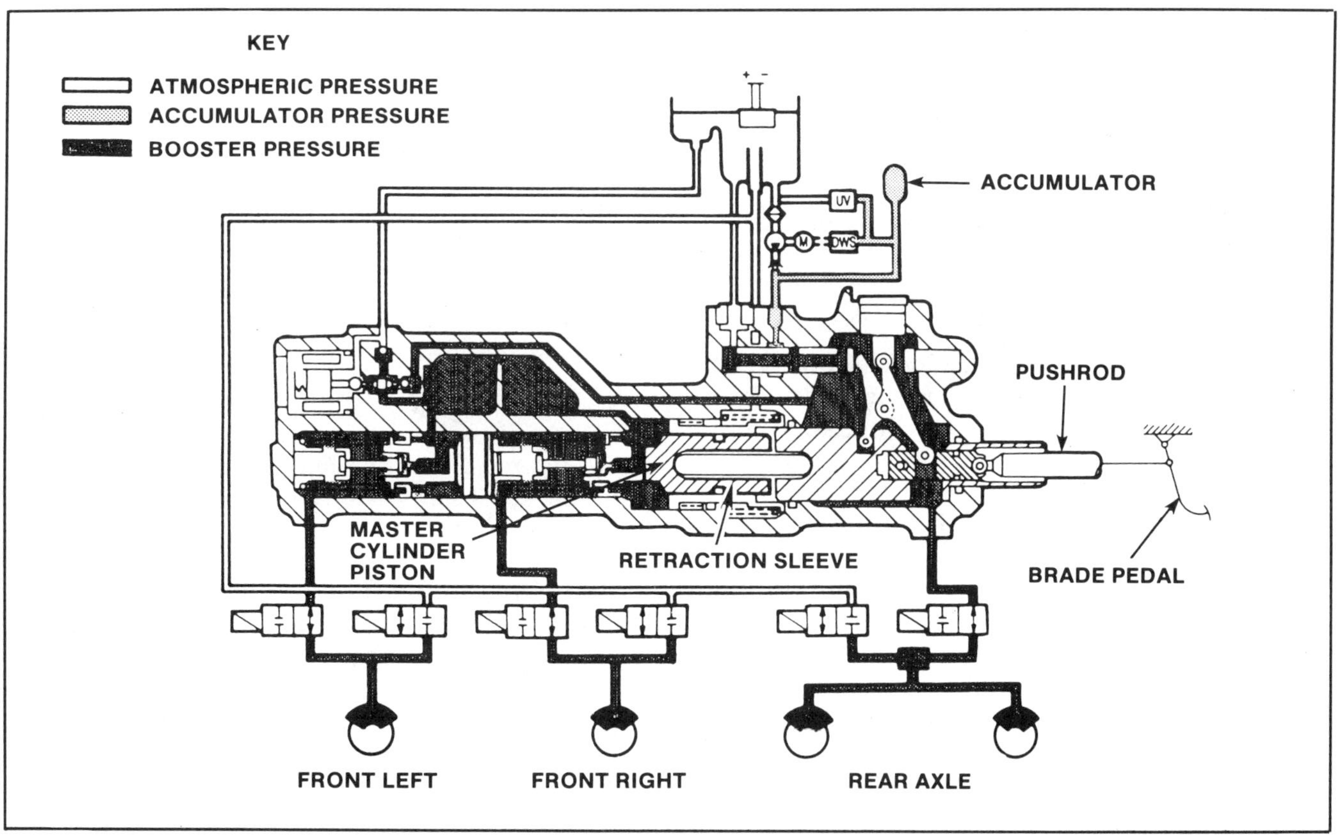

Figure 11-16 Anti-lock braking system in operation

When the solenoid valves are open, the master cylinder pistons supply hydraulic fluid to the front brakes and the boost pressure chamber provides hydraulic pressure to the rear. When the solenoid valves are closed, the hydraulic fluid from the master cylinder pistons and booster pressure chamber is cut off; the hydraulic fluid is returned from the brakes to the reservoir.

The electronic controller monitors the electromechanical components of the system. Malfunction of the ABS will cause the electronic controller to shut off or inhibit the system. However, normal power-assisted braking remains. Malfunctions are indicated by one or two warning lights inside the vehicle. Loss of hydraulic fluid or power booster pressure will disable the anti-lock brake system.

The four-wheel anti-lock brake system is self-monitoring. When the vehicle ignition switch is placed in the RUN position, the electronic controller will perform a preliminary self-check on the anti-lock electrical system indicated by a 3- to 4-second illumination of the amber CHECK ANTI-LOCK BRAKES light in the console. During vehicle operation, including normal and anti-lock braking, the electronic controller continually monitors all electrical anti-lock functions and many hydraulic performance characteristics.

In most malfunctions in the anti-lock brake system, the amber CHECK ANTI-LOCK BRAKES and/or BRAKE light will be illuminated. The sequence of illumination of these warning lights combined with the problem symptoms determine the appropriate diagnostic tests to perform. The diagnostic tests will then pinpoint the exact component needing service.

BRAKE SYSTEM BLEEDING

The front brakes of an anti-lock brake system can be bled in the conventional manner (see Chapter 3), with or without the accumulator being charged. However, bleeding the rear brakes requires a fully charged accumulator or a pressure bleeder attached to the reservoir cap opening, with a minimum of 35 PSI.

System Bleeding with Accumulator Charged

Once accumulator pressure is available to the system, the rear brakes can be bled by opening the rear brake caliper bleeder screws for 10 seconds at a time while holding the brake pedal in the applied position with the ignition switch in the RUN position. Repeat until an air-free flow of brake fluid has been observed at each caliper, then close the bleeder screws. Pump the brake pedal several times to complete the bleeding procedure. Adjust the brake fluid level in the reservoir to the MAX level with a fully charged accumulator.

WARNING: Care must be used when opening the rear caliper bleeder screws, due to the high pressure from a fully charged accumulator.

System Bleeding with Pressure Bleeder

When using a pressure bleeder, it must be attached to the reservoir cap opening, and a minimum of 35 PSI must be maintained on the system. With the brake pedal in the AT REST position and the ignition switch in the OFF position, open the rear caliper bleeder screws for 10 seconds at a time. Once an air-free flow of brake fluid has been observed at each caliper, close the bleeder screws and place the ignition switch in the RUN position. Pump the brake pedal several times to complete the bleeder procedure. Syphon off the excess fluid in the reservoir to adjust the level to the MAX level with a fully charged accumulator.

DIAGNOSIS AND TESTING

Diagnosis of anti-lock brake system malfunctions requires three basic steps that will isolate the fault accurately and quickly. The proper diagnostic procedure is as follows: prediagnosis inspection, light sequence determination, and component test charts.

CAUTION: Following the wrong sequence or bypassing steps will lead to unnecessary replacement of parts and/or incorrect resolution of the symptom. The information and procedures given in this chapter are typical of the various anti-lock systems on the market. For specific instructions, consult the vehicle's service manual.

To perform the diagnostic tests that are necessary for an anti-lock brake system, the equipment shown in Figure 11–17 is usually required. The proper use of this equipment is given in the owner's manual that accompanies each device.

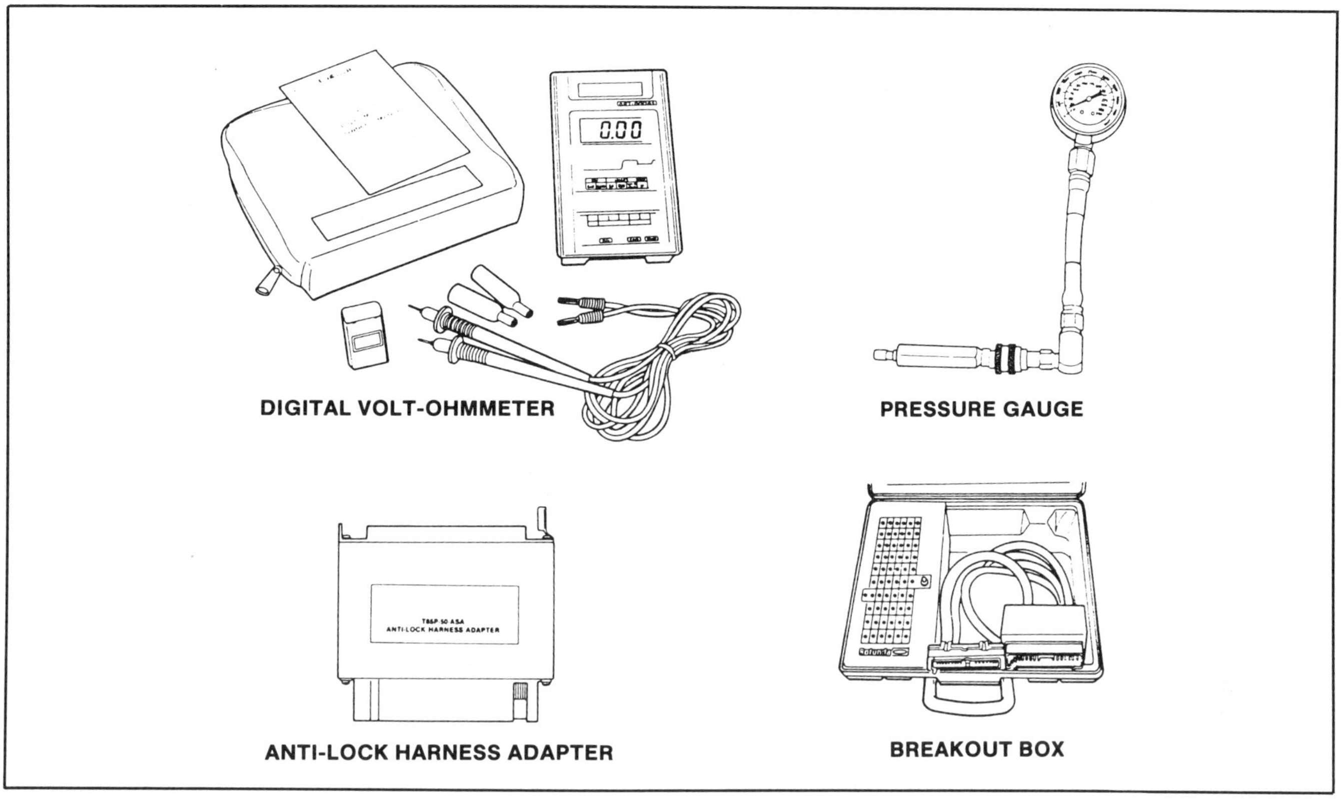

Figure 11-17 Equipment needed for checking ABS Systems

Prediagnosis Inspection

The prediagnosis inspection consists of a quick visual check of specific system components that could create an apparent anti-lock system malfunction. Inspecting the system before diagnosing specific symptoms can result in the detection of a simple defect, which could be the cause of an inoperative system. This inspection should be the first step in diagnosing a complaint.

If a malfunction in the ABS system occurs, perform the following checks:

1. Be sure the control module connector is securely connected.
2. Be sure that all ABS connectors are securely connected.
3. Check BRAKE and GAUGE for a blown fuse.
4. If the malfunction is still present after performing checks 1 through 3, refer to the Light Sequence procedure.

Light Sequence Determination

The second step in diagnosing this system is determining the warning light behavior. This system uses two warning lights, including a red BRAKE warning light and an amber ANTI-LOCK light. By observing when these lights come on, individual components and subsystems can be isolated as the cause of a malfunction.

Normal indicator light sequences are shown in Figures 11-18 and 11-19. In Figure 11-18, each light in a typical sequence is divided into seven areas that represent vehicle status, with the behavior of BRAKE and ANTI-LOCK lights represented in horizontal rows below the slanted vehicle status conditions at the top of the table. BRAKE and ANTI-LOCK light status is indicated by the shaded areas in the horizontal rows below each vehicle behavior condition. Right-hand diagonal shading indicates illumination of the ANTI-LOCK light. Left-hand diagonal shading indicates an illumination of the BRAKE light. A vehicle status block that is not entirely shaded means that the light is on only for part of the test period. In Figure 11-19, the Perform Test column suggests what system or tree chart to use. The following procedure illustrates how the light sequence should be used to determine the proper setup for further diagnosis. When evaluating the vehicle's condition, use caution if reduced braking ability is evident.

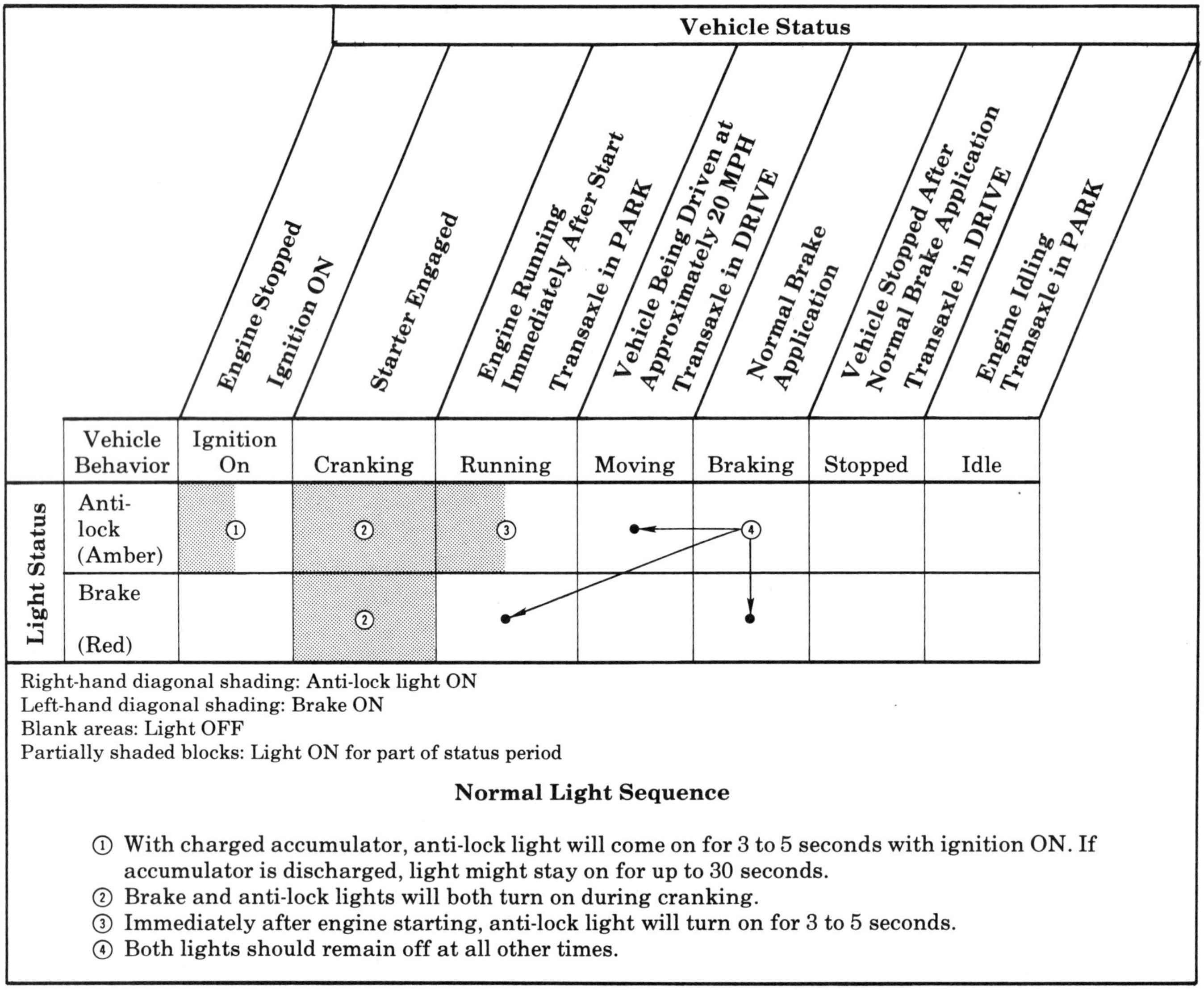

Figure 11-18 Chart of normal indicator light sequence

1. Turn the ignition switch to the OFF position for a minimum of 15 seconds.
2. Turn the ignition switch to the ON position, then observe the BRAKE and ANTI-LOCK lights, noting the following:
   - If the accumulator has been discharged, both indicator lights could remain on for approximately 30 seconds. If this occurs, allow the lights to turn off and turn the ignition switch to the OFF position for another 15 seconds, then proceed to step 3.
   - If the lights do not go out within 30 seconds, proceed to step 4.
3. Turn the ignition switch to the ON position and observe the warning lights. At this point, the BRAKE light should remain off. The ANTI-LOCK light should come on for 3 to 6 seconds, then go out.
4. Turn the ignition switch to the START position and observe the warning lights. Both warning lights should come on.
5. Start the engine, turn the key to the RUN position, and observe the warning lights. The BRAKE light should turn off immediately; the ANTI-LOCK light should remain on for 3 to 6 seconds, then go out.
6. Drive the vehicle at a minimum speed of 20 MPH for a short period of time and observe the warning lights. Both lights should remain off for this step and the remainder of the test. Note accurately the conditions under which either light comes on.

Sequence Number	Light Sequence	Symptom Description	Perform Test
1	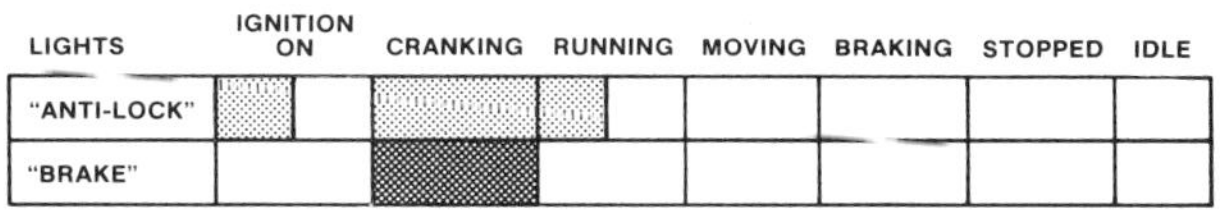	Normal light sequence with —Excessive pedal travel or spongy pedal —Anti-lock braking operation or valve cycling during normal stops on dry pavement —Poor vehicle tracking during anti-lock braking	H C D
2	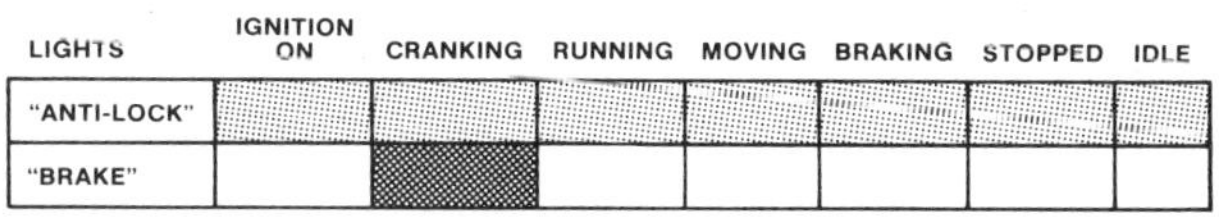	Continuous "anti-lock" light Normal "brake" light	A
3	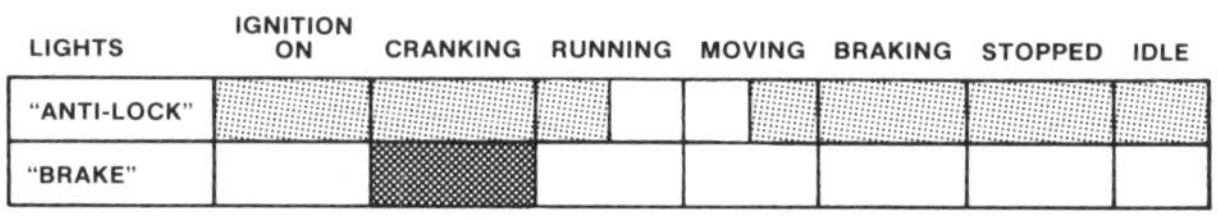	"Anti-lock" light comes on after vehicle starts moving. Normal brake light	C
4	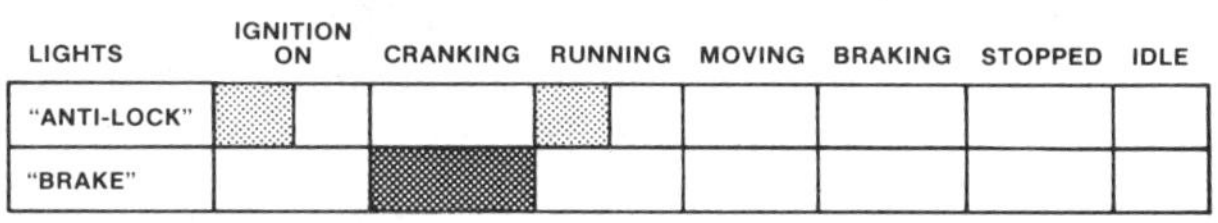	No "anti-lock" light while cranking. Normal "brake" light.	E
5	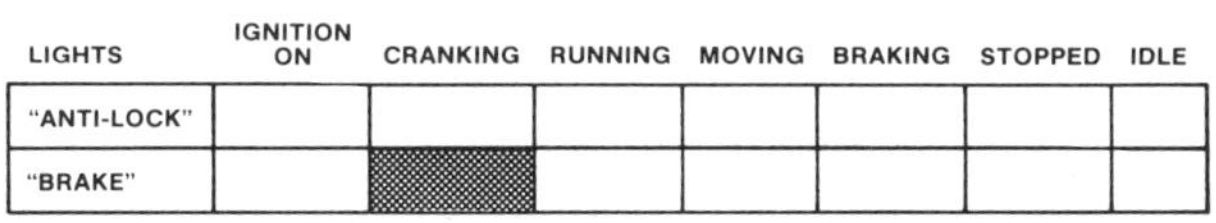	No "anti-lock" light Normal "brake" light	F
6	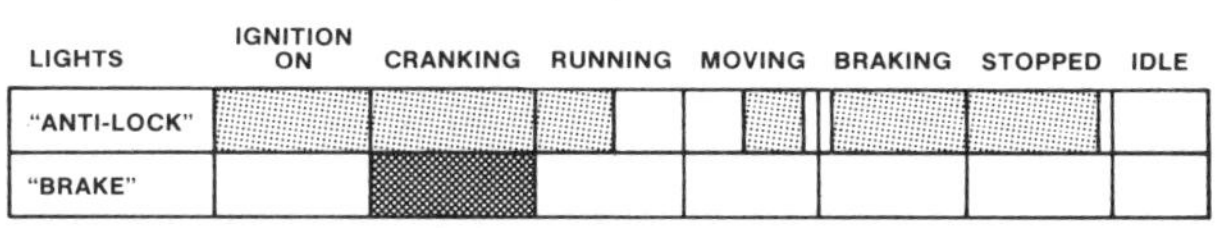	Intermittent "anti-lock" light while driving Normal "brake" light	G
7	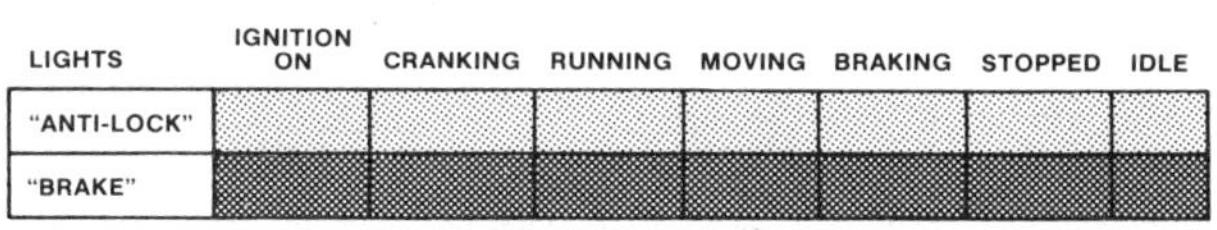	Continuous "anti-lock" light Continuous "brake" light	B
8	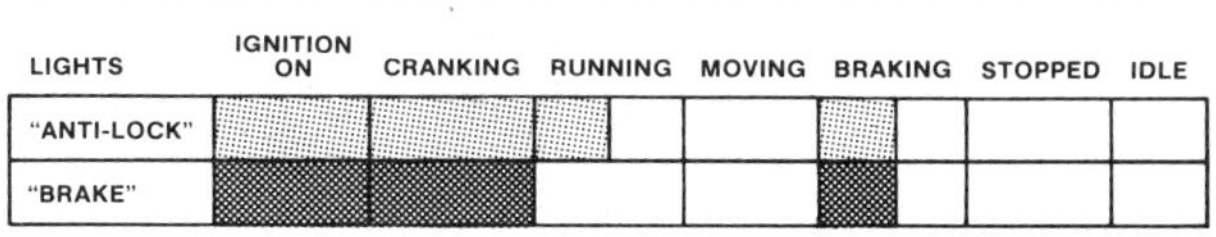	"Anti-lock" and "brake" lights come on while braking	B
9	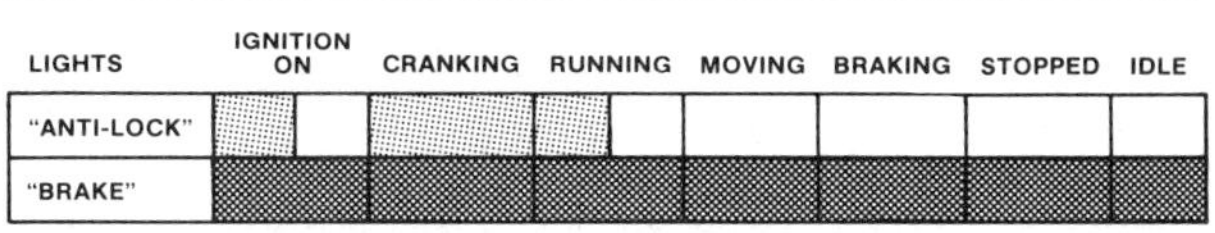	Normal "anti-lock" light Continuous "brake" light	B
10	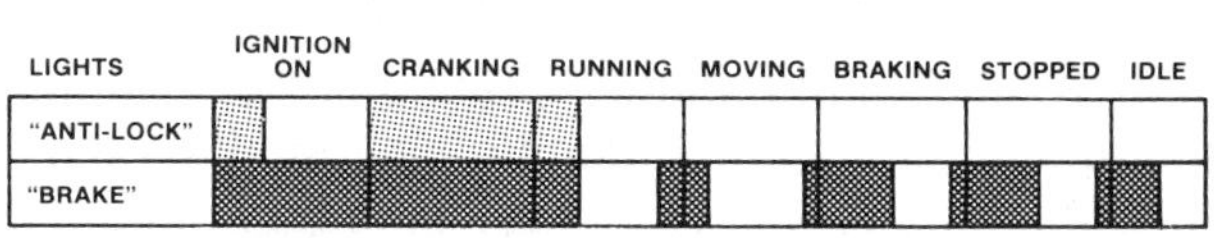	Normal or continuous "anti-lock" light Flashing "brake" light	B

Figure 11-19 Chart of indicator light sequence with a column of recommended tests

7. Stop the vehicle using normal brake application. Both lights should remain off.
8. Place the transaxle in PARK, then allow the vehicle to idle for several seconds and observe the warning lights. Both should remain off.
9. From the above tests, determine the light sequence and/or symptom that most closely matches the vehicle's condition, then perform the indicated tests.

If an intermittent fault cannot be diagnosed using the indicator light sequence chart, the problem will most probably be caused by a defective electrical connection or wiring. When an intermittent problem is encountered, check questionable circuits and electrical connectors for the following:

1. Poor mating of connector halves or terminals not fully seated in the connector body.
2. Improperly formed or damaged terminals. All connector terminals in the problem circuit should be carefully removed to increase contact tension.
3. Poor terminal wire connection.

Diagnosis Using Component Test Charts

Using the test equipment operating instructions for a description of the test sequences and display messages, perform the repair action as described in the symptom directory chart or trees. Three types of typical charts are illustrated in Figures 11-20, 11-21, and 11-22. It is interesting to note that the average ABS service manual contains sixty or more of these charts or trees.

ANTI-LOCK BRAKE SYSTEM SERVICING

As already noted, the Teves ABS units are continuously monitored by their built-in safety circuitry. If a problem develops, the ABS warning light will turn on and the system will cease operating without affecting the standard braking system. Thus, it is possible for the brake technician to diagnose the problem quickly and remove and replace a faulty component. At this time, internal master cylinder service is not possible; only external components (pump and motor, valve block assembly, accumulator, reservoir, and pressure switch) are serviceable following the proper procedures.

Normal brake repairs, such as replacing brake pads, caliper replacement, rotor machining or replacement, brake hose replacement, master cylinder or power booster replacement, or handbrake repair, can all be performed as usual. In other words, brake service on a Teves-equipped vehicle is similar to service on a conventional system with a few exceptions.

First, before beginning service, depressurize the accumulator to prevent personal injury from high-pressure fluid. This is accomplished by pumping the brake pedal (ignition off) until the pedal becomes hard. Air gaps between the speed sensors and toothed ring will not have to be adjusted after machining rotors if wheel bearing adjustment is made to manufacturer's specifications. Disc or drum brake service is performed as on a conventional vehicle.

After servicing, the front brakes can be bled with a pressure bleeder using the appropriate adapter. To bleed the rear brakes, first open the rear bleeder screws. Then have an assistant depress and hold the brake pedal down while turning on the ignition switch for at least 10 seconds. This will allow the high-pressure pump to push the air through the system. Repeat as necessary. But before opening a rear bleeder screw a second time, turn off the ignition and pump the brake pedal to relieve accumu-

G: PUMP MOTOR RELAY TEST		
Measure: RESISTANCE **At: MODULATOR VALVE (Disconnected)**		
• Ignition switch: OFF		
Measure Between	**Correct Resistance**	**For Diagnosis**
• Pump motor relay connected		
10 and 11	Approximately 40 ohms	See 1.
• Pump motor relay removed		
30 and 9	0 ohms	See 2.
9 and Ground	Approximately 0.3 ohms	See 3.

- If measurements in steps 1 and 2 are correct, but Pump Motor Relay does not operate, replace the Pump Motor Relay.

1. Check Pump Motor Relay Coil for an open. If the relay coil is good, replace the Modulator Valve.
2. Replace Modulator Valve.
3. Check for an open in Return Pump Motor or ground strap. If Return Pump Motor is good, replace the Modulator Valve.

Figure 11-20 Symptom directory chart

Symptom Message on Tester	Possible Cause	Repair Action
CHK LF SPD SENS (Check left front speed sensor.)	Open circuit in LF wheel speed sensor	Do Test F.
CHK RF SPD SENS (Check right front speed sensor.)	Open circuit in RF wheel speed sensor	Do Test F.
CHK LR SPD SENS (Check left rear speed sensor.)	Open circuit in LR wheel speed sensor	Do Test F.
CHK RR SPD SENS (Check right rear speed sensor.)	Open circuit in RR wheel speed sensor	Do Test F.
CHK ANTI-LOCK LT (Check anti-lock light.)	Anti-lock bulb burned out	Replace bulb. Check that bulb is properly connected.
	Connector pin not mated	Check that connectors C210 and C400 are securely mated.
	Open in circuit 915 or 639 between INJ 1 fuse, indicator, and control module	Check PNK/BLK wire (639) and WHT wire (915) for an open.
CHK MOTOR RELAY	Pump motor relay not closing	Do Test G: Pump Motor Relay Test.
	Open circuit to pump motor relay coil or pump ON input	Check for an open in the GRN/BLK wire (913). Check for open in the BLU/BLK wire (912).
CHK PUMP MOTOR	Pump motor inoperative	Do Test G. Table 2.
	Open ground circuit from pump motor to ground	Check BLK wire (150) from case of Modulator Valve to Ground at G320. Check that connector C448 is mated correctly.
CHK SPD SENS CKT (Check speed sensor circuit.)	Damaged wheel speed sensor	Check area of sensor for damage. Check sensor for damage.
	Misalignment of wheel speed sensor	Sensor should not contact toothed wheel.

Figure 11-21 Anti-lock system symptom directory chart

lator pressure. As with other types of anti-lock brake systems, use DOT 3 fluid only.

There are some general precautions and servicing techniques that must be kept in mind when working on Teves anti-lock brake system. They include:

1. If bodywork welding or any other welding is necessary, disconnect the electronic control unit or module plug to prevent overheating the unit. Be careful not to subject the control module to a heat of 200 degrees F or more for longer than 2 hours.

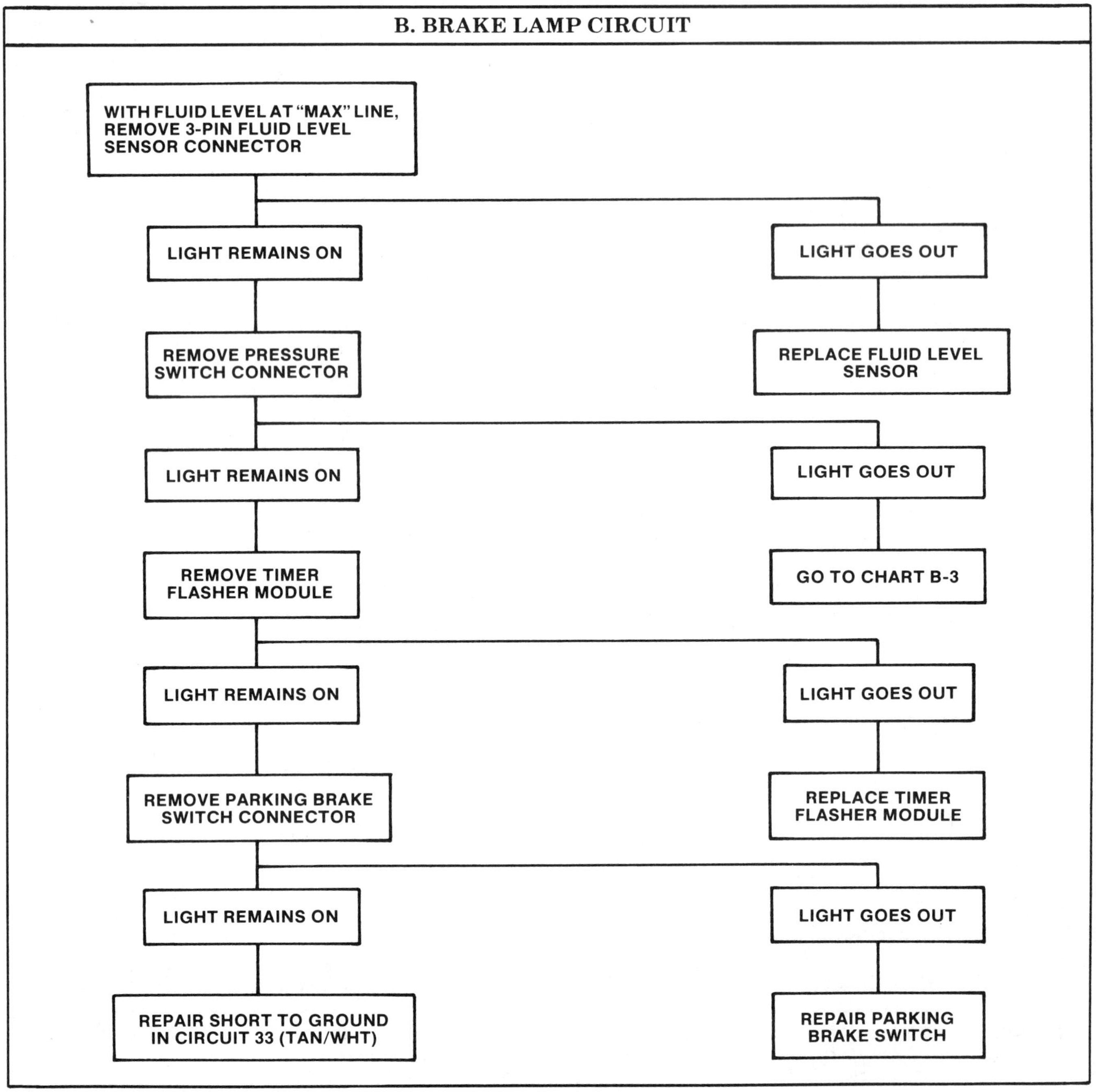

Figure 11-22 Brake light circuit using a tree type troubleshooting chart

2. When working on the pump, always check for special markings on the attaching screws and bolts. Some are not supposed to be tightened or loosened. If they are, they may require replacement.
3. If brake lines are disconnected, plug them immediately to prevent dirt from entering. And, if the lines are repeatedly disconnected and reconnected, double-check for leaks at their fittings. If leaks are detected, replace the brake lines as detailed in Chapter 4 and recheck connections.
4. Be sure the inlet and outlet brake lines are properly routed.
5. Do not use a fast charger for starting the engine.
6. Never disconnect the battery from the vehicle when the engine is running.

7. Do not connect or disconnect the control module electrical connector with the ignition switch in the ON position. Be sure all harness connectors are securely connected.
8. Avoid contact with the terminal of the control module. Static electricity can damage the unit.
9. Do not use a continuity tester to probe the ABS wire harness without referencing the manufacturer's diagnostic charts and test procedures. Damage to the control module may occur.
10. No screws on the modulator valve should be loosened. If the screws are loosened, it is no longer possible to get the brake circuit leak-tight.
11. When replacing the modulator valve, it should be removed through an access panel in the rear storage compartment. Be sure the vehicle interior and exterior are protected to avoid damage from brake fluid spillage. Do not remove the bottom of the storage compartment to replace the modulator valve.
12. Be sure that the rear storage compartment is wiped clean after replacing the modulator valve.
13. Do not support the suspension components by the wheel speed sensor wires.
14. Wheel speed sensors are a tight fit into the knuckle and should be pushed in by hand.
15. Do not hammer the sensors into position.
16. Each replacement wheel speed sensor is usually identified with a white tag, located approximately 3/4 inch from the neck of the sensor, labeled L (left) or R (right).
17. Tire and wheel assemblies should be removed when replacing wheel speed sensors.
18. Wheel speed sensors should be given anti-corrosion coating before installation to prevent corrosion. Do not use grease.
19. Be sure not to use silicone type brake fluid.
20. Faulty ABS components should be removed from the system and replaced with a new unit. Proper procedure for these servicing steps must be done as directed in the service manual or as directed on the replaced instruction sheet that comes with the part. Care must be taken when servicing anti-lock brake systems.

When checking the ABS operation, notice that the warning light will "light up" when the ignition is turned on, but it should turn off when the engine runs. The safety circuit is activated as soon as the engine runs and when the warning light turns off, it indicates that the ABS is fully operational.

When road-testing the car, the technician may notice a pulsating effect through the brake pedal. At low speeds he/she may also notice a noise from the hydraulic pump. These are normal and are actually "feedback" confirming that the ABS is operating.

WARNING: Before servicing any ABS component that contains high pressure, it is mandatory that the hydraulic pressure in the system be discharged. To do this, turn the ignition off and pump the brake pedal a minimum of twenty times until an increase in pedal force is clearly felt. Read the service manual instructions carefully before servicing any component of the anti-lock brake system.

SHOP TALK

After the problem condition has been repaired, repeat the entire test given in the symptom diagnosis chart or tree. This will reveal any other malfunction that might exist and should be corrected.

BOSCH ANTI-LOCK SYSTEMS

As with other types of ABS's, the three key elements—sensors, logic, and modulator—are required. In contrast to the Teves system, the Bosch system utilizes a conventional design master cylinder and vacuum booster. Brake fluid pressure from the master cylinder must first pass through the modulator valve before reaching the individual calipers. The system on the Corvette is illustrated in Figure 11–23.

The modulator valve assembly (Figure 11–24) consists of three solenoid-operated valves (one for each front wheel and one for both rear wheels), two accumulators (one for the front axle and one for the rear axle), and one pump and motor assembly (Figure 11–25).

Speed sensors (one per wheel) provide an alternating current signal to the control module to indicate the rotating speed of the respective wheels. These assemblies are illustrated in Figure 11–26.

As with the Teves ABS, a toothed ring (rotating with the wheel) in conjunction with the stationary sensor produces the signal. In order to provide accurate signals to the control module, the correct air

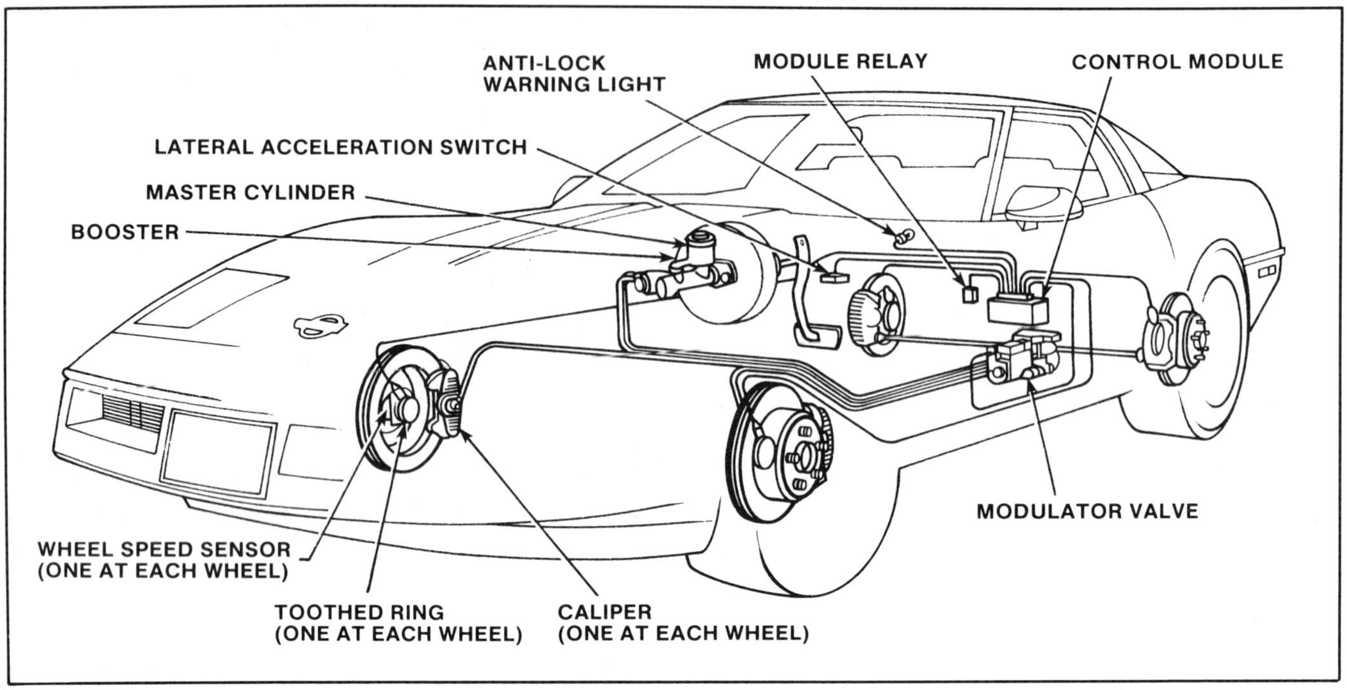

Figure 11-23 Main components of Bosch ABS

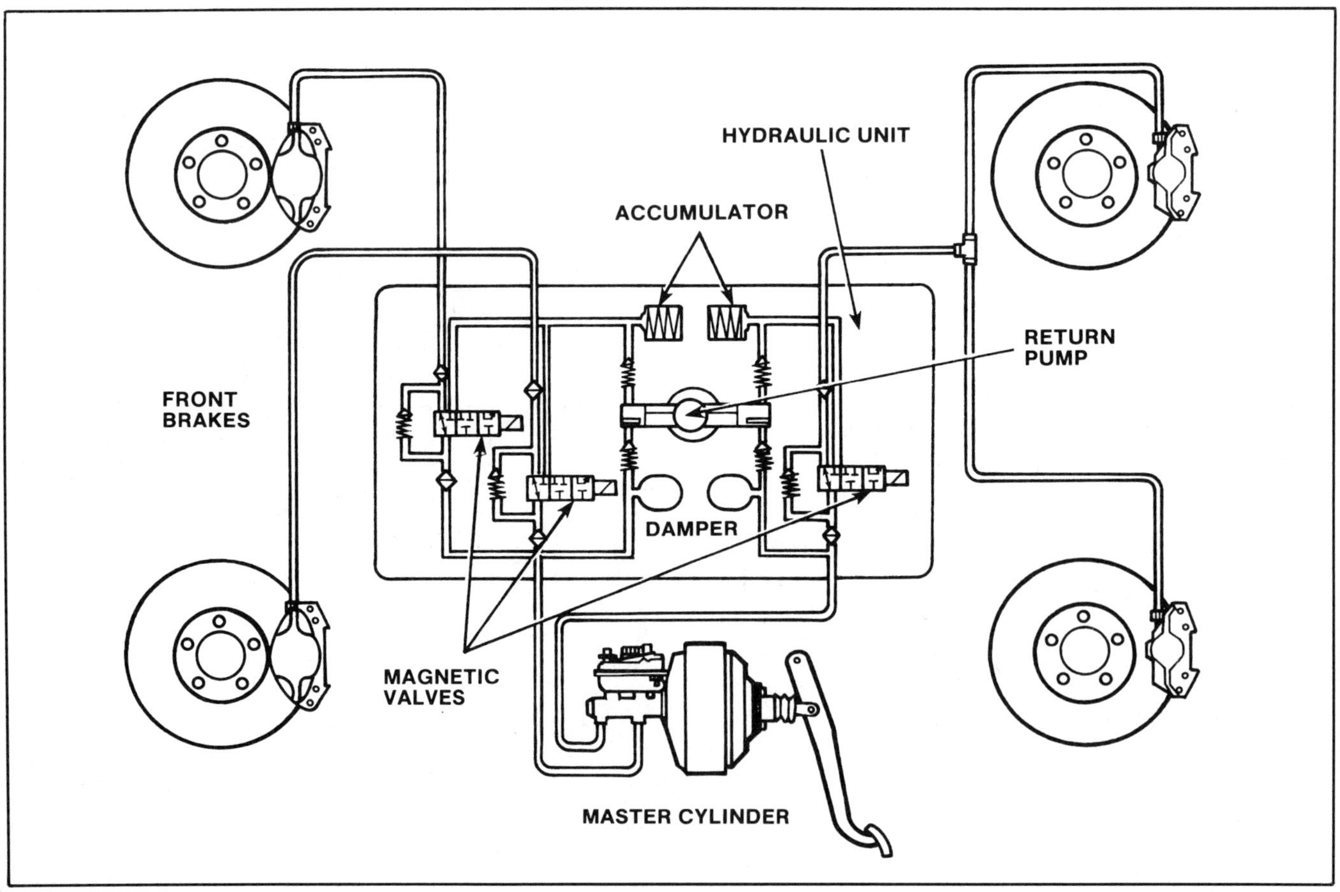

Figure 11-24 Modulator valve body in a Bosch ABS

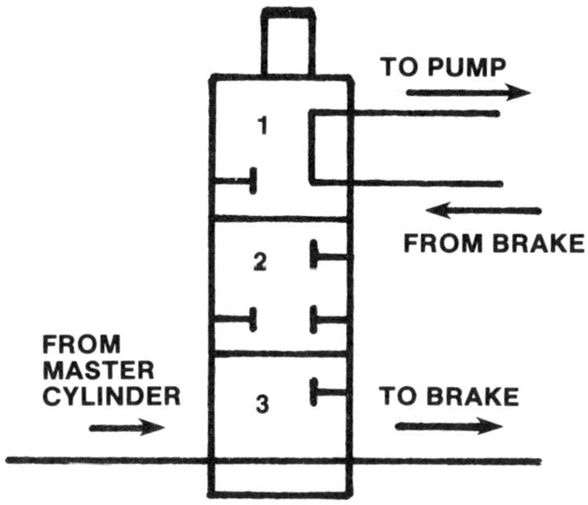

Figure 11-25 Three-position solenoid valve

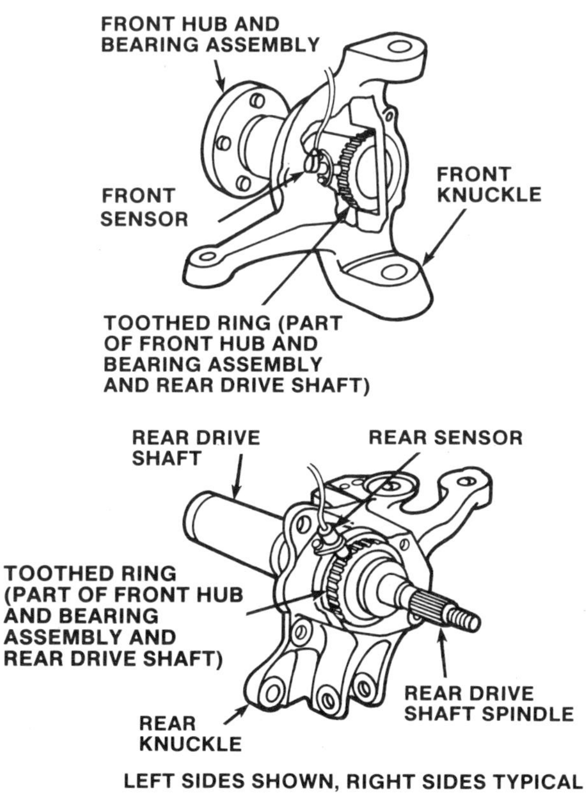

Figure 11-26 ABS sensor locations at front and rear wheels

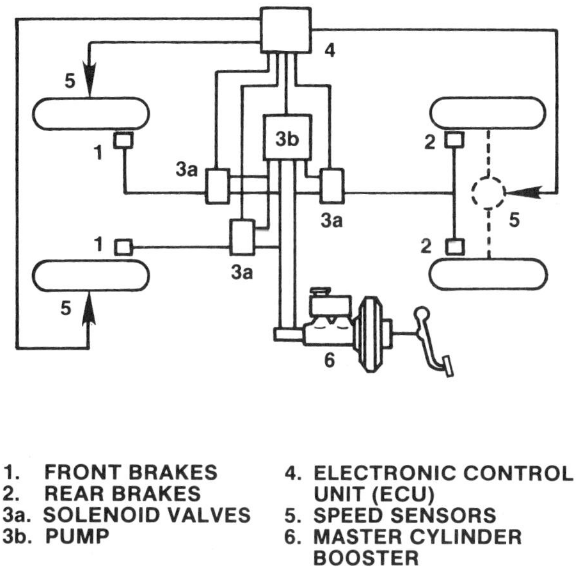

Figure 11-27 Three-circuit ABS schematic

gap between the sensor and the toothed ring is critical.

The control module receives information from the speed sensors (Figure 11-27), along with other input signals, processes the information and then sends output signals to the solenoid valves (located in the modulator valve), solenoid valve relay, pump motor relay, and anti-lock indicator as needed.

In the event of hard braking while cornering, a lateral acceleration switch (located in the storage compartment) will signal the control module and alter the anti-lock problem to compenste for braking while in a turn. The lateral acceleration switch is actually two mercury switches (one for left turns, one for right turns) in a common housing.

During normal braking, the solenoid valve in its "rest" (unpowered) position allows unrestricted hydraulic pressure from the master cylinder to pass through the valve and to the respective caliper(s) it serves (Figure 11-28), applying the brake.

In the event of hard braking or a lock-up-producing situation (snow, wet pavement, panic stop, etc.) causing one or more wheels to slow down too quickly, the speed sensor at that wheel(s) indicates this condition to the control module by its changing signal. The control module then sends a signal to the solenoid valve serving the caliper(s) (both rear wheels are served by one solenoid valve) where the lock-up is likely to occur (Figure 11-29). The solenoid valve then raises to its intermediate position, trapping existing hydraulic pressure at the caliper while also preventing increased hydraulic pressure from reaching the caliper.

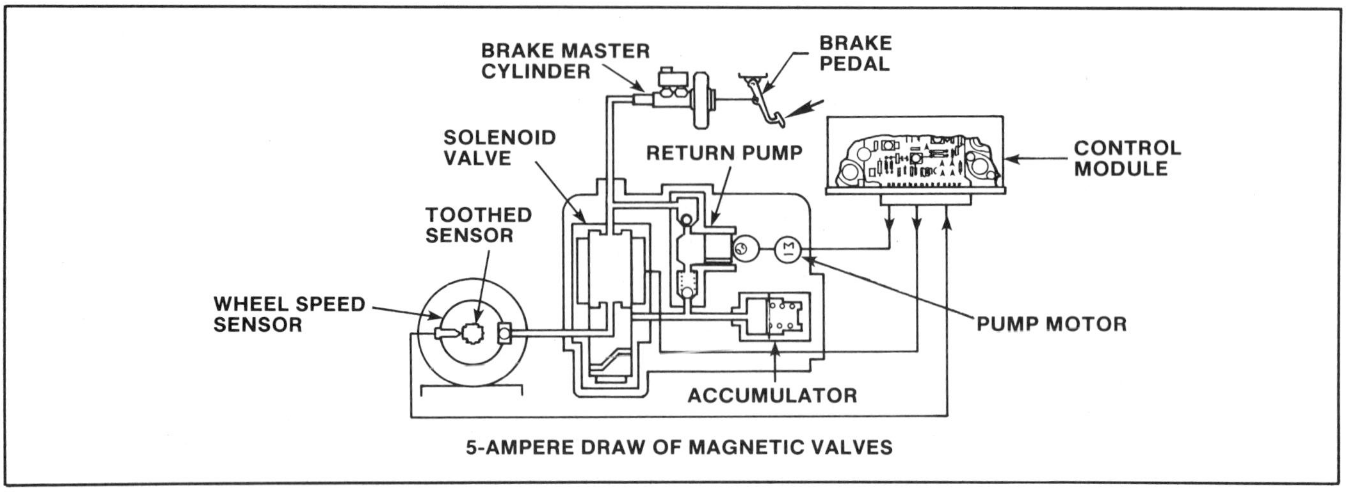

Figure 11-28 Bosch ABS during normal brake application

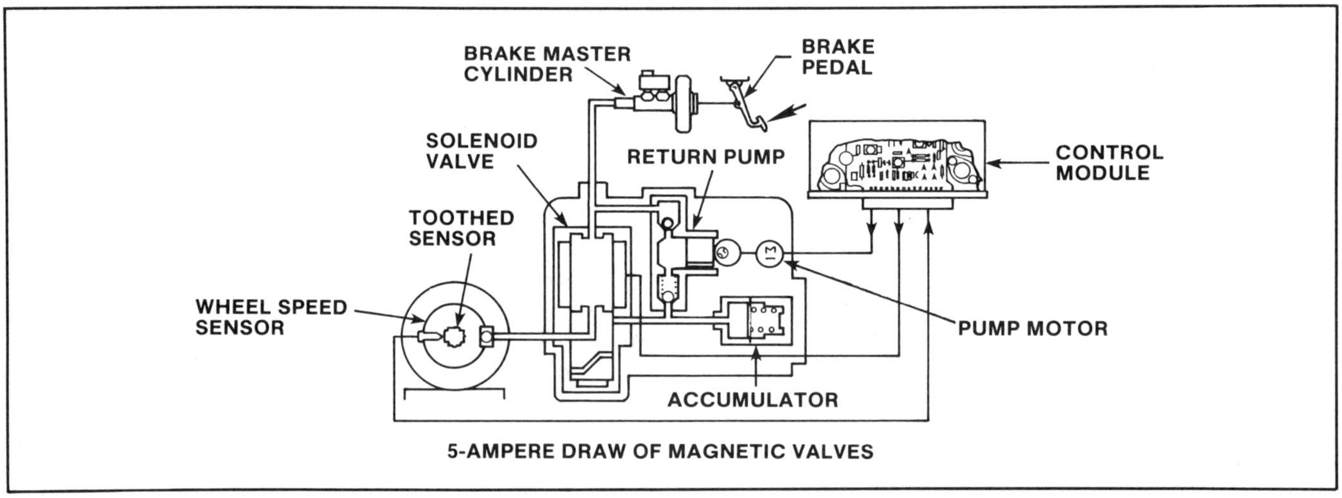

Figure 11-29 Maintaining hydraulic pressure at wheel

If the signal from the sensor indicates to the control module that lock-up is still likely to occur, the control module sends another signal to the solenoid valve, causing it to move to its fully raised position. With the solenoid valve in this position, existing hydraulic pressure (Figure 11-29) at the caliper is allowed to bleed off through the accumulator and return pump (Figure 11-30). The control module then sends a signal to the pump relay, actuating the return pump. As the brake fluid pressurizes, the accumulator and the pump turn on. Fluid also passes through the pump inlet ball check and into the pump.

With the pump now operating, the fluid is pressurized in the pump and discharged through the outlet ball check and returned to the inlet side of the solenoid valve. As this fluid is returned to the inlet side of the solenoid under considerable pressure, it will cause a pressure increase all the way back to the master cylinder. The brake pedal will raise each time this occurs. The fluid pressure will be unable to reach the caliper(s), however, because the position of the solenoid valve has isolated it from the rest of the system.

At this point in operation, hydraulic pressure to the caliper(s) has been fully released, allowing the wheel to rotate freely. When the speed sensor indicates to the control module that the speed of the wheel is increasing, the control module will deactivate the solenoid, allowing it to drop to the "apply" position (Figure 11-28) and shut off the return pump, completing a cycle. During anti-lock operation, the Bosch system can operate four to ten cycles per second as conditions require.

Routine brake service on a Bosch ABS-equipped vehicle is accomplished the same as on a vehicle without an anti-lock system, since none of the ABS components must be disturbed. After machining the front rotors, however, adjust the front wheel bearings per manufacturer's specifications to ensure correct air gap between the speed sensor and toothed ring.

Bleeding of the brakes may be done either manually or with an approved pressure bleeder providing 20 to 25 PSI supply pressure. In either case, the front of the vehicle must be raised above its rear to position the bleeder screw at the highest point of the caliper; this will allow all air to be bled from the system. And, if pressure bleeding, it will also be necessary to stroke the brake pedal during the bleeding operation to ensure that any air in the system is eliminated. The bleed sequence is left rear, right rear, left front, right front. As with the Teves system, use DOT 3 brake fluid only.

Which system is best? The Teves system is more compact and eliminates the need for a separate vacuum booster, while the Bosch system uses a conventional master cylinder and vacuum booster. The exotic master cylinder in the Teves system will be costly when it needs to be replaced, and its long-term durability may not be as good as a conventional master cylinder. The modulator valve assembly on the Bosch system, on the other hand, must be replaced rather than repaired if the unit is defective. In the United States, as already stated, most production model vehicles employ the Teves ABS.

Each system has built-in diagnostics to warn the driver in case of a failure, and each has a fail safe mode that allows the vehicle to revert to normal braking in case of a problem. For troubleshooting system problems. Bosch has a special microprocessor-based tester that can check out the ABS control module and the wheel speed sensor circuits.

Diagnosis and servicing of the Bosch ABS can only be accomplished with the appropriate service manual and diagnostic equipment.

LUCAS-GIRLING STOP CONTROL SYSTEM

A completely different approach to skid control has been taken by Lucas-Girling with their hydromechanical stop control system (SCS). Compared to the Bosch and Teves ABS, the Lucas-Girling system seems rather simple (Figure 11-31). Yet the utter simplicity of the system makes it a viable alternative for many low-cost applications.

To begin with, SCS has no computer or wheel speed sensors. The key to its operation is the two belt-driven mechanical modulator units, each controlling one front wheel and the opposing rear wheel (the brake system is split diagonally). The basic operating principle is still the same: to pulsate brake pressure if the wheels are in danger of locking up. But instead of using an electronic feedback system to monitor wheel speed and regulate pressure, the Lucas-Girling system relies on flywheel momentum and one-way clutches to open and close the pressure valves.

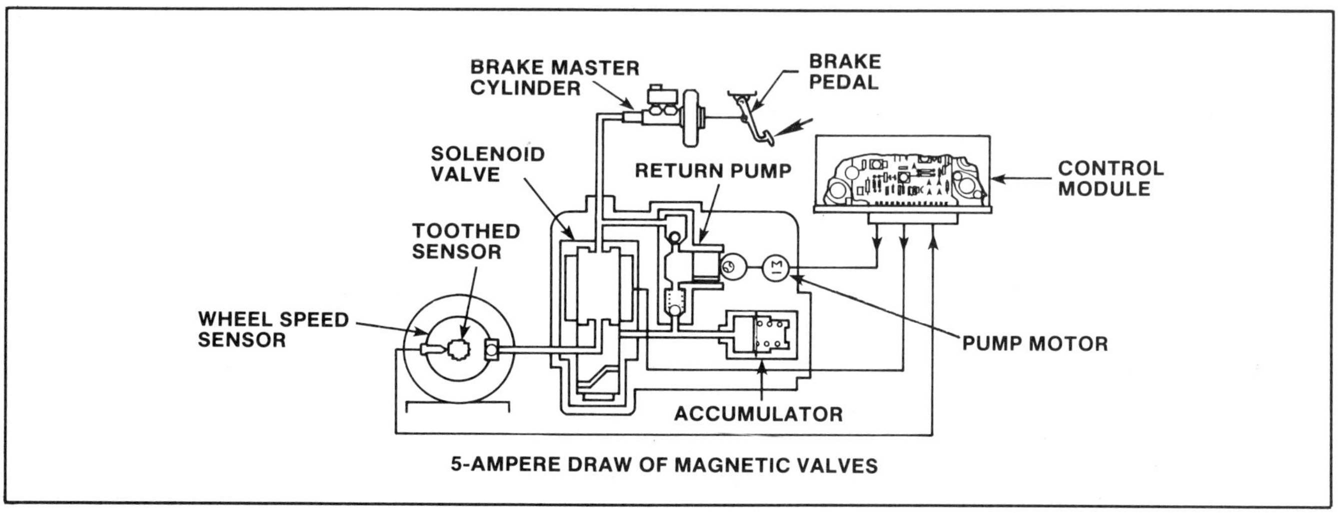

Figure 11-30 Reducing hydraulic pressure at a wheel

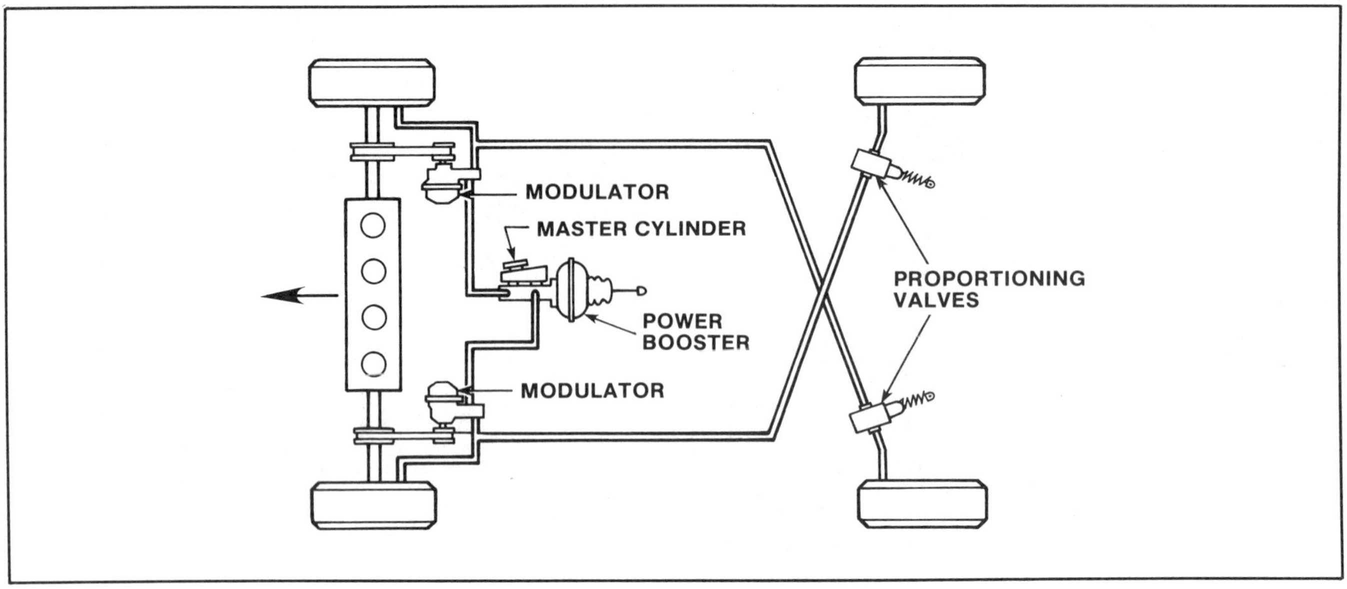

Figure 11-31 Lucas Girling stop control braking system (SCS)

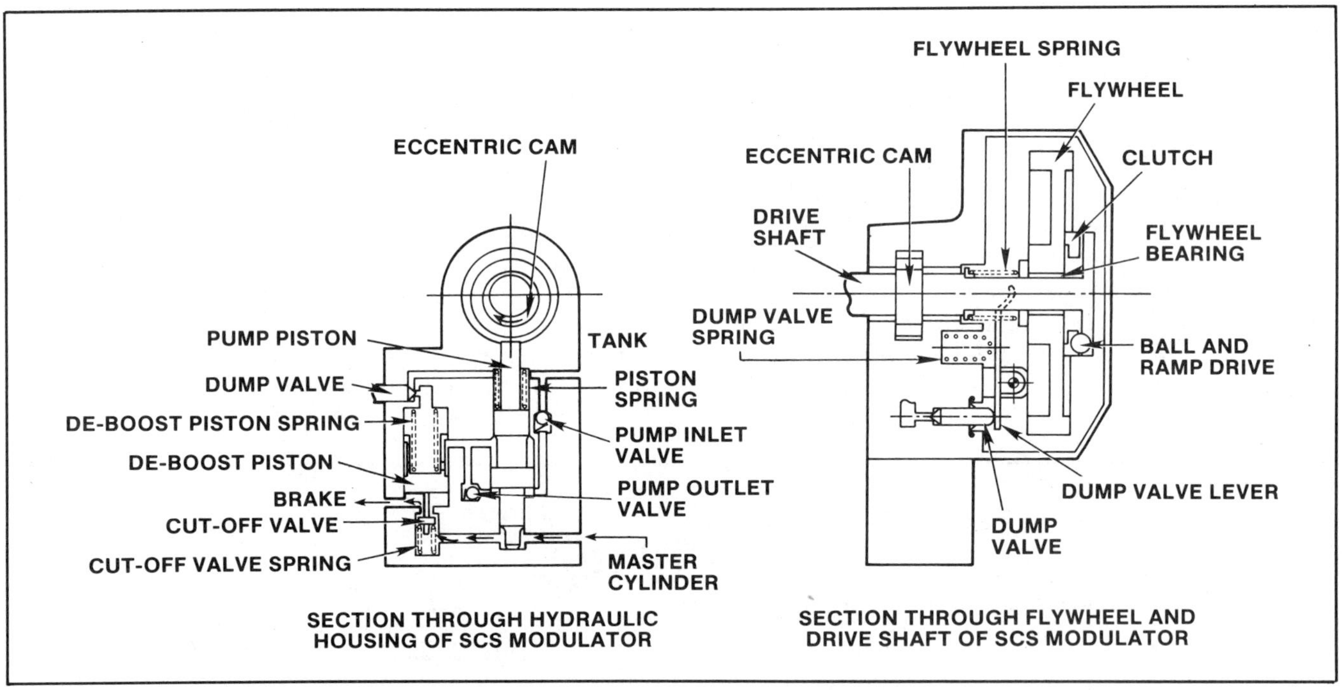

Figure 11-32 Functional sequence of Lucas Girling SCS

Driving down the road, the two modulator flywheels (which are driven off pulleys on each driveshaft) turn at the same speed as the wheels (Figure 11-32). When the brakes are suddenly applied, however, the momentum of the flywheels causes them to continue turning faster than the decelerating wheels.

This engages a one-way clutch mechanism that forces the flywheel against a lever that opens a pressure relief valve. This momentarily reduces brake pressure. The flywheel clutch then releases allowing an internal pump to generate fluid pressure to restore braking. The cycle continues to repeat until the tires cease to slip.

The diagonally opposite rear wheel to each front wheel is also controlled by the modulator through a conventional proportioning valve that regulates the braking effort between front and rear wheels. This operates such that when the right-hand wheels are running on a dry surface and the left-hand wheels on a slippery wet or icy surface, then, under hard braking, the front left-hand wheel's modulator will react to reduce the pressure in both the front left-hand and rear right-hand brakes thus ensuring that these wheels do not lock.

The design function and maintenance of the two-channel SCS are notably simple when compared with other known systems described earlier in this chapter. With an anti-lock braking system, which will be as widely fitted as this one, it is important that the brake technician will be able to service it competently, avoiding the added complication of sophisticated electronic test equipment. All the servicing details are given in the service manual.

The system obviously is not as sophisticated as the Bosch or Teves four-channel designs. But it does allow shorter stopping distances on slick surfaces, and it allows the driver to maintain steering control during panic stops. The design simplicity of Lucas Girling's SCS system coupled with the ease of its installation in production means that a price of well under half the cost of four-channel systems, makes it ideal for low to medium price FWD vehicles.

SHOP TALK

In order to reduce the cost of electronically controlled ABS units, manufacturers—in addition to the standard four-channel system (all four wheels have their own individual sensor)—have recently introduced a three-channel system. In this design, the front wheels are controlled individually, while the rear wheels are controlled as a pair. In general, the three-channel or three-sensor systems are used on "conventional" front-engine, rear-drive cars, where the lock-up sensing unit at the rear can be taken off the U-joint flange on the differential assembly. The SCS is still less costly and is well suited for front-wheel drive vehicles.

REVIEW QUESTIONS

Review the information in this chapter by answering the following questions.

1. Which of the following is *not* a component of an anti-lock brake system?
   a. power steering pump
   b. master cylinder
   c. solenoid valve body assembly
   d. wheel sensors
2. What is the name for the gas-filled pressure chamber that is part of the pump and motor assembly?
   a. controller
   b. accumulator
   c. sensor
   d. reservoir
3. Approximately how many control cycles are executed every second, depending upon road conditions?
   a. 10 to 14
   b. 14 to 20
   c. 20 to 24
   d. 4 to 10
4. What valve regulates the flow from the master cylinder pistons to the front brakes?
   a. controller valve
   b. solenoid valve
   c. sensor valve
   d. accumulator valve
5. What component monitors the electromechanical components of the system?
   a. electronic controller
   b. accumulator
   c. sensor
   d. master cylinder
6. When preparing to diagnose a malfunctioning anti-lock brake system, Mechanic A decides to visually inspect the system first. Mechanic B decides to inspect warning light behavior first. Who is right?
   a. Mechanic A
   b. Mechanic B
   c. both A and B
   d. neither A nor B
7. ABS systems
   a. can control front brakes individually or together.
   b. cannot control front brakes individually.
   c. always control rear brakes individually.
   d. can control rear brakes individually or together.
8. Mechanic A bleeds an anti-lock brake system with the accumulator charged. Mechanic B bleeds the system with a pressure bleeder without a fully charged accumulator. Who is right?
   a. Mechanic A
   b. Mechanic B

c. both A and B
d. neither A nor B

9. Under normal driving conditions, the anti-lock brake system
   a. does not operate.
   b. functions the same as a standard brake system.
   c. functions as a parking brake.
   d. none of the above.
10. Which of the following is *not* a process involved in a control cycle?
    a. eliminate pressure
    b. keep pressure constant
    c. reduce pressure
    d. increase pressure
11. Which warning light signals a possible malfunction in the anti-lock brake system?
    a. red BRAKE light
    b. amber ANTI-LOCK light
    c. either or both a and b
    d. neither a nor b
12. If an intermittent fault cannot be diagnosed using the indicator light sequence chart, the problem will most probably be caused by a defective.
    a. controller.
    b. electrical connection or wiring.
    c. chart.
    d. sensor.
13. Which of the following statements concerning ABS servicing is incorrect?
    a. Avoid contact with the terminal of the control module.
    b. Never disconnect the battery from the vehicle when the engine is running.
    c. Remove the tire and wheel assemblies when replacing wheel speed sensors.
    d. Use a fast charger for starting the engine.
14. What component is the heart of the anti-lock brake system?
    a. electronic controller
    b. reservoir
    c. booster
    d. accumulator
15. Rear wheel sensors can usually be removed from
    a. inside the engine compartment.
    b. inside the luggage compartment.
    c. underneath the vehicle.
    d. under the back seat.
16. What part of the anti-lock brake system does the job of charging the hydraulic accumulator?
    a. sensor
    b. hydraulic pump
    c. dedicated electric pump
    d. booster
17. Which of the following statements concerning anti-lock brake systems is incorrect?
    a. Loss of hydraulic fluid or power booster pressure will disable the system.
    b. The four-wheel anti-lock brake system is self-monitoring.
    c. The front brakes must be bled in the conventional manner with the accumulator charged.
    d. When this system is in use, the brake pedal can rise even as the brakes are being applied.
18. The Bosch ABS system
    a. eliminates the need for a separate vacuum booster.
    b. is more compact than the Teves system.
    c. has a special microprocessor-based tester for troubleshooting problems.
    d. all of the above.

CHAPTER 12

LIGHT-DUTY TRUCK BRAKES

OBJECTIVES

Upon completing this chapter, you should be able to

- Recognize and describe typical light-duty truck brake systems, including hydraulic brake systems, drum brakes, and disc brakes.
- Explain the procedure for replacing brake rotors in front-wheel, rear-wheel, and four-wheel drive models.
- Explain the procedure for replacing brake drums in the leading domestic light-duty trucks.
- Describe and service hub locks.
- Explain the operation of the rear anti-lock brake system.
- Recognize both electric and air brake systems.

Most light-duty trucks, like passenger vehicles on the market, are equipped with front-wheel disc brakes, while the drum type are used for the rear wheels. A few imports employ disc brakes on all four wheels.

LIGHT-DUTY TRUCK BRAKE SYSTEMS

Figure 12–1 illustrates a typical light-duty truck brake system.

HYDRAULIC BRAKE SYSTEM

Most light-duty truck hydraulic brake systems contain a dual master cylinder that performs in the following manner:

- *Rear Wheel Brakes*—The rear wheel brakes are connected to the secondary outlet port and are actuated by the secondary piston assembly.
- *Front Wheel Brakes*—The front wheel brakes are connected to the primary outlet port (farthest from dash panel) and are actuated by the primary piston assembly.

Both primary and secondary pistons function together. The proportioning valve, integral to the master cylinder, proportions pressure to the rear system. During heavy brake application, the proportioning valve reduces pressure to the rear brakes to avoid lockup.

Most modern light-duty truck hydraulic brake systems are assisted by a vacuum booster or hydro-boost (Figure 12–2). Many trucks are equipped with a vacuum pump, especially those vehicles with diesel engines. Complete operational and servicing data is given in Chapter 10.

The load- or height-sensing proportioning valve for the light-duty truck senses the vehicle load and proportions the braking between front and rear brakes in proportion to the load variations and degree of rear-to-front weight transfer during braking (Figure 12–3). The valve is located on the number four cross member and is activated through a linkage system that is connected to the rear axle housing cover. Pressure on the lever from the vehicle load will activate the height-sensing valve to regulate the hydraulic pressure to apply the rear brakes. For more information on the height-sensing proportioning valve, see Chapter 2.

Hydraulic brake system servicing that can be used for light-duty trucks is given in Chapter 4, while hydraulic brake bleeding techniques can be found in Chapter 5.

DRUM BRAKES

The rear brakes of most trucks are drum type with internal expanding shoes. The rear drum brakes are of the single anchor type, mounted to the same anchor, and actuated by one wheel cylinder. The wheel cylinder has two pistons. One piston exerts force against the upper end of the primary shoe; the other piston exerts force against the upper end of the secondary shoe.

The rear brake shoes are automatically adjusted when the vehicle is driven forward and reverse and the brakes applied several times sharply. Manual

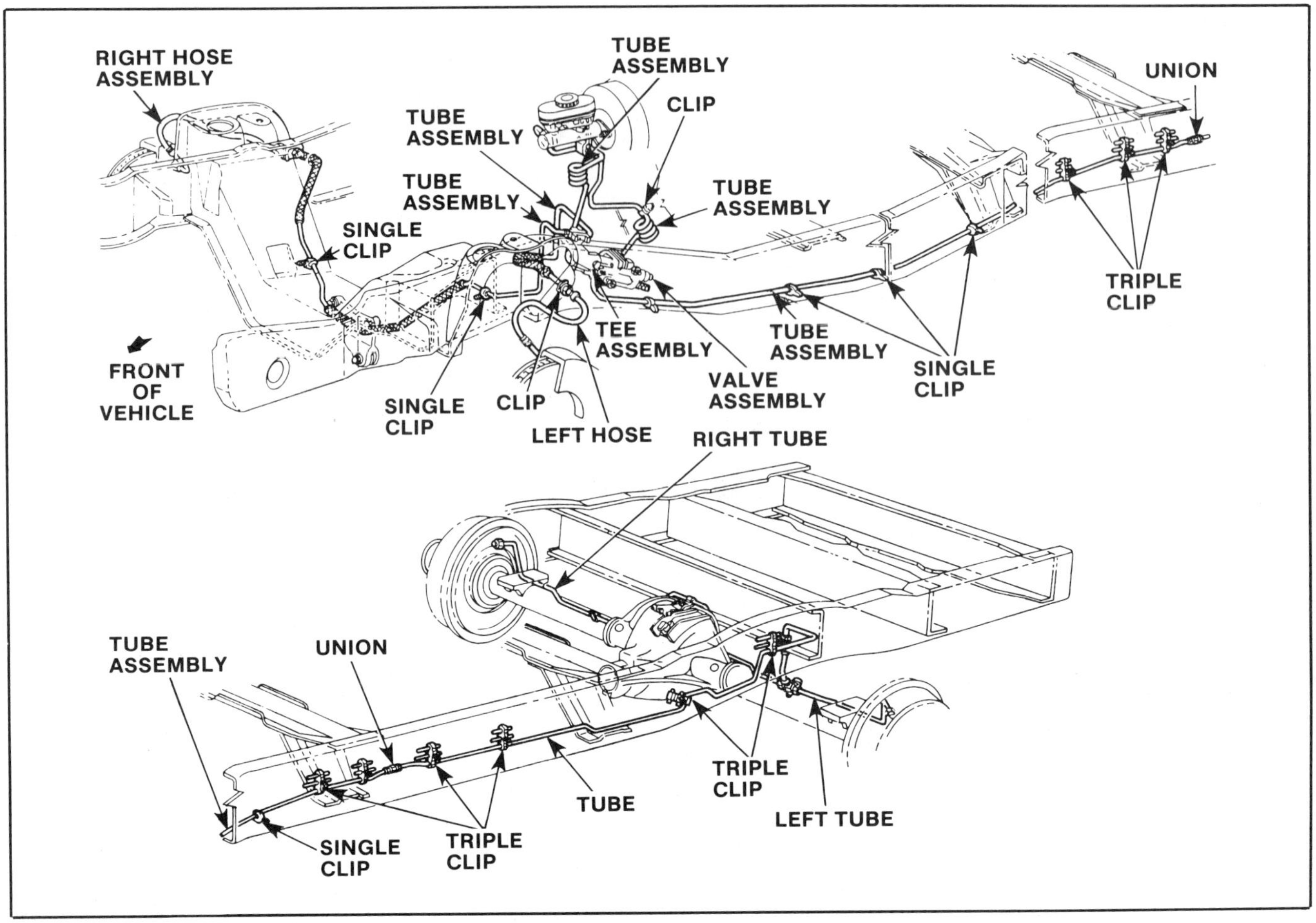

Figure 12-1 Typical light-duty truck brake system

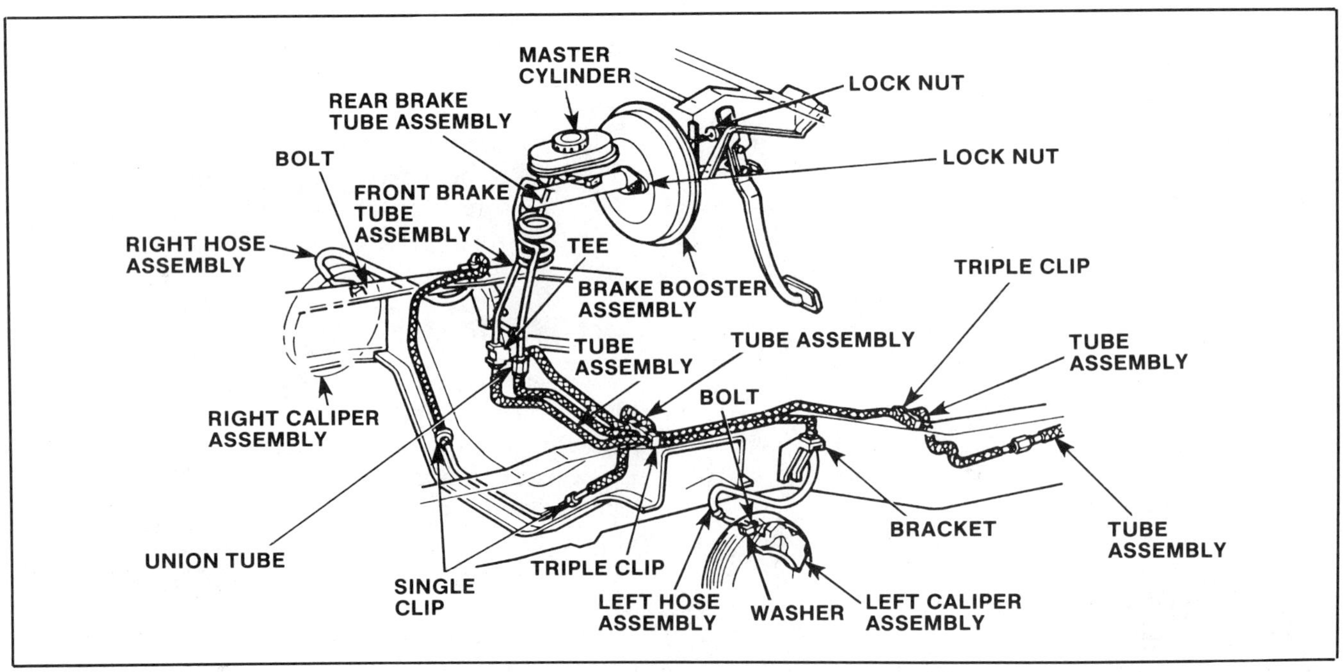

Figure 12-2 Hydro-boost on a light-duty truck hydraulic brake system

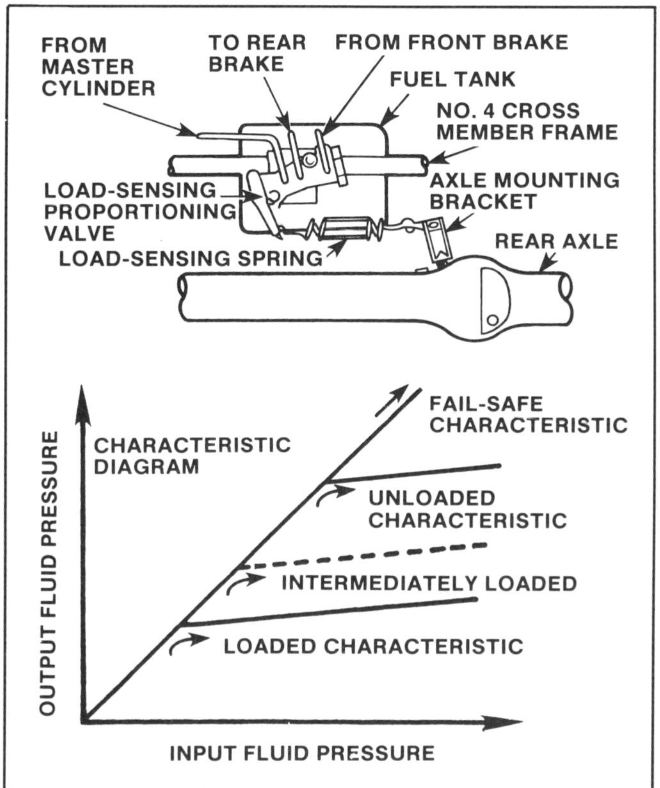

Figure 12-3 Operation of load-sensing proportioning valve

brake adjustment is required only after the brake shoes have been relined or replaced.

Figures 12-4 and 12-5 illustrate typical light and heavy drum brakes that are used on trucks. The primary brakes as with most passenger vehicles work in conjunction with the rear drum brakes.

The inspection and servicing of light-duty truck drums is essentially the same as for cars that were discussed in Chapter 6.

DISC BRAKES

Many so-called "heavy-duty" (HD) light-duty model trucks (4 × 2 or 4 × 4) are equipped with dual-piston sliding or floating caliper disc brakes. The floating caliper shown in Figure 12-6 has two pistons side by side in the housing. The caliper mounts on the anchor plate with mounting pins through bushings, protected by boot seals. The interchangeable pads attach to the caliper with two mounting pins. Two cap screws with washers attach the piston housing to the caliper. The anchor plate bolts directly to the spindle and the caliper sits forward. To service, procedure steps are as follows:

1. Remove pad mounting pins and anti-rattle springs.
2. Remove pads.

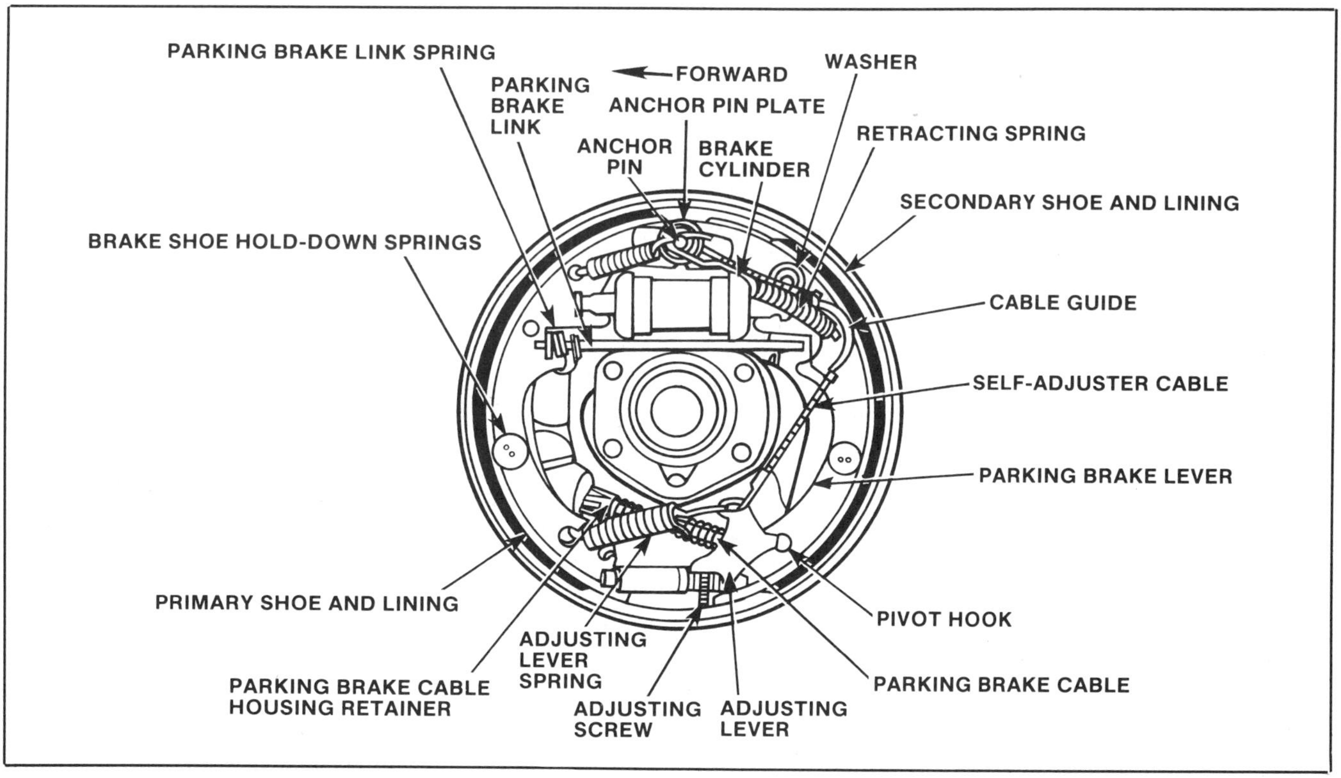

Figure 12-4 Typical light-duty truck self-adjusting brake assemblies

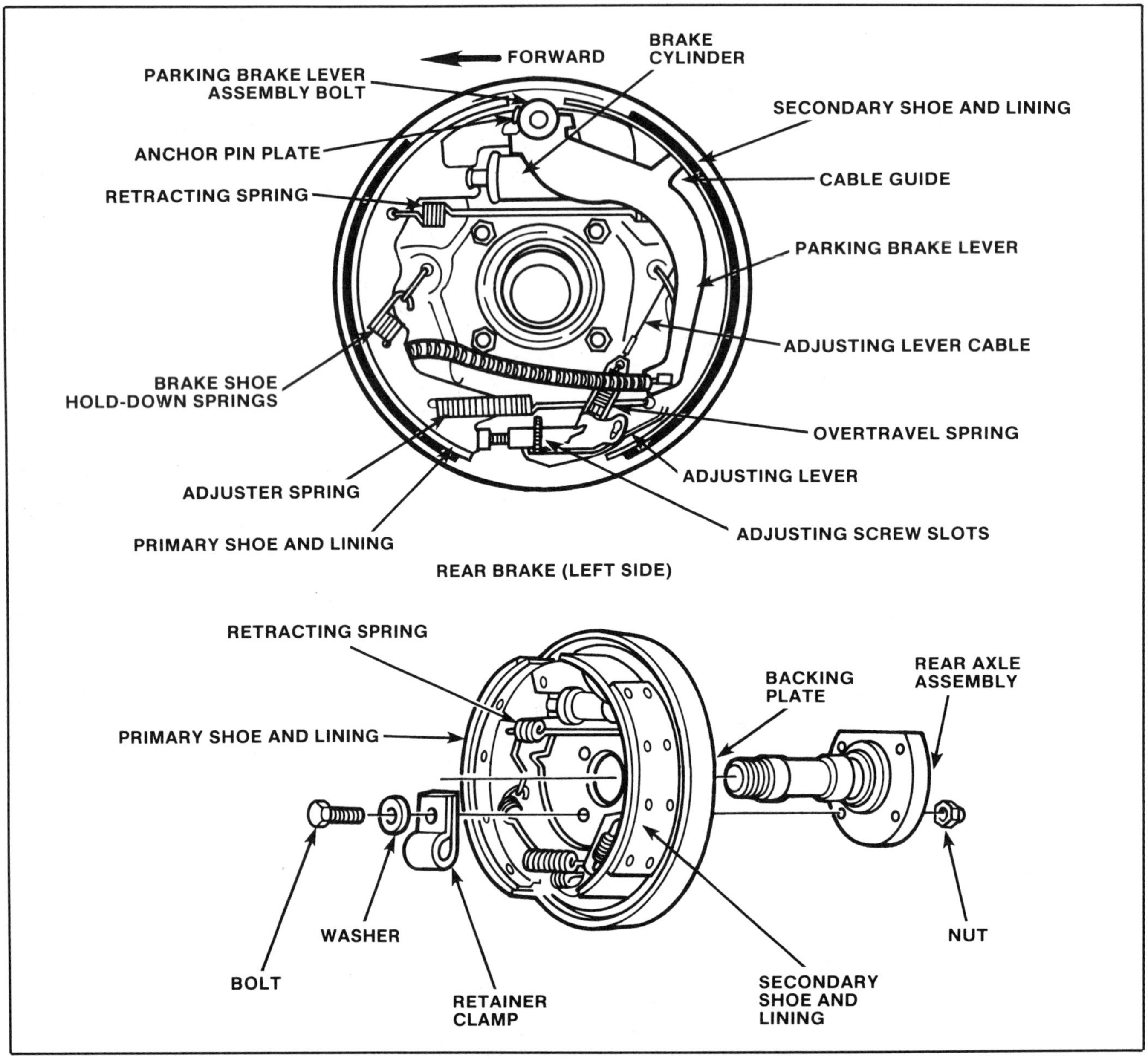

Figure 12-5 Heavy-duty self-adjusting brake assemblies

3. Disconnect brake hose.
4. Remove mounting pins and nuts holding caliper to the anchor plate.
5. Remove the caliper.
6. Crimp the chamfered end of the bushing, and remove with pliers.
7. Rebuild the caliper if necessary.
8. Assemble pin bushing boot onto caliper mounting pin bushing.
9. Insert pin bushing into anchor plate.
10. Peen the end of the bushing so that it conforms to the chamfered end of the anchor plate boss.
11. Insert the caliper mounting pin into the bushing from the chamfered end. Tap around the pin with a rubber mallet so that the bushing conforms to the anchor plate boss and is free in the bushing.
12. Insert the caliper mounting pin insulator into the large end of the anchor plate boss.
13. Lubricate the bushing.
14. Position the caliper over the rotor.
15. Install the caliper mounting pins.
16. Install the pads and their mounting pins and anti-rattle springs.
17. Install the brake hose.

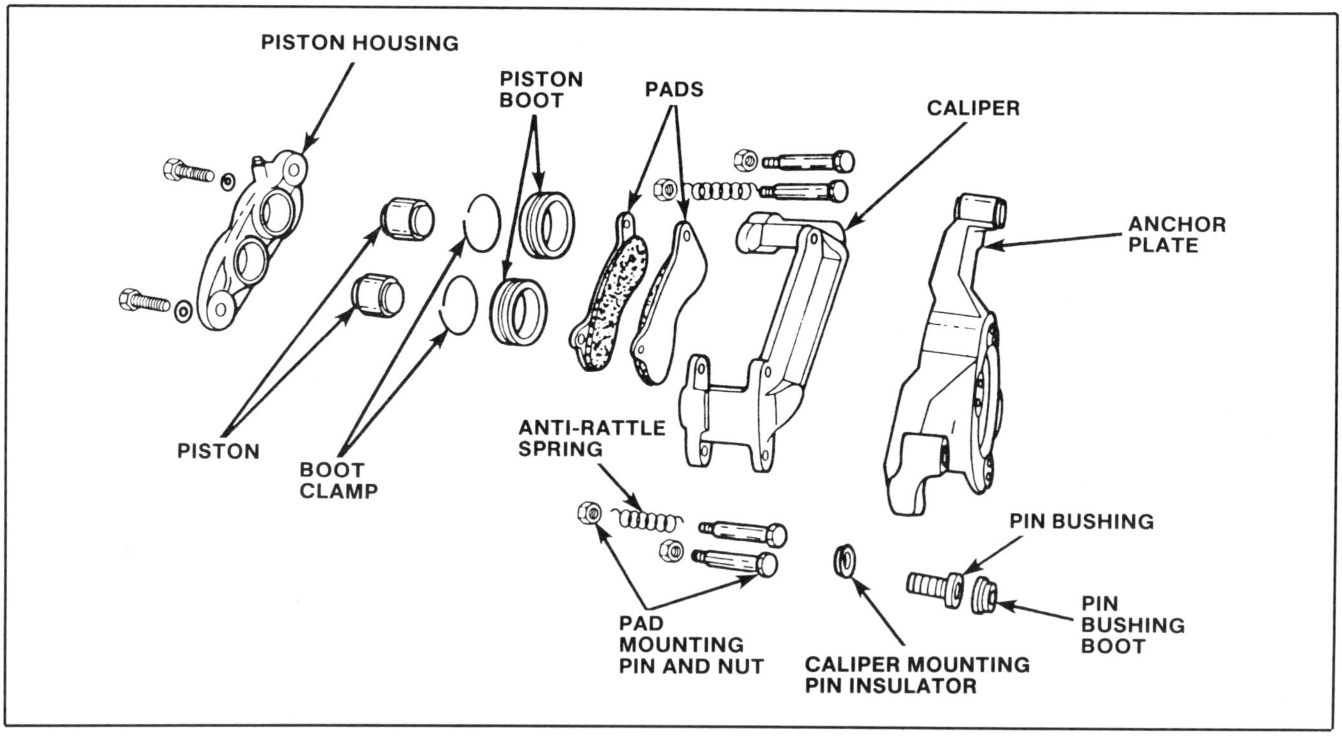

Figure 12-6 GM floating caliper disc brake

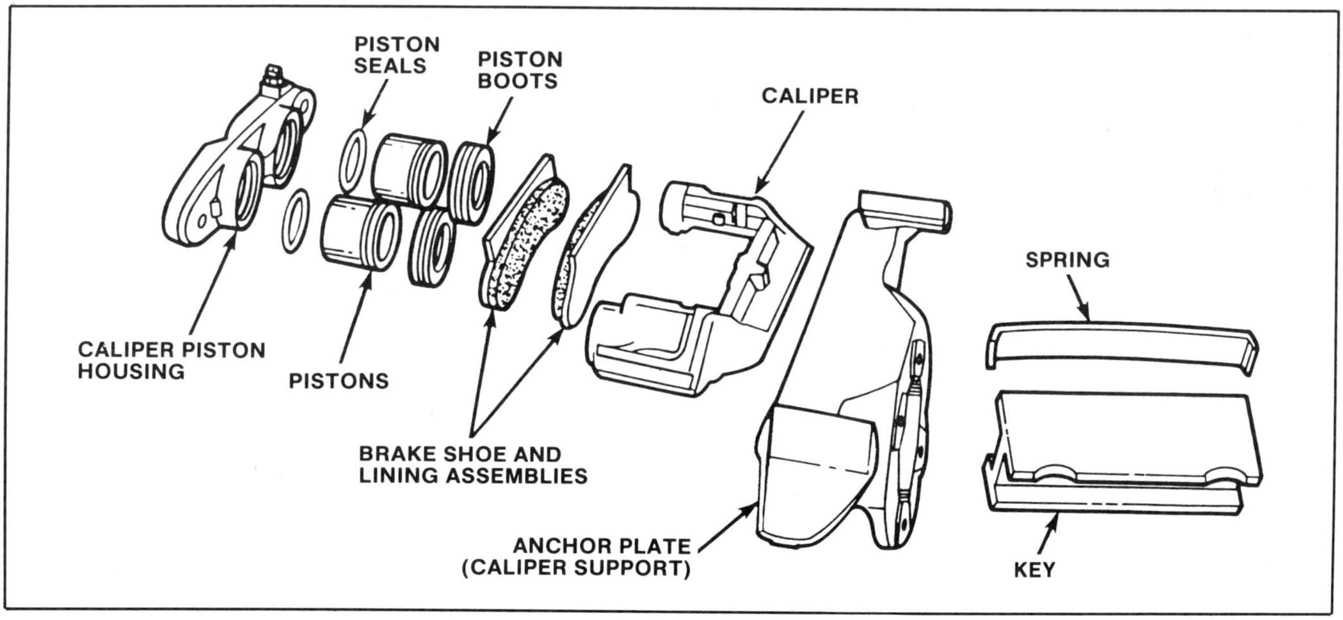

Figure 12-7 Ford dual-piston sliding caliper disc brake

18. Bleed the brakes.
19. Test the pedal and road test the vehicle.

The heavy-duty sliding caliper shown in Figure 12-7 has a layout similar to the floating two-piston, with two pistons on one side of the rotor. The caliper slides on and is supported by the anchor plate. It is held in place by a key and spring attached by a screw. One anti-rattle clip serves both pads, which slide on the caliper bridge. The pads are not interchangeable. The cylinder assembly attaches to the caliper with cap screws and

WHEEL TORQUE SPECIFICATIONS				
			Torque*	
Vehicle	Wheel	Bolt Size	N-m	Ft-Lbs
E-150, F-150, Bronco	5-Lug Wheel	1/2–20	135	100
E-250, E-350, F-250, F-350	8-Lug Wheel	9/16–18	190	140

*Torque specifications are for clean, dirt-and-paint free dry bolt and nut threads. Never use oil or grease on studs or nuts.

Figure 12-8 Typical manufacturer's disc brake torque and specifications

washers. The anchor plate and shield are bolted to the spindle with the caliper forward on the rotor.

1. Remove fluid from the master cylinder.
2. Remove the key retaining screw.
3. Use a brass drift and a hammer to drive out the key and spring.
4. Rotate the caliper away from the rotor.
5. Slide the opposite end of the caliper clear of the slide in the support.
6. Remove the caliper.
7. Remove the anti-rattle spring and pads.
8. Disconnect the brake hose.
9. Overhaul the caliper if necessary. When rebuilding, torque the caliper housing mounting bolts 155 to 185 foot-pounds.
10. Use a C-clamp to bottom the pistons.
11. Install the new pads.
12. Reconnect the brake hose.
13. Position the caliper rail into the slide on the support and rotate the caliper into position on the rotor.
14. Position the key and spring between the caliper and support. The spring should be between the caliper and the key with the tangs over the ends of the key.
15. Hold the caliper against the support with a screwdriver and hammer in the key and spring until the notch and hole correspond. Replace the screw and torque it 12 to 20 foot-pounds.

The disc brake assembly in a so-called "light" action light-duty truck is usually either single-piston floating or sliding caliper type. In fact, many of the brake designs described in Chapters 8 and 13 are suitable to "light" action trucks. Follow the servicing procedure given in these chapters.

Specific Disc Brake Rotor Replacement Procedures

The following information gives specific disc brake rotor replacement procedures for domestic light trucks:

FRONT-WHEEL DRIVE MODELS

1. Raise and support the front of the vehicle, then remove the wheel and tire assembly.
2. Remove the caliper from the mounting with brake hose attached and suspend it from the chassis using a piece of wire. Do not allow the caliper to hang by the brake hose.
3. Remove the disc brake rotor from the drive flange studs.
4. Reverse the procedure to install. Refer to the manufacturer's disc brake torque and specifications chart for caliper mounting bolt torque (Figure 12–8).

REAR-WHEEL DRIVE MODELS

1. Raise and support the front of the vehicle, then remove the wheel and tire assembly.
2. Remove the caliper from the mounting with brake hose attached and suspend it from the chassis using a piece of wire. Do not allow the caliper to hang by the brake hose.
3. Remove the dust cap, cotter pin, and wheel bearing adjusting nut and washer, then carefully remove the hub and brake rotor assembly from the spindle.
4. Reverse the procedure to install. Pack the wheel bearings with a suitable grease. Refer to the manufacturer's disc brake torque and specifications for wheel bearing adjustment and caliper mounting bolt torque.

FOUR-WHEEL DRIVE MODELS

1. Raise and support the front of the vehicle, then remove the wheel and tire assembly.
2. Remove the caliper mounting bolts, then lift the caliper from the mounting with brake hose attached (Figure 12–9), and suspend it from the chassis using a piece of wire.
3. Slide the disc brake rotor from the hub.
4. Reverse the procedure to install. Refer to the manufacturer's disc brake torque and specification chart for the correct caliper mounting bolt torque.

If rotor replacement is needed, most passenger car designs require only that the wheel and caliper be removed to allow the rotor to be pulled off the studs. On trucks, the rotor is often mounted behind the hub, or is part of the hub assembly. In this case, the hub must be removed. This might entail the removal of a locking hub or a drive gear and lock nuts and washers, spring cups, and bearings.

Care must be taken to see that these extra parts are assembled properly and are not damaged while off the vehicle. Reassemble according to factory recommended procedures.

Specific Drum Brake Replacement Procedures

The following are the brake drum replacement procedures for the leading domestic light-duty trucks. The drum installation procedure for Chevrolet and GMC trucks is as follows:

1. Raise and support the rear of the vehicle.
2. Remove the wheel and tire assembly.
3. Remove the bolts attaching the axle shaft flange to the hub.
4. On models equipped with Chevrolet or Spicer rear axles, tap on the axle shaft flange with a soft-faced mallet, then remove the axle shaft by twisting the shaft with locking pliers.
5. On models equipped with GMC or Timken (Rockwell) rear axles, remove the axle shaft using a suitable slide hammer.
6. Disengage the tang of the nut retainer from the lock nut, then remove the lock nut using a suitable tool.

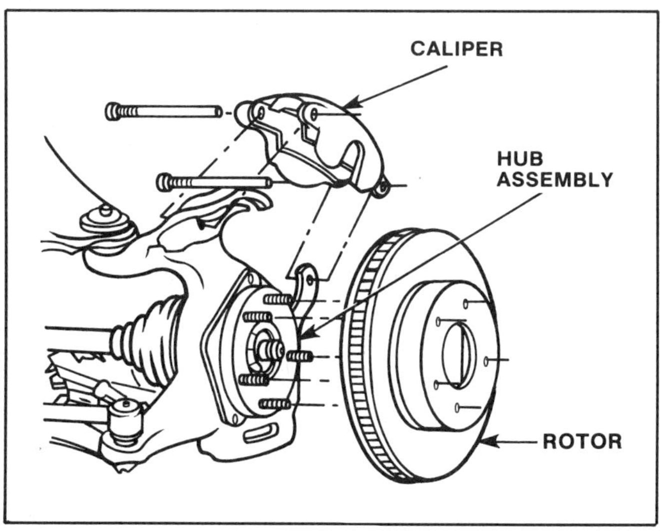

Figure 12-9 Removing caliper and rotor

7. Disengage the tang of the nut retainer from the adjusting nut, then remove the adjusting nut using a suitable tool.
8. Remove the thrust washer, then remove the brake drum and hub assembly from the axle housing.
9. Separate the brake drum from the hub. The brake drum can be retained to the hub by bolts, nuts, or wheel studs, depending on the model.
10. Reverse the procedure to install. Adjust the wheel bearing as follows:
    - Torque the inner adjusting nut to 50 foot-pounds while rotating the brake drum.
    - Back off the adjusting nut one-eighth turn.
    - Install the nut retainer, so that the short tang engages the adjusting nut slot.
    - Install and torque the outer lock nut to 250 foot-pounds, then bend the long end of the nut retainer into the slot of the lock nut.

Drum replacement procedure for Dodge and Plymouth trucks with Bendix heavy-duty 12- to 13-inch rear brakes is as follows:

1. Raise and support the rear of the vehicle.
2. Remove the wheel and tire assembly.
3. Remove the axle shaft retaining nuts, then remove the axle shafts. If the cones do not release, rap axle sharply at the center.
4. Using a suitable tool, remove the outer hub nut.
5. Straighten the lock washer, then remove the lock washer, inner nut, and bearing.
6. Remove the brake drum.
7. To install, position the brake drum on the axle housing.
8. Install the bearing and inner nut. Rotate the brake drum to seat the bearing rollers. Tighten the inner nut until a slight drag is at the drum, then back off the adjusting nut one-sixth turn. The brake drum should rotate freely with zero-to-slight end play.
9. Install the lock washer and outer nut. Bend the washer to lock the outer nut in place.
10. Position the gasket on the hub, then install the axle shaft, cone, and retaining nuts.
11. Adjust the rear drum brakes.

The drum installation procedure for Ford light-duty trucks is as follows:

1. Raise and support the rear of the vehicle.
2. Remove the wheel and tire assembly.
3. Loosen the brake shoe adjustment screw.
4. Remove the rear axle shaft retaining bolts and washers, then remove the rear axle shaft and gasket.

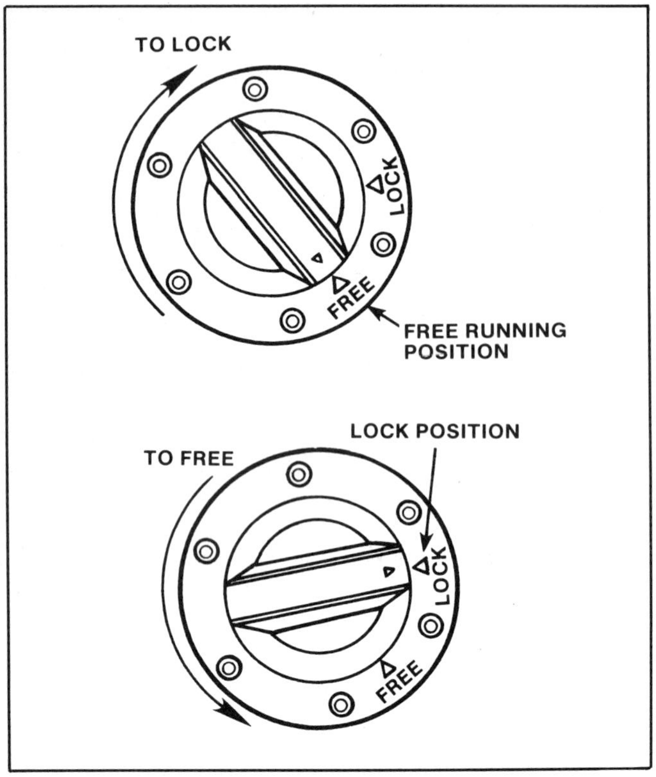

Figure 12-10 Four-wheel drive wheel hubs—manual locking

5. Remove the wheel bearing lock nut, lock washer, and adjusting nut.
6. Remove the brake drum-to-hub retaining bolts, then separate the brake drum from the hub.
7. To install, place the brake drum on the hub and install the attaching bolts.
8. Position the hub and drum assembly on the rear axle and start adjusting nut.
9. While rotating the drum, torque the adjusting nut 120 to 140 foot-pounds.
10. Back off the adjusting nut approximately one-eighth to three-eighths turn to obtain an end play of 0.001 to 0.010 inch. If the nut or locking wedge are damaged or do not allow for proper end play, they should be replaced.
11. Install the drive shaft and gasket, then install the torque attaching bolts 40 to 50 foot-pounds.
12. Install the wheel and tire assembly, then adjust the rear brakes.

LIGHT-DUTY TRUCK WHEEL HUBS AND BEARINGS

The hub locks on light-duty vehicles equipped with four-wheel (4 × 4) drive, either automatically or manually actuate the front driving axle (Figure 12–10). When actuated, the hub lock body assembly locks the hub and wheel and tire assembly to the front driving axle shaft. When released, the front driving axle shaft is disengaged from the hub body assembly and the hub and wheel-and-tire assembly rotate freely on the spindle.

Two tapered opposed roller bearings allow the hub and wheel-and-tire assembly to rotate on the

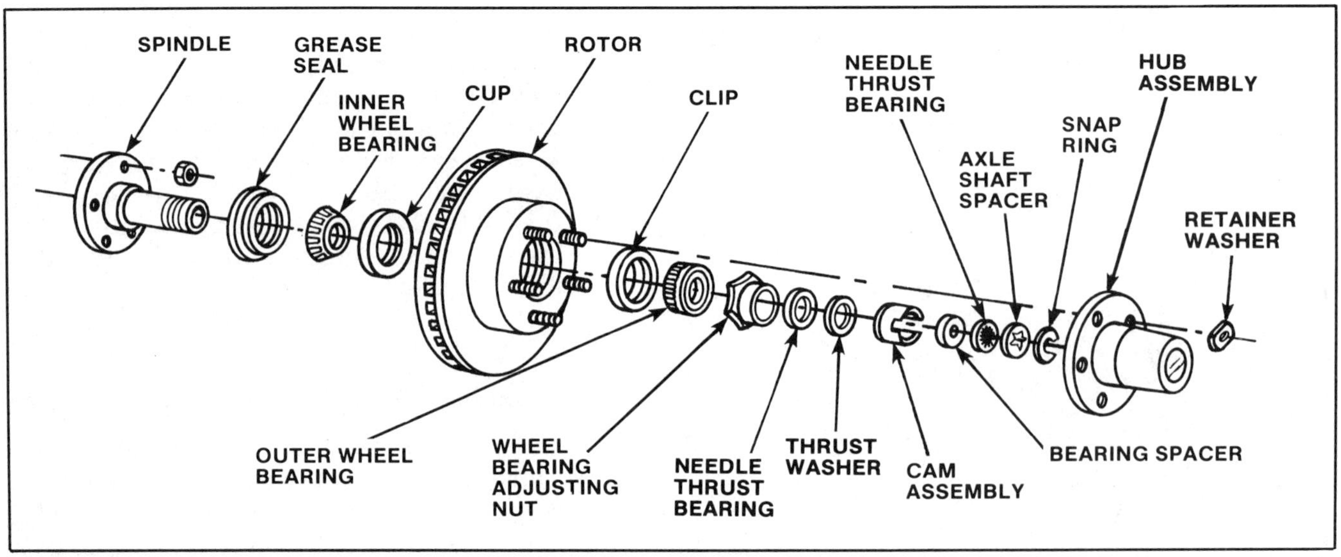

Figure 12-11 Automatic locking hub—disassembled

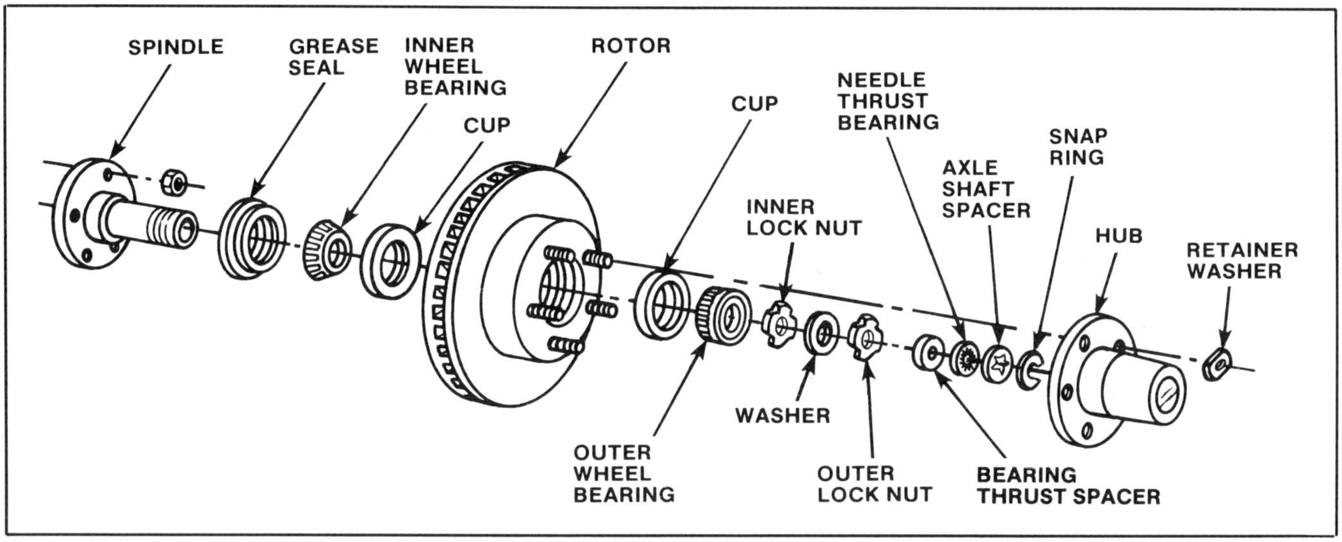

Figure 12-12 Manual locking hub—disassembled

spindle. A hub seal is installed inboard of the inner bearing to prevent wheel bearing lubricant from contaminating the brake caliper and rotor surfaces.

Some light-duty trucks (4 × 4) equipped with dual-wheel rear axles have the wheel-and-tire assembly attached to the drum with integral two-piece swiveling lug nuts.

AUTOMATIC LOCKING HUBS

The servicing procedure for a typical automatic locking hub, such as the one shown in Figure 2-11, is as follows:

1. Raise and support the front of the vehicle, then remove the wheel-and-tire assembly.
2. Remove the caliper from the mounting with brake hose attached and suspend it from the chassis using a piece of wire. Do not allow the caliper to hang by the brake hose.
3. Remove lug nut stud retainer washers, then the locking hub assembly.
4. Remove the snap ring from the end of the spindle shaft.
5. Remove the axle shaft spacer, needle thrust bearing, and bearing spacer.
6. Remove the cam assembly, thrust washer, and needle thrust bearing from wheel bearing adjusting nut.
7. Remove the wheel bearing adjusting nut, then slide the hub and brake rotor from the spindle.
8. Lubricate the wheel bearings with suitable grease, then install the brake rotor and hub assembly back onto the spindle.
9. Install the adjusting nut, then torque the nut to about 35 foot-pounds, while rotating the hub and rotor assembly.
10. Back off the adjusting nut one-quarter turn, then torque the nut to about 16 inch-pounds.
11. Align the nearest hole in the adjusting nut with the center of the spindle keyway slot. If necessary, advance the nut to the next hole.
12. Install the lock nut needle thrust bearing, thrust washer, and cam assembly.
13. Install the bearing thrust washer, needle thrust bearing, axle shaft spacer, and snap ring.
14. Install the lock hub assembly and retainer washers, then the wheel and tire assembly.
15. Ensure wheel end play on the spindle is 0.001 to 0.003 inch.
16. Install the brake caliper assembly. Refer to manufacturer's disc brake torque and specifications chart for caliper mounting bolt torque.
17. Install the wheel and tire assembly, then lower the vehicle.

MANUAL LOCKING HUBS

The servicing procedure for a typical manual locking hub such as the one shown in Figure 12-12 is as follows:

1. Raise and support the front of the vehicle, then remove the wheel and tire assembly.
2. Remove the caliper from the mounting with brake hose attached and suspend it from the chassis using a piece of wire. Do not allow the caliper to hang by the brake hose.

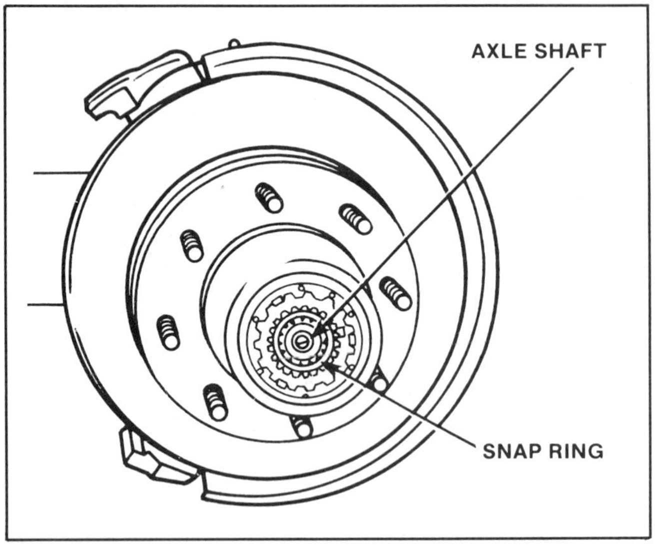

Figure 12-13 Snap ring removal from manual locking hub

3. Remove the lug nut stud retainer washers, then the locking hub assembly.
4. Remove the snap ring from the end of the spindle shaft (Figure 12-13).
5. Remove the axle shaft spacer, needle thrust bearing, and bearing thrust spacer.
6. Remove the outer wheel bearing lock nut using a suitable spanner wrench.
7. Remove the lock nut washer from the spindle (Figure 12-14).

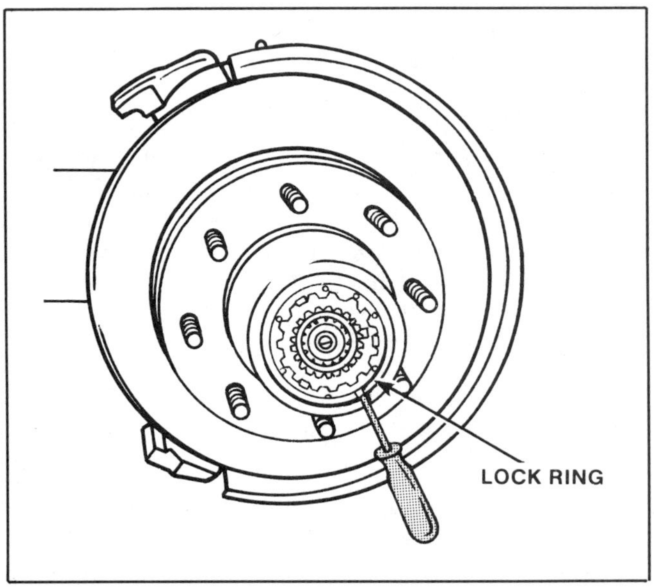

Figure 12-14 Lock nut washer removal from manual locking hub

8. Remove the wheel bearing inner lock nut, then slide the brake rotor and hub assembly from the spindle.
9. Lubricate the wheel bearing with a suitable grease, then position the brake rotor and hub assembly on the spindle.
10. Install the inner wheel bearing lock nut using a suitable spanner wrench, then torque the lock nut to approximately 35 foot-pounds.
11. Loosen the inner lock nut one-quarter turn, then install the lock washer on the spindle. If necessary, rotate the lock nut slightly to align the pin with closest hole in lock washer.
12. Install the outer bearing lock nut and torque to about 150 foot-pounds.
13. Install the bearing thrust spacer, needle thrust bearing, axle shaft spacer, and snap ring.
14. Install the locking hub assembly and retainer washers, then the wheel and tire assembly.
15. Ensure wheel end play on the spindle is 0.001 to 0.003 inch.
16. Install the caliper assembly. Refer to the manufacturer's disc brake torque and specification chart for caliper mounting bolt torque.
17. Install the wheel and tire assembly, then lower the vehicle.

CAUTION: Most hub nuts are right-hand (thread on the right spindle) and left-hand (thread on the left spindle). Each hub nut is stamped RH or LH to identify thread direction (Figure 12-15). Do not use power impact tools when performing operations on the hub nuts.

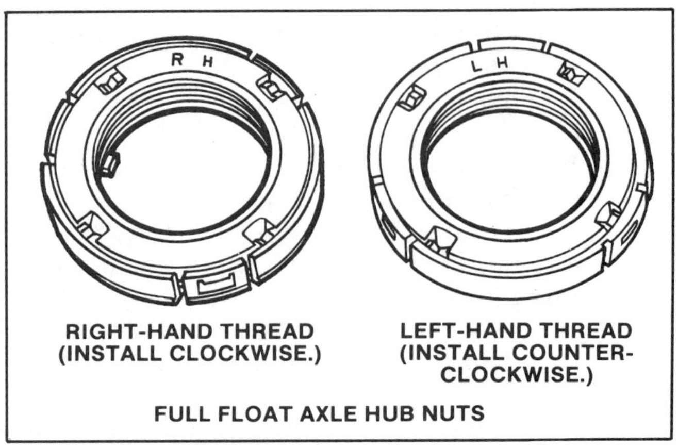

Figure 12-15 Hub nut installation

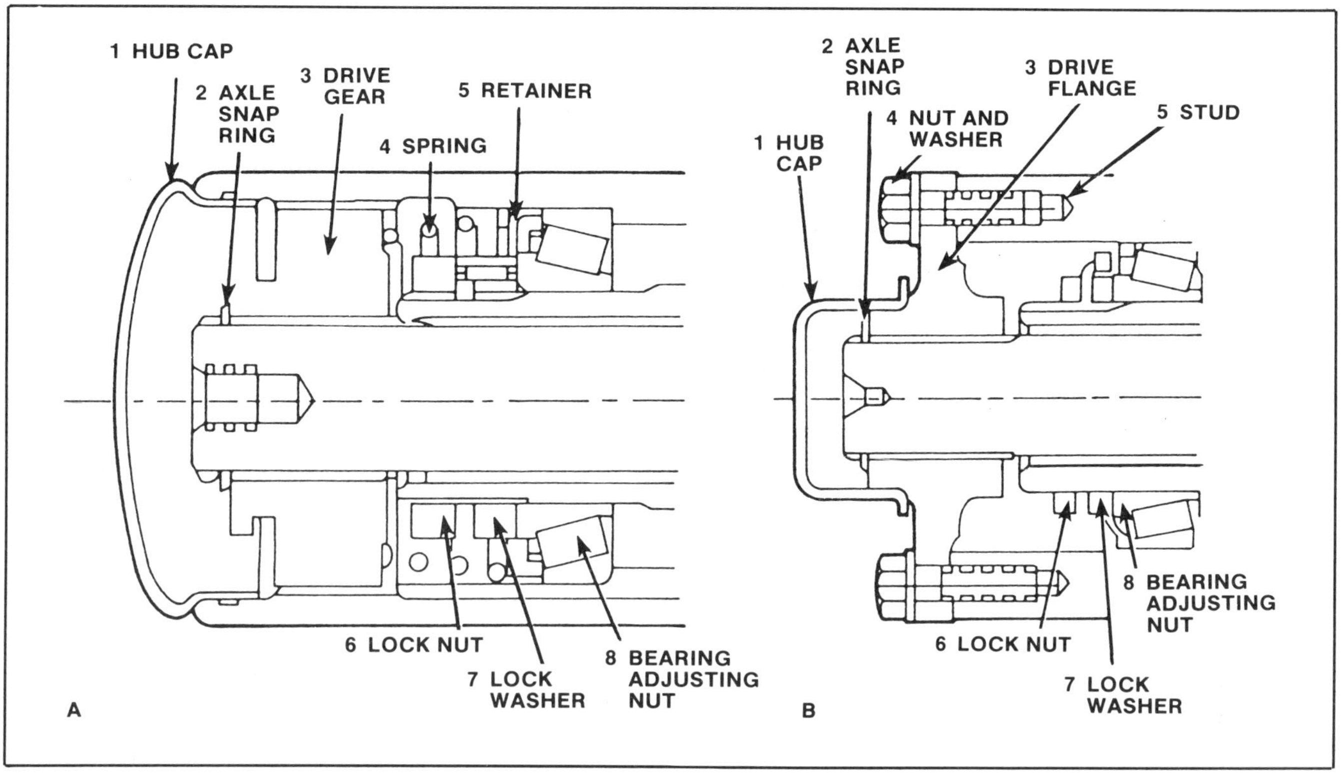

Figure 12-16 Front-wheel drive hub designs: (A) internal and (B) external

INSTALLING WHEEL LOCKOUT HUB

To install a wheel lockout hub, proceed as follows:

1. Select a hub lock to fit the hub design for the vehicle. Four-wheel drive vehicles use either external or internal type hubs on the front-drive axle; most recent models use internal locks.
2. Referring to Figure 12-16, remove items 1 through 5 on the internal hubs (1 through 4 on external hubs). Do not discard them; some will be reused. In some cases, item 5 in the external hub will require removal. (Hub lock installation kits contain instructions for specific applications.)
3. Carefully inspect the axle shaft splines to be sure they are free of nicks or burrs.
4. Separate the base and handle units of the hub lock assembly.
5. Apply a light coating of grease to the spline areas of the axle shaft and the hub lock base unit. Also apply a light coating of lubricant to the O-ring.
6. On external hubs, make sure the gasket surface is clean and smooth. Either the paper gasket furnished or RTV material can be used. Do not use a thick, soft gasket material.
7. Assemble the base unit to the wheel hub. O-rings make a snug fit on internal hubs and will usually require bumping with a soft mallet or the heel of the hand to get the base unit into place. Do not hit with a hammer. Lubricate O-rings.
8. Tighten mounting bolts or stud nuts on external hubs to 25 foot-pounds.
9. Reinstall the snap ring on the axle.
10. Push the hub lock handle into the base unit by hand. This will be a tight fit due to the O-ring.
11. Align screws holes "A" (Figure 12-17) with the groove in the base unit and install four screws. Do not force the screws; they will enter freely when proper alignment exists. Do not remove screws "B" from handle.

Hubs should be greased when wheel bearings are serviced. Do not pack the hub full. Also, lubricate O-rings. Hubs exposed to deep water should be disassembled, cleaned, inspected, and relubricated before reassembly.

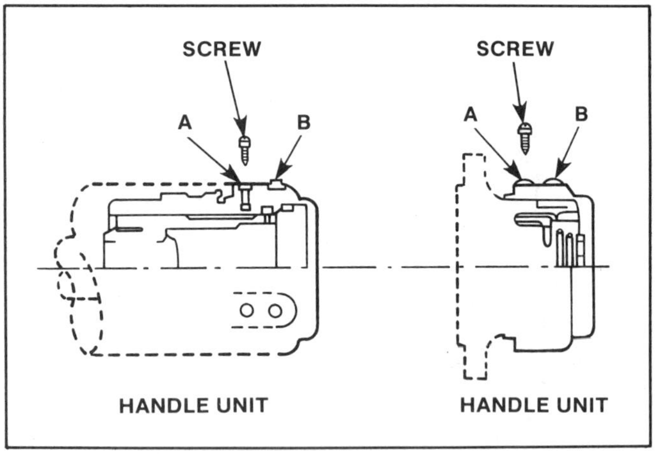

Figure 12-17 Installing screws in handle unit of wheel locknut hub

Generally, hub locks require no maintenance, but the hub should be serviced whenever brake work is done or wheel bearings are serviced. The wheel bearings should be serviced immediately after driving through sand or water to prevent any damage.

REAR ANTI-LOCK BRAKES

The rear anti-lock brake system (RABS) is being used more and more on light-duty trucks. Beginning with the 1987 model year, Ford trucks were available with anti-lock brakes on their rear axles.

Essentially, the Kelsey-Hayes rear anti-lock brake system (RABS) is an electronically controlled proportioning valve that enables modulation of the rear brakes to prevent lock-up, minimizing the tendency of a skid and reducing the overall stopping distance of the vehicle.

Components of the RABS, as illustrated in Figure 12-18, are a speed sensor and toothed (exciter) ring, a computer module, an electro-hydraulic brake proportioning valve and a wiring harness. The exciter ring (pressed onto the backside of the ring gear in the differential) and speed sensor operate together to generate an alternating current signal that indicates the speed of the rear wheels to the computer module. This signal is, in turn, evaluated by the control module. If corrective action is required, the control module sends an output signal

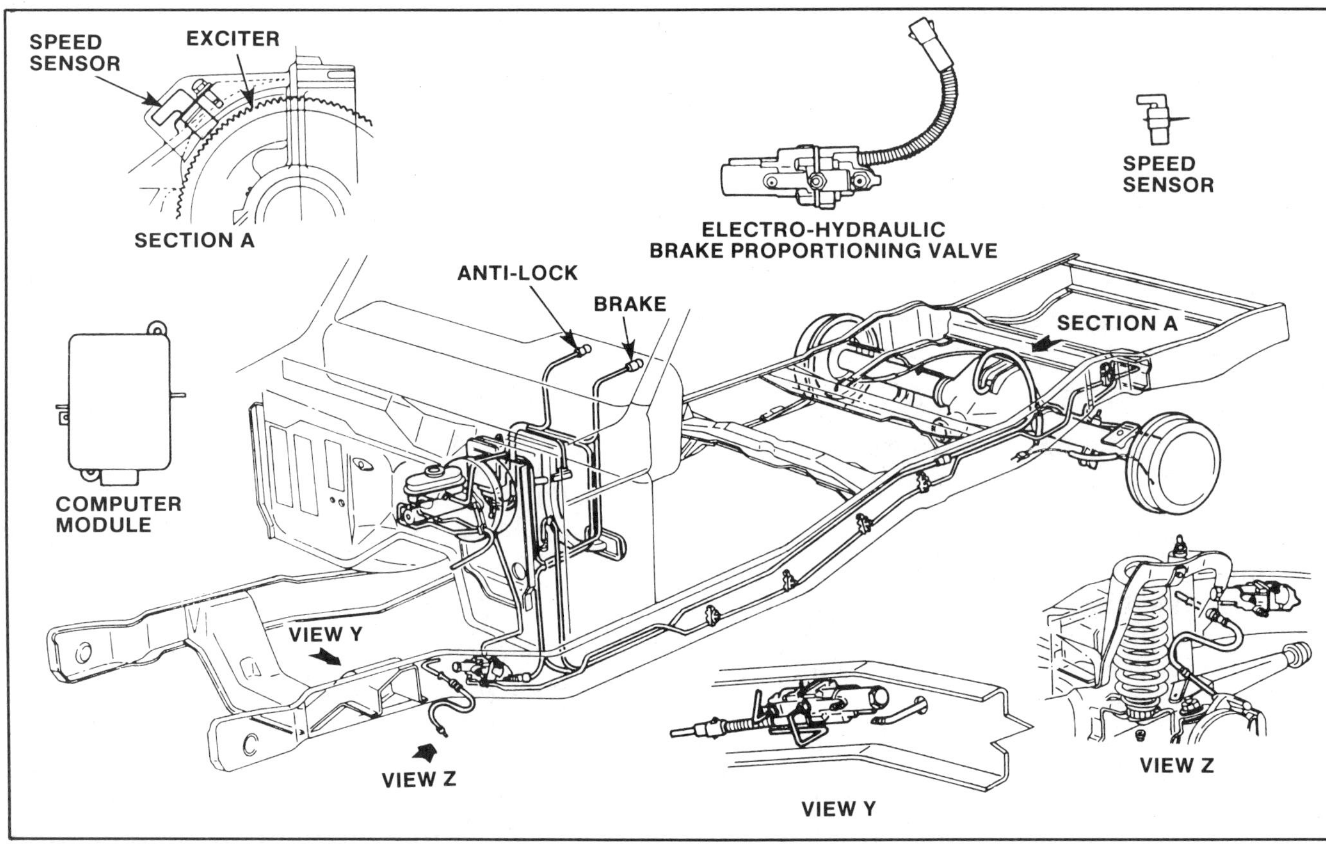

Figure 12-18 Main components of Kelsey-Hayes rear anti-lock brake system (*courtesy of Ford Motor Company)*

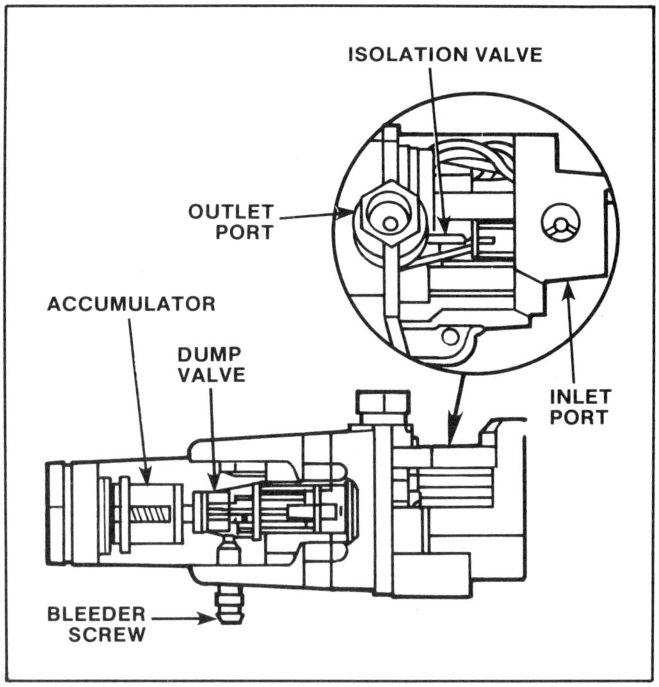

Figure 12-19 Electro-hydraulic proportioning valve

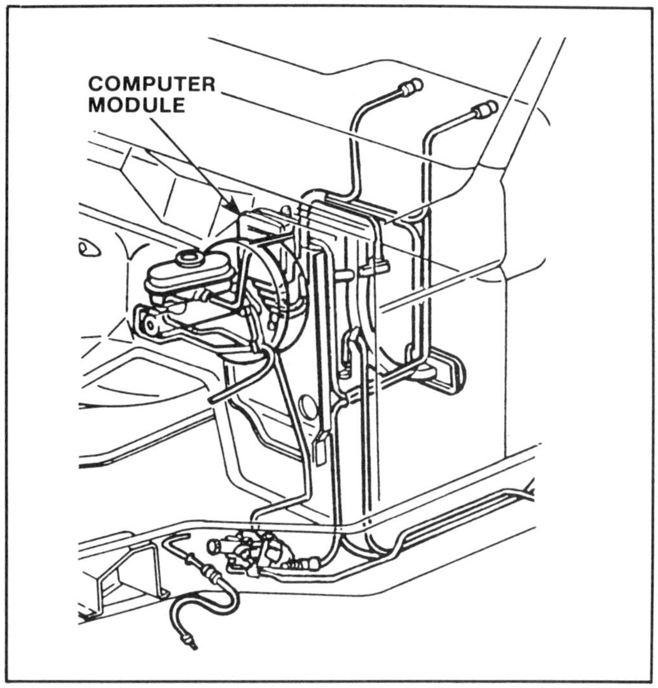

Figure 12-20 Computer module position on dash panel

through the wiring harness to the electrohydraulic (EH) brake proportioning valve (Figure 12-19).

The EH valve consists of a solenoid-operated isolation valve to isolate the rear brakes from the rest of the system, a solenoid-controlled dump valve to bleed pressure from the rear brakes, and an accumulator to store the fluid that has been bled off by the dump valve. During normal braking, brake fluid pressure from the master cylinder to the rear brakes passes unrestricted through the EH valve to the rear brakes. If, during hard braking, the control module senses that lock-up is likely to occur (reflected by the signal from the speed sensor), the control module (Figure 12-20) will energize the isolation valve located in the EH valve. This will cut off any further fluid pressure to the rear brakes and trap existing pressure to the rear brakes. If the signal from the speed sensor indicates that lock-up is going to occur, the control module will energize the dump valve located in the EH valve. The dump valve pulses rapidly to bleed off pressure from the rear brakes into the accumulator mounted on the EH valve, causing the hydraulic pressure to the rear brakes to decrease until lock-up no longer occurs.

In the event that the lock-up condition occurs only temporarily during braking, as soon as the control module determines that lock-up is no longer imminent, the control module will send a series of pulsing signals to the isolation valve to increase pressure to the rear brakes in order to minimize the stopping distance. Once the brake pedal has been released, the spring-loaded accumulator pushes the fluid back out and returns it to the master cylinder.

The rear anti-lock brake system has self-test capabilities much like those in other electronic control systems. Two warning lights are located in the instrument panel to alert the driver to a system malfunction. The red BRAKE warning light is used to indicate a low-fluid level condition, parking brake ON, or low vacuum (diesel only). During some conditions, the yellow warning light will go on only if the control module detects a malfunction in the anti-lock system that will affect the operation of the anti-skid system for approximately two seconds during ignition ON or engine cranking.

A typical self-test feature contains about 12 to 14 codes to indicate the area of the malfunction. When a malfunction is detected, the control module will shut down the system and the yellow anti-lock warning light will come on. This will permit normal braking. A code will be set in the system memory that can be retrieved by momentarily grounding the diagnosis pigtail and counting the flashes of the yellow light. If the key is turned off before the diagnostic is performed, the code will be lost. The light will remain out until the other mal-

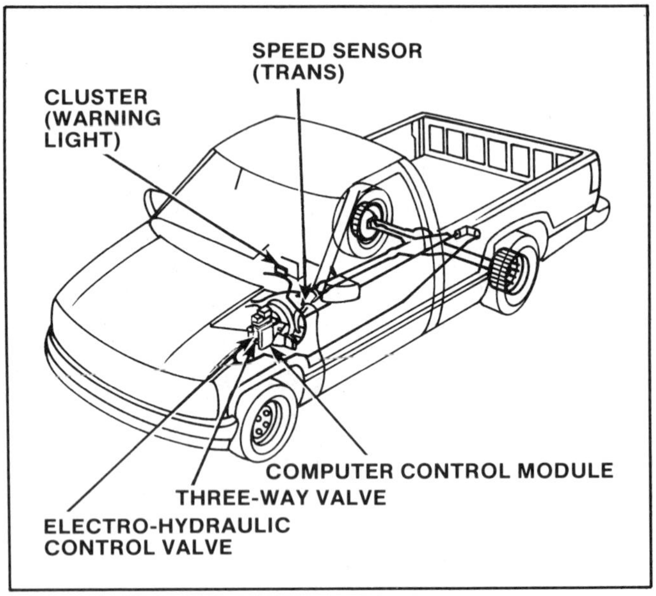

Figure 12-21 GM full-size light truck ABS from Delco

function is detected. If more than one fault exists, only the first code recorded will be displayed. Additional codes will output only after the first fault is corrected.

On vehicles with the Kelsey-Hayes RABS, normal brake service is performed as usual with one exception. After performing hydraulic service, begin pressure bleeding at the right rear, then left rear; then bleed the EH valve at the bleeder screw provided. Finish by bleeding the right front, then left front. As usual, use DOT 3 fluid only.

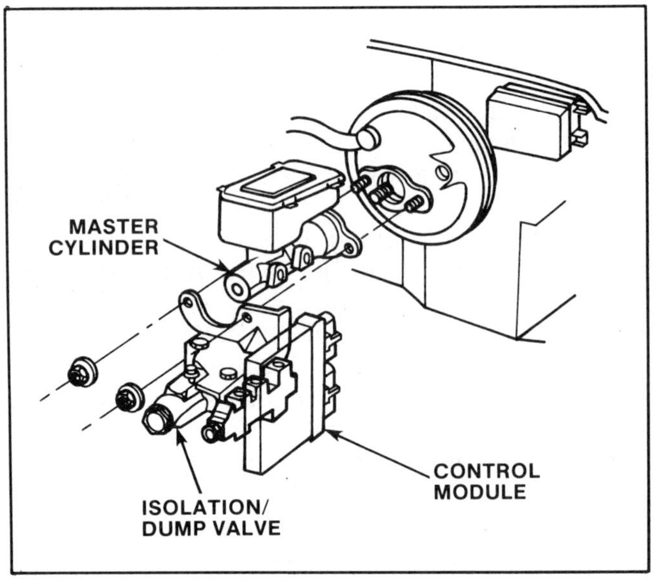

Figure 12-22 Valve and control module location on Delco ABS

Components of the RABS—EH valve, control module, and speed sensor—are non-serviceable and must be replaced rather than rebuilt if service is required. The procedures for the removal of faulty components and their replacement with new parts are given in the appropriate service manual.

DELCO ABS FOR GM LIGHT TRUCKS

In 1988 General Motors introduced the Delco anti-lock brake system for use on the rear wheels of the full-size light trucks. As can be noted in Figure 12-21, the components are similar to those used on the Kelsey-Hayes rear wheel anti-lock brakes introduced by Ford in 1987.

Operation of the Delco system is virtually identical to the operation of the Kelsey-Hayes system. The main difference between the two systems is in the location of the various components used in the system. The Delco system uses a transmission mounted speed sensor with the toothed ring on the transmission output shaft. The control module and EH valve are attached to the master cylinder assembly (Figure 12-22).

Brake service on the Delco system is identical to that on the Ford trucks. Diagnosis of an anti-lock malfunction also parallels that of the Ford system, although the Delco system has the ability to store trouble codes to be displayed through the warning light, even if the ignition system has been turned off. The Ford system loses all trouble codes once the ignition switch has been turned off.

ELECTRIC BRAKE SYSTEMS

Electric brakes are most often used on trailers. While the friction components are basically the same as for a hydraulic system, an electrical system is used to actuate the brakes instead of a hydraulic system. There are two basic electric brake systems with several variations.

In one system, the driver uses a foot- or hand-operated *rheostat*, or variable resistor, to control the amount of electrical current to an electromagnet. The electromagnet activates a cam that pushes the shoes against the drum. The braking force increases with increased current. When the current is stopped, the cam and brake shoes are pulled back to the release position by springs.

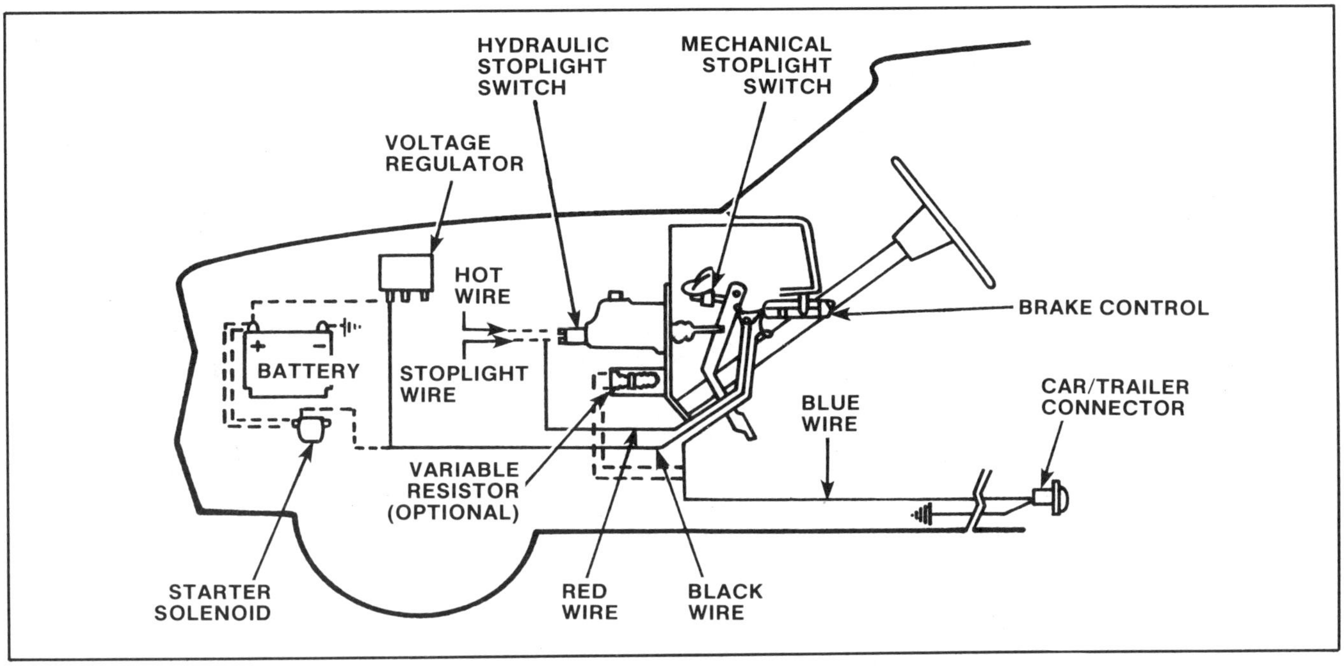

Figure 12-23 Hydraulically actuated electric brake control installation

An alternative to the foot- or hand-operated rheostat is one sensitive to brake hydraulic system pressure. When the brake pedal pressure is increased, hydraulic pressure actuates the rheostat that increases the current to the electromagnet. Figure 12-23 shows a hydraulically actuated electric brake control insulation.

The most common trailer brake system is the so-called "electric/electronic" arrangement. The key to this system is the braking sensor, which is composed of three parts: a weighted pendulum with a light separating tab, a light source, and a light sensitive transistor.

The braking sensor automatically senses the "G" force or stopping rate. This causes the pendulum to swing forward. The light-sensitive transistor can then "see" the light source that is normally blocked by the pendulum in its resting position. The amount of light reaching the transistor is controlled by the distance the pendulum swings. The output of the transistor is equal to the amount of light reaching it and is truly proportional to the vehicle's stopping rate. Once the amount of light is received, an electronic impulse is sent through the circuitry and the trailer brakes are applied in true proportion to the vehicle's stopping rate, bringing it to a safer and smoother stop. Figure 12-24 shows the wiring hookup for both 12-volt negative- and positive-ground electric/electronic hydraulic brake control systems.

ELECTRIC BRAKE SYSTEM DIAGNOSIS AND SERVICE PROCEDURE

The shoes, linings, drums, and attaching parts in the electric brake system are serviced in the same manner as those in a hydraulic brake system. All electrical connections must be clean and secure between the brake units of the wheels and the control devices. Poor connections or grounds will reduce braking effectiveness. If all connections are clean and secure and the circuit is unbroken, check the rheostat and/or brake control for proper operation. If either unit is faulty, replace it. A diagnostic table of electric brake system problems is given in Chapter 14.

SHOP TALK

When making the hydraulic connections, assemble the tube fittings to the brass connectors. The tube connections should be turned hand-tight, then with a wrench until a solid feeling is encountered. From that point, apply an additional one-sixth turn; do not overtorque because it may damage the fitting. Also do not run the tubing in a stright line. Form a small loop in the line before running it through the fire wall to permit flexing as shown in Figure 12-25.

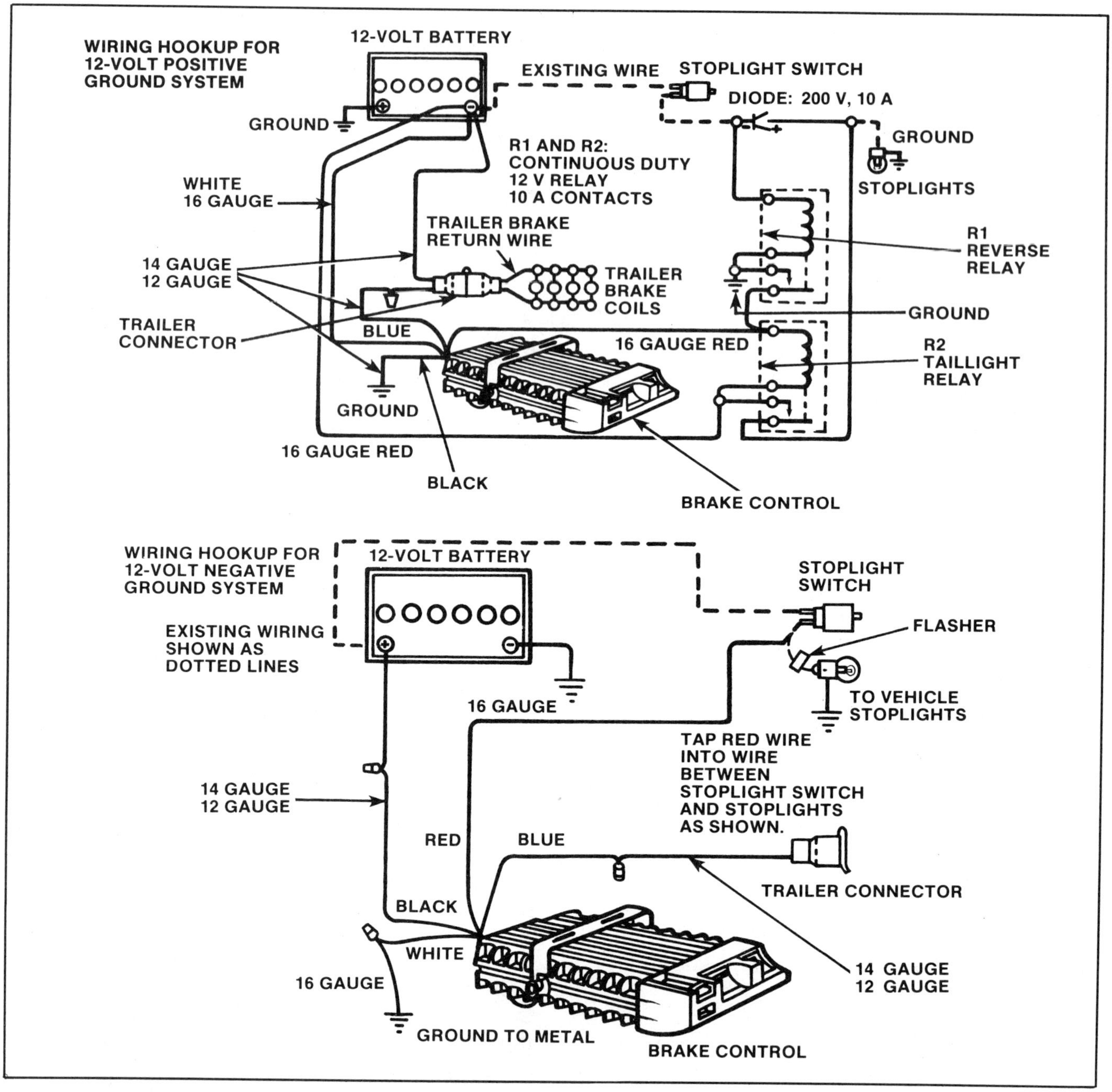

Figure 12-24 Electric/hydraulic brake control installation

> **CAUTION:** Do not crimp the line.

AIR BRAKE SYSTEMS

Air brakes and hydraulic brakes operate in a similar fashion. The difference is that air brakes are activated by compressed air instead of hydraulic pressure. Air brakes are used on heavier tracks and similar equipment. Figure 12-26 illustrates a typical air brake system's components.

When a hand-operated valve or a foot valve is operated, the brakes are applied. Air pressure is applied directly to the brake shoes through a diaphragm in the brake chamber through mechanical linkage. The compressor, driven by the engine, provides air pressure to the pressure tanks at ap-

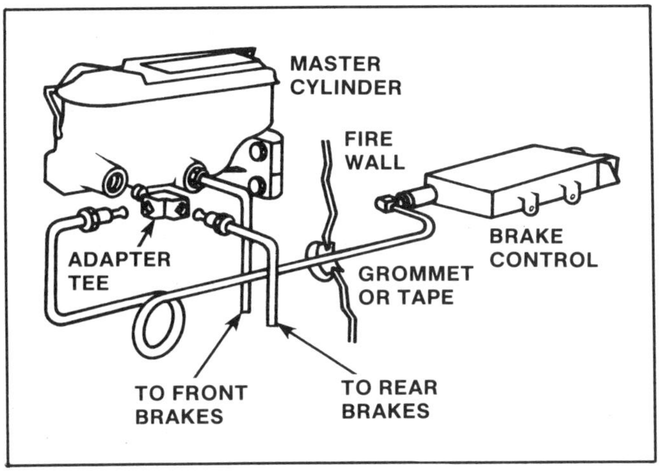

Figure 12-25 Tee installation

proximately 100 to 130 PSI. The governor cuts out the compressor operation when the pressure reaches about 130 PSI. The pressure drops when the brakes are applied. When the pressure drops to about 100 PSI, the governor again cuts in the compressor to raise the system pressure. When the system pressure drops to approximately 60 PSI, a warning buzzer sounds to alert the driver.

Basically there are three types of air brakes in use:

- Clam drum brake
- Wedge drum brake
- Disc brake

The *clam drum brake* (Figure 12-27) has been in use for many years. When air under pressure from

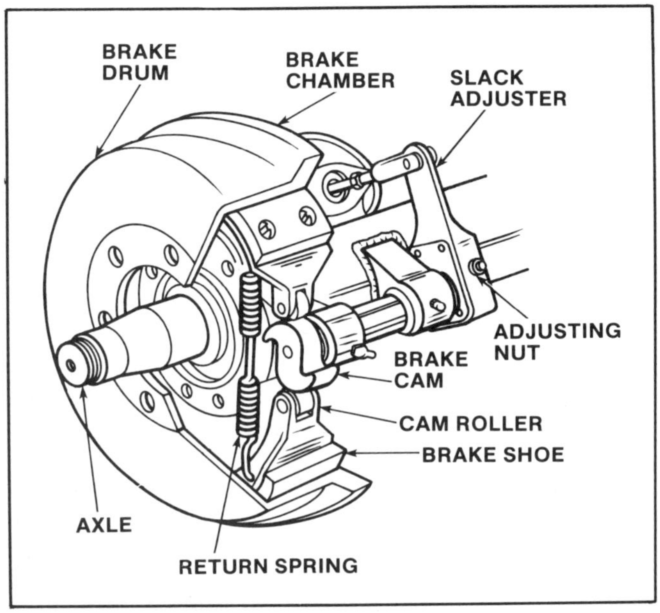

Figure 12-27 Brake shoe in contact with the drum in the clam-type brake

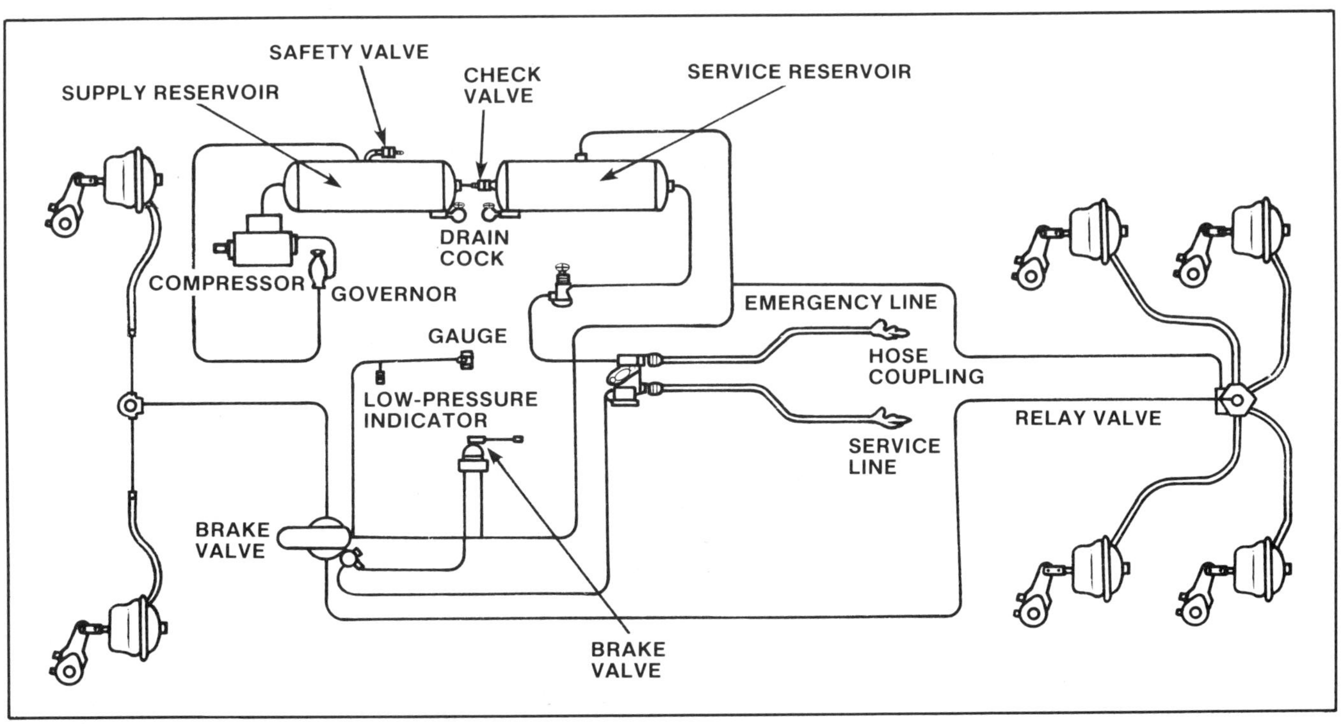

Figure 12-26 Typical air brake system components

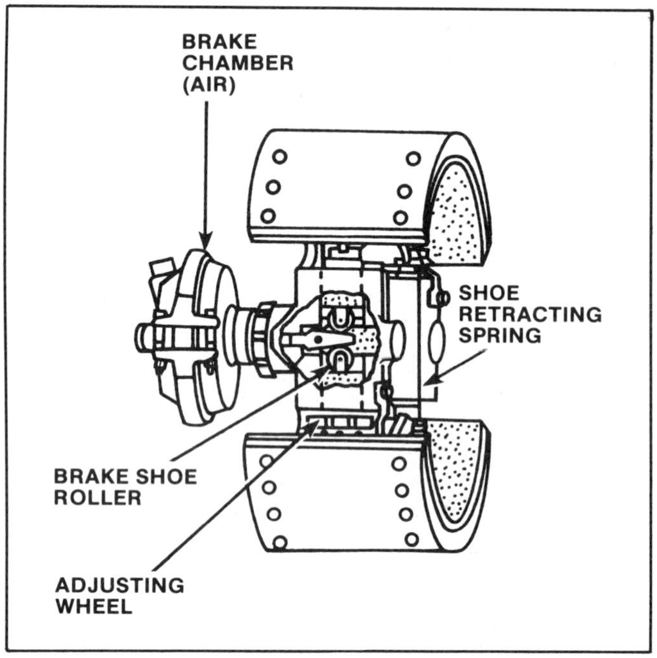

Figure 12-28 Typical single chamber wedge brake

the vehicle service tanks is directed to the brake chamber from the foot pedal by the driver, this air forces a rubber diaphragm, clamped around its outer circumference by the brake chamber, to invert itself. This action causes a pushrod to move forward against an internal return spring. Since the pushrod is attached to a lever (slack adjuster), this lever, which is splined to the brake camshaft, will cause it to rotate. Rotation of the camshaft forces the brake shoe rollers to be lifted as they run up the high point of the cam thereby forcing the brake shoes out against the rotating brake drum assembly.

With the *wedge design of air brake*, a wedge-shaped pushrod is forced by the action of the brake chamber pushrod between the brake shoe rollers, forcing the brake shoes against the drum.

The vehicle may be equipped with either a single chamber on each wheel (Figure 12-28) or with two chambers on each wheel, depending on the style and size of the vehicle. This type of brake arrangement is used a lot on vehicles with semiflotation tires. A good example is cement-mixer trucks.

Air-actuated disc brakes (Figure 12-29) have the following advantages:

- Better directional stability than with drum brakes because they do not depend on the servo action or wrap-up effect of drum brake shoes to increase the brake power. Therefore, straight line stops are assured with a disc brakes equipped vehicle.
- The clamping effect of disc brakes generates more braking power than a drum brake; therefore, shorter stopping distances are achieved with less tendency for the vehicle to pull, swerve, or weave.
- Disc brakes run cooler than drum brakes because of their design.
- Drum brakes are severely affected by water since the water is retained in the drum by centrifugal force as the drum rotates. With disc brakes, however, both water and contaminants are thrown off the brake surface of the rotor. In addition, the disc brake pads have a constant wiping action that assists in keeping the rotor clean and dry.

It is not the intention of this book to delve into the actual air brake system operation since there are a wide variety of schedules used on different classifications and applications of vehicles. Coupled with over 125 systems now in use in the United States, a separate book on air systems alone would be required to cover all of these to satisfaction; however, in Chapter 14, there is a simple diagnostic table of common air brake problems.

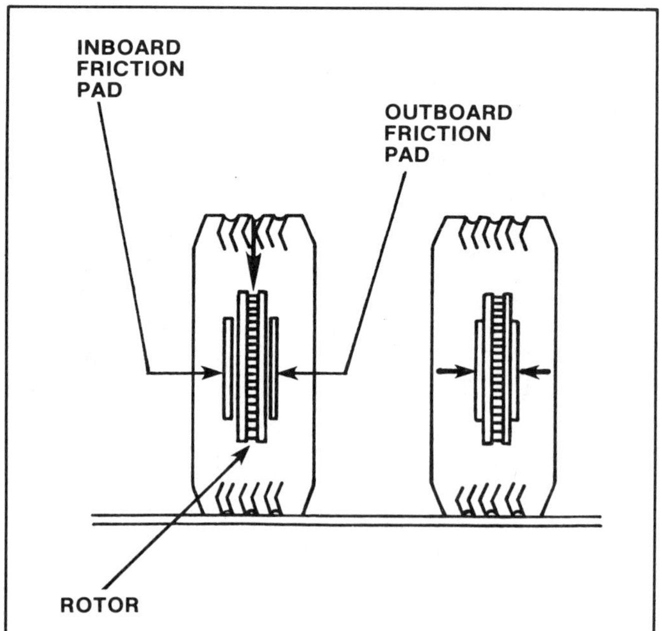

Figure 12-29 Air actuated disc brake. The basic diagram of disc brake whereby two friction pads are pushed toward the moving wheel brake rotor by air pressure as the brake pedal is activated.

REVIEW QUESTIONS

Review the information in this chapter by answering the following questions.

1. Most light-duty trucks on the market are equipped with
   a. disc brakes on front wheels and drum brakes on rear wheels.
   b. drum brakes on front wheels and disc brakes on rear wheels.
   c. drum brakes on all four wheels.
   d. none of the above.
2. What component coordinates the braking between front and rear brakes in proportion to the load variations and degree of rear-to-front weight transfer during braking?
   a. hydraulic valve
   b. master cylinder
   c. height-sensing proportioning valve
   d. vacuum booster
3. The rear brakes of most trucks are drum type with
   a. a single anchor.
   b. internal expanding shoes.
   c. one-wheel cylinder with two pistons.
   d. all of the above.
4. Many "heavy-duty" light-duty trucks are equipped with what types of disc brakes?
   a. dual-piston sliding
   b. floating caliper
   c. both a and b
   d. internal shoe
5. The disc brake assembly in a "light" action light-truck is usually
   a. single-piston floating.
   b. sliding caliper.
   c. both a and b.
   d. floating caliper.
6. On trucks, where is the rotor often mounted?
   a. behind the caliper
   b. behind the hub or part of the hub assembly
   c. under the wheel
   d. in front of the hub
7. What component on light-duty vehicles equipped with four-wheel drive actuates the front driving axle?
   a. hub locks
   b. axle shaft
   c. roller bearings
   d. hub seal
8. When servicing locking hubs on a light-duty truck, Mechanic A uses power impact tools when performing operations on the hub nuts. Mechanic B uses manual tools. Who is right?
   a. Mechanic A
   b. Mechanic B
   c. both A and B
   d. neither A nor B
9. After servicing the brakes and wheel bearings on a truck, Mechanic A services the hub. Mechanic B says hub locks require no maintenance so the hub need not be serviced. Who is right?
   a. Mechanic A
   b. Mechanic B
   c. both A and B
   d. neither A nor B
10. Electric brakes are used most often on
    a. passenger cars.
    b. high-performance cars.
    c. light-duty trucks.
    d. trailers.
11. On a heavy-duty sliding caliper, the caliper slides on and is supported by the
    a. anti-rattle clip.
    b. retainer clamp.
    c. retainer plate.
    d. anchor plate.
12. The types of air brakes in use are
    a. clam drum brake and wedge disc brake.
    b. wedge drum brake and clam disc brake.
    c. clam drum brake, wedge drum brake, and disc brake.
    d. drum, disc, and wedge.
13. What prevents wheel bearing lubricant from contaminating the brake caliper and rotor surfaces?
    a. exciter ring
    b. hub seal
    c. RABS valve
    d. none of the above
14. Most modern light-duty truck hydraulic brake systems are assisted by
    a. hydro-boost.
    b. a vacuum booster.
    c. both a and b.
    d. neither a nor b.
15. On most light-duty truck hydraulic brake systems with a dual master cylinder, which wheel brakes are connected to the primary outlet port and are actuated by the primary piston assembly?

a. front-wheel brakes
b. rear-wheel brakes
c. all four wheel brakes
d. none of the above

16. What valve is located on the number four cross member and is activated through a linkage system that is connected to the rear axle housing cover?
a. metering valve
b. proportioning valve
c. adjusting lever
d. anti-lock valve

17. In the air brake system, at what pressure will a warning buzzer alert the driver?
a. 130 PSI
b. 100 PSI
c. 0 PSI
d. 60 PSI

18. The heavy-duty sliding caliper has a layout similar to the
a. floating two-piston caliper.
b. floating single-piston caliper.
c. drum brake.
d. parking brake.

19. In the hub body assembly, two tapered opposed roller bearings allow the hub and wheel and tire assembly to rotate on the
a. rotor.
b. axle shaft.
c. spindle.
d. hub seal.

20. When servicing locking hubs, be sure that wheel end play on the spindle is
a. 0.001 to 0.003 inch.
b. 0.01 to 0.13 inch.
c. 0.1 to 0.3 inch.
d. 1 to 3 inches.

CHAPTER 13

IMPORT VEHICLE BRAKES

OBJECTIVES

Upon completing this chapter, you should be able to

- Recognize special features on import brake systems not found on domestic vehicles.
- Describe and service the Girling, ATE, Sumitomo, and Teves fixed calipers.
- Describe and service the ATE, Girling, Akebono, Renault/Fiat, and Lockheed/Akebono sliding calipers.
- Describe and service the ATE, Girling, Mazda/Toyota, Nissan (Datsun), and Sumitomo floating calipers.
- Explain the function and design of rear disc brakes, including Akebono, Girling, and ATE rear disc brakes.
- Explain the design of non-servo drum brakes on some European and Japanese cars.
- Explain the operation of two anti-lock braking systems currently used by imports: Teves and Bosch.

Import vehicle brakes have the same basic designs as the brakes of domestic cars: disc and drum. Import vehicles are also equipped with power brakes similar to the types described in Chapter 10. In addition, the types and servicing of the hydraulic systems given in Chapters 2 and 4 generally hold true for the imports (Figure 13-1). While the method of bleeding the hydraulic system given in Chapter 3 is basically correct for imports, the bleeding sequence should be checked in the vehicle's service manual since the procedure might vary from that of American cars.

The general operation of disc and drum is the same for both imports and domestics. However, import brake systems have special features that are not found in domestic vehicles. These features, as well as servicing data, are covered in this chapter.

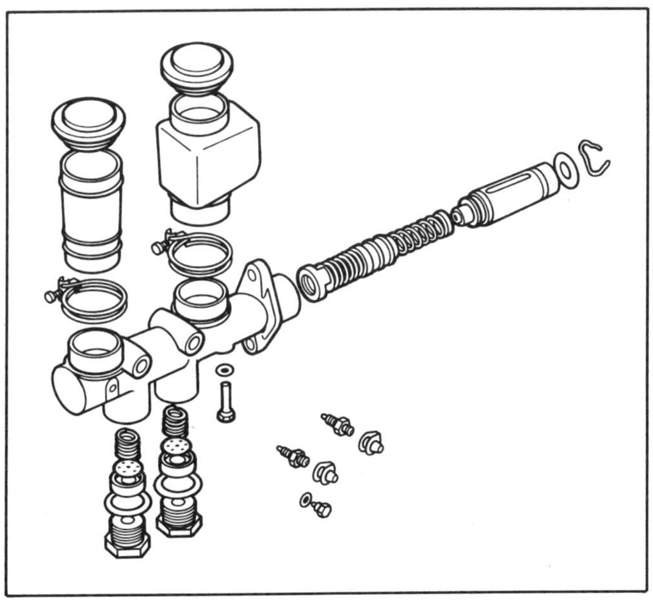

Figure 13-1 Import master cylinder

FIXED CALIPERS—GIRLING, ATE, SUMITOMO, TEVES

Fixed calipers by these manufacturers have been used on various models of imported cars, including BMW, Capri, Colt, Nissan, Mercedes-Benz, Saab, Toyota, Volkswagen, and Volvo.

GIRLING FOUR-PISTON CALIPER

The Girling four-piston (Figure 13-2)—similar to the other manufacturers' designs—is a fixed caliper brake. It has inner and outer halves that are

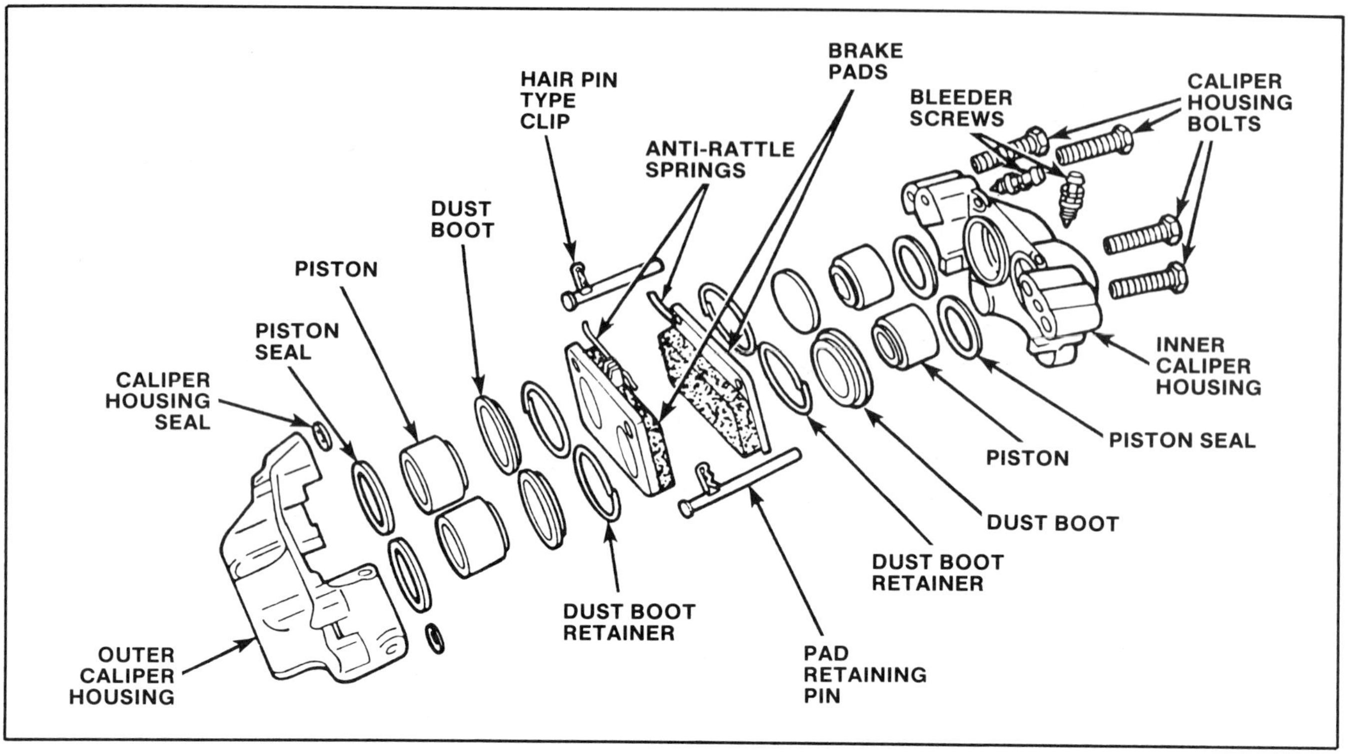

Figure 13-2 Girling four-piston caliper

bolted together with four mounting bolts and two different O-ring seals. The pads have anti-rattle springs attached on top. If one of the seals leaks or one of the pistons sticks, the entire caliper must be rebuilt. When disassembling this caliper, note the positions of damping shims and anti-rattle springs so that they can be replaced properly.

Before starting any servicing procedures on import disc brakes, remove half of the brake fluid from the master cylinder reservoir. Discard the fluid and reinstall the cover.

The final steps of any import brake servicing procedure, like those of domestic models, is to bleed the brake system, test the brake pedal, and road test the car.

When servicing a Girling four-piston caliper, proceed as follows:

1. Pull hairpin clips out of the pad retaining pins.
2. Pull out one of the retaining pins while holding the anti-rattle springs in place.
3. Remove the anti-rattle springs and the other pad retaining pin.
4. Force the pads away from the rotor and remove from the caliper.
5. Overhaul the caliper, if necessary.

CAUTION: Do not separate the caliper halves. Assembling the caliper halves requires special pressure-testing equipment, if the two halves are separated.

6. Position the damping shims with the directional arrow toward the forward rotor rotation.
7. Install a pad retaining pin.
8. Position the anti-rattle springs.
9. Install the outer pad retaining pin.
10. Install the hairpin clips.

ATE (TEVES) TWO-PISTON FIXED CALIPER

The ATE two-piston fixed caliper (Figure 13-3) is similar to the four-piston type with its four housing bolts and two O-ring seals. Pads are separated by a spreader spring. A retaining plate for the piston and a spring ring for the dust boot are used. The bleeder screw has a dust cap. This caliper does not have to be disassembled to replace the brake pads.

1. Disconnect the hydraulic hose and hose retainer clip.

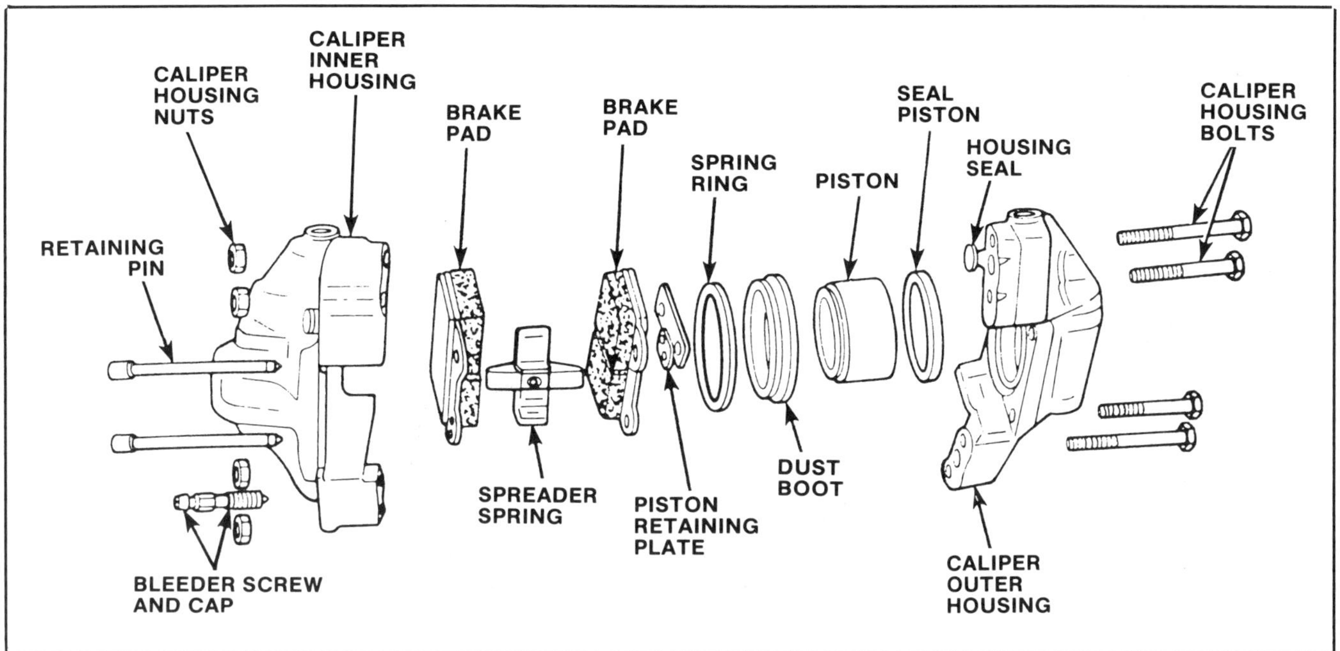

Figure 13-3 ATE two-piston fixed caliper

2. Remove the two mounting bolts from the mounting bracket.
3. Remove the brake hose.
4. Remove the spring locks and tap out the two retaining pins.
5. Lift out the spreader spring.
6. Force the pads away from the rotor and remove.
7. Rebuild the caliper one side at a time using a C-clamp to hold in the opposite piston.
8. Replace the caliper and torque the mounting bolts to 43 foot-pounds; secure the locking plate tabs.
9. Reconnect the hose and install the retaining clips at the bracket.
10. Bottom the pistons and install the new pads one at a time.
11. Replace the spreader spring, retaining pins, and spring locks, if used.

Volvo uses a triangulated hydraulic system. The front brakes are Girling four-piston fixed caliper type, and the rear service brakes are either Girling or ATE dual-piston fixed caliper. The master cylinder is a tandem type that feeds hydraulic pressure to two separate circuits.

One circuit applies pressure to the lower pistons on both front calipers and to both pistons on the right rear caliper. The other circuit applies pressure to the upper pistons on both front calipers and both pistons on the left rear caliper. If one circuit fails, the vehicle still has both front calipers working (only on half of their pistons, however) along with one of the rear calipers. Figure 13-4 illustrates the Volvo bleeding sequence.

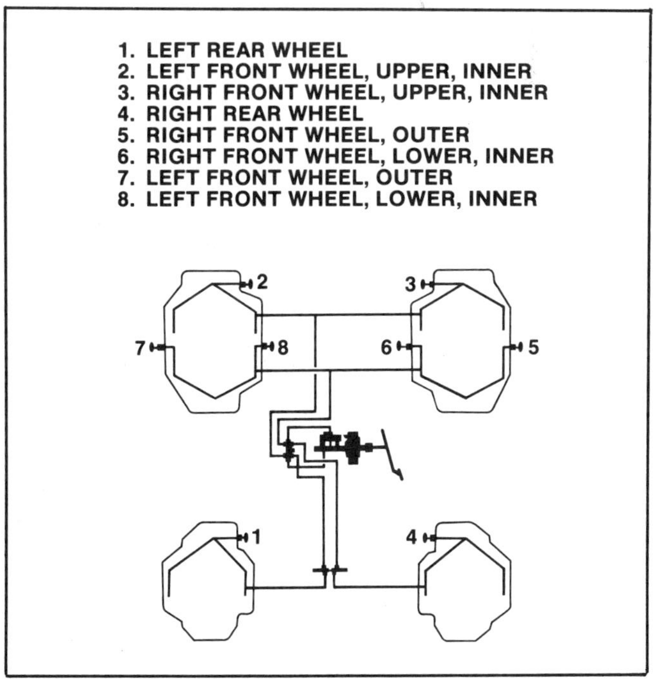

Figure 13-4 Volvo bleeding sequence

SLIDING CALIPER DISC BRAKES

Several different sliding caliper or yoke designs are used in import cars. The more common designs are described here.

ATE SINGLE-PISTON SLIDING YOKE CALIPER

This type of caliper is found mainly on various Audi and Volkswagen models. It can have one piston or two back-to-back pistons in a single cylinder. This design (Figure 13-5) has a fixed mounting frame bolted to the steering knuckle, which holds the pads. A floating frame or yoke slides on the fixed frame. The cylinder is attached to the yoke. The caliper has a cross spring on top of the pads. The outer pad is positioned in a notch. Though the inner pad comes out easily, the piston has a notch in it. This notch must be angled at 20 degrees to the top of the brake pad. A special tool is available for positioning the piston and another for pulling the pads. New pads can be installed without removing the caliper.

When servicing an ATE single-piston sliding yoke caliper, proceed as follows:

1. Remove the two mounting bolts.
2. Disconnect the brake hose and remove the bleeder screw. Drain the caliper of fluid.
3. Slide the mounting bracket away from the piston and off the guide spring.
4. Tap the caliper out of the floating frame.
5. Remove the guide spring.
6. Remove the retaining pin clip and tap out the retaining pins.
7. Remove the spreader.
8. Remove the inner pad with a tool.
9. Slide the floating frame outward to free the outer pad from the locating notch.
10. Remove the outer pad. Rebuild the caliper, if necessary.
11. Insert the new outer pad with the notch on the tab of the floating frame and slide the frame inward.
12. Bottom the piston with a C-clamp.
13. Install the inner pad and a retaining pin.
14. Position the spreader and insert the other retaining pin.
15. Install the retaining pin clip.
16. Replace the caliper guide spring.
17. Drive the caliper onto the floating frame by tapping the slotted ends.
18. Install the mounting bracket through the top of the floating frame.
19. Spread the guide spring. Slide the floating frame into the slots of the mounting bracket.
20. Locate the piston retaining plate over the cut out section of the piston.
21. Replace the caliper and the mounting bolts.

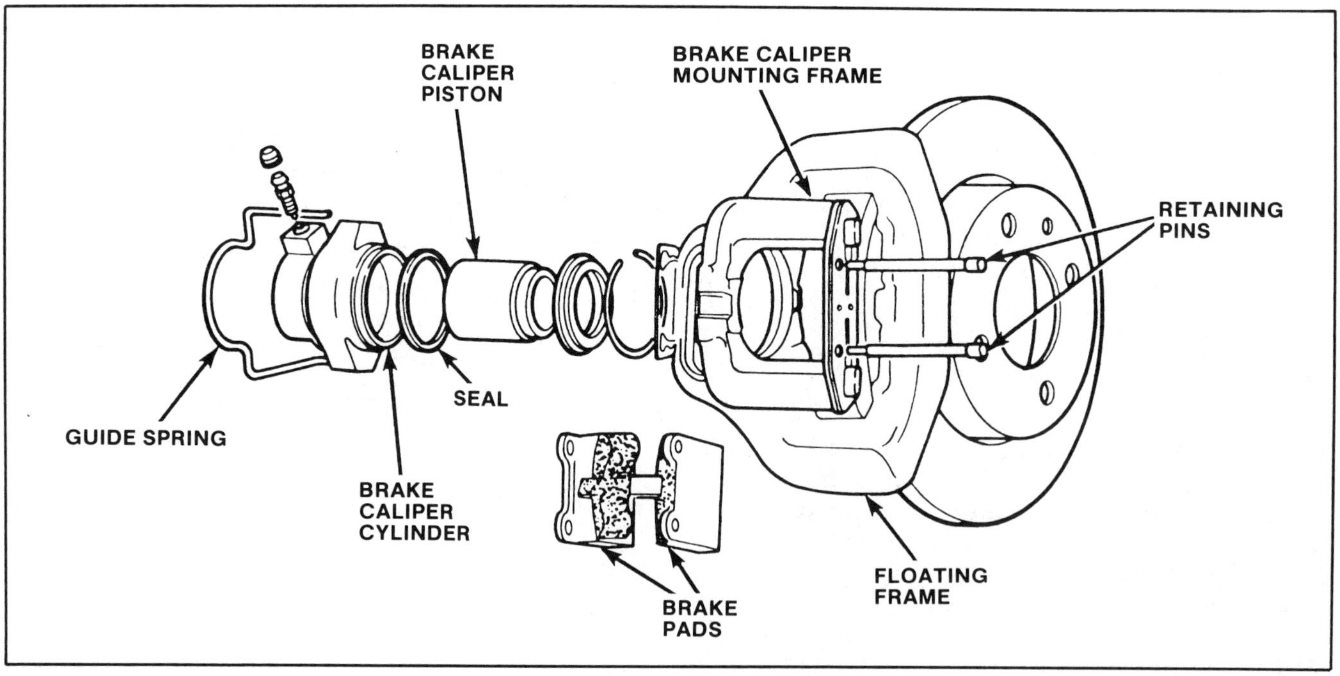

Figure 13-5 ATE single-piston sliding yoke caliper

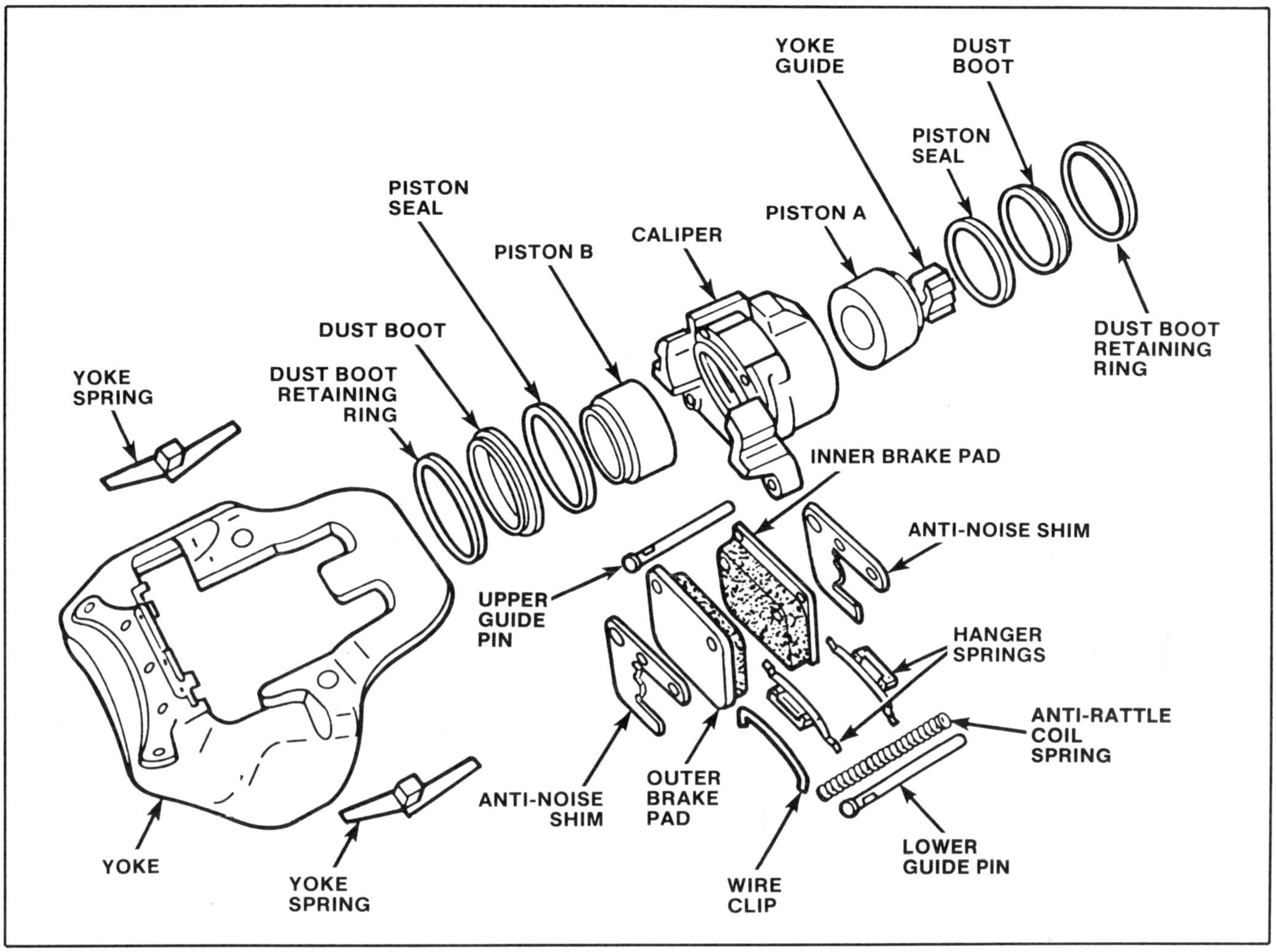

Figure 13-6 Girling sliding yoke caliper

22. Reconnect the brake hose and replace the bleeder screw.

GIRLING ANNETTE AND TOKICO DUAL-PISTON SLIDING YOKE CALIPER

This type of caliper is found on various models of Nissan (Datsun), Honda, and Mazda. Two back-to-back pistons are contained in one cylinder bore, and the cylinder body is bolted to the steering knuckle (Figure 13-6). When the brakes are applied, the piston that faces the rotor pushes on the inboard pad, and the other piston pushes on the yoke (which straddles the rotor and cylinder) moving it inward and applying the outboard pad.

When servicing a dual-piston sliding yoke caliper, proceed as follows:

1. Lift off the wire clip and remove the upper guide pin.
2. Remove the two hanger springs.
3. Tap out the lower guide pin. (Keep the anti-rattle coil spring.)
4. Slide the yoke outward. Remove the outer pad and anti-noise shim.
5. Slide the yoke inward. Remove the inner pad and shim.
6. Disconnect the brake hose.
7. Unbolt the caliper from the steering knuckle.
8. Remove and drain the caliper.
9. Push the pistons back into the cylinder.
10. Tap the cylinder body lightly at the outboard end to free it from the yoke.
11. Remove the retaining rings and boots from both ends of the cylinder.
12. Push out both pistons in one direction.
13. Overhaul the cylinder and pistons.
14. Do not mix the pistons. The inner piston has a plastic yoke guide.

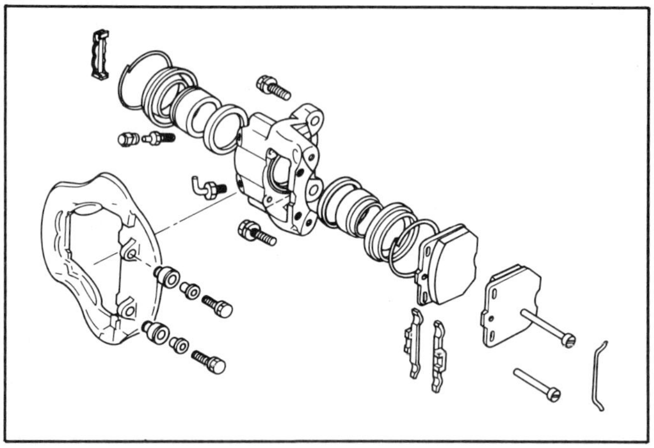

Figure 13-7 Yoke slides on guide pins

15. Turn the yoke guide so that it engages with the yoke.
16. Position the yoke with its springs inserted into the caliper groove.
17. Attach the caliper with mounting bolts to the steering knuckle.
18. Reconnect the hose.
19. Install new pads with anti-noise shims so that the triangle cutout points to the top of the caliper.
20. Replace the lower guide pin and anti-rattle coil spring.
21. Hook the hanger springs under the pin and over the pads.
22. Install the upper guide pin over the ends of the hanger springs.
23. Insert the wire clip lock in the holes in the guide pins.

In one variation of this design found on certain Nissans (Datsuns), the yoke floats on guide pins that thread into the cylinder body (Figure 13-7). The Nissan caliper has two different pistons in the same cylinder bore. Gripper pins bolt to the cylinder body and insert into two grippers on the yoke. The inner caliper piston has a special end and a yoke holder to make contact with the inner end of the yoke. The pads have anti-squeal springs and a clip for the pad pins. Service is similar on both types, and it is not necessary to remove the yoke on either type to replace the pads.

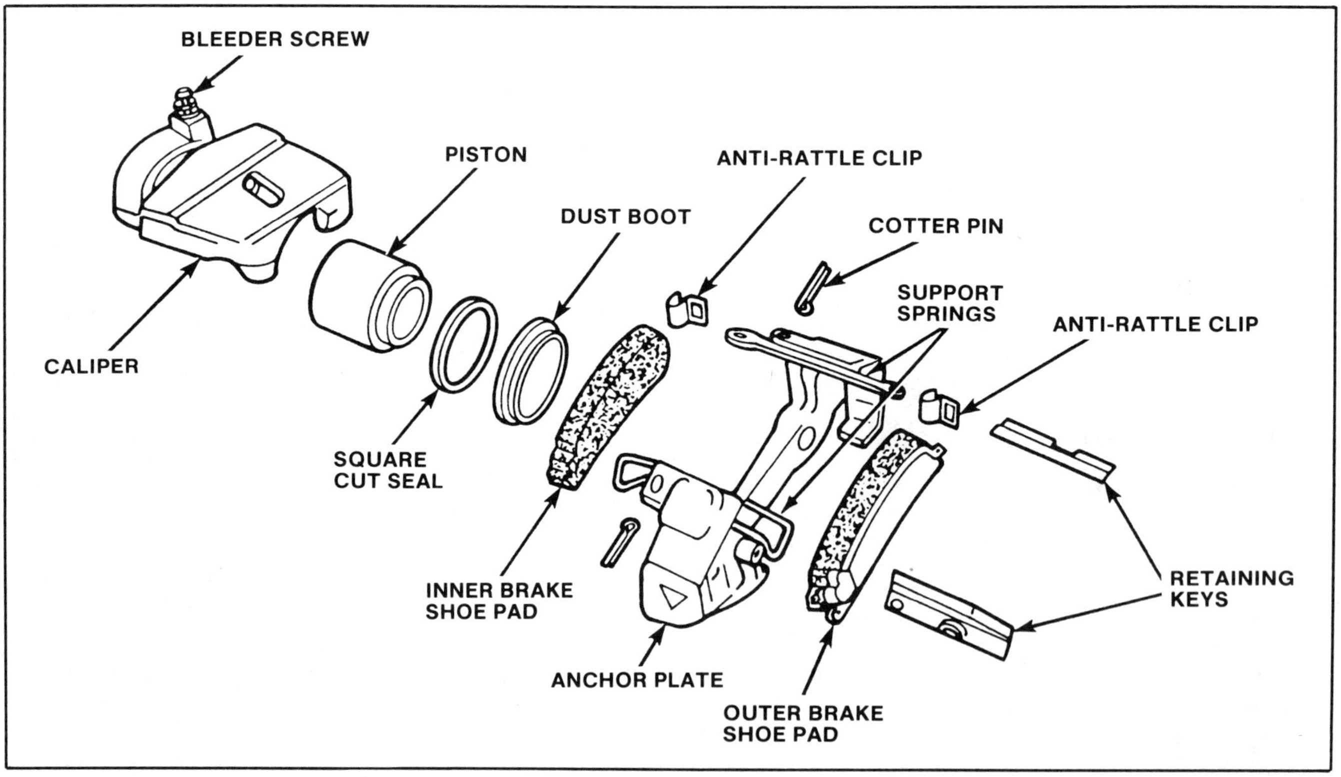

Figure 13-8 Girling/Akebono single-piston sliding caliper

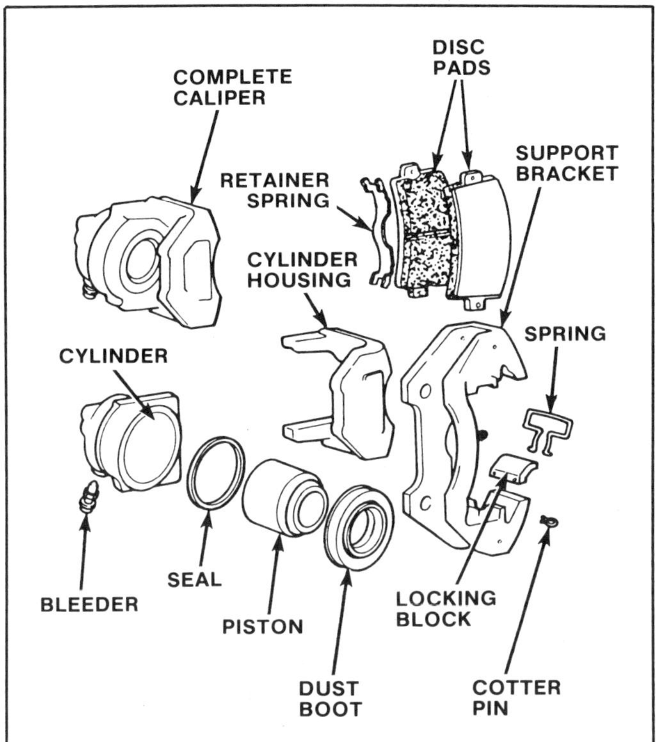

Figure 13-9 Renault/Fiat single-piston sliding caliper

GIRLING AND AKEBONO SLIDING CALIPER

The Girling and Akebono versions of this single-piston sliding caliper are quite similar (Figure 13-8), and both are very popular. They are found on certain models of Courier, Fiat, Fiesta, Honda, LUV, Mazda, Renault, Sapporo/Challenger, Subaru, Toyota, and other vehicles. The caliper is held to the mounting bracket or adapter by guides or keys and support plates are used under the pads to eliminate rattling.

When servicing a Girling or Akebono sliding caliper, proceed as follows:

1. Remove the cotter pins and tap out the two retaining keys.
2. Slide the brake pads out from under the support springs.
3. Keep the anti-rattle clips and do not remove the support springs.
4. Remove the brake hose and drain the caliper at the bleeder screw.
5. Rebuild the caliper, if necessary.
6. Clean all machined surfaces with an abrasive or wire brush.
7. Properly attach the anti-rattle clips onto the new pads.
8. Slide the new pads into the anchor plate.
9. Press the caliper down against the support springs on the anchor plate.
10. Slide in the retaining keys (upper and lower keys are not interchangeable).
11. Insert the cotter pins to secure the keys.
12. Connect the brake hose and bleed the system.

RENAULT/FIAT SLIDING CALIPER

This sliding single-piston caliper (Figure 13-9) has a cylinder held in a cylinder housing that mounts onto a support bracket. The housing is retained by a spring, locking block, and cotter pin. The pads are retained by a spring, which straddles the support bracket.

When servicing the Renault and Fiat sliding calipers, proceed as follows:

1. Disconnect the brake hose from the caliper.
2. Remove the cotter pins from the wedge-shaped locking blocks.
3. Tap out the caliper locking blocks with a punch.
4. Pull the caliper free.
5. Note the position of the tension springs when removing them.
6. Remove the pads.
7. Overhaul the caliper, if necessary.
8. If scoring or pitting of the rotor appears heavy, replace the rotor. Grinding Fiat rotors is not recommended.
9. Push the piston back into the caliper using a C-clamp.
10. Slide the caliper onto the mounting bracket.
11. Slide the locking blocks into position between the caliper and mounting bracket.
12. Drive locking blocks into place using a plastic hammer.
13. Position the springs properly.
14. Replace the cotter pins with new pins.
15. Connect the brake hoses.

LOCKHEED/AKEBONO PIVOTING CALIPER

This unique single-piston caliper is found on certain Nissan and Saab models (Figure 13-10). It is

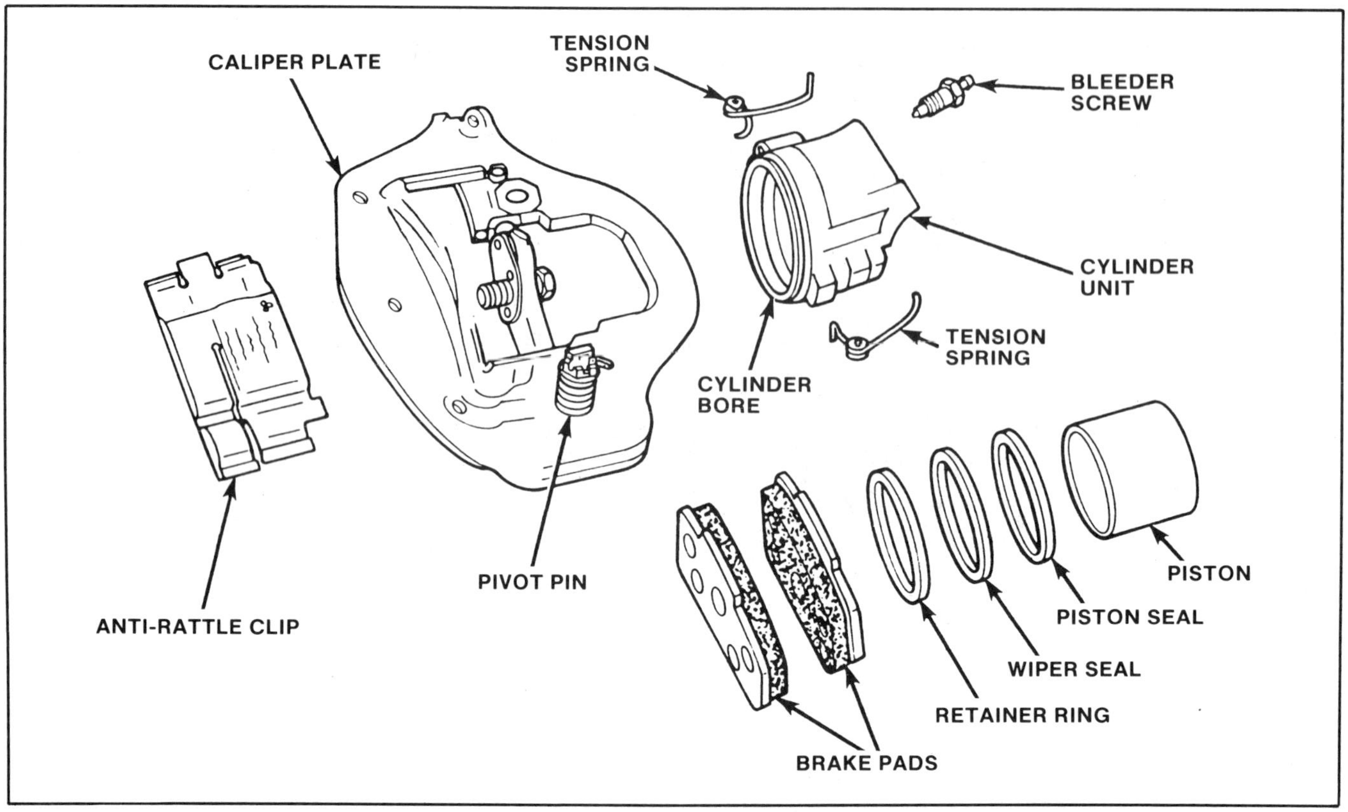

Figure 13-10 Lockheed/Akebono pivoting caliper

sometimes called a swing caliper. Instead of the caliper assembly sliding or floating sideways as on most other single-piston brakes, in this design, the caliper assembly pivots on a pin (Figure 13-11), which causes both pads to apply at an angle. This results in a slight self-energizing effect.

New pads are supplied with tapered linings (thicker at one end than at the other). Inboard and outboard pads are not interchangeable. Pad replacement does not require caliper removal.

When servicing a Lockheed/Akebono pivoting caliper, proceed as follows:

1. Remove the anti-rattle clip.
2. Swing the caliper away from the car.
3. Remove the outer pad, pushing the piston back into the bore, if necessary.
4. Move the caliper toward the car.
5. Remove the inner pad.
6. Disconnect the brake hose.
7. Remove the caliper mounting bolts and lift the caliper.
8. Release the tension springs.

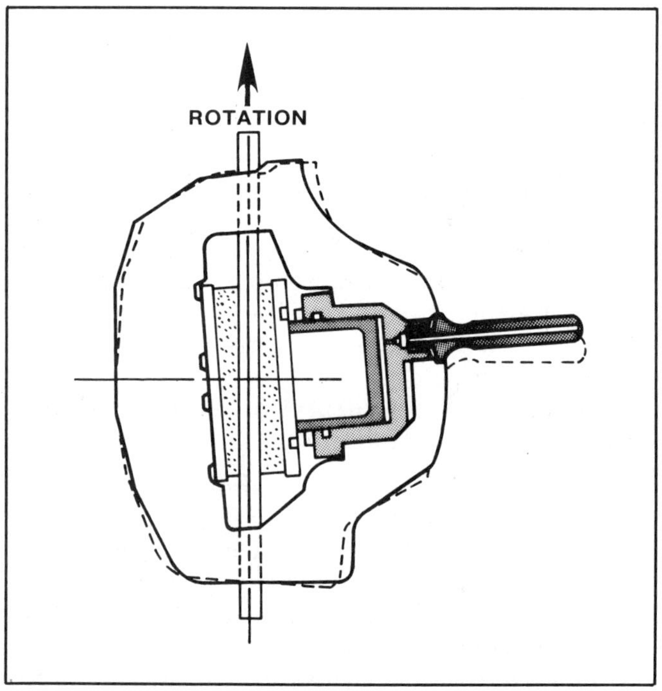

Figure 13-11 Action of a pivoting caliper disc brake

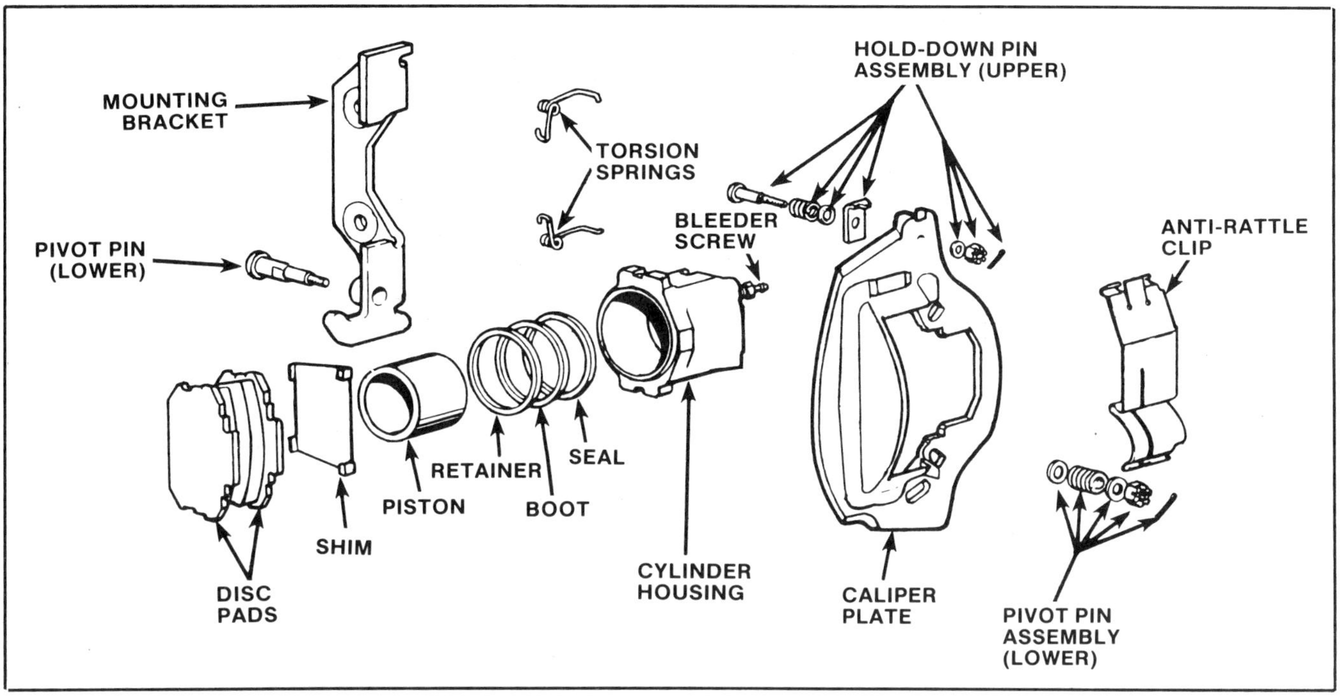

Figure 13-12 Nissan (Datsun) swing caliper

9. Remove the cylinder unit from the caliper plate.
10. Remove the piston, retainer ring, wiper seal, and piston seal.
11. Overhaul the caliper, if necessary.
12. Install the cylinder unit onto the caliper plate.
13. Replace the tension springs.
14. Install the caliper with the mounting bolts.
15. Move the caliper toward the car.
16. Install the inner pad. (Do not mix left and right pads.)
17. Rotate the caliper away from the car.
18. Install the outer pad.
19. Replace the anti-rattle clip.

Figure 13-12 illustrates another popular swing caliper disc brake. It is found on Nissan and other Japanese cars.

FLOATING CALIPER DISC BRAKES

Like sliding caliper disc brakes, there are several different floating caliper disc brakes used in import model vehicles. The more popular import designs include the following:

ATE AND GIRLING FLOATING CALIPER

These floating calipers of similar design are used on numerous small imported cars (Figure 13-13). They are made by ATE, Girling, and Japanese manufacturers for various models of Audi, Nissan, Honda, Mazda, and Toyota. These brakes

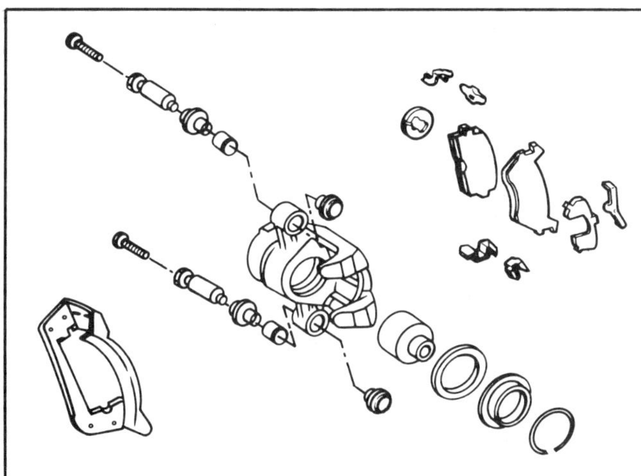

Figure 13-13 ATE/Girling single-piston floating caliper

are similar to many sliding caliper designs, but they float on guide pins and bushings threaded into a mounting bracket. The bracket is bolted to the steering knuckle.

Variations in pad retainers, shims, and anti-rattle and retaining springs will be encountered, but service procedures are basically the same. Note the position of all springs, clips, or shims when removing the pads. Work on one side at a time and use the other for reference.

When servicing an ATE or Girling floating caliper, proceed as follows:

1. Remove the caliper guide pins and anti-rattle springs or clips.
2. Remove the caliper from the rotor by slowly sliding the caliper assembly out and away from the rotor.
3. Support the caliper so it does not hang by the brake hose.
4. Slide the outboard brake pad out of the adapter.
5. Remove the inboard pad from the caliper or adapter. Also remove any shims behind the pads and note their positions.
6. Push the piston back into its bore using a large C-clamp.
7. Check the caliper for leaks or damage.
8. Remove the caliper. Rebuild or replace, if necessary.
9. Install the caliper. Install the anti-rattle hardware, then the new pads. Install any pad shims at this time.
10. Reposition the caliper over the rotor and line up the guide pin holes.
11. Install the guide pins and torque to specifications.
12. Fill the master cylinder with fresh brake fluid.

MAZDA/TOYOTA FLOATING CALIPER

This caliper (Figure 13-14) is very similar to the other small car floating calipers. The caliper slides on two caliper guide pins. Bushings provide a bearing surface upon which the caliper slides. Clips hold the pads in place and an anti-rattle spring reduces noise.

When servicing a Mazda/Toyota floating caliper, proceed as follows:

1. Remove the caliper guide pins.
2. Remove the anti-rattle spring and rotate the caliper up and off the rotor.
3. If you are only replacing the pads, tie up the caliper with wire and remove the pads, damping shims, and clips.
4. Rebuild the caliper if necessary.
5. Install new bushings, pads, and the anti-rattle spring.
6. Install the caliper.
7. Install the caliper guide pins.
8. Connect the brake hose if it was removed.

NISSAN (DATSUN) FLOATING-YOKE CALIPER

The Nissan floating-yoke caliper (Figure 13-15) has two different pistons in the same cylinder bore. Gripper pins bolt to the cylinder body and are inserted into two grippers on the yoke. The inner caliper piston has a special end and a yoke holder to make contact with the inner end of the yoke. The pads have anti-squeal springs and a clip for the pad pins.

When servicing a Nissan floating yoke caliper, proceed as follows:

1. Remove the pads.
2. Remove the caliper and drain it of fluid.
3. Remove the gripper pin attaching bolts.
4. Separate the yoke and cylinder body.
5. Remove the yoke holder from the piston.
6. Remove the retaining rings and the dust seals from the pistons.
7. Push out both pistons together. Remove the seals.

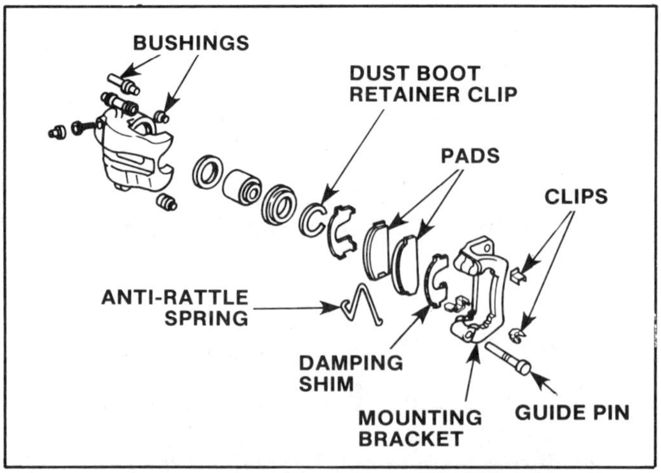

Figure 13-14 Mazda/Toyota floating caliper

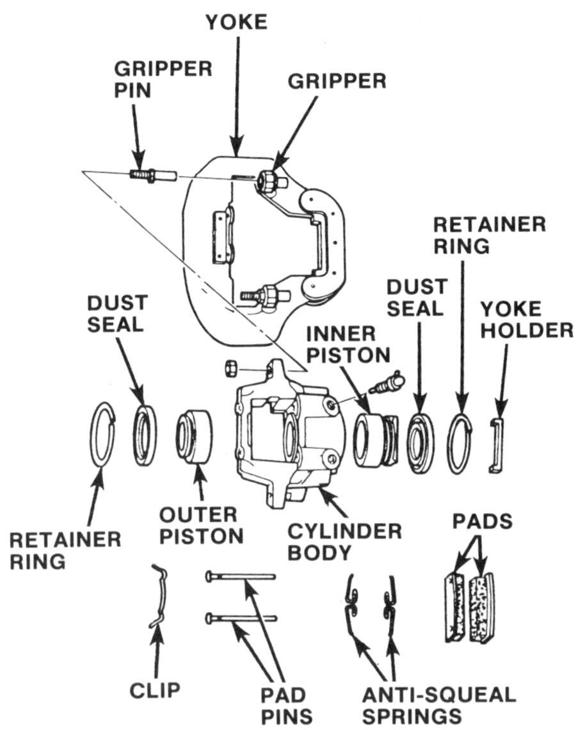

Figure 13-15 Nissan (Datsun) floating yoke caliper

8. Remove the gripper.
9. Clean all parts with brake fluid.
10. Push the outer piston into the cylinder. The end surface should correspond with the boot retaining ring, but do not force the piston groove inside the piston seal.
11. Apply brake grease to the dust boot sealing surface and install with the retaining ring.
12. Install the yoke holder to the inner piston.
13. Attach the gripper to the yoke.
14. Apply 1 percent soap solution to inner gripper wall.
15. Drive the gripper pin into position.
16. Support the outer piston end and press the yoke into the yoke holder so no clearance is present.
17. Replace the caliper.
18. Replace the pads.

SUMITOMO TORQUE PLATE FLOATING CALIPER

This single-piston floating caliper will be found on various Japanese cars including certain models of Chrysler/Mitsubishi (Arrow, Champ, Colt), Honda, Mazda, and Toyota.

In this design, a torque plate is bolted to the steering knuckle, and the caliper floats on the torque plate pins (Figure 13-16). The outer half of the caliper can be separated from the inner half for overhaul. The caliper does not have to be removed to replace the pads. There might be differences among variations in the type of pads, anti-rattle clips, and springs used, so do one side at a time and use the other for reference.

When servicing a Sumitomo torque plate floating caliper, proceed as follows:

1. Disconnect the brake hose.
2. Remove the two mounting bolts attaching the torque plate.
3. Remove and drain the caliper.
4. Pry the pad protector from the retaining pins using a screwdriver.
5. Remove the center of the M-clip from the hole in the outboard pad and its ends from the retaining pins. Note how the M-clip and K-spring are positioned.
6. Pull out the retaining pins and remove the K-spring from the inboard pad.
7. Remove the inner and outer pads.
8. Separate the caliper by removing the two bridge bolts.
9. Separate the torque plate from the inner half of the caliper and clean the pins.
10. Replace the torque plate bushing, rubber spacer, wiper sealer, and seal retainer.
11. Rebuild the caliper, if necessary.
12. Reassemble the torque plate to the inner half of the caliper with the rubber spacer installed onto the proper torque plate shaft.
13. Attach the inner and outer caliper halves and torque the caliper bridge bolts 58 to 69 foot-pounds.
14. Install the new pads.
15. Install one retaining pin.
16. Install the inboard pad K-spring with one end hooked under the retaining pin and the center over the top of the inner pad.
17. Insert the second retaining pin through the outer pad, over the K-spring, and through the inner pad.
18. Insert the ends of the M-clip into the holes in the retaining pins and press the center of the spring into the hole in the outboard pad.
19. Install the pad protector.

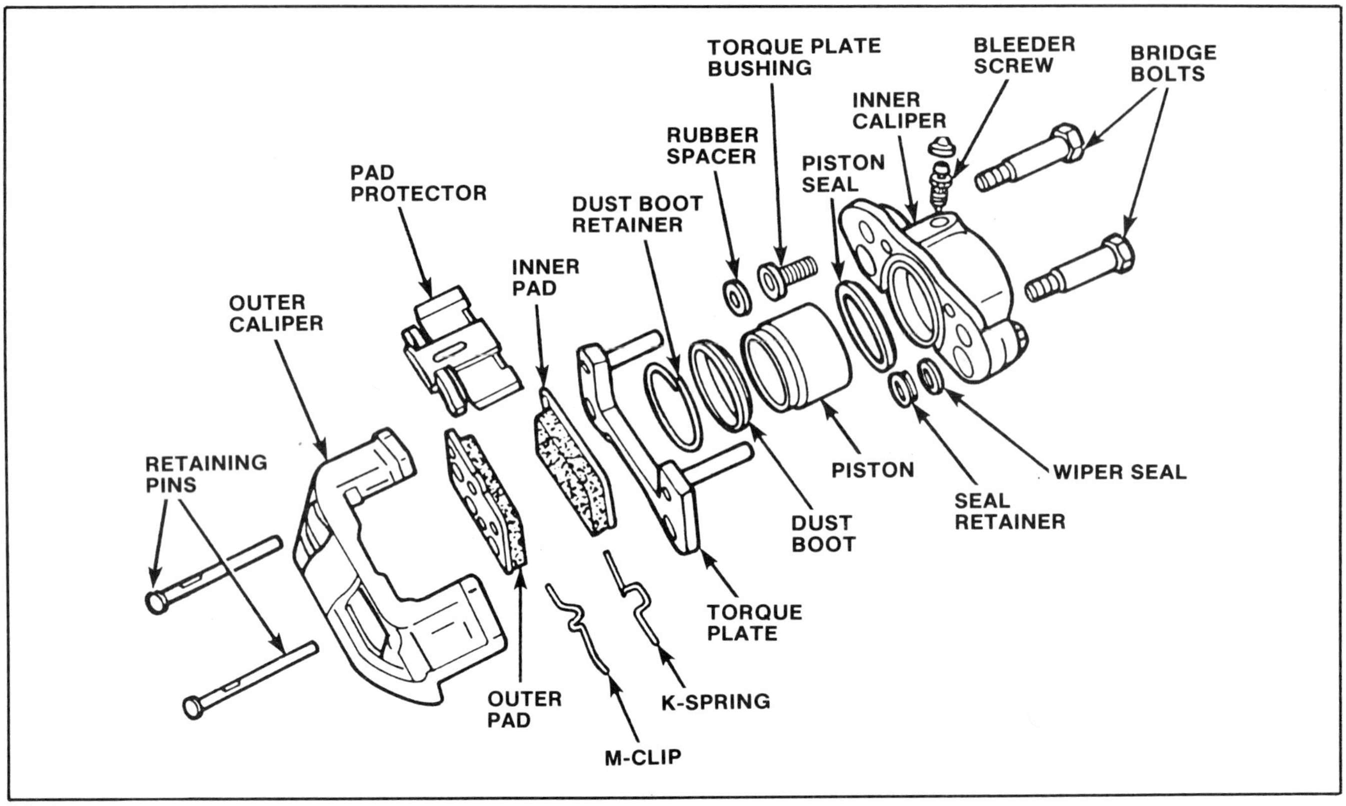

Figure 13-16 Sumitomo floating caliper

20. Install the caliper; torque the mounting bolts 51 to 65 foot-pounds.
21. Reconnect the hose and bleed the system.

GIRLING FLOATING (CLAM TYPE) CALIPER

Although this floating caliper (Figure 13-17) was designed for the Rabbit pickups, it is used by several import vehicle manufacturers in their cars. The guide pins have rubber dust boots, and the bleeder screw has a dust cap. The pads are mounted in a brake pad carrier. The caliper is attached with self-locking bolts that screw into the guide pins.

When servicing a Girling floating (clam type) caliper, proceed as follows:

1. Bottom the piston by pressing the caliper outward by hand.
2. Remove the lower caliper mounting bolt. Hold the guide pin head with a wrench while the bolt is loosened.
3. Rotate the caliper up.
4. Measure the brake pads. They should be replaced if they are less than 1/4-inch thick.

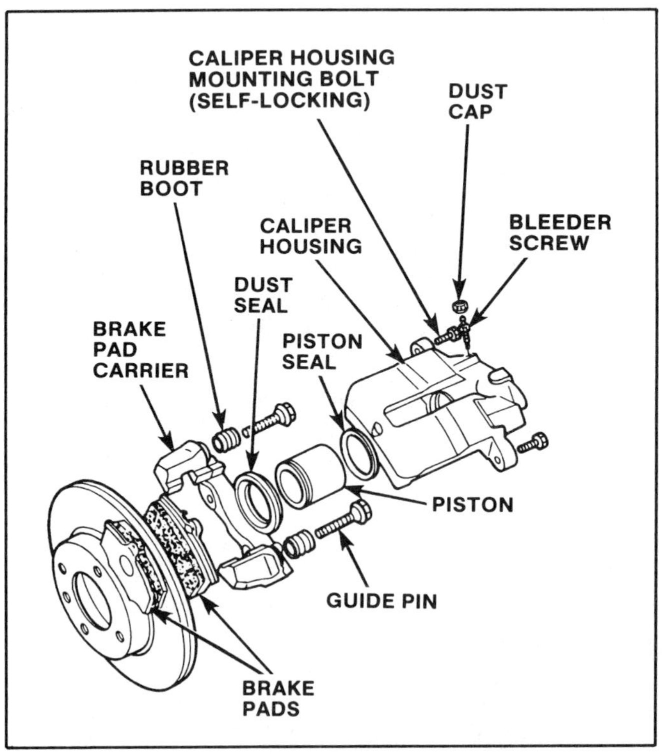

Figure 13-17 Girling floating (clam type) caliper

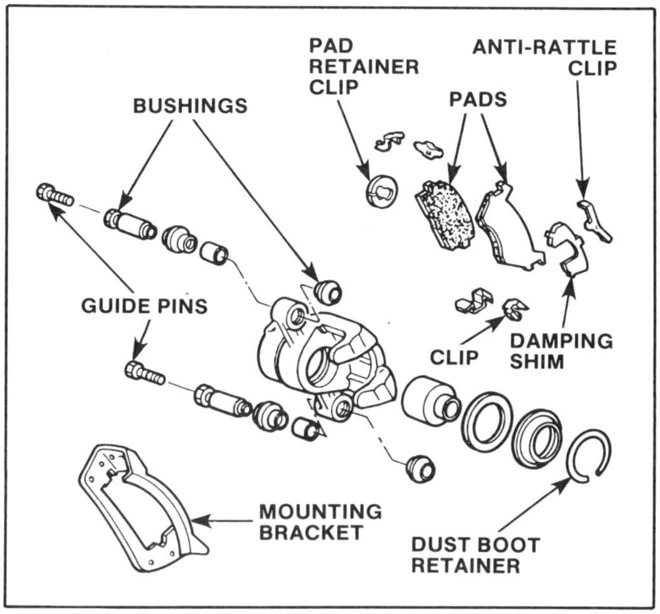

Figure 13-18 Toyota Tercel floating caliper

5. If pads are to be reused, mark their positions for reinstallation.
6. Remove the pads.
7. If the caliper must be rebuilt, remove the other mounting bolt.
8. Disconnect the brake hose and remove the caliper.
9. Rebuild the caliper.
10. Bottom the caliper piston and position the caliper over the rotor.
11. Install the pads.
12. Install new self-locking mounting bolts; tighten to 25 foot-pounds.

TOYOTA FLOATING CALIPER

This floating caliper (Figure 13-18) has guide pins and bushings. The inboard pad is held in place in the caliper by a retainer clip. The Toyota floating caliper was designed primarily for the Tercel models.

When servicing a Toyota floating caliper, proceed as follows:

1. Remove the anti-rattle clip from the caliper.
2. Remove the caliper guide pins.
3. Rotate the caliper up and off the rotor.
4. If you are only replacing the pads, tie up the caliper with wire and remove the pads and retainer clip. Note the location of any shims or clips.
5. Rebuild the caliper if necessary.
6. Install the caliper, shims, pads, clips, and anti-rattle spring.
7. Install the guide pins using new bushings.
8. Connect the brake hose if it was removed.

The Toyota Cressida single-piston disc brake is shown in Figure 13-19.

REAR DISC BRAKES

Like domestic cars, most small imports have front disc brakes and rear drum brakes. But more

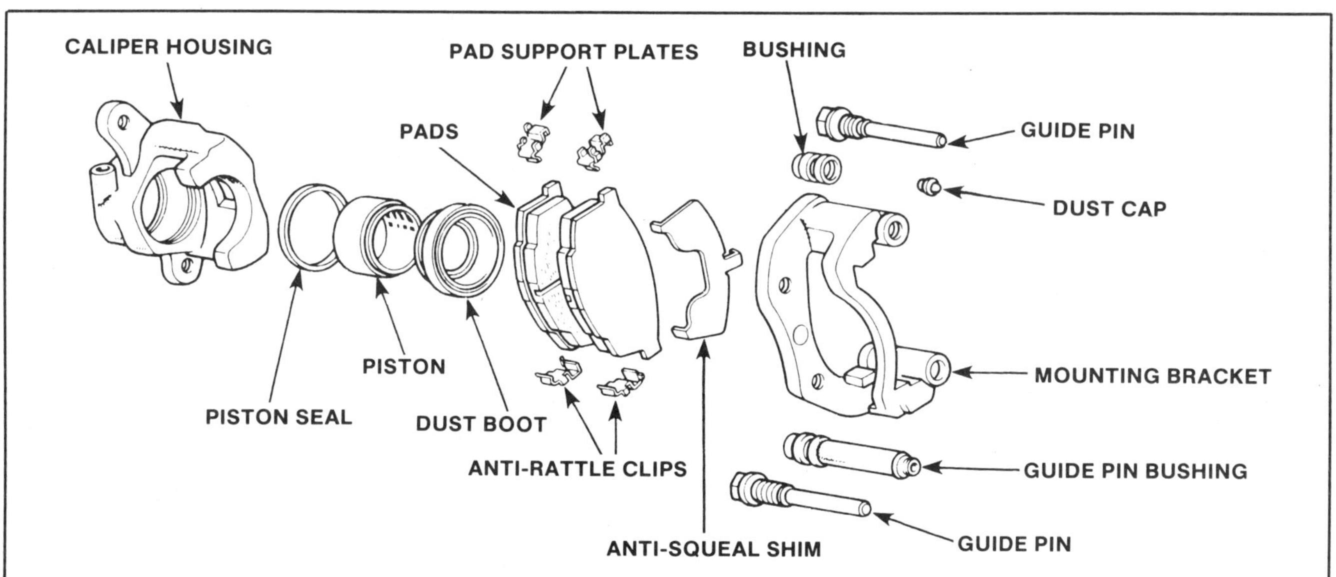

Figure 13-19 Toyota Cressida single-piston

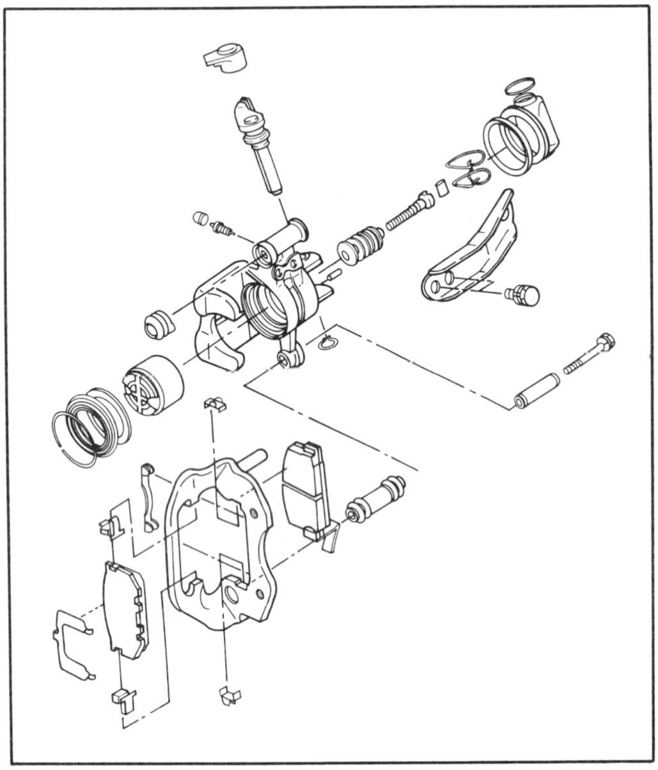

Figure 13-20 Akebono single-piston floating caliper with integral parking brake

and more imports have rear disc brakes that are similar to the front except for the addition of a parking brake mechanism. Some of the popular import rear disc brakes are given here.

AKEBONO SINGLE-PISTON CALIPER WITH PARKING BRAKE

This single-piston sliding caliper is found at the front of Subarus (Figure 13-20) and at the rear of various models of Chrysler/Mitsubishi (Arrow, Challenger, Sapporo), Renault, and Toyota.

The integral parking brake works as follows: When the parking brake is applied, the lever at the caliper rotates a cam against a wedge-shaped pawl. This pushes against a thrust screw, which is threaded into the back of the piston, and forces the piston outward just as hydraulic pressure does. The thrust screw rotates upon application so that automatic adjustment for pad wear is provided.

The service procedures given here are specifically for the Subaru front brake, but those for other applications are similar.

When servicing an Akebono rear disc single-piston brake, proceed as follows:

1. Release the parking brake.
2. Disconnect the parking brake cable from the lever at the caliper.
3. Unscrew the lower caliper lock pin. In some cases, it will be held in place by a bolt with a hex wrench.
4. Rotate the caliper up off the rotor so that it pivots on the upper pin.
5. Remove the pads.
6. Inspect the caliper for leaks and damage.
7. Remove the caliper. Rebuild or replace if necessary.
8. Rotate the piston clockwise, forcing it into its bore. A special wrench for this is available (Figure 13-21). Be certain the boot is not twisted or pinched.

CAUTION: If the piston is forced into its bore without being rotated, the self-adjusting mechanism will be damaged. The piston should be turned until the minimum clearance necessary for pad insertion is achieved. Leave the piston at the point of rotation that places its notches as shown in Figure 13-22.

9. Rebuild the caliper, if necessary.
10. Install the caliper. Install the new pads including any shims that might be present (no shim is used on the inboard pad).
11. Rotate the caliper back down over the rotor.
12. Install the lower caliper lock pin and the locking bolt, if present. Be careful not to twist or damage the pin's rubber boot.
13. Fill the master cylinder reservoir with fresh brake fluid.
14. Bleed the brakes.
15. Reattach the parking brake cable.
16. Obtain a high, hard pedal, refill the master cylinder reservoir, check the operation of the parking brake, and road test the car.

GIRLING SLIDING YOKE REAR DISC BRAKE CALIPER

This single-cylinder, dual-piston sliding yoke caliper with integral parking brake is found on certain models of a few European cars, such as at the front of Saabs (Figure 13-23) and the rear of

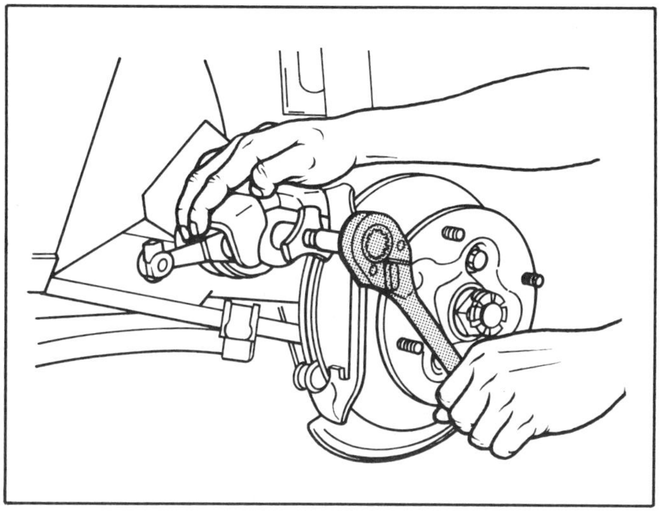

Figure 13-21 Pushing the piston back

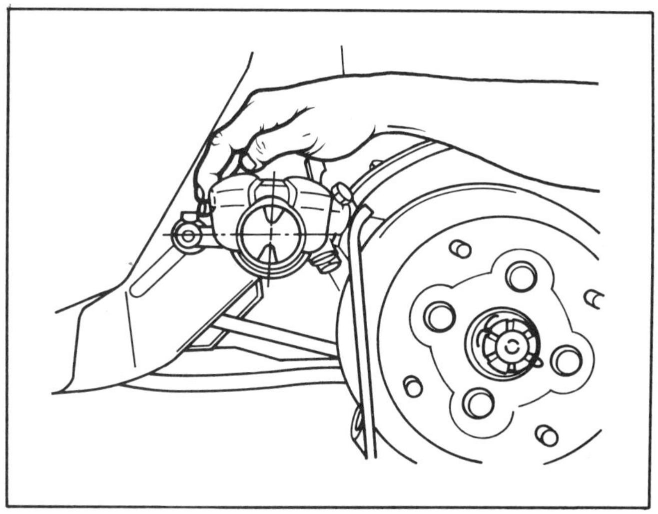

Figure 13-22 Specified position of piston notch

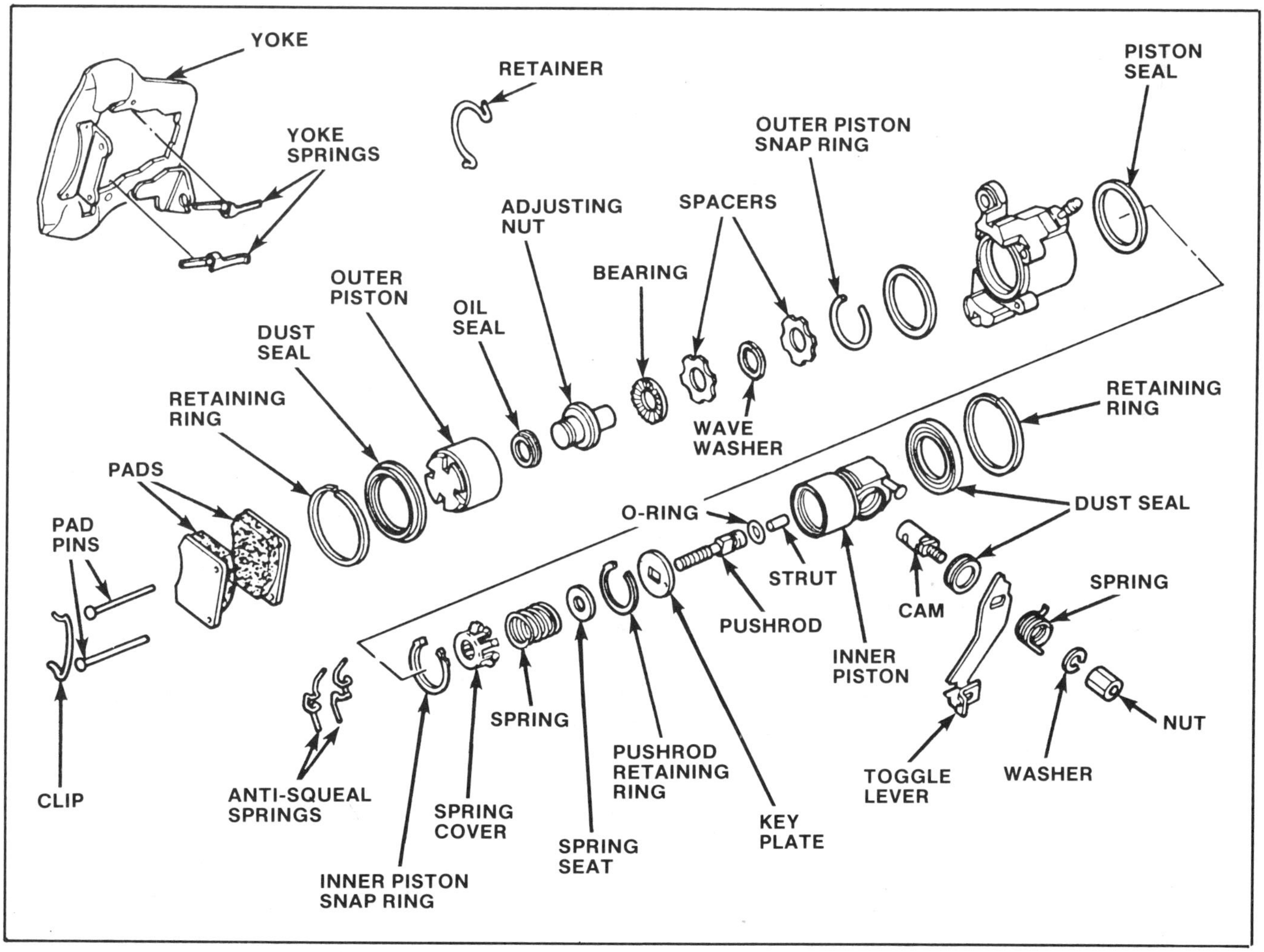

Figure 13-23 Girling sliding yoke rear disc brake caliper

Peugeots. This caliper has an inner and outer piston in a single-caliper body. It is a fixed caliper with a sliding yoke and parking brake mechanism.

When servicing a Girling sliding yoke, proceed as follows:

1. Drain the fluid.
2. Remove the pads.
3. Push in the caliper body to separate the caliper and yoke.
4. Remove the retaining rings and dust boots from both ends.
5. Push on the outer piston to remove the assembly from the caliper bore.
6. Remove the piston seals.
7. Turn the inner and outer pistons counterclockwise and disengage.
8. Disassemble the outer piston. Remove outer snap rings, spacers, wave washer, bearing, adjusting nut, and oil seal.
9. Disassemble the inner piston. Remove the inner snap ring, spring cover, spring, and spring gear.
10. Remove the pushrod retaining ring, key plate, pushrod, O-ring, and strut.
11. Hold the parking brake lever in a vise. Remove the nut, washer, return spring, lever, dust seal, and cam.
12. Clean and overhaul the caliper.
13. Begin reassembly by applying grease to the pushrod groove, O-ring, strut ends, oil seal, piston seals, and inside dust seals. Install new oil seal.
14. Slide the adjusting nut and seal into the outer piston.
15. Install the bearing, spacer, wave washer, and second spacer into the outer piston.
16. Secure all the internal parts with an outer piston snap ring.
17. Place the cam into the inner piston with the strut securely in the cam hole.
18. Install the O-ring onto the pushrod.
19. Align the square hole in the key plate with the pushrod and slide the assembly into the inner piston bore with the rounded part seated in the piston.
20. Install the pushrod retaining ring. Hold the spring seat, spring, and spring cover in position with a drift and install the inner piston snap ring.

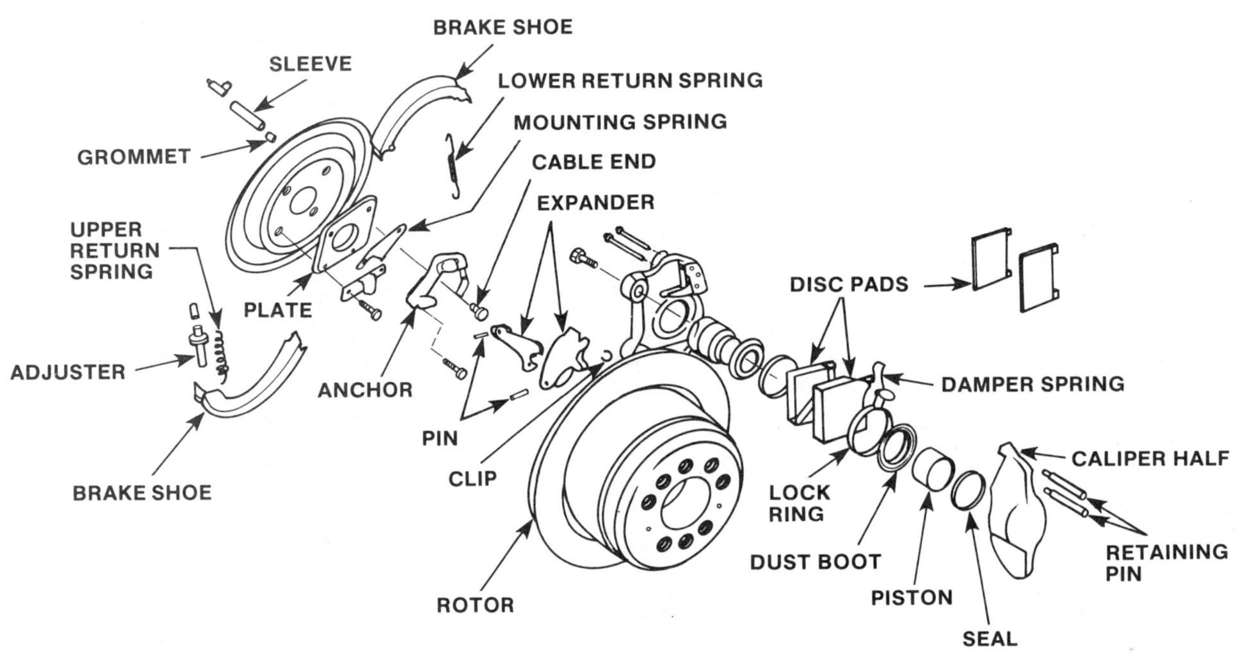

Figure 13-24 ATE/Girling disc brake with parking brake

21. Install the toggle lever dust seal, toggle lever with the square hole aligned on the cam, return spring, washer, and nut on the cam.
22. Turn the piston clockwise to reassemble. Coat with brake fluid.
23. Slide the piston assembly into the back of the caliper outer piston first.
24. Install the new dust boots and retaining rings.
25. Attach the yoke springs to the yoke.
26. Coat the yoke and caliper body pinholes with silicone.
27. Align the inner piston cutout with the yoke and reassemble.
28. Position the retainer wire securely in the piston groove.
29. Apply grease on clearances between the caliper body and pad, pad pin and pad, and pad pin to bracket.
30. Install the pads, springs, pins, and clip.
31. Replace the caliper.

Except for minor differences in the brake's anti-noise hardware, service procedures are identical to the Sumitomo sliding yoke caliper with parking brake (Figure 13-16).

ATE AND GIRLING REAR DISC WITH DRUM PARKING BRAKE

This combination of a two-piston fixed caliper and a small mechanically actuated drum type parking brake is found at the rear of certain European cars, notably Volvo (Figure 13-24).

Replacement of the disc brake pads is the same as that for the fixed caliper front brakes and is described at the beginning of Chapter 8.

When servicing an ATE and Girling rear disc with drum parking brake, replace the parking brake shoes as follows:

1. Unbolt the caliper and lift it off the rotor/drum. Suspend it by a wire.
2. Remove the two screws that hold the rotor/drum to the axle.
3. Remove the rotor/drum. It might be necessary to loosen the parking brake adjustment and/or back off the star wheel through the slot in the rotor/drum.
4. Remove the shoe return springs and the shoes. Loosen the parking brake cable further if necessary.
5. Remove the parking brake actuating lever spacer from the old trailing shoe.
6. Clean the backing plate in an approved manner, then inspect it for damage. Correct any axle seal leaks.
7. Inspect the rotor/drum, and machine or replace it as necessary.
8. Lubricate the rubbing surfaces on the backing plate. Install the parking brake actuating lever spacer and retaining spring on the new trailing shoe.
9. Install the new shoes, their return springs, and the star wheel adjuster (back it off all the way).
10. Mount the rotor/drum on the axle, and install the two screws.
11. Adjust the shoe-to-drum clearance by turning the star wheel with a tool inserted through the slot in the rotor/drum.
12. Adjust the parking brake cable.
13. Install the caliper.

DRUM BRAKES

Although servo brakes are found on some European and Japanese cars, the majority of import vehicles are of the non-servo design with a fixed shoe anchor. Those used at the rear of FWD vehicles usually require that the wheel bearings be removed in order to take the drum off the spindle or axle, and this operation is similar to that for the front drums found on older model RWD vehicles.

As can be seen in Figures 13-25, 13-26, and 13-27, there is little difference between the typical import drum brakes shown here and those illustrated in Chapters 5 and 6. Most drum brakes, whether domestic or import, are serviced basically in the same manner. There could be some differences in the brake shoes, adjusters, and parking brake linkage. Work on one wheel at a time and use the other wheel for reference.

ANTI-LOCK BRAKING SYSTEMS

Like domestic passenger cars, there are two basic anti-lock designs found on most imports: Bosch and Teves. They both operate in the same manner as the domestic ABS arrangement covered in Chapter 11.

As with domestic ABS, each wheel of the traction control system is constantly monitored via wheel sensors. These signals are monitored by the combined ABS/TCS controller, and it makes the distinction between brake slip and power-induced

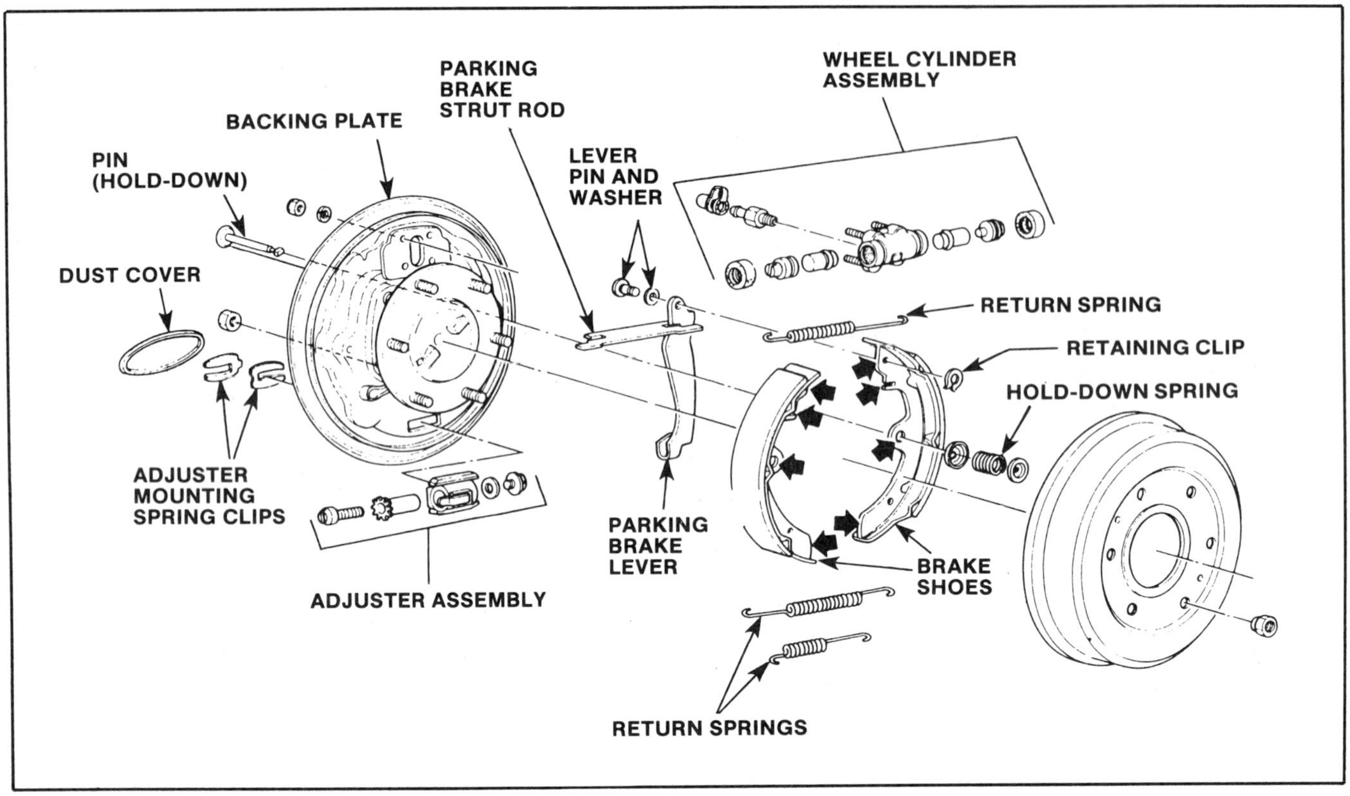

Figure 13-25 Nissan (Datsun) drum with dual star wheel adjustment

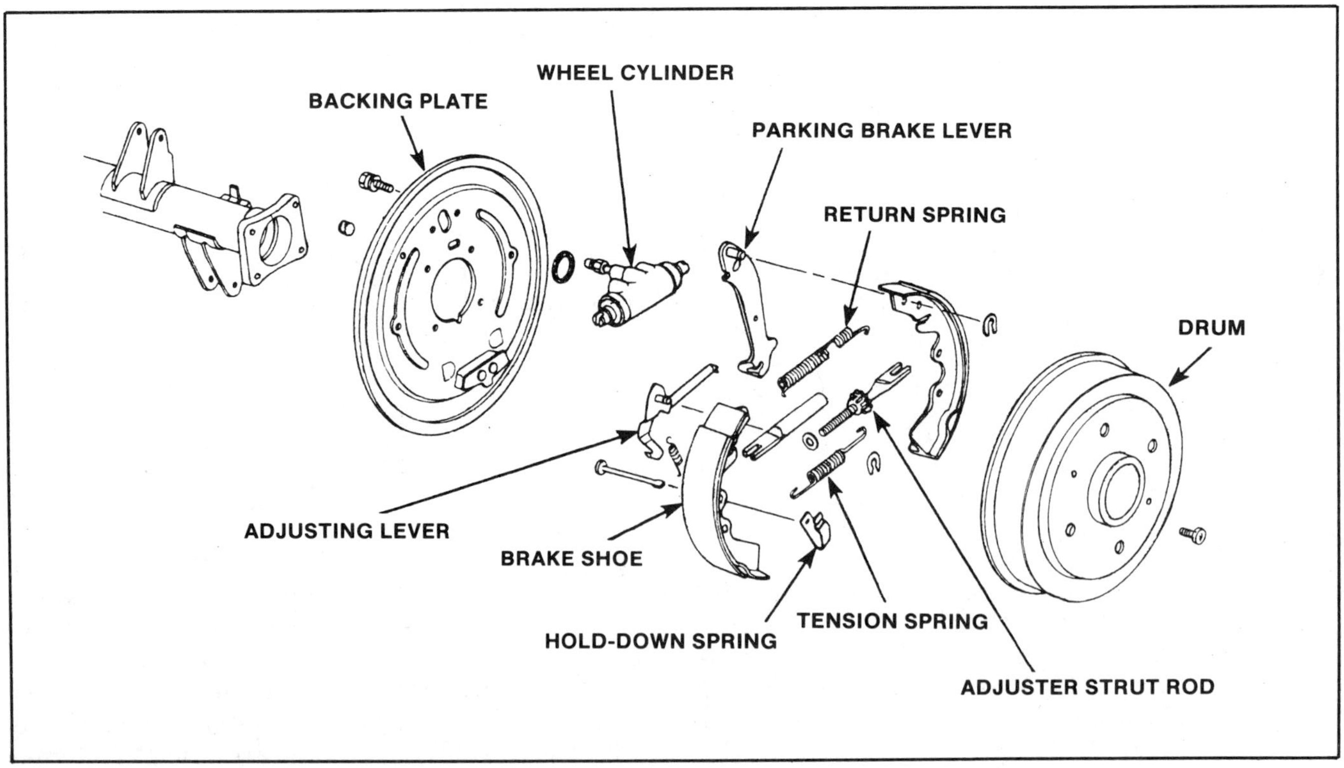

Figure 13-26 Mazda, Volkswagen with star wheel adjuster

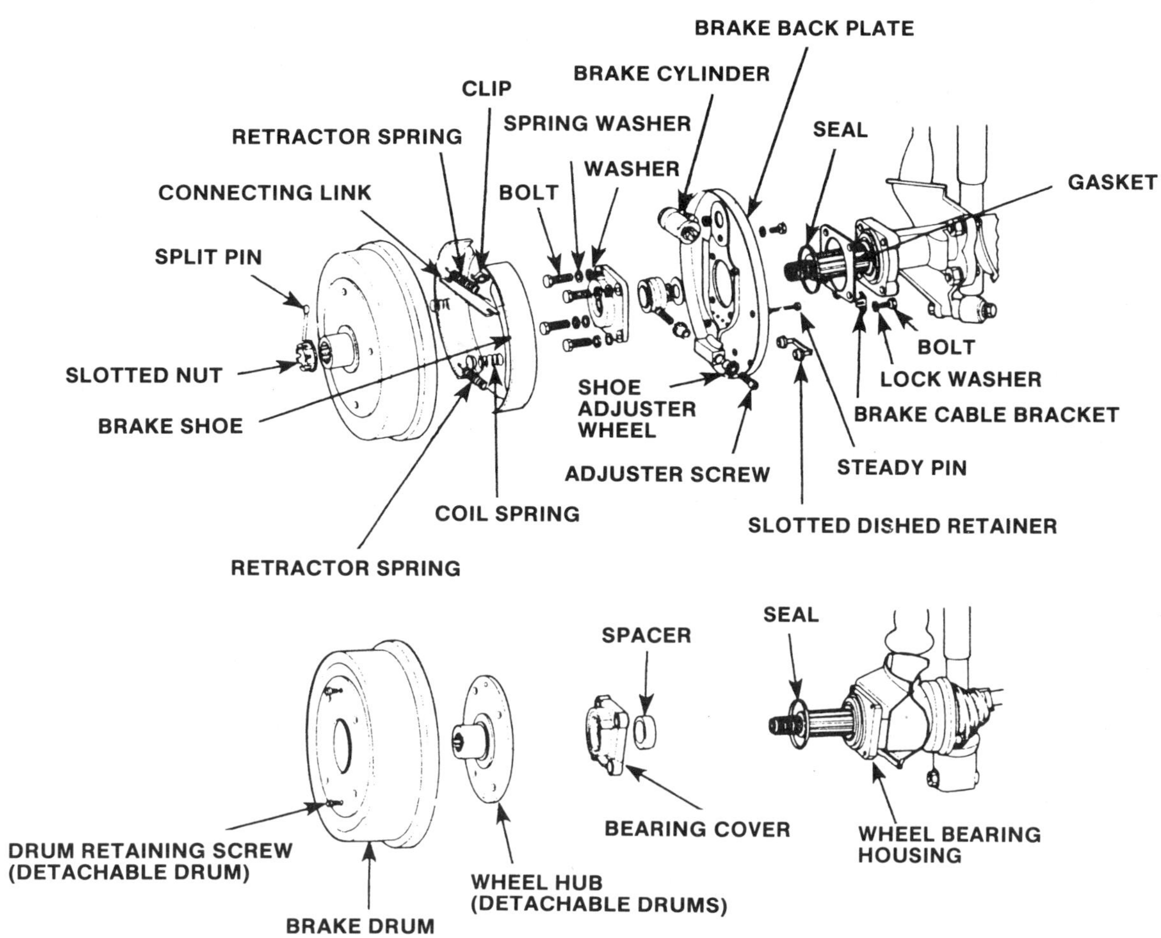

Figure 13-27 Volkswagen with dual star wheel adjuster

wheel spin at the drive wheels. If one or both of the drive wheels exceeds the available friction coefficient, the controller issues a command to increase brake pressure at the offending wheel until it slows down to the point known as brake intervention. At speeds in excess of 12.5 MPH, the brake intervention circuits are opened and the system disengaged to prevent brake overheating. All of this occurs without any action on the part of the driver.
power to the road. This happens within milliseconds, even if the throttle is abruptly opened wide.

With either system, should the slip coefficients vary from one side of the car to the other, the driver is able to make full use of the adhesion available at each drive wheel without compromising overall driving stability.

As mentioned in Chapter 11, the Lucas Girling stop control system can be found in some lower priced import passenger cars. In fact, rear anti-skid brake system is being offered in some import trucks.

REVIEW QUESTIONS

Review the information in this chapter by answering the following questions.

1. Which of the following is not a manufacturer of fixed calipers used on various models of imported cars?
   a. Girling
   b. ATE
   c. Sumitomo
   d. Tokico
2. What must be done before starting any servicing procedures on import disc brakes?
   a. Remove half of the brake fluid from the master cylinder reservoir.

b. Pull hair pin clips out of the pad retaining pins.
c. Bleed the brake system.
d. Test the brake pedal.

3. What import car model uses a triangulated hydraulic system?
   a. Volkswagen
   b. Volvo
   c. Fiat
   d. Toyota

4. The ATE single-piston sliding yoke caliper is found mainly on what import car model(s)?
   a. Audi
   b. Volkswagen
   c. both a and b
   d. neither a nor b

5. Which of the following car models does not use a Girling Annette and Tokico dual-piston sliding yoke caliper?
   a. Datsun
   b. Volkswagen
   c. Honda
   d. Mazda

6. What variation of the dual-piston sliding yoke caliper is found on certain Nissans (Datsuns)?
   a. The yoke floats on guide pins that thread into the cylinder body.
   b. The caliper has two different pistons in the same cylinder bore.
   c. Gripper pins bolt to the cylinder body.
   d. All of the above.

7. In the Girling and Akebono versions of the single-piston sliding caliper, what is used under the pads to eliminate rattling?
   a. support plates
   b. mounting brackets
   c. adapters
   d. cotter pins

8. The Renault sliding caliper is also found in what other car model?
   a. Fiat
   b. Volkswagen
   c. Nissan
   d. Saab

9. What is the unique single-piston caliper found on certain Nissan and Saab models sometimes called?
   a. fixed caliper
   b. Japanese caliper
   c. swing caliper
   d. disc caliper

10. When servicing floating calipers, Mechanic A works on one side at a time to use the other side for reference. Mechanic B works front to back. Who is right?
   a. Mechanic A
   b. Mechanic B
   c. both A and B
   d. neither A nor B

11. In the Sumitomo floating caliper design, where is the torque plate bolted?
   a. to the floor
   b. to the steering knuckle
   c. to the caliper
   d. to the clips

12. What floating caliper has guide pins and bushings and an inboard pad held in place in the caliper by a retainer clip?
   a. Girling
   b. Chrysler
   c. Akebono
   d. Toyota

13. In many imports, the rear disc brakes are similar to the front, except for the addition of
   a. drum brakes.
   b. the calipers.
   c. the parking brake mechanism.
   d. bushings.

14. When servicing rear disc brakes, Mechanic A forces the piston into its bore without being rotated. Mechanic B turns the piston. Who is right?
   a. Mechanic A
   b. Mechanic B
   c. both A and B
   d. neither A nor B

15. Which of the following is not an anti-lock braking system currently used by imports?
   a. ABS
   b. TCS
   c. ASC
   d. Teves traction control system

CHAPTER 14

DIAGNOSING BRAKE SYSTEM PROBLEMS

OBJECTIVES

Upon completing this chapter, you should be able to

- Fill out a prebrake job inspection checklist and a complete brake job checklist.
- Differentiate between various brake noises in order to diagnose problems in the braking system.
- Carry out low- and high-speed road tests in order to evaluate the performance of a vehicle's braking system.
- Recognize the external conditions that affect brake road-test performance.
- Recognize the symptoms, causes, and remedies for common drum brake conditions, including low pedal, springy or spongy pedal, pedal pulsation, dive, pedal fade, and decreasing pedal travel, among others.
- Recognize the symptoms, causes, and remedies for common disc brake conditions, including hard pedal, grabbing or uneven braking action, springy or spongy pedal, pedal pulsation, and dragging brakes, among others.
- Recognize the symptoms, causes, and remedies for common vacuum power brake conditions, including sensitive brakes, vibrating pedal, and pedal chatter, among others.
- Recognize the symptoms, causes, and remedies for common hydro-boost brake conditions, including excessive brake pedal effort, booster chatter, and grabby brakes, among others.
- Recognize the symptoms, causes, and remedies for common electric brake, air brake, and air disc brake conditions.
- Identify situations in which the way that a vehicle is driven can cause brake problems that otherwise would be blamed on shoddy work or defective parts.
- Explain the techniques required for complete brake service.

The first rule of quality brake service work is to do a *complete* job. For example, if new linings are installed without regard to the condition of the hydraulic system, the presence of a leaking or frozen caliper or wheel cylinder will cause the customer to come back complaining that the technician cannot be trusted to do good work, and the shop's reputation and schedule will be damaged. Worse yet, if a brake technician overlooks a weak brake hose or a condition that results in severe pulling or poor stopping power, the performed work could be the cause of an accident.

To aid in doing a complete inspection and diagnosis, Figures 14-1 and 14-2 illustrate two inspection report forms that are used in many brake shops. The use of such forms helps the brake technician avoid missing any brake tests and components.

All customer complaints must be carefully considered. Of course, the number one customer complaint is brake noise. But, when the driver hears brake noise for the first time, it is an advance notice of problems in the braking system. The severity of the problem and the ease with which it can be corrected depends on the correct diagnosis of the noise source. Rattles, clicking, grinding, and hammering from the wheels when the brake is in the unapplied position should be carefully investigated. Be sure the noise is not caused by the bearings or various suspension parts. If the noise is coming from the brake assembly, it is most likely caused by worn, damaged, or missing brake hardware or the poor fastening or mounting of brake components. Grinding noises usually occur when a stone or other object becomes trapped between the lining material and the rotor or drum.

When the brakes are applied, a clicking noise usually indicates play or hardware failure in the attachment of the pad or shoe. On recent brake jobs, the noise could be caused by the lining tracking cutting tool marks in the rotor or drum. A non-directional finish on rotors will eliminate this,

PREBRAKE JOB INSPECTION CHECKLIST

Owner ______________ (LAST) ______________ (FIRST) Phone __________ Date ___|___|___

Address ______________________________ License No. ______________

Make __________ Model __________ Mileage __________ Serial No. __________ Year __________

Special Key for Hubcaps/Wheels Location ______________ Owner Use Parking Brake Yes ☐ No ☐

4 Drum ☐ 4 Disc ☐ Disc/Drum ☐ P/B No ☐ Yes ☐ Vacuum ☐ Hydro ☐ ABS ☐

Owner Comments ____

1. CHECKS BEFORE ROAD TEST	Safe	Unsafe
Stoplight Operation		
Brake Warning Light Operation		
Master Cylinder Checks		
Fluid Level		
Fluid Contamination		
Under Hood Fluid Leaks		
Under Dash Fluid Leaks (No Power)		
Bypassing		

BRAKE PEDAL HEIGHT AND FEEL

Check One		Check One	
Low		Spongy	
Med		Firm	
High			

Power Brake Unit Checks

VACUUM	Safe	Unsafe	HYDRO	Safe	Unsafe
Vacuum Unit			Hydro Unit		
Engine Vacuum			P/S Fluid		
Vacuum Hose			P/S Belt Tension		
Unit Check Valve			P/S Belt Condition		
Reserve Braking			P/S Fluid Leaks		
			Reserve Braking		

Tire Pressure Specs	Front	Rear
Record Pressure Found		

RF ______ LF ______ RR ______ LR ______

Tire Condition

RF ______ LF ______ RR ______ LR ______

2. ROAD TEST	Yes	No	RF	LF	RR	LR
Brake Pull						
Brake Clunk						
Brake Scraping						
Brake Squeal						
Brake Grabby						
Brakes Lock Prematurely						
Wheel Bearing Noise						
Vehicle Vibrates						

STEERING WHEEL MOVEMENT WHEN STOPPING FROM 2-3 MPH YES/NO/RGT/LFT

Does ABS Work	YES	NO
Pedal Pulsation when Braking	YES	NO
Steering Wheel Oscillation when Braking	YES	NO
No Stopping Power	YES	NO
Warning Light Comes on when Braking	YES	NO
Difference in Pedal Height after Cornering	YES	NO
Nose Dive	YES	NO

3. In Shop Checks On Hoist	Yes	No	RF	LF	RR	LR
Brake Drag						
Intermittent Brake Drag						
Brake Pedal Linkage Binding						
Wheel Bearing Looseness						
Missing or Broken Wheel Fasteners						
Suspension Looseness						

	Front					Rear				
Mark Wheels and Remove	Right			Left		Right			Left	
Caliper/Piston Stuck RF LF RR LR										
Mark Drums and Remove	Spec	Safe	Unsafe	Safe	Unsafe	Spec	Safe	Unsafe	Safe	Unsafe
Measure Rotor Thickness or Drum Diameter.										
Measure Rotor Thickness Variation.										
Measure Rotor Runout.										
Lining Thickness										
Tubes and Hoses										
Fluid Leaks										
Broken Bleeders										
Leaky Seals										
Self-Adjuster Operation										
									Safe	Unsafe
Parking Brake Cables and Linkage										

Figure 14-1 Sample of Prebrake Job Inspection Checklist (*courtesy of Hennessy Industries, Inc.*)

COMPLETE BRAKE JOB CHECKLIST

Customer Name ______________________________

Address ____________________ Telephone Number: Work ____________ Home ____________

Vehicle Make ____________________ Year ____________ Mileage ____________

License Number ______________________________

Approximate Time Required for Service ______________________________

Inspected by ____________________ Date ____________________

The condition of the brake system of this vehicle is as follows:

Brake System Components	Required Services	$ Estimate Parts	$ Estimate Labor
Disc pads	☐ Semi-metallic ☐ Conventional		
Disc caliper	☐ Recondition ☐ Replace		
Disc hardware	☐ Replace		
Disc rotor	☐ Recondition ☐ Replace		
Grease seals	☐ Replace		
Front-wheel bearings	☐ Repack ☐ Replace		
Wheel cylinders	☐ Recondition ☐ Replace		
Rear brake hardware	☐ Replace		
Brake shoes	☐ Replace		
Brake drums	☐ Recondition ☐ Replace		
Parking brakes	☐ Adjust ☐ Lubricate ☐ Replace		
Power brake booster	☐ Service ☐ Replace		
Master cylinder	☐ Recondition ☐ Replace		
Brake fluid	☐ Drain old, and add new fluid		
Lines, hoses, and combination valve	☐ Replace		
Stoplight and other brake lights	☐ Replace Bulb ☐ Replace Switch ☐ Adjust ☐ Replace		
Other ____________			
	Total		

Figure 14-2 Sample of Complete Brake Job Checklist (*courtesy of Wagner Division, Cooper Industries Inc.*)

and so will a less pointed tip on the cutting tool used to refinish drums.

Grinding noises on application can mean metal-to-metal contact, either from badly worn pads or shoes, or from a serious misalignment of the caliper, rotor, wheel cylinder, or backing plate. Wheel cylinders and calipers that are frozen due to internal corrosion can also cause grinding or squealing noises.

Keep in mind that some disc brake pads have an audible sensor that signals when the lining has neared the end of its useful life. Other noise problems and their solutions are covered later in this chapter.

ROAD TESTING BRAKES

The road testing is very important; it allows the brake technician to evaluate brake performance under actual driving conditions. Whenever practical, perform the road test before beginning any work on the brake system. In every case, road test the vehicle after any brake work to make sure the brake system is working safely and properly.

> **WARNING:** Before test driving any car, first check the fluid level in the master cylinder, depress the brake pedal to be sure that there is adequate pedal reserve, and make a series of low-speed stops to be sure that the brakes are safe for road testing. Always make a preliminary inspection of the brake system in the shop before taking the vehicle on the road.

Brakes should be road tested on a dry, clean, reasonably smooth, and level roadway. A true test of brake performance cannot be made if the roadway is wet, greasy, or covered with loose dirt so that all tires do not grip the road equally. Testing will also be adversely affected if the roadway is crowned so as to throw the weight of the vehicle toward the wheels on one side, or if the roadway is so rough that wheels tend to bounce.

Test brakes at different speeds with both light and heavy pedal pressure; however, avoid locking the wheels and sliding the tires on the roadway. Locked wheels and sliding tires do not indicate brake efficiency since heavily braked, but turning wheels will stop the vehicle in less distance than locked wheels. More tire-to-road friction is present with a heavily braked turning tire than with a sliding tire. Also be careful not to induce fade unintentionally with continuous brake applications during road testing.

SHOP TALK

Fade is a temporary reduction of brake effectiveness, resulting from heat.

The following road tests will aid in the evaluation of the brake system's performance and the need for service:

- *Low-Speed Test for Effectiveness, Pulls, and Noise*—Make light and medium stops at 10 to 15 MPH. Bring the car to a complete stop each time. Observe the effort required to make each stop. Is it too light (grabby), ok, or too hard? Check for pulls. Unequal front brakes will cause pulls in the direction of the brake doing the most work. Unequal rear brakes might not cause noticeable pulling during low-speed stops. Check for noise. Open the windows and turn off all accessories and listen to determine the type of noise and the wheel from which it is coming. Driving aside a wall when applying brakes will make noises more audible.
- *High-Speed Test for Roughness or Pulsations*—Make *light stops* at the maximum legal speed. Check for roughness or pulsations by pedal feel and car vibration.
- *High-Speed Test for Effectiveness, Pulls, and Noise*—Make hard stops (just short of skid) at the maximum legal speed. Bring the car to a complete stop each time. Do not repeat stops within 2 miles of each other to avoid high brake temperatures. Observe the effort required to make each stop. Is it too light (grabby), ok, or too hard? Check for pulls. Unequal brakes will cause pulls in the direction of the brake doing the most work. Check for noise to determine the type and the wheel from which it is coming.
- *High-Speed Test for Fade*—Make three hard stops (just short of skid) from 60 MPH at 1/2-mile increments at the maximum legal speed. Check for pulls and observe the effort required to make the stops. These should be made without strenuous effort. Check pedal reserve after each stop.
- *High-Speed Test for Delayed Fade*—After making the previous fade test, let the car stand for 10 minutes. Then, accelerate quickly to maximum legal speed and make one hard stop, just short of a skid. Check for pulls and observe the effort required to make the stop. Then check the pedal reserve.

Keep in mind that there are external conditions that affect brake road-test performance. They are:

1. **Tires.** Tires having unequal contact and grip on the road will cause unequal braking. Tires must be equally inflated, and the tread pattern of right and left tires must be approximately equal.
2. **Vehicle Loading.** When the vehicle has unequal loading, the most heavily loaded wheels require more braking power than others. A heavily loaded vehicle requires more braking effort.
3. **Front-Wheel Bearings.** A loose front-wheel bearing permits the drum and wheel to tilt and have spotty contact with the brake linings, causing erratic brake action.
4. **Front-End Alignment.** Misalignment of the front end will cause the brakes to pull to one side. Also, a loose front-wheel bearing could permit the disc to tilt and have spotty contact with brake shoe linings, causing pulsations when the brakes are applied.
5. **Shock Absorbers.** Faulty shock absorbers that do not prevent the car from bouncing on quick stops can give the erroneous impression that the brakes are too severe.

After determining the condition of brakes either by road testing or based on the customer's description, refer to the troubleshooting guide. Most common problem conditions are listed by symptom in this diagnostic aid. This is to provide easy access to the information. Following the symptoms, the possible causes are listed along with brief corrective remedies. Many symptoms are similar to others and are sometimes found in groups, so one symptom will not necessarily give the complete answer. For complete service and repair procedures, always refer to the appropriate chapters of this book.

TROUBLESHOOTING GUIDE FOR DRUM BRAKES

CONDITION: LOW PEDAL (Pedal might go to floor.)

POSSIBLE CAUSE	*CORRECTIVE REMEDY*
1. Excessive clearance between drums and linings.	1. Adjust brakes.
2. Automatic adjusters not working properly.	2. Make several forward and reverse brake stops; if pedal does not come up, repair automatic adjusters.
3. Bent or distorted brake shoes.	3. Replace shoes and linings in axle sets.
4. Leak in hydraulic system.	4. Check master cylinder, wheel cylinders, tubes, and hoses for leakage; repair or replace faulty parts.
5. Air in hydraulic system.	5. Bleed hydraulic system.
6. Poor quality brake fluid (low boiling point).	6. Drain hydraulic system and fill with approved brake fluid.
7. Low brake fluid level.	7. Fill master cylinder and bleed hydraulic system.
8. Weak brake hoses that expand under pressure.	8. Replace defective hoses.
9. Residual pressure check valves not holding pressure in lines.	9. Repair or replace master cylinder.
10. Drum-to-shoe fit not right.	10. Grind shoes to fit drums exactly.
11. Shoes not properly aligned.	11. Examine backing plate bearing points and lining seal. Test drum for eccentricity.
12. Brake drum worn thin or out-of-round.	12. Replace drum.
13. Brakes out of adjustment.	13. If automatically adjusted, make several forward and reverse stops, or set and reset the parking brake, whichever actuates the automatic adjuster. If pedal does not come

CONDITION: LOW PEDAL (Pedal might go to floor.) (CONTINUED)

POSSIBLE CAUSE	*CORRECTIVE REMEDY*
	up, repair the automatic adjusters. On the star wheel type, check the pawl advancement. If manually adjusted, adjust as needed. Repair or replace adjusters if needed.
14. Leaking primary cup in master cylinder permitting fluid bypass.	14. Recondition the master cylinder or replace with a new cylinder. Bleed the hydraulic system.
15. Inferior brake fluid that evaporates or boils at low temperatures.	15. Flush and bleed the hydraulic system.
16. Leaking hydraulic system.	16. Check the entire hydraulic system and replace any defective parts. Bleed the hydraulic system.
17. Excessively oversize brake drums that are weakened and expand under heat and pressure.	17. Replace with new brake drums. Best results are obtained by replacing both drums on an axle.
18. Standard thickness brake lining with oversize drums causing heavy lining-to-drum contact at center of the lining, allowing the brake shoes to spring excessively at each end.	18. Use lining ground to the correct drum radius. Lay the brake shoe in the drum and check for the 0.005 to 0.010 inch clearance that should be present at the extreme ends for proper shoe-to-drum fit.
19. Wheel cylinder pushrod pins not properly located on slot in the brake shoes.	19. Relocate the pins properly and replace any damaged parts.
20. Improperly adjusted master cylinder pushrod.	20. Adjust the pushrod free play.
21. Master cylinder mounting loose.	21. Tighten nuts and bolts to specifications.
22. Glazed, worn, scored, or damaged brake linings.	22. Replace brake lining sets.

CONDITION: SPRINGY OR SPONGY PEDAL (Pedal has soft, springy, or spongy feel when depressed.)

POSSIBLE CAUSE	*CORRECTIVE REMEDY*
1. Air in the hydraulic brake system.	1. Refill and bleed. Check for possible fluid leaks at: —Master cylinder secondary cup —Wheel cylinders or disc calipers —Hose and line connections —Power brake vacuum line. Disconnect the vacuum line and check for the presence of fluid.
2. Fluid boil, vaporized by overheating.	2. Check by opening the bleeder screw. **CAUTION:** Cooling will restore braking, but condition will return when brakes are heated. Drain and replace the fluid. Check for cause of boiling:

CONDITION: SPRINGY OR SPONGY PEDAL
(Pedal has soft, springy, or spongy feel when depressed.) (CONTINUED)

POSSIBLE CAUSE	*CORRECTIVE REMEDY*
	—Fluid contaminated or wrong type for the system —Overheated brakes—thin drums or rotors, spotty lining contacts, severe overloading of the vehicle. Correct accordingly.
3. Shoes bent, distorted, or not centered in the drum.	3. Replace shoes and linings in axle sets. Adjust anchor pins (if provided). Check lining seating and drum eccentricity.
4. Brake drum cracked or too thin.	4. Replace drums in axle sets only.
5. Weak brake hoses that expand under pressure.	5. Replace defective hoses.
6. Brake tubing improperly positioned.	6. Check brake tubing for misposition near heat source. Hot fluid can boil and result in spongy pedal response.
7. Loss of system residual pressure might permit air to seep into wheel cylinder.	7. Replace the residual pressure check valve.
8. Improper use of parking brake (after the pedal is released) causes a pressure loss.	8. Set parking brake before releasing pedal.

CONDITION: PEDAL DROPS OR "FADES"
(Pedal travels—falls away—as brakes are held applied.)

POSSIBLE CAUSE	*CORRECTIVE REMEDY*
1. Fluid seepage at hose joints and line connections.	1. Replace faulty parts.
2. Fluid seepage (or sticking) at wheel cylinder piston cups.	2. Rebuild or replace wheel cylinder.
3. Excessive drum expansion.	3. Check for rebored thin drum, cracked drum, overheated conditions. Correct accordingly.
4. Caliper piston seal damage.	4. Replace seals.
5. Master cylinder main cup grooved.	5. Rebuild or replace master cylinder.
6. Bleeder screw open.	6. Close bleeder screws.
7. Brakes shoes improperly adjusted.	7. Readjust the brakes.
8. Poor lining-to-drum contact.	8. Use lining ground to the correct drum radius. Lay the brake shoes in the drum and check for 0.005 to 0.010-inch clearance that should be present at the extreme ends for proper shoe-to-drum fit.
9. Brakes adjusted too tightly.	9. Readjust brakes properly; wheels must spin freely.
10. Car driven with parking brake on.	10. Release parking brake and allow brakes to cool off.
11. Backing plate assemblies not lubricated, causing brake shoes to hang up and brakes to drag.	11. Lubricate all brake shoe-to-backing plate contact points.
12. Sluggish hydraulic system release, allowing the brakes to drag and heat up.	12. Check entire hydraulic system for freeness of moving parts. Replace any defective parts.
13. Brake shoe retracting springs weak, stretched, or wrong type.	13. Replace with proper springs in all wheels for best results.

CONDITION: PEDAL DROPS OR "FADES"
(Pedal travels—falls away—as brakes are held applied.) (CONTINUED)

POSSIBLE CAUSE	*CORRECTIVE REMEDY*
14. Thin or weakened brake drums expanding under pressure and heat.	14. Replace brake drums that do not meet the manufacturer's oversize limits.
15. Wrong type of brake lining used.	15. Use recommended brake shoe or lining set.
16. Power brake has poor release.	16. Overhaul or replace power unit.
17. Hydraulic system contaminated, piston rusted, sticky, or frozen.	17. Overhaul or replace hydraulic cylinders, etc.

CONDITION: PEDAL PULSATION
(Pedal chatter, fight, or vibration while braking.)

POSSIBLE CAUSE	*CORRECTIVE REMEDY*
1. Worn or damaged front-wheel bearings.	1. Replace bearings.
2. Out-of-round drums.	2. Recondition or replace the drums in axle sets.
3. Rotor has excessive lateral runout or non-parallelism.	3. Machine or replace.
4. Wheel bearing adjustment loose.	4. Adjust to specifications.
5. Bent rear axle.	5. Replace the axle.
6. Excessive drum runout due to faulty installation of new drums.	6. Remove and properly install the drums.
7. Loose brake drum on hub.	7. Reset the rivets or tighten bolts.

CONDITION: DIVE
(Excessive dipping of the vehicle front end upon brake application.)

POSSIBLE CAUSE	*CORRECTIVE REMEDY*
1. Metering valve faulty.	1. Replace valve; bleed system.
2. Rear half of system failure.	2. Completely check rear half or split system.

CONDITION: NO BRAKING EFFECT
(Pedal bottoms to floor with little or no braking.)

POSSIBLE CAUSE	*CORRECTIVE REMEDY*
1. Fluid reservoir low.	1. Refill reservoir. The disc reservoir drops more quickly than the drum reservoir.
2. Air in the hydraulic system.	2. Bleed the hydraulic system.
3. Vapor lock.	3. Flush out contaminated fluid. Replace with new.
4. Fluid leaking around primary cup of master cylinder.	4. Rebuild master cylinder with new seals. Repair or replace master cylinder.
5. Pistons bottomed in bores after caliper service or pads not properly in place.	5. Step on brakes lightly several times to move pistons out and pads into contact with rotor.
6. Caliper seal or piston damage.	6. Replace seals or piston.
7. Rear wheel cylinder leaking.	7. Repair or replace cylinder and install new shoes in axle sets.
8. Rear drum brake out of adjustment.	8. Adjust brakes. Check adjuster action.
9. Arms of brake pedal disconnected from pushrod.	9. Attach to pushrod.
10. Bleeder screw open.	10. Close bleeder screws; bleed system.

CONDITION: EXCESSIVE STOPPING DISTANCE AND/OR EFFORT
(Hard pedal; excessive pedal pressure is required to stop the car.)

POSSIBLE CAUSE	*CORRECTIVE REMEDY*
1. After extended use, linings worn out or poor quality brake linings.	1. Replace with approved shoes and linings in axle sets only.
2. "Faded" overheated condition, glazed linings, "blued" or heat-checked drums.	2. Lightly sand the lining surface or replace shoes and linings in axle sets. Recondition or replace drums.
3. Excessive lining clearance.	3. Check and correct pawl or link advancement on automatic adjusters. Adjust manual adjusters.
4. Grease—or brake fluid—soaked linings.	4. Repair grease seal or wheel cylinder, as necessary, and replace shoes and linings in axle sets only.
5. Low engine vacuum supply to power brake booster or defective power brake vacuum check valve.	5. Tune or repair the engine to obtain correct vacuum, or replace check valve.
6. Faulty power brake booster.	6. Replace the power booster.
7. Loose or leaking vacuum hose to power brake booster.	7. Tighten clamps, or replace hose, as required.
8. Wheel cylinder or master cylinder pistons stuck.	8. Rebuild or replace appropriate cylinder.
9. Brake drum bell-mouthed or barrel-shaped.	9. Refinish drum to proper limits, or replace (if beyond refinishing limits) in axle sets only.
10. Brake pedal binding on shaft.	10. Free up bind and lubricate the necessary parts.
11. Brake pedal interference.	11. Check pedal and linkage for any interference.
12. Insufficient clearance between the master cylinder pushrod and the piston, allowing hydraulic pressure buildup in the system.	12. Adjust the pushrod, allowing a 1/16-inch clearance between the end of rod and piston.
13. Replenishing port in the master cylinder plugged or primary cup swollen, allowing hydraulic pressure buildup in the system.	13. Recondition or replace with a new master cylinder, or adjust pushrod end play.
14. Frozen master or wheel cylinder pistons.	14. Recondition or replace all the cylinders.
15. Brake fluid contamination, causing sluggish hydraulic system operation.	15. Flush the hydraulic system.
16. Brake shoes with two different lining frictions installed in wrong location.	16. Install shoes in proper position or reline the brakes.
17. Cast-iron dust from surface of newly turned brake drum imbedded in brake linings.	17. Clean out dust and chips. Always wash the drum surface with a non-oil base cleaning fluid. Sand or regrind lining.
18. Wrong type of brake lining used.	18. Use recommended brake shoe or lining set.
19. Lack of lubrication at backing plate bosses, anchors, abutment blocks, shoe guides, etc.	19. Lubricate all points for free brake shoe movement.
20. Water film on brake lining surface.	20. Check backing plate for openings and be sure adjusting hole covers are in place.
21. Defective vacuum section in power brake unit or plugged vacuum line from engine.	21. Replace the power unit or vacuum line from intake manifold.

CONDITION: DECREASING BRAKE PEDAL TRAVEL

POSSIBLE CAUSE	*CORRECTIVE REMEDY*
1. Weak or broken shoe return springs.	1. Replace return springs.
2. Sticking wheel cylinder piston.	2. Repair or replace wheel cylinder.
3. Plugged master cylinder replenishing port(s).	3. Repair or replace master cylinder.
4. Swollen master cylinder cups.	4. Repair or replace master cylinder, flush hydraulic system, and fill with approved brake fluid.

CONDITION: CAR PULLS TO ONE SIDE

POSSIBLE CAUSE	*CORRECTIVE REMEDY*
1. Faulty suspension parts.	1. Repair suspension system.
2. Incorrect linings on shoes.	2. Replace with correct shoes and linings.
3. Brake fluid, oil, grease, water, mud, or dirt on linings.	3. Repair grease seal or wheel cylinder, as necessary, clean the affected parts, replace shoes and linings in axle sets only.
4. Distorted or incorrect shoes.	4. Replace with correct shoes and linings.
5. Shoes not centered in drums.	5. Adjust anchor pins (if provided).
6. Loose or distorted brake backing plate.	6. Tighten or replace the plate.
7. Weak shoe return springs or hold-downs.	7. Replace with new parts.
8. Drum bell-mouthed, barrel-shaped, threaded, cracked, or high hard spots on drum surface.	8. Recondition or replace drums, in axle sets only.
9. Sticking wheel cylinder piston.	9. Repair or replace wheel cylinder.
10. Abnormal deflection, wear, and improper alignment, including caster and camber, weak or broken springs, broken spring center bolt, loose U-bolts, loose steering gear or tie rods, worn shock absorbers, etc.	10. Check and correct front-end alignment. Repair or replace faulty parts.
11. Uneven tire pressure, tread wear or size, right to left.	11. Equalize to recommended pressures. Install correct size tires with good tread.
12. Faulty brake hose or obstruction in brake line.	12. Replace hose or line.
13. Charred or badly worn brake linings.	13. Replace linings (in axle sets) and other worn parts.

CONDITION: BRAKES DRAG, LOCK, OR OVERHEAT
(Either individual or all wheels drag or lock; severe reaction to pedal pressure.)

POSSIBLE CAUSE	*CORRECTIVE REMEDY*
1. Rough or corroded master cylinder bore. Master cylinder bypass port blocked by dirt or swollen primary cup.	1. Rebuild or replace the master cylinder. *(Do not rebuild* aluminum master cylinders.)
2. Binding brake pedal linkage; bind prevents the return of pedal against its stop.	2. Free up and lubricate.
3. Weak pedal return spring (if used).	3. Replace the spring.

CONDITION: BRAKES DRAG, LOCK, OR OVERHEAT (Either individual or all wheels drag or lock; severe reaction to pedal pressure.) (CONTINUED)

POSSIBLE CAUSE	*CORRECTIVE REMEDY*
4. Valve leaks, weak reaction springs, or bind in the power mechanism.	4. Repair or replace the power section of the power brake.
5. Loose or worn front-wheel bearings.	5. Adjust to specifications or replace.
6. Defective brake hose or lines, preventing the return of brake fluid.	6. Replace the faulty lines or hoses, as necessary.
7. Soft or swollen rubber parts caused by contaminated or incorrect brake fluid.	7. Replace all rubber parts (including hoses), flush the hydraulic system, and fill with approved brake fluid.
8. Incorrect, bent, or distorted shoes.	8. Replace shoes and linings in axle sets only.
9. Incorrect linings, or linings loose on shoes, or grease or brake fluid on linings.	9. Replace with correct shoes and linings in axle sets only. Repair grease seal or wheel cylinders.
10. Weak or broken shoe return springs.	10. Replace the shoe return springs.
11. Improper brake shoe adjustment.	11. Adjust shoes and repair the automatic adjuster, if necessary.
12. Shoes not centered in drums.	12. Adjust anchor pin (if provided).
13. Loose or distorted brake backing plate.	13. Tighten or replace backing plate.
14. Scored, hard-spotted, or out-of-round drums.	14. Refinish or replace the drums in axle sets.
15. Frozen parking brake cables.	15. Free up and lubricate cables, or replace.
16. Piston stuck in wheel cylinder.	16. Rebuild or replace wheel cylinder.
17. Plugged master cylinder vent port(s), and/or swollen or damaged master cylinder primary cups.	17. Rebuild or replace the master cylinder. Flush the hydraulic system and fill with approved brake fluid. *Do not rebuild* aluminum master cylinders.
18. Improper lining-to-drum clearance.	18. Adjust the brakes.
19. Brake pedal linkage binding.	19. Check for binding; relieve and lubricate.
20. Contaminated brake fluid causing rubber parts to swell in wheel cylinders or master cylinder.	20. Recondition or replace all the wheel cylinders. Flush and bleed the hydraulic system.
21. Hydraulic flex rubber hose line has torn inner fabric or foreign matter in line.	21. Replace all hydraulic flex hoses.
22. Grease, oil, or brake fluid on the lining surface.	22. Reline the brakes. Grinding or cleaning the brake lining surface will not give satisfactory results.
23. Brake shoes not releasing properly due to lack of backing plate lubrication at bosses, anchors, abutment blocks, etc., or deep grooves worn in backing plate bosses or worn shoe rims at boss locations.	23. Lubricate all contact points with lubricant. Replace worn backing plates or shoes.
24. Wheel cylinder-to-brake shoe pushrod pin not properly located.	24. Install the wheel cylinder-to-brake shoe pushrod pin properly and replace the damaged parts.
25. Loose or worn wheel and axle bearings.	25. Replace the worn bearings and adjust new bearings properly.
26. Loose backing plates anchors and wheel cylinders.	26. Tighten loose bolts.

CONDITION: BRAKES DRAG, LOCK, OR OVERHEAT (Either individual or all wheels drag or lock; severe reaction to pedal pressure.) (CONTINUED)

POSSIBLE CAUSE	*CORRECTIVE REMEDY*
27. Parking brake cables frozen or adjusted too tight.	27. Free up cables or replace them with new cables. Lubricate and adjust the cables properly.
28. Defective proportioning or combination valve (front disc/rear drum system).	28. Replace proportioning or combination valve.
29. Residual pressure check valve in outlet serving front disc brakes.	29. Remove residual pressure check valve; it is not needed in front disc brake system.

CONDITION: BRAKE NOISES (Brakes make many different types of noises; each has its own causes and remedies.)

SQUEAL (A high-pitched noise when brakes are used.)

POSSIBLE CAUSE	*CORRECTIVE REMEDY*
1. With new linings: Improper arc grind or eccentric drum causes the lining to "pinch" against the drum.	1. Reface or grind the drums, or regrind or replace linings, in axle sets only.
2. Dust or other particles imbedded in the linings.	2. Remove foreign matter. Replace the lining, if necessary, in axle sets only.
3. Loose lining rivets.	3. Reset the rivets or reline the shoes if the rivet holes in the linings are elongated.
4. Loose backing plate, anchor, drum, or wheel cylinder.	4. Tighten all loose parts and replace them if necessary.
5. Shoes scrape on support pads.	5. Clean and lubricate.
6. Weak or broken shoe hold-down springs.	6. Replace springs.
7. Support plate bent, shoes twisted.	7. Replace the plate and/or shoes.
8. Drums out-of-round or out-of-square.	8. Recondition or replace the drums, in axle sets only.
9. Drum silencer (screen door) spring missing or weak.	9. Replace the spring.
10. Lining of the wrong material.	10. Reline with new shoes.

GRIND (Metallic noise similar to a garbage disposal.)

POSSIBLE CAUSE	*CORRECTIVE REMEDY*
1. Lining worn enough to allow metal-to-metal contact. Rivets or shoe table rub on drum.	1. Replace shoes and linings in axle sets. Recondition or replace the drums, if damaged, in axle sets only.
2. Foreign material imbedded in lining.	2. Replace shoes and linings, in axle sets only.
3. Rough drum surface.	3. Recondition or replace the drums, in axle sets only.

THUMPS/CLUNKS (Deep noise that occurs on brake application.)

POSSIBLE CAUSE	*CORRECTIVE REMEDY*
1. Excessive lining clearance. Applying, linings thump drums; releasing, shoes clunk on anchors.	1. Adjust the brakes to proper clearance.

CONDITION: BRAKE NOISES (Brakes make many different types of noises; each has its own causes and remedies.) (CONTINUED)

CHATTER (A rapid, low-pitch noise.)

POSSIBLE CAUSE	*CORRECTIVE REMEDY*
1. Lining clearance too tight or too loose.	1. Adjust the linings to proper clearance.
2. Loose backing plate.	2. Tighten or replace the plate.
3. Grease, fluid, dust, or other contaminate on lining surface.	3. Recondition or replace the linings, in axle sets only.
4. Twisted brake shoe.	4. Replace the shoes.
5. Drum out-of-round, bell-mouthed, barrel-shaped, or cracked.	5. Recondition or replace the drum.
6. Weak or broken shoe return spring.	6. Replace the springs.
7. Loose wheel bearing.	7. Readjust wheel bearing.
8. Loose spindle and/or suspension parts.	8. Adjust or replace as needed.
9. Improper lining grind.	9. Replace linings in axle sets only.
10. Drum cracked.	10. Replace.
11. Dry adjuster threads.	11. Lubricate.

CLICK (Sharp metallic noise, like the ticking of a clock.)

POSSIBLE CAUSE	*CORRECTIVE REMEDY*
Shoes lift off of the support pads and snap back.	
1. Threaded drum surface.	1. Replace the drum, or recondition by machining.
2. Weak hold-downs.	2. Replace the hold-downs.
3. Grooves in brake shoe support pads.	3. Stone smooth and lubricate.
4. Hardware broken.	4. Replace hardware.
5. Excessive clearance between the lining and the brake drum.	5. Adjust the brakes to proper clearance.
6. Loose or worn front-end parts.	6. Tighten or replace the defective parts.
7. Excessively grooved backing plate bosses.	7. Replace the backing plate assemblies.
8. Lack of lubrication at backing plate bosses, abutment blocks, anchors, or any other points that require lubrication.	8. Lubricate all the necessary points.
9. Loose backing plates, wheel cylinders, and anchors.	9. Tighten the loose parts.
10. Cracked brake drums.	10. Replace drums in axle sets.
11. Baffle type cylinders not used on Falcon, Comet cars.	11. Replace with baffle type wheel cylinders.
12. Bent brake shoe hold-down pins.	12. Replace hold-down pins.
13. Drum lathe tool marks in newly turned drums caused by a chipped or improperly ground tool bit. This results in a threaded surface on which the shoes ride out, then snap back.	13. Regrind or replace the tool bit. Reface the brake drums and grind them if necessary.

GROAN (Deep, pulsating sound on brake application.)

POSSIBLE CAUSE	*CORRECTIVE REMEDY*
1. Unlubricated metal contacts or lands of backing plate.	1. Lubricate contacts or lands.

CONDITION: BRAKE NOISES (Brakes make many different types of noises; each has its own causes and remedies.) (CONTINUED)

SNAPPING (A sharp, cracking, or clicking noise on brake application.)

POSSIBLE CAUSE	*CORRECTIVE REMEDY*
1. Dry metal-to-metal contacts.	1. Lubricate.
2. Grooved shoe support pads.	2. Stone smooth and lubricate.
3. Loose backing plate or loose drum.	3. Tighten or replace the affected parts.
4. Loose or worn front-end components.	4. Tighten or replace the components.

CONDITION: ERRATIC BRAKE ACTION

POSSIBLE CAUSE	*CORRECTIVE REMEDY*
1. Loose or worn wheel bearings.	1. Adjust the bearings properly and replace the worn or damaged bearings.
2. Weak shock absorbers.	2. Replace the weak shock absorbers with new ones.
3. Worn front-end parts, such as ball joints, upper and lower control arms, bushings and pins, tie-rod ends, and steering arms, etc.	3. Overhaul the front end.
4. Improper front-end alignment.	4. Check the camber, caster, toe-in, etc.
5. Frozen or sluggish wheel cylinder pistons.	5. Recondition or replace the wheel cylinders.
6. Hydraulic rubber flex hoses plugged.	6. Replace all the flex hoses.
7. Backing plate assemblies not lubricated.	7. Lubricate all the necessary points.
8. Weak, distorted, or wrong type of brake shoe retracting springs.	8. Replace all the springs on an axle. Never replace springs on one side of axle only.
9. Lining not properly ground to fit drum.	9. Always check shoes in the drum for proper fit. A clearance of 0.005 to 0.010 inch should be present at toe and heel ends.
10. Standard diameter brake drum on one side and oversize drum on opposite side.	10. Replace the oversize drum. Most drums have to be replaced when they are 0.060 inch oversize.
11. Loose backing plates, brake anchors, and wheel cylinders.	11. Tighten loose parts.
12. Bent or distorted brake shoes, bent backing plates, or anchors.	12. Replace the defective parts.
13. Brake drum out-of-round, bell-mouthed, barrel-shaped, thin or loose on hub.	13. Reface and grind the drums. Replace the thin drums. Tighten the loose rivets or bolts.

CONDITION: LEAKING CYLINDER (Visual inspection reveals leaking fluid and/or contaminated linings.)

POSSIBLE CAUSE	*CORRECTIVE REMEDY*
1. Seal or piston damage.	1. Clean, inspect, and replace piston. Replace seals and boots.
2. Corroded or scored piston bore.	2. Disassemble caliper, hone bore; install new seals, boots, and pistons.
3. Piston improperly installed or seal cup over inlet/bleeder screw hole.	3. Reassemble correctly.

CONDITION: GRABBY BRAKES (Severe reaction to pedal pressure.)

POSSIBLE CAUSE	*CORRECTIVE REMEDY*
1. Incorrect or distorted shoes.	1. Replace with correct shoes and linings in axle sets.
2. Incorrect linings or loose linings on shoes.	2. Replace with correct shoes and linings in axle sets.
3. Grease or brake fluid on linings.	3. Repair grease seal or wheel cylinder, as necessary, and replace shoes and linings in axle sets.
4. Shoes not centered in drums.	4. Adjust anchor pins (if car has adjustable anchor pins).
5. Loose or distorted brake backing plate.	5. Tighten or replace backing plate.
6. Scored, hard-spotted, or out-of-round drums.	6. Refinish or replace drums in axle sets.
7. Rough or corroded master cylinder bore.	7. Repair or replace master cylinder.
8. Binding brake pedal linkage.	8. Free up and lubricate.
9. Faulty power brake.	9. Replace power section of power brakes.

CONDITION: GRAB OR PULL DUE TO ONE SENSITIVE BRAKE

POSSIBLE CAUSE	*CORRECTIVE REMEDY*
1. Improper brake shoe adjustment.	1. Readjust the brake shoes or replace with a new set.
2. Weak or broken springs.	2. Replace the defective parts.
3. Worn or loose front-end parts.	3. Tighten or replace any defective parts.
4. Improper front-wheel alignment.	4. Check and make any necessary corrections.
5. Brake drums eccentric, bell-mouthed, barrel-shaped or weak.	5. Reface or grind the drums, install new ones if they are oversize beyond the manufacturer's limits. It is good practice to replace both drums on the same axle. When refacing brake drums, keep the diameters to within 0.010 inch of each other.
6. Riveted lining loose on shoes.	6. Reset loose rivets or reline the brakes.
7. Loose backing plates, anchors, or wheel cylinder.	7. Tighten any loose parts.
8. High-friction lining on one side and lower friction lining on the opposite side.	8. Reline the brakes with proper lining friction in all wheels.
9. Primary and secondary or forward and reverse shoes and linings installed in wrong location on one side.	9. Install the brake shoes in their proper location and reline the brakes, if necessary.
10. Wheel cylinder or grease seals leaking, contaminating the brake lining.	10. Replace or overhaul the cylinder and replace grease seals. Fluid or grease cannot be removed from brake lining. Always reline brakes.
11. Improper size wheel cylinder on one side.	11. Replace with the correct size.
12. Self-adjuster not working on one side.	12. Free up and lubricate. Replace defective parts.

CONDITION: PEDAL RISING AND ALL BRAKES DRAGGING

POSSIBLE CAUSE	*CORRECTIVE REMEDY*
1. Blocked master cylinder replenishing port caused by foreign matter, brake fluid contamination, swollen primary cup, insufficient clearance between the master cylinder pushrod and the piston or a sluggish master cylinder piston return.	1. Recondition the master cylinder or replace it with a new cylinder. Adjust the master cylinder pushrod to the piston allowing at least a 1/16 inch clearance. Flush and bleed the hydraulic system.
2. Petroleum-based oil used in the hydraulic system damaging all internal rubber parts.	2. Replace all internal rubber parts and rubber flex hoses. Flush and bleed the hydraulic system.
3. Defective internal power brake unit parts.	3. Replace with a new power brake unit.
4. Internal master cylinder and power brake parts installed incorrectly.	4. Install all parts correctly or replace them with a new unit.
5. Master cylinder piston not returning to stop properly, thereby blocking replenishing port.	5. Check the master cylinder piston for proper return. Check the lock ring stop and washer for proper installation.

CONDITION: BRAKE SYSTEM WARNING LIGHT DOES NOT LIGHT

POSSIBLE CAUSE	*CORRECTIVE REMEDY*
1. Warning light bulb burned out.	1. Replace with new bulb.
2. Warning light switch has open circuit.	2. Check circuit and repair.
3. Warning light switch does not operate.	3. Replace with a new switch.

CONDITION: BRAKE SYSTEM WARNING LIGHT STAYS LIT

POSSIBLE CAUSE	*CORRECTIVE REMEDY*
1. One section of the dual-brake system inoperative.	1. Check hydraulic system for leaks and repair.
2. Differential pressure valve not centered.	2. Center the valve.
3. Wiring to warning light switch is grounded or shorted.	3. Locate problem wire and correct.
4. Warning light switch damaged or worn.	4. Replace switch.

TROUBLESHOOTING GUIDE FOR DISC BRAKES

CONDITION: DEPRESSING BRAKE PEDAL PRODUCES NO BRAKING EFFECT

POSSIBLE CAUSE	*CORRECTIVE REMEDY*
1. Reservoir fluid level low.	1. Check for cause of fluid leak, and repair as required. Refill the reservoir. Bleed system as needed.
2. Air in the hydraulic system.	2. Bleed the system.
3. Bleeder screw open.	3. Close the bleeder screw and bleed the system.

CONDITION: DEPRESSING BRAKE PEDAL PRODUCES NO BRAKING EFFECT (CONTINUED)

POSSIBLE CAUSE	*CORRECTIVE REMEDY*
4. Improperly positioned pads (while servicing the calipers, pistons may have been pushed back in cylinder bores).	4. Reposition pads (depress the pedal a second time).
5. Excessively tapered or warped pads.	5. Install new pads, in axle sets only.
6. Maladjusted rear brakes.	6. Adjust properly.
7. Leak in rear brake cylinder.	7. Hone the cylinder bore. Replace with new piston seals or replace caliper.
8. Piston seal damaged in one or more of the calipers.	8. Disassemble the caliper and replace the piston seals. Replace piston if damaged.
9. Leak past piston cups in master cylinder.	9. Recondition or replace the master cylinder.
10. Excessive rotor runout or bent rotor.	10. Check with dial indicator. Install new rotor if runout exceeds maximum specified.
11. Bad or excessively loose wheel bearings.	11. Adjust or replace as needed.
12. Pad and plate knock-back caused by loose wheel bearings or faulty front suspension.	12. Adjust or tighten parts or replace faulty parts, as necessary.
13. Leak in hydraulic system.	13. Check master cylinder, wheel cylinders, calipers, tubes, and hoses for leakage; repair or replace faulty parts. Bleed hydraulic system.
14. Weak brake hoses that expand under pressure.	14. Replace defective hoses.
15. Improperly adjusted manual master cylinder pushrod.	15. Adjust pushrod (if car has adjustable pushrod).
16. Rear brakes out of adjustment.	16. Check and adjust rear brakes.
17. Defective primary cup in the master cylinder.	17. Rebuild or replace master cylinder.
18. Brake shoe assemblies not properly seated or positioned.	18. Install new brake shoes in their correct position.

CONDITION: HARD PEDAL (Excessive pedal pressure required to stop vehicle.)

POSSIBLE CAUSE	*CORRECTIVE REMEDY*
1. Pads worn below minimum thickness.	1. Install new pads.
2. Master cylinder or power brake malfunction.	2. Check and correct the attached parts.
3. "Faded" overheated condition, glazed pads, "blued" or heat-checked rotors.	3. Replace the rotor and/or reface pads if sufficient lining remains.
4. Grease, oil, or brake fluid on linings.	4. Install new pads, in axle sets only.
5. Seized or frozen pistons.	5. Disassemble caliper and free up pistons, or replace caliper.
6. Pedal pushrod and linkage binding.	6. Free up and lubricate.
7. Lines, hoses, or connections dented, kinked, collapsed, or blocked.	7. Repair or replace defective parts.
8. Lining worn below specifications.	8. Replace with new lining.
9. Wrong lining friction or inferior grade lining.	9. Replace with recommended lining.
10. Frozen or sticking caliper piston.	10. Recondition the caliper.

CONDITION: SPRINGY OR SPONGY PEDAL
(Pedal has soft, springy, or spongy feel when depressed.)

POSSIBLE CAUSE	*CORRECTIVE REMEDY*
1. Poor quality brake fluid (low-boiling point).	1. Drain hydraulic system and fill with approved brake fluid.
2. Weak brake hoses that expand under pressure.	2. Replace defective hoses.
3. Air in hydraulic system.	3. Bleed hydraulic system.

CONDITION: PEDAL PULSATION (Brake roughness or chatter)

POSSIBLE CAUSE	*CORRECTIVE REMEDY*
1. Excessive lateral runout of brake rotor.	1. Check with dial indicator. Install new rotor if runout exceeds the maximum specified.
2. Excessive out-of-parallelism of brake rotor.	2. Check the parallelism (rotor thickness variation) with micrometer and remachine the rotor or install a new one, if the parallelism exceeds the maximum allowed.
3. Loose or worn steering or suspension parts.	3. Replace parts and realign.
4. Excessive front bearing clearance.	4. Readjust the bearings to specified torque.
5. Worn or damaged front-wheel bearing.	5. Replace bearing.
6. Pistons knocked back into caliper bore after violent cornering or travel over rough road.	6. Pump brake pedal several times to restore to normal height.
7. Loose caliper mounting bolts.	7. Torque caliper mounting bolts per specifications.
8. Loose wheel bearing adjustment.	8. Readjust wheel bearing per specifications.
9. Loose or worn front end parts. Front end out of alignment.	9. Replace defective parts. Check front end for caster and camber and correct.
10. Rear brake drums out-of-round.	10. Resurface or replace rear drums.
11. Brake fluid or grease on lining.	11. Replace with new lining on both calipers or rear brakes.

CONDITION: GRABBING OR UNEVEN BRAKING ACTION
(Severe reaction to pedal pressure.)

POSSIBLE CAUSE	*CORRECTIVE REMEDY*
1. Calipers loose on mounting.	1. Tighten caliper mounting bolts to torque specifications.
2. Pedal linkage binding.	2. Free up linkage and lubricate.
3. Incorrect tire pressure.	3. Inflate tires with correct pressure.
4. Worn front end parts, front end out of alignment, broken springs, etc.	4. Replace defective parts; check front end for caster and camber. Replace any broken springs.
5. Power brake unit defective.	5. See Chapter 6, Power Brakes and Parking Brakes.
6. Faulty metering or proportioning valve.	6. Replace metering or proportioning valve.
7. Lining contaminated with grease, oil, or fluid.	7. Replace with new brake pad assemblies on both calipers.
8. High-friction lining on one side and low-friction lining on the opposite side.	8. Reline with the same friction linings on all calipers.

CONDITION: GRABBING OR UNEVEN BRAKING ACTION (Severe reaction to pedal pressure.)

POSSIBLE CAUSE	*CORRECTIVE REMEDY*
9. Restricted or blocked lines, hoses, or passageways.	9. Unblock or replace lines or hoses. Recondition caliper and clear out all passageways.
10. Caliper pistons sluggish or frozen.	10. Recondition the caliper and replace the pistons, if necessary.
11. Soft or swollen seals in the caliper cylinder bores.	11. Flush the system with denatured alcohol and replace all seals.
12. Rough or corroded caliper cylinder bores.	12. Disassemble caliper and hone or sand bores with crocus cloth. Replace caliper if necessary.
13. Incorrect pads or loose pads on plate.	13. Replace with correct pads and plates in axle sets.
14. Rough or corroded master cylinder bore.	14. Repair or replace master cylinder.
15. Binding brake pedal linkage.	15. Free up and lubricate.

CONDITION: EXCESSIVE BRAKE PEDAL TRAVEL

POSSIBLE CAUSE	*CORRECTIVE REMEDY*
1. Pistons knocked back into caliper bore after violent cornering or travel over rough road.	1. Pump brake pedal several times to restore normal height.
2. Air in hydraulic system.	2. Bleed the hydraulic system.
3. Insufficient fluid in the hydraulic system.	3. Check and refill master cylinder and bleed.
4. Loose caliper attachment.	4. Torque caliper mounting bolts per specifications.
5. Loose wheel bearing attachment.	5. Readjust wheel bearing per specifications.
6. Improper power brake pushrod adjustment.	6. Readjust power brake pushrod.
7. Excessive lateral runout of the rotor.	7. Check rotor runout with dial indicator.
8. Rear brake adjustment required.	8. Check and adjust rear brakes.
9. Leaking hydraulic system.	9. Locate leak and repair.
10. Master cylinder malfunction.	10. Check and correct master cylinder operation.
11. Damaged or worn caliper piston seals.	11. Recondition the caliper.
12. Loose or worn wheel bearing, causing piston to be knocked back into bore.	12. Adjust or replace wheel bearing.
13. On rear disc brakes, excessive side-to-side play in axle, which knocks piston back into bore.	13. Reduce side-to-side axle play at wheel bearing or inside differential.

CONDITION: LEAKY CALIPER

POSSIBLE CAUSE	*CORRECTIVE REMEDY*
1. Cylinder bore surface scored or corroded.	1. Disassemble caliper, hone cylinder bore, replace seals and boots, if moving seal type. If fixed seal type, replace seals and boots.
2. Caliper piston seal damaged or worn.	2. Disassemble caliper and install new seals and boots.
3. Caliper piston damaged.	3. Replace piston.

CONDITION: BRAKES HEAT UP DURING DRIVING AND FAIL TO RELEASE

POSSIBLE CAUSE	*CORRECTIVE REMEDY*
1. Power brake malfunction.	1. Check and repair the power brake unit.
2. Residual check valve in disc brake circuit.	2. Remove residual pressure check valve.
3. Sluggish or sticking pedal linkage.	3. Free up and lubricate as necessary.
4. Frozen or seized pistons.	4. Disassemble the caliper and free up the pistons.
5. Vehicle driver habitually "rides" the brake pedal.	5. Instruct driver not to ride brake pedal.
6. Master cylinder does not release pressure.	6. Check master cylinder, pushrod free play, and fluid for contamination. Repair as required.
7. Residual pressure check valve installed in master cylinder outlet to disc brakes.	7. Remove residual pressure valve or replace master cylinder.
8. Brake assemblies not properly seated or positioned.	8. Replace with new brake shoes and position correctly.

CONDITION: CAR PULLS TO ONE SIDE

POSSIBLE CAUSE	*CORRECTIVE REMEDY*
1. Brake fluid, oil, or grease on linings.	1. Install new pads, in axle sets only.
2. Unmatched linings, uneven lining wear, or distorted pads.	2. Install new pads, in axle sets only.
3. Rough rotor surfaces on one rotor.	3. Resurface or replace rotors in axle sets.
4. Seized or frozen pistons.	4. Disassemble caliper and repair, or replace.
5. Loose caliper or caliper mounting bolts or guide pins. Collapsed stabilizers or worn bushings.	5. Tighten or replace and tighten bolts, if worn. Replace caliper hardware.
6. Uneven tire pressure, tread wear, or size—right to left.	6. Equalize to recommended pressures. Install correct size tires with good tread.
7. Excessive rotor parallelism or runout.	7. Resurface or replace rotor.
8. Restricted hose or line.	8. Examine the hoses and lines, and replace as necessary.
9. Caliper piston sticking.	9. Repair or replace caliper.
10. Front end out of alignment.	10. Reset alignment.
11. Faulty suspension parts.	11. Repair suspension system.

CONDITION: ONE FRONT BRAKE HAS EXCESSIVE DRAG

POSSIBLE CAUSE	*CORRECTIVE REMEDY*
1. Sticking caliper piston.	1. Repair or replace caliper.
2. Swollen caliper piston seal.	2. Repair caliper, flush hydraulic system, and fill with approved fluid.

CONDITION: ONE BRAKE DRAGS

POSSIBLE CAUSE	*CORRECTIVE REMEDY*
1. Loose or worn front-wheel bearings.	1. Adjust to specifications or replace.
2. Defective brake hose or hydraulic tube (preventing return of brake fluid).	2. Replace defective hose or tube, as necessary.

CONDITION: ALL BRAKES DRAG, BUT BRAKE ADJUSTMENT IS CORRECT

POSSIBLE CAUSE	*CORRECTIVE REMEDY*
1. Binding brake pedal.	1. Free up and lubricate.
2. Soft or swollen rubber parts caused by incorrect or contaminated brake fluid.	2. Replace all rubber parts, flush hydraulic system, and fill with approved brake fluid.
3. Plugged master cylinder compensating port(s).	3. Repair or replace master cylinder.

CONDITION: DECREASING BRAKE PEDAL TRAVEL

POSSIBLE CAUSE	*CORRECTIVE REMEDY*
1. Plugged master cylinder compensating port(s).	1. Repair or replace master cylinder.
2. Swollen master cylinder cups.	2. Repair or replace master cylinder, flush hydraulic system, and fill with approved brake fluid.

CONDITION: BRAKE NOISES (As with drum brakes, disc brakes make many different types of noises. Each has its own causes and remedies.)

CHATTER (A chatter when the pedal is pulsating during braking action.)

POSSIBLE CAUSE	*CORRECTIVE REMEDY*
1. Excessive lateral runout of rotor.	1. Check the runout with a dial indicator. Install new rotor if the runout exceeds the maximum specified.
2. Lack of rotor parallelism.	2. Check the parallelism with a micrometer. Remachine the rotor, or install a new one, as required.
3. Loose wheel bearing.	3. Readjust the bearings to specified torque.
4. Loose front spindle.	4. Adjust or replace as needed.
5. Loose rear suspension.	5. Adjust or repair as needed.
6. Lug bolt improperly torqued.	6. Retorque lug bolt to specifications.
7. Rear axle bent.	7. Replace axle.

SCRAPING (A lower pitched squeal, but not a pure tone.)

POSSIBLE CAUSE	*CORRECTIVE REMEDY*
1. Rust or mud buildup on edges of rotor and on caliper housing.	1. Clean or replace, as necessary.
2. Worn pad, metal tabs, or backing exposed to rotor.	2. Replace pads, in axle sets only. Repair or replace faulty parts.
3. Faulty caliper alignment, permitting rotor to scrape on housing.	3. Adjust the caliper alignment properly. Replace damaged or worn hardware.
4. Loose wheel bearing.	4. Readjust wheel bearings to correct specifications.
5. Mounting bolts too long.	5. Replace bolts with right length.

GROAN OR CREEP-GROAN (During slow release of the brakes.)

POSSIBLE CAUSE	*CORRECTIVE REMEDY*
1. Pressure on brake pedal too light.	1. Slightly increase the pedal effort to eliminate the noise.

CONDITION: BRAKE NOISES (As with drum brakes, disc brakes make many different types of noises. Each has its own causes and remedies.) (CONTINUED)

RATTLE (At low speeds on rough roads, front wheels only, brakes unapplied.)

POSSIBLE CAUSE	*CORRECTIVE REMEDY*
1. Excessive clearance between shoe and caliper.	1. Install new pads.
2. Shoe anti-rattle spring missing or not properly positioned.	2. Install new anti-rattle spring or position properly.
3. Caliper assembly frozen on mounts.	3. Clean and lubricate mounting points.
4. Loose caliper mounting bolts.	4. Torque caliper mounting bolts to specifications.

SQUEAL (A high-pitched sound when brakes are used.)

POSSIBLE CAUSE	*CORRECTIVE REMEDY*
1. Outboard pads loose in caliper.	1. Bend pad ears to tighten.
2. Glazed pads.	2. Resurface or replace pads, in axle sets only.
3. Insulator shim not in place between piston and pad plate.	3. Replace shim if originally supplied.
4. Worn or weak anti-rattle springs or clips.	4. Install new anti-rattle parts.
5. Pad wear indicators contacting rotor. (Squeal disappears when brakes are applied.)	5. Install new pads, in axle sets only.
6. Buildup of corrosion between caliper and pads.	6. Clean rotor with crocus cloth where pads make contact.
7. Outboard pad tabs loose.	7. Tighten tabs by bending.
8. Poor quality or contaminated linings.	8. Replace with brand-name linings.
9. If pads are equipped with wear sensors, linings worn out.	9. Replace pads.

CONDITION: GRINDING NOISE

POSSIBLE CAUSE	*CORRECTIVE REMEDY*
1. Linings worn down to rivets or metal backing plate.	1. Replace pads and machine rotors.

CONDITION: BRAKE SYSTEM WARNING LIGHT STAYS LIT

POSSIBLE CAUSE	*CORRECTIVE REMEDY*
1. One section of the dual brake system inoperative.	1. Check hydraulic system for leaks and repair.
2. Differential pressure valve not centered.	2. Center the valve.
3. Wiring to warning light switch is grounded.	3. Locate wiring that is grounded and correct.
4. Warning light switch damaged.	4. Replace switch.
5. Low vacuum.	5. Check vacuum.
6. Low-vacuum booster switch on relay damaged.	6. Replace switches.

CONDITION: BRAKE SYSTEM WARNING LIGHT DOES NOT LIGHT

POSSIBLE CAUSE	*CORRECTIVE REMEDY*
1. Warning light bulb burned out.	1. Replace with new bulb.
2. Warning light switch has open circuit.	2. Check circuit and repair.
3. Warning light switch does not operate.	3. Replace with a new switch.

TROUBLESHOOTING GUIDE FOR VACUUM POWER BRAKES

CONDITION: HARD PEDAL (Also see chapter for type of brake involved.)

POSSIBLE CAUSE	*CORRECTIVE REMEDY*
1. Insufficient engine manifold vacuum. There should be at least 15 inches at idle.	1. Tune engine and make any repairs necessary to generate proper vacuum (grind valves, replace timing chain or intake manifold gasket, etc.).
2. Faulty vacuum check valve.	2. Replace with new valve; do not repair or oil valve.
3. Plugged vacuum line fittings and vacuum hoses.	3. Clean out passages or replace with new fittings or replace vacuum hoses.
4. Auxiliary vacuum pump not working. This may be engine driven by the camshaft or by a belt. Or, it may be an electric pump.	4. See if pump generates vacuum by using a vacuum gauge. If not, replace pump. On electric type, check fuse and wiring.
5. External vacuum leaks in reserve tank and loose vacuum connections.	5. Replace with new tank; tighten all vacuum connections.
6. Internal vacuum leaks in power brake unit.	6. Install new power brake unit.
7. Plugged breather in power unit.	7. Clean or replace.
8. Dry or rusted vacuum piston.	8. Lubricate vacuum piston or replace power brake unit.
9. Wrong type of lining.	9. Reline with new brake shoe and lining sets for best results.
10. Restricted power brake air filter.	10. Wash filter in non-oil-base cleaning fluid. Allow to dry before replacing or install a new filter.
11. Broken counter reaction spring in power brake unit.†	11. Replace power brake unit.
12. Grease, oil, brake fluid, or foreign matter on lining surface.	12. Correct leaks; reline brakes with new brake shoe or lining sets for best results.
13. Frozen brake pedal linkage.	13. Free up pedal linkage and replace necessary parts.
14. Blocked air passage in the guide sleeve.	14. Replace power brake unit.
15. Jammed power cylinder air valve.	15. Replace power brake unit.
16. Collapsed vacuum hose or dented tubing from intake manifold to power brake unit.*	16. Replace with new hose or tubing.

†Only on Vacdraulic Power Unit.
*Only in Bellows Type Power Unit.

CONDITION: HARD PEDAL (Also see chapter for type of brake involved.) (CONTINUED)

POSSIBLE CAUSE	*CORRECTIVE REMEDY*
17. Bellows leaking in power brake unit.*	17. Replace power brake unit.
18. Defective vacuum valve in power brake unit.*	18. Replace power brake unit.
19. Valve operating rod assembly bent or binding in power brake unit.*	19. Replace power brake unit.
20. On hydraulic Hydro-Vac units, there is a low fluid level or slipping pump belt.	20. Fill pump reservoir with the proper fluid, or adjust or replace belt.

CONDITION: PEDAL GOES TO FLOOR (Also see chapter for type of brake involved.)

POSSIBLE CAUSE	*CORRECTIVE REMEDY*
1. Worn out brakes.	1. Replace shoes/linings or pads, in axle sets only.
2. Air in the system.	2. Bleed the hydraulic system.
3. Low fluid level, or external hydraulic fluid leaks.	3. Check for the leaks, and repair as necessary; replenish the fluid level.
4. Faulty residual check valve, primary cup, or contamination or rust.	4. Replace defective parts. Flush the system. Replace the power brake unit, if defective.
5. Booster dash unit mounting loose.	5. Tighten nuts and bolts to specifications.

CONDITION: PEDAL FAILS (OR SLOW) TO RELEASE (Also see chapter on disc or drum brakes.)

POSSIBLE CAUSE	*CORRECTIVE REMEDY*
1. Plugged breather in power brake unit.	1. Clean or replace.
2. Plugged compensator port in the power brake unit.	2. Adjust the pedal linkage, flush the system, or replace the power brake unit.
3. Piston stroke interference, or faulty residual check valve, or broken piston return spring, or sticky slide valve (do not oil) in the power brake unit.	3. Replace the power brake unit.
4. Excessive hydraulic seal friction in the power brake unit.	4. Replace the power brake unit. Replace brake fluid.

CONDITION: SENSITIVE BRAKES (Also see chapter for type of brake involved.)

POSSIBLE CAUSE	*CORRECTIVE REMEDY*
1. Grease, oil, or brake fluid on lining surface.	1. Correct leaks; reline brakes with new brake shoe or lining sets.
2. Burnt lining surface.	2. Replace lining.
3. Reaction diaphragm leakage.	3. Replace power brake unit.
4. Restricted or closed atmospheric passage in vacuum piston assembly.	4. Replace power brake unit.
5. Sticky slide valve action.	5. Replace power brake unit.
6. Brake pedal linkage binding.	6. Check for freeness; relieve and lubricate.
7. Rubber seals in power brake unit binding.	7. Replace power brake unit.

*Only in Bellows Type Power Unit.

CONDITION: PEDAL CHATTER
(Also see chapter for type of brake involved.)

POSSIBLE CAUSE	*CORRECTIVE REMEDY*
1. Power brake trigger collar damaged or missing.	1. Install a new rubber collar.
In bellows type power units:	
2. Power brake trigger's rubber collar damaged or missing.	2. Install a new rubber collar.
3. Power brake trigger bent or out of adjustment.	3. Replace the trigger, adjust the linkage.
4. Binding inside power brake unit.	4. Replace the power brake unit.
5. Improperly adjusted master cylinder pushrod.	5. Adjust the pushrod to proper clearance.

CONDITION: PEDAL VIBRATES AND POWER BRAKES CHATTER
(Also see chapter for type of brake involved.)

POSSIBLE CAUSE	*CORRECTIVE REMEDY*
1. Improperly adjusted brake pedal free play.	1. Readjust brake pedal free play.

TROUBLESHOOTING GUIDE FOR HYDRO-BOOST

CONDITION: EXCESSIVE BRAKE PEDAL EFFORT

POSSIBLE CAUSE	*CORRECTIVE REMEDY*
1. Loose or broken power steering pump belt.	1. Tighten or replace the belt.
2. Power steering reservoir empty.	2. Fill reservoir and check for external leaks.
3. Leaks in power steering, booster, or accumulator hoses.	3. Replace faulty parts.
4. Leaks at tube fittings, power steering, booster or accumulator connections.	4. Tighten fittings or replace tube seats, if faulty.
5. External leakage at accumulator.	5. Replace O-ring and retainer.
6. Faulty booster piston seal causing leakage at booster flange vent.	6. Overhaul with new seal or input rod and piston assembly (piston/accumulator assembly).
7. Faulty booster input rod seal with leakage at input rod end.	7. Overhaul with new seal kit.
8. Faulty booster cover seal and leakage between housing and cover.	8. Overhaul with new seal kit.
9. Faulty booster spool plug seal.	9. Overhaul with spool plug seal kit.
10. Spool valve bore scored or nicked.	10. Replace booster assembly.
11. Faulty spool action caused by contamination in system.	11. Flush steering system while pumping brake pedal.
12. Hydraulic lines routed incorrectly.	12. Reroute lines.
13. Hoses leaking.	13. Repair hoses.
14. Restriction pressure hose to booster.	14. Replace hose.
15. Binding in pedal linkage.	15. Repair and lubricate.
16. Internal leakage in hydro-boost.	16. Replace booster.

CONDITION: SLOW BRAKE PEDAL RETURN

POSSIBLE CAUSE	*CORRECTIVE REMEDY*
1. Excessive seal friction in booster.	1. Overhaul with new seal kit.
2. Faulty spool action.	2. Flush steering system while pumping brake pedal.
3. Broken power piston return spring.	3. Replace the spring.
4. Broken spool return spring.	4. Replace the spring.
5. Restriction in return line from booster to pump reservoir.	5. Replace the line.
6. Binding in linkage.	6. Repair and lubricate.

CONDITION: GRABBY BRAKES

POSSIBLE CAUSE	*CORRECTIVE REMEDY*
1. Faulty spool action caused by contamination in system.	1. Flush steering system while pumping brake pedal.
2. Broken spool return spring.	2. Replace spring.
3. Binding pedal linkage.	3. Repair and lubricate.

CONDITION: NOISE

POSSIBLE CAUSE	*CORRECTIVE REMEDY*
1. Power steering fluid low.	1. Fill reservoir.
2. Power steering fluid aerated.	2. Bleed system.
3. Blocked hoses.	3. Replace hose.
4. Power steering belt loose.	4. Tighten belt.

CONDITION: BOOSTER CHATTERS, PEDAL VIBRATES

POSSIBLE CAUSE	*CORRECTIVE REMEDY*
1. Power steering pump belt slips.	1. Tighten belt.
2. Low fluid level in power steering pump reservoir.	2. Fill reservoir and check for external leaks.
3. Faulty spool operation caused by contamination in system.	3. Flush steering system while pumping brake pedal.
4. Contamination in system.	4. Flush system and replace with new fluid.
5. Faulty power steering pump.	5. Repair or replace pump.

CONDITION: ACCUMULATOR LEAK DOWN (System does not hold charge.)

POSSIBLE CAUSE	*CORRECTIVE REMEDY*
1. Contamination in steering hydro-boost system.	1. Flush steering system while pumping brake pedal.
2. Internal leakage in accumulator system.	2. Overhaul unit using accumulator rebuild kit and seal kit, or replace power piston/accumulator assembly.
3. External leakage at accumulator welds.	3. Replace accumulator.
4. External leakage at accumulator fittings.	4. Tighten or replace fittings.

CAUTION: Power steering fluid and brake fluid should not be mixed. If brake seals contact steering fluid, or steering seals contact brake fluid, seal damage will result.

TROUBLESHOOTING GUIDE FOR ELECTRIC BRAKES

CONDITION: NO BRAKES

POSSIBLE CAUSE	*CORRECTIVE REMEDY*
1. Open circuit.	1. Check for broken wires, loose connections, improper grounding, faulty connector plug, between car and trailer, etc.
2. Improperly wired, or inoperative controller.	2. Rewire controller. Check controller operation.
3. Poor brake adjustment.	3. Adjust brakes.
4. Variable resistor defective.	4. Check resistor for loose connections.
5. Worn or defective magnet(s).	5. Replace magnet(s).
6. Short circuit.	6. Check electrical circuit.
7. Defective connector plug.	7. Check plug between trailer, and tow vehicle for loose connections, dirty, or corroded blades, or broken Bakelite insert in socket.
8. Burned out resistor.	8. Check resistor for continuity. Replace, if necessary.

CONDITION: INTERMITTENT OR SURGING BRAKES

POSSIBLE CAUSE	*CORRECTIVE REMEDY*
1. Out-of-round drums.	1. Turn, or replace drums.
2. Inadequate trailer ground.	2. Check for proper grounding. (NOTE: A ground through the trailer hitch is inadequate.)
3. Broken magnet lead wires.	3. Bench check magnets, and replace, if necessary.
4. Loose wheel bearings.	4. Check and adjust bearings.

CONDITION: WEAK BRAKES

POSSIBLE CAUSE	*CORRECTIVE REMEDY*
1. Loose connections.	1. Check that all connections are clean and tight.
2. Inadequate trailer ground.	2. Check for proper ground.
3. Short circuit.	3. Check electrical circuit.
4. Variable resistor setting incorrect.	4. Check for proper setting to avoid too much resistance.
5. Worn, or defective magnets.	5. Replace magnets.
6. Poor brake adjustment.	6. Adjust brakes.
7. Bent backing plate.	7. Check backing plate flange. Correct, if necessary.
8. Contaminated lining.	8. Check and replace badly contaminated linings.
9. Excessive load on trailer.	9. Check to be sure trailer is not underbraked. Also be sure to have brakes on every axle.
10. Using trailer brakes only.	10. Use of trailer brakes only can cause brake fade, or loss of friction due to excessive heat.
11. Inadequate gauge of wire.	11. See wiring recommendations.

CONDITION: WEAK BRAKES (CONTINUED)

POSSIBLE CAUSE	*CORRECTIVE REMEDY*
12. Stoplights connected in brake circuit.	12. Stoplights must not be connected in the brake circuit. The graduation of the current changes as it passes through the controller, resulting in weak or grabbing brakes. Wire a separate circuit for the stoplight switch.
13. Improper linings.	13. Replace with proper linings.
14. Worn out linings.	14. Reline with new linings.
15. Worn brake drums.	15. Inspect the brake drum surface. The surface should be free from scoring and excessive wear. Machine or replace drums as necessary.
16. Out-of-round drums.	16. Machine the brake drums.
17. Axle loose on springs or frames.	17. Inspect and make necessary repairs.
18. Lining loose on rivets.	18. Replace brake shoes.

CONDITION: GRABBING OR LOCKING BRAKES

POSSIBLE CAUSE	*CORRECTIVE REMEDY*
1. Flanges improperly installed.	1. Check flange location. Refer to axle manufacturer.
2. Contaminated lining.	2. Check and replace badly contaminated linings.
3. Controller not modulating.	3. Disconnect red wire on controller. Road test for braking modulation. If modulation is okay, check the red wire. Bench check controller, and replace, if necessary.
4. No variable resistor.	4. A variable resistor is required when brakes have greater power than is necessary for the weight on the axle. Install variable resistor, when necessary.
5. Weak or broken springs.	5. Check for weak or broken springs. Replace, if necessary.

CONDITION: DRAGGING BRAKES

POSSIBLE CAUSE	*CORRECTIVE REMEDY*
1. Brakes incorrectly adjusted.	1. Check brake adjustment.
2. Electrical defect in controller.	2. Insufficient gap between controller contactor pins and coil may cause brakes to drag.
3. Hydraulic defect in controller.	3. Excessive residual pressure in the tow car hydraulic system, or a "gummed up" controller cylinder may cause the controller to be held "on" slightly.
4. Flanges improperly installed.	4. Check flange location. Refer to axle manufacturer.
5. Badly corroded brake assembly.	5. Check brake assemblies for corrosion. Check to be sure magnet levers operate freely. Clean and lubricate brake assemblies.
6. Weak or broken springs.	6. Check for weak or broken springs. Replace, if necessary.
7. Worn or bent magnet lever arm.	7. Replace magnet and lever arm.

CONDITION: NOISY BRAKES

POSSIBLE CAUSE	*CORRECTIVE REMEDY*
1. Lining excessively worn.	1. Check and replace shoes, if necessary.
2. Weak or broken springs.	2. Check for weak or broken springs. Replace, if necessary.
3. Flange improperly located; bent backing plate.	3. Check and repair, if necessary.
4. Contaminated lining.	4. Check and replace badly contaminated linings.
5. Improper bearing adjustment.	5. Check and adjust bearings. Check for worn or damaged bearings. Replace, if necessary.
6. Brakes incorrectly adjusted.	6. Check brake adjustment.
7. Improperly adjusted shoes.	7. Adjust star wheel until there is a heavy drag, and back off adjuster slightly.
8. Grease on lining.	8. Replace leaky seal and linings.
9. Worn magnets.	9. Check for excessive or uneven wear. Replace, if necessary.

CONDITION: BREAKAWAY SWITCH FAILS TO FUNCTION

POSSIBLE CAUSE	*CORRECTIVE REMEDY*
1. Weak or dead battery.	1. Replace with new 12-volt battery. If brakes fail to function, replace breakaway switch.
2. Faulty breakaway switch wiring.	2. Check breakaway switch circuit for broken or frayed wires. Replace wire, where necessary, with proper wire. All splices must have a good connection.
3. Faulty breakaway switch.	3. Check breakaway switch by pulling pin and attempting to tow trailer. If switch works, brakes will engage.
4. Only one brake working.	4. Check the amperage at each brake. Where no amperages are indicated, check the wires leading to brake. If no defect is found, remove magnet from backing plate, and check for amperage capacity, ground, or short. If brake is okay electrically, check for mechanical defects.
5. Poor electrical connections.	5. Check wiring for loose connections, broken wires, or worn insulation. Rewire as necessary.

TROUBLESHOOTING GUIDE FOR AIR BRAKE SYSTEMS

TRUCKS, TRACTORS, BUSES

CONDITION: INSUFFICIENT BRAKES

POSSIBLE CAUSE	*CORRECTIVE REMEDY*
1. Brakes out of adjustment.	1. Adjust brakes.
2. Worn linings.	2. Replace with approved shoes and linings in axle sets only.

CONDITION: INSUFFICIENT BRAKES (CONTINUED)

POSSIBLE CAUSE	*CORRECTIVE REMEDY*
3. Lack of lubrication at backing plate bosses, anchors, abutment blocks, shoe guides, etc.	3. Lubricate all points for free brake shoe movement.
4. Wrong type of brake lining used.	4. Use recommended brake shoe or lining set.
5. Poor fit between lining and drum.	5. Grind shoes to fit drum exactly.
6. Low air pressure.	6. Check valves and lines for air leaks. Check governor and compressor.
7. Incorrect angle between slack adjuster and brake chamber pushrod.	7. Adjust for proper angle.

CONDITION: BRAKES RELEASE TOO SLOWLY

POSSIBLE CAUSE	*CORRECTIVE REMEDY*
1. Brakes out of adjustment.	1. Adjust brakes.
2. Binding in brake linkage.	2. Relieve and lubricate.
3. Lack of lubrication at backing plate bosses, anchors, abutment blocks, shoe guides, etc.	3. Lubricate all points for free brake shoe movement.
4. Brake valve not returning to fully released position.	4. Repair or replace valve.
5. Restricted tubing or hose line.	5. Replace tubing or hose line.
6. Exhaust port of brake valve or quick-release valve restricted or plugged.	6. Clean plugged valve.
7. Defective brake valve or quick-release valve.	7. Replace malfunctioning valve.
8. Binding in camshaft or anchor pins.	8. Lubricate or replace.

CONDITION: BRAKES APPLY TOO SLOWLY

POSSIBLE CAUSE	*CORRECTIVE REMEDY*
1. Brakes out of adjustment.	1. Adjust brakes.
2. Lack of lubrication at backing plate bosses, anchors, abutment blocks, shoe guides, etc.	2. Lubricate all points for free brake shoe movement.
3. Low air pressure.	3. Check valves and lines for leaks. Check governor and compressor.
4. Excessive leakage when brakes applied.	4. Check entire system and correct problem.
5. Restricted tubing or hose line.	5. Replace tubing or hose line.
6. Binding in camshaft or anchor pins.	6. Lubricate or replace.
7. Binding in brake linkage.	7. Relieve and lubricate.

CONDITION: BRAKES GRAB

POSSIBLE CAUSE	*CORRECTIVE REMEDY*
1. Grease on brake lining.	1. Replace shoes and linings in axle sets.
2. Brake drum out-of-round.	2. Refinish or replace drums in axle sets.
3. Defective brake valve.	3. Replace valve.
4. Brake rigging binding.	4. Check entire system and make necessary repairs.
5. Incorrect brake linings.	5. Replace with correct linings.

CONDITION: UNEVEN BRAKES

POSSIBLE CAUSE	*CORRECTIVE REMEDY*
1. Brakes out of adjustment.	1. Adjust brakes.
2. Worn linings.	2. Replace with approved shoes and linings in axle sets only.
3. Lack of lubrication at backing plate bosses, anchors, abutment blocks, shoe guides, etc.	3. Lubricate all points for free brake shoe movement.
4. Grease on brake lining.	4. Replace shoes and linings in axle sets.
5. Brake shoe return spring or brake chamber spring weak or broken.	5. Replace defective spring.
6. Brake drum out-of-round.	6. Refinish or replace drums in axle sets.
7. Leaking brake chamber diaphragm.	7. Replace as necessary.

TRAILERS

CONDITION: INSUFFICIENT BRAKES
(See same under "Trucks" or the following)

POSSIBLE CAUSE	*CORRECTIVE REMEDY*
1. Defective relay-emergency valve.	1. Replace valve.
2. Restricted tubing (service line).	2. Replace tubing.

CONDITION: BRAKES RELEASE TOO SLOWLY
(See same under "Trucks" or the following)

POSSIBLE CAUSE	*CORRECTIVE REMEDY*
1. Exhaust port of relay-emergency valve restricted or plugged.	1. Replace valve.

CONDITION: BRAKES DO NOT RELEASE

POSSIBLE CAUSE	*CORRECTIVE REMEDY*
1. Brake system no properly connected to towing vehicle.	1. Secure connections properly.
2. Relay-emergency valve in emergency position.	2. Correct or replace valve.
3. Tractor protection valve malfunctioning.	3. Replace valve.

CONDITION: BRAKES APPLY TOO SLOWLY
(See same under "Trucks" or the following)

POSSIBLE CAUSE	*CORRECTIVE REMEDY*
1. Excessive air leakage with brakes applied.	1. Check entire system and make necessary repairs.

CONDITION: BRAKES DO NOT APPLY

POSSIBLE CAUSE	*CORRECTIVE REMEDY*
1. Brake system not properly connected to towing vehicle.	1. Secure connections properly.
2. Tractor protection valve malfunctioning.	2. Replace valve.

CONDITION: BRAKES DO NOT APPLY (CONTINUED)

POSSIBLE CAUSE	*CORRECTIVE REMEDY*
3. No air pressure.	3. Check valves and lines for air leaks. Check governor and compressor.
4. Plugged tube or hose.	4. Replace tube or hose.

CONDITION: BRAKES GRAB
(See same under "Trucks" of the following)

POSSIBLE CAUSE	*CORRECTIVE REMEDY*
1. Defective relay-emergency valve.	1. Replace valve.

CONDITION: UNEVEN BRAKES
(See same under "Trucks")

TROUBLESHOOTING GUIDE FOR AIR DISC BRAKES*

CONDITION: BRAKES INEFFECTIVE

POSSIBLE CAUSE	*CORRECTIVE REMEDY*
1. "Green" lining.	1. Burnish lining.
2. Grease or oil on linings.	2. Burnish or replace shoe and linings as necessary.
3. Incorrect shoe and linings.	3. Replace with approved Kelsey-Hayes shoe and linings.
4. Lining worn below minimum recommended thickness.	4. Replace shoe and linings.
5. Air leak.	5. Check system for leaks.
6. Anti-lock false cycling.	6. Refer to anti-lock manual.
7. Spindle has reached maximum adjustment.	7. Replace shoe and linings.

CONDITION: BRAKE ROUGHNESS OR CHATTER

POSSIBLE CAUSE	*CORRECTIVE REMEDY*
1. Excessive rotor thickness variation.	1. Refinish or replace as required.
2. Excessive lateral rotor runout.	2. Same as above.
3. Excessive front wheel bearing clearance.	3. Adjust wheel bearing to manufacturer's specifications.
4. Grease or oil on linings.	4. Burnish or replace shoe and linings as necessary.
5. Loose anchor plate.	5. Torque mounting bolts 185 to 215 ft-lbs torque.
6. Loose caliper.	6. Check retainers and springs.
7. Excessively worn caliper rails.	7. Replace caliper and/or anchor plate as necessary.
8. Excessively worn shoe ends.	8. Replace shoe and linings.
9. Worn or broken suspension.	9. Repair.

*Courtesy of Kelsey-Hayes Company

CONDITION: SENSITIVE BRAKES

POSSIBLE CAUSE	*CORRECTIVE REMEDY*
1. Malfunctioning treadle valve.	1. Replace valve. tight.
2. Malfunctioning relay valve.	2. Replace valve.
3. Incorrect lining.	3. Replace with approved Kelsey-Hayes shoe and linings.

CONDITION: SIDE PULL

POSSIBLE CAUSE	*CORRECTIVE REMEDY*
1. Air imbalance.	1. Check valves and lines for obstructions and leaks.
2. Loose caliper.	2. Check retainers and springs.
3. One side not adjusting.	3. Check for boot or seal damage; overhaul if necessary.
4. Loose or seized wheel bearing.	4. Adjust or replace as necessary.
4. Improper air chamber.	4. Adjust or replace as necessary.
5. Improper air chamber.	5. Replace with correct assembly.
6. Loose wheel clamp nuts.	6. Tighten to manufacturer's specifications.
7. Incompatible wheel rim widths.	7. Ensure wheel rims on axle are same.
8. Grease or oil on linings.	8. Burnish or replace shoe and linings as required.
9. Incompatible tires.	9. Ensure tires on axle are same.
10. Incorrect tire pressures.	10. Adjust to correct pressures.
11. Air chamber diaphragm ruptured.	11. Replace as necessary.
12. Worn or broken suspension.	12. Repair.
13. Incorrect steering geometry.	13. Align.

CONDITION: SQUEALING BRAKES

POSSIBLE CAUSE	*CORRECTIVE REMEDY*
1. Incorrect lining.	1. Replace with approved Kelsey-Hayes shoe and linings.
2. Foreign material embedded in lining.	2. Replace shoe and linings.
3. Loose anchor plate.	3. Torque mounting bolts 185 to 215 ft-lbs torque.
4. Rotor distortion.	4. Check rotor with dial indicator; replace or refinish rotor.
5. Excessively worn shoe ends.	5. Replace shoe and linings.

CONDITION: RATTLE OR OTHER NOISES

POSSIBLE CAUSE	*CORRECTIVE REMEDY*
1. Brake not adjusting; excessive shoe clearance.	1. Overhaul caliper.
2. Loose caliper.	2. Check retainers and springs.
3. Loose anchor plate.	3. Torque mounting bolts 185 to 215 ft-lbs torque.
4. Excessively worn shoe ends.	4. Replace shoe and linigs.
5. Loose wheel bearing.	5. Tighten to manufacturer's specifications.
6. Lining worn below minimum recommended thickness.	6. Replace shoe and linings.
7. Yoke bolts too long.	7. Replace with correct bolts.

CONDITION: BRAKE DRAG

POSSIBLE CAUSE	*CORRECTIVE REMEDY*
1. Caliper mechanism not fully returning.	1. Air chamber and caliper. • Check air chamber if it is a recommended part. • Check if air chamber diaphragm is free (*not* iced up). Boot and/or air dryer is recommended if vehicle is continuously used in very wet cold environment. OR • Overhaul caliper.
2. Heavily corroded retainers and springs.	2. Decontaminate or replace as necessary.
3. Seized wheel bearing.	3. Replace.

CONDITION: BRAKE LOCKS ON AFTER STANDING WITH PARKING BRAKE APPLIED

POSSIBLE CAUSE	*CORRECTIVE REMEDY*
1. Caliper mechanism not fully returning.	1. Check if air chamber diaphragm is free (*not* iced up). Boot and/or air dryer is recommended if vehicle is continuously used in very wet cold environment.

SERVICE FREQUENCY

It is important that the brakes be inspected at definite intervals. Many brake shops have customer reminder programs as to when the vehicle's brakes should be checked. In addition, many states have vehicle inspection laws that include brake inspections, and, of course, the inspection intervals contained in these laws should be observed.

The interval between required brake servicings can vary widely depending upon many factors, such as:

- Individual driving habits
- Type of driving (stop-and-go city driving or highway driving)
- Road and climate conditions
- Vehicle load

It is important, if possible, for the brake mechanic to know how the client operated the vehicle. Of course, every brake mechanic should be concerned about customer comebacks. The customer usually blames the quality of work and the mechanic blames defective parts. Getting to know the driving habits of the customer can be the solution to the comeback problem. Certain habits, such as the following, can cause brake problems that might appear to be defective parts or shoddy work but are, in fact, due to the way the car is driven.

Left-Foot Braker. Due to the popularity of the automatic transmission, drivers using their left foot to apply the brakes are commonplace. This forces the brake system to overcome the power of the engine in addition to its normal function of stopping the vehicle. Left-foot brakers rest their foot on the brake pedal. The effect of this slight pressure is a constant braking action. Differences in individual brake parts used cause one wheel's return springs, seals, pistons, etc., to have less mechanical resistance than the others. The slight brake application results in that wheel's brakes dragging. When this vehicle comes into the shop, symptoms appear in one of two ways:

1. If the car is normally driven in stop-and-go traffic, one wheel will be severely worn with signs of extreme heat buildup (blued drums, burnt shoes, springs, rubber, etc.). This condition can appear on any wheel and repairing it can only be a stop-gap measure. The problem could show up on other wheels in a relatively short period of time after you have done the work.

2. If the car is normally driven at highway speeds over relatively long distances, the heat created by one wheel dragging, as described above, causes the constant brake application to be more severe. This results in all four wheels appearing severely worn and burnt. The danger of this situation is that the heat created can cause vaporization of the brake fluid, which results in complete brake failure.

While the problem might not show up for 6 months or more from the time of the original job, it usually happens within a much shorter time period. The driver should be asked if this has ever happened before and the pedal inspected for signs of wear on the left side. If the left-foot brake condition is apparent, let the customer know that he/she is the cause of the problem, not the parts used or the work done.

Delivery Route Driver. A driver that delivers newspapers, magazines, packages, etc., can also cause extreme and unusual wear to the brakes. This is caused by overloading and close multiple stops, which do not allow the brakes to cool. Many times the engine idle speed is set at a very high rate so that the brakes have to overcome the power of the engine as well as stop the vehicle. When the two rear brakes are severely worn, this is a telltale sign of this condition. You might also see the two right-hand brakes worn. This type of vehicle overloading causes the rear wheels to do the bulk of the braking instead of the front wheels, which were designed for it.

Let the driver know that his/her vehicle is inadequate for the job.

Long Stopper or One-Wheel Braker. The long stopper applies the brakes so gradually and gently that only one wheel is braking. This results in under use of the other three wheels, which causes them to freeze up. Usually, the customer will not have a complaint about this unless he/she has tried to make a panic stop. Inform the person that his/her braking habits are creating a dangerous condition. If the driver ever gets into a situation where he/she must to make a rapid stop, the system will not work properly.

It is important to keep in mind that the use of automatic adjusters does not remove the need for service, and as a general rule, brakes should be inspected at least every 12,000 miles under normal operating conditions. Under severe operating conditions or near the end of the useful lining life, the brakes should be inspected more frequently.

IMPORTANT TIPS FOR COMPLETE BRAKE SERVICE

As stated in the opening paragraph of this chapter, the basic concept in good brake service work is to do a complete brake job. For instance, a sticky wheel cylinder or a frozen caliper piston can function improperly, causing the customer to return his/her car after a few miles complaining about the quality of the technician's work. A few minutes spent at the time of reline to inspect all parts, functional and non-functional, should also create additional business and profits. A complete brake reline should return the braking system to a "like-new" condition and give the customer the feeling that he/she has better performing brakes than he/she ever had. Quality parts are absolutely essential to reach this point, and without them one can only expect substandard performance. Performance must be measured at all times, under all conditions, and with all drivers.

The following is a list of parts that should be considered for replacement:

1. Lined brake shoes and disc brake pads
2. Brake drums and rotors
3. Wheel cylinders, calipers, pistons, and rubber parts
4. Brake hoses
5. Backing plates
6. Retracting springs and hold-down springs
7. Disc hardware
8. Parts or all of the automatic adjuster mechanism
9. Bearings and bearing races
10. Grease seals
11. Various front-end parts
12. Parking brake parts such as levers, lock washers, and brake cables
13. Assorted other parts that can vary from car to car

These items vary as to the need for replacement, but each part directly or indirectly affects braking. Remember that if a damaged component or malfunction is discovered while making any of the following brake system checks, repairs must be made before attempting to continue with additional diagnosis.

As a review of the various servicing techniques explained in the previous chapter of the book, here are some very important hints necessary for better complete brake servicing that all brake technicians should keep in mind:

1. First, check the power brake unit on cars equipped with power brakes. Check the vacuum reading, look for blocked or collapsed vacuum hoses or lines, a carbonized vacuum intake manifold fitting, or a defective vacuum check valve.
2. Prior to removing wheels and drums or rotors, apply the brake pedal firmly and check the master cylinder, power brake unit, hydraulic lines, rubber flex hoses, and connections for brake fluid leaks while the brakes are cold. Tighten, replace, or overhaul any of these items found leaking.

SHOP TALK

Heat generated by high-speed stops will expand brake drums and increase shoe clearance, thereby permitting the pedal pad to go closer to the toeboard than when the brakes are cold.

3. **Drum Brakes.** After the brake shoes are removed from the backing plate, check for freeness of the wheel cylinder pistons and for wheel cylinder leakage. Always overhaul or replace wheel cylinders prior to relining brakes, even if the cylinders are not leaking. This is good insurance for better brake performance and eliminates the possibility of leaking wheel cylinders after brakes are relined. Use hydraulic brake cylinder assembly fluid when overhauling or replacing cylinders. Coat wheel cylinder pistons, rubber sealing cups, and cylinder bore with the assembly fluid. This prevents frozen or sticky pistons, rusting of cylinder bores, erratic brake action, sluggish release, lining fade, poor lining life, glazed linings, hard pedal action, etc.

   Disc Brakes. After disc brake pads are removed from the caliper assembly, check for freeness of the piston in the caliper bore. Check for hydraulic leakage. Replacement of rubber piston seals and boots will assure that performance will be up to a "like-new" condition. Use hydraulic assembly fluid when overhauling the caliper.
4. Do not attempt to wash the master cylinder bore, the wheel cylinder bore, the disc brake caliper bore or piston, and power brake parts with anything but clean denatured alcohol or fluid recommended by the car manufacturer. Never permit the caliper assembly to hang with the weight on the brake hose; support it on the suspension or hang it by a piece of wire.
5. Do not attempt to salvage used brake fluid.
6. Do not attempt to use mineral oil in the hydraulic system. If mineral oil has been used, replace all rubber parts, including the rubber flex hose, and flush the entire hydraulic system with denatured alcohol or recommended fluid. For best results, refill the system with brake fluid that meets or exceeds SAE and DOT specifications. When it is necessary to disconnect a hydraulic hose, plug the hose to prevent any foreign material from entering it.
7. Flush the hydraulic system in order to remove sludge and moisture from the system. Use brake fluid and a pressure tank for flushing and bleeding.
8. **Drum Brakes.** Thoroughly clean the backing plate, anchors, bosses, and so on. Clean the brake components in either denatured alcohol, clean brake fluid, or a cleaning fluid designated for use in cleaning brake components. With a wire brush, wipe away any oil, grease, or brake fluid. Apply a light coat of lubricant to all contact points on the backing plate, anchors, cams, bosses, etc. This allows the brake shoes to move and release freely. Use a high-temperature lubricant.

   Disc Brakes. As in drum brakes, all metal-to-metal contact points should be cleaned with a wire brush and polished with emery cloth so that all parts will move freely. Keep oil and brake fluid off the caliper rotor and the brake pads.
9. **Drum Brakes.** Always replace damaged springs for proper brake performance. If springs are replaced on one side of the axle, the other side should also be changed; otherwise, erratic brake action will occur. Springs will lose tension due to excessive heat in brake and drum assemblies and in many cases the springs will be stretched and distorted. If not replaced, erratic brake action, poor shoe release, and brake squeals will occur.

   Disc Brakes. All disc brake hardware should be inspected to make sure that parts function properly. Bent and damaged bolts, pin guides, sleeves, bushings, positioners, anti-rattle springs, caliper support springs, etc., must be replaced. Work on one wheel at a time to avoid "popping" pistons out of the other caliper and to allow the other caliper assembly to be used as a guide.

10. Inspect the caliper for leaks. If leakage is present the caliper must be overhauled.
11. Use crocus cloth to remove rust, corrosion, pitting, and scratches from the piston bore. If the bore cannot be cleaned up with crocus cloth, a light honing is permitted. Do not hone a plated bore. When using compressed air to remove caliper piston(s), avoid high pressures. A safe pressure to use is 30 PSI.
12. Before the brake pads are installed, apply a disc brake noise suppressor to the back of the pads to prevent brake squeal. For best results, follow the directions on the container.
13. Disconnect parking brake cables at the cross shaft; check for cable freeness. Wash cables, apply lubricant, then work cable in and out of the conduit. If cables are frozen or wire strands broken, replace with new cables. When replacing cables, install replacement parts that are of correct length for the proper operation of the brake. Cables too long or too short will give trouble as will cables with weak or collapsing conduits.
14. Loose or worn front-end parts, weak shock absorbers, weak or broken coil springs, sheared center bolts in leaf type springs, broken spring leafs, weak springs, low tire inflation, incorrect wheel alignment, tire balance, loose backing plates, loose wheel cylinders, and loose anchor pins will cause erratic brake performance, clunking noises, vibrations, brake squeals, brake chatter, etc. It is most important to check the items mentioned when servicing or relining brakes and make necessary repairs.
15. New grease seals are relatively inexpensive and should be used with every reline. Both bearings, inner and outer, should be repacked with clean, fresh axle grease of high quality. The grease should be similar to that recommended by the car manufacturer. If grease or oil gets into a lining, it destroys its effectiveness, causing erratic performance. Never attempt to clean lining surface with gasoline, various cleaners, etc. Replace lining if it has become contaminated. Do not attempt to reline one wheel without relining its opposite on the same axle.
16. Many times adjusting hole covers are not replaced on the backing plate and brake drum after brake adjustments. Rain and road dust get into the brake assemblies through these holes, causing erratic brake action. It is recommended that the holes be covered at all times.
17. Always torque the lug nuts when installing a wheel on a vehicle with disc brakes. Never use an impact gun to tighten the lug nuts. Warpage of the rotor could result if an impact gun is used.
18. Water in the brake assembly causes hard pedal and fade because water acts as a lubricant. The recovery time for a water-soaked brake depends upon the capacity of the brake lining to absorb water. There is less of a tendency for disc brakes to be affected by water because the disc pads tend to wipe the rotor clean and the spinning rotor tends to throw the water off.

REVIEW QUESTIONS

Review the information in this chapter by answering the following questions.

1. Which of the following is a possible cause of low pedal in a drum brake system?
   a. caliper piston seal damage
   b. open bleeder screw
   c. air in the hydraulic system
   d. all of the above.
2. Excessive pedal pressure is required to stop a given vehicle. Upon inspection, a water film is found on the drum brake lining surface. Mechanic A checks the backing plate for openings and makes sure the adjusting hole covers are in place. Mechanic B flushes the hydraulic system and adjusts the master cylinder pushrod. Who is right?
   a. Mechanic A
   b. Mechanic B
   c. both A and B
   d. neither A nor B
3. Which of the following statements concerning brake drums is incorrect?
   a. It is good practice to replace both drums on the same axle.
   b. An out-of-round drum can cause brake chatter.
   c. A loose drum is irreparable; it must be replaced.
   d. When refacing drums, keep the diameters to within 0.010 inch of each other.
4. Which of the following is a possible cause of hard pedal in a disc brake system?
   a. poor quality brake fluid
   b. heat-checked rotor
   c. excessive lateral runout of rotor
   d. excessive out-of-parallelism of rotor

5. The disc brakes on a given vehicle are heating up during driving and failing to release. Upon inspection, it is discovered that the master cylinder is not releasing pressure. Mechanic A says the master cylinder must be replaced. Mechanic B says the answer could simply be contaminated brake fluid. Who is right?
   a. Mechanic A
   b. Mechanic B
   c. both A and B
   d. neither A nor B
6. Which of the following is a possible cause of a car with disc brakes pulling to one side?
   a. rough surfaces on one rotor
   b. scored cylinder bore surface
   c. swollen master cylinder cups
   d. all of the above
7. In a hydro-boost brake system, excessive seal friction in the booster is a possible cause of
   a. slow brake pedal return.
   b. grabby brakes.
   c. vibrating pedal.
   d. noise.
8. Under normal operating conditions, brakes should be inspected at least every
   a. 20,000 miles.
   b. 5,000 miles.
   c. 25,000 milles.
   d. 12,000 miles.
9. Heat generated by high-speed stops
   a. expands the brake drums and increases shoe clearance.
   b. contracts the brake drums and decreases shoe clearance.
   c. expands the brake drums but does not affect shoe clearance.
   d. contracts the brake drums but does not affect shoe clearance.
10. When using compressed air to remove caliper pistons, use a pressure of no more than
    a. 50 PSI.
    b. 30 PSI.
    c. 20 PSI.
    d. 10 PSI.

APPENDIX A

GLOSSARY

Access Slots Openings in the backing plates or drums that allow access to the star wheel adjusters.

Accumulator A reserve system component in a hydraulic brake booster unit that retains a small amount of pressurized fluid.

Adapter On disc brakes, a bracket on which the caliper is rigidly mounted, or slides or floats. It is bolted to the spindle or steering knuckle, or the rear axle or control arm.

Adjusting Cams Eccentric bolts that adjust the shoe-to-drum clearance. Positioned in the backing plate of drum brakes, the cam positions the shoe(s) closer to the drum. Some adjust automatically and some manually.

Air Brake System System utilizing compressed air to activate the brakes.

Aligning Cups Devices used in mounting floating drums on the arbor of a brake drum lathe.

Anchor Pin A steel pin rigidly attached to the backing plate of drum brakes. Return springs are attached to the anchor pin and to the brake shoes to hold the shoes against the anchor pin in a non-applied position. In an applied position, it prevents the shoes from rotating with the drums.

Anodize An electrochemical process that coats and hardens the surface of aluminum.

Anti-rattle Springs, Clips, or Washers Disc or drum brake parts (hardware) that prevent pad or shoe vibration.

Arbor The rotating shaft of a lathe on which a drum or rotor is mounted for machining.

Arcing Shoes On drum brakes, when shoes are supplied with ungrounded linings, the lining material must be cut down on a special machine (arch grinder) so that it matches the diameter of the drum. This practice is less common than in the past, the new shoes today are usually supplied with pre-arced or preground linings.

Asbestos A fibrous substance that does not burn or conduct heat. Used in combination with bonding materials such as asphalts and resins to make brake linings. Airborne asbestos is reported to be a health hazard.

Automatic Adjuster A drum brake mechanism that adjusts the lining clearance as wear occurs. It is commonly actuated on reverse stops or when the parking brake is set.

Auxiliary Drum Parking Brake Incorporates an auxiliary parking brake drum inside a rear rotor on some four-wheel drive disc brake systems.

Axle A rod or bar on which wheels turn. In automotive use, it is also a shaft transmitting driving torque to wheels.

Backing Plate A metal plate that serves as the foundation for the brake shoes and other drum brake hardware.

Bail The spring-wire loop used to secure the cover on most master cylinder reservoirs.

Ball Bearing An anti-friction bearing that uses a series of steel balls held between inner and outer bearing races.

Barrel Shape A drum defect caused by excessive wear at the center of the friction surface.

Bearing Cone The inner race for a ball or roller bearing.

Bearing Cup The outer race for a ball or roller bearing.

Bearing Race The inner or outer ring that provides the smooth, hard contact surface for the balls or rollers in a bearing. *See also* **Cup.**

Bell-Mouthed Drum A brake drum that, through wear or warping, is deformed so that its open end has a larger diameter than its closed end.

Bellows Seal An expanding diaphragm used as a seal between the master cylinder reservoir and the reservoir cover. It prevents air from contacting the fluid, yet it allows the fluid to change in volume.

Bench Bleeding A procedure used to bleed the air from a new or rebuilt master cylinder before installation in a vehicle.

Bleeder Hose A length of rubber tubing used in bleeding brakes.

Bleeder Jar A glass or transparent plastic container used to detect the escape of air while bleeding brakes.

Bleeder Screw A small hollow screw or valve found at drum brake wheel cylinders, in disc brake calipers, and adjacent to the outlet ports of some master cylinders. It is opened to release pressure and bleed air and fluid from the hydraulic system after reconditioning or in the event air has gotten into the system.

Bleeder Tubes Short curved pieces of tubing used to facilitate the bleeding of master cylinders.

Bleeder Valve A valve placed in a hydraulic system where it can be opened to allow the release of air.

Bleeder Wrench A tool used to open bleeder valves.

Bleeding Expelling air from a brake hydraulic system so that it can function properly.

Bleeding Sequence Order of bleeding brake system components.

Boiling Point The temperature at which liquid changes to vapor.

Bonding Agent The cement used to secure bonded lining to a brake shoe.

Bonding Linings Lining attached to a brake shoe with an oven-cured adhesive.

Boot A flexible rubber or plastic cover used over the open ends of master cylinders and wheel cylinders to keep out water and other foreign matter.

Bore The area of a cylinder through which a piston travels back and forth. May also refer to the walls or the inside diameter of a cylinder.

Bore Diameter The diameter of a cylinder.

Bosses Raised ledges, or platforms, on a backing plate, that support the brake shoes.

Brake An energy conversion mechanism used to slow, stop, or hold a vehicle.

Brake Fade Occurs when friction surfaces become hot enough to cause the coefficient of friction to drop to a point where the application of severe pedal pressure results in little actual braking.

Brake Feel A feeling transmitted from the brake system back to the driver during braking.

Brake Fluid A special non-petroleum fluid used to transfer hydraulic pressure through the brake system.

Brake Flushing Cleaning the brake system by pumping through denatured alcohol or clean brake fluid to remove contaminants.

Brake Hose A flexible conductor of hydraulic pressure in the brake system.

Brake Lathe A machine used to refinish brake drums and disc brake rotors.

Brake Light Switch An electrical switch, operated either mechanically or hydraulically depending on the car model, that causes the brake lights to come on when the brakes are activated.

Brake Line Small-diameter rigid steel tubing used to channel hydraulic pressure from the master cylinder to the wheel cylinder or caliper in a brake system. May have either a double or ISO flaring at fittings.

Brake Lining A special friction material with which brake shoes or brake pads are lined. It withstands high temperatures and pressure. The molded material is either riveted or bonded to the brake shoe, with a suitable coefficient of friction for stopping a vehicle.

Brake Pad The friction lining and plate assembly that is forced against the rotor to cause braking action in a disc brake.

Brake Pedal The foot-actuated lever the operator depresses to cause brake application.

Brake Shoe The curved metal part, faced with brake lining, which is forced against the brake drum to produce braking action.

Brake Shoe Grinder A machine used to grind the lining on a brake shoe so that it will fit a particular drum.

Brake Shoe Hold-Downs Spring-loaded retainers that hold brake shoes against the backing plate.

Brake Spoon A tool used to turn star wheel adjusters and, thus, to adjust the brake lining-to-drum clearance.

Brake System The vehicle system that slows or stops a vehicle. A combination of brakes and a control system.

Brake Warning Light An instrument light that warns of brake system malfunction.

Bridge Bolt The bolt(s) that hold together the two halves of a fixed caliper.

Caliper A disc brake component that changes hydraulic pressure into mechanical force and uses that force to press the brake pads against the rotor and stop the vehicle. Calipers come in three basic types: fixed, floating, and sliding, and can have one or more pistons.

Car Stands Pedestal-type supports for holding up a car once the car has been raised.

Castellated Nut A nut having slots through which a cotter pin can be passed to secure the nut to its bolt or stud.

Centering Cones Devices used to mount floating drums on the arbor of a brake drum line.

Center of Gravity The point about which the weight of a car is evenly distributed; the point of balance.

Coefficient of Friction A relative measurement of the friction developed between two objects in contact with each other.

Coil Spring A length of spring-steel wire wound in the shape of a spiral.

Combination Brake System A dual brake system that uses disc brakes at the front wheels and drum brakes at the rear wheels.

Combination Valve A pressure control valve within a brake hydraulic system that combines two or more valves in a common housing. It might contain a metering valve, a proportioning valve, and a pressure differential valve that controls the brake warning light.

Concave Drum A brake drum deformed in such a way that the diameter at the center of the friction surface is larger than that at the ends.

Convex Drum A brake drum deformed in such a way that the diameter at the center of the friction surface is smaller than that at the ends.

Corrosion The gradual wearing away of material by chemical reaction.

Cotter Pin A round locking pin formed by a folded semicircular steel wire. The pin is locked by spreading the paired ends of the wire.

Cup The common hydraulic piston seal in master and wheel cylinders. By design, hydraulic pressure assists the sealing action.

Cup Expanders Metal discs formed to fit inside piston cups and to keep the lips of the cups in tight contact with the cylinder walls while the hydraulic system is not pressurized.

Cylinder Hone A rotating instrument fitted with abrasive material used to remove roughness and deposits, and to polish the bores of wheel cylinders and master cylinders.

Dampening Belt A rubber belt that is wound around the outside of a brake drum or rotor before the drum or rotor is machined. The belt dampens out vibrations that might affect the quality of the finished surface.

Denatured Alcohol Ethyl alcohol that contains methanol to render it unfit for consumption. Used to clean brake system.

Diagonal Brake System A dual brake system with separate hydraulic circuits connecting diagonal wheels together (RF to LR and LF to RR).

Dial Indicator A precision instrument that indicates linear measurement on a dial face.

Diaphragm A flexible, impermeable membrane upon which pressure acts to produce mechanical movement.

Disc A flat, round iron or steel plate that rotates with the wheel hub and against which the brake pads press when the brakes are applied. Also called a rotor.

Disc Brake A type of brake that generates stopping power by the application of pads against a rotating disc (rotor).

Disc Brake Pad An assembly consisting of friction material and its steel backing.

DOT Department of Transportation. A government agency that establishes vehicle standards.

Double Flare A type of flare used on the ends of brake lines for extra strength. The tubing flared end is doubled over.

Drum The part of a drum brake that rotates with the axle hub. The brake shoes press against the inside of the drum when the brakes are applied.

Drum Brake A type of brake in which stopping friction is created by shoes pressing against the inside of the rotating drum.

Dual Brake System A brake system that uses two hydraulic circuits. Should one circuit fail, the other remains operational.

Dual Master Cylinder Type of master cylinder associated with dual brake systems.

Duo-Servo A drum brake design with increased stopping power due to the servo or self-energizing effect of the brake.

Dust Boot Flexible rubber or plastic covers that seal out foreign matter from brake components.

Dust Covers Small plugs made of rubber or metal. Used to cover the access holes in backing plates and drums.

Eccentric Off center. A drum defect caused by unequal wear, drum distortion, or both.

Edge Code A manufacturer's code used to identify friction materials.

Electric/Electronic Brake System An electrical or electronic system is used to actuate the brakes.

Energy The ability to do work.

EPA Environmental Protection Agency

Equalizer A device used in parking brake systems to equalize the pull of both rear brake cables.

Feeler Gauge A thin strip of metal of known thickness used to measure the clearance between two parts.

Filler Port A large hole connecting the master cylinder reservoir with the master cylinder bore. The filler port permits fluid to flow from the reservoir into the hydraulic system.

Fixed Anchor A non-adjustable anchor pin. It can be riveted or welded to the backing plate, or it can pass through the backing plate and attach to a part of the suspension system.

Fixed Caliper A disc brake caliper that is rigidly mounted to the steering knuckle or spindle or control arm. It can have one or two pistons on each side of the disc and does not float or slide.

Fixed Seal (Disc Brakes) A caliper piston seal fixed in the caliper. The piston moves through this seal.

Flaring Tool A tool used to give the ends of tubing a flared shape.

Floating Caliper A disc brake caliper that has a piston or pistons on only one side of the disc. The caliper bore moves away from the rotor in order to press the pad on the other side against the disc when the brakes are applied. The caliper "floats" on pins or bolts.

Floating Drum A brake drum that is not secured to a hub.

Flushing A method of cleaning a hydraulic system by pumping alcohol or brake fluid through the system.

Flushing Hydraulic System The procedure for replacing old brake fluid with new fluid.

Foot-Pound A unit of measurement for torque. In tightening, 1 foot-pound is the torque obtained by a pulling force of 1 pound applied to a wrench handle 12 inches long.

Force A push or pushing effort measured in pounds.

Four Piston Caliper A disc brake caliper of the fixed type that has two pistons on each side of the disc.

Friction The resistance to motion between two bodies in contact with each other.

Gravity Bleeding A method of purging air from a hydraulic system by allowing the fluid to force air out of an opened bleeder valve by its own weight.

Guide Pins or Bolts Floating calipers slide on these pins to center over the rotor when braking action occurs.

Hard Spots Areas in the friction surface of a brake drum or rotor that have become harder than the surrounding metal; sometimes called "islands of steel."

Heat Dissipation The transfer of heat. In brake systems the heat produced by braking is transferred to the air.

Heel The end of the brake shoe nearest the anchor pin.

Herringbone Pattern The characteristic pattern cut by a tool bit when a brake drum is machined without the proper use of a dampening belt.

High Lead Screw A screw with a low pitch so that an engaged nut would move an appreciable distance with a partial revolution.

Hold-Down A spring tension mechanism that holds a brake shoe against a backing plate.

Hold-Down Spring A spring used in a hold-down.

Hone To remove metal with a fine abrasive stone.

Hub A rotating component at the center of a wheel that contains the wheel bearings. It is attached to, or integral with, the brake drum or rotor.

Hubbed Drum A brake drum that is mounted on a hub.

Hydraulic Using fluids to transmit force and motion.

Hydraulic Brake System A system utilizing the properties of fluids under pressure to activate brakes.

Hydraulic Circuit The patch of hydraulic fluid from the master cylinder to the wheels through all valves, lines, hoses, and fittings.

Hydraulic Piston A piston in a cylinder that is acted upon by or acting on a hydraulic fluid.

Hydraulic Pressure The force per unit area exerted in all parts of a hydraulic system. Usually pounds per square inch (PSI).

Hydraulic Service Bench Bench equipment used when disassembling, assembling, and testing brake calipers.

Hydraulics The science of the use of fluids to transmit force and motion.

Hydro-Boost A type of brake power booster that uses hydraulic pressure provided by the power steering pump to reduce pedal effort. Its small size and freedom from a vacuum source offers some advantages over a "conventional" vacuum booster.

Hygroscopic Readily absorbing and retaining moisture.

Intake Port The passage through which brake fluid flows in the master cylinder from the reservoir to refill the low-pressure area ahead of the cup on the return stroke.

Integral Made in one piece.

Integrally Molded Linings Disc brake pad sets that have the linings cured integrally with the pad plates.

ISO Flare The type of brake line flare made popular by the International Standards Organization.

Kinetic Energy The energy of motion.

Lack of Parallelism A measurement of brake rotor thickness variation at various points around a rotor. Usually no greater than 0.0005 inch is acceptable.

Lateral Runout A measurement of the lateral change in position of the rotor surface during one revolution. Usually no greater than 0.005 inch is acceptable.

Leading Shoe The shoe toward the front of the vehicle in a non-servo drum brake.

Lining The actual friction material that contacts the brake disc or drum when the brakes are applied. Attached to a drum brake shoe or a disc brake, it can be made of asbestos or a semi-metallic material. Various friction, wear, noise, and heat-resisting characteristics are available, depending on the cost, quality, or application.

Lockup The point at which braking power overcomes the traction of the vehicle's tires and skidding occurs. The most efficient stopping occurs just before lockup is reached. Locked wheels cause loss of control, long stopping distances, and flat-spotting of the tires.

Low-Drag Caliper A caliper designed to reduce pad drag on the rotor.

Low Pedal The condition in which the brake pedal must travel very far or very close to the floor before the brakes are applied.

Lubricant Any material, usually liquid or semiliquid, that reduces friction when placed between two moving parts.

Machining In brakes, the practice of turning a drum or a rotor on a special lathe to remove surface imperfections such as scoring, and to eliminate runout and other dimensional problems. Drums must not be machined to a larger diameter nor rotors to a lesser thickness than the dimension stamped or cast on them.

Manual Bleeding A two-man technique for bleeding brakes. One pumps the brakes, the other opens and closes the bleeder screws.

Master Cylinder The device in a brake system that converts mechanical force to hydraulic pressure.

Maximum Diameter The largest diameter to which a brake drum can be machined or worn before it becomes unsafe. This dimension is usually stamped or cast into the drum near the hub. In most cases, this is 0.060 inch over original diameter.

Mechanical Brakes A brake system that is actuated mechanically, usually by rods or cables.

Metering Valve A valve used to delay application of front disc brakes.

Minimum Thickness The least amount of thickness to which a brake disc can be machined or worn before it becomes unsafe. This dimension is usually stamped or cast into the disc.

NIOSH National Institute for Occupation Safety and Health

Nondirectional Finish The recommended machine finish on a face of a rotor.

Non-Servo A type of drum brake in which the shoes work independently of each other.

OSHA Occupational Safety and Health Administration.

Out-of-Round In drum brakes, a drum that is deformed in such a way that its diameter varies at different points. It can be egg-shaped, for example. The drum must be machined or replaced to prevent brake grabbing and pulsation.

Pad A disc brake lining and metal back riveted or molded together.

Pad Wear Indicator A mechanical or electrical warning device on disc brakes that indicates need for pad replacement.

Parallelism On disc brakes, the thickness variation of a rotor's wear surface as measured with a micrometer. In most cases, if the variation exceeds 0.001 inch, the rotor should be machined or replaced to prevent increased pedal travel and pulsation.

Parking Brake A mechanically applied brake used to prevent a parked vehicle's movement.

Parking Brake Cable Cable(s) that transmit brake actuating force in the parking brake system.

Parking Brake Equalizer A device to equalize pull between the parking brake actuator and two wheels.

Pascal's Law A basic law of hydraulics. When pressure is exerted on a confined fluid, the pressure is transmitted equally and in all directions.

Penetrating Oil A very thin oil that is used to penetrate rust and corrosion, and to free rusted parts.

Phenolic This refers to wheel cylinder or caliper pistons made of plastic (phenolic) material rather than metal.

Piston A moveable plug that fits in a cylinder.

Piston Cup A rubber cup-shaped part that seals a cylinder and eliminates leakage between the piston and the cylinder walls.

Piston Stops Tabs, or protrusions, on a backing plate positioned to prevent the wheel cylinder pistons from leaving the wheel cylinder.

Pitting Surface irregularities resulting from corrosion.

Play Movement between two parts.

Power Brake A system utilizing energy from the vehicle's engine to reduce the amount of brake pedal pressure that must be exerted by the driver to stop the car.

Pressure The amount of force applied to a definite area. It is measured in pounds per square inch (PSI).

Pressure Bleeder A container with a brake fluid and air compartment separated by a diaphragm. Used with adapters to supply a constant, clean pressurized source of brake fluid for bleeding.

Pressure Bleeding Bleeding the brake system using a pressure bleeder to charge the system. One person opens and closes the bleeder screws as required.

Primary Piston In a dual or tandem master cylinder, the piston against which the brake pedal or power booster pushrod bears.

Primary Shoe The brake shoe toward the front of the vehicle in a duo-servo drum brake.

Proportioning Valve A pressure reduction valve used in rear brake circuit.

Pull The tendency of a car to pull or lead to one side when the brakes are applied.

Pushrod (Master Cylinder) The rod that transmits the movement and force of the driver from the brake pedal lever to the master cylinder piston.

Pushrod (Wheel Cylinder) The rod that transmits the movement and force of the wheel cylinder piston to the brake shoe.

Quick Take-Up Master Cylinder A type of master cylinder used to provide additional fluid volume required by low-drag calipers.

Radial Load A load that is applied at 90 degrees to an axis of rotation.

Radii Adapter A mounting device that is used to center a drum or rotor on the arbor of a lathe. A radii adapter centers the drum or rotor through contact with the bearing races.

Reservoir A storage area for extra brake fluid. Usually integral with the master cylinder.

Reservoir Diaphragm Gasket The gasket under the master cylinder reservoir cap that separates the fluid from the atmosphere. Moves with the fluid to allow venting above the fluid.

Residual Pressure Check Valve This valve retains slight positive pressure in the hydraulic system lines and wheel brakes.

Retaining Key A metal plate used to retain some sliding calipers in their anchor plates.

Retaining Spring A spring that is used to connect the lower ends of a pair of brake shoes and hold them in contact with the star wheel adjuster.

Retracting Spring A spring that is used to pull the brake shoes away from the drum when the brake pedal is released. The retracting spring also pushes the wheel cylinder piston back into its bore and thus returns the brake fluid to the master cylinder.

Return Springs Springs used on drum brakes to pull the brake shoes away from the drums when the brakes are released.

Reverse Bleeding A method of purging air from a hydraulic system by forcing fluid into the system at a bleeder valve and allowing the air to escape at the master cylinder.

Riveted Linings On drum brake shoes or disc brake pads, linings that are attached to their steel backing with rivets.

Roller Bearing An anti-friction bearing that uses a series of steel rollers held between inner and outer bearing races.

Rotor A disc that is attached to a wheel or hub to provide a friction surface for a brake system.

Runout On disc brakes, this refers to the amount of wobble in the disc as it rotates. Measured with a dial indicator, it must be corrected by machining or replacement if it exceeds the specified limit, or increased pedal travel and pulsation can occur.

Scoring On brakes, this refers to grooves worn into the friction surface of a drum or rotor. If scoring depth exceeds specified limits, the drum or rotor must be machined or replaced.

Seals On brakes, this refers to the rubber rings or cups that retain fluid in the hydraulic system or to the grease seal that keeps wheel bearing grease from migrating into a drum or onto a disc.

Secondary Piston In a dual or tandem master cylinder, it is the piston that is normally pushed forward by fluid trapped between it and the primary piston.

Secondary Shoe The brake shoe located toward the rear of the vehicle in a duo-servo drum brake.

Self-Adjusting Any brake capable of automatically compensating for brake lining wear.

Self-Energizing In drum brakes, this refers to the way most designs increase friction between the drum and the lining by means of a wedging action so that stopping power exceeds what is available from hydraulic pressure.

Semi-Metallic Type of brake lining using metallic components in place of asbestos in its formulation. Used in applications where heat renders organic components ineffective.

Servo Action Describes how one brake shoe adds to the application force of the other.

Shoe On drum brakes, the lining and its steel backing. The shoe and lining press against the inside of the brake drum to provide stopping power.

Shoe Anchor The point in a drum brake system where the braking forces are transmitted to the chassis.

Shoe Return Spring Drum brake spring that retracts the brake shoe when brakes are released.

Single Brake System A brake system using only one hydraulic circuit for all wheels.

Single Master Cylinder A master cylinder with only one pressure chamber. Used in single brake systems.

Single-Piston Caliper This refers to a disc brake caliper that has one hydraulic cylinder and piston as opposed to calipers with two or four pistons.

Skid Control System A system designed to respond to a locking wheel by relieving hydraulic pressure to the locking brake.

Sliding Caliper A disc brake caliper that has a piston or pistons on one side of the disc only. The caliper moves sideways on machined "ways" or keys to press the pad on the other side against the disc.

Snap Ring A split ring that is held in a groove by its own tension. Internal split rings are used in grooves cut around the bore of a hole. External snap rings are used in grooves cut around a shaft.

Spindle The part of the front suspension system around which a front wheel rotates. A shaft or pin around which another part rotates.

Splash Shield A metal deflector used to protect a disc brake rotor from road splash. It can be formed to improve the flow of air over the rotor.

Spongy Pedal A condition where the brake pedal is not solid when depressed, but bounces softly.

Star Wheel Adjuster A drum brake mechanism that separates the brake shoes. A star wheel rotates a nut on a threaded link to adjust brake lining clearance.

Stroking Seal (Disc Brakes) A disc brake hydraulic seal that is set in a piston groove and moves in the caliper bore with the piston.

Surge Bleeding A bleeding technique designed to dislodge air from wheel cylinders by creating turbulence in the wheel cylinder.

Tandem Master Cylinder A master cylinder that consists of a single bore with two pistons and separate fluid compression chambers. Used with "split" braking systems, this design insures that there will be some braking power in one braking circuit even if there is a severe fluid loss in the other. It has been mandatory on cars sold in the US since 1967.

Taper A lack of parallelism. A drum or rotor defect in which the thickness of the drum or rotor at the outer edge differs from its thickness at the inner edge.

Tapered Roller Bearing An anti-friction bearing that uses a series of tapered steel rollers held between tapered inner and outer races. Tapered roller bearings are used where both radial and thrust loads must be handled.

Three-Way Valve A hydraulic system component that incorporated metering, proportioning, and warning light activation.

Thrust Load A load that is applied in line with an axis of rotation.

TIR Total Indicator Reading. In disc brakes, this refers to a rotor runout reading of the entire swing of the dial indicator's needle, both above and below zero.

Tire The rubber and cord donut on the wheel rim that is filled with pressurized air and transmits vehicle forces (including braking forces) to the road.

Tire Rotation Moving the wheel and tire assemblies to different locations to equalize wear patterns.

Tire Tread The portion of a tire that contacts the road surface.

Toe The end of a brake shoe that is not adjacent to its anchor.

Tolerance A permissible variation, usually stated as extremes of a specification.

Tool Bit The hardened steel or carbide blade that cuts away metal during machining.

Torque A force that tends to produce a twisting or turning motion.

Torque Sequence The order in which a series of bolts or nuts should be tightened.

Torque Wrench A wrench or handle that indicates the amount of torque applied to a bolt or nut. A tool used to tighten bolts and nuts to a specific torque.

Trailing Shoe The shoe toward the rear of the vehicle is a non-servo drum brake.

Tube Seat An insert or machined face, against which a flared tube end seals.

Tubing Bender A tool used to bend tubing without kinking or deforming its walls.

Tubing Cutter A tool used to cut tubing. In operation, the tubing is held between a pair of rollers and a sharp wheel. The tool is moved around the tubing, and the wheel cuts the tubing cleanly and without distortion.

Tubing Wrench A wrench used to turn fittings on tubing. A tubing wrench distributes the turning forces evenly around the fitting and minimizes the possibility of damage.

Union A hydraulic coupling used to connect two brake lines.

Vacuum Booster A power brake actuating mechanism that uses vacuum on one side of a diaphragm as a power source.

Ventilated Rotor A disc brake rotor whose friction surfaces are separated by cooling fins.

Viscosity The thickness or fluid body of a liquid.

Warning Light Valve A valve in the hydraulic circuits of a dual brake system that switches on a dash warning light when one system fails.

Ways On disc brake single-piston calipers, machined surfaces on the adapter and caliper on which the caliper slides sideways.

Wear Sensor A projection on an inboard brake pad that causes a squealing sound when the brake pads are worn thin.

Wheel Bearing Seals Rubber or leather seals used at wheel hubs to keep wheel bearing grease from migrating into a brake drum or onto a brake disc.

Wheel Cylinder A device in the drum brake that converts hydraulic pressure to mechanical force to apply to the brakes.

APPENDIX B

INCHES AND MILLIMETER CONVERSION CHART

DECIMAL AND METRIC EQUIVALENTS					
Fractions	Decimal (in.)	Metric (mm)	Fractions	Decimal (in.)	Metric (mm)
1/64	.015625	.397	33/64	.515625	13.097
1/32	.03125	.794	17/32	.53125	13.494
3/64	.046875	1.191	35/64	.546875	13.891
1/16	.0625	1.588	9/16	.5625	14.288
5/64	.078125	1.984	36/64	.578125	14.684
3/32	.09375	2.381	19/32	.59375	15.081
7/64	.109375	2.778	39/64	.609375	15.478
1/8	.125	3.175	5/8	.625	15.875
9/64	.140625	3.572	41/64	.640625	16.272
5/32	.15625	3.969	21/32	.65625	16.669
11/64	.171875	4.366	43/64	.671875	17.066
3/16	.1875	4.763	11/16	.6875	17.463
13/64	.203125	5.159	45/64	.703125	17.859
7/32	.21875	5.556	23/32	.71875	18.256
15/64	.234275	5.953	47/64	.734375	18.653
1/4	.250	6.35	3/4	.750	19.05
17/64	.265625	6.747	49/64	.765625	19.447
9/32	.28125	7.144	25/32	.78125	19.844
19/64	.296875	7.54	51/64	.796875	20.241
5/16	.3125	7.938	13/16	.8125	20.638
21/64	.328125	8.334	53/64	.828125	21.034
11/32	.34375	8.731	27/32	.84375	21.431
23/64	.359375	9.128	55/64	.859375	21.828
3/8	.375	9.525	7/8	.875	22.225
25/64	.390625	9.922	57/64	.890625	22.622
13/32	.40625	10.319	29/32	.90625	23.019
27/64	.421875	10.716	59/64	.921875	23.416
7/16	.4375	11.113	15/16	.9375	23.813
29/64	.453125	11.509	61/64	.953125	24.209
15/32	.46875	11.906	31/32	.96875	24.606
31/64	.484375	12.303	63/64	.984375	25.003
1/2	.500	12.7	1	1.00	25.4

INDEX

Complete Brake Systems

Student Technician's Shop Manual

Delmar Publishers Inc.®

NOTICE TO THE READER

Publisher does not warrant or guarantee any of the products described herein or perform any independent analysis in connection with any of the product information contained herein. Publisher does not assume, and expressly disclaims, any obligation to obtain and include information other than that provided to it by the manufacturer.

The reader is expressly warned to consider and adopt all safety precautions that might be indicated by the activities described herein and to avoid all potential hazards. By following the instructions contained herein, the reader willingly assumes all risks in connection with such instructions.

The publisher makes no representations or warranties of any kind, including but not limited to, the warranties of fitness for particular purpose or merchantability, nor are any such representations implied with respect to the material set forth herein, and the publisher takes no responsibility with respect to such material. The publisher should not be liable for any special, consequential or exemplary damages resulting, in whole or in part, from the readers' use of, or reliance upon, this material.

Text, Design, and Production by
Scharff Associates, Ltd.

Scharff Staff
Production Manager: Marilyn Strouse-Hauptly
Staff Writer: Dave Caruso
Cover Design: Eric Schreader
Logo Design: Ed Foulk

Delmar Staff
Editor-in-Chief: Mark W. Huth
Associate Editor: Joan Gill

For information, address Delmar Publishers Inc.
3 Columbia Circle, PO Box 15015
Albany, New York 12212-5015

Printed in the United States of America
Published simultaneously in Canada
by Nelson Canada,
a division of International Thomson Limited

10 9 8

Library of Congress Cataloging-in-Publication Data

Complete brake systems.

Includes index.
1. Automobiles—Brakes I. Scharff, Robert.
II. Delmar Publishers.
TL269.C66 1989 629.2'46 88-28517
ISBN 0-8273-3539-3 (pbk.)
ISBN 0-8273-3540-7 (instructor's guide)
ISBN 0-8273-3541-5 (shop manual)

CONTENTS

PREFACE

Complete Brake Systems Student Technician's Shop Manual is designed to reinforce important brake tools, equipment, and procedures through written and hands-on testing. Today's automotive braking system is very complex; in order to diagnose and correct problems that arise, a great deal of information must be learned. Thus, the need becomes apparent for a textbook and accompanying workbook containing the latest information, products, and procedures.

SCOPE OF THE BOOK

Chapter 1 is an introduction to brake principles that stresses safety procedures.

Chapters 2, 3, and 4 discuss the hydraulic system, including components, bleeding, and service procedures.

Drum brakes are covered in Chapters 5 and 6, and disc brakes in Chapters 7 and 8.

Chapters 9 through 13 discuss parking brakes, power brakes, anti-lock brakes, light-duty truck brakes, and import vehicle brakes, respectively.

Finally, Chapter 14 provides a comprehensive diagnosis of brake problems.

CHAPTER 1

BRAKE PRINCIPLES AND FUNDAMENTALS

PRACTICE QUESTIONS

1. With 25 PSI force applied to the brake pedal and a four-square-inch piston area as shown in Figure 1-1, how many pounds of pressure would exit the master cylinder?
   a. 6.25 lb pressure
   b. 100 lb pressure
   c. 250 lb pressure
   d. 625 lb pressure

2. Mechanics should wear an approved respirator when working with brake lining and other automotive friction materials because of the hazards of ______________.
   a. flying pieces
   b. fiberglass dust
   c. asbestos dust
   d. fire

3. Figure 1-2 shows a measuring tool called a ______________.
   a. dial indicator
   b. micrometer
   c. clamp
   d. brake adjuster

4. Figure 1-3 shows another measuring tool called a ______________.
   a. dial indicator
   b. drum micrometer
   c. safe-set gauge
   d. caliper

5. When grinding a rotor from a vehicle using semi-metallic brake linings, Mechanic A creates a directional finish. Mechanic B creates a nondirectional, swirl-like finish. Who is right?
   a. Mechanic A
   b. Mechanic B
   c. Both A and B
   d. Neither A nor B

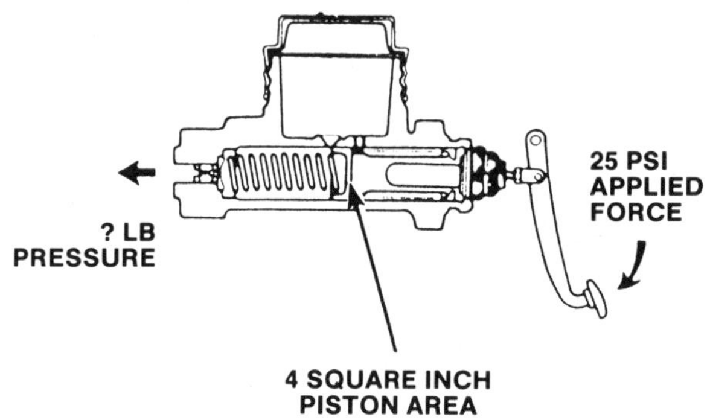

Figure 1-1

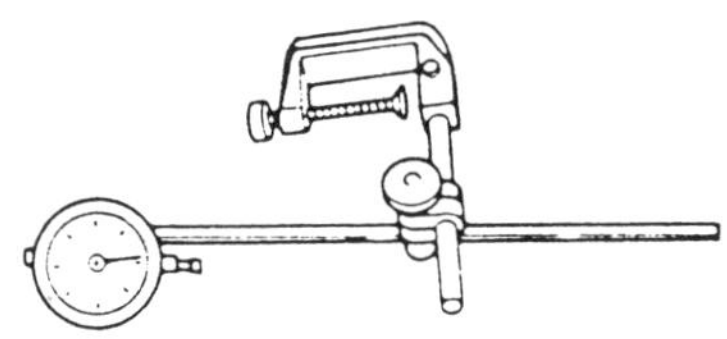

Figure 1-2

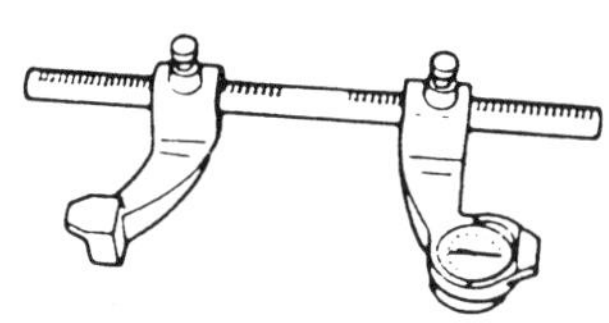

Figure 1-3

6. Approximately what percentage of the braking effort must be provided by the front brakes when stopping a vehicle?
   a. 75 percent
   b. 100 percent
   c. 50 percent
   d. 25 percent

7. Most cars manufactured in the United States use what type of brakes on the front wheels?
   a. drum brakes
   b. parking brakes

c. combination brakes
d. disc brakes

8. The invention of what component has greatly reduced the occurrence of rear wheel lockup?
   a. combination valve
   b. metering valve
   c. proportioning valve
   d. pressure sensors

9. What component of a typical disc brake is indicated in Figure 1-4?
   a. caliper
   b. rotor
   c. pad
   d. disc

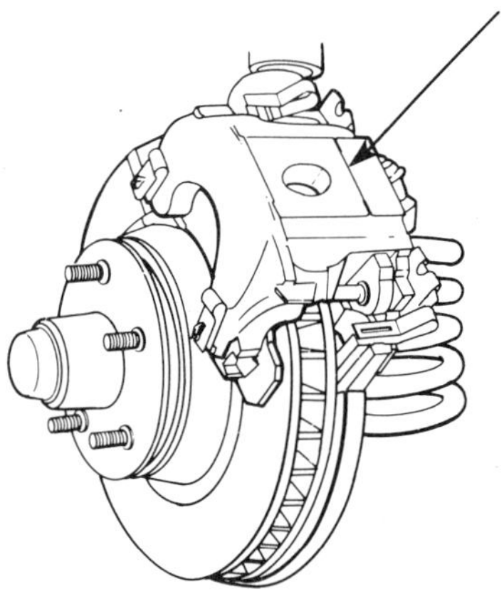

Figure 1-4

10. What component of a typical hydraulic brake system is indicated in Figure 1-5?
    a. combination valve
    b. power brake
    c. parking brake
    d. master cylinder

SHOP ASSIGNMENT 1

Examine a typical drum brake and locate the following parts: primary shoe, wheel cylinder, backing plate, secondary shoe, parking brake strut, and automatic adjuster. Wear respirator or dust mask and safety goggles or glasses with side shields.

SHOP ASSIGNMENT 2

Examine a typical disc brake and locate the following parts: caliper, rotor, and pad. Wear respirator or dust mask and safety goggles or glasses with side shields.

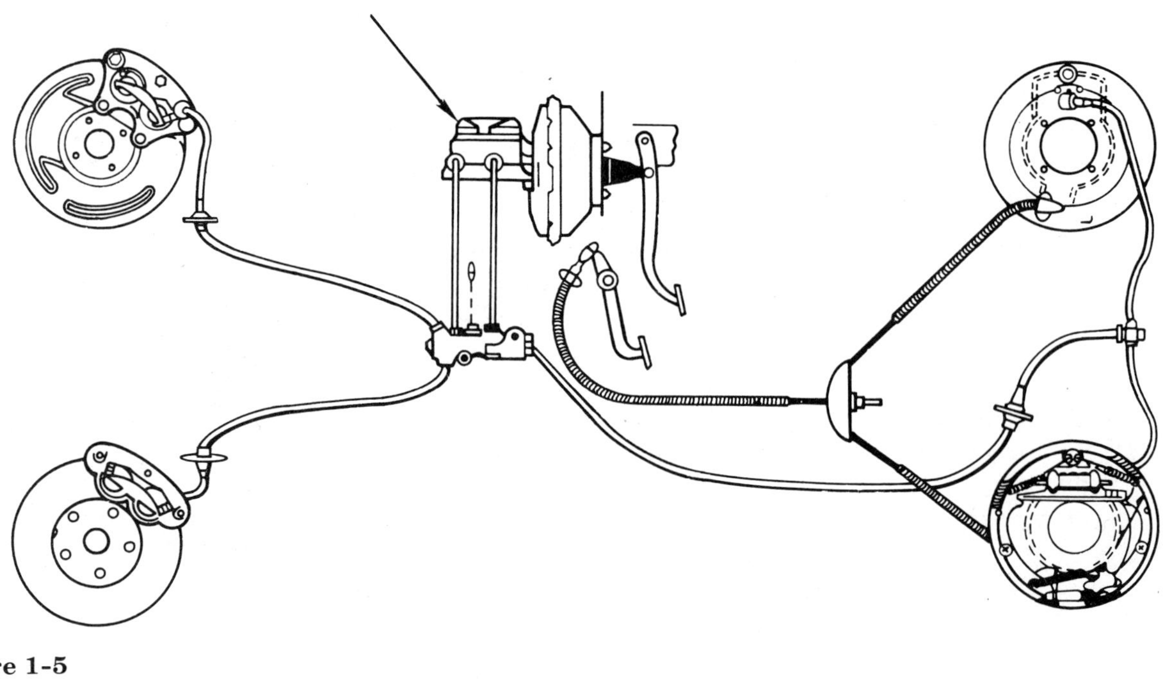

Figure 1-5

CHAPTER 2

HYDRAULIC SYSTEM—COMPONENT OPERATIONS

PRACTICE QUESTIONS

1. A vehicle's aluminum master cylinder has pitting and scoring in the cylinder bore. Mechanic A decides to rebuild the master cylinder. Mechanic B installs a new unit. Who is right?
   a. Mechanic A
   b. Mechanic B
   c. Both A and B
   d. Neither A nor B

2. Which of the following components serves the same function?
   a. replenishing port and compensating port
   b. vent port and by-pass port
   c. vent port and relief passage
   d. all of the above

3. What component of a dual-system master cylinder is indicated by the letter A in Figure 2-1?
   a. reservoir diaphragm
   b. secondary reservoir
   c. vent port
   d. piston stop bolt

4. In Figure 2-1, what component is indicated by the letter B?
   a. reservoir diaphragm
   b. primary reservoir
   c. primary cups
   d. vent port

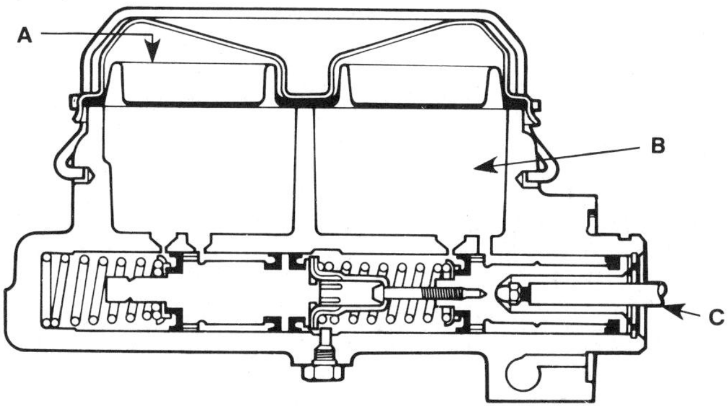

Figure 2-1

5. In Figure 2-1, what component is indicated by the letter C?
   a. piston stop bolt
   b. vent port
   c. hydraulic pushrod
   d. primary piston

6. A master cylinder under repair has a tapped hole in the reservoir, but no bolt had been installed in production. Mechanic A plugs the hole. Mechanic B installs a piston stop bolt. Who is right?
   a. Mechanic A
   b. Mechanic B
   c. Both A and B
   d. Neither A nor B

7. When beginning to reassemble a master cylinder, Mechanic A inspects the rubber grommets that the reservoir outlets fit into, and replaces them if they are cracked or torn. Mechanic B lubricates suitable grommets with special silicone brake assembly lubricant or fresh brake fluid. Who is right?
   a. Mechanic A
   b. Mechanic B
   c. Both A and B
   d. Neither A nor B

8. When replacing brake line tubing, Mechanic A uses tubing made of steel. Mechanic B uses tubing made of copper. Who is right?
   a. Mechanic A
   b. Mechanic B
   c. Both A and B
   d. Neither A nor B

9. What type of wheel cylinder is pictured in Figure 2-2?
   a. step bore wheel cylinder
   b. single piston fixed wheel cylinder
   c. double piston wheel cylinder
   d. baffle type piston wheel cylinder

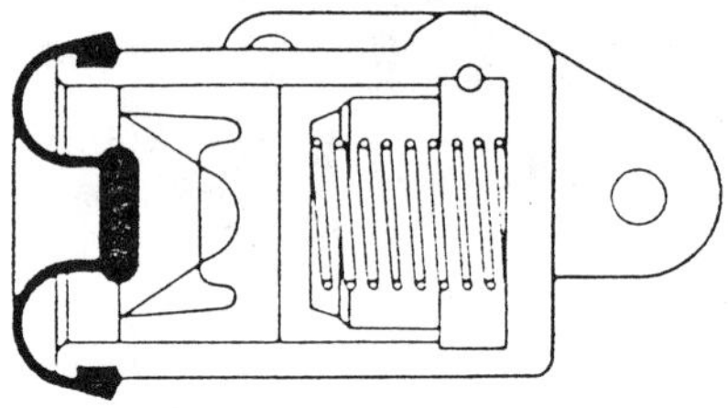

Figure 2-2

10. The brake warning light in the instrument panel of a vehicle is activated by the
    _______________.

    a. proportioning valve
    b. stoplight switch
    c. metering valve
    d. pressure differential switch

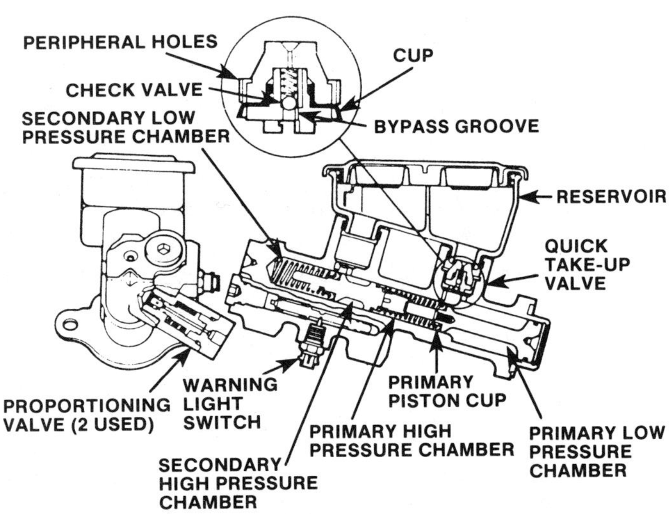

Figure 2-3

SHOP ASSIGNMENT 3

Examine a GM quick take-up master cylinder and locate the following parts: reservoir, quick take-up valve, primary and secondary low-pressure chambers, primary and secondary high-pressure chambers, primary piston cup, warning light switch, and proportioning valve. Examine the following parts of the quick take-up valve: cup, bypass groove, peripheral holes, and ball check valve. Refer to Figure 2-3 as needed.

SHOP ASSIGNMENT 4

Examine the differences in design between mechanically operated stoplight switches (used on most recent model cars) and hydraulically operated switches (used on some older model cars). Refer to Figure 2-4 as needed.

SHOP ASSIGNMENT 5

Fill a master cylinder with clean, specified brake fluid. Cast-iron master cylinders should be filled to within 1/4 inch of the top of the reservoir to allow for brake expansion when heated. Aluminum master cylinders with transparent plastic reservoirs are marked with "FILL TO" or "FULL" to show the proper level for the brake fluid. Brake fluid level should be checked twice each year.

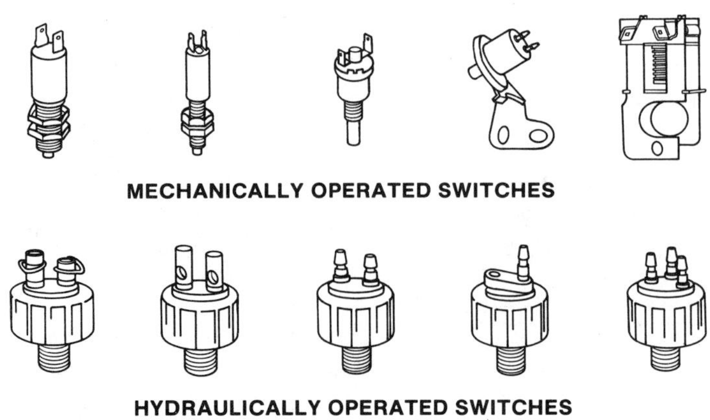

Figure 2-4

CHAPTER 3

HYDRAULIC SYSTEM—BLEEDING

PRACTICE QUESTIONS

1. The first step in the bleeding procedure is ____________.
   a. to clean dirt from the filler cap before removing it from the master cylinder
   b. to slip a hose over the end of the bleeder screw
   c. to bleed at each wheel cylinder and caliper in succession
   d. to refill the reservoir with brake fluid

2. Upon encountering rear calipers that are difficult to bleed, Mechanic A lightly taps them with a hammer to regroup the air bubbles. Mechanic B expels the air bubbles from a different point. Who is right?
   a. Mechanic A
   b. Mechanic B
   c. Both A and B
   d. Neither A nor B

3. When beginning the bleeding sequence, Mechanic A bleeds at the master cylinder bleeder screws. On a different vehicle, Mechanic B cracks the outlet tube nuts at the master cylinder while a helper applies pressure to the brake pedal. Who is right?
   a. Mechanic A
   b. Mechanic B
   c. Both A and B
   d. Neither A nor B

4. How many people are required to manually bleed a brake hydraulic system?
   a. one
   b. two
   c. three
   d. four

5. After attaching a bleeder hose to a bleeder screw, Mechanic A holds the other end of the hose below the level of the screw. Mechanic B submerges the other end of the hose in a glass jar partially filled with brake fluid. Who is right?
   a. Mechanic A
   b. Mechanic B
   c. Both A and B
   d. Neither A nor B

6. How many people are required to pressure bleed a brake hydraulic system?
   a. one
   b. two
   c. three
   d. four

7. Why should the pressure bleeder tank remain still throughout the bleeding operation?
   a. to maintain adequate pressure
   b. to allow the procedure to be completed as quickly as possible
   c. to prevent any spilling of brake fluid
   d. to prevent air in the tank from getting into the lines

8. What method of bleeding may be used when routine manual or pressure bleeding methods prove inadequate?
   a. final touch bleeding
   b. flushing
   c. surge bleeding
   d. vacuum bleeding

9. Figure 3–1 shows a special tool used to hold open what type of valve?
   a. proportioning valve
   b. metering valve
   c. valve port
   d. height-sensing valve

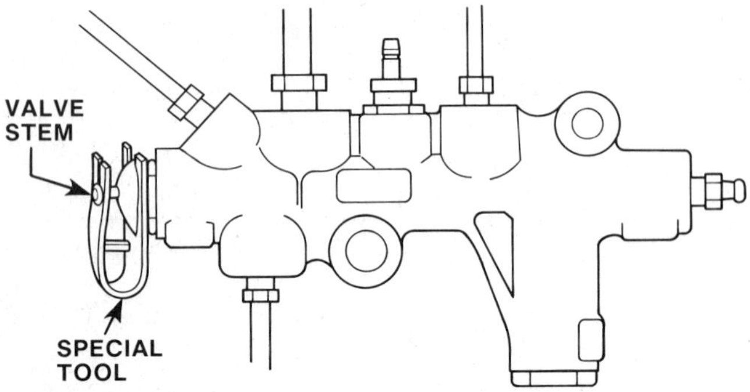

Figure 3-1

10. When beginning to bench bleed a reconditioned master cylinder, Mechanic A first mounts the master cylinder in a vise. Mechanic B first fills the reservoirs with fresh brake fluid. Who is right?
    a. Mechanic A
    b. Mechanic B
    c. Both A and B
    d. Neither A nor B

JOB SHEET

SHOP ASSIGNMENT 6
MANUALLY BLEED A HYDRAULIC SYSTEM

NAME ____________ STATION ____________ DATE ____________

Tools and Materials

Bleeder hose
Wrench
Glass jar
Brake fluid

Protective Clothing

Safety goggles or glasses with side shields

Procedure

1. Attach one end of the bleeder hose to a bleeder screw.

   Task completed ____

   a. Place the other end of the hose in a glass jar partially filled with brake fluid.

   Task completed ____

2. Apply moderate steady pressure on the brake pedal.

   Task completed ____

   a. Open the bleeder screw with a wrench.

   Task completed ____

3. Is the fluid coming from the submerged end of the hose clear and free from air bubbles?

   Yes ____ No ____

   a. If no, continue until answer is yes.

   Task completed ____

   b. If yes, close the bleeder screw and release the brake pedal.

   Task completed ____

4. Does the brake pedal reach the floor without removing all of the air bubbles?

   Yes ____ No ____

   a. If no, go to step 5.

   Task completed ____

   b. If yes, close the bleeder screw and release the pedal slowly.

   Task completed ____

c. Repeat steps 2 and 3.

Task completed ____

5. Repeat the procedure at all other bleed points.

Task completed ____

6. Make sure reservoir fluid is 1/4 inch from the top.

Task completed ____

PROBLEMS ENCOUNTERED: ________________________________

____

____

INSTRUCTOR'S COMMENTS: ________________________________

____

____

JOB SHEET

SHOP ASSIGNMENT 7
PRESSURE BLEED A HYDRAULIC SYSTEM

NAME ______________ STATION ______________ DATE ______________

Tools and Materials

Pressure bleeder and adapter
Brake fluid
Clean rag
Bleeder hose
Glass jar
Wrench

Protective Clothing

Safety goggles or glasses with side shields

Procedure

1. Is there enough brake fluid in the pressure bleeder to complete the operation?

   Yes ____ No ____

   a. If no, correct.

   Task completed ____

   b. Bring the unit to a working pressure of 15 to 20 PSI.

   Task completed ____

2. Use a rag to clean the master cylinder cover.

   Task completed ____

   a. Remove the cover.

   Task completed ____

3. Remove the reservoir diaphragm gasket.

   Not applicable ____ Task completed ____

   a. Clean gasket seal.

   Not applicable ____ Task completed ____

   b. Fill reservoir with brake fluid.

   Task completed ____

4. Attach the adapter to the reservoir.

   Task completed ____

5. Attach one end of the bleeder hose to the adapter fitting.

Task completed ____

   a. Is the coupling sleeve fully engaged?

Yes ____ No ____

   b. If no, then correct.

Task completed ____

6. Place the other end of the hose in a glass jar partially filled with brake fluid.

Task completed ____

7. Open the fluid supply valve.

Task completed ____

   a. Bleed until all air bubbles have been removed from the brake fluid coming from the submerged end of the hose.

Task completed ____

8. Repeat the procedure at all other bleed points.

Task completed ____

9. To disconnect the pressure bleeder, first close the supply valve.

Task completed ____

   a. Wrap the coupling in a rag to protect the car's finish from drips.

Task completed ____

   b. Undo the coupler sleeve above the master cylinder.

Task completed ____

10. Is the reservoir fluid level within 1/4 inch of the top?

Yes ____ No ____

   a. If no, then correct.

Task completed ____

11. Replace the gasket and cap.

Task completed ____

PROBLEMS ENCOUNTERED: ____

____

____

INSTRUCTOR'S COMMENTS: ____

____

____

JOB SHEET

SHOP ASSIGNMENT 8
BENCH BLEED A MASTER CYLINDER

NAME ________________ STATION ________________ DATE ________________

Tools and Materials

Vise
Tubing
Brake fluid
Wooden dowel or smooth, rounded rod
Wrench
Clean rag

Protective Clothing

Safety goggles or glasses with side shields

Procedure

1. Mount the cylinder in a vise.

   Task completed ____

   a. Is the bore horizontal?

   Yes ____ No ____

   b. If no, then correct.

   Task completed ____

2. Connect short lengths of tubing to the outlet ports.

   Task completed ____

   a. Bend them.

   Task completed ____

3. Fill the reservoirs with brake fluid.

   Task completed ____

   a. Is the level above the ends of the tubes?

   Yes ____ No ____

   b. If no, then correct

   Task completed ____

4. Use a wooden dowel or smooth, rounded rod to slowly pump the master cylinder until bubbles stop appearing at the tube ends.

   Task completed ____

5. Using the same tool, push the piston 1/4 to 1/2 inch from the fully released position.

   Task completed ____

a. Allow it to return about 1/8 inch.

Task completed ____

b. Push it back to the initial 1/4- to 1/2-inch position.

Task completed ____

c. Move the piston back and forth rapidly in this manner until air bubbles stop coming from the ports located inside the reservoirs.

Task completed ____

6. Remove the tubing.

Task completed ____

7. Refill the reservoirs.

Task completed ____

a. Secure the master cylinder cap.

Task completed ____

8. Install the master cylinder on the vehicle.

Task completed ____

a. Attach the lines, but do not tighten tube connections.

Task completed ____

9. Depress the brake pedal slowly several times to force any trapped air out of the connections.

Task completed ____

a. Tighten the nut slightly before releasing the pedal.

Task completed ____

b. Loosen the nut before depressing the pedal.

Task completed ____

c. Is a rag handy to catch the fluid to protect the vehicle's finish?

Yes ____ No ____

d. If no, correct.

Task completed ____

10. Have all air bubbles been removed?

Yes ____ No ____

a. If no, continue procedure.

Task completed ____

11. Tighten the connections to manufacturer's specifications.

Task completed ____

12. Are the master cylinder reservoirs filled with brake fluid?

Yes ____ No ____

a. If no, correct.

Task completed ____

PROBLEMS ENCOUNTERED: ____

____

____

INSTRUCTOR'S COMMENTS: ____

____

____

JOB SHEET

SHOP ASSIGNMENT 9
BENCH BLEED A MASTER CYLINDER USING A SPECIAL BLEEDER SYRINGE

NAME ____________________ STATION ____________________ DATE ______________

Tools and Materials

Plugs
Vise
Brake fluid
Bleeder syringe

Protective Clothing

Safety goggles or glasses with side shields

Procedure

1. Plug the master cylinder outlet holes.

   Task completed ____

   a. Gently clamp it in a vise with the pushrod end slightly elevated.

   Task completed ____

2. Fill the master cylinder halfway with fluid.

   Task completed ____

3. Remove an outlet plug from the cylinder.

   Task completed ____

   a. Depress the syringe plunger fully.

   Task completed ____

   b. Place the rubber tip firmly against the outlet hole to form a seal.

   Task completed ____

4. Slowly pull the plunger out until the syringe is half full of fluid.

   Task completed ____

   a. Remove the syringe from the outlet.

   Task completed ____

5. Invert the syringe and slowly depress its plunger to expel all of the air.

   Task completed ____

6. Hold the syringe (with fluid) against the same outlet.

   Task completed ____

a. Slowly depress the plunger.

Task completed ____

b. Have the air bubbles in the reservoir disappeared?

Yes ____ No ____

c. If no, continue procedure.

Task completed ____

d. If yes, remove the syringe and plug the outlet.

Task completed ____

7. Repeat this procedure at the other outlets.

Task completed ____

a. Plug them tightly.

Task completed ____

8. Reclamp the master cylinder in the vise with the pushrod end tilted downward slightly.

Task completed ____

9. Slide the master cylinder pushrod back and forth slowly about 1/8 inch until no air bubbles are seen in the reservoirs.

Task completed ____

10. Again mount the master cylinder in the vise with the pushrod end up.

Task completed ____

11. Fill the syringe with brake fluid.

Task completed ____

a. Expel the air as was done in step 5.

Task completed ____

12. Remove one outlet plug at a time and repeat steps 5, 6, and 7.

Task completed ____

PROBLEMS ENCOUNTERED: ____________________

____

____

INSTRUCTOR'S COMMENTS: ____________________

____

____

HYDRAULIC SYSTEM—SERVICING

PRACTICE QUESTIONS

1. Excessive side movement in the brake pedal indicates that ____________.
   a. nothing is wrong
   b. a leak exists in the lines
   c. pedal mounting parts are worn
   d. the master cylinder should be checked

2. Mechanic A tests both regular brakes and power brakes in the same manner with the engine not running. Mechanic B tests power brakes with the engine running. Who is right?
   a. Mechanic A
   b. Mechanic B
   c. Both A and B
   d. Neither A nor B

3. What are usually the first indicators that something could be wrong in the brake system?
   a. changes in brake pedal feel or travel
   b. brakes locking up
   c. no braking power
   d. noise when brakes are applied

4. In Figure 4-1, what does measurement A indicate?
   a. pedal free height
   b. pedal assembly
   c. pedal pad
   d. pedal travel

5. In Figure 4-1, what does measurement B indicate?
   a. pedal free height
   b. pedal assembly
   c. pedal pad
   d. pedal travel

6. Mechanic A notices a trace of brake fluid on the booster shell below the master cylinder mounting flange and begins looking for a leak. Mechanic B notices the fluid but considers it to be normal. Who is right?
   a. Mechanic A
   b. Mechanic B
   c. Both A and B
   d. Neither A nor B

7. To remove a frozen bleeder screw, Mechanic A uses a special wrench to exert pressure on the screw and strikes it with a hammer. Mechanic B uses heat from a welding torch. Who is right?
   a. Mechanic A
   b. Mechanic B
   c. Both A and B
   d. Neither A nor B

8. What is the most common symptom of fluid leaking past the cups in the master cylinder?
   a. spongy pedal that goes all the way to the floor
   b. no braking ability
   c. the need to pump the pedal to achieve proper braking ability
   d. puddle under the vehicle

9. After installing a new master cylinder, Mechanic A first bleeds the hydraulic system. Mechanic B first tests the the vehicle's braking ability to determine if bleeding is necessary. Who is right?
   a. Mechanic A
   b. Mechanic B

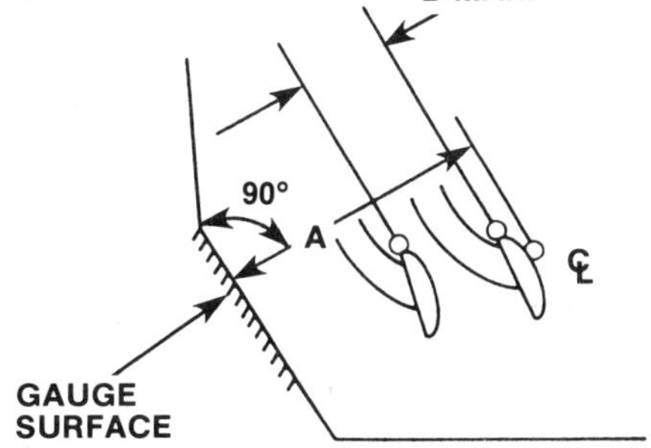

Figure 4-1

c. Both A and B
d. Neither A nor B

10. Which of the following cannot be repaired but must be replaced?
    a. metering valve
    b. proportioning valve
    c. combination valve
    d. all of the above

JOB SHEET

SHOP ASSIGNMENT 10
TEST FOR AIR TRAPPED IN A HYDRAULIC SYSTEM

NAME ______________ STATION ______________ DATE ______________

Tools and Materials

Brake fluid
Clear plastic wrap or similar covering
Towels and water

Protective Clothing

Safety goggles or glasses with side shields

Procedure

1. Are the master cylinder reservoirs filled with brake fluid.

   Yes ____ No ____

   a. If no, fill them.

   Task completed ____

2. Loosely place the gasket and cover assembly on the master cylinder.

   Task completed ____

3. Cover the reservoirs with clear plastic wrap or similar covering to keep brake fluid off the vehicle's finish.

   Task completed ____

   a. Are towels and water on hand just in case?

   Yes ____ No ____

   b. If no, correct.

   Task completed ____

4. Watch the master cylinder while a helper pumps the brakes rapidly 20 times and holds down the brake pedal on the last application?

   Task completed ____

5. Observe the fluid in the reservoirs while the other person releases the brake pedal quickly.

   Task completed ____

a. Did the brake fluid spout as high as 6 inches out of the reservoir?

Yes ____ No ____

b. If yes, air is trapped in the system; bleed the system that is spouting the fluid to remove the air.

Task completed ____

PROBLEMS ENCOUNTERED: ____________________

____

____

INSTRUCTOR'S COMMENTS: ____________________

____

____

JOB SHEET

SHOP ASSIGNMENT 11
TEST FOR OPEN REPLENISHING PORTS IN A MASTER CYLINDER

NAME ______________ STATION ______________ DATE ______________

Tools and Materials

Wrenches
Brake fluid

Protective Clothing

Safety goggles or glasses with side shields

Procedure

1. Remove the cover from the master cylinder.

   Task completed ____

   a. Are the reservoirs filled with fluid?

   Yes ____ No ____

   b. If no, fill with fluid.

   Task completed ____

2. Observe the reservoirs while a helper pumps the brake pedal.

   a. Can a small ripple or geyser be seen in the reservoirs?

   Yes ____ No ____

   b. If yes, go to step 3b.

   c. If no, loosen the master cylinder attaching bolts about 1/4 inch.

   Task completed ____

   d. Pull the master cylinder out and hold it in position.

   Task completed ____

3. Again observe the reservoirs while a helper pumps the brake pedal.

   a. Is turbulence present in the reservoirs?

   Yes ____ No ____

   b. If yes, adjust the brake pedal pushrod length.

   Task completed ____

   c. If no, replace the master cylinder.

   Task completed ____

PROBLEMS ENCOUNTERED: ______________________________

____

____

INSTRUCTOR'S COMMENTS: ______________________________

____

____

JOB SHEET

SHOP ASSIGNMENT 12
REMOVE AND DISASSEMBLE A TYPICAL DUAL-SYSTEM MASTER CYLINDER

NAME ______________ **STATION** ______________ **DATE** ______________

Tools and Materials

Wrenches
Vise, if needed
Screwdrivers
Clean denatured alcohol
Brake fluid
Crocus cloth or cylinder hone, if necessary
Feeler gauge
New hardware

Protective Clothing

Safety goggles or glasses with side shields

Procedure

1. Disconnect hydraulic tubing.

Task completed ____

   a. Remove nuts or bolts that secure master cylinder to firewall or power brake.

Task completed ____

   b. Remove master cylinder from the vehicle.

Task completed ____

2. Remove reservoir cover.

Task completed ____

   a. Drain brake fluid from the reservoirs.

Task completed ____

3. Remove the piston stop bolt from the bottom or side of the master cylinder.

Task completed ____

4. Press in on pushrod or a rod inserted in primary piston, and remove the lock ring from the cylinder bore or piston retainer from mounting flange.

Not applicable ____ Task completed ____

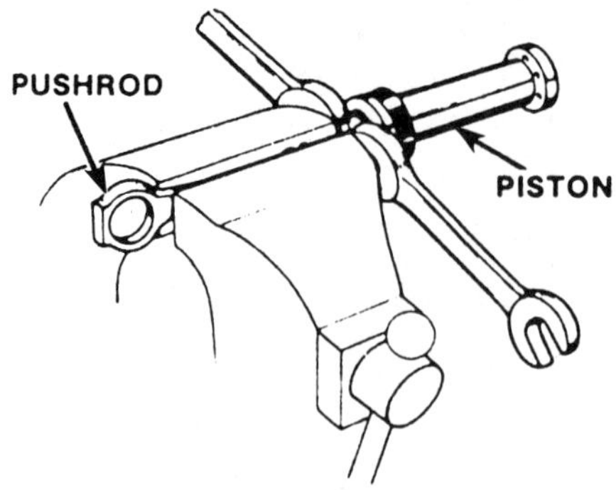

Figure 4-2

a. If the pushrod is joined to the piston with a plastic or rubber retainer, clamp the pushrod in a vise as shown in Figure 4-2.

Not applicable ____ Task completed ____

b. Then use open end wrenches to pry the piston from the pushrod.

Not applicable ____ Task completed ____

5. Remove the parts from the cylinder bore.

Task completed ____

a. Disassemble the secondary piston.

Task completed ____

6. Clamp the master cylinder in a vise with the outlet ports facing up.

Task completed ____

a. Does an outlet port contain a residual pressure check valve?

Yes ____ No ____

b. If yes, go to step 7.

Not applicable ____ Task completed ____

c. If no, is the tube seat damaged?

Yes ____ No ____

d. If no, go to step 9.

Not applicable ____ Task completed ____

7. Place a flat washer on a 6/32 self-tapping screw.

Task completed ____

a. Thread the screw into the tube seat insert.

Task completed ____

8. Use two screwdrivers inserted under the flat washer to pry upward and remove the tube seat.

Task completed ____

a. Remove the check valve.

Not applicable ____ Task completed ____

b. Remove the check valve spring from the outlet port.

Not applicable ____ Task completed ____

9. Thoroughly clean the master cylinder in clean denatured alcohol.

Task completed ____

a. Clean any other parts to be reused (except rubber components) in clean denatured alcohol.

Task completed ____

PROBLEMS ENCOUNTERED: ______________________________

____

____

INSTRUCTOR'S COMMENTS: ______________________________

____

____

JOB SHEET

SHOP ASSIGNMENT 13
REASSEMBLE AND INSTALL A BENDIX TYPE DUAL-SYSTEM MASTER CYLINDER

NAME ____________ STATION ____________ DATE ____________

Tools and Materials

Brake fluid
Dull scribe
Wrenches

Protective Clothing

Safety goggles or glasses with side shields

Procedure

1. Dip the secondary piston parts in clean brake fluid and assemble.

   Task completed ____

2. Lubricate the cylinder bore and cups with brake fluid.

   Task completed ____

3. Guide the secondary piston assembly, spring end first, into the cylinder bore. Use a dull scribe to ease the cups into the bore.

   Task completed ____

4. Guide the primary piston into the bore in the same manner.

   Task completed ____

5. While pressing in the primary piston, assemble the lock ring in the cylinder bore or the retainer on the mounting flange. Still pressing in the primary piston, assemble the piston stop bolt.

   Task completed ____

6. Insert the check valve spring in the check valve, then place them in the vent port, spring end first.

   Task completed ____

   a. Is the spring seated in the bottom of the port?

   Yes ____ No ____

   b. If no, correct.

   Task completed ____

7. Position the tube seat insert in the port. Press the tube seat into place by threading a tube nut into the port.

   Task completed ____

8. If applicable, assemble the boot on the master cylinder hub.

Not applicable ____ Task completed ____

9. If applicable, assemble the pushrod retainer on the pushrod.

Not applicable ____ Task completed ____

   a. Insert the pushrod to the primary piston, pressing in until the retainer snaps into place.

Not applicable ____ Task completed ____

10. Install the master cylinder on the vehicle.

Task completed ____

   a. Connect the hydraulic tubing.

Task completed ____

11. If it was disconnected, reattach the pushrod to the brake pedal.

Not applicable ____ Task completed ____

PROBLEMS ENCOUNTERED: ____

____

____

INSTRUCTOR'S COMMENTS: ____

____

____

JOB SHEET

SHOP ASSIGNMENT 14
REMOVE AND INSTALL A HEIGHT-SENSING PROPORTIONING VALVE ASSEMBLY

NAME ____________ STATION ____________ DATE ____________

Tools and Materials

Hoist
Socket wrench assortment
Torque wrench
Screwdriver assortment
Groove joint pliers
Bolts and washers

Protective Clothing

Safety goggles or glasses with side shields

Procedure

1. Raise the vehicle on a hoist with the rear suspension and rear axle hanging in the fully released position.

   Task completed ____

2. Remove the rear wheels and tires.

   Task completed ____

3. Remove the linkage assembly from the valve.

   Task completed ____

4. Remove the brake hose and discard the old washers.

   Task completed ____

5. Remove the valve from the mounting bracket.

   Task completed ____

   a. Replace with a new valve.

   Task completed ____

6. Attach the sensing valve to the mounting bracket with two bolts.

   Task completed ____

   a. Torque the bolts 14 to 22 foot-pounds.

   Task completed ____

7. Squeeze the plastic bushing of the linkage assembly onto the serrated section of the sensing valve shaft, using groove joint pliers.

   Task completed ____

8. Install the nut securing link assembly to the valve shaft.

Task completed ____

    a. Torque the nut 8 to 10 foot-pounds.

Task completed ____

9. Install the rear brake hose assembly to the sensing valve using the existing bolt and two new washers.

Task completed ____

    a. Torque the bolt 17 to 25 foot-pounds.

Task completed ____

10. Bleed the brakes as per recommendations.

Task completed ____

    a. Are any leakage problems evident?

Yes ____ No ____

    b. If yes, correct.

Task completed ____

11. Replace the wheels and tires.

Task completed ____

    a. Torque the lug nuts to the proper specifications.

Task completed ____

12. Lower the vehicle.

Task completed ____

    a. Are any problems encountered in a road test?

Yes ____ No ____

    b. If yes, explain below and correct.

Task completed ____

PROBLEMS ENCOUNTERED: ____

____

____

INSTRUCTOR'S COMMENTS: ____

____

____

CHAPTER 5

DRUM BRAKES—OPERATION

PRACTICE QUESTIONS

1. What drum brake component maintains the correct shoe position and clearance?
   a. backing plate
   b. wheel cylinder
   c. retainer
   d. return springs and hold-down parts

2. After installing new brake lining, Mechanic A cautions the customer to avoid high-speed stops and severe brake applications during the first 50 miles. Mechanic B cautions the customer to avoid severe usage during the first 500 miles. Who is right?
   a. Mechanic A
   b. Mechanic B
   c. Both A and B
   d. Neither A nor B

3. What type of shoe return spring alignment is pictured in Figure 5–1?
   a. cable automatic adjuster with auxiliary spring
   b. separate spring guide
   c. separate spring link
   d. manual adjuster

4. What type of shoe anchor is pictured in Figure 5–2?
   a. bolted to backing plate
   b. swaged to backing plate
   c. threaded into steering arm
   d. self-centering (sliding) anchor

5. The adjuster system on most vehicles is installed on the ____________.
   a. secondary shoe
   b. star wheel
   c. primary shoe
   d. anchor

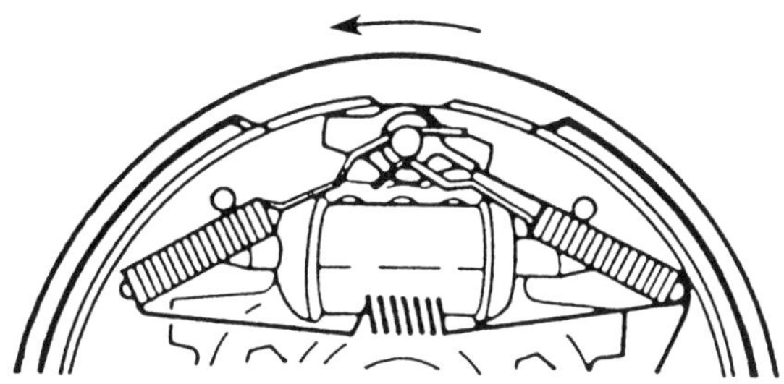

Figure 5-1

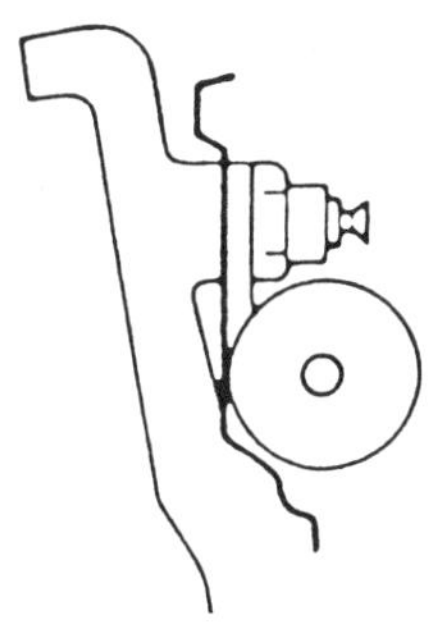

Figure 5-2

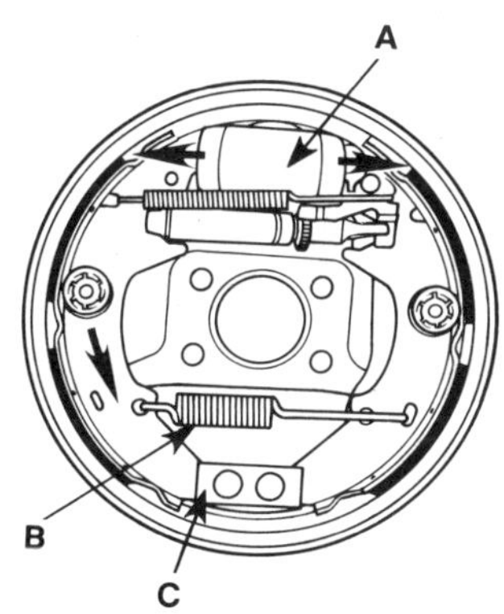

Figure 5-3

6. What component of a non-servo drum brake is indicated by the letter A in Figure 5–3?
   a. primary shoe
   b. wheel cylinder
   c. anchor plate
   d. retaining spring

7. What component of a non-servo drum brake is indicated by the letter B in Figure 5-3?
   a. primary shoe
   b. wheel cylinder
   c. anchor plate
   d. retaining spring

8. What component of a non-servo drum brake is indicated by the letter C in Figure 5-3?
   a. primary shoe
   b. wheel cylinder
   c. anchor plate
   d. retaining spring

9. Brake shoes are usually held in position by ________________.
   a. spring tension
   b. the backing plate
   c. adjusters
   d. the star wheel

10. What type of drum brake is known as a leading-trailing shoe type?
    a. duo-servo
    b. non-servo
    c. automatically adjusted
    d. manually adjusted

SHOP ASSIGNMENT 15

Examine the components of a duo-servo brake and locate the following components: wheel cylinder, return springs, anchor pin, primary and secondary shoes, connecting spring, and adjusting screw assembly. Wear respirator and safety goggles or glasses with side shields. Refer to Figure 5-4 as needed.

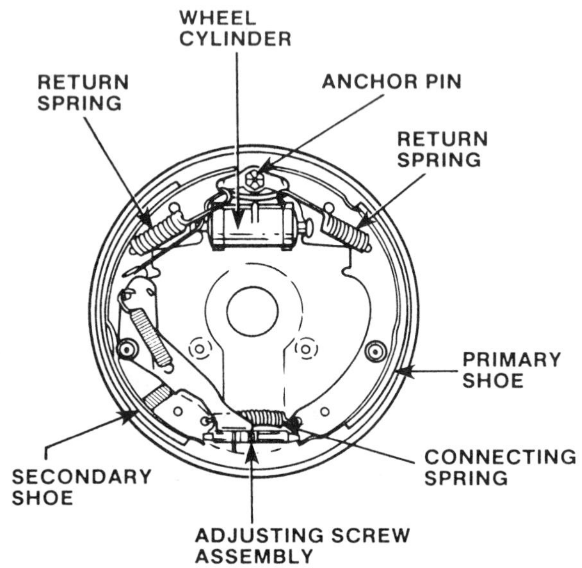

Figure 5-4

JOB SHEET

SHOP ASSIGNMENT 16
ADJUST AN
ECCENTRIC TYPE SHOE ANCHOR

NAME ____________ **STATION** ____________ **DATE** ____________

ADJUSTABLE FIXED ANCHORS

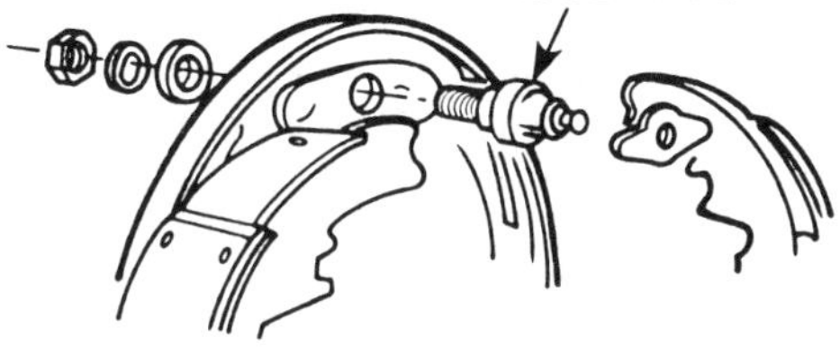

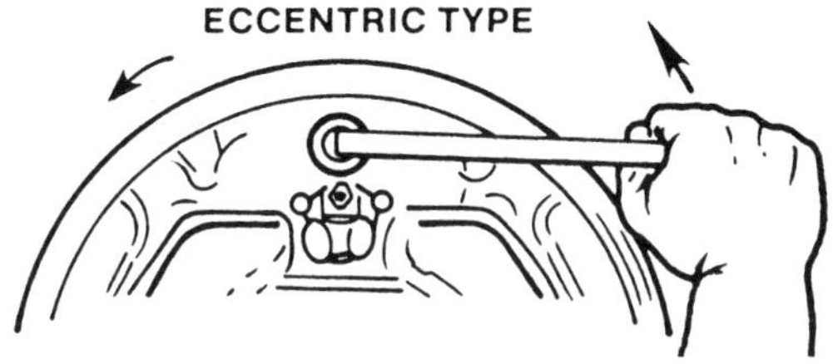

Figure 5-5

Tools and Materials

Wrenches

Protective Clothing

Safety goggles or glasses with side shields
Respirator

Procedure

1. Tighten the star wheel to heavy brake drag.

   Task completed ____

2. Rotate the eccentric anchor in the direction that frees the brake until drag is encountered again.

   Task completed ____

3. Return the eccentric to the position between the two points of drag where the drum rotates freely.

   Task completed ____

4. Repeat this sequence until drag cannot be relieved.

   Task completed ____

5. Tighten the anchor nut.

   Task completed ____

6. Back off the star wheel to a normal manual adjustment.

   Task completed ____

PROBLEMS ENCOUNTERED: ____

____

____

INSTRUCTOR'S COMMENTS: ____

____

____

CHAPTER 6

DRUM BRAKES—INSPECTING AND SERVICING

PRACTICE QUESTIONS

1. When inspecting a drum brake, Mechanic A checks for lining thickness and wheel cylinder leaks. Mechanic B checks for leaking axle seal and contaminated lining. Who is right?
   a. Mechanic A
   b. Mechanic B
   c. Both A and B
   d. Neither A nor B

2. When servicing wheel brake parts, Mechanic A cleans up with a dry brush. Mechanic B cleans with an OSHA-approved washer. Who is right?
   a. Mechanic A
   b. Mechanic B
   c. Both A and B
   d. Neither A nor B

3. When preparing a typical duo-servo drum brake for inspection and servicing, Mechanic A uses a brake spring tool only to unhook the springs from the anchor pin or anchor plate. Mechanic B uses the brake spring tool to also remove the adjusting link from the anchor pin or anchor plate. Who is right?
   a. Mechanic A
   b. Mechanic B
   c. Both A and B
   d. Neither A nor B

4. The mechanic in Figure 6-1 is removing what brake component?
   a. springs
   b. adjusting levers
   c. anchor plates
   d. shoes

5. After inspecting brake linings, Mechanic A replaces only the worn lining. Mechanic B replaces the worn lining and the lining on the wheel on the opposite side of the car. Who is right?
   a. Mechanic A
   b. Mechanic B
   c. Both A and B
   d. Neither A nor B

6. When a drum is removed for inspection, brake fluid drips out of the wheel cylinders when the rubber dust boots are pulled away. Mechanic A concludes that the cylinder needs rebuilding or replacing. Mechanic B concludes that the amount of fluid dripping is normal. Who is right?
   a. Mechanic A
   b. Mechanic B
   c. Both A and B
   d. Neither A nor B

7. What method of brake drum mounting is depicted in Figure 6-2?
   a. swaged stud mounting
   b. rear axle shaft flange mounting
   c. riveted mounting
   d. tire rim and lug nut mounting

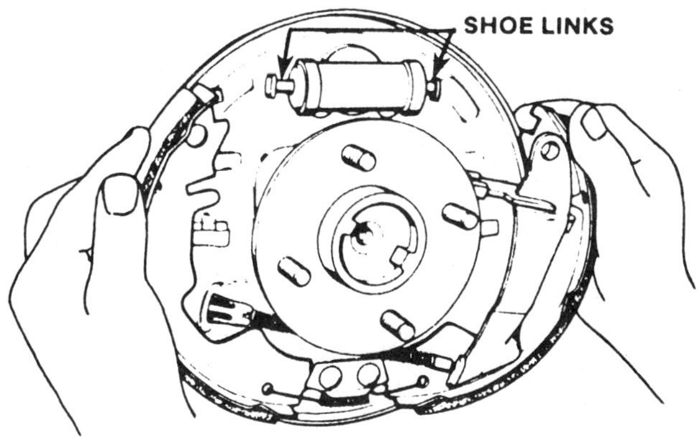

Figure 6-1

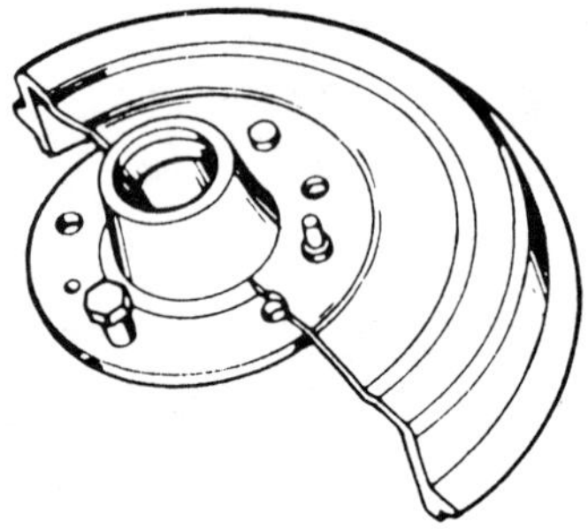

Figure 6-2

8. What is the most common method of brake drum mounting?
   a. swaged stud mounting
   b. rear axle shaft flange mounting
   c. riveted mounting
   d. tire rim and lug nut mounting

9. When disassembling an unfamiliar brake assembly, Mechanic A works on both front or both rear wheels at the same time to make sure each step is done identically on each side. Mechanic B works on one wheel at a time and uses the other wheel as a reference. Who is right?
   a. Mechanic A
   b. Mechanic B
   c. Both A and B
   d. Neither A nor B

10. What tool is used to measure a drum?
    a. drum micrometer
    b. drum gauge
    c. brake lathe
    d. depth gauge

JOB SHEET

SHOP ASSIGNMENT 17
PREPARE A DUO-SERVO DRUM BRAKE FOR INSPECTION AND SERVICING

NAME ______________ STATION ______________ DATE ______________

Tools and Materials

Screwdriver assortment
Brake spring tool
Wheel cylinder clamps (if applicable)
Wood block
Pliers
Socket wrench assortment

Protective Clothing

Respirator
Safety goggles or glasses with side shields

Procedure

1. Disconnect the cable from the parking brake lever.

Task completed ____

2. Back off the adjusting screw or release the shoe adjusting cams to provide ample lining-to-drum clearance.

Task completed ____

   a. Is the lining dragging on the drum?

Yes ____ No ____

   b. If yes, back off the brake adjustment by rotating the adjustment screw.

Task completed ____

3. If necessary, install wheel cylinder clamps on the wheel cylinders to prevent fluid or air leakage into the system.

Not applicable ____ Task completed ____

4. Note the location, position, and color of all brake shoe return springs so that they can be reinstalled later easily.

Task completed ____

   a. Use a brake spring tool to remove the shoe return springs.

Task completed ____

5. Block up the brake pedal with a wood block so that it cannot be depressed.

Task completed ____

6. Remove the shoe retaining or hold-down cups and spring.

Task completed ____

7. Remove the self-adjuster parts, including the actuating link, lever and pivot assembly, sleeve (through the lever), and return spring.

Not applicable ____ Task completed ____

8. On rear brakes, spread the shoes slightly to free the parking brake strut.

Not applicable ____ Task completed ____

   a. Remove the strut with its spring.

   Task completed ____

   b. Disconnect the parking brake lever from the secondary shoe.

   Task completed ____

9. Slip the anchor plate off the pin.

Task completed ____

   a. Spread the anchor ends of the shoes and disengage them from the wheel cylinder links, if used.

   Not applicable ____ Task completed ____

10. Remove the shoes connected at the bottom as an assembly by adjusting the screw and spring.

Task completed ____

11. Overlap the anchor end of the shoes to relieve spring tension.

Task completed ____

   a. Unhook the adjusting screw spring.

   Task completed ____

   b. Remove the adjusting screw assembly.

   Task completed ____

PROBLEMS ENCOUNTERED: ____

____

____

INSTRUCTOR'S COMMENTS: ____

____

____

JOB SHEET

SHOP ASSIGNMENT 18
REPLACE A WHEEL CYLINDER

NAME ____________ STATION ____________ DATE ____________

Tools and Materials

Socket wrench assortment
Tubing wrench
Small picks (if applicable)
Cleaning solvent
Torque wrench
Screwdriver assortment

Protective Clothing

Respirator
Safety goggles or glasses with side shields

Procedure

1. Remove the brake shoe assemblies from the backing plate.

   Task completed ____

2. On front axles, remove the small, U-shaped lock clip that holds the brake hose to the chassis bracket.

   Not applicable ____ Task completed ____

3. Disconnect the hydraulic hose from the steel line located on the chassis.

   Task completed ____

   a. On solid rear axles, use the appropriate tubing wrench to disconnect the hydraulic line where it enters the wheel cylinder.

   Not applicable ____ Task completed ____

4. Remove the plates, shims, and bolts holding the wheel cylinder to the backing plate.

   Task completed ____

   a. On some late models, the wheel cylinder is attached with a retaining ring that can be removed with two small picks.

   Not applicable ____ Task completed ____

5. Remove the wheel cylinder.

   Task completed ____

   a. Clean the area with a good cleaning solvent.

   Task completed ____

6. On front brakes, install a new hose.

Not applicable ____ Task completed ____

   a. Pass the hose through the hole in the backing plate.

   Not applicable ____ Task completed ____

   b. Reinstall the anchor bolts or retaining ring.

   Not applicable ____ Task completed ____

7. Connect the hose to the steel tubing, starting the tubing nut into the hose by hand to avoid stripping the threads.

Task completed ____

   a. Tighten with sufficient torque to prevent leakage.

   Task completed ____

8. Replace the small horseshoe lock that clamps the hose to the chassis bracket.

Task completed ____

9. Replace the brake shoes.

Task completed ____

10. Bleed the system.

Task completed ____

PROBLEMS ENCOUNTERED: ______________________________

____

____

INSTRUCTOR'S COMMENTS: ______________________________

____

____

CHAPTER 7

DISC BRAKES—OPERATION

PRACTICE QUESTIONS

1. Heavier automobiles, high-performance cars, and trucks generally have what type of rotor?
   a. ventilated rotor
   b. lightweight rotor
   c. solid rotor
   d. heavyweight rotor

2. When servicing a rotor, Mechanic A always replaces the splash shield. When servicing a rotor, Mechanic B replaces the splash shield only when it has been damaged or when the spindle is replaced. Who is right?
   a. Mechanic A
   b. Mechanic B
   c. Both A and B
   d. Neither A nor B

3. What type of caliper mounting hardware is pictured in Figure 7-1?
   a. support key assembly
   b. bushings
   c. guide pins
   d. anti-rattle or support clips

4. What type of caliper mounting hardware is used in conjunction with guide bolts to cushion caliper movement and eliminate brake noise?
   a. support key assembly
   b. bushings
   c. guide pins
   d. anti-rattle or support clips

5. When replacing disc brake pads, Mechanic A replaces semi-metallic pads with asbestos because it has a higher coefficient of friction. Mechanic B replaces disc brake pads with pads made of the same material as previously used. Who is right?
   a. Mechanic A
   b. Mechanic B
   c. Both A and B
   d. Neither A nor B

6. Two-piston or three-piston fixed caliper brakes are used primarily on ____________.
   a. domestic trucks
   b. import vehicles
   c. both a and b
   d. neither a nor b

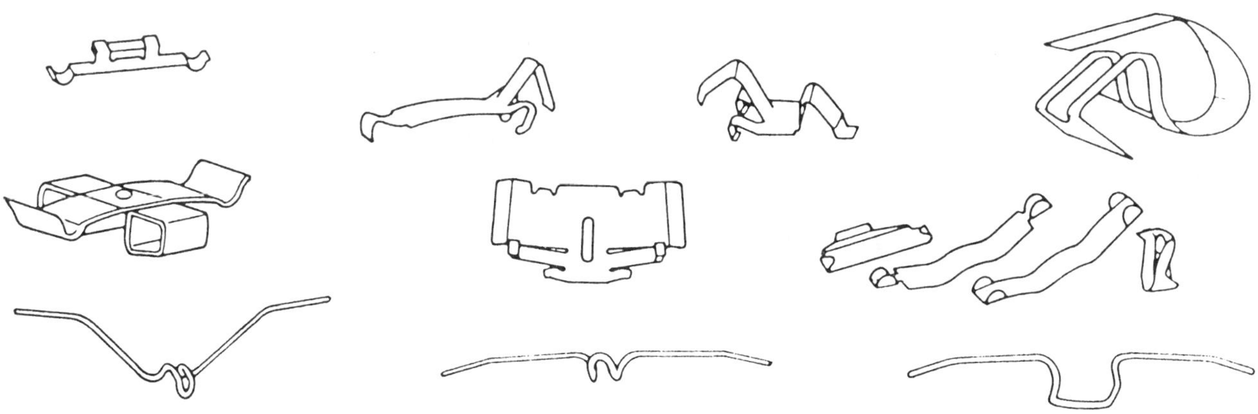

Figure 7-1

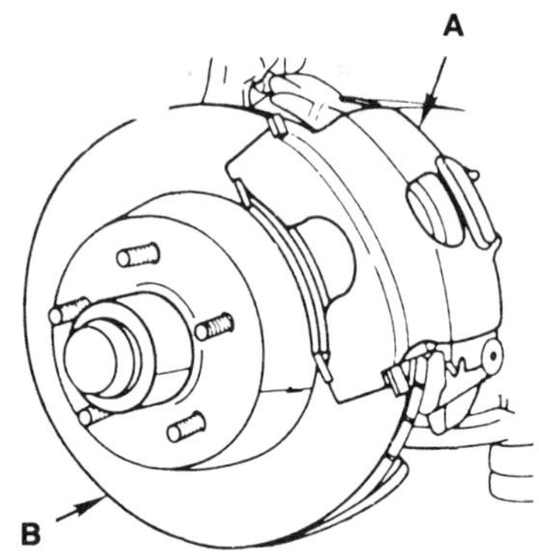

Figure 7-2

7. What component of a sliding caliper disc brake is indicated by letter A in Figure 7-2?
   a. caliper housing
   b. splash shield
   c. anchor plate
   d. rotor

8. What component of a sliding caliper disc brake is indicated by letter B in Figure 7-2?
   a. caliper housing
   b. splash shield
   c. anchor plate
   d. rotor

9. When servicing pin-slider disc brakes, Mechanic A does not remove the anchor plate. Mechanic B always removes the anchor plate. Who is right?
   a. Mechanic A
   b. Mechanic B
   c. Both A and B
   d. Neither A nor B

10. What component is included on rear disc brakes that is not included on front disc brakes?
    a. splash shield
    b. parking brake mechanism
    c. automatic adjusters
    d. drum brake

SHOP ASSIGNMENT 19

Examine a caliper and locate the following components: anchor plate, caliper housing, piston, dust boot, seal, inner shoe and lining, outer shoe

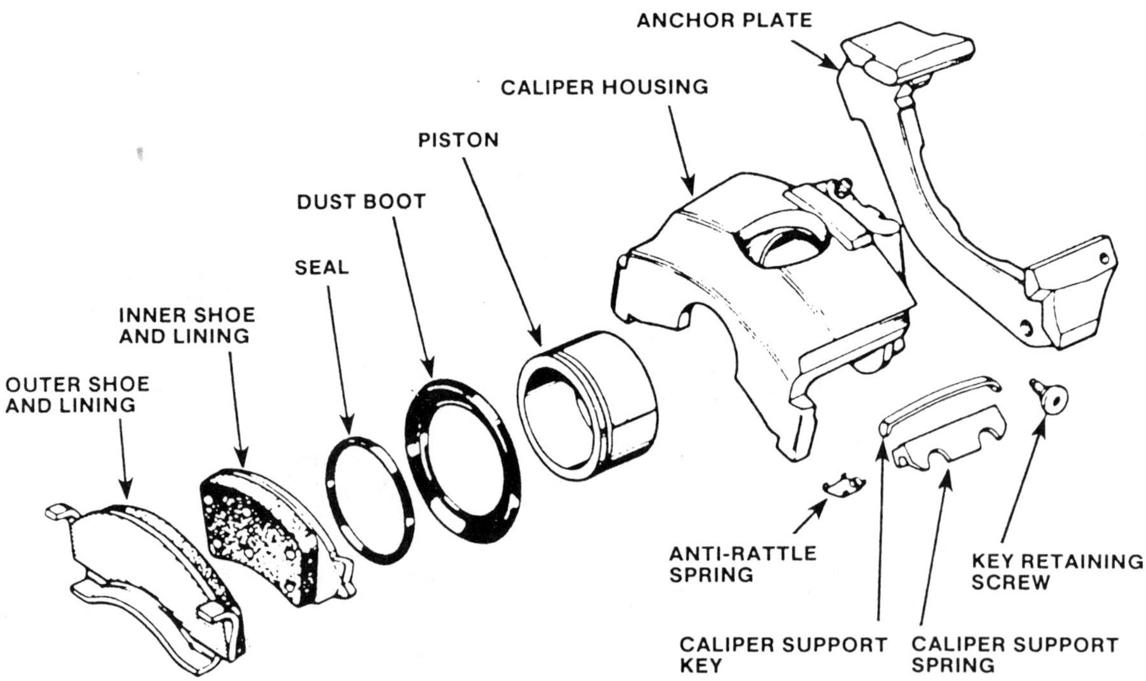

Figure 7-3

and lining, anti-rattle spring, caliper support key, caliper support spring, and key retaining screw. Refer to Figure 7-3 as needed. Wear a respirator and safety goggles or glasses with side shields.

SHOP ASSIGNMENT 20

Examine rear disc brakes and locate the following components: boot, piston, seal, housing, and bleeder screw. Examine the parking brake mechanism and locate the following components: thrust screw, anti-rotation pin, balls, operating shaft, thrust washer, end retainer, operating lever, and seals. Refer to Figure 7-4 as needed. Wear a respirator and safety goggles or glasses with side shields.

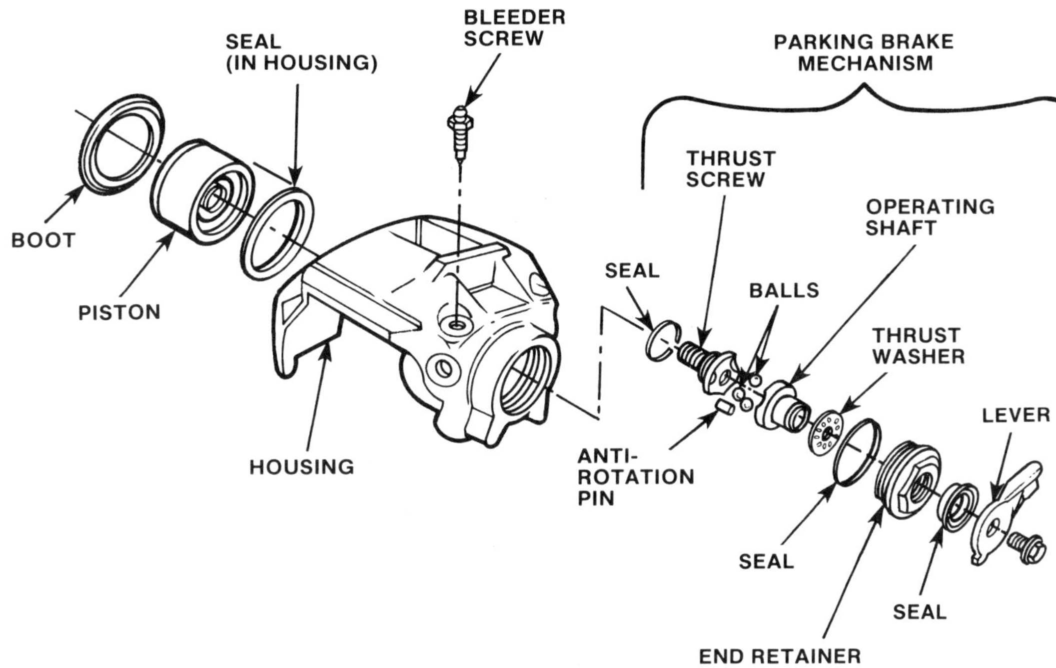

Figure 7-4

CHAPTER 8

DISC BRAKES—INSPECTING AND SERVICING

PRACTICE QUESTIONS

1. What is the first task that should be done when servicing a disc brake system?
   a. Secure the vehicle on stands or a hoist.
   b. Road test the vehicle.
   c. Disconnect the battery ground cable.
   d. Check brake fluid level.

2. Before beginning brake work, Mechanic A fills the reservoir(s) with brake fluid. Before beginning brake work, Mechanic B removes two-thirds of the fluid. Who is right?
   a. Mechanic A
   b. Mechanic B
   c. Both A and B
   d. Neither A nor B

3. Mechanic A replaces disc brake pads if they measure 1/32-inch thick or less at the thinnest part. Mechanic B replaces pads if they measure 1/8-inch thick or less at the thinnest part. Who is right?
   a. Mechanic A
   b. Mechanic B
   c. Both A and B
   d. Neither A nor B

4. Mechanic A replaces all pistons with even a partial crack. Mechanic B replaces only those pistons with cracks that go all the way across. Who is right?
   a. Mechanic A
   b. Mechanic B
   c. Both A and B
   d. Neither A nor B

5. What component is the mechanic removing in Figure 8-1?
   a. dust boot
   b. caliper
   c. brake lining
   d. piston seal

6. What tool can be used to check piston-to-bore clearance?
   a. screwdriver
   b. wooden or plastic stick
   c. feeler gauge
   d. micrometer

7. When applying sealer between the dust boot ring and the caliper housing, Mechanic A also applies sealer between the piston and the cylinder. Mechanic B makes sure no sealer gets between the piston and cylinder. Who is right?
   a. Mechanic A
   b. Mechanic B
   c. Both A and B
   d. Neither A nor B

8. Overtorquing or uneven torquing the lug nuts on the wheel with an impact wrench can cause ______________.
   a. warpage
   b. lateral runout
   c. overheating
   d. scoring

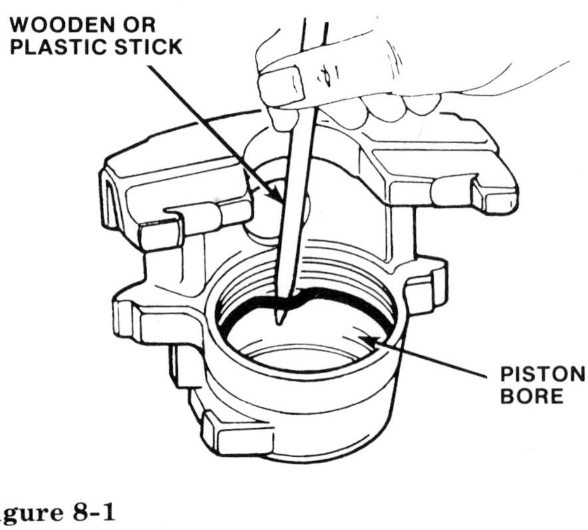

Figure 8-1

9. Bluing and/or glazing on the lining surface are signs of ______________.
   a. warpage
   b. lateral runout
   c. overheating
   d. scoring

10. When servicing disc brakes, Mechanic A cleans and repacks the wheel bearings before reinstalling the rotor. Mechanic B reinstalls the rotor before cleaning and repacking the wheel bearings. Who is right?
    a. Mechanic A
    b. Mechanic B
    c. Both A and B
    d. Neither A nor B

JOB SHEET

SHOP ASSIGNMENT 21
REMOVE A TYPICAL HUB AND ROTOR ASSEMBLY

NAME ______ **STATION** ______ **DATE** ______

Tools and Materials

Wrenches
Wire
Screwdrivers

Protective Clothing

Safety goggles or glasses with side shields

Procedure

1. Remove the wheel and tire assembly.

   Task completed ____

2. Remove the caliper assembly. (Do not disconnect the brake hose unless the caliper is to be removed for servicing.)

   Task completed ____

   a. Does the caliper require servicing?

   Yes ____ No ____

   b. If no, suspend the caliper from a wire hook or loop on the front suspension.

   Not applicable ____ Task completed ____

   c. If yes, disconnect the brake hose.

   Task completed ____

3. Is the rotor integral with the wheel bearing hub?

   Yes ____ No ____

   a. If no, go to step 4.

   Not applicable ____ Task completed ____

   b. If yes, remove the grease cap from the hub.

   Task completed ____

   c. Remove the cotter pin.

   Task completed ____

   d. Remove the adjusting nut.

   Task completed ____

e. Remove the thrust washer from the spindle.

Task completed ____

4. Remove the outer wheel bearing cone.

Task completed ____

a. Remove the roller assembly from the hub.

Task completed ____

5. Remove the hub and rotor assembly from the spindle.

Task completed ____

PROBLEMS ENCOUNTERED: ______________________________

____

____

INSTRUCTOR'S COMMENTS: ______________________________

____

____

JOB SHEET

SHOP ASSIGNMENT 22
INSTALL A TYPICAL HUB AND ROTOR ASSEMBLY

NAME ______________ STATION ______________ DATE ______________

Tools and Materials

Grease seal
Socket wrench assortment
Torque wrench

Protective Clothing

Safety goggles or glasses with side shields

Procedure

1. Install a new grease seal.

   Task completed ____

2. Place the hub and rotor assembly over the spindle.

   Task completed ____

   a. Push it in as far as it will go.

   Task completed ____

3. Install the outer bearing over the spindle.

   Task completed ____

   a. Install the thrust washer over the spindle.

   Task completed ____

   b. Install the nut over the spindle.

   Task completed ____

4. Tighten the bearing adjustment nut while rotating the hub.

   Task completed ____

   a. Back it off and adjust it to the proper torque as specified in the vehicle service manual.

   Task completed ____

5. Install the lock washer.

   Task completed ____

   a. Install the cotter pin.

   Task completed ____

   b. Install the grease cup.

   Task completed ____

6. Install the caliper over the rotor.

Task completed ____

7. Install the wheel.

Task completed ____

a. Replace the tire.

Task completed ____

8. Road test the vehicle.

Task completed ____

PROBLEMS ENCOUNTERED: ____

____

____

INSTRUCTOR'S COMMENTS: ____

____

____

JOB SHEET

SHOP ASSIGNMENT 23
SERVICE A SINGLE-PISTON SLIDING CALIPER

NAME ____________ **STATION** ____________ **DATE** ____________

Tools and Materials

Appropriate screwdrivers
Wire brush
C-clamp
Wood block

Protective Clothing

Safety goggles or glasses with side shields

Procedure

1. Remove the screw holding each of the anti-rattle springs and caliper retaining clips.

   Task completed ____

2. Rotate and lift the caliper from the anchor plate.

   Task completed ____

3. Remove the outer shoe from the caliper housing.

   Task completed ____

   a. Remove the inner shoe from the anchor plate.

   Task completed ____

4. Check for leaks.

   Task completed ____

5. Overhaul the caliper.

   Task completed ____

6. Clean the machined guides with a wire brush.

   Task completed ____

7. Seat the piston using a C-clamp and wood block.

   Task completed ____

8. Position the new inner pad on the anchor plate, placing the ears of the shoe on the machined abutments.

   Task completed ____

9. Install the new outer pad.

Task completed ____

a. Do the metal flanges fit the caliper snugly?

Yes ____ No ____

b. If no, correct.

Task completed ____

10. Bend the flanges if necessary.

Not applicable ____ Task completed ____

a. Snap them into place by hand or with a C-clamp.

Task completed ____

PROBLEMS ENCOUNTERED: ______________________________

____

____

INSTRUCTOR'S COMMENTS: ______________________________

____

____

JOB SHEET

SHOP ASSIGNMENT 24
DO A COMPLETE CALIPER OVERHAUL ON A DELCO-MORAINE FOUR-PISTON FIXED CALIPER

NAME ______________________ STATION ______________________ DATE ______________

Tools and Materials

Appropriate screwdrivers
Brake fluid
Piston ring compressor
Silastic sealant
Boot seal installer
O-rings
Torque wrench

Protective Clothing

Safety goggles or glasses with side shields

Procedure

1. If working on a front caliper, remove the brake hose.

   Not applicable ____ Task completed ____

   a. If working on a rear caliper, remove the brake line.

   Not applicable ____ Task completed ____

2. Remove the two caliper mounting bolts.

   Task completed ____

   a. Remove the caliper from the mounting bracket.

   Task completed ____

3. Remove the shoe retaining pin.

   Task completed ____

   a. Remove the brake shoes.

   Task completed ____

4. Remove the two caliper bolts.

   Task completed ____

   a. Separate the caliper halves.

   Task completed ____

5. Remove the O-rings from around the fluid transfer hose.

   Task completed ____

6. Push the pistons down in the bore, using them as a fulcrum while prying out the dust boots with a screwdriver.

Task completed ____

7. Remove the pistons.

Task completed ____

   a. Remove the piston springs.

Task completed ____

8. Remove the boots and seals from the pistons.

Task completed ____

9. Service the caliper bores.

Task completed ____

   a. Clean all components.

Task completed ____

10. Install the seal in the groove closest to the flat end of the piston with the lip on the seal facing the large end of the piston.

Task completed ____

11. Place the piston spring in the bottom of the bore.

Task completed ____

12. Lubricate the piston seal with brake fluid.

Task completed ____

   a. Install the piston in its bore using a piston ring compressor to protect the seal as it goes in.

Task completed ____

13. Place a bead of silastic sealant in the boot groove.

Task completed ____

   a. Assemble the boot in the groove.

Task completed ____

14. Is the piston movement in the bore satisfactory?

Yes ____ No ____

   a. If no, check the location of the piston seal and piston spring.

Task completed ____

15. Using a boot installer, seat the steel boot retaining ring in its counterbore.

Task completed ____

16. Depress the piston to place a bead of silastic sealant between the boot retainer ring and the housing.

Task completed ____

17. Perform the above procedure for each piston in both caliper halves.

Task completed ____

18. Place new O-rings in the grooves around the fluid transfer holes in the outboard caliper.

Task completed ____

19. Dip the caliper bolts in clean brake fluid.

Task completed ____

a. Fit the caliper halves together.

Task completed ____

20. Install the caliper bolts.

Task completed ____

a. Torque them to 130 foot-pounds for front calipers.

Not applicable ____ Task completed ____

b. Torque them to 60 foot-pounds for rear calipers.

Not applicable ____ Task completed ____

21. Position the caliper over the rotor.

Task completed ____

a. Depress the pistons using two screwdrivers.

Task completed ____

22. Install the two caliper mounting bolts.

Task completed ____

a. Torque them to 70 foot-pounds.

Task completed ____

23. Install the shoes.

Task completed ____

a. Install the brake hose.

Task completed ____

b. Install the brake line.

Task completed ____

24. Bleed the brakes.

Task completed ____

25. Test the brake pedal.

Task completed ____

a. Road test the vehicle.

Task completed ____

PROBLEMS ENCOUNTERED: ____

____

____

INSTRUCTOR'S COMMENTS: ____

____

____

CHAPTER 9

PARKING BRAKES—OPERATION AND SERVICING

PRACTICE QUESTIONS

1. What component of a typical integral parking brake is indicated by the letter A in Figure 9-1?
   a. parking brake strut
   b. equalizer
   c. parking brake pedal
   d. parking brake cable

2. What component of a typical integral parking brake is indicated by the letter B in Figure 9-1?
   a. parking brake strut
   b. equalizer
   c. parking brake pedal
   d. parking brake cable

3. In order to remove the rear cable, Mechanic A first raises the vehicle and then removes the rear wheels. Mechanic B raises the vehicle and then detaches the retaining clip from the brake cable retainer bracket. Who is right?
   a. Mechanic A
   b. Mechanic B
   c. Both A and B
   d. Neither A nor B

4. Applying air pressure to the parking brake vacuum release system could damage ____________.
   a. the vacuum tester gauge
   b. the hoses
   c. the actuator diaphragm in the parking brake vacuum motor
   d. all of the above

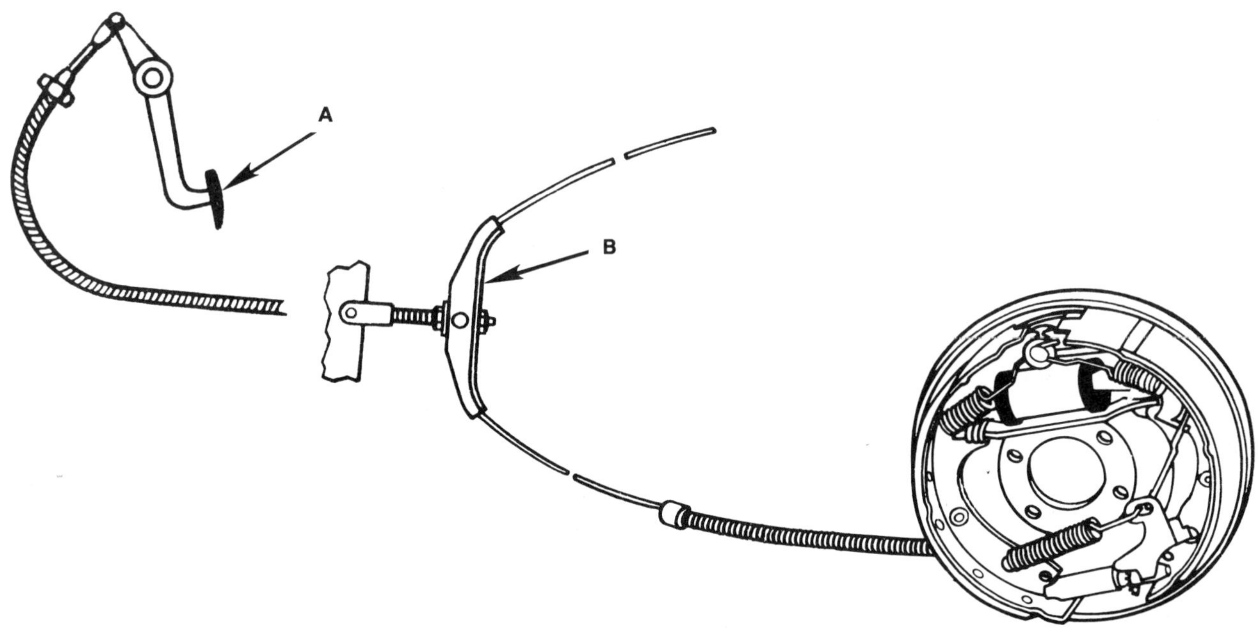

Figure 9-1

5. When testing a parking brake vacuum release system, what tool is used to check minimum pressure at all points where vacuum is applied?
   a. actuator diaphragm
   b. feeler gauge
   c. guide wire
   d. vacuum tester gauge

6. When servicing a disc/drum parking brake, Mechanic A discovers damaged wear shims and replaces them. Mechanic B discovers damaged wear shims and replaces the entire mounting plate. Who is right?
   a. Mechanic A
   b. Mechanic B
   c. Both A and B
   d. Neither A nor B

7. What component of an internal shoe parking brake is indicated by the letter A in Figure 9-2?
   a. shoe
   b. cable
   c. band
   d. adjusting screw

8. What component of an internal shoe parking brake is indicated by the letter B in Figure 9-2?
   a. shoe
   b. cable
   c. band
   d. adjusting screw

9. What component of an internal shoe parking brake is indicated by the letter C in Figure 9-2?
   a. shoe
   b. cable
   c. band
   d. adjusting screw

10. When testing parking brakes, Mechanic A parks the vehicle facing up on an incline of 30 degrees or less. When testing parking brakes, Mechanic B parks the vehicle facing down on an incline of 30 degrees or less. Who is right?
    a. Mechanic A
    b. Mechanic B
    c. Both A and B
    d. Neither A nor B

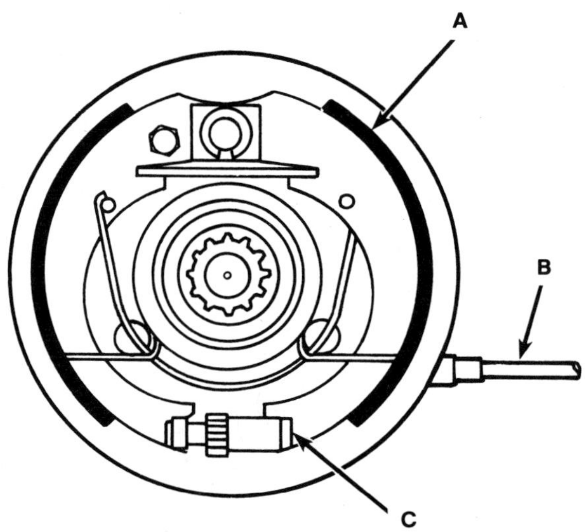

Figure 9-2

JOB SHEET

SHOP ASSIGNMENT 25
REMOVE A FRONT PARKING BRAKE CABLE

NAME ______ STATION ______ DATE ______

Tools and Materials

Appropriate screwdrivers
Wire, about 50 inches

Protective Clothing

Safety goggles or glasses with side shields

Procedure

1. Disengage the front parking brake cable from its connector.

   Task completed ____

2. Use a screwdriver to force the cable housing and attaching clip forward out of the body cross member.

   Task completed ____

3. Fold back the left front edge of the floor covering.

   Task completed ____

   a. Pry out either the rubber grommet or rubber cable cover from the floor pan.

   Task completed ____

   b. Remove the floor pan clip.

   Task completed ____

4. Engage the parking brake.

   Task completed ____

5. Work the brake cable up and out of the clevis linkage or guide bracket.

   Task completed ____

6. Use a screwdriver to force the cable assembly and clip down and out of the pedal assembly bracket.

   Task completed ____

   a. If necessary, attach a substantial wire about 50 inches long to the lower end of the cable housing to be used as a guide for cable installation.

   Not applicable ____ Task completed ____

7. Work the cable and housing assembly up through the floor pan.

Task completed ____

a. If used, disconnect the guide wire but leave in place for new front cable installation.

Not applicable ____ Task completed ____

PROBLEMS ENCOUNTERED: ____________________

____

____

INSTRUCTOR'S COMMENTS: ____________________

____

____

JOB SHEET

SHOP ASSIGNMENT 26
INSTALL A FRONT
PARKING BRAKE CABLE

NAME ______________ STATION ______________ DATE ______________

Tools and Materials

Appropriate screwdrivers
Appropriate wrenches

Protective Clothing

Safety goggles or glasses with side shields

Procedure

1. If applicable, pull the guide wire connected to the new cable through the floor pan.

   Not applicable ____ Task completed ____

   a. If not applicable, insert the new cable through the floor pan.

   Not applicable ____ Task completed ____

2. Insert the retainer into the hole in the bottom of the parking brake pedal assembly bracket.

   Task completed ____

3. Insert the cable into the hole.

   Task completed ____

   a. Place the cable end fitting into the clevis linkage or guide bracket.

   Task completed ____

4. Force the upper cable housing into the retainer until it is firmly seated against the pedal assembly bracket.

   Task completed ____

5. Insert the retainer into the hole in the cross member or bracket.

   Task completed ____

6. Insert the cable through the hole in the cross member.

   Task completed ____

   a. Force the cable housing into the retainer until it is firmly seated.

   Task completed ____

7. If applicable, push the cable seal into the hole in the dash panel with enough force to seat the locking tabs.

Not applicable ____ Task completed ____

   a. Otherwise, reinstall the rubber cable cover and floor pan clip.

Not applicable ____ Task completed ____

8. Attach the cable to the connector.

Task completed ____

9. Adjust the rear service brakes.

Task completed ____

   a. Adjust the parking brake.

Task completed ____

10. Test for free-wheel rotation by applying and releasing the parking brake several times.

Task completed ____

PROBLEMS ENCOUNTERED: ____

____

____

INSTRUCTOR'S COMMENTS: ____

____

____

JOB SHEET

SHOP ASSIGNMENT 27
REMOVE AND INSPECT A GIRLING-LOCKHEED DISC/DRUM PARKING BRAKE

NAME ____________ STATION ____________ DATE ____________

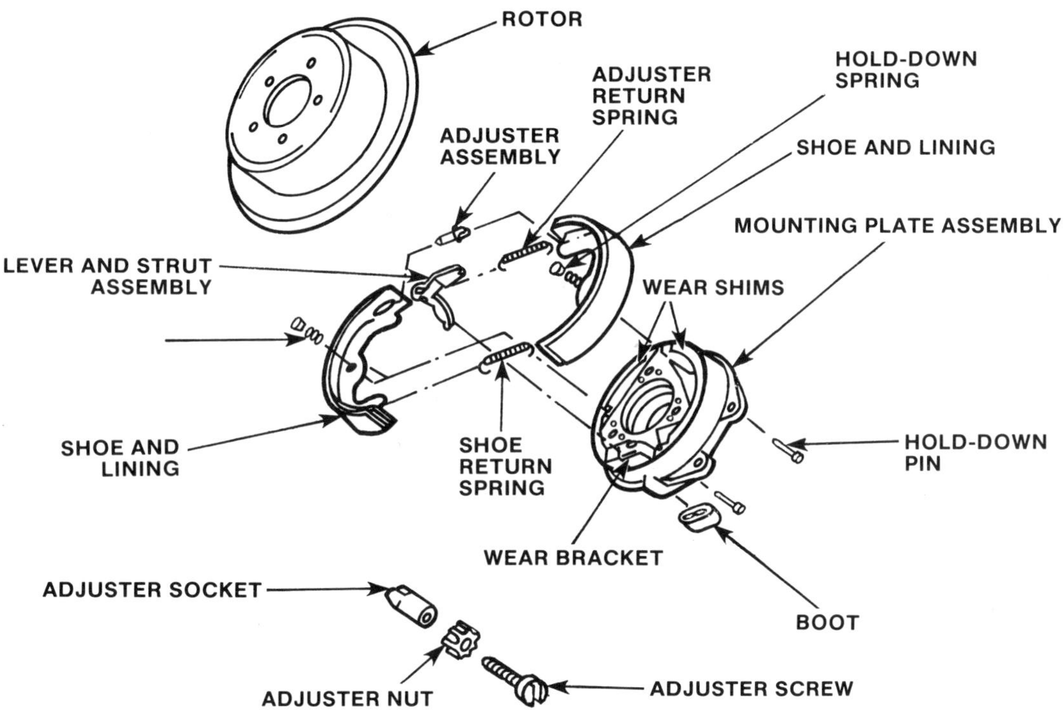

Figure 9-3

Tools and Materials

Socket wrench assortment
Solvent
Brush or filtered vacuum cleaner
Appropriate screwdrivers
Lubricating oil

Protective Clothing

Safety goggles or glasses with side shields

Procedure

1. Release the parking brake.

Task completed ____

   a. Remove the wheel and tire assembly.

Task completed ____

2. Remove the disc/drum combination.

Task completed ____

a. If necessary, turn the star wheel through the slot in the disc/drum hub to back off the adjustment of the shoes.

Not applicable ____ Task completed ____

3. Clean with solvent and a brush or an approved filtered vacuum cleaner.

Task completed ____

4. Spread the shoes.

Task completed ____

a. Remove the adjuster assembly.

Task completed ____

5. Remove the adjuster springs.

Task completed ____

6. Remove the hold-down springs by pushing down on the spring and sliding the pin over until it can be removed through the hole at the end of the slot.

Task completed ____

7. Remove the shoe return springs.

Task completed ____

a. Remove both shoes.

Task completed ____

8. Clean and inspect the adjuster assembly.

Task completed ____

a. Lubricate the threads.

Task completed ____

9. Clean and inspect the wear bracket.

Task completed ____

a. Clean and inspect the wear shims.

Task completed ____

b. Are the bracket and shims serviceable?

Yes ____ No ____

c. If no, replace the entire mounting plate assembly.

Not applicable ____ Task completed ____

10. Clean and inspect the rubber boot.

Task completed ____

a. Replace the boot, if necessary.

Not applicable ____ Task completed ____

11. Clean and inspect all remaining hardware.

Task completed ____

12. Clean all parts.

Task completed ____

a. Lubricate the wear bracket.

Task completed ____

b. Lubricate the wear shims.

Task completed ____

PROBLEMS ENCOUNTERED: ______________________________

____

____

INSTRUCTOR'S COMMENTS: ______________________________

____

____

CHAPTER 10

POWER BRAKES—OPERATION AND SERVICING

PRACTICE QUESTIONS

1. Mechanic A uses only specified hydraulic fluid in hydro-boost systems. Mechanic B uses a mixture of hydraulic fluid and clean brake fluid. Who is right?
   a. Mechanic A
   b. Mechanic B
   c. Both A and B
   d. Neither A nor B

2. A spongy pedal with power brakes indicates ______________.
   a. grease on the linings
   b. the presence of a leak
   c. air in the system
   d. a low fluid level

3. When servicing power brakes, Mechanic A checks the power brake circuit first. Mechanic B checks for faults in the basic hydraulic system first. Who is right?
   a. Mechanic A
   b. Mechanic B
   c. Both A and B
   d. Neither A nor B

4. What feature is designed into the power brake system to give the brake pedal "feel" to the operator?
   a. diaphragm
   b. power unit
   c. accumulator
   d. reaction control

5. Figure 10-1 shows brake system components in what position?
   a. unapplied
   b. partially applied
   c. applied
   d. released

6. When bleeding a pressure-multiplying type power unit with more than one bleeder screw, Mechanic A bleeds the one at the pressure cylinder first. Mechanic B bleeds the bleeder screw at the control valve first. Who is right?
   a. Mechanic A
   b. Mechanic B
   c. Both A and B
   d. Neither A nor B

7. What tool is used to check engine vacuum?
   a. feeler gauge
   b. vacuum gauge
   c. vacuum pump
   d. vacuum tester

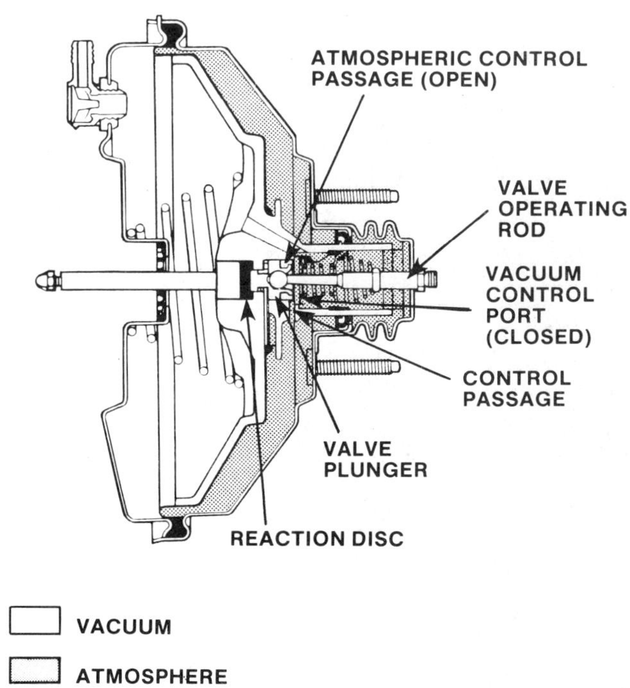

Figure 10-1

8. A pushrod that is too long will cause ______________.
   a. excessive brake pedal travel
   b. groaning noises when the brake is applied
   c. excessive servicing
   d. the master cylinder to close off the replenishing port

9. What hydro-boost component is indicated by the letter A in Figure 10-2?
   a. accumulator
   b. power steering gear
   c. master cylinder
   d. power steering pump

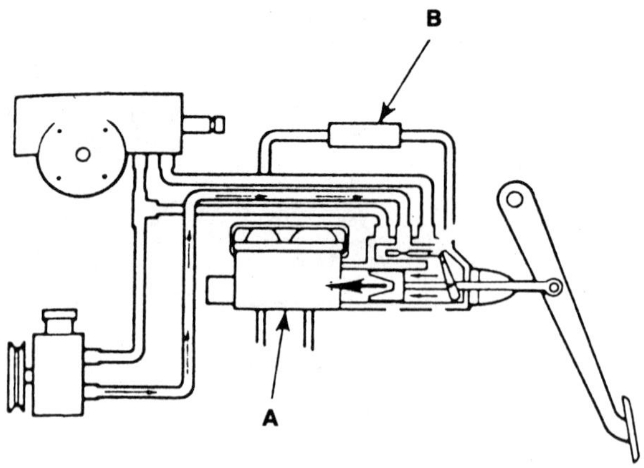

Figure 10-2

10. What hydro-boost component is indicated by the letter B in Figure 10-2?
    a. accumulator
    b. power steering gear
    c. master cylinder
    d. power steering pump

JOB SHEET

SHOP ASSIGNMENT 28
CHECK A VACUUM SYSTEM FOR LEAKS

NAME ____________ STATION ____________ DATE ____________

Tools and Materials

Vacuum gauge

Protective Clothing

Safety goggles or glasses with side shields

Procedure

1. Turn the engine on and let it idle.

   Task completed ____

2. Attach a vacuum gauge to an intake manifold port.

   Task completed ____

   a. Is the reading low?

   Yes ____ No ____

   b. If yes, write "engine problems" in the section for Problems Encountered below.

   Not applicable ____ Task completed ____

3. Disconnect the vacuum tube or hose running from the intake manifold to the booster unit, placing a thumb over it quickly so the engine does not stall.

   Task completed ____

4. Is a strong vacuum present at the end of the tube or hose?

   Yes ____ No ____

   a. If yes, move to the bottom of page.

   Task completed ____

   b. If no, shut off the engine.

   Task completed ____

   c. Is the tube or hose collapsed, crimped, or clogged?

   Yes ____ No ____

   d. If yes, replace the tube or hose if necessary.

   Not applicable ____ Task completed ____

PROBLEMS ENCOUNTERED: ____

____

____

INSTRUCTOR'S COMMENTS: ____

____

____

JOB SHEET

SHOP ASSIGNMENT 29
REMOVE A SINGLE-DIAPHRAGM OR TANDEM-DIAPHRAGM BOOSTER

NAME ______________ STATION ______________ DATE ______________

Tools and Materials

Appropriate wrenches

Protective Clothing

Safety goggles or glasses with side shields

Procedure

1. Remove the nuts attaching the master cylinder to the power brake unit.

   Task completed ____

2. Carefully slide the master cylinder off the mounting studs.

   Task completed ____

   a. Rest it against the fender shield.

   Task completed ____

3. Disconnect the vacuum hose from the power brake unit.

   Task completed ____

PROBLEMS ENCOUNTERED: ______________________________

____

____

INSTRUCTOR'S COMMENTS: ______________________________

____

____

JOB SHEET

SHOP ASSIGNMENT 30
INSTALL A SINGLE-DIAPHRAGM OR TANDEM-DIAPHRAGM BOOSTER

NAME ______________ STATION ______________ DATE ______________

Tools and Materials

Appropriate wrenches
Torque wrench
Lubricating oil
Retainer clip

Protective Clothing

Safety goggles or glasses with side shields

Procedure

1. Rotate the linkage as needed to position the power brake booster in the dash panel.

   Task completed ____

2. Install the mounting nuts and washers.

   Task completed ____

   a. Tighten to specifications.

   Task completed ____

3. Position the master cylinder on the power brake unit.

   Task completed ____

   a. Install the mounting nuts.

   Task completed ____

   b. Tighten to specifications.

   Task completed ____

4. Attach the vacuum hose.

   Task completed ____

5. Coat the bearing surface of the pedal pin with a lubricant.

   Task completed ____

6. Connect the linkage to the pedal pin.

   Task completed ____

   a. Install a new retainer clip.

   Task completed ____

7. Coat the bushings and sleeve with lubricant.

Task completed ____

a. Install them in the linkage.

Task completed ____

8. Position the linkage.

Task completed ____

a. Install the lower pivot bolt.

Task completed ____

b. Tighten to specifications.

Task completed ____

PROBLEMS ENCOUNTERED: ________________________________

____

____

INSTRUCTOR'S COMMENTS: ________________________________

____

____

JOB SHEET

SHOP ASSIGNMENT 31
DISASSEMBLE A TYPICAL
HYDRO-BOOST UNIT

NAME ______________________ **STATION** ______________________ **DATE** ______________

Tools and Materials

Vise
Small punch
Appropriate wrenches
Appropriate screwdrivers
Drip pan

Protective Clothing

Safety goggles or glasses with side shields

Procedure

1. Secure the hydro-boost in a vise with the master cylinder end up.

   Task completed ____

2. Depress the spool valve plug.

   Task completed ____

   a. Use a small punch to unseat the retainer.

   Task completed ____

   b. Remove and discard the retainer.

   Task completed ____

3. Remove the spool valve plug.

   Task completed ____

   a. Remove and discard the O-ring.

   Task completed ____

4. Remove the spool valve sleeve.

   Not applicable ____ Task completed ____

   a. Remove the spool valve return spring.

   Task completed ____

5. Remove and discard the spring retainer from the power piston bore.

   Not applicable ____ Task completed ____

6. Remove the return spring from the power piston bore.

   Task completed ____

7. Unscrew the return line fitting.

Task completed ____

a. Discard the O-ring.

Task completed ____

8. Remove the housing-to-cover bolts.

Task completed ____

a. Holding the unit over a drip pan, separate the cover from the housing.

Task completed ____

b. Allow the unit to drain.

Task completed ____

9. Remove and discard the housing-to-cover seal.

Task completed ____

10. Detach the spool valve from the ratio lever.

Task completed ____

11. Remove the input rod from the cover.

Task completed ____

a. Remove and discard the seals.

Task completed ____

12. Remove the power piston/accumulator assembly.

Task completed ____

13. Remove and discard the power piston seal from the bore.

Task completed ____

14. Remove and discard the tube seat from the hose ports.

Task completed ____

PROBLEMS ENCOUNTERED: ____

____

____

INSTRUCTOR'S COMMENTS: ____

____

____

JOB SHEET

SHOP ASSIGNMENT 32
REASSEMBLE A TYPICAL HYDRO-BOOST II TYPE UNIT

NAME ________________ STATION ________________ DATE ________

Tools and Materials

Clean power steering fluid
Tube nut and tube seats
Plugs
O-rings
Torque wrench and other wrenches, as needed
Special tool for seal installation
New seals

Protective Clothing

Safety goggles or glasses with side shields

Procedure

1. Lubricate all parts and bores with clean power steering fluid.

   Task completed ____

2. Does the unit use tube seats?

   Yes ____ No ____

   a. If not, go to step 2h.

      Not applicable ____ Task completed ____

   b. If yes, set the new seats in position in the ports.

      Task completed ____

   c. Use a spare tube nut to reseat the new tube seat.

      Task completed ____

   d. Remove the tube nut.

      Task completed ____

   e. Inspect the seat for metal flakes or foreign material.

      Task completed ____

   f. Plug the ports until it is time to connect the lines.

      Task completed ____

   g. Go to step 3.

      Task completed ____

   h. Replace the O-rings on the hose lines.

      Task completed ____

3. Place the new O-ring on the return hose fitting.

Not applicable ____ Task completed ____

a. Install it into the housing.

Task completed ____

b. Torque to 7 foot-pounds.

Task completed ____

4. Use the special tool shown in Figure 10-3 to install the new seal or seals onto the input rod.

Task completed ____

a. Insert the rod and seals into the cover.

Task completed ____

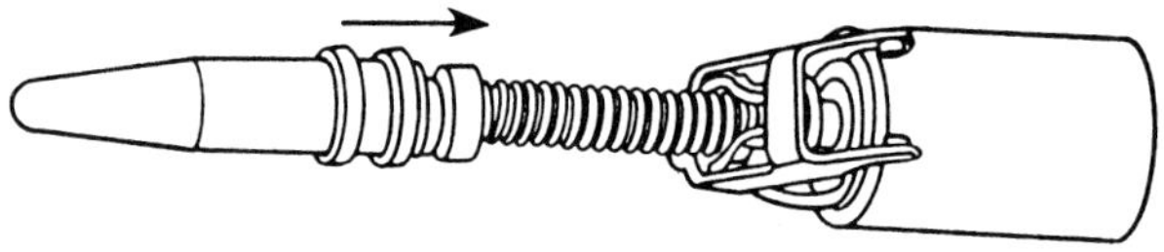

Figure 10-3

5. Install the new power piston seal.

Task completed ____

6. Install the new housing-to-cover seal.

Task completed ____

7. Align the power piston into the bore.

Task completed ____

a. Attach the spool valve to the ratio lever.

Task completed ____

b. Install the power piston/spool valve into place.

Task completed ____

8. Install the housing-to-cover bolts.

Task completed ____

a. Torque to 20 foot-pounds.

Task completed ____

9. Install the spool valve sleeve.

Not applicable ____ Task completed ____

a. Install the spool valve return spring.

Task completed ____

10. Place the new O-ring on the spool plug and install.

Task completed ____

a. Depress the plug.

Task completed ____

b. Insert the new retainer.

Task completed ____

11. Insert the power piston return spring.

Task completed ____

a. Is the spring seated on the ledge or bore?

Yes ____ No ____

b. If no, correct.

Not applicable ____ Task completed ____

12. Mount the hydro-boost on the vehicle.

Task completed ____

13. Connect the hydraulic hoses.

Task completed ____

a. Torque to 20 foot-pounds.

Task completed ____

14. Install the master cylinder.

Task completed ____

PROBLEMS ENCOUNTERED: ____

____

____

INSTRUCTOR'S COMMENTS: ____

____

____

JOB SHEET

SHOP ASSIGNMENT 33
REMOVE, INSPECT, AND INSTALL AN ALTERNATOR-DRIVEN VACUUM PUMP

NAME ______ STATION ______ DATE ______

Tools and Materials

Appropriate screwdrivers
Appropriate wrenches
Feeler gauge
Engine oil

Protective Clothing

Safety goggles or glasses with side shields

Procedure

1. Disconnect the oil lines from the vacuum pump.

Task completed ____

   a. Disconnect the vacuum hose from the pump.

Task completed ____

2. Remove the vacuum pump attaching bolts.

Task completed ____

3. While holding the center plate, remove the vacuum pump from the rotor shaft by turning in the direction of the shaft.

Task completed ____

4. Carefully remove the vanes from the pump housing.

Task completed ____

5. Measure the length of the vanes.

Task completed ____

   a. Do they meet manufacturer's specifications?

Yes ____ No ____

   b. If no, replace them.

Not applicable ____ Task completed ____

6. Measure the inside diameter of the housing.

Task completed ____

   a. Does it meet manufacturer's specifications?

Yes ____ No ____

b. If no, replace the vacuum pump.

Not applicable ____ Task completed ____

7. Apply light pressure to the check valve with a screwdriver.

Task completed ____

a. Does it operate smoothly?

Yes ____ No ____

b. If no, replace it.

Task completed ____

8. Is oil leakage evident on the inner face of the rear housing cover?

Yes ____ No ____

a. Does the inner face of the oil seal show signs of damage or wear?

Yes ____ No ____

b. If yes to either question, replace the oil seal.

Not applicable ____ Task completed ____

9. Position the rotor on the center plate housing with the serrated boss turned up.

Task completed ____

10. Align the holes in the center plate and the rotor.

Task completed ____

11. Install the vanes in the rotor with the chamfered side of the vanes turned outward.

Task completed ____

12. Install the housing.

Task completed ____

a. Does the O-ring protrude past the slot in the center?

Yes ____ No ____

b. If yes, correct.

Task completed ____

13. Install the vacuum pump on the alternator.

Task completed ____

a. Secure it with mounting bolts.

Task completed ____

14. Connect the oil lines to the pump.

Task completed ____

a. Connect the vacuum hose to the pump.

Task completed ____

15. Add 5 cubic centimeters of clean engine oil to the filler port on the pump.

Task completed ____

16. Does the alternator pulley turn freely by hand?

Yes ____ No ____

a. If no, correct.

Not applicable ____ Task completed ____

PROBLEMS ENCOUNTERED: ____

____

____

INSTRUCTOR'S COMMENTS: ____

____

____

CHAPTER 11

ANTI-LOCK BRAKE SYSTEMS—OPERATION AND SERVICING

PRACTICE QUESTIONS

1. The Bosch and Teves systems keep track of wheel deceleration rates with ____________.
   a. solenoid valves
   b. magnetic sensors
   c. the modulator
   d. a vacuum

2. When servicing the brakes on a vehicle with ABS, Mechanic A does not service ABS components, but replaces them as a complete unit. Mechanic B services ABS components in the field. Who is right?
   a. Mechanic A
   b. Mechanic B
   c. Both A and B
   d. Neither A nor B

3. When the engine of a particular vehicle is first started, both the ANTI-LOCK and the BRAKE warning lights stay on for about 20 seconds. Mechanic A looks for problems in the hydraulic brake system. Mechanic B looks for problems in the ABS system. Who is right?
   a. Mechanic A
   b. Mechanic B
   c. Both A and B
   d. Neither A nor B

4. Mechanic A bleeds the front brakes of an anti-lock brake system in the conventional manner with the accumulator fully charged. Mechanic B bleeds the front brakes of an anti-lock brake system in the conventional manner without the accumulator being charged. Who is right?
   a. Mechanic A
   b. Mechanic B
   c. Both A and B
   d. Neither A nor B

5. When beginning to diagnose an anti-lock brake system malfunction, Mechanic A refers to component test charts first. Mechanic B conducts a prediagnosis inspection first. Who is right?
   a. Mechanic A
   b. Mechanic B
   c. Both A and B
   d. Neither A nor B

6. What tool used for checking ABS systems is pictured in Figure 11-1?
   a. volt-ohmmeter
   b. anti-lock harness adapter
   c. breakout box
   d. pressure gauge

7. In the Bosch anti-lock system, where is the lateral acceleration switch located?
   a. in the storage compartment
   b. at each wheel
   c. under the rear seats
   d. in the master cylinder

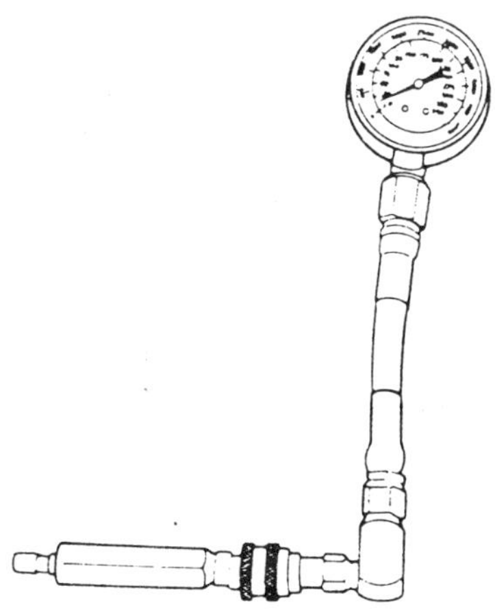

Figure 11-1

8. What two important components in the Bosch and Teves ABS systems are not used in the Lucas-Girling stop control system?
   a. computer and master cylinder
   b. wheel sensors and pressure valves
   c. computer and wheel sensors
   d. vacuum booster and solenoid valve

9. When servicing ABS brakes, Mechanic A removes wheel sensors for inspection. Mechanic B does not disturb the wheel sensors. Who is right?
   a. Mechanic A
   b. Mechanic B
   c. Both A and B
   d. Neither A nor B

10. What happens if the ABS actuation system has a malfunction?
    a. It shuts itself off.
    b. The brakes lock.
    c. Braking power is reduced.
    d. Nothing.

SHOP ASSIGNMENT 34

Examine and locate the following key components of the Teves anti-skid brake system: electronic controller, rear axle, anti-lock sensors, front spindle, and disc brake. Refer to Figure 11-2 as needed.

SHOP ASSIGNMENT 35

Examine and locate the following key components of the Bosch anti-lock brake system: control module, module relay, anti-lock warning light, lateral acceleration switch, master cylinder, booster, wheel speed sensors, toothed rings, calipers, and modulator relay. Refer to Figure 11-3 as needed.

SHOP ASSIGNMENT 36

Examine and locate the following key components of the Lucas-Girling stop control system: modulator, master cylinder, power booster, and proportioning valve. Refer to Figure 11-4 as needed.

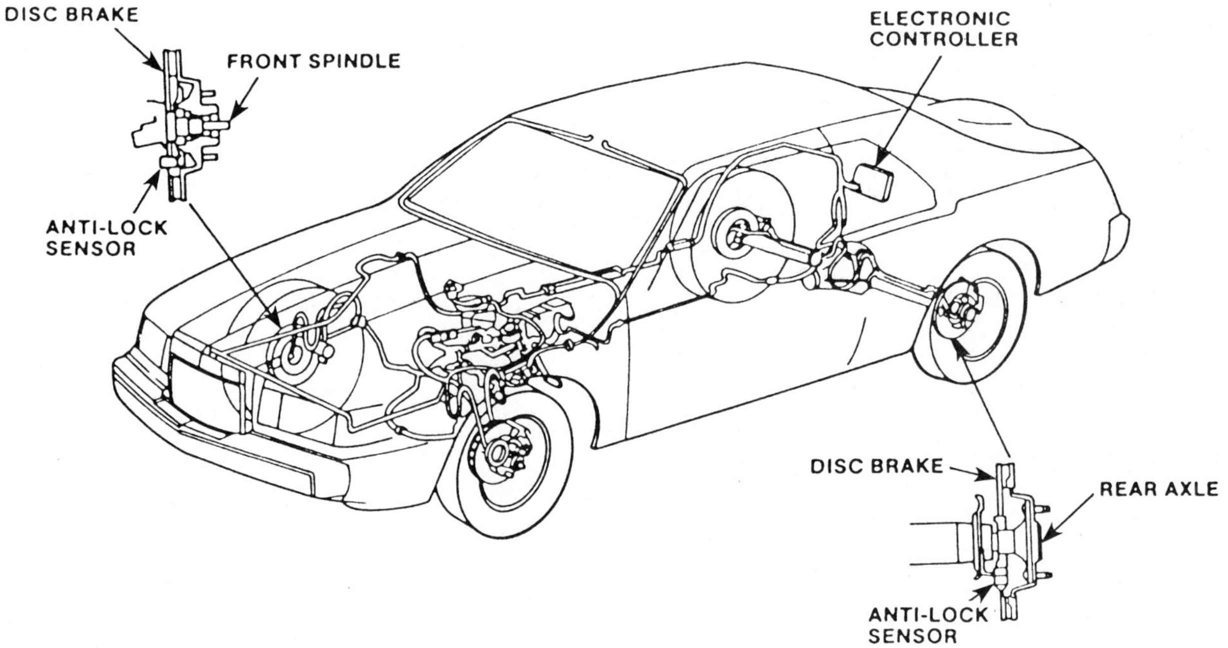

Figure 11-2

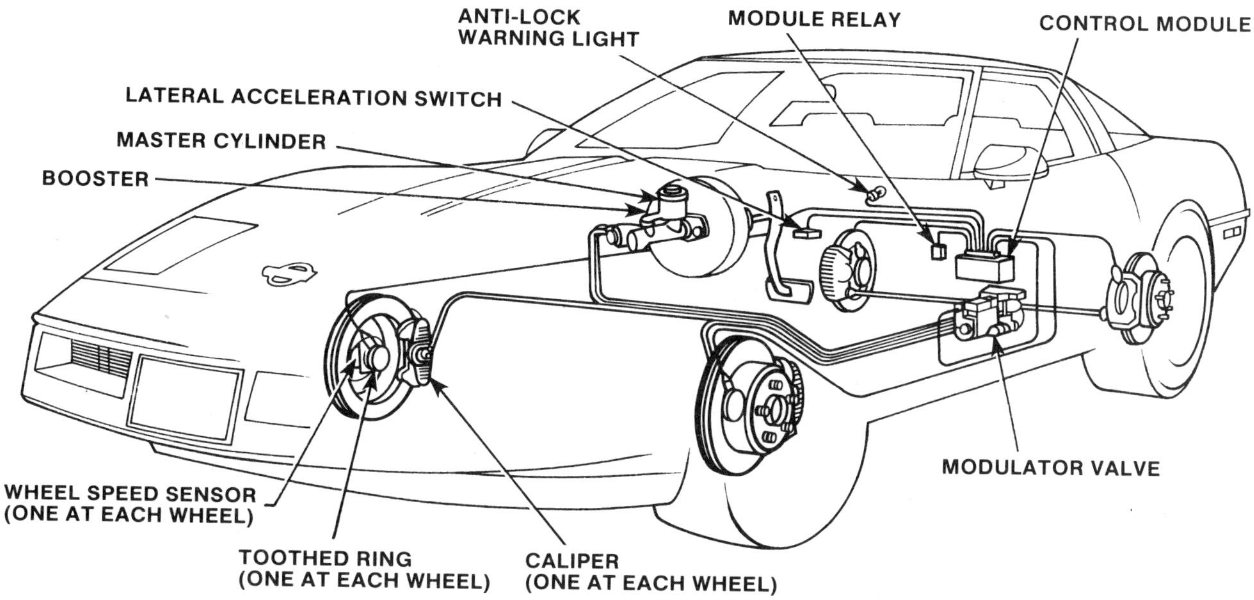

Figure 11-3

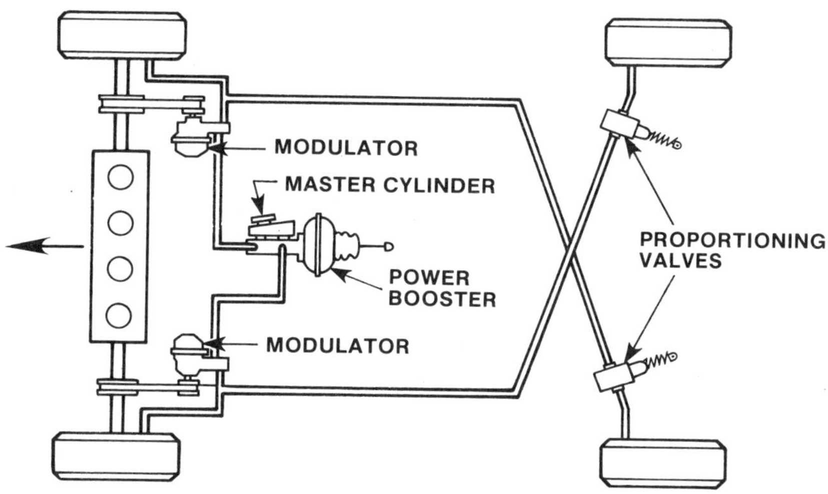

Figure 11-4

CHAPTER 12

LIGHT-DUTY TRUCK BRAKES

PRACTICE QUESTIONS

1. What component of a typical light-duty truck self-adjusting brake assembly is indicated by the letter A in Figure 12-1?
   a. self-adjuster cable
   b. anchor pin
   c. brake cylinder
   d. primary shoe and lining

2. What component of a typical light-duty truck self-adjusting brake assembly is indicated by the letter B in Figure 12-1?
   a. self-adjuster cable
   b. anchor pin
   c. brake cylinder
   d. primary shoe and lining

3. What component of a typical light-duty truck self-adjusting brake assembly is indicated by the letter C in Figure 12-1?
   a. self-adjuster cable
   b. anchor pin
   c. brake cylinder
   d. primary shoe and lining

4. To replace a disc brake rotor in a domestic light truck, Mechanic A first raises the vehicle and then removes the caliper mounting bolts. Mechanic B first raises the vehicle and then removes the wheel and tire assembly. Who is right?
   a. Mechanic A
   b. Mechanic B
   c. Both A and B
   d. Neither A nor B

5. How does rotor replacement differ between most passenger cars and many trucks?
   a. The rotor on trucks is easier to remove.
   b. The rotor on trucks requires less hardware.

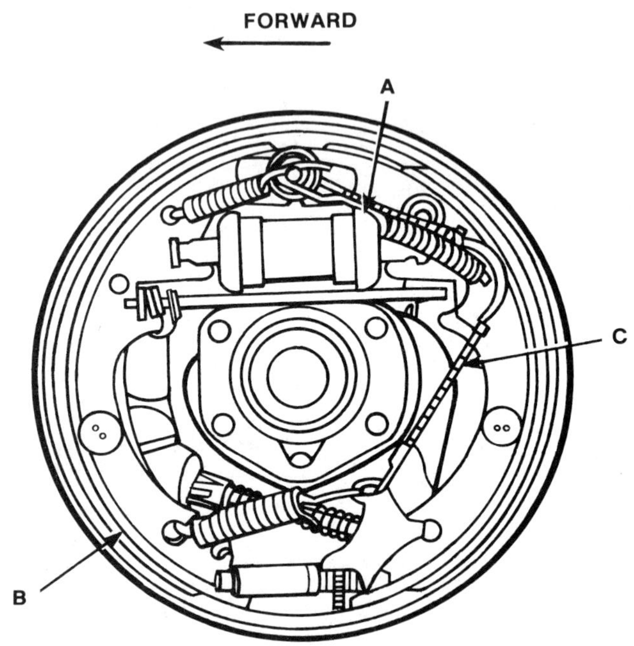

Figure 12-1

   c. The rotor on trucks is often mounted behind the hub or is part of the hub assembly.
   d. All of the above.

6. Mechanic A installs some hub nuts clockwise and some counterclockwise. Mechanic B installs all hub nuts in the same direction. Who is right?
   a. Mechanic A
   b. Mechanic B
   c. Both A and B
   d. Neither A nor B

7. When servicing wheel bearings, Mechanic A greases the hubs. Mechanic B does not disturb the hubs. Who is right?
   a. Mechanic A
   b. Mechanic B
   c. Both A and B
   d. Neither A nor B

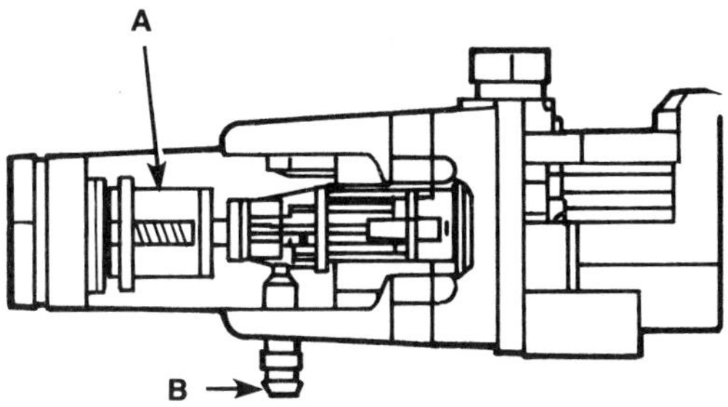

Figure 12-2

8. What component is indicated by the letter A in Figure 12-2?
   a. isolation valve
   b. accumulator
   c. outlet port
   d. bleeder screw

9. What component is indicated by the letter B in Figure 12-2?
   a. isolation valve
   b. accumulator
   c. outlet port
   d. bleeder screw

10. What component in some electric brake systems acts as a variable resistor to control the amount of electrical current to an electromagnet?
    a. braking sensor
    b. weighted pendulum
    c. rheostat
    d. transistor

JOB SHEET

SHOP ASSIGNMENT 37
REPLACE A DISC BRAKE ROTOR
ON A REAR WHEEL DRIVE VEHICLE

NAME ______________________ STATION ______________________ DATE ______________

Tools and Materials

Jack stands (2)
Socket wrench assortment
Wire
Grease
Torque wrench

Protective Clothing

Safety goggles or glasses with side shields

Procedure

1. Raise the front of the vehicle on jack stands.

Task completed ____

   a. Remove the wheel and tire assembly.

Task completed ____

2. Remove the caliper with the brake hose still attached.

Task completed ____

   a. Suspend it from the chassis with a piece of wire.

Task completed ____

3. Remove the dust cap.

Task completed ____

   a. Remove the cotter pin.

Task completed ____

   b. Remove the wheel bearing adjusting nut and washer.

Task completed ____

4. Carefully remove the hub and brake rotor assembly from the spindle.

Task completed ____

5. Pack the wheel bearings with grease.

Task completed ____

6. Reverse the procedure to install the new rotor assembly.

Task completed ____

7. Refer to the manufacturer's disc brake torque and specifications for the correct wheel bearing adjustment.

Task completed ____

   a. Torque the caliper mounting bolt to manufacturer's specifications.

Task completed ____

PROBLEMS ENCOUNTERED: ______________________________

____

____

INSTRUCTOR'S COMMENTS: ______________________________

____

____

JOB SHEET

SHOP ASSIGNMENT 38
REPLACE A BRAKE DRUM
ON A FORD LIGHT-DUTY TRUCK

NAME ____________________ STATION ____________________ DATE ______________

Tools and Materials

Jack stands (2)
Socket wrench assortment
Appropriate screwdrivers
Torque wrench

Protective Clothing

Safety goggles or glasses with side shields
Respirator

Procedure

1. Raise the rear of the vehicle on jack stands.

Task completed ____

a. Remove the wheel and tire assembly.

Task completed ____

2. Loosen the brake shoe adjustment screw.

Task completed ____

3. Remove the rear axle shaft retaining bolts and washers.

Task completed ____

a. Remove the rear axle shaft and gasket.

Task completed ____

4. Remove the wheel bearing lock nut, lock washer, and adjusting nut.

Task completed ____

5. Remove the brake drum-to-hub retaining bolts.

Task completed ____

a. Separate the brake drum from the hub.

Task completed ____

6. Begin installation by placing the brake drum on the hub.

Task completed ____

a. Install the attaching bolts.

Task completed ____

7. Position the hub and drum assembly on the rear axle.

Task completed ____

a. Start the adjusting nut.

Task completed ____

8. While rotating the drum, torque the adjusting nut 120 to 140 foot-pounds.

Task completed ____

9. Back off the adjusting nut approximately one-eighth to three-eighths turn.

Task completed ____

a. Is there an end-play of 0.001 to 0.010 inch?

Yes ____ No ____

b. If no, correct.

Not applicable ____ Task completed ____

10. Install the drive shaft and gasket.

Task completed ____

a. Install the attaching bolts.

Task completed ____

b. Torque the bolts 40 to 50 foot-pounds.

Task completed ____

11. Install the wheel and tire assembly.

Task completed ____

a. Adjust the rear brakes.

Task completed ____

PROBLEMS ENCOUNTERED: ____________________

____

____

INSTRUCTOR'S COMMENTS: ____________________

____

____

JOB SHEET

SHOP ASSIGNMENT 39
SERVICE A TYPICAL AUTOMATIC LOCKING HUB

NAME ____________________ STATION ____________________ DATE ____________

Tools and Materials

Jack stands
Wire
Grease
Torque wrench and other appropriate wrenches

Protective Clothing

Safety goggles or glasses with side shields

Procedure

1. Raise and support the front of the vehicle.

Task completed ____

   a. Remove the wheel and tire assembly.

Task completed ____

2. Remove the caliper from the mounting with the brake hose attached.

Task completed ____

   a. Suspend it from the chassis with a piece of wire.

Task completed ____

3. Remove the lug nut retainer washers.

Task completed ____

   a. Remove the locking hub assembly.

Task completed ____

4. Remove the snap ring from the end of the spindle shaft.

Task completed ____

5. Remove the axle shaft spacer.

Task completed ____

   a. Remove the needle thrust bearing.

Task completed ____

   b. Remove the bearing spacer.

Task completed ____

6. Remove the cam assembly.

Task completed ____

a. Remove the thrust washer.

Task completed ____

b. Remove the needle thrust bearing from the wheel bearing adjusting nut.

Task completed ____

7. Remove the wheel bearing adjusting nut.

Task completed ____

a. Slide the hub and brake rotor from the spindle.

Task completed ____

8. Lubricate the wheel bearings with suitable grease.

Task completed ____

a. Install the brake rotor and hub assembly back onto the spindle.

Task completed ____

9. Install the adjusting nut.

Task completed ____

a. Torque to approximately 35 foot-pounds while rotating the hub and rotor assembly.

Task completed ____

10. Back off the adjusting nut one-quarter turn.

Task completed ____

a. Torque the nut to approximately 16 foot-pounds.

Task completed ____

11. Align the nearest hole in the adjusting nut with the center of the spindle keyway slot.

Task completed ____

a. If necessary, advance the nut to the next hole.

Not applicable ____ Task completed ____

12. Install the lock nut needle thrust bearing.

Task completed ____

a. Install the thrust washer.

Task completed ____

b. Install the cam assembly.

Task completed ____

13. Install the bearing thrust washer.

Task completed ____

a. Install the needle thrust bearing.

Task completed ____

b. Install the axle shaft spacer.

Task completed ____

c. Install the snap ring.

Task completed ____

14. Install the lock hub assembly.

Task completed ____

a. Install the retainer washers.

Task completed ____

b. Install the locking hub assembly.

Task completed ____

15. Is the wheel end-play on the spindle 0.001 to 0.003 inch?

Yes ____ No ____

a. If no, correct.

Not applicable ____ Task completed ____

16. Install the brake caliper assembly.

Task completed ____

a. Torque the caliper mounting bolt to manufacturer's specifications.

Task completed ____

17. Install the wheel and tire assembly.

Task completed ____

18. Lower the vehicle.

Task completed ____

PROBLEMS ENCOUNTERED: ____

____

____

INSTRUCTOR'S COMMENTS: ____

____

____

CHAPTER 13
IMPORT VEHICLE BRAKES

PRACTICE QUESTIONS

1. When servicing Girling four-piston calipers, Mechanic A separates the caliper halves. Mechanic B does not separate the caliper halves. Who is right?
   a. Mechanic A
   b. Mechanic B
   c. Both A and B
   d. Neither A nor B

2. What component of a Girling four-piston caliper is indicated by the letter A in Figure 13-1?
   a. dust boot
   b. caliper housing
   c. piston
   d. brake pads

3. What component of a Girling four-piston caliper is indicated by the letter B in Figure 13-1?
   a. dust boot
   b. caliper housing
   c. piston
   d. brake pads

4. What component of an ATE single-piston sliding yoke caliper is indicated by the letter A in Figure 13-2?
   a. retaining pins
   b. guide spring
   c. brake caliper cylinder
   d. brake caliper piston

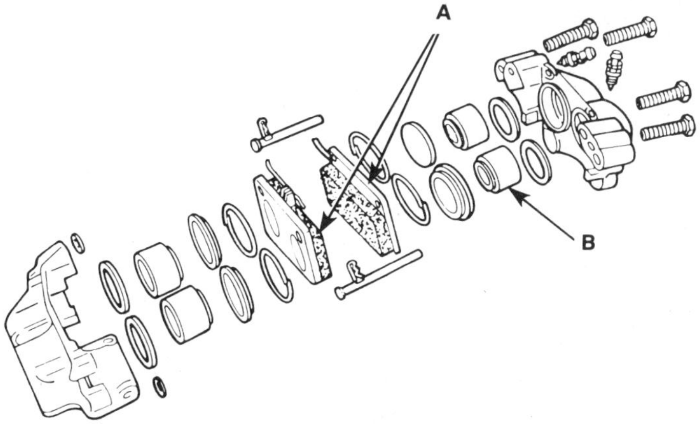

Figure 13-1

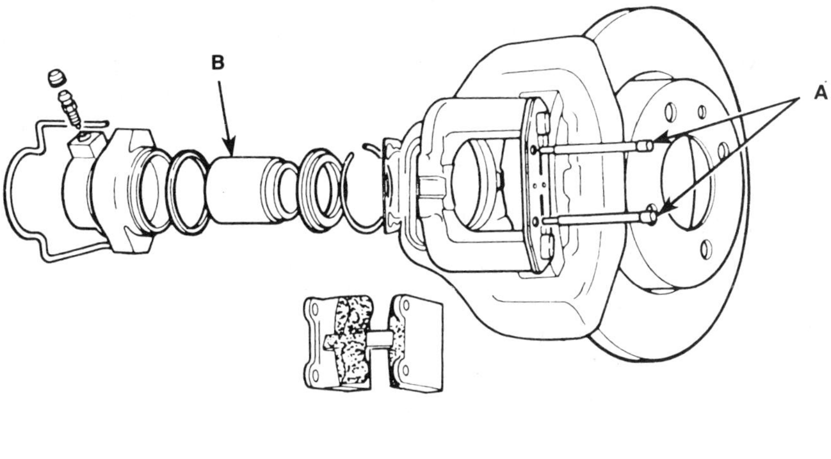

Figure 13-2

5. What component of an ATE single-piston sliding yoke caliper is indicated by the letter B in Figure 13-2?
   a. retaining pins
   b. guide spring
   c. brake caliper cylinder
   d. brake caliper piston

6. What component of a Nissan (Datsun) swing caliper is indicated by the letter A in Figure 13-3?
   a. disc pads
   b. cylinder housing
   c. caliper plate
   d. mounting bracket

7. What component of a Nissan (Datsun) swing caliper is indicated by the letter B in Figure 13-3?
   a. disc pads
   b. cylinder housing
   c. caliper plate
   d. mounting bracket

8. What component of a Mazda, Volkswagen drum brake with star wheel adjuster is indicated by the letter A in Figure 13-4?
   a. parking brake lever
   b. drum
   c. brake shoe
   d. wheel cylinder

9. What component of a Mazda, Volkswagen drum brake with star wheel adjuster is indicated by the letter B in Figure 13-4?
   a. parking brake lever
   b. drum
   c. brake shoe
   d. wheel cylinder

10. What component of a Mazda, Volkswagen drum brake with star wheel adjuster is indicated by the letter C in Figure 13-4?
    a. parking brake lever
    b. drum
    c. brake shoe
    d. wheel cylinder

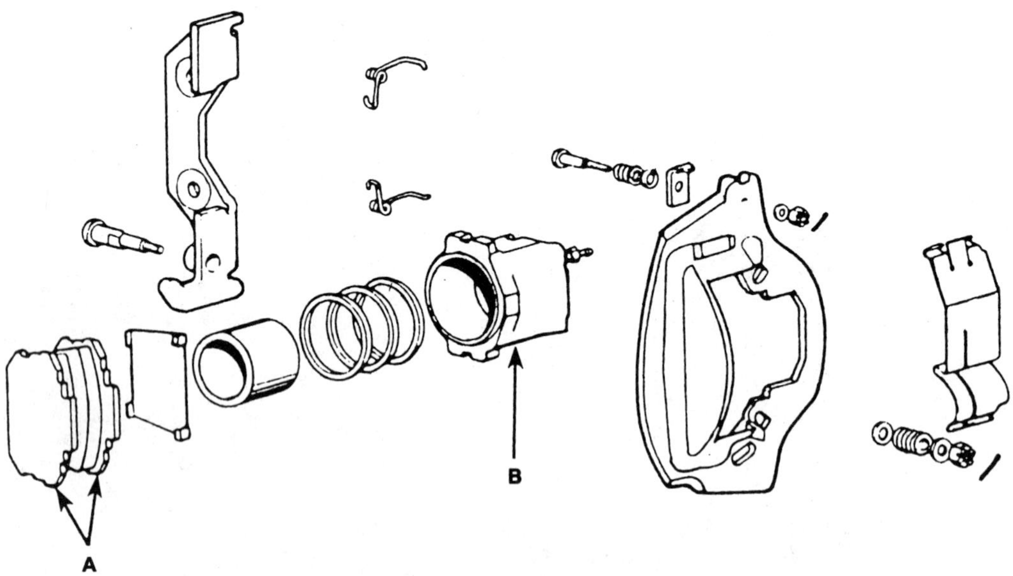

Figure 13-3

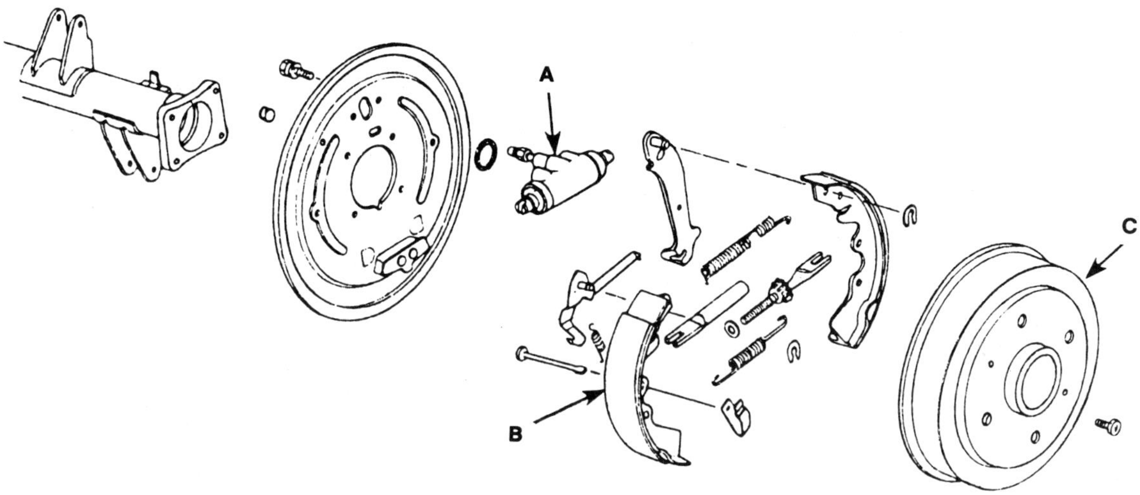

Figure 13-4

JOB SHEET

SHOP ASSIGNMENT 40
SERVICE AN ATE
TWO-PISTON FIXED CALIPER

NAME ______________________ STATION ______________________ DATE ______________

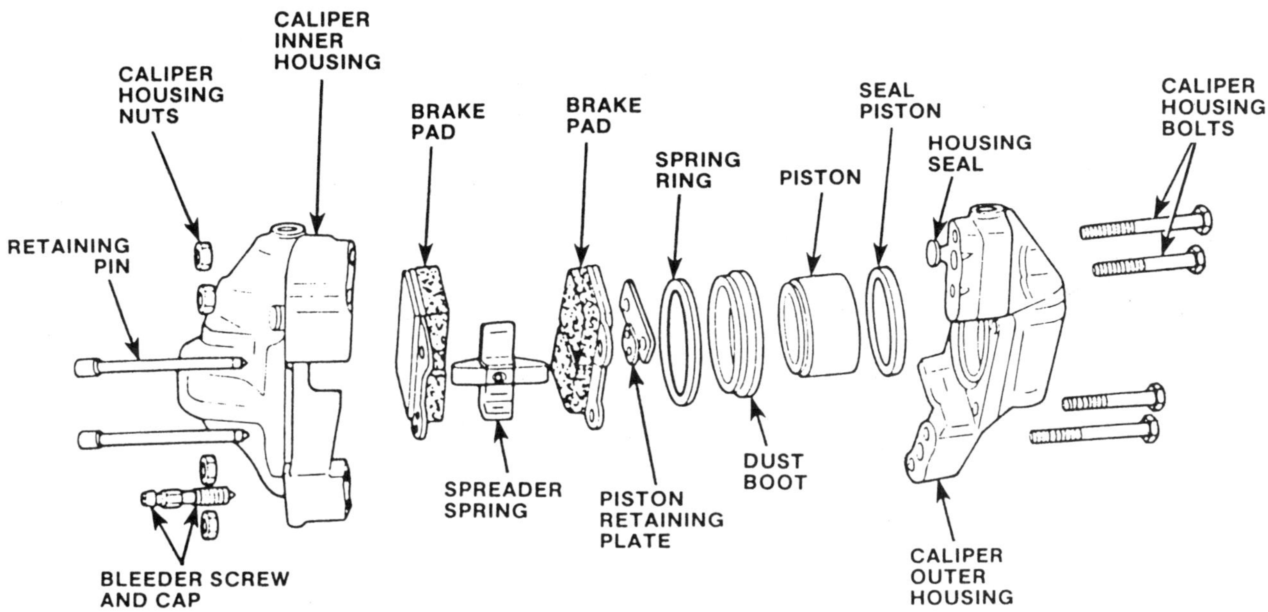

Figure 13-5

Tools and Materials

Appropriate screwdriver
Appropriate wrenches
Mallet
C-clamp
Torque wrench

Protective Clothing

Safety goggles or glasses with side shields

Procedure

1. Disconnect the hydraulic hose and hose retainer clip.

Task completed ____

2. Remove the two mounting bolts from the mounting bracket.

Task completed ____

3. Remove the brake hose.

Task completed ____

4. Remove the spring locks.

Task completed ____

a. Tap out both retaining pins.

Task completed ____

5. Lift out the spreader spring.

Task completed ____

6. Force the pads away from the rotor.

Task completed ____

a. Remove the pads.

Task completed ____

7. If applicable, rebuild the caliper one side at a time.

Not applicable ____ Task completed ____

a. Use a C-clamp to hold in the opposite piston.

Not applicable ____ Task completed ____

8. Replace the caliper, if necessary.

Not applicable ____ Task completed ____

9. Torque the mounting bolts to 43 foot-pounds.

Task completed ____

a. Secure the locking plate tabs.

Task completed ____

10. Bottom the pistons.

Task completed ____

a. Install the new pads one at a time.

Task completed ____

11. Replace the spreader spring.

Not applicable ____ Task completed ____

a. Replace the retaining pins.

Not applicable ____ Task completed ____

b. Replace the spring locks.

Not applicable ____ Task completed ____

PROBLEMS ENCOUNTERED: ______________________________

____

____

INSTRUCTOR'S COMMENTS: ______________________________

____

____

JOB SHEET

SHOP ASSIGNMENT 41
SERVICE A GIRLING
SINGLE-PISTON SLIDING CALIPER

NAME ______________ STATION ______________ DATE ______________

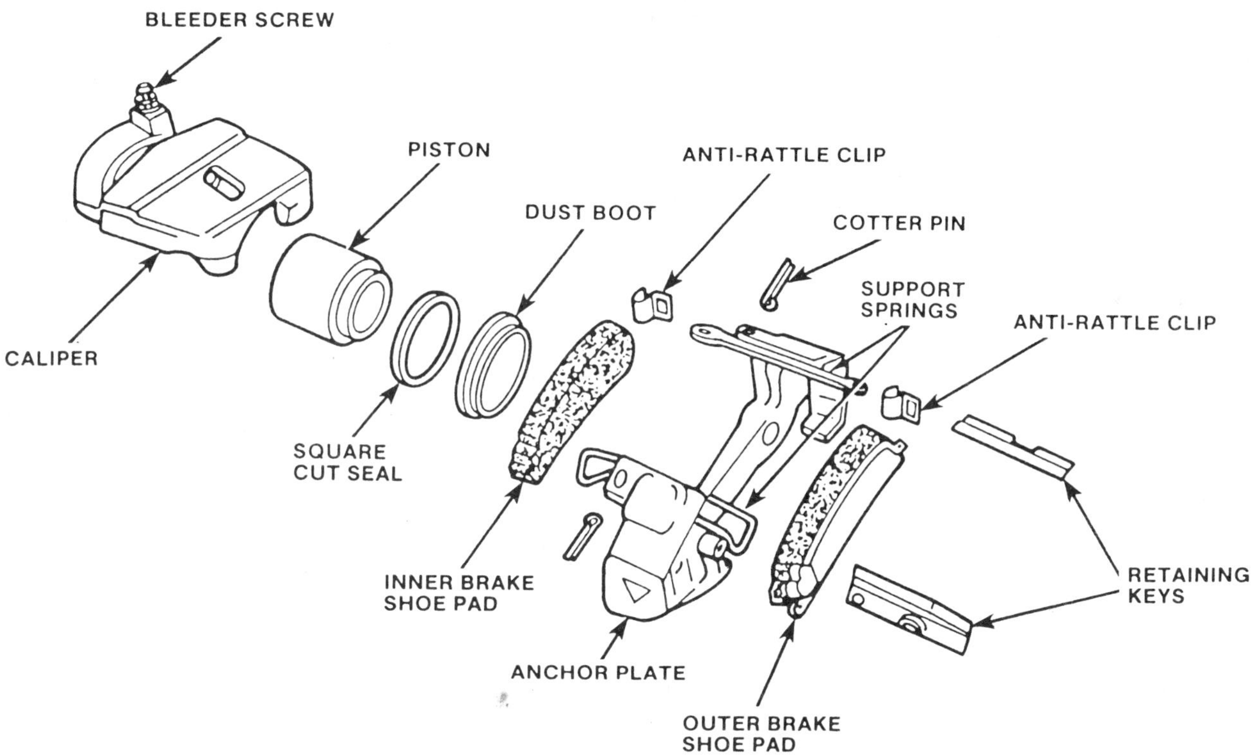

Figure 13-6

Tools and Materials

Mallet
Abrasive or wire brush
Appropriate screwdrivers

Protective Clothing

Safety goggles or glasses with side shields

Procedure

1. Remove the cotter pins.

Task completed ____

   a. Tap out the two retaining keys.

Task completed ____

2. Slide the brake pads out from under the support springs.

Task completed ____

3. Remove the anti-rattle clips. (Do not remove support springs.)

Task completed ____

4. Remove the brake hose.

Task completed ____

   a. Drain the caliper at the bleeder screw.

Task completed ____

5. If necessary, rebuild the caliper.

Not applicable ____ Task completed ____

6. Use an abrasive or wire brush to clean all machined surfaces.

Task completed ____

7. Attach the anti-rattle clips onto the new pads.

Task completed ____

8. Slide the new pads into the anchor plate.

Task completed ____

9. Press the caliper down against the support springs on the anchor plate.

Task completed ____

10. Slide the retaining keys in place.

Task completed ____

   a. Secure the retaining keys with the cotter pins.

Task completed ____

11. Connect the brake hose.

Task completed ____

12. Bleed the system.

Task completed ____

PROBLEMS ENCOUNTERED: ____

____

____

INSTRUCTOR'S COMMENTS: ____

____

____

JOB SHEET

SHOP ASSIGNMENT 42
SERVICE AN AKEBONO REAR DISC SINGLE-PISTON BRAKE

NAME ______________________ STATION ______________________ DATE ______________

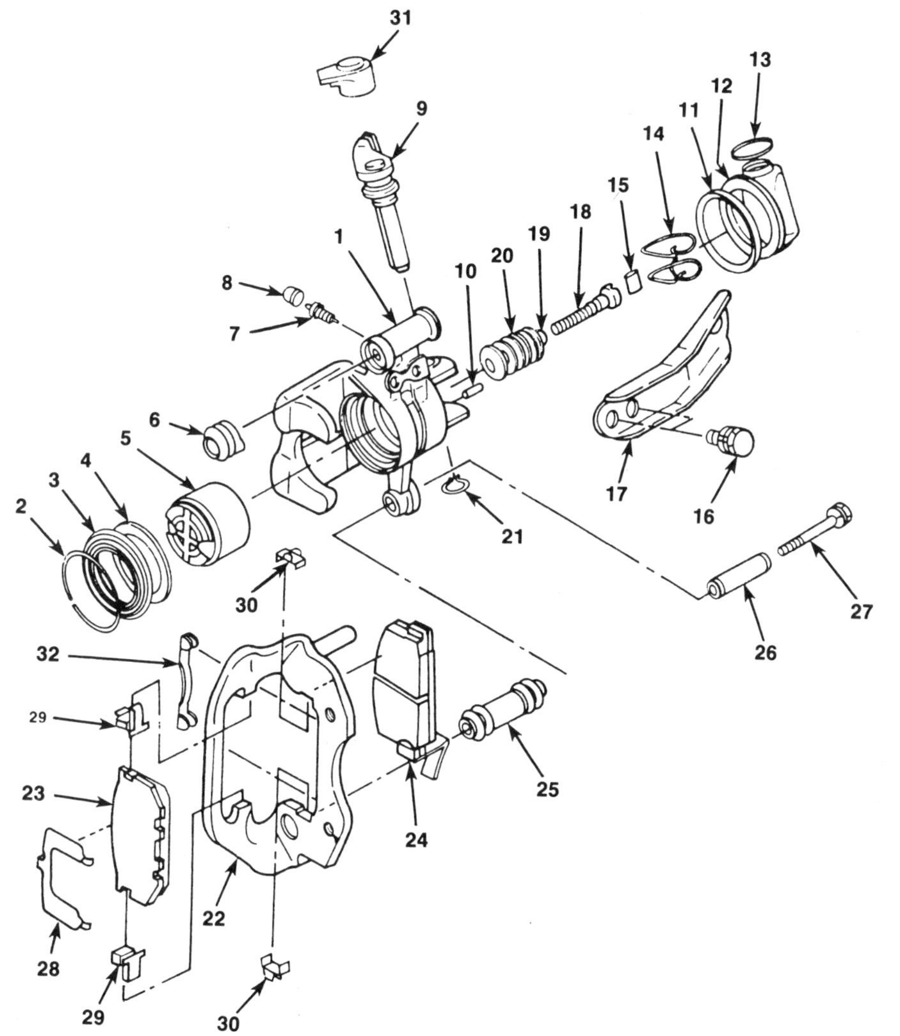

1	CALIPER BODY	12	LEVER CAP	23	OUTER PAD
2	BOOT RING	13	GARTER SPRING	24	INNER PAD
3	PISTON BOOT	14	RETURN SPRING	25	LOCK PIN BOOT
4	PISTON SEAL	15	CONNECTING LINK	26	SLEEVE
5	PISTON	16	BOLT ASSEMBLY	27	LOCK PIN
6	GUIDE PIN BOOT	17	BRACKET	28	SHIM
7	AIR BLEEDER SCREW	18	SPINDLE	29	OUTER PAD CLIP
8	AIR BLEEDER CAP	19	O-RING	30	INNER PAD CLIP
9	LEVER AND SPINDLE	20	CONE SPRING WASHERS	31	LEVER CAP (UPPER)
10	SPRING PIN	21	SNAP RING	32	INNER PAD CLIP
11	CAP RING	22	SUPPORT		

Figure 13-7

Tools and Materials

Appropriate screwdrivers
Hex wrench, if applicable
Special piston wrench
Brake fluid
Appropriate wrenches

Protective Clothing

Safety goggles or glasses with side shields

Procedure

1. Release the parking brake.

Task completed ____

2. Disconnect the parking brake cable from the lever at the caliper.

Task completed ____

3. Unscrew the lower caliper lock pin.

Task completed ____

4. Rotate the caliper up off the rotor so that it pivots on the upper pin.

Task completed ____

5. Remove the pads.

Task completed ____

6. Inspect the caliper for leaks or damage.

Task completed ____

7. Remove the caliper.

Task completed ____

   a. Rebuild or replace as needed.

Not applicable ____ Task completed ____

8. Use the special wrench to rotate the piston clockwise, forcing it into the bore, being careful not to twist or pinch the pin's rubber boot. (See Figure 13-21 on page 317 in book.)

Task completed ____

9. Install the caliper.

Task completed ____

10. Install the new pads, including any shims that might be present.

Task completed ____

11. Rotate the caliper back down over the rotor.

Task completed ____

12. Install the lower caliper lock pin.

Task completed ____

   a. Install the locking bolt.

Not applicable ____ Task completed ____

13. Fill the master cylinder reservoir with fresh brake fluid.

Task completed ____

14. Bleed the brakes.

Task completed ____

a. Reattach the parking brake cable.

Task completed ____

15. Obtain a high, hard pedal.

Task completed ____

a. Refill the master cylinder reservoir.

Task completed ____

16. Check the operation of the parking brake.

Task completed ____

a. Road test the vehicle.

Task completed ____

PROBLEMS ENCOUNTERED: ____

____

____

INSTRUCTOR'S COMMENTS: ____

____

____

JOB SHEET

SHOP ASSIGNMENT 43
REPLACE THE PARKING BRAKE SHOES ON A GIRLING REAR DISC WITH DRUM PARKING BRAKE

NAME ____________ STATION ____________ DATE ____________

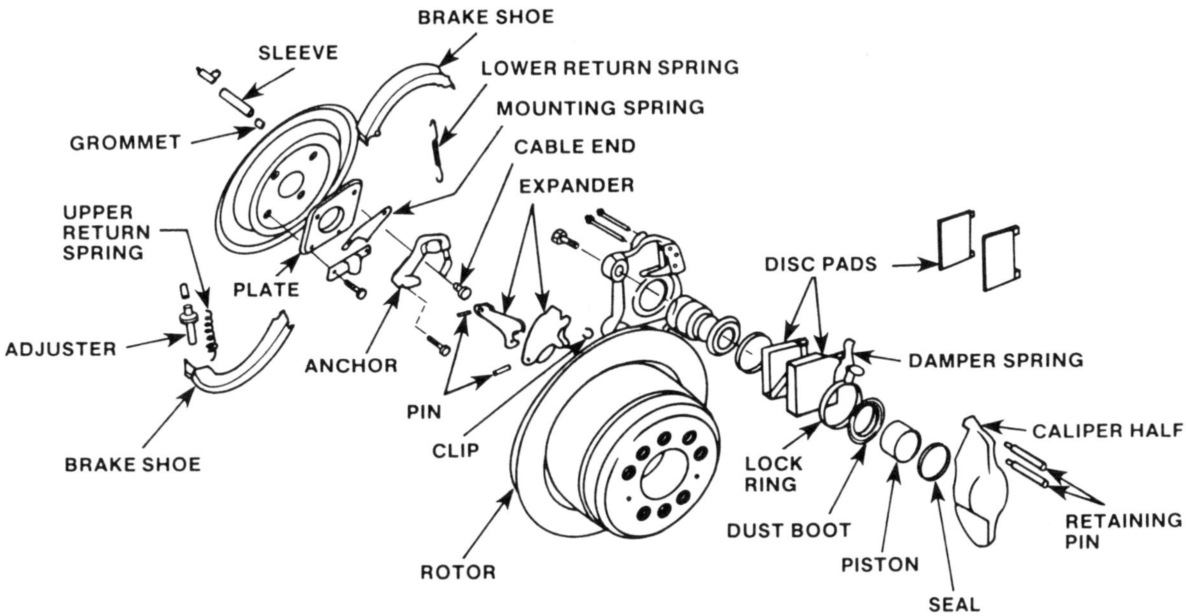

Figure 13-8

Tools and Materials

Appropriate wrenches
Wire
Appropriate screwdrivers
Cleaner
Lubricant
Brake adjusting tool, if necessary

Protective Clothing

Safety goggles or glasses with side shields

Procedure

1. Unbolt the caliper.

Task completed ____

a. Lift it off the rotor/drum.

Task completed ____

b. Suspend it by a wire.

Task completed ____

2. Remove the two screws holding the rotor/drum to the axle.

Task completed ____

a. Remove the rotor/drum.

Task completed ____

3. Remove the shoe return springs.

Task completed ____

a. Remove the shoes.

Task completed ____

4. Remove the parking brake actuating lever spacer from the old trailing shoe.

Task completed ____

5. Clean the backing plate.

Task completed ____

a. Inspect the backing plate.

Task completed ____

b. Do any axle seal leaks exist?

Yes ____ No ____

c. If yes, correct them.

Not applicable ____ Task completed ____

6. Inspect the rotor/drum.

Task completed ____

a. Machine or replace as necessary.

Not applicable ____ Task completed ____

7. Lubricate the rubbing surfaces on the backing plate.

Task completed ____

8. Install the parking brake actuating lever spacer on the new trailing shoe.

Task completed ____

a. Install the retaining spring on the new trailing shoe.

Task completed ____

9. Install the new shoes.

Task completed ____

a. Install the return springs.

Task completed ____

b. Install the star wheel adjuster.

Task completed ____

c. Back off the star wheel adjuster all the way.

Task completed ____

10. Mount the rotor/drum on the axle.

Task completed ____

a. Install the two screws.

Task completed ____

11. Adjust the shoe-to-drum clearance by turning the star wheel with a tool inserted through the slot in the rotor/drum.

Task completed ____

12. Adjust the parking brake cable.

Task completed ____

13. Install the caliper.

Task completed ____

PROBLEMS ENCOUNTERED: ____

____

____

INSTRUCTOR'S COMMENTS: ____

____

____

CHAPTER 14

DIAGNOSING BRAKE SYSTEM PROBLEMS

PRACTICE QUESTIONS

1. A vehicle's drum and wheel are tilting, and they have spotty contact with the brake linings, causing erratic brake action. Mechanic A looks for a loose front-wheel bearing. Mechanic B looks for worn lining. Who is right?
   a. Mechanic A
   b. Mechanic B
   c. Both A and B
   d. Neither A nor B

2. Before test driving a vehicle, Mechanic A makes a preliminary inspection of the brake system in the shop. Mechanic B checks the fluid level in the master cylinder. Who is right?
   a. Mechanic A
   b. Mechanic B
   c. Both A and B
   d. Neither A nor B

3. What causes brake fade?
   a. heat
   b. water
   c. both a and b
   d. neither a nor b

4. What faulty component can give the misleading impression that brakes are too severe?
   a. tires with low air pressure
   b. faulty shock absorbers
   c. loose wheel bearings
   d. front end misalignment

5. The high-speed road test for roughness or pulsations involves ______________.
   a. making light stops at the maximum legal speed
   b. making hard stops at the maximum legal speed
   c. making light stops at various speeds
   d. making hard stops at various speeds

6. A vehicle comes into the shop often with the two rear brakes severely worn. Mechanic A suspects the driver to be a left-foot braker. Mechanic B suspects the driver is a delivery route driver. Who is right?
   a. Mechanic A
   b. Mechanic B
   c. Both A and B
   d. Neither A nor B

7. A leaking caliper is discovered on a vehicle. Mechanic A overhauls the caliper. Mechanic B replaces the caliper. Who is right?
   a. Mechanic A
   b. Mechanic B
   c. Both A and B
   d. Neither A nor B

8. Water in the brake assembly causes ______________.
   a. hard pedal
   b. brake fade
   c. loss of braking ability
   d. all of the above

9. Mechanic A relines only those wheels that need new lining. Mechanic B always also relines the opposite wheel on the same axle. Who is right?
   a. Mechanic A
   b. Mechanic B
   c. Both A and B
   d. Neither A nor B

10. Mechanic A uses a crocus cloth to remove rust, corrosion, pitting, and scratches from a plated piston bore. Mechanic B hones a plated piston bore. Who is right?
    a. Mechanic A
    b. Mechanic B
    c. Both A and B
    d. Neither A nor B

JOB SHEET

SHOP ASSIGNMENT 44
CONDUCT A PREBRAKE JOB INSPECTION

Examine a sample vehicle and complete the checklist that follows on the next page.

NAME ______________________ STATION ______________________ DATE ______________

PREBRAKE JOB INSPECTION CHECKLIST

Owner ______________________ (LAST) ______________________ (FIRST) Phone ______________ Date ___|___|___

Address ____ License No. ______________

Make __________ Model __________ Mileage __________ Serial No. __________ Year __________

Special Key for Hubcaps/Wheels Location ______________ Owner Use Parking Brake Yes ☐ No ☐

4 Drum ☐ 4 Disc ☐ Disc/Drum ☐ P/B No ☐ Yes ☐ Vacuum ☐ Hydro ☐ ABS ☐

Owner Comments ____

1. CHECKS BEFORE ROAD TEST	Safe	Unsafe
Stoplight Operation		
Brake Warning Light Operation		
Master Cylinder Checks		
Fluid Level		
Fluid Contamination		
Under Hood Fluid Leaks		
Under Dash Fluid Leaks (No Power)		
Bypassing		

BRAKE PEDAL HEIGHT AND FEEL

Check One		Check One		
Low		Spongy		
Med		Firm		
High				

Power Brake Unit Checks

VACUUM	Safe	Unsafe	HYDRO	Safe	Unsafe
Vacuum Unit			Hydro Unit		
Engine Vacuum			P/S Fluid		
Vacuum Hose			P/S Belt Tension		
Unit Check Valve			P/S Belt Condition		
Reserve Braking			P/S Fluid Leaks		
			Reserve Braking		

Tire Pressure Specs	Front	Rear
Record Pressure Found		
RF ______ LF ______ RR ______ LR ______		
Tire Condition		
RF ______ LF ______ RR ______ LR ______		

2. ROAD TEST	Yes	No	RF	LF	RR	LR
Brake Pull						
Brake Clunk						
Brake Scraping						
Brake Squeal						
Brake Grabby						
Brakes Lock Prematurely						
Wheel Bearing Noise						
Vehicle Vibrates						

STEERING WHEEL MOVEMENT WHEN STOPPING FROM 2-3 MPH YES/NO/RGT/LFT

Does ABS Work	YES	NO
Pedal Pulsation when Braking	YES	NO
Steering Wheel Oscillation when Braking	YES	NO
No Stopping Power	YES	NO
Warning Light Comes on when Braking	YES	NO
Difference in Pedal Height after Cornering	YES	NO
Nose Dive	YES	NO

3. In Shop Checks On Hoist	Yes	No	RF	LF	RR	LR
Brake Drag						
Intermittent Brake Drag						
Brake Pedal Linkage Binding						
Wheel Bearing Looseness						
Missing or Broken Wheel Fasteners						
Suspension Looseness						

	Front					Rear				
Mark Wheels and Remove	Right			Left		Right			Left	
Caliper/Piston Stuck RF LF RR LR	Spec	Safe	Unsafe	Safe	Unsafe	Spec	Safe	Unsafe	Safe	Unsafe
Mark Drums and Remove										
Measure Rotor Thickness or Drum Diameter.										
Measure Rotor Thickness Variation.										
Measure Rotor Runout.										
Lining Thickness										
Tubes and Hoses										
Fluid Leaks										
Broken Bleeders										
Leaky Seals										
Self-Adjuster Operation										
									Safe	Unsafe
Parking Brake Cables and Linkage										

JOB SHEET

SHOP ASSIGNMENT 45
DIAGNOSE AND CORRECT THE REASON FOR BRAKE DRAG ON A VEHICLE WITH DISC BRAKES

NAME ________________ STATION ________________ DATE ________________

Tools and Materials

Jack stands (2)
Screwdrivers
Wrenches
New hardware, if needed

Protective Clothing

Safety goggles or glasses with side shields

Procedure

1. Raise the vehicle.

   Task completed ____

2. Is the caliper mechanism fully returning?

   Yes ____ No ____

   a. If yes, go to step 2.

      Task completed ____

   b. If no, is the air chamber diaphragm free and not iced up?

      Not applicable ____ Yes ____ No ____

   c. If yes, go to step 1g.

      Task completed ____

   d. If no, correct.

      Not applicable ____ Task completed ____

   e. Is a boot and/or air dryer being used?

      Yes ____ No ____

   f. If no, correct.

      Not applicable ____ Task completed ____

   g. Overhaul the caliper.

      Task completed ____

3. Is corrosion present on the retainers and springs?

Yes ____ No ____

   a. If no, go to step 3.

Task completed ____

   b. If yes, decontaminate or replace, as necessary.

Task completed ____

4. Is a wheel bearing seized?

Yes ____ No ____

   a. If yes, replace.

Task completed ____

5. Road test the vehicle.

Task completed ____

   a. Is brake drag evident?

Yes ____ No ____

   b. If yes, repeat steps 1 through 5.

Task completed ____

PROBLEMS ENCOUNTERED: ____________________

____

____

INSTRUCTOR'S COMMENTS: ____________________

____

____

ANSWER KEYS

CHAPTER 1

1. b	3. a	5. b	7. d	9. a
2. c	4. b	6. a	8. c	10. d

CHAPTER 2

1. b	3. a	5. c	7. c	9. b
2. d	4. b	6. a	8. a	10. d

CHAPTER 3

1. a	3. c	5. b	7. d	9. b
2. a	4. b	6. a	8. c	10. a

CHAPTER 4

1. c	3. a	5. d	7. c	9. a
2. b	4. a	6. b	8. a	10. d

CHAPTER 5

1. d	3. a	5. a	7. d	9. a
2. b	4. c	6. b	8. c	10. b

CHAPTER 6

1. c	3. a	5. b	7. c	9. b
2. b	4. d	6. a	8. d	10. a

CHAPTER 7

1. a	3. d	5. b	7. a	9. a
2. b	4. b	6. c	8. d	10. b

CHAPTER 8

1. b	3. a	5. d	7. b	9. c
2. b	4. b	6. c	8. a	10. a

CHAPTER 9

1. c	3. a	5. d	7. a	9. d
2. b	4. c	6. b	8. b	10. c

CHAPTER 10

1. a	3. b	5. c	7. b	9. c
2. c	4. d	6. a	8. d	10. a

CHAPTER 11

1. b	3. d	5. b	7. a	9. b
2. a	4. c	6. d	8. c	10. a

CHAPTER 12

1. c	3. a	5. c	7. a	9. d
2. d	4. b	6. a	8. b	10. c

CHAPTER 13

1. b	3. c	5. d	7. b	9. c
2. d	4. a	6. a	8. d	10. b

CHAPTER 14

1. a	3. c	5. a	7. c	9. b
2. c	4. b	6. b	8. d	10. a